WITHDRAWN
UTSA LIBRARIES

New Pathways for Organic Synthesis

Practical Applications of Transition Metals

New Pathways
for Organic Synthesis

Practical Applications of Transition Metals

H. M. Colquhoun
J. Holton

Imperial Chemical Industries plc
New Science Group
Runcorn, United Kingdom

D. J. Thompson

Imperial Chemical Industries plc
Organics Division
Huddersfield, United Kingdom

and

M. V. Twigg

Imperial Chemical Industries plc
Agricultural Division
Billingham, United Kingdom

Plenum Press • New York and London

Library of Congress Cataloging in Publication Data

Main entry under title:

New pathways for organic synthesis.

  Bibliography: p.
  Includes index.
  1. Chemistry, Organic—Synthesis. 2. Organometallic compounds. 3. Transition
metal compounds. I. Colquhoun, H. M.
QD262.N477   1983   547′.0569   83-16085
ISBN 0-306-41318-3

LIBRARY
The University of Texas
At San Antonio

© 1984 Plenum Press, New York
A Division of Plenum Publishing Corporation
233 Spring Street, New York, N.Y. 10013

All rights reserved

No part of this book may be reproduced, stored in a retrieval system, or transmitted
in any form or by any means, electronic, mechanical, photocopying, microfilming,
recording, or otherwise, without written permission from the Publisher

Printed in the United States of America

Foreword

The continually growing contribution of transition metal chemistry to synthetic organic chemistry is, of course, widely recognized. Equally well-known is the difficulty in keeping up-to-date with the multifarious reactions and procedures that seem to be spawned at an ever-increasing rate. These can certainly be summarized on the basis of reviews under the headings of the individual transition metals. More useful to the bench organic chemist, however, would be the opposite type of concordance based on the structural type of the desired synthetic product. This is the approach taken in the present monograph, which presents for each structural entity a conspectus of the transition metal-mediated processes that can be employed in its production. The resulting comparative survey should be a great help in devising the optimum synthetic approach for a particular goal. It is presented from an essentially practical viewpoint, with detailed directions interspersed in the Houben–Weyl style. The wide scope of the volume should certainly encourage synthetic organic chemists to utilize fully the range and versatility of these transition metal-mediated processes. This will certainly be a well-thumbed reference book!

Cambridge University   R. A. RAPHAEL

Preface

In recent years an enormous amount of work has been done on the catalysis of organic reactions by various transition metal species and on the organic reactivity of organo-transition-metal compounds. Many new, facile, and potentially useful pathways to organic products have been discovered and it is unfortunate that, because the majority of this work has been carried out in "inorganic" laboratories, many practicing organic chemists are probably unaware of these developments.

The objective of the present book is to help remedy this situation by providing those engaged in the preparation of pharmaceuticals, natural products, herbicides, dyestuffs, and other organic chemicals, with a practical guide to the application of transition metals in organic synthesis. Here we have collected a considerable number of transition-metal-based procedures that have genuine applications in synthesis, arranged according to the nature of the organic product or synthetic transformation being carried out. Many illustrative examples of actual practical procedures are given, though extensive reference to the original literature is also provided if more detailed information is required.

The procedures described are often straightforward and generally require only commercially available or readily prepared inorganic materials. However details for the preparation of catalysts are not always easily accessible, being widely scattered in the primary literature, and a short chapter is therefore devoted to the preparation and handling of transition metal catalysts.

The reaction pathways offered by transition metal reagents and catalysts have much to offer in organic synthesis and it is the hope of the present authors that this book will help to encourage their wider routine application.

Contents

Chapter 1

Introduction

Organometallic reagents have been used in preparative organic chemistry since the beginning of this century. Some main-group organometallic compounds such as the organomagnesium and organolithium reagents are particularly useful and have found widespread application. Although transition elements such as the precious metals palladium and platinum, and also nickel (in the form of Raney nickel), are much used as hydrogenation catalysts, other transition metals and their compounds have not been widely applied in organic chemistry. Organocopper species have, however, been employed in a number of synthetic methods that now find routine application, and these methods have been the subject of a number of books and reviews.[1-5]

With the growth of interest in organo-transition-metal chemistry during recent years it has become clear that many transition metals, their compounds, and in particular their organometallic derivatives, are potentially very important catalysts and reagents for organic synthesis. Since the characterization of ferrocene (dicyclopentadienyl iron) in the early 1950s, organo-transition-metal chemistry has developed to become a distinct, and very large, subject in its own right. During this period much effort has been directed toward the preparation, isolation, and characterization of novel compounds, while theoretical studies have concentrated on their structural and bonding patterns. The stability and properties of organo-transition-metal compounds depend on the metal concerned, its oxidation state, and the nature of the groups bound to the metal. Many such compounds are very stable, and, having physical properties similar to conventional organic compounds, can be handled in the same way. Others, however, are air or water sensitive, and require special techniques for their manipulation.

1

The reactivity of an organic molecule (or residue) bound to a transition metal can be significantly different from that of the free molecule, because organo-transition-metal compounds often have available reaction pathways that are not found in conventional organic chemistry. Participation of the metals' *d* orbitals in bonding to the organic group is usually responsible for this behavior, and initially these reactions may appear "unnatural" to the traditional organic chemist. For example, nucleophilic attack on an alkyl-substituted arene may seem "unnatural," but this and many other apparently improbable reactions take place when transition metals are involved, and they can provide efficient routes to desirable products. In order to exploit these dramatic changes in reactivity, it may be necessary to prepare and isolate an organometallic compound, carry out a reaction, and then remove the transition metal to liberate the final product. However, in many instances it is possible to form the organometallic compound *in situ* and exploit its reactivity in a catalytic cycle. This can be particularly convenient in practice if, as is often the case, the intermediate can be generated from readily available metal salts or easily prepared simple complexes. For example, ethanolic solutions of rhodium trichloride produce, on warming, a rhodium hydride that is a powerful catalyst for double-bond migration reactions—that is, isomerization of alkenes via hydrogen migration. In this way, using only very small quantities of rhodium trichloride in the presence of ethanol, remote double bonds can be brought into conjugation with one another, and compounds having an exocyclic double bond can be smoothly converted to isomers with an endocyclic double bond. A particularly attractive feature of these reactions, which in many instances are difficult to achieve at all by other means, is that they take place under very mild conditions (typically 50–100°C), and reaction times can be as short as an hour. This isomerization procedure therefore seems likely to find considerable application in areas such as natural product synthesis. There are a number of other useful transformations that are possible using transition metal catalysts or reagents that cannot be achieved easily in other ways. The asymmetric hydrogenation of prochiral alkenes to optically active products using chiral transition metal complexes as catalysts is a good example, and with catalyst systems now available this reaction proceeds with excellent selectivity.

With the advantages that accompany the use of transition metal species in organic synthesis it is not surprising that industrial laboratories have maintained an active interest in the area, particularly in catalytic applications with a view to large-scale processes. An impressive number of industrial processes have in fact been introduced that are based on homogeneous transition metal catalysts,[6–8] and some of them are listed in Table 1.1. Perhaps the most innovative of these commercial processes is the manufacture of *l*-DOPA via asymmetric hydrogenation of a prochiral alkene to produce a single optical isomer [Eq. (1)]. Among the very large-scale

$$\text{RCH=C(NHCOCH}_3)\text{CO}_2\text{R} \xrightarrow[\text{chiral rhodium catalyst}]{\text{hydrogen}} l\text{-RCH}_2\overset{*}{\text{C}}\text{H(NH}_2)\text{CO}_2\text{H} \quad (1)$$

industrial processes using transition metal catalysts, some of the more important involve direct carbonylation of organic substrates, that is, the introduction of a carbonyl group into a molecule by insertion of carbon monoxide. Although extrusion of carbon monoxide from a molecule is well known in organic synthesis, direct carbonylation of, say, an alkyl halide to form the acyl halide may appear a distinctly "unnatural" reaction. Nonetheless, this reaction can be achieved in the laboratory under relatively mild conditions with the appropriate transition metal catalyst (Chapter 6), and a variety of related carbonylation reactions are carried out on a large scale in industry. Acetic acid, for example, is now manufactured by the direct carbonylation of methanol in the presence of a homogeneous rhodium catalyst, and the linear alcohols used in the synthesis of biodegradable detergents are derived from terminal alkenes via hydroformylation with carbon monoxide and hydrogen in the presence of cobalt catalysts.

Transition-metal-based procedures clearly have considerable potential in laboratory scale synthesis and some examples of catalytic reactions are collected in Table 1.2, but as yet transition-metal-based reagents and catalysts have not become widely integrated into the repertoire of synthetic organic chemistry. A considerable number of books and reviews concerned with organo-transition-metal chemistry are available, and many of these mention potential applications in organic synthesis, but relatively few emphasize synthetic aspects or indicate the procedures necessary to make use of these reactions in the laboratory. The present book is directed toward the practicing organic chemist, and its major objective is to familiarize the reader with the more important transformations that can be conveniently brought about in the laboratory by transition metal reagents and catalysts.

The reactions that have been collected together in this book were chosen principally for their general usefulness in synthesis. Coverage, of necessity, is selective rather than comprehensive. It is not possible in a work of this size to include all the reactions involving transition metals that may have some potential use in organic synthesis. Emphasis has therefore been given to nontrivial transformations, which offer advantages over conventional organic reactions. Practical details are given, and where possible illustrative procedures have been selected that do not require the use of special techniques or complex and expensive equipment. An inert atmosphere may sometimes be needed, but even then, simple means of obtaining this are normally satisfactory. Sufficient detail is generally given about

Table 1.1. Selected Examples of Industrial Processes Based on Homogeneous Transition Metal Catalysts

Process	Reaction	Metal catalyst	Comments	Ref.
Acetic acid from methanol	$CH_3OH + CO \rightarrow CH_3CO_2H$	Cobalt, rhodium	Cobalt catalysts require high pressure and temperature, whereas the rhodium-catalyzed reaction can be operated even at atmospheric pressure.	9
Hydroformylation of alkenes	$RCH=CH_2 + CO/H_2 \rightarrow RCH_2CH_2CHO$	Cobalt, rhodium	Oldest large-scale process using homogeneous transition metal catalysts. Normally the aldehyde products are hydrogenated to alcohols.	10
Oxidation of alkenes to aldehydes	$RCH=CH_2 + O_2 \rightarrow RCH_2CHO$	Palladium with copper	The Wacker process, important when acetylene-based processes were being replaced, now almost obsolete.	11
Adiponitrile from butadiene	$CH_2=CHCH=CH_2 + 2HCN \rightarrow NC(CH_2)_4CN$	Nickel	Several catalysts are available for the first addition. Regio-selectivity appears important, but nickel arylphosphite complexes catalyze isomerization of branched intermediates as well as the addition reactions.	12
Polymerization of alkenes	$RCH=CH_2 \rightarrow [-CRH-CH_2-]_n$	Titanium, zirconium, vanadium	Very large-scale production of high-molecular-weight polymers and rubbers. Some of the catalysts are heterogeneous.	13

Process	Reaction	Catalyst	Description	No.
Hydrogenation of alkenes and aromatics	$RCH=CHR + H_2 \rightarrow RCH_2CH_2R$	Rhodium, cobalt	Very many transition metal complexes catalyze alkene hydrogenation. $RhCl(PPh_3)_3$ is the most studied catalyst; use of chiral ligands can afford high-purity optically active products, e.g., *l*-DOPA.	14
Codimerization of ethylene and butadiene to *trans*-1,4-hexadiene	$CH_2=CH_2$ $+ CH_2=CHCH=CH_2$ $\rightarrow$	Rhodium, nickel	A special case of alkene oligomerization. Oligomerization of alkenes and dienes to form dimers, trimers, etc., is used extensively.	15
Oxidation of *p*-xylene to terephthalic acid	$+ 3O_2 \rightarrow$ ... $+ 2H_2O$	Cobalt, manganese	Original oxidation process used nitric acid at 150–250°C; use of air (or oxygen) in the presence of cobalt and promoters such as bromine is more economical.	16
Propylene epoxidation	$CH_3CH=CH_2 + PhCHCH_3$ (OOH) $\rightarrow CH_3CHCH_2$ (O) $+ PhCHOHCH_3$	Molybdenum	Ethyl benzene is easily converted to its hydroperoxide, and in the presence of molybdenum naphthenate it reacts with propylene at 90°C.	17

Table 1.2. Catalytic Applications of Some Selected Transition Metal Compounds—Illustrative Examples of Laboratory Scale Reactions

Transition metal catalyst	Reaction type	Typical conditions	Comments
$Co(C_5H_5)_2$	$2CH \equiv CH + PhCN$ → [2-Ph pyridine]	150°C, 9 atm., 9 h	Provides an efficient route to bipyridines from cyanopyridines.
$NiCl_2(Ph_2PCH_2CH_2PPh_2)_2$	$ArCl + MgRX$ (or LiR) → $ArR + MgXCl$ (or LiX)	25°C	Activation of arylhalides in reactions with Grignard reagents and lithium alkyls.
$Ni(COD)_2{}^a$	$3CH_2CHCHCH_2$ → cyclododecatriene	20–100°C, 1 atm	Specific product obtained in good yield. Presence of phosphine or phosphite allows controlled formation of other products.
$RhCl(PPh_3)_3$	$RCH=CH_2 + H_2$ → RCH_2CH_3	25°C, 1 atm	Excellent selectivity, insensitive to functional groups. Related catalysts with optically active ligands permit asymmetric hydrogenation of double bonds.
$RhCl_3 \cdot 3H_2O$	[cyclohexenone with pentenyl side chain → 4-pentylphenol]	50–100°C, 1 atm, 2 h	$RhCl_3 \cdot 3H_2O$ in the presence of ethanol is a powerful catalyst for double-bond isomerizations forming conjugated systems.

I–C$_6$H$_4$–CO$_2$CH$_3$ + PhCH=CH$_2$ → *trans*- C$_6$H$_4$(CH=CHPh)(CO$_2$CH$_3$)	Pd(CH$_3$CO$_2$)$_2$[b]	100°C, 2 h	One mole of base (e.g., NEt$_3$) is needed to remove the HI. With bromides it is necessary to use PdCl$_2$(PPh$_3$)$_2$.
(CH$_3$)$_2$CHNO$_2$ + CH$_2$=CHCH=CH$_2$ → (CH$_3$)$_2$CCH$_2$CH$_2$CH=CH$_2$ NO$_2$	PdCl$_2$(PPh$_3$)$_2$	50°C, 1 atm	General reaction of activated hydrogens, catalyzed by a number of palladium compounds.
ArX + CO + ROH → ArCO$_2$R + HX	PdCl$_2$(PPh$_3$)$_2$	60–100°C	One mole of base (e.g., NEt$_3$) needed to remove HX. Replacing the alcohol by an amine leads to formation of amides.

[a] COD, 1,5-cyclo-octadiene.
[b] Pd(CH$_3$CO$_2$)$_2$ is in fact Pd$_3$(CH$_3$CO$_2$)$_6$.

reaction conditions, sensitivity of functional groups, etc., to allow preliminary evaluation of procedures for particular applications. The experimental details that are provided in many examples are helpful in this respect, and extensive references to the original literature may be given so that further information can be obtained when necessary. In most cases the procedures described use transition metal compounds that are either available commercially or are easily prepared. Details of the preparation of many of these transition metal compounds are given in Chapter 9. These laboratory preparations are all relatively straightforward and, in view of the cost of commercial reagents and catalysts, their use can often result in a considerable financial saving.

The use of procedures involving particularly toxic reagents has been avoided wherever possible, although some reactions using volatile metal carbonyls (iron pentacarbonyl and nickel tetracarbonyl) are mentioned. Alternative methods that do not require their use are, however, usually available. The extremely high toxicity of nickel tetracarbonyl (TLV 0.001 ppm) must be fully taken into account by anyone considering its use. *The use of nickel carbonyl must be supervised by a person experienced in handling it.*

This book emphasizes the practical aspects of organic synthesis using transition metals, and it is appropriate that some chapters are concerned with the preparation of a particular class of compound (e.g., the preparation of carbonyl compounds), while others deal with a particular type of reaction (e.g., hydrogenation, isomerization). In this way each chapter has its own distinct character. The cross-references given in the text and the extensive indexes are intended to unify the material and to make easily accessible all of the relevant information that is available on each topic.

Chapter 2

Formation of Carbon–Carbon Bonds

2.1. Formation of Carbon–Carbon Single Bonds

Despite the fundamental importance of carbon–carbon bond formation in organic synthesis, there are only a very limited number of "classical" methods available which proceed selectively, in high yield, under mild conditions. However, transition-metal-mediated addition, substitution, dimerization, oligomerization, and coupling reactions often *do* proceed in this way, and the present chapter is concerned with applications of such chemistry to organic synthesis.

The ability of transition metal reagents and catalysts to promote carbon–carbon bond formation has been recognized for many years. The "Kharasch" reactions, involving copper-catalyzed addition of Grignard reagents to activated alkenes, and cobalt-catalyzed substitution of organic halides by Grignard reagents were, for example, described in 1941, but it is only relatively recently that the full potential of transition metals in this area has been appreciated. The involvement of transition metals in organic reactions, via d-orbital participation, can in fact give access to a wide range of novel reaction pathways leading to carbon–carbon bond formation. Simple alkenes, for example, react readily with electrophiles and are not normally susceptible to nucleophilic attack, but coordination of the alkene to a transition metal inverts this reactivity and allows nucleophilic addition of carbanions to the double bond [Eq. (1)]. Other pathways more or less specific to transition metal chemistry include reductive elimination of two organic fragments from the metal, with C–C bond formation [Eq. (2)], alkene "insertion" into a metal–carbon or metal–hydrogen bond [Eq. (3)], and alkene elimination from a metal center by β-hydrogen abstraction [Eq. (4)].

$$\left[\begin{array}{c} R^{\ominus} \\ \diagdown \\ M \end{array}\right]^{n+} \longrightarrow \left[\begin{array}{c} R \\ \diagup \\ M \end{array}\right]^{(n-1)+} \tag{1}$$

$$R{-}M{-}R \quad \rightarrow \quad [M] + R{-}R \tag{2}$$

$$M{-}R \xrightarrow{\quad} \underset{M-R}{\diagup\!\diagdown} \longrightarrow \underset{R}{\overset{M}{\diagup\!\diagdown}} \tag{3}$$

$$\underset{M}{\diagdown} \!\! \diagup\!\! R \longrightarrow \overset{H}{\underset{M}{|}} + R \!\! \diagup\!\! \diagup \tag{4}$$

These elementary steps can lead to catalysis of known reactions, e.g., alkene dimerization, by providing low-energy pathways, but perhaps more importantly, they can also generate entirely new syntheses such as the catalytic arylation of alkenes described in Section 2.1.4.2.

2.1.1. Preparation of Substituted Alkenes

2.1.1.1. Dimerization of Alkenes

The nonoxidative dimerization of alkenes is catalyzed by a variety of transition metal complexes, zero-valent nickel complexes and rhodium and palladium halides being among the most commonly used. The range of substrates that has been examined tends to be limited to simple alkenes like ethylene and propylene, but of potential commercial interest[1] is the rhodium-catalyzed dimerization of methyl acrylate [Eq. (5)].

$$CH_2{=}CHCO_2Me \xrightarrow[\text{MeOH}]{\text{RhCl}_3} MeO_2CCH{=}CHCH_2CH_2CO_2Me \tag{5}$$

Nickel catalysts convert alkenes into a mixture of dimers, trimers and higher oligomers, the rate being generally first order in nickel, second order in alkene, and decreasing in the order

$$CH_2{=}CH_2 > CH_3CH{=}CH_2 > \text{cycloalkene} > MeCH{=}CHMe$$

The catalyst can be prepared *in situ* from various nickel(II) complexes by reduction in the presence of a Lewis acid, using diethylaluminum ethoxide or ethylaluminum sesquichloride. The postulated active catalyst is a hydridonickel species and the dimerization can be envisaged as a series of sequential alkene insertions into Ni–H and Ni–C bonds followed by β-elimination (Scheme 1).

Scheme 1

$$HNi(C_2H_4) + C_2H_4 \rightarrow CH_3CH_2Ni(C_2H_4)$$

$\uparrow$ $\qquad\qquad$ $\downarrow C_2H_4$

$$HNi(C_2H_4) + CH_3CH_2CH{=}CH_2 \leftarrow CH_3(CH_2)_3Ni(C_2H_4)$$

$$[Ni = Ni(II)X \text{ (phosphine)}_2]$$

In practice isomerization occurs so that a mixture of 1-butene and *cis*- and *trans*-2-butenes is obtained.[2]

The dimerization of propylene produces hexenes, methylpentenes, and 2,3-dimethylbutenes, the relative amounts of which are determined by the regioselectivity of the additions. In the first step the addition of Ni–H is kinetically controlled and favors the *n*-propylnickel derivative which rearranges to the more stable isopropyl nickel species. In the second step the regiospecificity of addition is highly dependent on the structure of the phosphine ligand (Scheme 2).[3]

Scheme 2

CH_3	CH_3
$CH_3{-}CH{-}Ni$	$CH_2{-}CH{-}Ni$
$CH_2{=}CH{-}CH_3$	$CH_3{-}CH{=}CH_2$
Preferred attack in presence of nonbulky ligands	Preferred attack in presence of bulky ligands

The mechanism for the conversion of ethylene to butene in the presence of a rhodium catalyst has been studied.[4] In this scheme an ethylrhodium(III) species, formed by oxidative addition of HCl to a π-ethylenerhodium(I) complex, undergoes a rate-limiting migratory insertion of ethylene into the ethyl–rhodium bond followed by β-elimination of the butyl derivative. 1-Butene formed initially is again rapidly isomerized to 2-butene. The absence of higher oligomers is attributed to butene release and isomerization occurring faster than insertion.

The dimerization of ethylene in the presence of a variety of palladium complexes has been studied and again yields a mixture, consisting mostly of *cis-* and *trans-*2-butenes.[5]

2.1.1.2. Cross-Coupling

A useful method for the synthesis of unsaturated compounds involves the selective cross-coupling of Grignard reagents with vinyl halides in the presence of various transition metal catalysts [Eq. (6)].

$$R-Mg-Br \; + \; \diagdown\!\diagup^{Br} \; \xrightarrow{\text{[catalyst]}} \; \diagdown\!\diagup^{R} \tag{6}$$

The most commonly used catalysts are based on either iron or nickel. Although ferric chloride has been used as the catalyst precursor, tris(dibenzoylmethido)iron(III), Fe(DBM)$_3$, appears to be more effective.[6] The iron-catalyzed reaction is, moreover, stereospecific, *trans*-1-bromopropene producing only *trans*-2-butene in good yield (70%–80%). The reaction even works with secondary and tertiary alkyl magnesium halides.

Preparation of Propenylcyclohexane[6]

To a solution of cyclohexylmagnesium bromide (45 mmol) in tetrahydrofuran (THF) is added Fe(DBM)$_3$ (0.15 mmol) in THF. After stirring for 5 min, 1-bromopropene (12 mmol, 10 ml) is added and the solution cooled in an ice bath to prevent the THF from boiling. After 1 h the mixture is filtered and the dark filtrate reduced by distillation to half its original volume. The residual liquid is then extracted with large amounts of 5% hydrochloric acid and pentane. Separation of the organic layer followed by distillation gives the product in 65% yield.

The Fe(DBM)$_3$ catalyst is prepared as follows: To an aqueous solution of ferric chloride (0.6 g) is added an ethanolic solution of dibenzoylmethane (1.85 g). An immediate reaction produces a red solid which is completely precipitated by the addition of 50% aqueous ammonia. After filtration the solid is washed with water, dried, and recrystallized from benzene–hexane to give the product in 70% yield as red needles, mp 240° with decomposition.

The selective cross-coupling of Grignard reagents with vinyl halides is also catalyzed by various nickel complexes.[7] Typically the reaction is catalyzed by a dihalodiphosphinenickel complex (**1**) and the reaction generally gives a high yield using vinyl chloride.[8]

Preparation of α-Vinylnaphthalene[8]

Vinyl chloride (6 ml) is condensed at −78°C in a 100-ml glass bomb tube containing dichloro[1,2-bis(diphenylphosphino)ethane]nickel(II), [NiCl$_2$(dpe)] (0.18 g, 0.35 mmol) and ether (10 ml). To this is added, at the same temperature, a solution of α-naphthylmagnesium bromide (55 mmol) in 50 ml of a mixture of ether:benzene:THF (2:3:1). The resulting homogeneous solution, after standing at room temperature for 20 h, is hydrolyzed with dilute hydrochloric acid. The organic layer plus the ether extracts of the aqueous layer are combined, washed with water, dried over calcium chloride, and concentrated *in vacuo*. Distillation of the residue under reduced pressure gives the product 6.8 g, 80% yield (based on the Grignard reagent), bp 89°C (5 mm Hg).

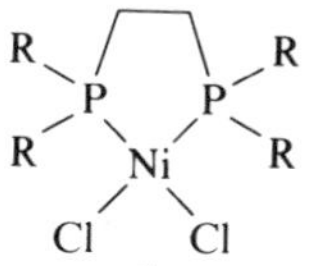

1

An extension of the nickel-catalyzed reaction uses a chiral phosphine–nickel complex, for example, Ni[(−)DIOP]Cl$_2$, for the asymmetric cross-coupling of secondary alkyl Grignard reagents with vinyl halides [Eq. (7)].[9]

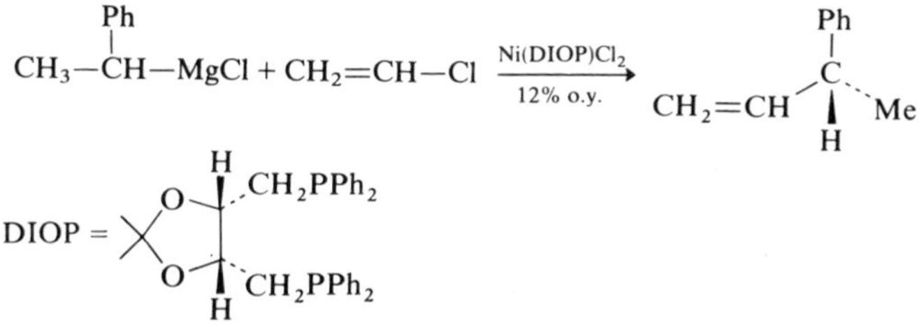

$$\text{(7)}$$

Using the more complex chiral ligands based on ferrocenyl phosphines, much higher optical yields (o.y.) have been achieved.[10] For example, in the nickel-catalyzed cross-coupling of PhCHMeMgCl with vinyl bromide, optical yields of up to 63% have been obtained. The optical yield is very dependent on the structure of the ligand, the presence of a polar NMe$_2$ group being essential for high optical yields. With the ligand EPPF (**2**; X = H) the optical yield was only 4%, whereas using the ligand FePN (**2**; X = NMe$_2$) the optical yield is around 60%. This striking increase in optical

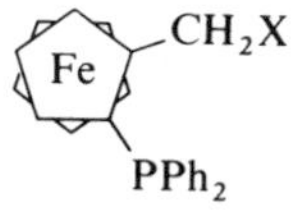

2

yield on going from the simple to the polar functional group-substituted phosphine is presumably related to the extra rigidity imposed on the transition state by coordination of the NMe_2 group to the magnesium atom of the Grignard reagent during halogen/alkyl exchange at nickel. This reaction is remarkable in that it requires both kinetic resolution and rapid racemization of the Grignard reagent by the chiral nickel catalyst in order to achieve the high chemical and optical yields.

A recent extension of the nickel-catalyzed cross-coupling reaction involves the formation of substituted alkenes from Grignard reagents and silyl enol ethers (**3**).[11] The readily available complex nickel(II) acetylacetonate is the most active catalyst, the reaction [Eq. (8)] going to completion

$$
\begin{array}{c}
R^1 \\ \diagdown \\ \quad C{=}C \quad \\ R^2 \diagup \diagdown OSiMe_3 \\ \mathbf{3}
\end{array}
\quad + R^4MgX \xrightarrow{[Ni]}
\begin{array}{c}
R^1 R^3 \\ \diagdown \diagup \\ \quad C{=}C \quad \\ R^2 \diagup \diagdown R^4
\end{array}
\qquad (8)
$$

in 3–6 h in refluxing ether. By using $NiCl_2(PPh_3)_2$ as the catalyst, the cross-coupling proceeds in a regio- and stereospecific manner, but the catalytic activity is lower and a higher boiling solvent is required.

Another group of transition metal complexes which are useful for the synthesis of substituted alkenes are the π-allylnickel(II) halides.[12] These are valuable reagents since they can be prepared by a number of methods and are easily purified and stored for several weeks in the absence of oxygen. Yields of 75%–90% can be obtained by heating allyl halides with nickel carbonyl in benzene, but a more easily handled synthesis in the laboratory is the reaction of bis(1,5-cyclooctadiene)–nickel with allyl halides at −10°C. A third method, but one of less value, is the reaction of bis(π-allyl)nickel(II) with hydrobromic acid. In polar coordinating solvents these complexes react with a range of organic halides to produce substituted alkenes [Eq. (9)].[13] The reaction proceeds equally well for aryl, vinyl, or

$$
R^1{-}\!\!\left\langle\!\!\diagup{-}Ni \begin{array}{c} X \\ \diagup \diagdown \\ \diagdown \diagup \\ X \end{array} Ni \diagdown\!\!\right\rangle{-}R^1 + 2R^2X \longrightarrow 2R^2\diagup\!\!\diagdown\!\!\diagup_{R^1}^{\diagup}
\qquad (9)
$$

alkyl halides and in the presence of hydroxy, ester, and other common functional groups. For example, the complex **4** reacts with 1-iodo-3-chloropropane [Eq. (10)] to give **5**.[14]

$$
EtO_2C{-}\!\!\left\langle\!\!\diagup{-}Ni \begin{array}{c} Br \\ \diagup \\ \diagdown \\ )_2 \end{array}\right. + I(CH_2)_3Cl \longrightarrow
\begin{array}{c} CO_2Et \\ \diagup\diagdown\diagup\diagdown\diagup\diagdown Cl \end{array}
\qquad (10)
$$

$$
\mathbf{4} \qquad\qquad\qquad\qquad\qquad\qquad \mathbf{5}
$$

Preparation of 4-(2-methylallyl)cyclohexanol[12a]

π-(2-methylallyl)nickel bromide (0.55 g, 1.42 mmol) is weighed under nitrogen into a flask equipped with a three-way stopcock and rubber serum stopper. Purified dimethyl formamide (DMF) (4.0 ml) is added via a syringe, followed by a solution of *trans*-4-iodocyclohexanol (0.643 g, 2.84 mmol) in DMF (2.0 ml). After addition the solution is stirred at 23°C for 22 h under a positive pressure of nitrogen. The green reaction mixture is poured into ether and the organic layer washed with water (four times) and dried over magnesium sulfate. Removal of the solvent gives the product (388 mg, 89%) as a mixture of epimers (38:62).

2.1.1.3. From Alkynes

Hydrometallation of alkynes [Eq. (11)] produces intermediates of the type **6** which are particularly useful for the synthesis of substituted alkenes and dienes.

$$R^1C\equiv CR^2 \xrightarrow{\text{MH}} \begin{matrix} R^1 \\ \diagdown \\ H \diagup \end{matrix} C=C \begin{matrix} R^2 \\ \diagup \\ \diagdown M \end{matrix} \tag{11}$$

6

The *trans*-alkenylaluminum compounds (**7**), for example, which are readily prepared by the addition of aluminum hydrides to alkynes, react with aryl bromides or iodides in the presence of a nickel catalyst to give the corresponding *trans*-arylalkene (**8**) in high yield.[15] The reaction [Eq. (12)] proceeds under mild conditions and is highly stereoselective giving 99% of the *trans*-product.

$$R^1C\equiv CH \xrightarrow{\text{HAlR}_2^2} \begin{matrix} R^1 \\ \diagdown \\ H \diagup \end{matrix} C=C \begin{matrix} H \\ \diagup \\ \diagdown AlR_2^2 \end{matrix} \xrightarrow[\text{Ni(PPh}_3)_4]{\text{ArX}} \begin{matrix} R^1 \\ \diagdown \\ H \diagup \end{matrix} C=C \begin{matrix} H \\ \diagup \\ \diagdown Ar \end{matrix} \tag{12}$$

7 **8**

Preparation of trans-Hex-1-enylnaphthalene[15]

To Ni(PPh$_3$)$_4$, prepared by the reaction of anhydrous Ni(acac)$_2$ (0.25 mmol) with HAlBu$_2^i$ (0.25 mmol) in the presence of PPh$_3$ (1.0 mmol) in THF (5 ml), is added *trans*-hex-1-enyldi-isobutylalane (10 mmol) in hexane (10 ml) followed by 1-bromonaphthalene (5 mmol) at 25°C. After stirring for 3 h the mixture is quenched with 3 M hydrochloric acid. Isolation of the product in the normal way gives *trans*-hex-1-enyl naphthalene in 93% yield.

Hydroalumination is, however, not without complication and does not only produce the alkenylalane (**7**), but by substitution also produces the

alkenylalanate in varying amounts. Moreover, hydroalumination of substituted alkynes can sometimes prove difficult and, as an alternative, hydrozirconation of alkynes [Eq. (13)] using Cp_2ZrHCl has been studied.[16] The

$$R^1C\equiv CR^2 \xrightarrow{\eta^5\text{-}(C_5H_5)_2ZrCl(H)} \underset{\underset{\textbf{9}}{H}\,\,\,\,\,\,ZrCp_2Cl}{\overset{R^1\,\,\,\,\,\,R^2}{\diagup\!\!\!\diagdown}} \qquad (13)$$

reaction proceeds in high yield with highly stereospecific *cis*-addition of
the metal hydride to the alkyne, and with good regioselectivity when
unsymmetrically substituted alkynes are used.[17] The metal becomes
attached to the carbon atom bearing the least bulky group. These alkenylzir-
conium compounds (**9**) couple with aryl halides in the presence of nickel(0)
catalyst to give the corresponding *trans*-aryl alkenes in a similar fashion
to the reaction of alkenylalanes.[18] The complex **9** will also undergo conjugate addition to α,β-enones in the presence of $Ni(acac)_2$ at 0°C.[16] High
yields of the conjugate addition product **10** are obtained with no isomerization [Eq. (14)]. A better catalyst for the reaction can be obtained by
reducing the nickel compound prior to the reaction by addition of 1
equivalent of diisobutyl aluminum hydride.

$$\textbf{9} + \underset{R^5}{\overset{O}{\underset{\diagdown}{R^4\diagdown\!\!\!/\!\!\!\diagup}}} \xrightarrow{Ni(II)} \underset{\textbf{10}}{R^4\diagdown\!\!\!/\!\!\!\diagup\overset{O}{\diagdown}\diagup\overset{R^5}{\diagdown}\!\!\!/\!\!\!\underset{R^2}{\overset{}{\diagup}}\!\!=\!\!\diagdown R^1} \qquad (14)$$

Preparation of 3-(trans)-4,4-Dimethylbuten-1-yl) Cyclohexanone
(a) Chlorobis(η^5-cyclopentadienyl)-3,3-dimethylbutyl zirconium: To 3.96 g
     (15.4 mol) of the zirconium hydridochloride suspended in benzene
     (25 ml) is added 3,3-dimethylbutene (2 ml; 15.4 mmol). After stirring
     for several hours the reaction mixture is filtered, and evaporation of the
     solvent gives the product as yellow crystals (88% yield).
(b) Conjugate addition: The above complex (1.92 g, 5.67 mmol) and 2-
     cyclohexen-1-one (0.623 g, 6.48 mmol) are dissolved in THF (30 ml)
     and cooled to 0°C. $Ni(acac)_2$ (0.151 g, 0.59 mmol) is added and after
     stirring at 0°C for 6.5 h the mixture is allowed to warm to room tem-
     perature. Saturated aqueous ammonium chloride is added, and the
     mixture extracted with ether. After washing with saturated aqueous
     sodium bicarbonate solution and brine, the ethereal layer is dried
     (Na_2SO_4) and distilled to give the product, which is purified by prepara-
     tive liquid chromatography (73% yield).

A related series of alkenyl metal complexes (**11**) is formed in high
yield by reaction of alkynes with organoalanes [Me_3Al, for example, Eq.

the nucleophile generally attacks the carbon exocyclic to the ring, but with six-membered rings variation of the ligands around the palladium can affect the position of attack. Use of tri-*o*-tolylphosphine, for example, produces predominant attack at the endocyclic position.[28] Moreover, the use of optically active phosphine ligands can lead to optically active allylic alkylation products, albeit in low optical yields.[34]

Scheme 4

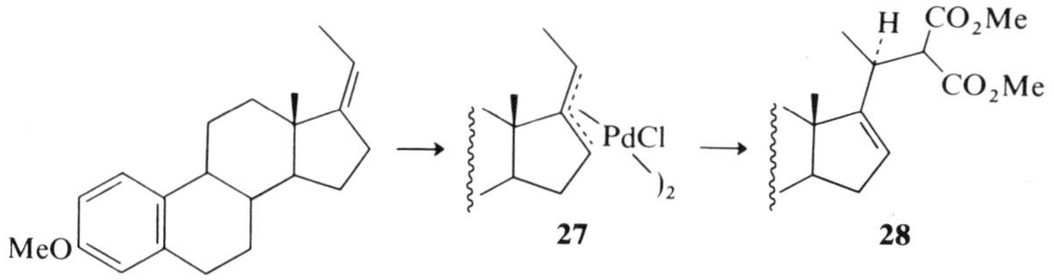

26 → **27** → **28**

Preparation of Bis[chloro-(16,17,20-η³-3-methoxy-19-norpregna-1,3,5(10), 17(20)-tetraene)palladium(II)] (27)[31]

A mixture of cupric chloride dihydrate (2.66 g, 14.4 mmol) PdCl$_2$ (0.75 g, 4.25 mmol), sodium acetate (2.24 g, 27.3 mmol), sodium chloride (1.70 g, 29.0 mmol) and acetic anhydride (2 ml) in glacial acetic acid (32 ml) is heated for 3 h at 100°C. The reaction mixture is cooled to 60°C and the ethylidene compound **26** (2.0 g, 6.78 mmol) added. After heating at 70°C for 60 h the mixture is filtered and partitioned between ether and water. The ethereal layer is washed with saturated aqueous sodium carbonate, water, and dried over MgSO$_4$. Removal of the solvent *in vacuo* gives the crude product (2.64 g) which is purified by column chromatography (silica/chloroform/hexane) to give the yellow complex (**27**) (1.25 g, 67%).

Alkylation of the Complex 27 with Malonate

To sodium hydride (27.7 mg, 1.16 mmol) in THF (4 ml) is added dimethylmalonate (15.2 mg, 1.16 mmol) and the mixture stirred for 45 min at room temperature. The palladium complex **27** (202 mg, 0.323 mmol) and 1,2-bis(diphenylphosphino) ethane (184.7 mg, 0.46 mmol) are stirred in THF (4 ml) for 45 min at room temperature and this solution is then added to the former and the mixture stirred at room temperature for 43 h. The reaction mixture is partitioned between ether and water, extracted with ether (4 × 3 ml), and then dried over Na$_2$SO$_4$. After removal of the solvent *in vacuo* the product (**28**) is isolated by preparative TLC (3:1 hexane/ethyl acetate) as a yellow oil (160 mg, 81%).

Although the stoichiometric allylic reactions already described are useful, and the palladium can be recovered and recycled, it would be more desirable to have processes which were catalytic in palladium. One successful approach to this problem is outlined in Scheme 5. In this sequence the

Scheme 5

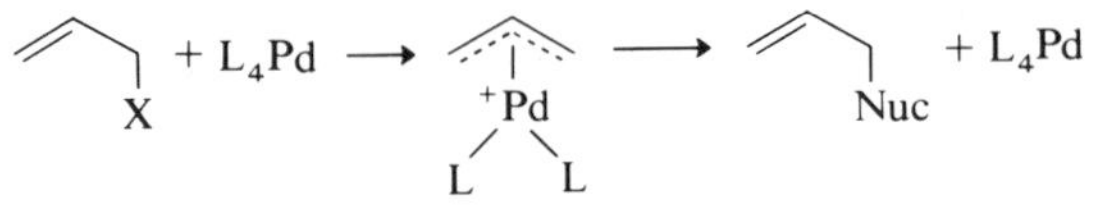

allylic position is activated by a potential leaving group and palladium(0) complexes initiate ionization to form the same intermediate as in the stoichiometric process. Nucleophilic attack then occurs to give the allylic alkylation product and regenerate the palladium catalyst. Typical catalysts include tetrakis-(triphenyl phosphine)palladium and bis[1,2-bis(diphenyl-phosphino)ethane]palladium. The catalyst can also be generated by *in situ* reduction of palladium(II) salts in the presence of phosphine ligands but this method often gives lower yields.[28]

In displacement reactions stereochemical inversion of configuration is usual, but this palladium-assisted alkylation proceeds with net retention of configuration, for example, the conversion of **29** to **30** [28] [Eq. (24)]. Other leaving groups, such as acyl ether or hydroxy, have been studied, but generally acetoxy is preferred.[33]

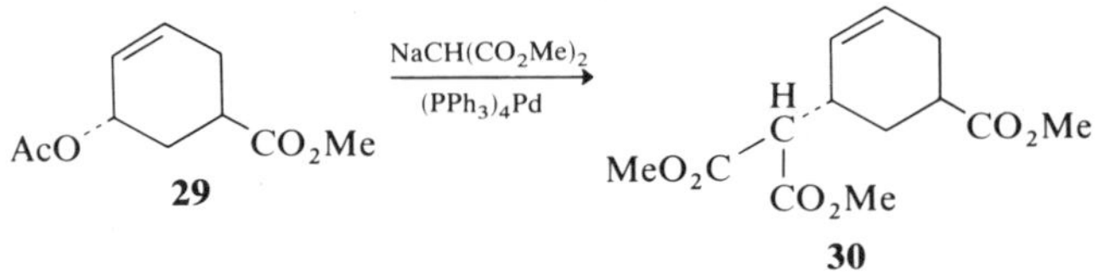

$$(24)$$

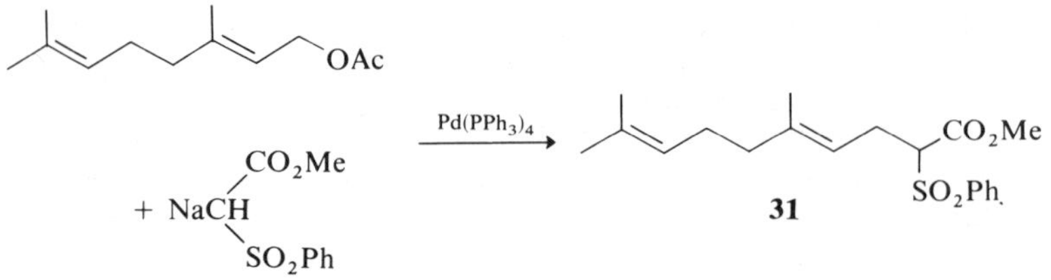

$$(25)$$

Preparation of 5,9-Dimethyl-2-phenylsulfonyldeca-(E)-4,8-diene (**31**[32])
[*Eq.* (25)]

A mixture of geranyl acetate (230 mg, 1.27 mmol), triphenyl phosphine (30.4 mg, 0.17 mmol) and Pd(Ph₃P)₄ (48 mg, 0.04 mmol), in dry THF (2 ml) is stirred at room temperature for 15 min. A solution of the sodium salt of methyl phenylsulfonylacetate in THF (8 ml), generated from methyl phenyl-sulfonylacetate (948 mg, 4.42 mmol) and sodium hydride (168.5 mg of 57% mineral oil dispersion, 4.0 mmol), is added and the mixture refluxed for 36 h. The reaction mixture is partitioned between ether and water and the aqueous layer further extracted with ether. The combined ethereal extracts are dried and evaporated *in vacuo* to give an oil which is chromatographed on silica gel (hexane/ethyl acetate) to give the product **31** (345 mg, 84%).

Using a chiral-phosphine palladium catalyst, asymmetric induction can be achieved in this type of alkylic alkylation.[34] The allylic acetate **32**, for example, undergoes asymmetric alkylation with optical yields in the range 20%–46%, the higher optical yields being obtained with bulkier nucleophiles, for example, the sodium salt of methyl phenylsulfonyl acetate [Eq. (26)].

$$ \text{optical yield:} \quad 20\%\text{--}46\% \tag{26} $$

32

2.1.2. Dienes and Polyenes

The methods that will be discussed for the synthesis of dienes and polyenes closely resemble those discussed earlier for the synthesis of sub-stituted monoenes, so as far as possible the topic will be split into the same sections and, where appropriate, reference can then be made back to the relevant section.

2.1.2.1. Dimerization, Oligomerization, and Telomerization of Dienes

Reaction of conjugated 1,3-dienes, particularly butadiene and isoprene, to form linear and cyclic oligomers and telomers in the presence of a variety of transition metal complexes has been a very active area of research over the last 20 years. Three types of reaction can be broadly defined as (a) polymerization, (b) cyclization, and (c) linear oligomerization or telomerization. Polybutadiene and polyisoprene are produced and used mainly as synthetic rubber on an industrial scale, usually using titanium or nickel catalysts. Cyclization reactions will be discussed in Chapter 3, so this section will concentrate on linear oligomerization and telomerization.

Unlike nickel catalysts, which generally form cyclic dimers and trimers, palladium compounds catalyze linear dimerization of conjugated dienes, 1,3-butadiene, for example, being converted into 1,3,7-octatriene in the presence of catalytic amounts of $[(Ph_3P)_2Pd(maleic anhydride)_2]$.[35] A variety of other metal complexes have been studied[12b] and by a careful choice of catalyst very high selectivity can be obtained. Isoprene, for example, in the presence of a (1,4-diaza-1,3-diene)chromium complex (**33**) dimerizes in a "tail to tail" fashion to give (E)-2,7-dimethylocta-2,4,6-triene (**34**) in 80% yield[36] [Eq. (27)].

$$\left(\begin{array}{c} \overset{R}{\underset{\|}{N}} \\ \overset{\|}{\underset{N}{N}} \\ \overset{|}{R} \end{array} \right)_2 Cr \qquad R = (Me_2CH)_2CH-$$

33

(27)

34

Trimerization of 1,3-dienes can also be observed as the major reaction pathway by a careful choice of catalyst. Above 50°C butadiene is catalytically trimerized by π-allylpalladium acetate to give 1,3,6,10-dodecatetraene (**35**) with 77% selectivity[37] [Eq. (28)]. A small amount of 1,3,7-octatriene is also formed and this becomes the major product when triphenylphosphine is added to the reaction (Pd/PPh$_3$ = 1).

(28)

35

Preparation of 1,3,6,10-Dodecatetraene (**35**)[37]

A 200-ml autoclave is charged with π-allylpalladium acetate (2.10 g, 5 mmol), benzene (60 ml), and butadiene (30.6 g, 0.565 mol) and then heated at 50°C for 22 h. After venting off the excess butadiene, thd autoclave is discharged and the palladium metal which has formed is removed by filtration. Removal of the solvent *in vacuo* gives a residue (15.7 g) from which $C_{12}H_{18}Pd_2(OAc)_2$ (0.41 g) is precipitated by addition of pentane. After evaporation of the pentane *in vacuo* the product (**35**) (7.1 g) is obtained by fractional distillation at reduced pressure (bp 88–94°C, 15 mm Hg).

Although the simple dimerization and oligomerization of dienes is interesting, a more useful reaction to the synthetic organic chemist is the telomerization of conjugated dienes. It is well known that simple alkenes, in the presence of palladium(II) compounds, undergo either nucleophilic addition or substitution, and examples of this can be found in other parts of this book. Similarly, nucleophiles react with butadiene to form dimeric telomers in which the nucleophile is introduced mainly at the terminal position to form 8-substituted 1,6-octadiene (**36**), with 3-substituted 1,7-octadiene (**37**) being produced as a minor product [Eq. (29)]. The range

$$2 \diagup\!\!\!\diagdown\!\!\!\diagup\!\!\!\diagdown \ + \ YH \ \xrightarrow{\text{[Pd]}} \ \diagup\!\!\!\diagdown\!\!\!\diagup\!\!\!\diagdown\!\!\!\diagup\!\!\!\diagdown\!\!\!\diagup Y \quad \text{(Major)}$$

36

+ \hfill (29)

$$\diagup\!\!\!\diagdown\!\!\!\diagup\!\!\!\diagdown\!\!\!\overset{\overset{\displaystyle Y}{|}}{\diagup}\!\!\!\diagdown \quad \text{(Minor)}$$

37

of nucleophiles which will undergo this type of reaction includes water, carboxylic acids, primary and secondary alcohols, phenols, ammonia, primary and secondary amines, enamines, active methylene compounds, and nitroalkanes. Several excellent reviews[38] have been written on this topic and hence a few examples will suffice to illustrate the scope of the reaction.

Phenol reacts readily with butadiene to give octadienyl phenyl ether (**36**; Y = OPh) in high yield, with the branched isomer (**37**; Y = OPh) a minor product. A variety of palladium species have been used to catalyze this reaction, the best of which appears to be palladium chloride–sodium phenoxide.[39] Similarly primary alcohols react readily to form ethers. The reaction of acetic acid with butadiene has been extensively studied,[38b] not least because it appears a promising industrial route to *n*-octanol. Initially a complex mixture of products was obtained but, by varying the conditions, high selectivity to 8-acetoxy-1,6-octadiene (**36**; Y = OAc)[40] was achieved. Optimum conditions include the incorporation of phosphites into the catalytic system and using molar amounts of amine.

Compounds with methylene groups attached to two electronegative groups, such as carbonyl, alkoxycarbonyl, cyano, nitro, or sulfonyl, react smoothly with butadiene, the acidic hydrogens being replaced by the 2,7-octadienyl group to give mono- and disubstituted products. Selectivity to the monosubstituted product can be obtained by using a palladium(II) salt in the presence of a bidentate phosphine ligand.[41]

In general, methylene groups activated by only one electron-withdrawing group are not reactive, the exceptions being nitroalkanes which react smoothly with butadiene. Careful selection of reaction conditions, however, is essential to achieve high yields of the desired nitroalkene. Simple oligomerization of butadiene can be the predominant reaction but by using a solvent such as isopropanol and a palladium phosphine catalyst in the presence of base, high yields of the 1,6-octadienyl substituted compound (**38**) can be obtained[42] [Eq. (30)].

$$\text{(30)}$$

38

Preparation of 9-methyl-9-nitro-1,6-decadiene[42]

$PdCl_2(PPh_3)_2$ (0.42 g), KOH (1.68 g), 2-nitropropane (26.7 g), butadiene (35 g), and isopropanol (150 ml) are heated in a 500 ml autoclave at 50°C for 4 h. After venting off the excess butadiene the reaction mixture is poured into ether and the organic layer extracted thoroughly with water. After drying and removal of the ether *in vacuo* the residue is distilled to give 1,3,7-octatriene (1.9 g) and 9-methyl-9-nitro-1,6-decadiene (49 g).

2.1.2.2. Cross-Coupling

The direct coupling of two unlike alkenyl groups by reaction of an alkenyl metal derivative with an alkenyl halide is generally difficult. In the presence of a catalytic amount of $Pd(PPh_3)_4$, however, alkenyl iodides react stereospecifically with Grignard reagents of the type **39** to produce the diene **40** in high yield (~80%) and under very mild conditions[43] [Eq. (31)].

$$\text{(31)}$$

39   **40**

As in the synthesis of substituted alkenes, π-allylnickel halides have proved to be useful intermediates. As early as 1951 it was observed[44] that 1-chloro-2-butene or 3-chloro-1-butene, in the presence of $Ni(CO)_4$, dimerized to produce the 1,5-dienes **41** and **42** in high yield [Eq. (32)]. Corey[45] demonstrated that these allylic coupling reactions take place via π-allylnickel(II) halides. The complex **43** is stable at 25°C for several days

$$\begin{array}{ccc}
\text{ClCH=CHMe} & & \text{MeCH=CHCH}_2\text{CH}_2\text{CH=CHMe} \\
\text{or} & \xrightarrow{\text{Ni(CO)}_4} & \text{or} \qquad \textbf{41} \\
\text{CH}_2\text{=CHCHClMe} & & \text{CH}_2\text{=CHCH(Me)CH}_2\text{CH=CHMe} \\
& & \textbf{42}
\end{array} \qquad \text{(32)}$$

but on addition of allyl bromide it is quantitatively converted to biallyl in a few minutes [Eq. (33)].

$$\tag{33}$$

43

The reactions of π-allylnickel halides with allyl bromides have been used for the synthesis of a number of terpenoid compounds. Geranyl acetate (**44**) and farnesyl acetate (**45**) have been obtained from the nickel complex **46** and the appropriate allylic halide.[46] Similarly the complex **47** and the bromide **48** react to form geranyl acetate (**44**)[47] [Eqs. (34)–(36)].

$$\tag{34}$$

46   **44**

$$\tag{35}$$

46 +   **45**

$$\tag{36}$$

47   **48**

One problem with this type of reaction of course is the tendency for homocoupling rather than cross-coupling between the two different allylic moieties. It has been observed that electron-withdrawing groups on the coordinated allyl group favor cross-coupling[48] and a change in the leaving group from halide to dithiocarbamate can also be used to advantage.[49]

Preparation of Geranyl Acetate (**44**)[47]

A solution of 1,1-dimethyl-π-allyl nickel bromide, prepared from prenyl bromide (7.5 g, 50 mmol) and nickel carbonyl (12 g, 70 mmol) (CAUTION!) in benzene (80 ml), is mixed with 4-acetoxy-1-bromo-2-methyl-2-butene (6.2 g, 30 mmol) in benzene (100 ml) and the mixture stirred for 50°C for 6 h. After cooling, the solution is poured into ice-water, acidified with hydrochloric acid and the aqueous layer extracted with ether. The combined organic layer is washed with water, dried (MgSO₄), and fractionally distilled to give geranyl acetate (**44**) (3.5 g, 60%), bp 92–94°C (3.5 mm Hg).

2.1.2.3. From Alkynes

Trans-alkenyl-aluminum compounds (**49**), which are readily prepared by the addition of aluminum hydrides to alkynes as described earlier, react with alkenyl halides [Eqs. (37) and (38)] to produce the conjugated (E, E)

(37)

(38)

or (E, Z) dienes **50** and **51** in good yield in the presence of palladium or nickel catalysts.[50] The palladium-catalyzed reaction in each case is highly stereospecific (97%) and produces only very small amounts (2%) of the homocoupled product. Stereospecificity using nickel catalysts is slightly lower and although only a limited amount of work has been reported on functionalized alkenes, ester groups can be tolerated.

(E)-Alkenyl zirconium complexes, prepared by the reaction of alkynes with [H(Cl)ZrCp$_2$], react with alkenyl halides [Eq. (39)] in the presence of

(39)

a palladium catalyst to form conjugated dienes of high isomeric purity (97%) and in good yield.[51] The complexes (**52**) are as good as the corresponding alkenylalanes in terms of yield and stereoselectivity and, moreover, the presence of certain oxygen functionalities, for example, ether and ketone, can be tolerated in the alkenyl halide.

Related alkyne-derived intermediates include the dialkenylchloro-boranes (**53**), which undergo cross-coupling with organic halides in the presence of methyl copper[22] [Eq. (40)], and (E)-alkenyl pentafluorosilicates (**54**), which undergo cross-coupling in the presence of a palladium catalyst.[52] The latter reagents are readily prepared from alkynes by platinum-catalyzed hydrosilylation followed by treatment with KF. The cross-coupling takes

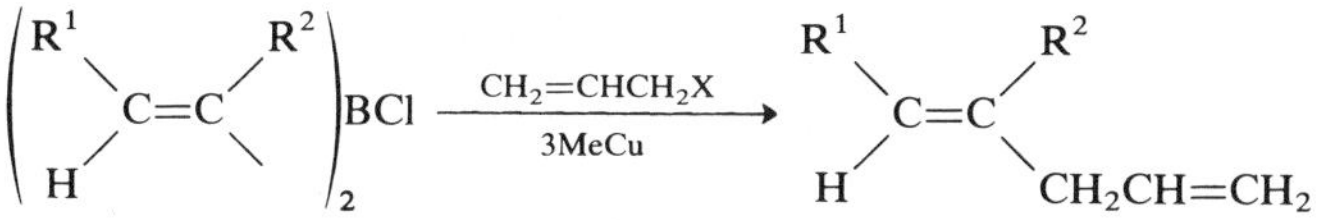

$$\left(\begin{array}{c} R^1 \\ \diagdown \\ C=C \\ \diagup \quad \diagdown \\ H \end{array} \begin{array}{c} R^2 \\ \diagup \\ \diagdown \end{array} BCl\right)_2 \xrightarrow[\text{3MeCu}]{CH_2=CHCH_2X} \begin{array}{c} R^1 \qquad R^2 \\ \diagdown \quad \diagup \\ C=C \\ \diagup \quad \diagdown \\ H \qquad CH_2CH=CH_2 \end{array} \qquad (40)$$

53

place in the presence of $Pd(OAc)_2$ and the reaction (Scheme 6) can tolerate certain functional groups, for example esters, which are incompatible with hydroalumination.

Scheme 6

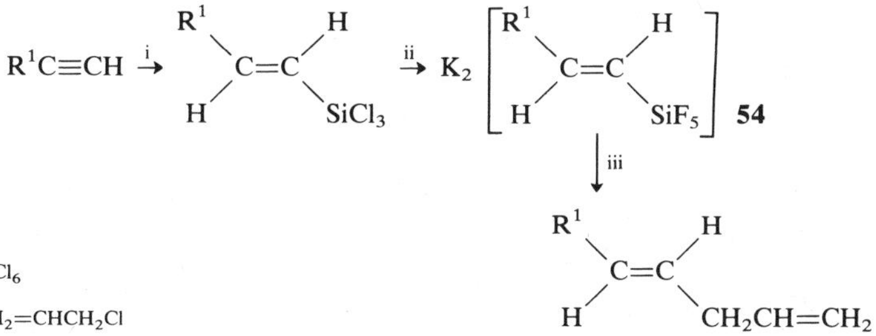

$$R^1C\equiv CH \xrightarrow{i} \begin{array}{c} R^1 \qquad H \\ \diagdown \quad \diagup \\ C=C \\ \diagup \quad \diagdown \\ H \qquad SiCl_3 \end{array} \xrightarrow{ii} K_2 \left[\begin{array}{c} R^1 \qquad H \\ \diagdown \quad \diagup \\ C=C \\ \diagup \quad \diagdown \\ H \qquad SiF_5 \end{array}\right] \ \mathbf{54}$$

$$\downarrow iii$$

$$\begin{array}{c} R^1 \qquad H \\ \diagdown \quad \diagup \\ C=C \\ \diagup \quad \diagdown \\ H \qquad CH_2CH=CH_2 \end{array}$$

(i) $HSiCl_3/H_2PtCl_6$
(ii) KF
(iii) $Pd(OAc)_2/CH_2=CHCH_2Cl$

A simple one-step procedure for the synthesis of 1,4-pentadienyl halides (**55**) involves the addition reaction [Eq. (41)] of allyl halides to alkynes in the presence of $PdCl_2(PhCN)_2$.[53] 1-Hexyne, for example, reacts with allyl chloride to give 5-chloro-1,4-nonadiene in almost quantitative yield.

Preparation of 5-Chloro-1,4-nonadiene[53]

To a mixture of $PdBr_2(PhCN)_2$ (1.81 g, 4 mmol) and allyl bromide (80 ml) is added, dropwise, 1-hexyne (6.65 g, 80 mmol) with stirring over 1 h, and the resulting mixture is stirred for a further 1 h. After removal of the excess allyl bromide *in vacuo* the residue is chromatographed on an alumina column (80 g) with petroleum ether to give the product (12 g), bp 53°C (3.5 mm Hg).

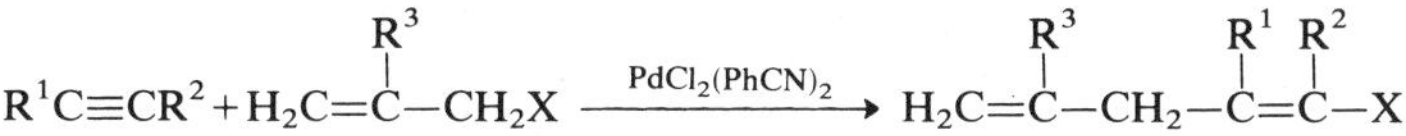

$$R^1C\equiv CR^2 + H_2C=\overset{\displaystyle R^3}{\overset{\displaystyle |}{C}}-CH_2X \xrightarrow{PdCl_2(PhCN)_2} H_2C=\overset{\displaystyle R^3}{\overset{\displaystyle |}{C}}-CH_2-\overset{\displaystyle R^1}{\overset{\displaystyle |}{C}}=\overset{\displaystyle R^2}{\overset{\displaystyle |}{C}}-X \qquad (41)$$

55

A further coupling reaction [Eq. (42)] which leads to the formation of the conjugated enyne (**57**) is brought about by reaction of the alkynyl

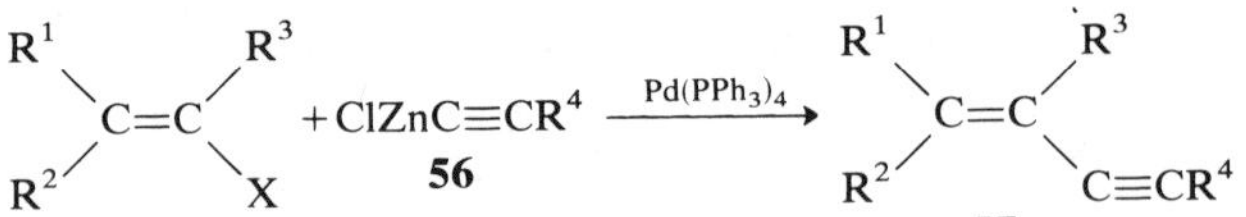

$$R^4C\equiv CZnCl \quad 56 \qquad \xrightarrow{\text{Pd(PPh}_3)_4}$$

$$\tag{42}$$

zinc reagent (**56**) with alkenyl iodides or bromides in the presence of catalytic amounts of Pd(PPh$_3$)$_4$.[54] The reagent **56** is readily obtained from the corresponding alkynyl–lithium and zinc chloride and the reaction gives about an 80% yield with 97% stereoselectivity. This reaction generally goes under very mild conditions and appears to be specific to palladium, Ni(PPh$_3$)$_4$ giving only very low yields. The reaction is applicable to the synthesis of both internal and terminal conjugated enynes, a unit which appears in a number of natural products.

2.1.3. Substituted Alkynes

An interesting use of a coordinated ligand in organic synthesis is demonstrated by the reaction of the cobalt-stabilized propargyl cation (**58**) (Scheme 7). This complex, which is generated *in situ* by protonation of the corresponding hydroxy complex selectively monoalkylates–dicarbonyl systems to give the complex (**59**) in good yield (65%–95%).[55] Demetallation with ferric nitrate then releases the organic ligand in virtually quantitative yield. Furthermore the cation **58** readily and selectively α-alkylates ketones

Scheme 7

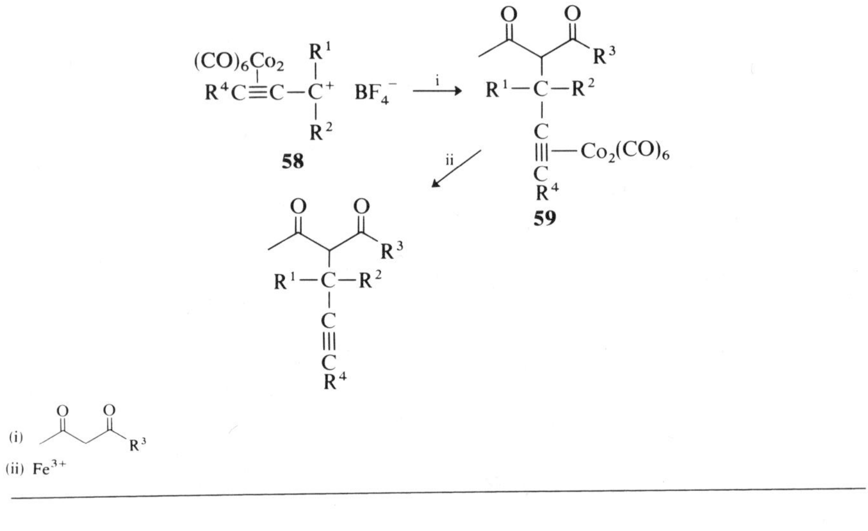

(i)

(ii) Fe^{3+}

and ketone derivatives, including enol ethers and enol acetates, to produce keto-alkynes **60** [55a] [Eq. (43)].

$$\mathbf{58} + \text{(structure)} \xrightarrow{\text{(ii) } Fe^3} \text{(structure)} \quad \mathbf{60} \tag{43}$$

Preparation of 5-Undecyn-2-one (**60**; R^1, R^2, R^3 = H; R^4 = C_5H_{11})[55]

$Co_2(CO)_8$ (12 g, 35 mmol) is added to a stirred solution of 2-octyn-1-ol (3.2 g, 25 mmol) in benzene (50 ml) at room temperature under N_2. After stirring for 2 h the dark red solution is filtered through a column of Celite and alumina under N_2 and removal of the solvent gives the hydroxy-alkyne complex as a red oil (9.0 g, 86%). This complex (2.3 g, 5.6 mmol) in propionic anhydride (2 ml) is treated with 40% aqueous HBF_4 (4.0 ml, 23 mmol) at $-45°C$ under N_2 and then stirred at this temperature for 10 min. Trituration of the mixture with ether (3 × 100 ml) gives a residual red oil which is dissolved in dry CH_2Cl_2 (20 ml) containing benzene (2 ml). Isopropenyl acetate (4.0 ml, 34 mmol) is added to this solution at 0°C and stirring continued for a further 15 min. The solution, after treatment with excess $NaHCO_3$ and $MgSO_4$ and filtration through Celite, is evaporated *in vacuo* to give a dark red oil (2.5 g, 98%). A solution of this complex (1.9 g, 4.2 mmol) in 95% ethanol (20 ml) is cooled to 0°C and excess ferric nitrate added with stirring over 1 h. After CO evolution ceases the mixture is stirred for a further 2 h then poured into water (4.00 ml) and extracted with ether (4 × 50 ml). The combined ether extracts give, after washing with water, drying over $MgSO_4$, and removal of the solvent *in vacuo*, the product as an oil (0.65 g, 94%).

Terminal and internal arylalkynes have been synthesized[56] by the palladium-catalyzed reaction of alkynyl-zinc reagents with aryl halides [Eq. (44)]. The reaction is complete within minutes at room temperature using

$$RC{\equiv}CZnCl + ArX \xrightarrow{Pd(PPh_3)_4} RC{\equiv}CAr \tag{44}$$

aryl iodides or activated aryl bromides, and yields are in the region of 80%. The alkynyl-zinc reagent is claimed to be superior to the corresponding Grignard or organoalkali metal reagent and the most convenient catalyst appears to be $Pd(PPh_3)_4$. $Ni(PPh_3)_4$ leads to the same reaction but yields are much lower.

Alkylated alkynes can be synthesized in good yield from 1-bromoalk-1-ynes and trialkylalanes in the presence of catalytic amounts of the nickel complex, bis(*N*-methyl-salicylaldimino)nickel, $[Ni(mesal)_2]$[57] [Eq. (45)]. In

$$R_3^2Al + R^1C{\equiv}CBr \xrightarrow{[Ni(mesal)_2]} R^1C{\equiv}CR^2 \tag{45}$$

the absence of the nickel catalyst the reaction is very slow. One drawback of this method is that, as in related reactions, only one of the alkyl groups of the trialkylalane is used.

The conjugate addition of an alkynyl group to an α,β-unsaturated ketone is a useful transformation, but not one that is readily accomplished. Organocuprates, the most commonly used reagents for 1,4 addition of alkyl and alkenyl groups to α,β-enones, cannot be employed in alkynylation reactions owing to the strong bonding of the alkynyl ligand to copper. Acetylenic alanes do conjugatively add their alkynyl unit to α,β-enones, but only under certain circumstances. In the presence of catalytic amounts of the complex formed by reaction of Ni(acac)$_2$ with diisobutylaluminum hydride (DiBAH) however, alkynylaluminum dialkyls undergo conjugate addition to both S-*cis*- and S-*trans*-enones.[58] Only the alkynyl group of the mixed alane is transferred and only 1,4 addition is observed. In cases where two stereochemical isomers are possible, conjugate addition of alkynyl units proceeds to give only one of these. For example, the alkoxy cyclopentenone **61** gives the compound **62** [Eq. (46)], possessing *anti*-stereochemistry, in 85% isolated yield. Other alkynyls, those of magnesium and zinc for example, do not react and other nickel complexes, NiCl$_2$(PEt$_3$)$_2$ for example, are ineffective.

$$R^1C{\equiv}CAlMe_2 \; + \quad \text{(structure 61)} \quad \xrightarrow{[Ni]} \quad \text{(structure 62)} \tag{46}$$

61   **62**

Preparation of Trans-3-(3,3-dimethyl-1-butynyl)-4-cumyloxy)-cyclopen-tanone (**62**; $R^1 = CMe_3$, $R^2 = CMe_2\,Ph$)[58]

To a solution of Ni(acac)$_2$ (0.09 g, 0.36 mmol) in ether (5 ml) at 0°C is added 0.60 ml (0.32 mmol) of a 0.53 M solution of DiBAH in toluene. To this mixture is then added 8 ml (3.6 mmol) of a 0.45 M solution of dimethyl-neohexynyl aluminum in ether and the mixture is cooled to -5°C when 4-(cumyloxy)2-cyclopentenone (0.36 g, 1.65 mmol) in ether (10 ml) is added dropwise over 15 min. After stirring at -5°C for 1.5 h the mixture is hydro-lyzed with saturated KH$_2$PO$_4$ and enough 10% aqueous H$_2$SO$_4$ is added to dissolve the aluminum salts. The organic layer is separated, extracted with ether, washed with saturated aqueous NaHCO$_3$ and saturated aqueous NaCl and dried (Na$_2$SO$_4$). Removal of the solvent gives a residue which is purified by liquid chromatography to give the product (0.42 g, 85%).

2.1.4. Substituted Arenes

2.1.4.1. Via π-arene and π-cyclohexadienyl complexes

The complexation of an aromatic molecule "face-on" to a transition metal is generally considered to involve donation of electrons from the filled π orbitals of the arene into vacant metal orbitals of σ and π symmetry, together with significant back-donation from filled metal d orbitals of δ symmetry into the lowest energy π-antibonding orbitals of the arene[59] (**63**).

63

By far the most intensively studied complexes of this type (from the point of view of organic synthesis) are those containing the tricarbonylchromium group[60] (**64**), since these are often available[61] by direct reaction of the arene with $Cr(CO)_6$ or with a simple derivative such as $[Cr(pyridine)_3(CO)_3]$. The effects of such complexation on the reactivity of an aromatic molecule can be briefly summarized as *an enhanced ability to stabilize a positive or negative charge located either on the aromatic nucleus or on a side chain*; as a result, certain reaction pathways normally forbidden to unactivated arenes become readily accessible. In nucleophilic displacements, for example, the $[Cr(CO)_3]$ group has a similar effect to that of a p-nitro substituent, and in both cases it is the ability of the activating group to stabilize a charged intermediate (**65** or **66**) that facilitates the reaction.

64   **65**   **66**

(X, leaving group; Nu, nucleophile)

A carbonionic center α to the ring is similarly stabilized (**67**) by delocalization of charge into the $[arene\text{-}Cr(CO)_3]$ moiety; the acidity of protons on an α-carbon is in fact increased by approximately two orders of magnitude as a result of π complexation to tricarbonylchromium.[62]

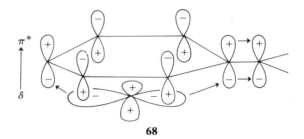

67

Remarkably, however, a *cationic* center in the α-position is also stabilized, presumably by electron donation from the metal, either directly to the α-carbon or, more probably, via the $\delta \rightarrow \pi^*$ interaction (**68**) noted above.[63]

68

A further consequence of π complexation is that, due to the large steric blocking effect of the $[Cr(CO)_3]$ group, reactions occur only on the *exo* (noncomplexing) face of the aromatic ring or side chain, allowing a high degree of stereoselectivity in syntheses using this type of complex. This blocking effect can be used as a synthetic tool in its own right, so that for example the indanone complex **69** is alkylated specifically on the *exo* face to give, after decomplexation, the optically active indanol in 100% enantiomeric excess[64] (Scheme 8).

Once a desired reaction has been carried out on the arene, it is of course necessary to remove the $[Cr(CO)_3]$ group, and a variety of reagents are available for this purpose. The most widely used are mild oxidizing agents such as Ce(IV), I_2, O_2, amine oxides, and peracids, but in certain favorable cases the product arene can be displaced, by pyridine [to give $Cr(Py)_3(CO)_3$],[65] or directly by another arene.[66]

Other π-arene complexes which have found some use in organic synthesis are based on iron[67] (**70, 71**) and manganese[68] (**72**). These cationic complexes are even more susceptible to nucleophilic attack than the neutral chromium analogs, but because the preparations of the cations involve relatively harsh conditions, they have been much less widely exploited in organic synthesis.

Scheme 8

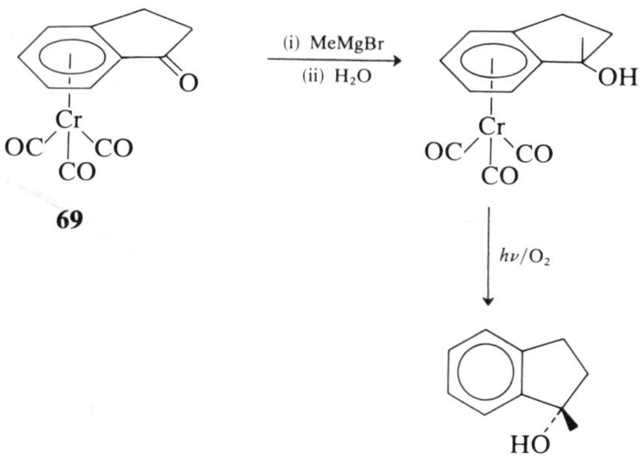

Substitution on the Ring. The classical functional groups which activate the aromatic nucleus to nucleophilic attack (nitro, cyano, carbonyl) are generally difficult to introduce selectively and reversibly, whereas, as we have seen, the tricarbonylchromium group is very convenient in this respect. Arene tricarbonylchromium complexes have thus been extensively studied as reagents for the arylation of carbanions. Chlorobenzene tricarbonyl chromium (**73**), for example, reacts with moderately nucleophilic carbanions, including $LiC(Me)_2CN$ which gives, after decomplexation with iodine, phenylisobutyronitrile (**74**) in high yield[69] (Scheme 9). The corresponding fluorobenzene complex is even more reactive, and it seems that the classical order of reactivity for activated nucleophilic aromatic substitution (F > Cl > Br > I) is retained in reactions of π-arene complexes. Less reactive nucleophiles, including $LiCuMe_2$, MeMgBr, and $PhCOCH_2HgCl$, fail to react with the above chlorobenzene complex, whereas much more nucleophilic reagents such as LiMe, LiPh, and $LiCH_2CO_2Me$, add to the

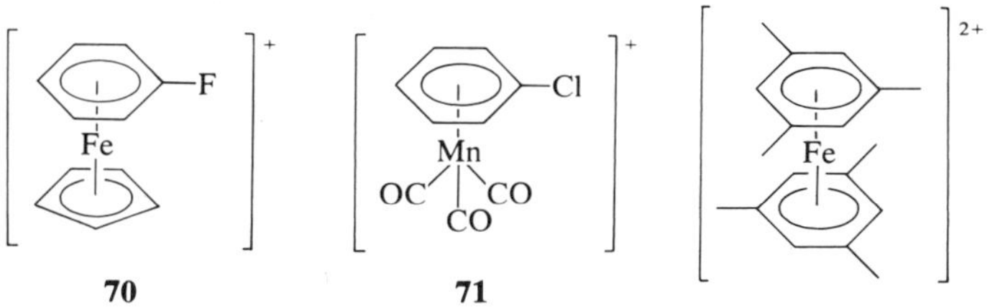

Scheme 9

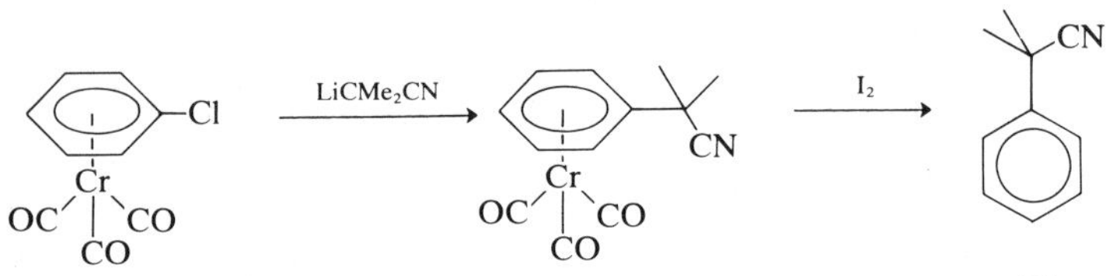

73   **74**

aromatic ring *without* displacing chloride ion. The latter effect results from kinetic control of the site of attack, which is initially *ortho* or *meta* to the chlorine substituent [Eq. (47)]. This addition is irreversible for strongly

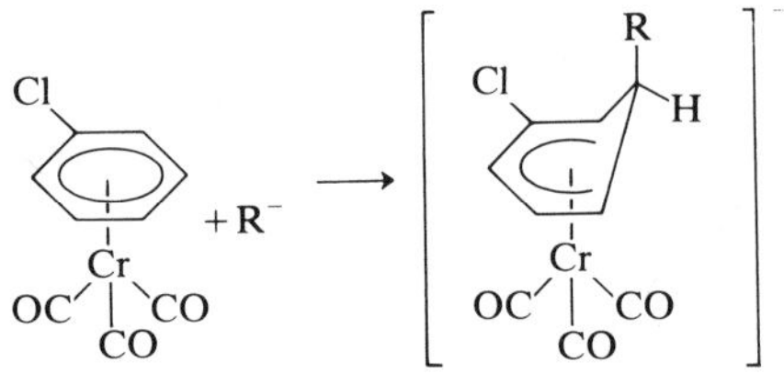

$$(47)$$

nucleophilic carbanions, but with those of moderate nucleophilicity, dissociation is possible, so that it is only when the kinetically less favored [C–Cl] site is eventually attacked that irreversible loss of chloride leads to the observed product.[70] An alternative procedure for alkylation of coordinated arenes dispenses with the requirement for a leaving group altogether. Instead, the intermediate addition product (**75**) is treated *in situ* with an oxidizing agent such as iodine, which both abstracts hydride from the intermediate and releases the product arene from the metal[71] (Scheme 10).

Scheme 10

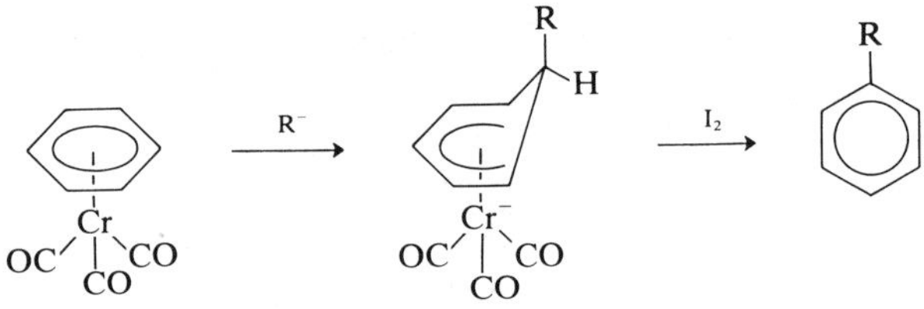

75

Benzene Chromiumtricarbonyl[72]

In a 500-cm^3 round-bottomed flask are placed 2-methylpyridine (100 cm^3), benzene (100 cm^3), a few boiling chips, and hexacarbonyl-chromium (8.80 g). The flask is fitted with a reflux condenser, evacuated, and refilled with nitrogen several times to ensure an oxygen-free atmosphere, and the reaction mixture is refluxed for 96 h. After cooling, the excess reactants [including Cr(CO)$_6$] are removed on the rotary evaporator. [CAUTION: This must be carried out in a fume cupboard as Cr(CO)$_6$ is toxic.] The yellow-green residue is extracted repeatedly with hot diethyl ether, until the extracts after filtration are colorless. The filtered extracts are combined, evaporated to dryness, washed with pentane, and dried, giving 7.75 g of yellow crystals (91%), mp 161–162°C.

An alternative procedure for this, and a range of other arenetricar-bonylchromium complexes, has recently been given by Mahaffy and Pauson.[61]

Phenylisobutyronitrile[71]

Lithium diisopropylamide is prepared from *n*-butyl lithium (12.8 cm^3 of a 1.95 *M* solution in hexane, 25.2 mmol) and diisopropylamine (3.84 cm^3, 27.5 mmol) in 50 cm^3 of dry THF, by mixing the reagents at −78°C under argon and allowing the mixture to stir at 0°C for 15 min. The solution is cooled to −78°C and isobutyronitrile (2.20 cm^3, 25 mmol) is added dropwise over 5 min. The mixture is warmed to 0°C for 15 min, cooled again to −78°C, and a solution of benzenetricarbonylchromium (5.20 g, 25 mmol) in THF (25 cm^3) is added dropwise over 5 min. The yellow solution is held at 0°C for 30 min, cooled to −78°C, and a solution of iodine (25 g, 98 mmol) in 75 m^3 of THF is added dropwise over 15 min. (CAUTION: Carbon monoxide is evolved.) After stirring at room temperature for 3 h, 5% aqueous sodium bisulfite (100 cm^3) is added, and the solution is extracted with diethyl ether (200 cm^3). The ether layer is washed successively with water and aqueous sodium chloride, dried over magnesium sulfate, and concentrated to a yellow oil. Flash distillation (25°C, 0.001 mm) gives colorless phenylisobutyronitrile (3.65 g, 98%).

Substituent-directing effects in this type of "oxidative alkylation" reaction are apparently related to the nature of the intermediate addition product, which may be regarded as a complexed cyclohexadienyl anion. Electron-donating groups (X) are least destabilizing in the *meta* position (**76**), and a methoxy group, for example, directs incoming nucleophiles (R$^-$) almost exclusively to this position.[73]

76

Oxidative alkylation has been applied to the synthesis of carbocyclic compounds[73] [Eq. (48)] and to a novel preparation of phenyl mesitylene,[67a] which provides the best synthesis so far reported for this compound (Scheme 11).

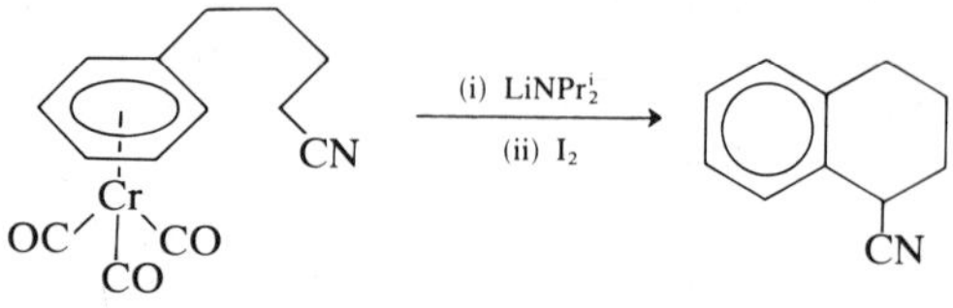

$$\text{(48)}$$

77

Scheme 11

(95%)

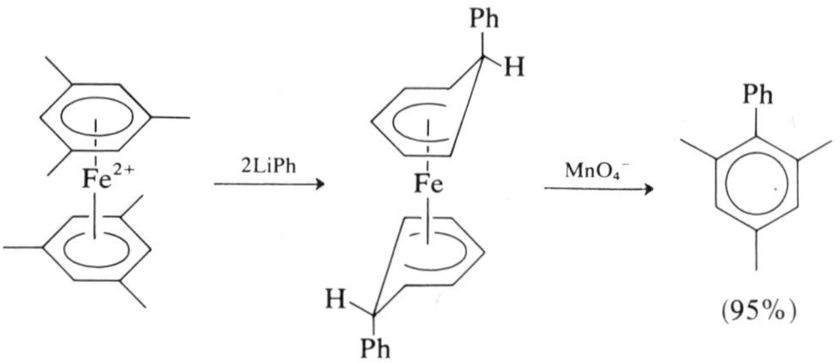

As an alternative to direct oxidation, the intermediate addition product may be protonated with trifluoracetic acid to give a substituted cyclo-hexadiene which is liberated from the metal by iodine oxidation (Scheme 12).[74] Cationic π-cyclohexadienyl complexes are similarly susceptible to nucleophilic attack by carbanions [Eq. (49)]. The resulting cyclohexadiene complexes may be oxidatively cleaved by a variety of reagents (R_3NO,

Scheme 12

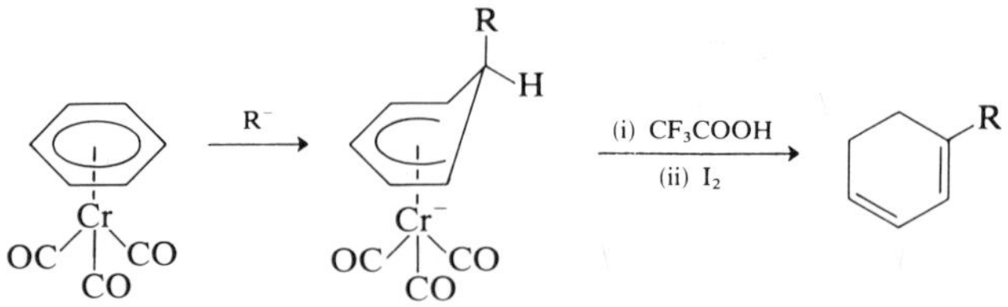

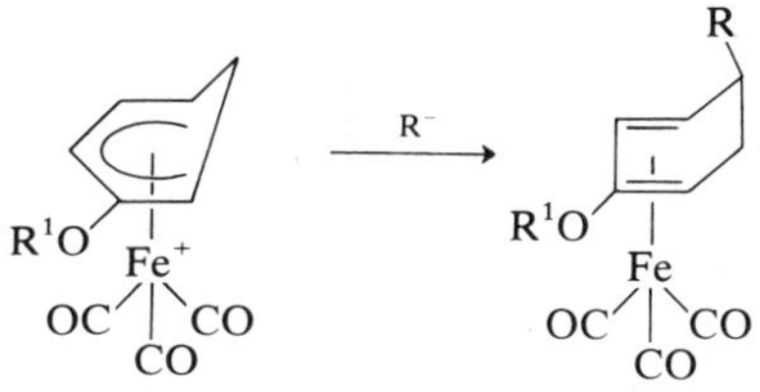

$$(49)$$

CrO$_3$, Pd/C, etc.) to give substituted arenes, cyclohexadienes, or cyclohexenones, depending on the reagent and conditions used. This type of chemistry, originally discovered by Birch,[75] has recently been extensively applied by Pearson and coworkers[76] in the synthesis of a wide variety of natural products such as the sceletium alkaloid (**78**) shown in Scheme 13. The alkoxy-substituted cyclohexadienyl complexes are of particular value, as the substituent exerts a powerful *para*-directing effect on the incoming nucleophile, allowing highly selective transformations to be achieved.

Scheme 13

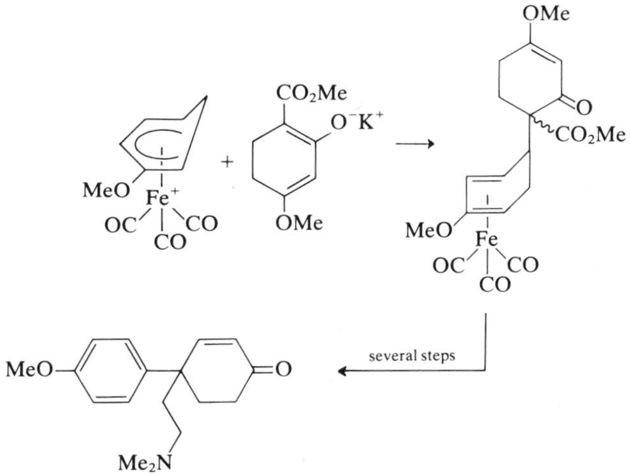

Substitution on the Side Chain. Although methyl phenylacetate is unaffected by treatment with sodium hydride and iodomethane in dmf at 25°C, when it is complexed to tricarbonylchromium the increased acidity of its α-protons allows rapid dialkylation under these conditions[77] [Eq. (50)]. The organic product is liberated by air–oxidation in sunlight.

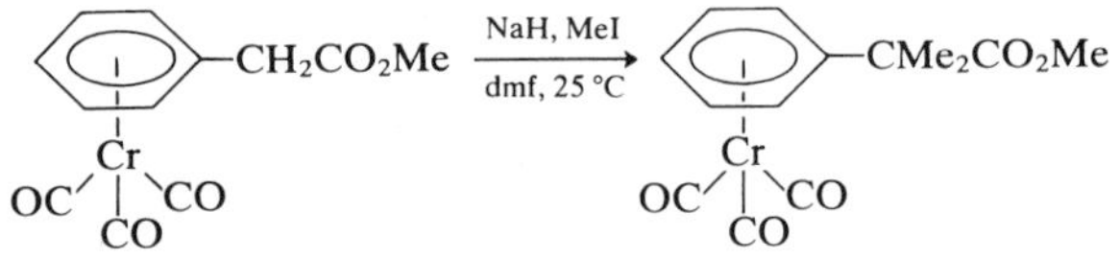
CH2CO2Me
NaH, MeI
dmf, 25 °C
CMe2CO2Me
Cr
OC
CO
CO

(50)

The corresponding ethylbenzene complex may be methylated at the α-position using potassium *t*-butoxide and iodomethane in dmso, but in the absence of the activating methoxycarbonyl group only monoalkylation is now observed. If, however, one of the carbonyl groups on chromium is replaced by the stronger acceptor-ligand thiocarbonyl,[78] then sufficient stabilization of the second intermediate anion **79** is achieved to allow dialkylation to occur (Scheme 14).[60]

Scheme 14

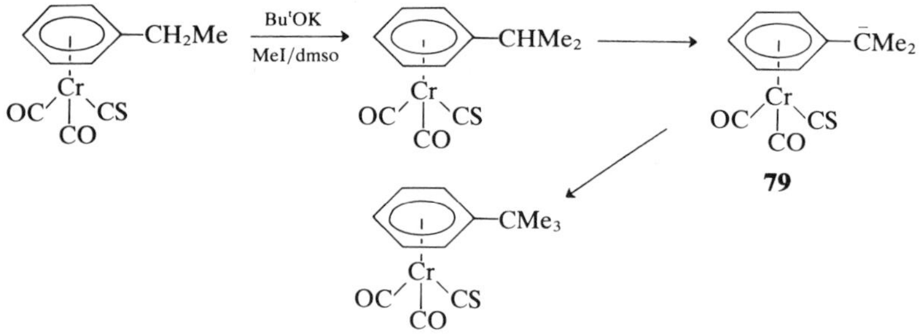
CH2Me
BuᵗOK
MeI/dmso
CHMe2
C̄Me2
CMe3
Cr
OC
CS
CO
79

The opposite result is found if one carbonyl group is replaced by the weaker π-acceptor triphenylphosphine. Even the methyl phenylacetate complex is now inert to alkylation at the α-carbon,[77] and such results clearly illustrate the "fine-tuning" possibilities of π-arene complexation in synthesis.

The combined effects of α-proton acidification and *exo*-only attack have led to the development of a novel, stereospecific annulation reaction (Scheme 15) in which a proton α to the complexed arene is lost in preference to one α to a carbonyl group.[79]

Scheme 15

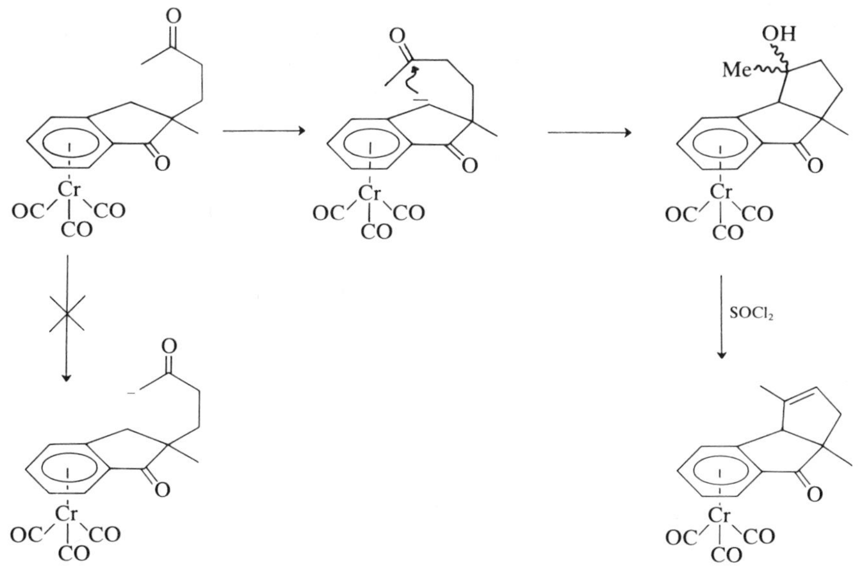

Methyl Phenylisobutyrate Tricarbonylchromium[77]

The tricarbonylchromium complex of methyl phenylacetate (0.3 g, 1.07 mmol) is dissolved in dmf (50 cm³) and treated with sodium hydride (0.5 g, 20 mmol). The mixture is shaken for 10 min at room temperature, and iodomethane (0.6 g, 4.2 mmol) is added. After stirring for a further 5 min the solution is poured into 500 cm³ of water (CAUTION: sodium hydride reacts vigorously) and the mixture extracted with diethyl ether. After drying the ether extract over magnesium sulfate, evaporation gives 0.32 g of methyl phenylisobutyrate tricarbonylchromium (97%), mp 55°C.

2.1.4.2. Via σ-Aryl Complexes

Alkyl Arenes. As discussed in the previous section, aryl halides which are normally inert to nucleophilic substitution by carbanions can be activated by π complexation to transition metals. This technique, though potentially valuable, is generally stoichiometric in transition metal and involves separate complexing and decomplexing steps, so that for many simple substitutions, a catalytic reaction would be desirable.

It has in fact been shown in recent years that catalytic quantities of certain nickel or palladium complexes [particularly $NiCl_2(Ph_2PCH_2CH_2PPh_2)$ [8]

and Pd(PPh$_3$)$_4$ [80]] promote such substitutions with high efficiency, though where divalent catalyst precursors are used it seems likely that the true catalyst is formed by reduction to the zerovalent state *in situ*. The catalytic cycle (Scheme 16) then involves oxidative addition of the aryl halide, substitution of halide by the carbanion, and reductive elimination of the product.

Scheme 16

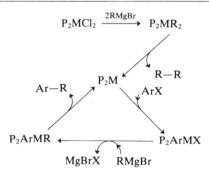

(M = Ni or Pd; P = tertiary phosphine)

1,3-Di-n-butyl Benzene [8]

To a stirred mixture of [NiCl$_2$(Ph$_2$PCH$_2$CH$_2$PPh$_2$)] (0.208 g, 0.39 mmol), 1,3-dichlorobenzene (7.93 g, 53.9 mmol) and diethyl ether (50 cm^3) at 0°C, is added, under nitrogen, *n*-butyl magnesium bromide (120 mmol) in diethyl ether (50 cm^3) over 10 min. The resulting black mixture is heated at reflux for 20 h, cooled, and hydrolyzed with dilute hydrochloric acid. The organic layer and ether extracts from the aqueous layer are combined, washed with water, dried over CaCl$_2$, and concentrated *in vacuo*. Distillation at reduced pressure gives pure di-*n*-butyl benzene as a colorless liquid (9.6 g, 94%), bp 80°C at 5 mm Hg.

The catalyst necessary in this type of reaction depends principally on the nature of the halide, so that aryl iodides react with carbanions even in the presence of palladium metal[81] (produced *in situ* by reduction of the chloride or acetate), whereas bromides require a nickel or palladium–phosphine complex and chlorides react only in the presence of nickel–phosphine catalysts. The carbanion may be in the form of an organolithium, Grignard, or organozinc[80,82] reagent, and even alkali metal cyanides react with aryl halides in the presence of Ni(PPh$_3$)$_4$ [83] or Pd(PPh$_3$)$_3$ [84] under very mild conditions to give nitriles in high yield. Terminal alkynes, which are relatively acidic and thus easily converted to carbanions, may be catalytically arylated at 100°C in the presence of $(Ph_3P)_2Pd(OAc)_2$ or $Pd(PPh_3)_4$ simply by using a tertiary amine to deprotonate the alkyne[85] [Eq. (51)].

$$RC\equiv CH + ArX + NR_3 \xrightarrow{Pd(PPh_3)_2(OAc)_2} RC\equiv CAr + R_3NHX \qquad (51)$$

Addition of copper(I) iodide to this type of system allows the reaction to be carried out at room temperature, and in particular facilitates the synthesis of diaryl alkynes from acetylene itself.[86] The role of copper is not well understood, but it may be significant that, under more vigorous conditions, copper(I) alkynyls react directly with aryl iodides to give aryl alkynes in high yield.[87] This reaction has been used in the synthesis of a variety of heterocyclic compounds[88] (Scheme 17).

Scheme 17

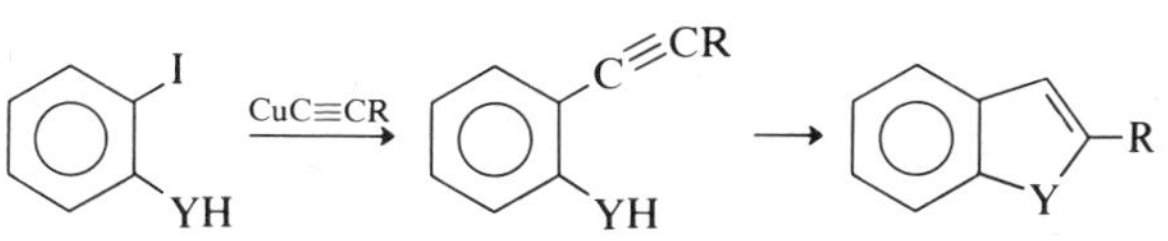

$(Y = -O-, -CO_2-, -S-, \text{ or } -NR-)$

Organocopper compounds in general can alkylate aryl halides,[89] but harsh conditions are often required, the reaction is usually limited to aryl iodides, and the copper reagent must be prepared in a separate step. Direct alkylation by an organolithium or Grignard reagent, using a nickel or palladium catalyst, would therefore seem to be the method of choice for alkylation of aryl halides.

Organocopper reagents are of more value when a reaction in the opposite sense is required. Lithium diphenyl cuprate, for example, phenyl-ates alkyl halides and tosylates in high yield, with predominant inversion of configuration at the saturated carbon atom.[90] The order of effectiveness of leaving groups in such reactions is $OTs > I \sim Br > Cl$.[91] Acyl halides are also phenylated by phenyl cuprates,[92] and moreover the carbonyl group of the resulting ketone is not attacked, a marked advantage over conventional organolithium reagents which yield significant quantities of tertiary carbinols.

The relative inability of organocopper compounds to attack carbonyl groups has been widely exploited in the 1,4 addition of carbanions to the

C=C double bonds of α,β-unsaturated carbonyl compounds (Scheme 18) (this topic is fully discussed in Section 2.1.5). Although catalytic quantities of copper(I) salts are sometimes used in conjunction with lithium or Grignard reagents, preformed organocopper compounds such as LiCuPh$_2$ are generally preferred owing to their greater selectivity.[93]

Scheme 18

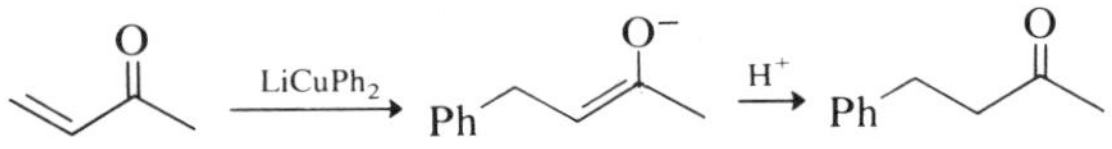

3-Methyl-5-phenylcyclohexanone[93]

    (*Note*: The preparation and reaction of the organocuprate reagent must be carried out under dry nitrogen, using dry distilled diethyl ether.)

    To a cold solution (0°C) of lithium diphenylcuprate, prepared from 124 mmol of phenyllithium and 8.90 g (62.0 mmol) of anhydrous copper(I) bromide in 118 cm^3 of ether, is added a solution of 6.64 g (60.3 mmol) of 5-methyl-2-cyclohexan-1-one in 50 cm^3 of ether. The mixture is stirred for 15 min at 0°C, and an excess of aqueous NH$_4$Cl (containing ammonia to pH 8) is then added, and air is passed through the solution to oxidize insoluble copper(I) species to a soluble copper(II)–ammonia complex. The organic layer is separated, combined with an ether extract of the aqueous phase, and then washed with aqueous NH$_4$Cl, dried, and concentrated. The residual yellow oil is distilled (96–110°C at 0.09 mm Hg) in a short-path still, giving 7.62 g of crude *trans*-3-methyl-5-phenylcyclohexanone. This material, which contains some *cis* compound, may be purified by conversion to the oxime (mp 92–93°C), followed by hydrolysis and distillation (90–95°C at 0.05 mm Hg) to give the product as a colorless oil: infrared shows $V(C{=}O)$ at 1721 cm^{-1}.

    A technique for selective *ortho*-alkylation of aromatic compounds depends on the phenomenon of *ortho*-metallation, whereby direct palladation of an aryl group in a donor ligand gives a chelated organopalladium complex (**80**) [Eq. (52)]. Compounds which can be metallated in this way

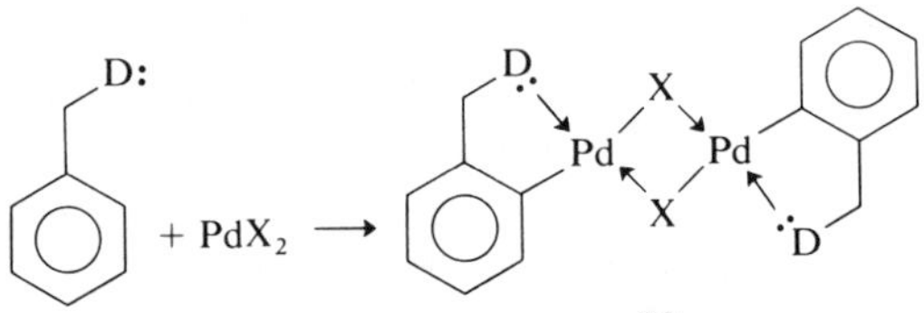

(52)

80

include *N,N*-dialkyl benzylamines, and azobenzenes. Reaction of the resulting complexes with alkyllithium or Grignard reagents, in the presence of triphenylphosphine, affords *o*-alkyl substitution products in high yield[94] (Scheme 19).

Scheme 19

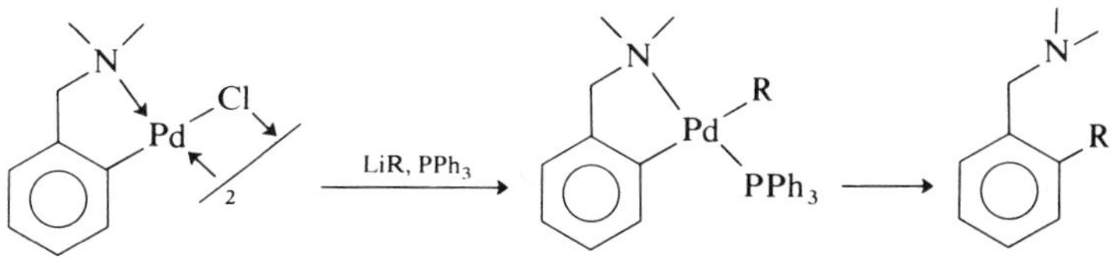

Since the reaction is stoichiometric in palladium, it is clearly unsuitable for large-scale preparative work, but because of its specificity it should prove valuable when only small quantities of material are required.

Vinyl Arenes. The vinylic hydrogen atoms of a wide variety of alkenes may be substituted using aryl or heterocyclic halides in the presence of a palladium catalyst[95] [Eq. (53)]. The reaction is thought to proceed via

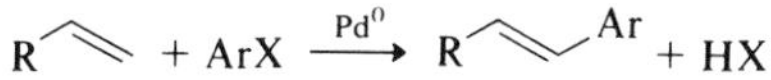

$$R\diagup\diagdown + ArX \xrightarrow{Pd^0} R\diagup\diagdown\diagup^{Ar} + HX \qquad (53)$$

oxidative addition of the halide to give a palladium aryl which "inserts" alkene and then undergoes a β-hydride elimination to release the product and regenerate the catalyst (Scheme 20).

Scheme 20

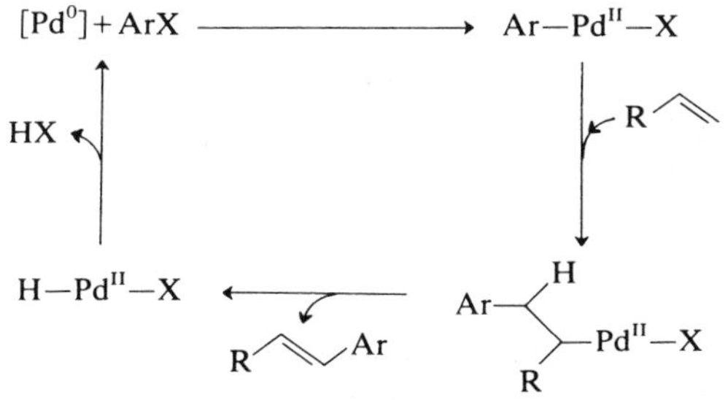

With very reactive halides such as aryl iodides, palladium acetate may be used as the catalyst precursor, since it is reduced to an active Pd^0 species under the conditions of the reaction.[96] Bromoarenes, however, also require addition of a tertiary phosphine such as PPh_3 or $P(o\text{-tolyl})_3$ to stabilize the catalyst, or alternatively a preformed complex may be used $[(PPh_3)_2Pd(OAc)_2$ is convenient].[97] A stoichiometric amount of tertiary amine is used to remove the acid formed in the reaction, and under these conditions the catalytic activity of palladium is normally very high. The reaction is also generally stereospecific, so that the aryl group appears almost exclusively at the less substituted carbon atom of the double bond and, in the case of a terminal alkene as starting material, *trans* to the other substituent [Eq. (54)].

$$\text{(54)}$$

81

Trans-4-carbomethoxystilbene (**81**)[96]

A mixture of 0.045 g (0.2 mmol) of palladium acetate, 5.24 g (20 mmol) of methyl-p-iodobenzoate, 2.86 cm^3 of styrene (25 mmol), and 4.74 cm^3 (20 mmol) of tri-n-butylamine, in a 25-cm^3 round-bottomed flask fitted with a reflux condenser, is stirred magnetically at 100°C for 2 h. The reaction mixture is cooled and the solid mass then broken up with a spatula and triturated with 100 cm^3 of water. The resulting solid is filtered off, and recrystallized from 225 cm^3 of boiling ethanol, with hot filtration through Celite to remove palladium metal. The yield is 3.50 g (74%) of shining plates: mp 158–160°C.

This reaction has been applied to the synthesis of heterocyclic compounds, so that quinoline, for example, is obtained by reaction of 2-iodoaniline with acrolein dimethyl acetal (Scheme 21). 2-Quinolones (**82**) were similarly obtained from 2-iodoaniline and dimethyl maleate [Eq. (55)], but attempts to prepare coumarin derivatives by an analogous route from 2-iodophenol were unsuccessful.[98]

Palladium-catalyzed vinylic arylation is an extremely general reaction. It is tolerant, for example, of carboxylic acid groups in both alkene and aryl halide,[99] and proceeds normally in the presence of aldehyde, cyano, nitro, methoxy, thiomethyl, amide, and pyridyl groups. The presence of a carboxylic acid function can in fact be a definite advantage as it allows extraction of the product by aqueous base and thus simplifies the work-up

Scheme 21

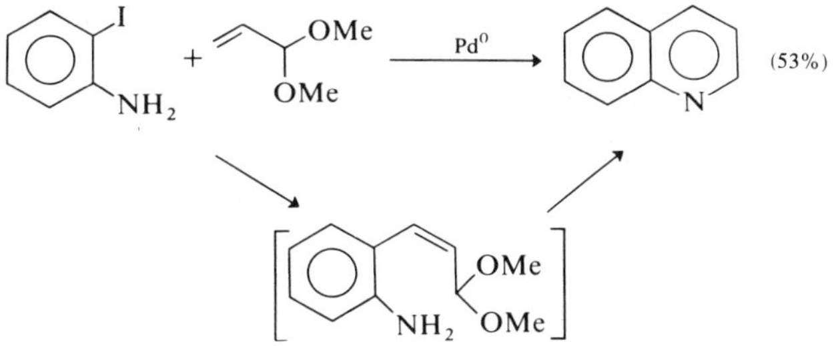

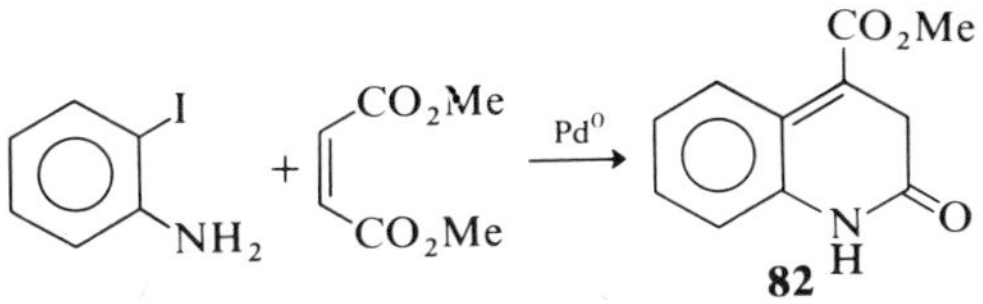

(55)

procedure. When the aryl halide contains a strongly electron-donating substituent, however, such as —OH or —NH₂, use of the "standard" palladium acetate–triphenylphosphine catalyst gives only very low yields (2%–4%) due to rapid catalyst deactivation. This is caused by reductive elimination of aryltriphenylphosphonium salt from the aryl palladium intermediate (**83**) [Eq. (56)], but can be prevented by use of the sterically hindered tri(*o*-tolyl)phosphine, which does not undergo quaternization under normal reaction conditions.[100]

$$PPh_3 - \underset{\underset{X}{|}}{\overset{\overset{Ar}{|}}{Pd}} - PPh_3 \rightarrow [PPh_3Ar]^+ X^- + PPh_3 + Pd \qquad (56)$$

83

A variant of vinylic arylation involves an allylic alcohol as the unsaturated reactant, and this reaction provides a convenient preparation of 3-aryl aldehydes and ketones, since palladium catalyzes both vinylic substitution and subsequent isomerization (Scheme 22).[101]

Scheme 22

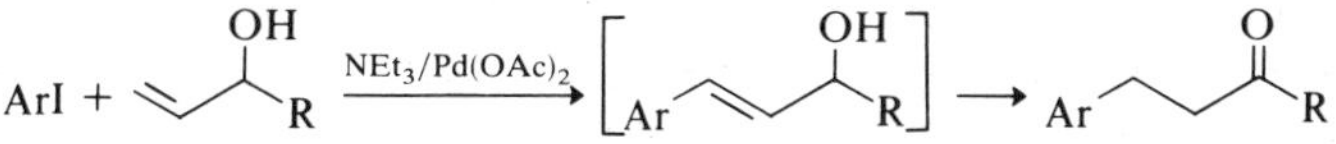

Recent reports indicate that arenediazonium salts,[102] and even aromatic amines,[103] can function as arylating agents in vinylic substitution, but yields are very variable and aryl iodides or bromides remain the reagents of choice for this reaction.

Since aromatic compounds can be metallated directly by palladium(II) salts, the alkene arylation reaction may in certain cases be carried out using simple unsubstituted arenes (Scheme 23).[104] As with aryl halide substrates,

Scheme 23

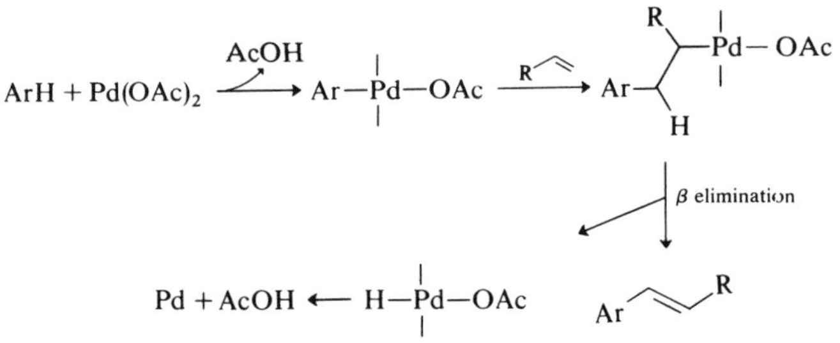

cyclization reactions are possible and indone derivatives (**84**), for example, are formed by intramolecular coupling of 3-phenylpropenes[105] [Eq. (57)].

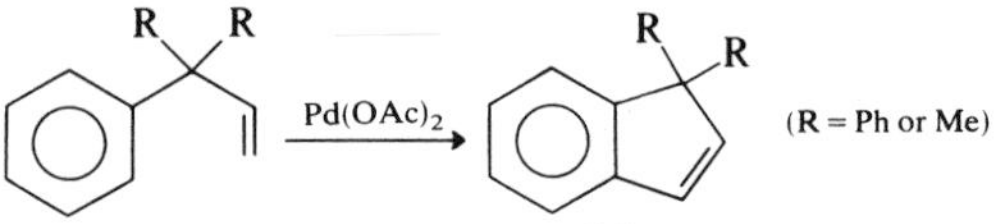

(R = Ph or Me)   (57)

84

The major disadvantage of direct palladation is of course that the reaction is no longer catalytic, but is stoichiometric in palladium. Some success has been achieved in reoxidizing the palladium *in situ* by addition of copper(II) salts and/or by operating under oxygen pressure,[106] but only a limited range of alkenes are suitable for such procedures. Finally, the initial palladation reaction, being essentially an electrophilic aromatic substitution of H^+ by Pd^{2+}, tends to give a mixture of isomers, so that overall the direct coupling of arene and alkene is not as generally useful in synthesis as the very versatile reaction between an alkene and an aryl halide.

Biaryls. The oxidative substitution of π-arene complexes with aryl lithium reagents provides a novel route to symmetrical or unsymmetrical

an ice-bath for 3 min, and then oxidized by passing oxygen over the surface of the cold solution with vigorous stirring for 30 min. The mixture is partitioned between Et_2O and aqueous NH_4Cl/NH_3, and the ether phase is dried and concentrated. Chromatography on silica-gel (hexane) yields 2.96 g of crude 1,8-diphenylnaphthalene, which is recrystallized from hexane to give 2.77 g (65.4%) of pure product as white needles, mp 149–151°C.

An alternative coupling technique involves direct reaction of an aryllithium or Grignard reagent with copper(II) chloride, an approach used successfully in the synthesis of a number of cyclic polyarenes[116] [Eqs. (66) and (67)].

$$(66)$$

$$(67)$$

The uses of organocopper reagents in coupling reactions have been surveyed by Posner (1975),[117] Kauffmann (1974),[116] and Normant (1972),[118] and the Ullmann reaction has been reviewed by Fanta.[113]

2.1.5. Substituted Alkanes (By Conjugate Addition of Organocopper Reagents to Activated Alkenes)

Although stabilized carbanions such as diester enolates and β-diketonates undergo almost exclusive 1,4 addition to α,β-unsaturated carbonyl compounds (Michael reaction), more strongly nucleophilic species such as alkyl-lithium or -magnesium reagents tend to add preferentially to the carbonyl group, giving carbinols. In 1941, however, Karasch and Tawney[119] discovered that addition of a catalytic amount of cuprous chloride to the reaction between methylmagnesium and isophorone (**88**) increased the selectivity toward 1,4 addition by a factor of 50 (Scheme 27).

Scheme 27

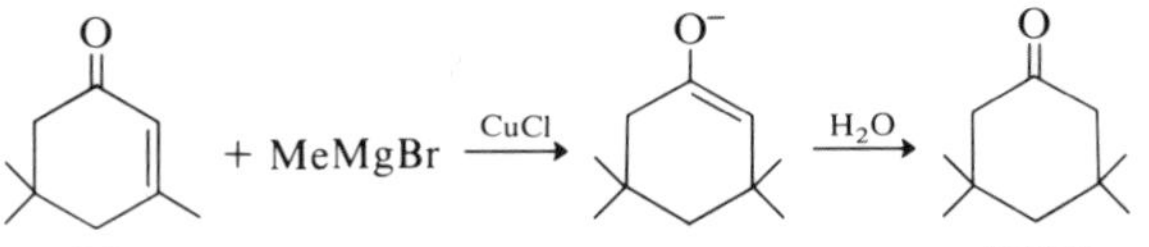

88   (82%)

The catalytic technique was then used, despite some limitations, until 1966, when it was demonstrated[120] that organocopper compounds were the active intermediates, and that preformed methylcopper and lithium dimethylcuprate underwent extremely effective 1,4 addition to *trans*-3-penten-2-one. Since this report, it has been confirmed that selective conjugate addition is often achieved much more effectively with stoichiometric organocopper reagents than by copper catalysis of Grignard or organolithium reactions.[93] The enone **89**, for example, gives a 56:34 mixture of 1,2- and 1,4-addition products on reaction with methylmagnesium iodide in the presence of cuprous acetate catalyst, whereas lithium dimethylcuprate gives a quantitative yield of the 1,4 adduct **90**[121] [Eq. (68)].

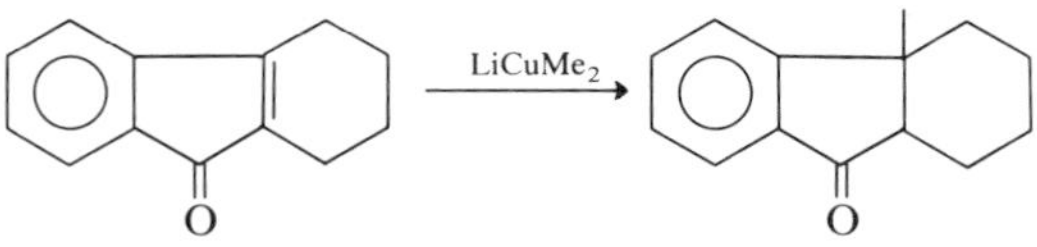

89   **90**

(68)

Reasons for the effectiveness of organocopper reagents in conjugate addition have been difficult to determine, since little structural information is available. Alkyl copper(I) compounds such as CuMe are certainly oligomeric, and although dialkyl cuprates are conventionally written as monomers ($Li^+R_2Cu^-$), it is likely that here too we are dealing with polynuclear systems containing bridging alkyl or aryl groups.[122] A variety of mechanisms for copper-promoted conjugate addition have been suggested, and probably the most widely accepted (though still controversial) pathway is shown in Scheme 28. In this mechanism,[123] the first step is a one-electron transfer from an anionic copper(I) species to the activated alkene, followed by combination of the resulting copper(II) alkyl and radical anion to give a transient copper(III) species which reductively eliminates the product. (*N.B.* Since we are dealing with polynuclear systems, the quoted oxidation states for copper should not be taken too literally.)

Scheme 28

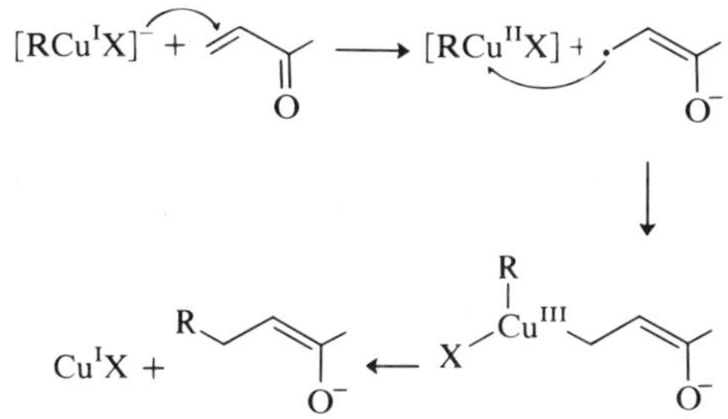

The one-electron reduction potentials of a range of enones correlate well with the ease of conjugate addition by organocuprates, providing good evidence that electron transfer plays a significant role in the reaction.[124] In agreement with this, no such correlation was found for conventional (Michael) additions of stabilized carbanions.[123] Anionic copper(I) alkyls, which are better able to transfer an electron to the substrate than neutral alkyls, are certainly required for conjugate addition, since isolated, salt-free alkyl copper compounds RCu are inert in this reaction.[125] In the presence of halide ions, however, which are invariably present when such reagents are prepared *in situ*, the equilibrium

$$RCu + LiX \rightleftharpoons Li^+RCuX^-$$

apparently provides a sufficient concentration of cuprate anion for reaction to occur.

For synthetic purposes, diorganocuprate reagents R_2CuLi, being soluble, are generally preferred to the less reactive and less selective heterogeneous mixtures of lithium halides and insoluble copper alkyls RCu, though the latter can be solubilized by addition of a neutral ligand such as a tertiary phosphine[126] or secondary sulfide.[127] An alternative method of solubilizing RCu, avoiding the use of such ligands which can be troublesome to separate afterwards, involves formation of a mixed-cuprate species RR^1CuLi, in which the second component R^1 is nontransferable in conjugate addition. Ethynyl groups, for example, as noted in Section 3, are inert to such transfer, and *t*-butyl ethynyl[124] and 3-dimethylamino propynyl[128] have been recommended as readily available, nontransferable groups for mixed-cuprate reagents $Li(RC\equiv C-Cu-R^1)$.

Since the first-formed product of conjugate addition is an enolate anion, the latter may be protonated by aqueous work-up (as in most reactions), alkylated at the enolate carbon, e.g., by alkyl iodides,[129] or may undergo further intramolecular reaction. Particularly elegant examples of the latter are shown in Scheme 29, where ring closure of an intermediate carbanion provides a route to bicyclic or spirocyclic compounds.[130]

Scheme 29

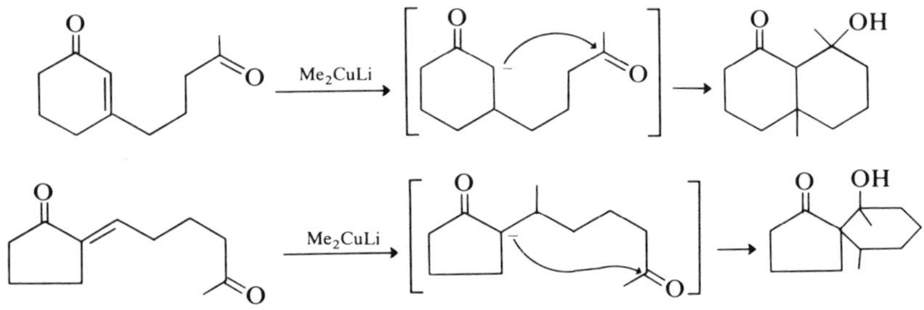

Lithium bis(α-trimethylsilylvinyl)cuprate (**91**) and lithium di(α-methoxyvinyl)cuprate (**92**) behave as "acetyl anion equivalents" in conjugate addition,[131] and examples of the use of these reagents are given in Schemes 30 and 31, and in the following experimental section.

Scheme 30

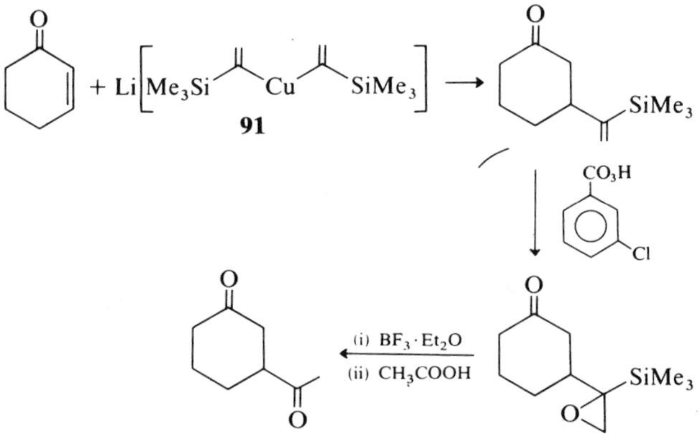

Scheme 31

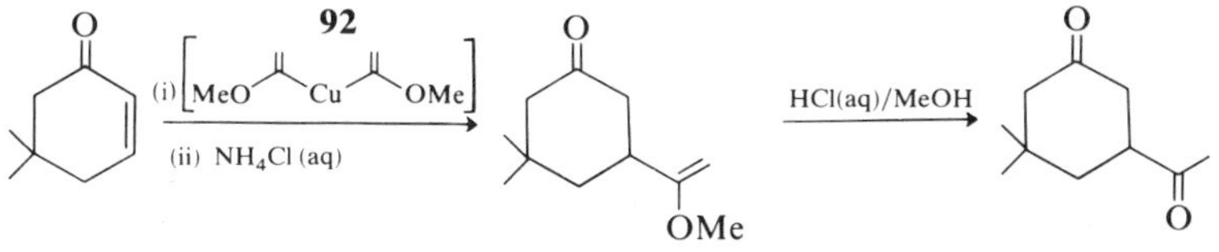

5,5-Dimethyl-3-(α-methoxyvinyl)cyclohexanone[131]

A solution of α-methoxyvinyllithium, prepared from 5.24 cm³ (69.6 mmol) of methyl vinyl ether and 54.4 cm³ (43.5 mmol) of 0.8 *M* t-butyl lithium in 30 cm³ of dry THF, is cooled to below −40°C and added over 15 min via a glass-jacketed dropping funnel containing dry ice/isopropanol to a solution of 4.17 g (21.9 mmol) of purified† copper(I) iodide and 6.30 mmol of dimethyl sulfide in 30 cm³ THF at −40°C under dry nitrogen. After stirring for 30 min, 1.80 g of 5,5-dimethyl-2-cyclohexen-1-one (14.5 mmol) in 50 cm³ of THF is added over 10 min to the yellow mixture and stirred at −40°C for 10 min, and then at −10°C for 45 min. The red mixture is quenched with 20% NH₄Cl, extracted with ether, and the crude product isolated and distilled (73–76°C, 0.25 mm Hg), affording 1.76 g (67%) of 5,5-dimethyl-3-(α-methoxyvinyl)cyclohexanone. Ir shows bands at 1824 (v, C=O) 1661 and 1631 cm⁻¹. Hydrolysis with 0.025 *M* HCl in methanol–water (4 : 1) yields 5,5-dimethyl-3-acetylcyclohexanone in 80% distilled yield.

In conjugate additions to enones (**94**) alkyl substituents at the α- or α-positions do not significantly affect the reaction, and even two substituents at the β-position hinder addition only slightly. With the less reactive enoate esters (**93**), however, substitutent effects become important, and a single

93   **94**

alkyl group at the β-position severely retards the reaction with conventional organocuprate reagents. Two β-substituents prevent addition of such reagents completely, but a novel group of complexes, formulated as [RCuBF₃], are reported to be so reactive that they will undergo conjugate addition even with β-disubstituted enoate esters.[132]

The actual nature of "RCuBF₃" and related copper–boron reagents RCuBR₃ [133] is by no means clear, but polynuclear structures involving copper–alkyl–boron bridges seem likely.

† Recrystallized from aqueous potassium iodide and then extracted (Soxhlet) with hot THF.

The literature of copper-promoted conjugate addition, up to the end of 1971, has been exhaustively reviewed by Posner,[93] and more recent results have been surveyed by Noyori.[134] Unlike the vast majority of reactions promoted or catalyzed by transition metals, conjugate addition is now firmly established as a routine tool for organic synthesis.

2.2. Formation of Carbon–Carbon Double Bonds

Compared with the variety of transition-metal-mediated transformations that lead to the formation of carbon–carbon single bonds, there are relatively few such reactions yielding carbon–carbon multiple bonds. However, during recent years two alkene forming reactions involving transition metals have come to prominence, though despite considerable work their scope and full potential have probably not yet been realized. These synthetically significant reactions are (a) catalytic metathesis of alkenes and (b) reductive coupling of aldehydes or ketones to alkenes. In some situations, however, it may also be desirable to obtain an alkene by deoxygenation of the corresponding epoxide and a number of efficient transition metal-mediated reactions which are now available for doing this are briefly mentioned in this section.

2.2.1. Catalytic Alkene Metathesis

The transition-metal-induced dismutation of alkenes via a formal exchange of alkylidene groups [Eq. (69)] is referred to as alkene metathesis.

$$R^1HC{=}CHR^1 + RHC{=}CHR \rightleftharpoons 2RHC{=}CHR^1 \tag{69}$$

The thermal reaction is symmetry forbidden according to Woodward–Hoffman rules, and indeed, extremely high temperatures are required to induce the thermal reaction.[135] The catalytic reaction therefore represents one of the most remarkable transformations of organic chemistry, particularly as in some situations equilibrium can be established within minutes at room temperature.

A number of catalyst systems are available, both homogeneous and heterogeneous. The most active homogeneous catalysts are generally derived from a molybdenum or tungsten compound, a strongly reducing and/or alkylating agent such as an aluminum or tin alkyl, together with an additive to inhibit side reactions and improve selectivity. Additives that have been employed include pyridine, acetonitrile, ethanol, and triphenylphosphine. Typical catalysts are $WCl_6/EtAlCl_2$, $WCl_6/SnMe_4$,

$Mo(PPh_3)_2(NO)_2Cl_2/Al_2Me_3Cl_3$, $PhWCl_3/AlCl_3$, and $WCl_6/LiBu$, while commonly used solvents include chlorobenzene and trichloroethylene. Reactions must be çarried out under an inert atmosphere, though there is some evidence suggesting *traces* of oxygen are necessary.

During the last decade extensive efforts have been directed toward elucidating the mechanism of alkene metathesis and, although several different pathways have been proposed, almost all available evidence is consistent with the metal–carbene chain process outlined in Scheme 32. Space does not permit further discussion of mechanistic aspects, and interested readers are directed to the many excellent reviews that are available[136] on this topic.

Scheme 32. The Basic Cycle of the Carbene Chain Mechanism for Alkene Metathesis

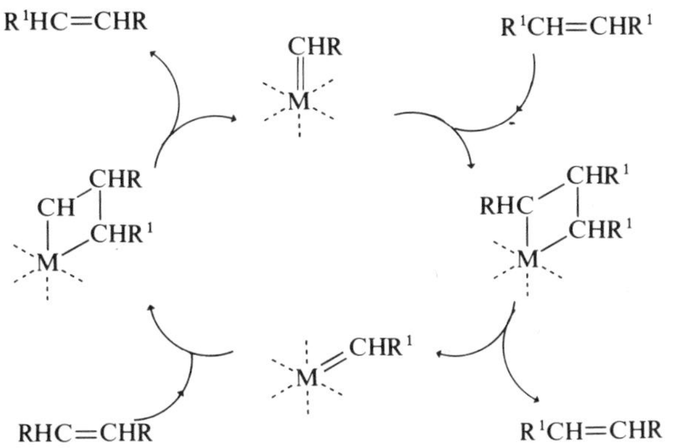

In general alkene metathesis is a reversible equilibrium-controlled process, and with simple alkenes the enthalpy of reaction is essentially zero (making and breaking similar carbon–carbon double bonds); consequently the product distribution is determined by entropy differences and a statistical distribution of alkyldiene groups is obtained. Accordingly, the metathesis of 2-pentene can afford a maximum 50% conversion to 2-butene and 3-hexene [Eq. (70)]. With more complex alkenes, electronic and steric

$$2CH_3CH_3CH=CHCH_3 \rightleftharpoons CH_3CH=CHCH_3 + CH_3CH_2CH=CHCH_2CH_3 \quad (70)$$

factors play important roles in determining their thermodynamic stabilities and in some instances quite good yields of a particular alkene can be obtained. If thermodynamic data are available potential yields may be estimated.

The "degenerate metathesis" of α-alkenes illustrated in Eq. (71) is a facile process that may have application in preparing some deuterated

$$RCH{=}CH_2 + R^1CH{=}CD_2 \;\rightleftharpoons\; RCH{=}CD_2 + R^1CH{=}CH_2 \qquad (71)$$

α-alkenes, but normal metathesis of such alkenes forming ethylene is slow compared with that of unhindered internal alkenes. However, owing to its volatility, loss of ethylene from the metathesis of α-alkenes drives the equilibrium in the desired direction, and this reaction has been employed in several syntheses. For instance, the plant growth regulator triacontanol has been prepared by isomerization and oxidation of 15-triacontene [Eq.

$$CH_3(CH_2)_{13}CH{=}CH_2 \xrightarrow[\text{80°C, 5 h}]{WCl_6,\ Bu_4Sn} CH_3(CH_2)_{13}CH{=}CH(CH_2)_{13}CH_3 + C_2H_4 \quad (72)$$

(72)] obtained from the metathesis of 1-hexadecene.[137] A number of insect *Pheromones* have been synthesized from readily available alkenes via catalytic metathesis. Yields of the desired alkene are not high, typically about 10%, but this route has the advantage of being a single-stage process using relatively cheap reactants, and requiring only short reaction times (sometimes as short as 5 min). In this way a 10% yield of 9-tricosene, the *cis*-isomer of which is a pheromone of the house fly (*Musca domestica*), is obtained after 5 min from the metathesis of 9-octadecene (4 mol) and 2-hexadecene (1 mol) [Eq. (73)] at 50°C in the presence of a catalyst derived from WCl_6, $EtAlCl_2$, and ethanol in a molar ratio of $1:3:5$, with an alkene-to-tungsten molar ratio of about a thousand.[138–140]

$$CH_3(CH_2)_7CH{=}CH(CH_2)_7CH_3 + CH_3(CH_2)_{12}CH{=}CHCH_3$$
$$\xrightarrow[\text{EtAlCl}_2/\text{EtOH}]{WCl_6} CH_3(CH_2)_7CH{=}CH(CH_2)_{12}CH_3 + CH_3(CH_2)_7CH{=}CHCH_3 \qquad (73)$$

15-Triacontene from 1-Hexadecene [Eq. (72)][137]

1-hexadecene (8.6 ml, 30 mmol), tungsten hexachloride in tri-chloroethylene (22 ml, 1.4 mmol), acetonitrile (0.05 ml, 0.96 mmol), and tetrabutyltin in trichloroethylene (6.0 ml, 2.8 mmol) are injected under argon into a 50-ml flask equipped with a reflux condenser and a magnetic stirrer. The mixture is maintained at 80°C under argon for 5 h, before filtering through alumina to remove tungsten compounds. Distillation under reduced pressure (172–174°C, 0.015 mm Hg) affords 15-triacontene (2.5–3.8 g, 40%–60%). Recrystallization from benzene gives the product as white crystals (mp 53–55°C).

Further metathesis of initially formed products is a potential problem that is reduced by restricting reaction times. Other side reactions include isomerization (double-bond migration), oligomerization, and alkylation of aromatic solvent. These may be minimized by employing low temperatures as well as short reaction times, by careful choice of catalyst system, and by selective "poisoning" with an additive as discussed above. Chlorinated solvents are resistant to alkylation, and chlorobenzene and chloroform are often used as inert solvents. However, the most serious limitation to the widespread use of alkene metathesis is the sensitivity of the catalysts to functional groups, most of which (other than hydrocarbon residues) are strong inhibitors unless they are remote from the double bond. Thus 1-chloro-9-octadecene undergoes metathesis over rhenium-based heterogeneous catalysts,[141] as do long-chain unsaturated fatty acid esters.[142] Similarly *p*-halogenoallylbenzenes undergo normal metathesis, but the presence of a methoxy group in the ring completely inhibits the reaction.[143] Interestingly, although unsaturated amines will not react, quaternary ammonium derivatives do, presumably due to the removal of the basic character of the nitrogen atom.[144] It is to be hoped that means of protecting a range of functional groups will be devised to increase the general usefulness of alkene metathesis.

An interesting application of alkene metathesis is the ring-opening polymerization of cyclic alkenes. As expected, the cyclooctene derivative in Eq. (74) gives an unusual alkenamer.[145] Reactions of this kind have been

$$n \quad \rightarrow \quad \left[=CH(CH_2)_2CH-CH(CH_2)_2CH= \right]_n \qquad (74)$$

widely studied but have found few synthetic applications owing to their lack of selectivity. By carefully controlling conditions, metathesis of cycloalkenes offers the possibility of forming macrocyclic polyenes as well as alkenamers. For example, the dimer of cyclooctene, 1,9-cyclohexadecadiene, was obtained in some 20% yield from cyclooctene [Eq. (75)].[146]

$$2 \quad \rightarrow \qquad (75)$$

2.2.2. Dehalogenation of vic-Dihalides and Coupling of gem-Dihalides, Aldehydes, and Ketones

During recent years titanium- and to a smaller extent vanadium chloride-based reagents have been shown to be particularly effective in the reductive coupling of dihalides and carbonyl compounds to alkenes. Thus, the reduced species derived from $TiCl_3$ or $TiCl_4$ and a strong reducing agent such as $LiAlH_4$ are very effective for dehalogenating *vic*-dihalides and for coupling *gem*-dihalides to substituted ethylenes[147] as illustrated in Eqs. (76) and (77). Related vanadium reagents sometimes induce these reactions, though generally they are less effective than titanium reagents, which have the advantages of being more reliable than older methods based on zinc metal and less hazardous than sodium in liquid ammonia.

$$2Ph_2CCl_2 \xrightarrow[96\%]{TiCl_3/LiAlH_4} Ph_2C{=}CPh_2 \qquad (76)$$

$$\text{\raisebox{0pt}{\includegraphics}} \xrightarrow[78\%]{TiCl_3/LiAlH_4} \qquad (77)$$

Cyclohexene from vic-dibromocyclohexane [Eq. (77)]
    Lithium aluminum hydride (1.9 g, 0.05 mol) is added in small portions to a stirred slurry of $TiCl_3$ (15.4 g, 0.1 mol) in THF (200 ml) under a nitrogen atmosphere. Stirring of the hydrogen-evolving black slurry is continued for 30 min. The *vic*-dibromocyclohexane (24.2 g, 0.1 mol) in THF (50 ml) is added over 30 min before refluxing the mixture for 8 h. Addition of aqueous ammonium chloride followed by ether extraction and distillation affords the product in 78% yield.

Before the introduction of low-oxidation-state titanium reagents, no general, direct method was available for reductively coupling aldehydes and ketones to alkenes. The reagent derived from zinc and $TiCl_4$ is convenient and effective for coupling *aromatic* aldehydes and ketones[148] [Eq. (78)].

$$2Ph_2CO \xrightarrow[reflux,\,97\%]{Zn/TiCl_3/THF} Ph_2C{=}CPh_2 \qquad (78)$$

More strongly reducing agents are, however, required to couple *saturated* ketones and aldehydes. The reagent obtained from $LiAlH_4$ and $TiCl_3$ (or $TiCl_4$) works particularly well for saturated and conjugated reactants, and affords high yields of alkyl-substituted alkenes. Because double bonds are not rapidly reduced by these systems a high degree of unsaturation in the reactant can be tolerated. This is illustrated by the coupling of retinal to

β-carotene shown in Eq. (79).[149] In some instances the more reactive,

$$(79)$$

though more hazardous, reagent formed from potassium metal and $TiCl_3$ may be used with otherwise-resistant alkyl-substituted reactants.[151] With this reagent intramolecular coupling of dicarbonyls to give good yields of cyclic alkenes is possible.[152] However, large ring systems have been prepared from saturated dialdehydes and diketones using the safer reagent derived from a zinc–copper couple and $TiCl_3$ in dimethoxyethane[153] [Eq. (80)]. The yields of large cycloalkenes prepared in this way are surprisingly

$$CHO(CH_2)_{12}CHO \xrightarrow[\text{reflux 1 h, 71\%}]{\text{Zn/Cu, TiCl}_3\text{, DME}} \qquad (80)$$

Table 2.1. Formation of Alkenes via Reductive Coupling with Titanium Chloride Reagents

Reactants	Product	Reagent	Refs.
Ar_2CO	$Ar_2C{=}CAr_2$	$TiCl_4/Zn$	148
R_2CO	$R_2C{=}CR_2$ (R = alkyl)	$TiCl_3/LiAlH_4$	149
	(R = alkyl/aryl)	$TiCl_3/Li$ or K	151, 152
(ring)COR / COR	(ring with R, R)	$RiCl_3/Zn/Cu$	150, 152
$ArRCX_2$	*trans*-$ArRC{=}CRAr$	$TiCl_3/LiAlH_4$	147
(ring X, X)	(ring)	$TiCl_3/LiAlH_4$	147

high, and this type of coupling reaction is now a preferred route to such compounds (see Chapter 3). Unsymmetric aryl/alkyl tetrasubstituted alkenes have been obtained in good yield by coupling two different ketones with TiCl$_3$/Li [Eq. (81)]. With alkyl ketones an excess of one ketone must

$$CH_3(CH_2)_4CHO + PhCHO \xrightarrow[84\%]{TiCl_3/Li} PhHC{=}CH(CH_2)_4CH_3 \qquad (81)$$

be used to suppress self-coupling.[154] When preparing an unsymmetric dimethyl alkene, acetone is used in excess, which has the further advantage that its self-coupling product, tetramethylethylene, is volatile and easily removed from the product mixture.[155]

In all of the titanium-mediated reactions discussed above, readily reducible functional groups such as nitro, sulfoxide, etc., are not expected to survive the reaction conditions. However, in most cases aromatic aldehydes and ketones with acyloxy, carbomethoxy, and tosyloxy groups are unaffected under normal reaction conditions.[150]

2.2.3. Deoxygenation of Epoxides

Deoxygenation of epoxides is useful for the preparation of certain alkenes, in structure determination, and when epoxidation is employed to protect an alkene. A number of strongly reducing metals and metal salts such as zinc,[156] magnesium amalgam,[157] low-oxidation-state iron,[158] chromium,[159] and titanium compounds[160,161] are effective. Thus, ferric chloride in THF when treated with three equivalents of *n*-butyl lithium under an inert atmosphere gives a black solution[158] that reduces a variety of epoxides to alkenes at room temperature, though refluxing is advantageous with sterically hindered epoxides. Similarly the reagent prepared in THF from TiCl$_3$ and LiAlH$_4$ smoothly reduces epoxides at reflux temperature [Eq. (82)]. The optimum TiCl$_3$/KiAlH$_4$/epoxide ratio was shown[160,161] to be $4:1:2$. Reductions of this type usually take place non-stereospecifically, though the FeCl$_3$/BuLi reagent affords a high yield of *cis-* or *trans*-stilbene from the corresponding isomeric epoxide.[158]

$$CH_3(CH_2)_7\overset{\displaystyle O}{\overset{\diagup\,\diagdown}{CHCH_2}} \xrightarrow[\text{reflux 3 h, } 68\%]{TiCl_3/LiAlH_4} CH_3(CH_2)_7CH{=}CH_2 \qquad (82)$$

Deoxygenation of an Epoxide with TiCl$_3$/LiAlH$_4$

To a stirred slurry of TiCl$_3$ (3.1 g, 20 mmol) in dry THF (60 ml) under nitrogen is added in small portions LiAlH$_4$ (0.2 g, 5 mmol). Hydrogen is evolved, and after stirring for 15 min a solution of the epoxide (10 mmol) in

THF (10 ml) is added to the mixture. After refluxing for 3 h the cooled reaction mixture is diluted with water and extracted several times with ether. The combined extract is washed with brine, dried ($MgSO_4$), and solvent removed under reduced pressure to afford the desired alkene. Typically, yields are in the range 40%–96%.

Deoxygenation of epoxides to alkenes with retention of stereochemistry may be achieved[162] by using the complex anion $C_5H_5Fe(CO)_2^-$ in a sequence of reactions that do not require isolation of intermediates [Eq. (83)]. However, if the alkoxide is decomposed by

$$\text{(83)}$$

refluxing, rather than by using iodide, the alkene produced in high yield has inverted stereochemistry relative to the starting epoxide.[163] These procedures appear well suited for diaryl and dialkyl epoxides and are not complicated by unsaturation in the substrate. A more convenient means of deoxygenating ester-bearing epoxides (and no doubt epoxides in general) with inversion of stereochemistry makes use of commercially available $Co_2(CO)_8$ (0.15 equivalents) at room temperature.[164] After 18 h almost quantitative yields of the inverted alkene are obtained [Eq. (84)]. With

$$\text{(84)}$$

larger amounts of $Co_2(CO)_8$ (1 equivalent) the reaction is faster though less stereospecific. Alkene-forming reactions related to deoxygenation of epoxides include the ferrous-iodide-induced deamination of 2-benzoyl-aziridines [Eq. (85)], which proceeds with retention of stereochemistry,[165] and the reduction of bromohydrins with $TiCl_3/LiAlH_4$ [Eq. (86)], which has the advantage over the more conventional zinc/acetic acid method in being nonacidic.[161,166]

$$\text{(85)}$$

$$\text{(86)}$$

Chapter 3

Formation of Carbocyclic Compounds

3.1. Introduction

Numerous general methods exist in organic chemistry for the synthesis of cyclic compounds.[1] The catalytic hydrogenation of a benzenoid substrate, for example, provides a convenient route to cyclohexane derivatives (Section 7.4). Generally, though, ring synthesis is achieved by intramolecular cyclization of a difunctional substrate. Typical is the synthesis of cycloalkanones, which is readily achieved by a number of methods (Table 3.1). These methods require that two functional groups in the same molecule react together, i.e., intramolecularly rather than intermolecularly; therefore high dilution and small-to-medium ring sizes are preferred.

Transition metals have been used to promote intramolecular cyclization, e.g., nickel-promoted coupling of allylic dibromides or the reductive

Table 3.1. General Methods for the Formation of Cycloalkanones

Reaction	Method	Preferred ring size
Dieckmann condensation[2]	Dicarboxylic ester + sodium alkoxide	5,6,7 ·
Thorpe Ziegler condensation[2]	Dinitrile + sodium alkoxide	7,8,15
Ruzicka synthesis[3]	Distillation of a dicarboxylic acid metal salt	5,6,7
Cyclic acyloin condensation[4]	Dicarboxylic ester + sodium	9–14

coupling of diketones, but only in a few cases do they offer advantages over more conventional organic methods. The oxidative coupling of terminal alkynes promoted by copper catalysts (Sondheimer method)[5] is perhaps one of the best known transition-metal-catalyzed reactions in organic synthesis and has been used to synthesize cyclic products containing as many as 54 carbon atoms.

The real advantage of transition metals in forming cyclic products lies in their ability to promote the cycloaddition of unsaturated substrates, a reaction not easily achieved by conventional methods. Alkynes, alkenes, 1,2, and 1,3 dienes coordinate to transition metals and undergo cycloaddition reactions to produce 3- to >30-carbon atom rings. In many cases these reactions are catalytic and require only small amounts of organometallic reagent to afford large quantities of cyclic product. The noble metals Os, Rh, Pd, and especially Co and Ni are the preferred transition metals for this reaction.

The cycloaddition reaction is exemplified by the nickel-catalyzed cyclooligomerization of butadiene (Scheme 1).[6] By carefully controlling the conditions, divinylcyclobutane, 2-vinyl methylene cyclopentane, 4-vinyl cyclohexene, cyclooctadiene, and cyclododecatriene can all be obtained in good-to-excellent yield from butadiene. By incorporating other unsaturated substrates, e.g., an alkene or alkyne, cyclodecadienes and cyclodecatrienes, respectively, can be obtained. Many of these products require complex alternative procedures for their preparation, yet by using organometallic reagents they are readily derived, catalytically, from simple starting materials in one step. The butadiene co- and homo-oligomerizations often do not require the use of pressure equipment but are performed at atmospheric pressure by bubbling butadiene through the reaction. Only when very volatile comonomers are employed, e.g., ethylene, acetylene, or allene, is pressure equipment necessary.

Although many of the reactions described are catalytic, a number of stoichiometric reactions are also discussed. These reactions generally require expert air-sensitive handling techniques and therefore only selected examples are included. The reactions of stable metal carbene complexes, for instance, are numerous[7] but are not included; similarly iron cyclobutadiene reactions[25] are only briefly mentioned.

3.2. Formation of Three-Membered Rings

Transition-metal-promoted carbene addition to alkenes or alkynes, and promoted cyclization provide practical routes to three-membered car-

Scheme 1

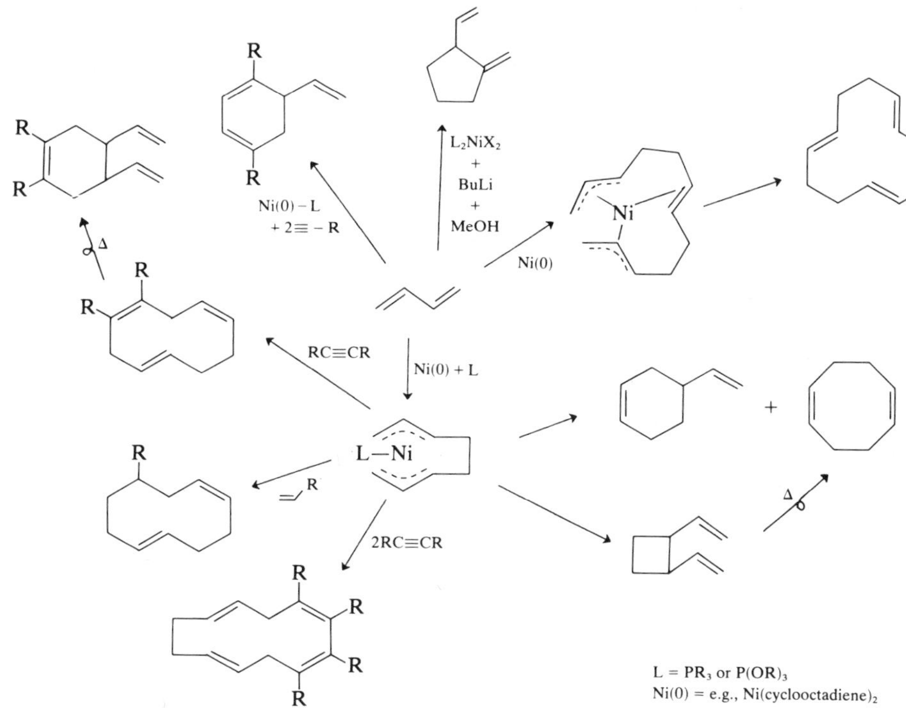

bocyclic rings. The former method is the most studied because a three-membered ring incorporates considerable strain energy, and therefore intramolecular cyclization to form a cyclopropanoid derivative is not a favored process. Carbene generation is possible both thermally and chemically without the use of transition metals, but transition metals have the ability to stabilize the very reactive carbene moiety. Many examples of stable transition metal–carbene complexes are known,[8] but little is known of the transient metal–carbene species described below and their existence is mainly by inference. Although stable, well-characterized organometallic carbene complexes react with alkenes to produce cyclic derivatives, this is not always so and numerous other reactions have been observed.[7] The exact role of the transition metal is therefore not well understood.

3.2.1. Carbene Addition to Alkenes or Alkynes

3.2.1.1. Using gem-Dihalo Compounds

The organozinc reagent obtained from methylene iodide and a copper–zinc couple reacts with alkenes to form a three-membered ring (the Simmons–Smith reaction)[9] [Eq. (1)].

$$\overset{\diagdown}{\underset{\diagup}{\overset{\displaystyle C}{\underset{\displaystyle C}{\|}}}} \ + \ CH_2I_2 \ + \ Zn(Cu) \ \xrightarrow[\text{ether}]{\text{reflux}} \ \overset{\diagdown}{\underset{\diagup}{\overset{\displaystyle C}{\underset{\displaystyle C}{\Big|}}}}\!\!CH_2 \ + \ ZnI_2 \ + \ (Cu) \tag{1}$$

The zinc–copper couple is simply refluxed in a solution of methylene iodide, alkene, and ether. An iodine crystal may be added to promote formation of the zinc reagent. A large excess of zinc reagent (16 : 1 Zn–CH_2I_2 : alkene) is used to ensure high conversion of alkene (up to 80%), and reaction times vary from one hour to two days depending on the substrate. A wide variety of alkenes have been examined and full experimental details are available.[9] Alkynes have also been studied but the reaction is complex and cyclopropenes are not obtained. A development of the Simmons–Smith reaction uses copper powder in place of zinc–copper couple[10] [Eq. (2)]. Trihalomethanes have been examined with this reagent and produce monohalocyclopropanes [Eq. (3)].

$$\text{(octagon)} \ + \ Br_2CHCO_2Me \ \xrightarrow[55°C,\ 50\ h]{Cu} \ \text{(fused ring)}\ \underset{CO_2Me}{\overset{H}{C}} \tag{2}$$

71%
exo : endo 1.3 : 1

$$\text{(hexagon)} \ + \ CHClI_2 \ \xrightarrow[2.5\ h]{Cu,\ 70°C} \ \text{(fused ring)}\ \underset{H}{\overset{Cl}{C}} \tag{3}$$

48%
exo : endo 1 : 2

These reactions are thought to proceed through organometallic intermediates and not via free carbenes, hence a preference for one isomer is often observed. Stereoselectivity is readily understood on the basis of a crowded transition state and prediction of the predominant isomer can often be made from simple model considerations.

3.2.1.2. Using Diazoalkanes

Many transition metals catalyze cyclopropanation with diazoalkanes [Eq. (4)]. For example, copper and silver salts, ligand-modified copper,

$$R \overset{\diagdown}{\underset{R'}{}} C{=}N{=}N + \underset{\diagup C \diagdown}{\overset{\diagdown C \diagup}{\|}} \xrightarrow{\text{Catalyst}} \overset{R}{\underset{R'}{}} \overset{\diagdown}{C} \overset{\diagup C \diagdown}{\underset{\diagdown C \diagup}{|}} + N_2 \qquad (4)$$

cobalt, palladium, and rhodium catalysts as well as Lewis acid metal halides have all been studied. Copper salts have received the most attention.[11] Alkenes, allenes, and alkynes are all suitable substrates [Eqs. (5), (6) and (7)] and *cis* addition is often observed. When two geometric isomers are possible from *cis* addition, the isomer that requires the least crowded transition state predominates (cf. the Simmons–Smith reaction, Section 3.2.1.1). The mechanism and stereochemistry of these reactions have been extensively surveyed.[12]

$$\bigcirc\!\!\!\!/ \ + \ N_2CHCO_2Et \ \xrightarrow[\text{Et}_2\text{O, reflux}]{\text{Cu}} \quad (5)$$

30%
exo : endo, 85 : 15

$$Bu^n CH{=}C{=}CH_2 + N_2CHCO_2Et \xrightarrow{CuCl_2}$$

$$\underset{\substack{Bu^n \\ 10\%}}{\overset{CO_2Et}{\triangle}} CH_2 \ + \ \underset{Bu^nCH \ 4\%}{\overset{CO_2Et}{\triangle}} \ + \ \underset{\substack{Bu^n \\ 12\%}}{\overset{CO_2Et}{\triangle}} CO_2Et \qquad (6)$$

$$Pr^n C{\equiv}CPr^n + N_2CHCO_2Et \xrightarrow[70°C]{CuSO_4} \underset{\substack{Pr^n \quad Pr^n \\ 85\%-95\%}}{\overset{H \quad CO_2Et}{\triangle}} \qquad (7)$$

Full experimental details and a tabular survey of known reactions have been given.[11] Yields vary considerably due to a number of side reactions such as C–H insertion, carbene dimerization, and alkyl radical dimerization [Eq. (8)].

$$\bigcirc\!\!\!\!/ \ + \ N_2C(CO_2Me)_2 \xrightarrow{\text{copper catalyst}} \underset{CO_2Me}{\overset{CO_2Me}{C}} \ + \ \overset{\substack{CO_2Me \\ H{-}C{-}CO_2Me}}{\bigcirc\!\!\!\!/} \qquad (8)$$

$+ (MeO_2C)_2C{=}C(CO_2Me)_2$
$+ (MeO_2C)_2HCCH(CO_2Me)_2$

General Procedure—Preparation of Ethyl trans-2,3-di-n-propylcyclopropane-1-carboxylate

Ethyl diazoacetate (19.38 g, 180 mM) is added slowly to a rapidly stirred mixture of *trans*-4-octene (38.08 g, 360 mM) and anhydrous cupric sulfate at 90°C. After complete reaction (2 h) the mixture is filtered to remove the catalyst and the filtrate distilled under reduced pressure to give *trans*-4-octene (21 g) plus a mixture of ethyl *trans*-2,3-di-*n*-propylcyclopropane-1-carboxylate and diethyl fumarate (20.8 g). The second fraction is extracted with cold 2% KMnO₄ (aqueous), until the aqueous solution retains a permanent purple color (equivalent to 1.2 g KMnO₄) to remove the by-product. The remaining cyclopropyl ester is redistilled (19 g, 62%, based on octene used).

The problem with the above reactions is the instability of the diazo compound, some of which are explosive when dry. This seriously limits the type of carbene species that can be generated easily.

Palladium salts can also be used to prepare cyclopropane derivatives from diazocompounds. Diazomethane in the presence of palladium acetate effectively adds a methylene group to α,β-unsaturated ketones[13] [Eq. (9)].

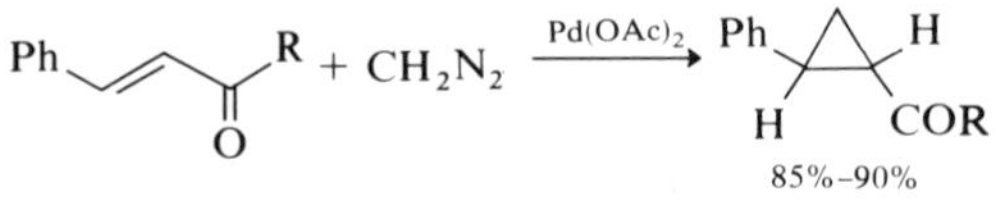

$$\text{Ph} \diagup\diagdown R + CH_2N_2 \xrightarrow{Pd(OAc)_2} \quad (9)$$

85%–90%

High yields are only achieved with (a) disubstituted alkenes, (with tri- or tetra-substitution no reaction occurs) and (b) diazomethane (for instance, with ethyl diazoacetate yields fall to 50%). This procedure was developed for use with α,β-unsaturated steroidal ketones and produces good results providing the double bond is not trisubstituted [Eq. (10)].

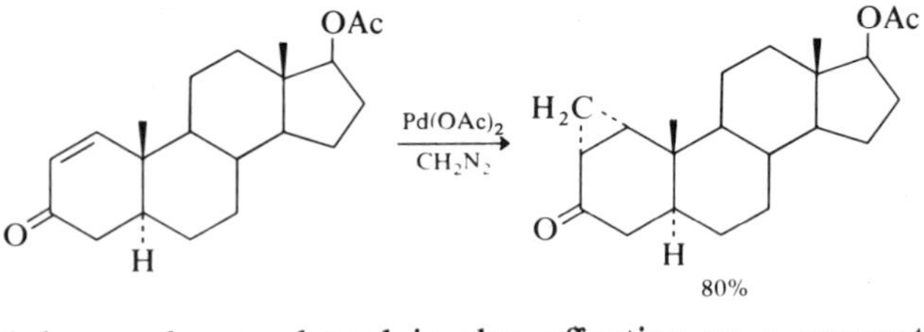

$$\xrightarrow[CH_2N_2]{Pd(OAc)_2} \quad (10)$$

80%

Molybdenum hexacarbonyl is also effective as a promoter for the addition of diazo compounds to alkenes.[14] α,β-Unsaturated esters and nitriles readily add diazocarbonyl compounds at 65°C in the presence of a catalytic amount of molybdenum(0) [Eq. (11)]. Yields of 50%–70% are reported.

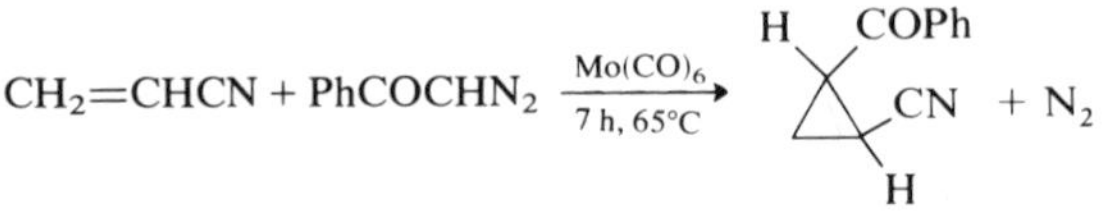

$$CH_2{=}CHCN + PhCOCHN_2 \xrightarrow[7\,h,\,65°C]{Mo(CO)_6} \quad + N_2 \quad (11)$$

Cyclopropanation reactions using copper or palladium catalysts are sensitive to substitution at the double bond and, with these crowded alkene substrates, rhodium(II) carboxylates have proven more effective catalysts.[15] The efficiency of the rhodium complex depends on its solubility, and freely soluble butanoate or pivaloate derivatives are preferred. Best results are obtained using butyldiazoacetate as the carbene source [Eq. (12)].

$$
\begin{array}{c}
R^1 \\
 \\
R^2
\end{array}
C{=}C
\begin{array}{c}
R^3 \\
 \\
R^4
\end{array}
+ N_2CHCO_2R \xrightarrow{Rh(O_2CBu^t)_2}
\begin{array}{c}
H \quad CO_2R \\
R^1 \quad \quad R^3 \\
R^2 \quad \quad R^4
\end{array}
\qquad (12)
$$

$R^1 = R^3 = Et, R^2 = R^4 = H; R = Me(68\%), Bu(100\%);$
$R^1 = R^2 = R^3 = R^4 = Me; R = Me(75\%);$
$R^1 = Me, R^2 = R^3 = H, R^4 = n\text{-}C_5H_{11}; R = Me(77\%)$
$R^1 = R^4 = Pr^n, R^2 = R^3 = H; R = Bu^n(85\%)$

3.2.1.3. Asymmetric Cyclopropanation

The main commercial use of catalyzed cyclopropanation reactions is the preparation of insecticides based on the chrysanthemate esters (**1**). These are among the most potent pesticides known. Four possible isomers of **1** exist; *cis* and *trans* geometrical isomers about the cyclopropyl ring

1

and the R and S form of each of these geometrical isomers. Only one of these isomers has high biological activity and work has therefore been directed to selective formation of this isomer.[16] Asymmetric cyclopropanation is important in this context. Because the transition metal is involved in the transition state, the ligand environment around the metal center can influence the stereochemistry of the reaction. The Simmons–Smith reaction employing chiral substrates and the cyclopropanation with diazo compounds under the influence of chiral copper complexes generally gives poor optical yields (<10%). Even so, by carefully matching the catalyst, ligand, and diazo compound, excellent enantiomeric excess of one isomer has been achieved using copper systems.[17] The chiral copper catalyst **2**, prepared

2   (R = 5-*t*-butyl-2-octyloxyphenyl)

from an optically active α-amino acid, with ethyldiazoacetate gives reasonable optical yields (60%–70%) of the chrysanthemate ester. But, by using a chiral diazoalkane (e.g., l-menthyldiazoacetate), the optical yield is greatly improved (80%–90%) [Eq. (13)].

$$\text{(structure)} + N_2CHCO_2R \xrightarrow{2} \underset{trans}{\text{(structure) CO}_2\text{R}} + \underset{cis}{\text{(structure) CO}_2\text{R}} \qquad (13)$$

R = l-menthyl; chemical yield = 67%, optical yield = 87% (*trans*), 25% (*cis*)

The matching of all reactants required to obtain high optical yields with copper catalysts prompted investigation for a more general catalyst, and chiral cobalt systems were found to be useful.[18] Air-sensitive cobalt(II) complexes of camphorquinonedioximes (3) are active in the cyclopropanation of alkenes with diazoacetates[19] [Eqs. (14) and (15)]. The optical yield

3 (cqdH)

$$PhCH=CH_2 + N_2CHCO_2Et \xrightarrow{Co(cqd)_2 \cdot H_2O} \underset{\text{optical yield } 67\%}{\text{(structure) Ph CO}_2\text{Et}} + \underset{\text{optical yield } 75\%}{\text{(structure) CO}_2\text{Et}} \qquad (14)$$

chemical yield 92%

$$Ph_2C=CH_2 + N_2CHCO_2Et \xrightarrow{Co(cqd)_2 \cdot H_2O} \underset{\substack{\text{optical yield } 70\% \\ \text{chemical yield } 95\%}}{\text{(structure) Ph CO}_2\text{Et}} \qquad (15)$$

and *trans : cis* ratio improve with increasing bulk of the diazoacetate. For example, in Eq. (14) using ethyl diazoacetate the optical yield is 75% (*trans* isomer) ($t:c$ 0.85), using iso-butyl diazoacetate the optical yield is 80% (t) ($t:c$ 0.92), and using neopentyldiazoacetate the optical yield is 88% (t) ($t:c$ 2.34). This catalyst has two other interesting features. With conjugated dienes regioselectivity is observed for cyclopropanation of the less hindered terminal double bond [Eq. (16)] and, secondly, whereas

$$Ph\text{(structure)} + N_2CHCO_2Et \xrightarrow{Co(cqd)_2} PhCH=CHCH-CHCO_2Et \atop \underset{H_2}{\overset{C}{\diagdown \diagup}} \qquad (16)$$

92%

attempted cyclopropanation of activated alkenes (e.g., methyl acrylate) generally results in 1,3-dipolar addition to form pyrazolines (**4**), in the presence of $Co(cqd)_2$ a cyclopropane is produced with accompanying optical induction, although yields are low (11%) [Eq. (17)]. In common with other cyclopropanation reactions there are numerous restrictions; e.g.,

$$\text{(17)}$$

optical yield 33%

alkylalkenes, cycloalkenes or dienes, sterically hindered dienes, vinyl ethers, allenes, and alkynes cannot be cyclopropanated. Interestingly, palladium catalysts containing chiral ligands do not produce asymmetric induction.[20]

3.2.1.4. Formation of Cyclopropenes

In contrast to the copper or cobalt catalysts, and the Simmons–Smith reagent, rhodium carboxylates catalyze the cyclopropenation of terminal alkynes with diazoacetates [Eq. (18)].[21] The steric bulk of the substituent

$$\text{(18)}$$

at the triple bond does not significantly influence the yield, but polar groups do decrease it drastically (Table 3.2), owing to the reduced stability of the

Table 3.2. Rhodium-Catalyzed
Cyclopropenation [Eq. (18)]

R	Yield, %
Bu^n	84
Bu^t	86
Cyclohexyl	80
$MeCO_2CH_2$	40
CH_3OCH_2	46
$(EtC{\equiv}CEt)$	68

corresponding cyclopropene, which rearranges easily. Acetylenic alcohols also undergo cyclopropenation but insertion into the O—H bond [Eq. (19)] can also occur.

$$\tag{19}$$

	5, %	6, %
Propargyl alcohol	60	12
3-Butyn-1-ol	50	19
1-Pentyn-3-ol	54	21
3-Methyl-1-pentyn-3-ol	37	36

General Procedure—Rhodium-Catalyzed Cyclopropenation of Alkynes[21]

Methyl diazoacetate (4 mM) is added slowly to a stirred suspension of $Rh_2(O_2CMe)_4$ (0.05 mM) in the alkyne (10 mM) (substrate:catalyst, 200) at 25°C under a nitrogen atmosphere. After complete reaction (10 h) the mixture is filtered and the filtrate distilled to give the cyclopropene in up to 86% yield.

3.2.2. Cyclization Reactions

Intramolecular α,ω-eliminations have been widely used to prepare cyclopropane derivatives.[22] The reaction is effected by a variety of reagents including, for example, $AlCl_3$, $HgCl_2$, or halogens. One convenient method employs $TiCl_4$ to promote the cyclodesilylation of butenyl silanes with acid chlorides [Eq. (20)]. A variety of aliphatic, aromatic, and olefinic acid

$$\tag{20}$$

R = Me(38%), Pr^i(55%), Bu^n(65%), 1-propene(51%), or Ph(50%)

chlorides produce cyclopropylmethylketones in reasonable yields. Low temperatures (−78°C) are required because the product can react further to give 2-chlorobutylketones. Several alkyl-substituted-3-butenylsilanes also react with pivaloyl chloride to afford the corresponding cyclopropyl derivatives, e.g., **7**, 67%, and **8**, 65%.

A second cyclization method for forming three-membered rings involves the intramolecular reductive coupling of 1,3 diols. $TiCl_3$–$LiAlH_4$ reductively cyclizes 1,3-diphenyl-1,3-propanediol to form a mixture of *cis*- and *trans*-1,2-diphenylcyclopropane[23] [Eq. (21)]. Moderate yields are generally obtained, but long reaction times (2 days at reflux in glyme) are required.

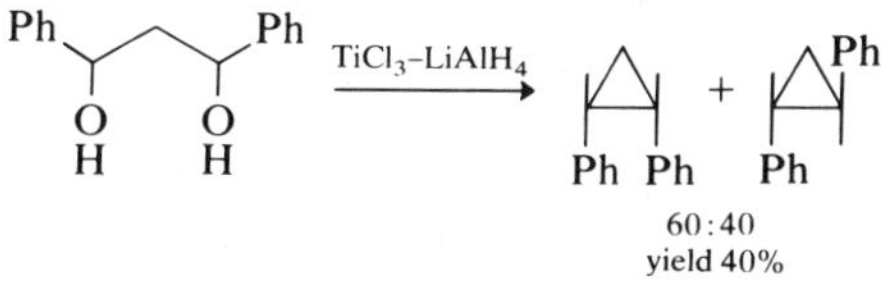

$$\text{(21)}$$

The use of trimethylsilyl as a leaving group during cyclization has already been mentioned [Eq. (20)]; ironcyclopentadienyl dicarbonyl can serve a similar function. 3-Methylbut-3-enyldicarbonylcyclopentadienyl-iron reacts regiospecifically with free radicals and with electrophiles at the δ carbon of the butenyl ligand, resulting in displacement of the metal and formation of cyclopropyl compounds in high yield[24] (Scheme 2). These reactions require a preformed iron complex, but this then reacts rapidly (a few seconds) at ambient temperature.

Scheme 2

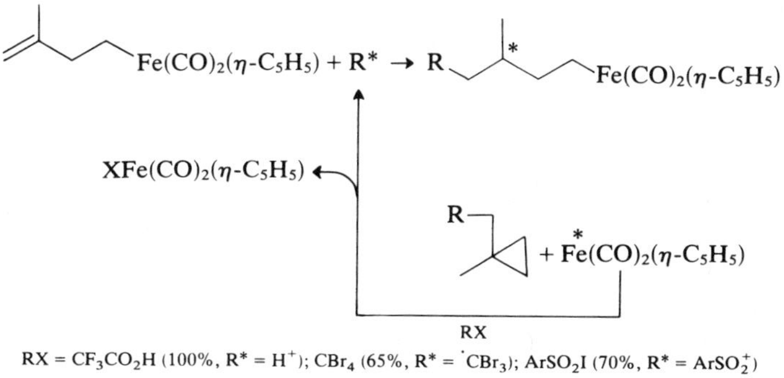

$RX = CF_3CO_2H$ (100%, $R^* = H^+$); CBr_4 (65%, $R^* = {}^\cdot CBr_3$); $ArSO_2I$ (70%, $R^* = ArSO_2^+$)

3.3. Formation of Four-Membered Rings

Transition-metal-catalyzed reactions provide a variety of methods for the formation of four-membered rings. Examples include the $2\pi + 2\pi$ cycloaddition of alkenes and the cyclization of a 1,4-functionalized compound.

A series of metal-stabilized, substituted cyclobutadiene complexes has been prepared including chromium, molybdenum, tungsten, iron, and cobalt derivatives. Cyclobutadienyl iron tricarbonyl has received most attention,[25] and although free cyclobutadiene is extremely labile, an extensive range of organic chemistry on the coordinated ligand has been developed. The cyclobutadienyl complex is not usually produced by the cycloaddition of two alkyne units but by reaction of 3,4-dichlorocyclobutene with an iron complex such as $Na_2Fe(CO)_4$. Many reactions can be performed on the cyclobutadienyl iron complex to produce a wide range of substituted rings. Unfortunately subsequent removal of the substituted cyclobutadiene from the iron center is not easy and the four-membered ring is rarely isolable. The reactions of coordinated cyclobutadiene will not be discussed further because they require specialized organometallic handling techniques and do not represent general synthetic methods.

3.3.1. 2π–2π Cycloaddition of Alkenes

2π–2π Cycloadditions are thermally forbidden and photochemically allowed[26] except when the alkenes are coordinated to a metal center; then photochemical initiation is not required. This type of four-membered ring synthesis usually requires activated alkenes (or alkynes), typically strained alkenes, but the dimerization of butadiene to divinylcyclobutane has been investigated in depth.

3.3.1.1. Dimerization of 1,3-Dienes

The reaction of butadiene with a variety of zero-valent nickel catalysts produces a range of cyclooligomers (see Scheme 1, Section 3.1), including divinylcyclobutane, vinylcyclohexene, and cyclooctadiene. To obtain maximum yield of the four-membered ring requires careful choice of catalyst, catalyst concentration, and conversion. Yields of divinylcyclobutane up to 40% have been achieved using Ni(cyclooctadiene)$_2$ plus tris(2-biphenylyl)phosphite [Ni(COD)$_2$–P(OC$_6$H$_4$—C$_6$H$_5$)$_3$, 1:1] and stopping reaction at 85% conversion[27,28] [Eq. (22)]. The major problem with this reaction is the subsequent Cope rearrangement of the cyclobutane to cyclooctadiene [Eq. (23)]. This reaction severely limits the overall yield of divinylcyclobutane, but, although yields are only moderate, the butadiene

$$2 \quad \xrightarrow[\text{50°C}]{\text{Ni(COD)}_2\text{-P(OC}_6\text{H}_4\text{-}o\text{-C}_6\text{H}_5)_3} \qquad (22)$$

40%   2%   57%

overall yield 85%

$$\text{(graphic)} \xrightarrow[\text{or (b) } 80°C]{\text{(a) NiL}_n, 20°C} \text{(graphic)} \qquad (23)$$

oligomerization is catalytic and can easily produce large quantities of product.

Procedure—Preparation of Divinylcyclobutane[27]
    Liquid butadiene is added to an equimolar mixture of Ni(COD)$_2$ and tris(2-biphenylyl)phosphite under nitrogen in an autoclave (typically butadiene:Ni, 70:1). The mixture is stirred at 50°C. After a suitable time (e.g., 5 h) the reaction is cooled to 10°C and the mixture slowly transferred to a flask at 40°C, under a vacuum (10^{-1}–10^{-2} mm Hg) so as to flash-distill the products from the catalyst. This method prevents excessive rearrangement to cyclooctadiene. The distillate is redistilled under vacuum and at low temperature to yield divinylcyclobutane (40%), vinylcyclohexene (2.5%), and cyclooctadiene (57%), plus other oligomeric material (1%).

    Substituted butadienes have also been dimerized using nickel(0) catalysts[29] [Eq. (24)], and a two-stage synthesis of the monoterpene gradisol has been developed using the dimerization of isoprene[30] [Eq. (25)]. Although the yield of the cyclobutane derivative (**9**) is low (~15%), the reaction has the advantage of being simple to perform on a large scale.

$$\text{R} \diagdown \diagup \xrightarrow[\text{R = Me or Et}]{\substack{\text{Ni(COD)}_2 \\ \text{P(OC}_6\text{H}_4\text{-}o\text{-C}_6\text{H}_5)_3}} \text{(graphic)} \qquad (24)$$

$$2 \diagup \diagdown \xrightarrow[\text{P(OC}_6\text{H}_4\text{-}o\text{-C}_6\text{H}_5)_3]{\text{Ni(COD)}_2} \underset{\substack{\textbf{9} \\ 12\%\text{–}15\%}}{\text{(graphic)}} \xrightarrow[\text{2. OH}^-]{\text{1. (Me}_2\text{CHCHMe)}_2\text{BH}} \text{(graphic)} \text{OH} \qquad (25)$$

    The dimerization of butadiene to divinylcyclobutane has been achieved using palladium salts containing noncoordinating anions such as ClO_4^- or BF_4^-.[31] No experimental details or yields are provided but this represents an unusual catalytic reaction of palladium, which usually produces linear structures from butadiene.

3.3.1.2. *Codimerization of Butadiene and Ethylene*

    Butadiene has also been codimerized with ethylene to form vinyl-cyclobutane[32] [Eq. (26)]. It is unusual for ethylene to be involved in a reaction leading to a four-membered ring; generally linear products

$$\| \; + \; \diagup\!\!\!\diagdown\!\!\!\diagup \xrightarrow[\text{135–150°C, 1200 psi}]{\text{Ti(CH}_2\text{Ph)}_4\text{–bipy, 30 min}} \quad \boxed{}\!\!\!\diagup\!\!\!\diagdown \tag{26}$$

$$\underset{74\%}{}$$

are formed. With nickel catalysts, for example, 1,4-hexadiene is produced. However a variety of titanium-based catalysts, including $Ti(CH_2Ph)_4$–bipy and $Ti(\eta\text{-}C_5H_5)(CH_2Ph)_3$, catalyze the cyclization reaction.

3.3.1.3. Cycloaddition of Strained Alkenes

As described earlier, cyclobutane derivatives can be prepared from strained ring alkenes using a variety of catalysts. Norbornene and norbornadiene have received the most attention.[33] The iron catalyst $Fe(acac)_3$–$AlEt_2Cl$ in the presence of $Ph_2PCH_2CH_2PPh_2$ gives a 20% yield of the four-membered ring species **10** from butadiene and norbornadiene.[34] Similarly, norbornene and butadiene in the presence of $Ni(Bu_3P)_2Br_2$–$NaBH_4$ produce a moderate yield of **11**.[35] Maintaining a solution of norbornadiene (4 mM), methylenecyclopropane (4 mM), $Ni(COD)_2$ (0.2 mM), and

$$\underset{\textbf{10}}{} \qquad\qquad\qquad \underset{\textbf{11}}{}$$

triphenylphosphine (0.22 mM) in benzene (10 ml) at 20°C for 24 h under nitrogen produces a single 1 : 1 adduct (**12**) in 86% yield[36] [Eq. (27)]. The formation of the individual dimers is suppressed in this reaction but they are known to form smoothly via individual catalyzed reactions. Thus methylenecyclopropane is dimerized by $Ni(COD)_2$ at $-15°C$ to give **13** in 10% yield[37] [Eq. (28)] and norbornadiene is dimerized by a variety of catalysts including nickel complexes.[33]

$$\underset{}{} + \underset{}{} \xrightarrow[\text{PPh}_3]{\text{Ni(COD)}_2} \underset{\textbf{12}}{} \tag{27}$$

$$2 \; \underset{}{} \xrightarrow[-15°C,\, 1\,h]{\text{Ni(COD)}_2} \underset{\textbf{13}}{} \tag{28}$$

The reaction of norbornene with allyl acetate catalyzed by nickel tetrakis(tri-isopropylphosphite) affords a mixture containing the cyclobutane derivative **14** and also **15** [Eq. (29)].[38] The intermediate **16** can either eliminate the nickel moiety to give **14** or rearrange prior to elimination to give **15**. At higher temperatures rearrangement is favored (**14**:**15**, 3:7).

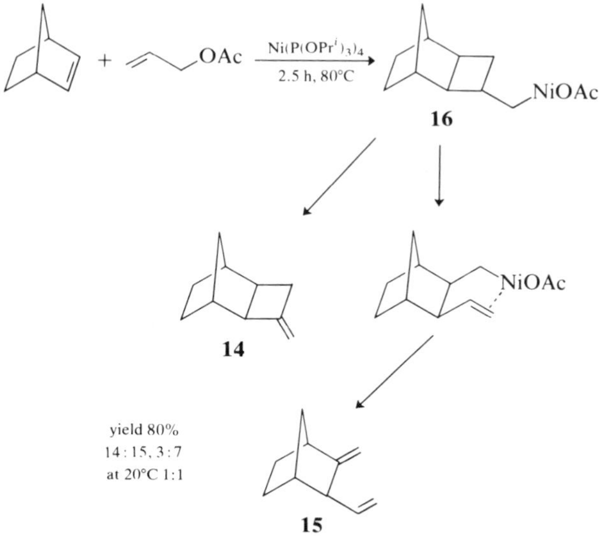

(29)

Procedure—Allylation of Norbornene

Allyl acetate (5 g, 50 mM), 2-norbornene (4.7 g, 50 mM), and $Ni(P(OPr^i)_3)_4$ (0.89 g, 1 mM), are heated to 80°C in THF (20 ml) for $2\frac{1}{2}$ h under an inert atmosphere. The solution is filtered and fractionally distilled, collecting the fraction boiling at 62–64°C/140 mm Hg. The two components of this fraction are separated by preparative gas liquid chromatography (glc) on a SE30 (5% silicone) column to give **14**, 1.6 g, and **15**, 3.75 g. Overall yield 80%.

Allyl alcohol also reacts with norbornene by a copper(I)-catalyzed photocycloaddition reaction[39] [Eq. (30)]. Using norbornadiene and allyl

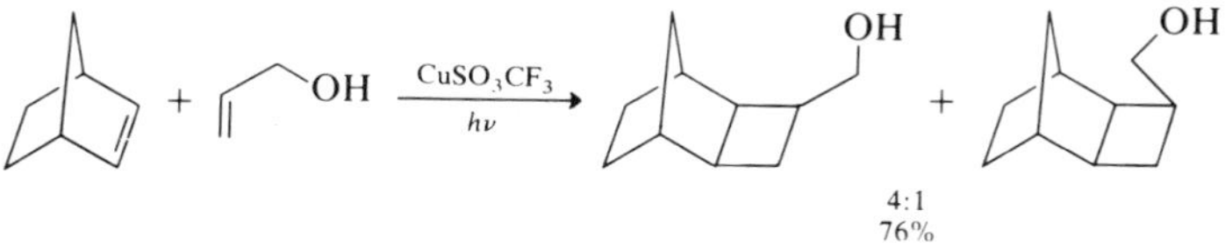

(30)

alcohol a mixture of two ethers is formed, **17** and **18**, probably as a result of an initial catalytic intramolecular photocyclization of the diene, to produce **19**, which then reacts with allyl alcohol. Copper(I)-catalyzed photocycloadditions are known for many cyclic alkenes though not for acyclic

17   18   19

alkenes.[40] Predictably, therefore, the dimerization of allyl alcohol is not a side reaction in the formation of **17** and **18**. Cyclopentene, cyclohexene, and cycloheptene all dimerize to give as the major product the $2\pi + 2\pi$ cyclodimer. Copper triflate is the favored catalyst for these reactions because it is able to coordinate alkenes more strongly than simple copper(I) halides.

General Procedure—Copper (I)-Catalyzed Photocycloaddition[40] *of*
Cyclohexene
    A solution of copper triflate (0.2 g) in cyclohexene (4 ml) and THF (1 ml) (sealed in a quartz tube) is irradiated at 254 nm for $4\frac{1}{2}$ days. The resulting mixture is dissolved in pentane, washed with aqueous potassium cyanide (30 ml, 2.5 M) to remove copper(I) salts, and dried over $MgSO_4$. Solvent and unreacted cyclohexene are removed by distillation and the residual oil is distilled under reduced pressure to yield 1.7 g (81%) of product, bp 96–98°C (10 mm Hg). The distillate consists of three dimers (**20, 21, 22**) which may be separated by column chromatography.

60%   10%   30%
20   **21**   **22**

3.3.1.4. Cyclodimerization of Alkenes, Alkynes, and Allenes

    The cyclodimerization of acyclic alkenes has been reported for the electron-rich, aromatic alkene, phenylvinylether in the presence of iron(III) or manganese(II) complexes.[41] Irradiation (300-W mercury arc) of an acetonitrile solution of phenylvinylether and $Fe(bipy)_3(ClO_4)_3$ for 40 h gives a mixture of cyclodimers **23** and **24** [Eq. (31)]. Unfortunately, no yields

(31)

23   24

1:1

are reported. Although 1,1-dimethylindene gives the same $2\pi + 2\pi$ cyclo-addition reaction, 1,1-diphenylethylene yields the six-membered ring **25**

25

by α-hydrogen abstraction. The dimerization of methylene cyclopropane has been mentioned earlier and related cyclopropene complexes have also been reported to dimerize.[42] Interestingly, palladium catalyzes the dimerization of 1-methyl and 1,3,3-trimethyl-cyclopropene[42] [Eq. (32)] as well as the cyclotrimerization of 1,1-dimethylcyclopropene[43] [Eq. (33)]. Linear

(32)

100%

(33)

oligomerization is usually observed with palladium catalysts. Dimerization of 1,1-dimethylcyclopropene is readily achieved using a nickel(0) phosphine catalyst[44] [Eq. (34)].

(34)

The cyclodimerization of allene has been reported but uses a rather inefficient catalytic method.[45] Allene is passed over the catalyst system $[Ph_2PC_6H_4PPh_2Ni(CO)_2]$ [prepared from $Ni(CO)_4$ and bis(diphenylphosphino)benzene] at 200°C. Volatile products consisting of 1,2- (8%) and 1,3-dimethylcyclobutane (36%) and trimers (16%) are isolated, but rapid catalyst deactivation occurs due to tar formation.

Allene has also been shown to react with strained alkenes such as norbornadiene[46] [Eq. (35)]. Heating (120°C) a mixture of allene and norbornadiene in the presence of bis(triphenyl)phosphine maleic anhydride palladium affords **26** in 25% yield. The product **26** can conceivably arise from the cycloaddition of methyl acetylene, formed *in situ* from allene, to norbornadiene; however, replacing methylacetylene for allene in the reaction does not produce **26**.

$$= \cdot = \; + \; \text{(norbornadiene)} \quad \xrightarrow[\text{120°C}]{\text{Pd(PPh}_3)_2\text{(MA)}} \quad \text{(product)} \tag{35}$$

26
25%

The $2\pi + 2\pi$ cycloaddition of alkynes and strained alkenes is known.[33,47] $RuH_2(PPh_3)_4$ catalyzes the addition of dimethylacetylenedicarboxylate to norbornene derivatives with moderate efficiency[48] [Eq. (36)]. Unfortunately only dimethylacetylenedicarboxylate is successful; other alkynes are either unreactive or decompose.

$$\text{(norbornene)} + MeO_2CC{\equiv}CCO_2Me \quad \xrightarrow[\text{80°C, 24 h}]{RuH_2(PPh_3)_4} \quad \text{(product)} \begin{array}{l} CO_2Me \\ CO_2Me \end{array} \tag{36}$$

27
52%

Procedure—Cycloaddition of Dimethylacetylenedicarboxylate to Norbornene[48]
A mixture of norbornene (10 mM), dimethylacetylenedicarboxylate (10 mM), and benzene (10 ml) in the presence of $RuH_2(PPh_3)_4$ (0.2 mM) is heated in a sealed tube at 80°C for 24 h. Distillation of the reaction mixture affords 1.2 g (52%) of **27**, bp 83–86°C, 0.25 mm Hg.

An interesting cyclobutene derivative is formed by the reaction of diphenylacetylene with a cobalt isocyanide complex.[49] Cobalt carbonyl reacts with, for example, 2,6-xylylisocyanide at 80–90°C over 1 h to form the dimeric red-brown complex $Co_2(C_9H_9N)_8$ (**28**) which is air sensitive. Reaction with diphenylacetylene at reflux gives two products, a cyclobutene (**29**) and a cyclopentene (**30**). The reaction is noncatalytic but provides a simple route to an interesting four- (and five-) membered ring.

$$Co_2(CO)_8 + RCN \rightarrow \underset{\textbf{28}}{Co_2(RCN)_8} \quad \xrightarrow{PhC{\equiv}CPh} \quad \underset{\textbf{29}}{\begin{array}{c} Ph \quad NR \\ Ph \quad NR \end{array}} + \underset{\textbf{30}}{\begin{array}{c} Ph \quad NR \\ {=}NR \\ Ph \quad NR \end{array}} \tag{37}$$

3.3.2. Cyclization Reactions

The number of cyclization reactions forming four-membered rings is not extensive. Although metallacyclopentadiene complexes are common in organometallic chemistry, liberation of a cyclobutadiene derivative (by C–C bond formation) does not readily occur. Further reactions of the metallacycle are numerous leading to five- and six-membered rings; these are discussed later (Sections 3.4.3.1, 3.5.1, 3.5.2).

A very simple and useful four-membered ring synthesis uses the ability of transition metals to readily catalyze the trimerization of alkynes to benzene derivatives.[50] For example, ring closure of a 1,5-hexadiyne with an alkyne in the presence of cyclopentadienylcobaltdicarbonyl is facile and produces a very useful cyclobutene derivative[51] (**31**) [Eq. (38)]. The strain

$$
\begin{bmatrix} -C \equiv CR \\ -C \equiv CR' \end{bmatrix}
+
\begin{matrix} R'' \\ C \\ \|\| \\ C \\ R''' \end{matrix}
\xrightarrow{\text{Co}(\eta\text{-}(C_5H_5)(CO)_2}
\text{(31)}
\tag{38}
$$

incorporated in forming the four-membered ring might be expected to lead to side reactions, but often reasonable yields of product are obtained (Table 3.3). The process is most efficient using bis(trimethylsilyl)acetylene, which does not homotrimerize (a problematic side reaction) in the presence of the cobalt catalyst. The preparation of substituted benzocyclobutenes is hampered by halogen, acetyl, or alkoxide substituents. Alkoxide substituents, for example, lead to the formation of naphthalene derivatives[52] [Eq. (39)]. This reaction is thought to proceed via an intermediate benzocyclobutene (**32**), followed by opening of the four-membered ring,

Table 3.3. *Cobalt-Catalyzed Formation of Benzocyclobutenes [Eq. (38)]*

R	Diyne R'	Alkyne R''	R'''	Product yield, %
H	H	CO_2Me	CO_2Me	14
H	H	Ph	H	17
H	H	Ph	Ph	48
H	H	$SiMe_3$	$SiMe_3$	65
H	H	CH_2OMe	$SiMe_3$	55
Me	Me	Ph	Ph	20
Me	Me	CO_2Me	CO_2Me	28
CH_2OMe	CH_2OMe	Me_3Si	H	25
Me_3Si	H	Me_3Si	CH_2OMe	16

(39)

incorporation of a second alkyne molecule, and loss of alkoxide. A series of organic syntheses have been developed based on these cobalt-catalyzed reactions. The trimethylsilyl group, for example, is a good leaving group in electrophilic substitution reactions and provides an entry into a variety of derivatized benzocyclobutenes. Further examples of cobalt-catalyzed alkyne cyclizations and the related chemistry are described in Section 3.5.1.

General Procedure—Preparation of 4-Trimethylsilyl-5-methoxymethyl-benzocyclobutene[53]

Two cyclization methods have been reported. The first uses bis(trimethylsilyl)acetylene as a solvent and reactant to yield bis(trimethylsilyl)-substituted products. This method is described later (Section 3.5.1). An alternative procedure uses an *n*-octane solvent and can be used to prepare other disubstituted derivatives.

To a refluxing solution of $Co(\eta-C_5H_5)(CO)_2$ (20 μl) in *n*-octane (60 ml) under nitrogen is slowly added, over 48 h, a mixture of 1,5-hexadiyne (780 mg, 10 mM) and trimethylsilylpropargyl methyl ether (1.42 g, 10 mM) containing $Co(\eta-C_5H_5)(CO)_2$ (30 μl) in *n*-octane (12 ml), using a syringe pump. After the addition is complete volatiles are removed under vacuum, leaving a brown oil that is chromatographed on silica eluting with pentane:ether (92:8). The pale yellow oil is redistilled (60°C/0.01 mm Hg) to yield 4-trimethylsilyl-5-methoxymethyl benzocyclobutene (1.2 g, 55%).

3.4. Formation of Five-Membered Rings

Five-membered carbocyclic rings are relatively strain free and therefore cyclization reactions play an important part in their preparation. Cycloaddition reactions, on the other hand, require an odd-numbered carbon fragment. No 2 + 2 + 1 cycloadditions involving two alkenes or alkynes plus a carbene moiety have been reported, but a few stoichiometric reactions using carbon monoxide or an isonitrile as the one carbon fragment are known. Numerous three-carbon fragments have been generated for the 2 + 3 cycloaddition reaction, which is often very efficient. The dimerization of dienes, especially butadiene, can also lead to five-membered ring formation.

3.4.1. Cycloaddition Reactions Involving Alkenes, Alkynes, and Dienes

3.4.1.1. Butadiene Cyclooligomerization

The cyclodimerization of butadiene using nickel–phosphine catalysts favors formation of mixed products: divinylcyclobutane, vinylcyclohexene, and cyclooctadiene. By careful choice of catalyst and reaction conditions, each of these products may be selectively obtained in good yield (see Sections 3.3.1.1, 3.5.4, 3.7.1). Using a different nickel–phosphine catalyst, exclusive formation of a five-membered carbocyclic compound has been achieved[54] [Eq. (40)]. The catalyst produced by reaction of bis(trialkylphosphine)nickel dihalide and butyl lithium in the presence of alcohol (25–40 molar amounts) dimerizes butadiene to 2-methylene vinyl cyclopentane.

$$2 \; \diagup\!\!\!\diagdown\!\!\!\diagup \quad \xrightarrow[\substack{\text{BuLi, MeOH, C}_6\text{H}_6 \\ 60\text{–}65°\text{C, 50 h}}]{\text{Ni(Bu}_3\text{P)}_2\text{X}_2} \quad \text{90\%} \tag{40}$$

The choice of an alkyl phosphine, especially tributylphosphine, is important; to produce the five-membered ring, strong σ-donating phosphines are preferred. More bulky phosphines, for example, tricyclohexylphosphine, cause linear rather than cyclodimerization. The dramatic change in reaction pathway is emphasized by the 90% yield of the vinyl cyclopentane.

Procedure—Preparation of 2-methylenevinylcyclopentane[54]

Bis(tri-*n*-butylphosphine)dichloronickel (0.67 g, 1.3 mM) is placed under an inert atmosphere in a heavy glass tube, fitted with a three-way stopcock. At room temperature benzene (5 ml) and butyl lithium (1.3 mM) in hexane (0.54 ml) are added, causing a slight exotherm and a dark red coloration. The mixture is stirred for 30 min, and methanol (1.5 ml, 40 mM) added, followed by butadiene (3.9 g, 70 mM) at −78°C. The tube is sealed and heated to 65–70°C for 40 h. Distillation without separation of the catalyst gives 2.7 g (70%) of 2-methylene vinyl cyclopentane as a colorless liquid, bp 121–122°C. By using bis(tri-*n*-butylphosphine)dibromonickel, yields of 90% are reported.

A related reaction of butadiene, $(PPh_3)_2NiCl_2$ (5 mM), and *n*-propyl magnesium bromide (250 mM) at reflux (i.e., room temperature using a dri-kold condenser) for 24 h affords, after hydrolysis, *cis*-2-methyl vinyl cyclopentane (**33**) (67%).[55] Work-up using acetone yields the *cis* alcohol **34**. Heating the mixture to 110°C for 24 h (sealed tube) gives the thermally more stable *trans* compound **35** (*trans* : *cis* 11 : 1). This thermal rearrangement reaction is not nickel catalyzed. These nickel-catalyzed butadiene cyclooligomerizations offer simple, effective routes to cyclopentane derivatives containing functionalities for further reaction.

Scheme 3

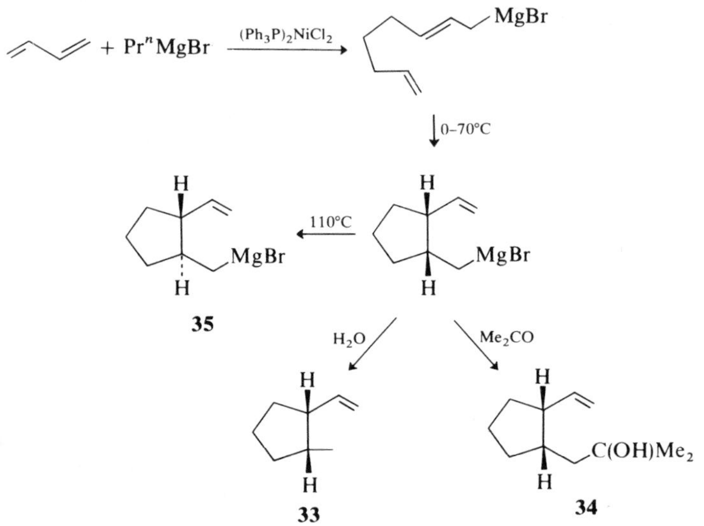

3.4.1.2. *Modified Diels–Alder Reactions*

In the Diels–Alder reaction the presence of a transition metal is known to alter the course of the reaction or the product distribution. Dienophiles such as maleic anhydride and *p*-benzoquinone coordinated to a $Pd(PR_3)_2$ moiety resist Diels–Alder reaction because of the considerable electron drift from palladium to the dienophile. Hence the complex $Pd(Ph_3P)_2$ (*p*-benzoquinone) does not undergo the normal Diels–Alder reaction with butadiene; instead a novel cyclization occurs[56] [Eq. (41)]. The complex **36** can be isolated in 75% yield after reaction at 60°C. Treatment with Ph_3Sb

liberates the organic product **37**. The reaction is general for other quinones; for example, acetyl-*p*-benzoquinone, methoxycarbonyl-*p*-benzoquinone, and 1,4-naphthoquinone coordinated to palladium all afford adducts analogous to **37**.

3.4.1.3. 3 + 2 Cycloaddition Reactions

A number of 3 + 2 cycloaddition reactions which form cyclopentanoid derivatives have been reported. Three methods are described here, two of which use iron complexes; the third requires the use of palladium to promote ring formation. The first method involves the use of Fe(η-C$_5$H$_5$)·(σ-C$_3$H$_4$R)(CO)$_2$ (**38**), which is prepared by metallation of an allyl halide or tosylate with Na[Fe(η-C$_5$H$_5$)(CO)$_2$].[57] Reaction of **38** with the electron-deficient alkene tetracyanoethylene at room temperature leads to the rapid formation of the cyclopentyl complex **39** in good yield [Eq. (42)]. Various substituted allyls (e.g., Me,[57] Ph, COCHMe$_2$, or OMe[58]) have also been examined and undergo a similar reaction sequence. The organic ring is easily liberated from the iron moiety by reaction with Ce(IV) in carbon-monoxide-saturated methanol to form the ester **40** or in the case of the methoxide derivative the acetal **41**.[58] A heterocyclic five-membered ring may be formed by reaction of the iron complex **38** with an isocyanide. The use of propargyl or allenyl fragments in place of allyl produces a similar reaction giving cyclopentenes rather than cyclopentanes[59] [Eq. (43)]. Alkenes of similar reactivity to tetracyanoethylene have been used, for example, 1,2-dicyano-1,2-bis(trifluoromethyl)ethylene, but the reaction is not general for a wide variety of alkenes.

$(\eta\text{-}C_5H_5)(CO)_2FeCH_2C\equiv CPh + (CN)_2C=C(CN)_2 \rightarrow$

$$(\eta\text{-}C_5H_5)(CO)_2Fe \qquad\qquad (43)$$

General Procedure—Formation of a Cyclopentane via an Iron Allyl Complex [57]

Under a nitrogen atmosphere [Fe(η-C$_5$H$_5$)(CO)$_2$]Na (50 mM) [prepared by the sodium amalgam reduction of {Fe(η-C$_5$H$_5$)(CO)$_2$}$_2$ in THF (150 ml)] is cooled to $-78°C$ and allyl chloride (50 mM) added slowly. The reaction is allowed to warm to room temperature over 1 h and the solution filtered under nitrogen. A 10% excess of tetracyanoethylene (55 mM) in THF is added to the iron complex at room temperature and the yellow crystalline product is either crystallized directly or purified by chromatography on neutral alumina (yield 80%). Reaction with (NH$_4$)$_2$Ce(NO$_3$)$_6$ in carbon-monoxide-saturated alcohol liberates the ester **40** quantitatively.

Related to the iron allyl reactions described above is the 3 + 2 cyclo-addition reaction of $\alpha\alpha'$-dibromoketones and aryl alkenes aided by iron carbonyl.[60] The iron carbonyl reacts with $\alpha\alpha'$-dibromoketone to form an allyl iron intermediate (**42**), which then further reacts with aryl alkenes to form a cyclopentanone[61] [Eq. (44)]. Although iron nonacarbonyl is most

$$(CO)_n Fe-O$$

42

$$(44)$$

65%–70%

R' = Me, Pr^i; R'' = H, Me and R^a = H, Me, Ph; R^b = H, Ph; R^c = H effective, iron pentacarbonyl may also be used. The reaction is sensitive to both electronic and steric factors, hence only aryl alkenes react to form cyclopentanones. Using β-substituted alkenes or ketones with increased substitution at the α and α' positions, lower yields of the cyclic product are obtained. Replacing the aryl alkene by an enamine extends the range of this reaction and provides a synthesis of cyclopentenones. Reaction of secondary α,α'-dibromo ketone and a morpholino enamine in the presence of iron carbonyl affords the corresponding 3-morpholinocyclopentanone. The derived adduct then eliminates morpholine giving the 2-cyclopentenone[62] [Eq. (45)]. A wide range of cyclopentenones have been prepared from secondary dibromoketones in high yield (Table 3.4). Unfortunately,

(45)

Table 3.4. Formation of Cyclopentenones from α,α'-Dibromoketones[62]

Dibromoketone (43)	Alkene (44)			Cyclopentenone (45)
R	R^a	R^b	R^c	% Yield
Me	H	Me	Me	79
Me	Ph	H	H	91
Me	Et	H	Me	73
Me	$+CH_2+_3$		H	74
Me	$+CH_2+_4$		H	100
Me	$+CH_2+_5$		H	100
Me	$+CH_2+_{10}$		H	90
Me	H	$+CH_2+_5$		71
Me	H	$+CH_2+_{11}$		65
Et	Ph	H	H	64
Et	Et	H	Me	70
Et	$+CH_2+_4$		H	89
Pr^i	Ph	H	H	72
Pr^i	$+CH_2+_4$		H	73
Ph	$+CH_2+_3$		H	66

primary or tertiary analogs do not undergo this reaction. The attractive features of this synthesis include the ready availability of the starting materials, operational simplicity, and high yields.

These reactions offer a method for the preparation of bicyclic ketones (e.g., 5/7 fused bicyclic ketones) and hence a simple route into azulene chemistry. Various other synthetic applications have been devised for this reaction, for example, the synthesis of (±)α-cuparenone (46) [Eq. (46)][63] (an earlier seven-step route gave only 3.3% yield) and the synthesis of spiro (n, 4) alkenone systems [Eq. (47)].

(46)

(47)

General Procedure—Preparation of 2,5-dimethyl-3-phenylcyclopent-2-enone[63]

2,4-Dibromopentan-3-one (2.44 g, 10 mM), α-morpholinostyrene (3.78 g, 21 mM) and $Fe_2(CO)_9$ (4 g, 11 mM) in benzene (25 ml) are stirred under an inert atmosphere at 30°C for 20 h. The reaction mixture is then diluted with ethyl acetate (50 ml) and washed with a saturated sodium bicarbonate solution followed by a solution of potassium nitrate (or brine). The organic layer is separated and dried (Na_2SO_4). Concentration under vacuum gives an orange oil (3.0 g) that is purified by column chromatography (silica gel 150 g, 1 : 10 ethyl acetate : hexane) to yield the title product (1.68 g, 91%) as a semisolid (after recrystallization from hexane, mp 57–59°C). When an undeaminated adduct is formed, the whole reaction mixture is treated with 3% EtOH–NaOH at room temperature for 5–30 min to liberate the cyclopentenone.·

A third 3 + 2 cycloaddition reaction uses alkenes containing electron-withdrawing groups; these have been successfully cyclocoupled to a three-carbon fragment to produce a variety of cyclopentanoid products.[64] The three-carbon fragment is derived from 2-acetoxymethyl-3-allyltrimethyl-silane (47), which is easily prepared in good yield (60%) from α-methyl-allylalcohol by reaction with two equivalents of BuLi in ether and two

Table 3.5. Formation of Substituted Cyclopentanones [Eq. (48)]

Alkene	Product yield, %	Reaction time, h
Methyl acrylate	68	43
Methyl methacrylate	50	85
Methyl crotonate	38	60
Dimethyl maleate	60	42
Methyl cinnamate	70	4.5
Methyl benzyl idenemalonate	65	5
Chalcone	85	11
Methyl vinyl ketone	30	42

equivalents of TMEDA at 0°C, followed by THF, warming to room temperature, reaction with Me_3SiCl, hydrolysis, and acetylation with CH_3COCl/pyridine at 0°C. In the presence of tetrakis(triphenylphosphine) palladium and diphos (1–4 mol %), (**47**) adds to electron-deficient alkenes [Eq. (48) and Table 3.5]. Simple alkyl-substituted alkenes, e.g., norbornene

$$Me_3Si\diagdown\diagup OAc \; + \; \overset{Y}{\underset{Z}{\diagup\diagdown}} \quad \xrightarrow[\substack{THF, \, reflux \\ 3 \, days}]{Pd(PP_3)_4, \, diphos} \quad \underset{Y \quad Z}{\bigpentagon} \qquad (48)$$

47

or electron-rich alkenes such as enamines, failed to react. The overall reaction represents the addition of trimethylenemethane to alkenes. The synthetic usefulness is enhanced by the exocyclic methylene group, which can serve as a protected carbonyl as well as a site for further reaction. For example, cyclopropanation followed by hydrolysis leads to a *gem*-dimethyl group as found in coriolins.

General Procedure—Formation of Substituted Cyclopentanones[64] *[Eq. (48)]*
A mixture of an alkene containing electron-withdrawing substituents (15–36 mM), 2-acetoxymethyl-3-allyltrimethylsilane (10 mM), Pd(PPh$_3$)$_4$ (3–9 mol %), and diphos (1–4 mol %) is refluxed in toluene or THF for up to 3 days (depending upon the alkene; see Table 3.5). Work-up simply involves removal of the volatiles followed by chromatography. Yields are generally in the range 50%–85%.

3.4.1.4. Miscellaneous Cycloaddition Reactions

Five-membered carbocyclic products have been obtained in a number of alkene cycloaddition reactions. Methylenecyclopropane affords 2-vinylmethylenecyclopentane in the presence of bis(tri-*n*-butylphosphine) nickel dibromide–butyllithium in methanol.[65] This reaction is misleading because the cyclopropane acts only as a source of butadiene, which is known to form **48** in the presence of this catalyst[54] (see Section 3.4.1.1). Dimerization of methylene cyclopropane by nickel bis(cyclooctadiene) at −15°C produces a mixture (45% overall yield) of **49**, 65%, and **50**, 20%,[37] but by using Ni(COD)$_2$–maleic anhydride (1 : 10), a 68% conversion with 97% selectivity to the cyclopentane product is achieved[66] [Eq. (50)]. These species are

$$ \text{(49)} $$

48
91%

$$ \text{(50)} $$

49   **50**
97%   3%

derived directly from methylene cyclopropane and not via butadiene, although an extensive side reaction is formation of higher molecular weight oligomers (mol wt 1100). Nickel bisacrylonitrile catalyzes the cycloaddition of methylene cyclopropane to electron-deficient alkenes, e.g., methyl acrylate,[67] methyl vinyl ketone, and acrylonitrile[68] [Eq. (51)]. Disubstituted

$$ \text{(51)} $$

70%

alkenes react more slowly giving mixed isomeric products in low yields. These reactions are suggested to proceed through a transition-metal-stabilized trimethylene methane intermediate (see Section 3.4.1.3), which is generated from methylene cyclopropane, and therefore represent further examples of the 3 + 2 cycloaddition reaction.

Diphenyl cyclopropenone reacts with ketenes in the presence of catalytic amounts of nickel carbonyl to give a 1,2 cyclic dione[69] [Eq. (52)].

$$ \text{(52)} $$

85%

This reaction also occurs in the presence of iron pentacarbonyl, but in this case equimolar amounts of reagent are required.

Quadricyclane (**51**) reacts with electron-deficient alkenes in the presence of nickel(0) creating a five-membered ring[76] (**52**) [Eq. (53)]. Although

(53)

X = CN, yield 50%
X = CO₂Me, yield 58%

the conversion of quadricyclane into norbornadiene is catalyzed by nickel(0), this does not occur here because reaction of the latter with electron-deficient alkenes gives a very poor yield of the five-membered ring compound. Norbornadiene does react with alkynes in a Diels–Alder fashion to produce a cyclopentene analog of **52** [71] [Eq. (54)]. The catalyst for this reaction is prepared by reaction of cobalt trisacetylacetonate with diethylaluminum chloride in the presence of 1,2-bis(diphenylphosphino)ethane. This reaction is unusual because the alkynes do not contain electron-withdrawing groups, which are generally required to produce these cycloaddition products.

(54)

R = R′ = H(90%); R = R′ = Ph(86%); R = H,R′ = Me(16%), Et(25%), or Ph(41%)

3.4.2. Cycloaddition of Two Unsaturated Species with a Third Substrate

This type of reaction is exemplified by the coupling of alkynes or alkenes in the presence of carbon monoxide to produce a five-membered cyclic ketone. For example, alkyne cobalt complexes react stoichiometrically with ethylene, terminal alkenes, and cycloalkenes (C_5H_8 to C_8H_{14}) in refluxing toluene, giving single-product cyclopentenones [Eqs. (55) and (56)].[72] A related stoichiometric reaction uses nickel carbonyl and two alkynes to give moderate yields of the cyclopentenone analogs[73] [Eq. (57)].

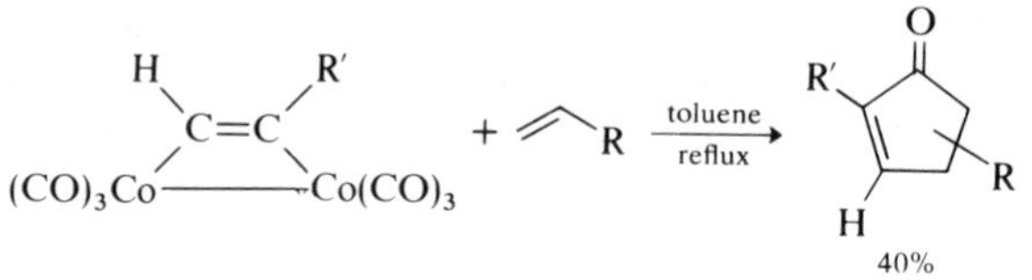

$$\text{(55)}$$

R' = H, R = Me or Ph
R' = R = Ph
R' = Me, R = n-C$_5$H$_{11}$

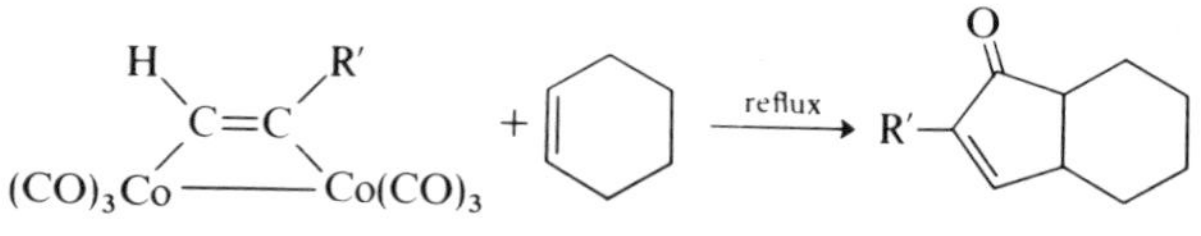

40%

$$\text{(56)}$$

$$2\, RC\equiv CR \xrightarrow[\text{HCl}]{\text{Ni(CO)}_4}$$

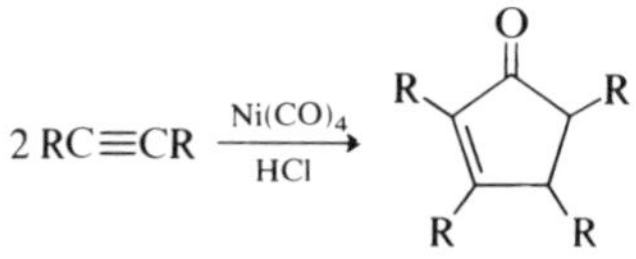

R = H, 15%
R = Et, 70%

$$\text{(57)}$$

Similarly a butadiene complex can also produce a cyclic ketone in the presence of carbon monoxide. Tricarbonyliron diene complexes react with anhydrous aluminum chloride to produce cyclopentenones on work-up[74] [Eq. (58)]. Interestingly the nonconjugated isomer is formed in contrast to

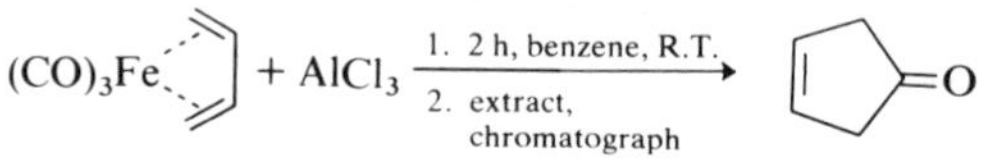

$$\text{(58)}$$

the previous examples. Reactions involving CO are discussed in more detail in Chapter 6.

Isocyanides have also been incorporated in a cycloaddition reaction to form an interesting five-membered ring.[75] Co(PPh$_3$)(η-C$_5$H$_5$)-(PhC$\equiv$CPh) reacts with 2,6-dimethylphenylisocyanide at room temperature to give air-stable Co(η-C$_5$H$_5$)(PhC$\equiv$CPh)(2,6-Me$_2$C$_6$H$_3$NC)$_2$. Reaction with a further molecule of 2,6-dimethylphenylisocyanide at 130°C affords tris(2,6-dimethylphenylimino)diphenylcyclopentene (**53**) [Eq. (59)]. The reaction is not catalytic.

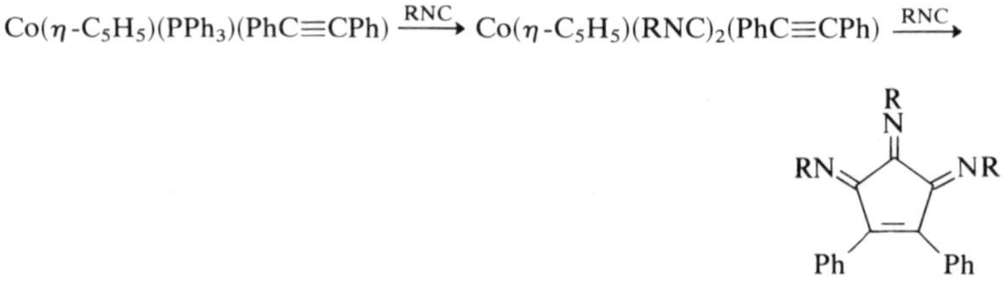

$$Co(\eta\text{-}C_5H_5)(PPh_3)(PhC\equiv CPh) \xrightarrow{\text{RNC}} Co(\eta\text{-}C_5H_5)(RNC)_2(PhC\equiv CPh) \xrightarrow{\text{RNC}}$$

$$\text{(59)}$$

53

The reactions described so far in this section have restricted synthetic applications because they are not catalytic and require specialized organometallic techniques. A more useful stoichiometric reaction uses an air-stable palladium reagent for the cyclization of hexadiene in acetic acid. Generally, linear oligomers and telomers are obtained from palladium-catalyzed reactions but 1,5-hexadiene gives 3-acetoxymethylenecyclopentane in good yield[76] [Eq. (60)]. Other dienes, 1,6-heptadiene, 2,6-octadiene, or 1,7-octadiene give the expected linear telomers.

$$\text{(60)}$$

Procedure—Preparation of 3-Acetoxymethylenecyclopentane[76]

1,5-Hexadiene (43 mM) is added to palladium acetate (1 g, 4.3 mM) in acetic acid (40 ml) and stirred for 5–7 h at 50°C until precipitation of palladium metal is complete. After centrifuging, the metal residue is filtered off, ether (300 ml) is added, and the solution washed in turn with water, sodium bicarbonate, and brine. The solution is then dried and ether and excess diene removed by distillation. The residue is distilled under reduced pressure to give 3-acetoxymethylenecyclopentane (64% by glc, based on Pd(OAc)$_2$).

3.4.3. Cyclization of Difunctional Substrates

3.4.3.1. Cyclization of α,ω-Dihalo Compounds

Cyclopentanoid derivatives can be prepared by the transition-metal-catalyzed cyclization of a functionalized linear chain. α,ω-Dibromobutanes form metallacyclopentanes (54) in the presence of a nickel(0) complex.[77] These metallacycles are not readily cyclized to form the five-membered cyclic product (e.g., by reaction with carbon monoxide), although reaction with alkenes or alkynes does lead to cyclohexane or -hexene derivatives. The nickel complex 54 does react with *gem*-dihalides to liberate the cyclopentane, and similarly substituted cyclopentanes are formed by reaction with MeCHBr$_2$ (methylcyclopentane 59%) or CHCl$_3$ (chlorocyclopentane 27%) (Scheme 4). Larger rings, such as cyclohexane (37%) or cycloheptane (17%), are formed from the appropriate dihalide but in much lower yield. Reaction of the nickel metallacyclopentane 54 with oxygen leads interestingly to ring closure and formation of cyclobutane.

Scheme 4

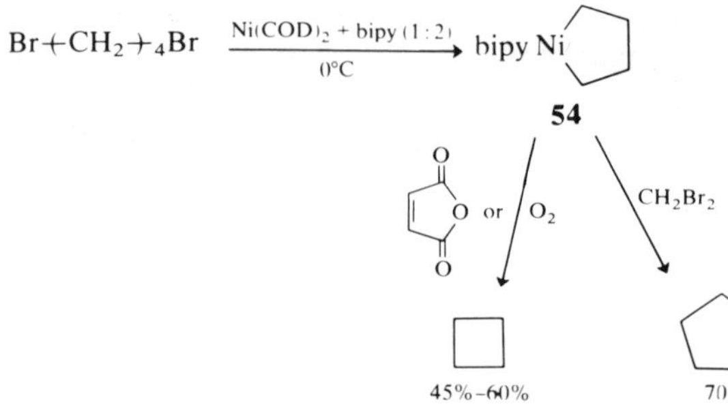

$$54$$

45%–60%   70%

General Procedure—Formation of Cyclopentane from an α,ω-Dibromide [77]

The nickel metallacycle **54** (Scheme 4) is prepared *in situ* by reaction of $Ni(COD)_2$ (2 mM) with 2,2'-bipyridyl (6 mM) and 1,4-dibromobutane (1 mM) in THF (10 ml) at 0°C under nitrogen. Methylene dibromide (2 mM) is added to the complex, still at 0°C, and the mixture allowed to warm to room temperature with stirring (4 h). Filtration and distillation gives cyclopentane (70%).

3.4.3.2. Cyclization of Unsaturated Halo Compounds

Disodium tetracarbonyl ferrate, $Na_2Fe(CO)_4$, reacts with β-allenic bromides to produce cyclopentenones in moderate yield[78] [Eq. (61)]. In contrast, reaction of $Na_2Fe(CO)_4$ with a linear alkyl halide followed by allene does not yield the cyclic derivative but rather an α,β unsaturated ketone [Eq. (62)]. Using γ-ethylenic bromides produces cyclohexanones.

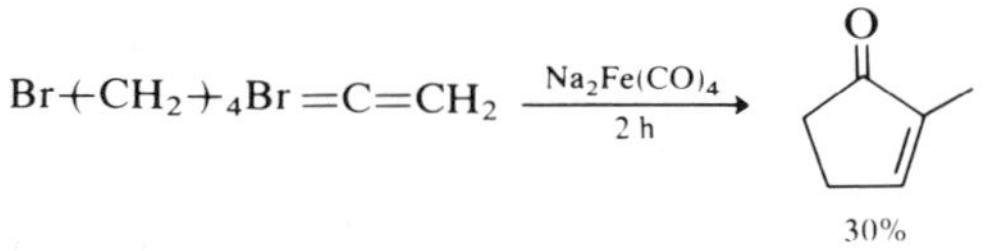

$$\text{Br} + \text{CH}_2 +_4 \text{Br} = \text{C} = \text{CH}_2 \xrightarrow[2\text{ h}]{Na_2Fe(CO)_4} \qquad (61)$$

30%

$$Na_2Fe(CO)_4 + RX \longrightarrow NaFe(CO)_4R \xrightarrow[2. \ H^+, Me_3NO]{1. \ =\cdot=} \qquad (62)$$

The iodoalkyne $Bu^tC\equiv C(CH_2)_4I$ can be cyclized by $Ni(CO)_4$–$KOBu^t$ to give a mixture of cyclic esters, t-butyl-2-(1-cyclopentenyl) hexanoate and t-butyl-2-cyclopentylidenehexanoate, in good yield[79] [Eq. (63)]. When shorter- or longer-chain iodoalkynes are used linear products predominate. Although the reaction is simple to perform the toxicity of nickel carbonyl limits its applicability; however, a similar cyclization is promoted by lithium dialkyl cuprates, which are free of the toxicity problem[80] [Eq. (64)]. The

$$Bu^tC\equiv C(CH_2)_4I \xrightarrow[\text{6 h, R.T.}]{Ni(CO)_4, KOBu^t} \quad \underset{41\%}{\overset{Bu^t}{\diagdown}} \overset{CO_2Bu^t}{\diagup} \quad + \quad \underset{56\%}{\overset{Bu^t}{\diagdown}} \overset{CO_2Bu^t}{\diagup} \tag{63}$$

$$PhC\equiv C(CH_2)_nX \xrightarrow[\text{2. H}_2O]{1. R_2CuLi} \quad \underset{\mathbf{55}}{\overset{Ph}{\diagdown}} \overset{H}{\diagup} \quad + \quad \underset{\mathbf{56}}{\overset{Ph}{\diagdown}} \overset{R}{\diagup} \tag{64}$$

$$+ PhC\equiv C(CH_2)_nR + PhC\equiv C(CH_2)_nH$$
$$+ PhC\equiv C(CH_2)_{n-2}CH=CH_2$$

reaction of $PhC\equiv C(CH_2)_4Br$ with a fivefold excess of lithium di-*n*-butyl-cuprate in pentane : ether (10 : 1) initially at $-30°C$, then at reflux for 6 h, gives a mixture containing **55** (79%) and **56** (13%), together with small amounts of linear product. Use of the iodide increases the yield of cyclic products **55** and **56** (91 : 8) to 99%. Hydrolysis with D_2O leads to 91% incorporation of deuterium at the double bond of **55**, demonstrating that this hydrocarbon is derived from a stable organometallic precursor. Because the initial cyclic product is a stable organometallic complex there is synthetic potential in these cyclizations. For example, the organometallic intermediate can be reacted with a variety of reagents rather than simply hydrolyzed [Eqs. (65)–(68)]. Four- and six-membered rings may also be formed using

$$\underset{}{\overset{Ph}{\diagdown}} \overset{CuBu^tLi}{\diagup} \xrightarrow{\text{MeI}} \underset{}{\overset{Ph}{\diagdown}} \overset{Me}{\diagup} \quad 53\% \tag{65}$$

$$\underset{}{\overset{Ph}{\diagdown}} \overset{CuBu^tLi}{\diagup} \xrightarrow{BrCH_2CH=CH_2} \underset{}{\overset{Ph}{\diagdown}} \overset{CH_2CH=CH_2}{\diagup} \quad 66\% \tag{66}$$

$$\underset{}{\overset{Ph}{\diagdown}} \overset{CuBu^tLi}{\diagup} \xrightarrow{I_2} \underset{}{\overset{Ph}{\diagdown}} \overset{I}{\diagup} \quad 83\% \tag{67}$$

$$\underset{}{\overset{Ph}{\diagdown}} \overset{CuBu^tLi}{\diagup} \xrightarrow{NBS} \underset{}{\overset{Ph}{\diagdown}} \overset{Br}{\diagup} \quad 69\% \tag{68}$$

the appropriate haloalkyne but cycloheptanes and larger rings are not formed. Although not transition-metal-catalyzed, it is interesting to note that haloalkenyls are cyclized by magnesium via a Grignard reagent to produce a five-membered carbocycle[81] [Eq. (69)].

$$X \overset{Mg}{\longrightarrow} \qquad \qquad (69)$$

3.4.3.3. *Reductive Coupling*

Dibenzylidenacetone undergoes reductive coupling followed by cyclization in the presence of K_2OsCl_6 and zinc dust to give the five-membered ring **57**.[82] Other α,β-unsaturated ketones react with K_2OsCl_6–Zn in the same manner, and hence this reaction may be used as a preparative method for the synthesis of cyclic ketols.

$$PhCH{=}CHCCH{=}CHPh \xrightarrow[\text{Zn, MeOH}]{K_2OsCl_6} \qquad (70)$$

57 100%

General Procedure—Formation of Cyclic Ketols[82]

A solution of dibenzylidenacetone ($0.1\ M$) in methanol (200 ml) is stirred with a mixture of K_2OsCl_6 ($0.1\ M$) and zinc dust ($0.3\ M$) at reflux temperature. After 3 h the reaction is complete. A white precipitate of the product and unreacted K_2OsCl_6–Zn forms upon cooling. This is filtered, washed with methanol, and extracted with boiling chloroform (soxhlet). The white crystalline compound obtained from the chloroform extraction is recrystallized from chloroform to yield the cyclic product **57** (90%). Only zinc is consumed in the reaction; K_2OsCl_6 remains unchanged and can be reused.

A related substrate, an alkyne–alkene, is cyclized by a palladium catalyst to produce the bicyclic system **58** [Eq. (71)].[83] Heating the inter-

$$RCH{=}CHC{-}C{\equiv}CR' \xrightarrow[\text{2. }\Delta]{\text{1. Pd(PhCN)}_2Cl_2\text{, benzene}} \qquad (71)$$

58

R, R' = Ph, Ph; Me, Ph; Ph, Me;
Me, Me; Me, Et.

mediate palladium complex in methanol as opposed to benzene produces a different product (**59**).

59

Intramolecular reductive coupling has been used to prepare five-membered rings. Several reagents have been reported for the pinacolic coupling of ketones and aldehydes.[84] One of the most versatile is $CpTiCl_3/LiAlH_4$, although $Mg(Hg)/TiCl_4$ may be used in less difficult cyclizations [Eqs. (72) and (73)].

(72)

90%

(73)

49%

General Procedure—Cyclic Pinacolic Coupling[84]

To 70–80-mesh magnesium (0.195 g, 8 mM) in THF (3 ml) under an inert atmosphere is added mercuric chloride (0.06 g, 0.22 mM) and the mixture is stirred for 15 min at room temperature. The solvent is removed and the amalgam washed with THF (three times) followed by the addition of THF (4 ml). The reaction is cooled to −10°C and $TiCl_4$ (0.33 ml, 3 mM) added dropwise. A solution of the ketoaldehyde (1 mM) in THF (4 ml) is added and stirred at 0°C for 75 min. The reaction mixture is then treated with aqueous K_2CO_3 (0.5 ml) at 0°C for 15 min and the resulting blue-black slurry diluted with ether and filtered through Celite. The filtrate is washed with saturated brine, dried ($MgSO_4$), and concentrated to give the product in good yield. The other catalyst is prepared by slowly adding a solution of $LiAlH_4$ (0.75 mM) to $CpTiCl_3$ (0.99 mM) in THF under an inert atmosphere and stirring the mixture for 1 h at 50°C. The procedure is then identical with the above.

3.4.3.4. Intramolecular Cyclization of Alkenes and Alkynes

Intramolecular coupling of an alkyne and alkene can lead to cyclopentene derivatives. 2-Alkenyl-1-alkynyl benzenes cyclize in the presence of Pd(PhCN)$_2$Cl$_2$.[85] The reaction proceeds through a palladium allyl intermediate (**60**) and only requires stirring at room temperature for ~10 min.

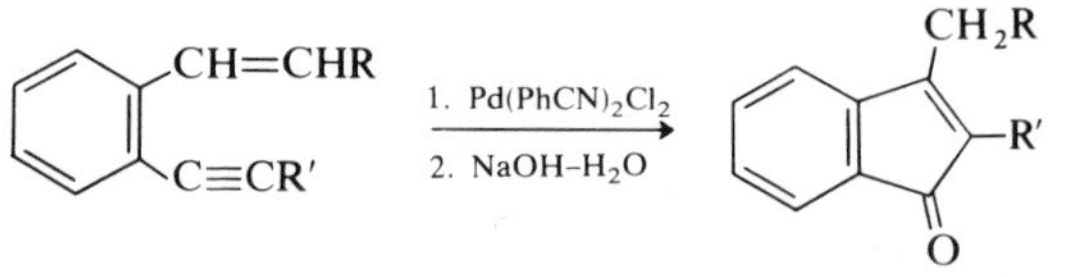

$$
\text{(74)}
$$

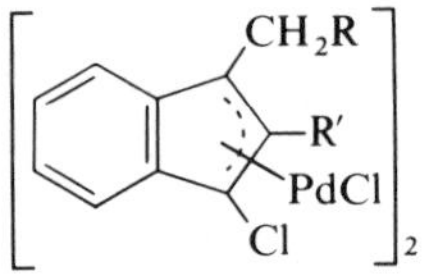

R = R' = Ph (54%)
R = R' = Me (30%)
R = Bu^t, R' = Ph (35%)
R = Me, R' = Ph (15%)

Wilkinson's catalyst [Rh(PPh$_3$)$_3$Cl] has been shown to be effective for the cyclization of 4,4-disubstituted 1,6-dienes to methylene cyclopentanes [Eq (75)].[86] Simply refluxing for 8–72 h in redistilled chloroform saturated

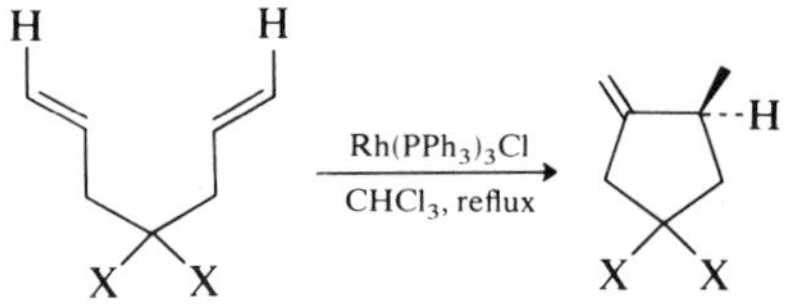

$$
\text{(75)}
$$

X = COMe (87%), COPh (75%), CO$_2$Et (90%) or X$_2$ = 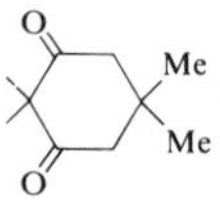 (60%)

with hydrogen chloride gives high yields of product (60%–90%). It is interesting to compare the effect of rhodium(I) with that of palladium(II) which, with similar substrates, catalyzes the formation of cyclopentenes in high yield[87] [Eq. (76)]. Although the rhodium-catalyzed cyclization does not occur with substituted alkenes, by using palladium the reaction is tolerant of substitution. These two simple catalytic reactions allow the formation of both cyclopentane and cyclopentene derivatives from the same substrate.

$$ (76) $$

R = H, R' = Me   X = Y = COMe (93%)
R = R' = Me   X = Y = COMe (82%)

R = R' = H   XY = (65%)

General Procedure—Cyclization of 4,4 Disubstituted 1,6 Dienes[86,87]

The substrate (100 mM) and catalyst (1–3 mM) [i.e., either Rh(PPh$_3$)$_3$Cl or Pd(OAc)$_2$] are heated in boiling chloroform (redistilled) through which hydrogen chloride has been bubbled slowly for 6 min prior to heating. The mixture is refluxed for 8–72 h (Rh) or 5–10 h (Pd), and after work-up high yields of the cyclopentane (Rh) or the cyclopentene (Pd) are obtained.

3.4.3.5. Intramolecular Cyclization of Unsaturated Ethers

The palladium-promoted intramolecular cyclization of silyl enol ethers of alkenylmethyl ketones leads to the formation of cyclic α,β-unsaturated ketones[88] [Eq. (77)]. The reaction has been used successfully to produce high yields of 3- and 4-substituted cyclopentenones, but cyclohexenones and cycloheptenones [Eq. (78)] are only formed in low yields (<40%), and

$$ H_2C=C(CH_2)_2C=CH_2 \xrightarrow[\text{MeCN}]{\text{Pd(OAc)}_2} \quad (77) $$

OSiMe$_3$ 87%

$$ \xrightarrow{\text{Pd(OAc)}_2} \quad (78) $$

OSiMe$_3$ 25%

large rings are not formed. Initial results were based on a stoichiometric palladium-promoted reaction but the indications are that the reaction may be performed catalytically. The active palladium(II) species is regenerated by cupric acetate and oxygen. This regeneration method has not yet been reported for the reaction described above but it has been successfully applied to a related reaction—the palladium acetate-catalyzed dehydrosilylation of silyl enol ethers—to yield α,β-unsaturated carbonyl compounds.[88]

General Procedure—Cyclization of Silyl Enol Ethers of Alkenyl Ketones[88]

To a stirred solution of $Pd(OAc)_2$ (1 mM, 0.225 g) in acetonitrile (4 ml) under nitrogen is added slowly 2-trimethylsilyloxy-4-phenyl-1,5-hexadiene (1 mM). The mixture is stirred at room temperature for 10 h. As the reaction progresses, deposition of metallic palladium occurs. The reaction mixture is concentrated under vacuum and extracted with hexane. Purification by glc gives pure 3-methyl-4-phenyl-2-cyclopentenone (83%).

The catalytic reaction is performed using a mixture of palladium acetate (0.1 mM), and copper acetate (0.2 mM) in acetonitrile (8 ml), stirring by bubbling oxygen through the mixture. In this way reduction to palladium(0) is avoided.

The nickel(II) catalyst $Ni(NO_3)_2(PBu_3)_2–2KOBu^t$ (**61**) in ethanol cyclizes 2,7-octadienyl isopropyl ether into 2-methylenevinylcyclopentane in moderate yield[89] [Eq. (79)].

$$\text{(79)}$$

A palladium-catalyzed 1,3 shift has provided a novel cyclopentanone synthesis[90] [Eq. (80)]. Alkylidenetetrahydrofurans (**62**) are known to

$$\text{(80)}$$

undergo thermal rearrangement to cycloheptanones, whereas heating in DMSO (50–120°C) in the presence of palladium(0) produces cyclization to the cyclopentanone **63**. This reaction has been used to prepare the prostaglandin A_2 intermediate (**64**) in excellent yield (69%) [Eq. (81)].

$$\text{(81)}$$

3.4.3.6. Miscellaneous Reactions

A simple five-membered ring synthesis uses 1,3-diiodopropane with an α,β-unsaturated ester in the presence of metallic copper and cyclohexylisonitrile.[91] The reaction is thought to proceed through a 3-iodopropylcopper isonitrile complex, followed by addition of the α,β-unsaturated carboxyester and finally cyclization by intramolecular elimination of a copper–halide–isonitrile complex.

$$I\!+\!CH_2\!+\!_3I + RCH\!=\!CHR' \xrightarrow[\text{110°C, 12 h}]{\text{Cu, C}_6\text{H}_{11}\text{NC}} \quad (82)$$

trans

R = R' = CO$_2$Et (90%)
R = H, R' = CO$_2$Et (58%)

3,3-Bis(iodomethyl)oxetane ($\overline{OCH_2C(CH_2I)_2CH_2}$) can also be used in place of 1,3-diiodopropane to yield the corresponding oxaspirane carboxylate. The preferred halide for these reactions is iodide; bromo and chloro compounds give only low yields of cyclic product. Using diiodobutane, the cyclohexane carboxylates are readily formed but the yield is low (37%–55%).

3.5. Formation of Six-Membered Rings

The formation of six-membered rings is relatively easy because they are strain free; even so, the synthesis of cyclohexanoid derivatives using organometallic catalysts has not received the extensive study given to the forming of five-membered rings. There may be many reasons for this, but it is worth noting that conventional ring forming reactions work well for six-membered ring synthesis, and that the chemistry of benzenoid compounds is extensive and more easily controlled. General methods involving transition metals for six-membered ring synthesis include the trimerization of alkynes to form aromatic derivatives and the cycloaddition reaction of alkenes, alkynes, and dienes to yield a variety of saturated and unsaturated rings.

3.5.1. Alkyne Trimerization—Formation of Aromatic Rings

The trimerization of alkynes to form aromatic compounds is highly exothermic yet, surprisingly, in the absence of metal catalysts high temperatures are required for this transformation. In contrast, the transition-metal-catalyzed reaction occurs at room temperature and since its discovery in the 1940s has been intensively investigated.[92] The reaction is typically catalyzed by nickel, cobalt, rhodium, or palladium, although Ziegler systems (Ti–Al compounds) have been employed.

Nickel(0) catalysts [e.g., $Ni(Ph_3P)_2(CO)_2$] have received the most attention and have been examined with a wide variety of substrates.[92e,93] Studies using cobalt, rhodium, and palladium systems [e.g., $CpCo(CO)_2$, $CpRh(CO)_2$, $PdCl_2(PhCN)_2$] have allowed the isolation of intermediates and have demonstrated that the reaction proceeds through a metallacyclopentadiene intermediate (**65**).[94] Insertion of a third alkyne produces a metallacycloheptatriene that undergoes reductive elimination to liberate the aromatic ring (Scheme 5). A second mechanism can operate if the third alkyne is a good dienophile; a Diels–Alder reaction with the metallacycle then takes place followed by reductive decyclization (Scheme 5).

Scheme 5

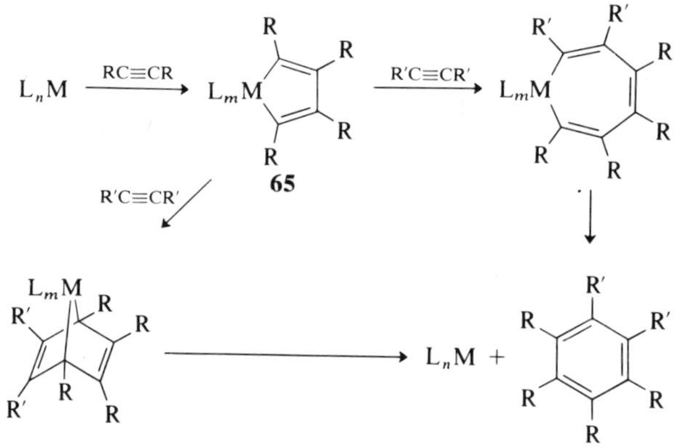

Generally, disubstituted alkynes give the symmetrical hexasubstituted product and monosubstituted alkynes give 1,2,4- and 1,3,5-trisubstituted benzenes; the relative amounts of each depend on the reaction conditions, catalyst, and nature of the substituents [Eq. (83)]. Alkynes containing a

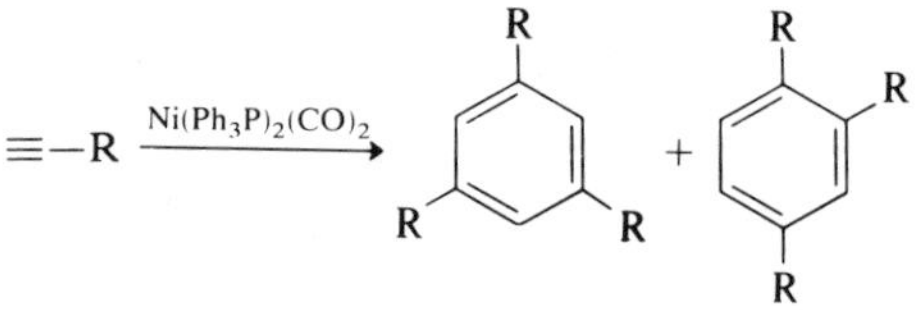

$$\equiv\!-R \xrightarrow{\text{Ni}(Ph_3P)_2(CO)_2} \tag{83}$$

wide variety of functional groups (e.g., alkyl, aryl, alkene, alkyne, ether, alcohol, ester, carboxylic acid, amine, or silyl) have been successfully cyclized (Table 3.6).[92e] Not all alkynes are readily cyclotrimerized; for example, propargyl chloride cannot be cyclized and propargyl aldehyde gives only low yields of benzenetricarbaldehyde. Similarly, alkynediols and their derivatives are also prone to give poor yields of aromatics.

Typically, for the cyclization of alkynes using a nickel(0) catalyst, the alkyne is slowly added to a benzene solution of the catalyst [e.g., $Ni(PPh_3)_2(CO)_2$] at 70°C under nitrogen. The reaction, which usually begins immediately, can be controlled by carefully adding the alkyne at such a rate to maintain the reaction temperature at 70–80°C without external heating. Despite the simplicity of this reaction it has not been used extensively in organic synthesis, although recently alkyne trimerization has been adapted and a series of useful organic syntheses produced.

The extension of the trimerization reaction employs an α,ω-diyne which is cooligomerized with an alkyne [Eq. (84)].[50] This reaction is

$$\tag{84}$$

catalyzed by commercially available cyclopentadienyl cobalt dicarbonyl [$CpCo(CO)_2$]. To suppress the oligomerization and cyclotrimerization of the monoalkyne, which can be serious side reactions, bis(trimethylsilyl)acetylene is preferred. The trimethylsilyl groups are readily removed at a later stage. Using this alkyne, slow addition of the diyne to refluxing solvent ($Me_3SiC{\equiv}CSiMe_3$) containing catalytic amounts of $CpCo(CO)_2$

Table 3.6. *Examples of Alkyne Cyclization Catalyzed by Ni(0) [Eq. (83)]*

Substrate	Product	% Yield (Ratio)
Propargyl alcohol	1,3,5-Tris(hydroxymethyl)benzene 1,2,4-Tris(hydroxymethyl)benzene	95 (1:1)
3-Butyn-2-ol	1,3,5-Tris(α-hydroxyethyl)benzene 1,2,4-Tris(α-hydroxyethyl)benzene	89 (1:1)
Methyl propiolate	Trimesic acid Trimellitic acid	85
N,N-Dimethyl-2-propinyl-1-amine	1,3,5-Tris(dimethylamino)benzene	67
Propargyl ethyl ether	1,3,5-Tris(ethoxymethyl)benzene 1,2,4-Tris(ethoxymethyl)benzene	77 (1:1)

results in the formation of *o*-bis(trimethylsilyl) substituted aromatic products in high yield. The alkyne solvent can be reused without build-up of harmful impurities and therefore its use in large excess presents no economic problems. Using this method, a general route to indans, tetralins,[95] and anthraquinones[53] has been developed [Eqs. (85), (86), and (87)].

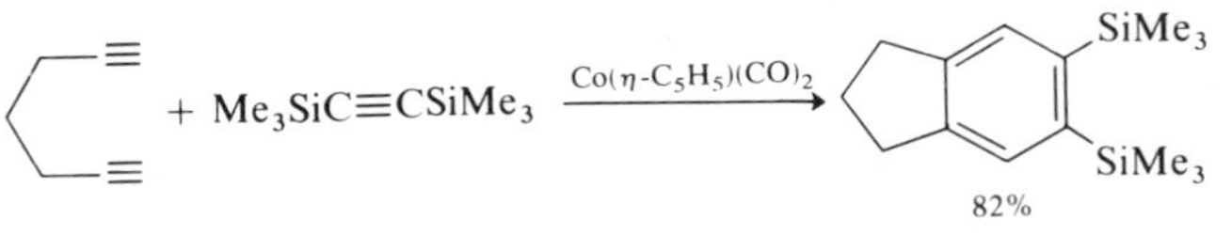

(85)

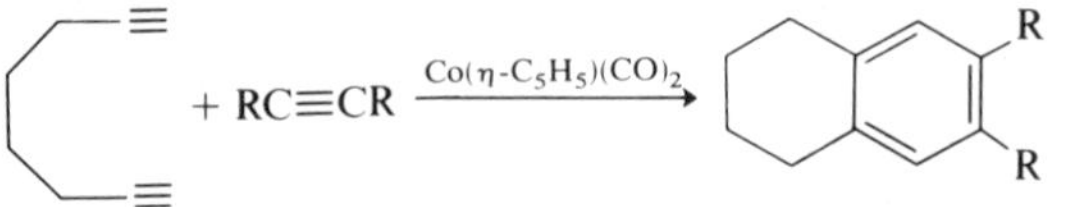

(86)

R = Me_3Si, 85%
= Ph, 21%
= CO_2Me, 26%

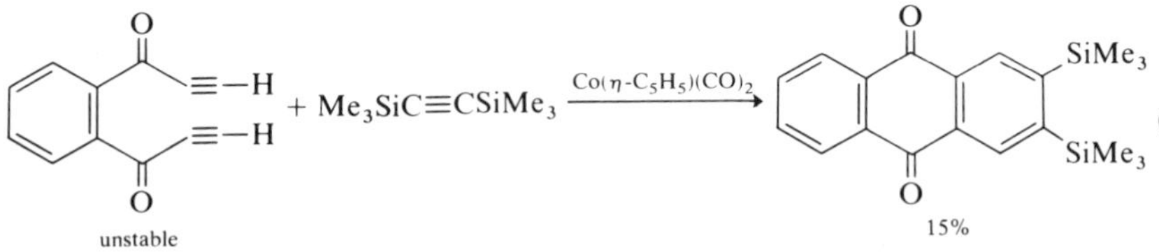

(87)

General Procedure—Preparation of 5,6-Bis(trimethylsilyl)indan[95]

1,6-Heptadiyne (1 g, 12.8 mM) in bis(trimethylsilyl)acetylene (5 ml) containing $CpCo(CO)_2$ (20 μl) is added to refluxing (oil bath 140°C) bis(trimethylsilyl)acetylene (35 ml) containing $CpCo(CO)_2$ (30 μl; 72 mg, 0.4 mM total) under nitrogen. The mixture is magnetically stirred for 72 h. The reaction is then cooled and all volatiles vacuum distilled to give recovered solvent ($Me_3SiC{\equiv}CSiMe_3$) that is usable, as such, in further cyclizations. The dark oily residue is chromatographed on silica, diluting with pentane (200-ml fractions). Fractions 2 and 3 contain the product as a yellow oil that crystallizes on standing. Recrystallization from ether–methanol at −60°C gives the title indan (80%, mp 68–69°C).

The synthesis of benzocyclobutenes has already been described (Section 3.3.2), but an extension of this reaction offers a route into substituted benzene derivatives. The trimethylsilyl group is a good leaving group in electrophilic substitution reactions and hence a large variety of derivatized compounds can be prepared (Scheme 6).[50] Benzocyclobutenes are in equilibrium with *o*-xylylenes, and, using the ether-substituted diyne **66**, the xylylene is trapped by a further molecule of bis(trimethylsilyl)acetylene in

Scheme 6

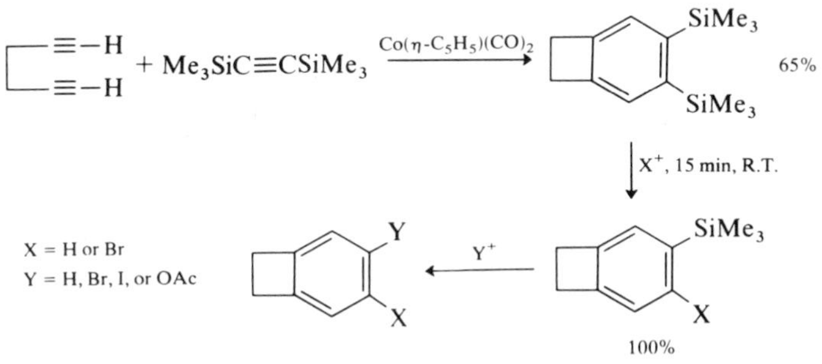

a Diels–Alder reaction to produce the substituted naphthalene **67** (Scheme 7).[96] This reaction can be developed into a polycyclic synthesis by including an alkene, carbonyl, or imine in the side chain leading to an intramolecular

Scheme 7

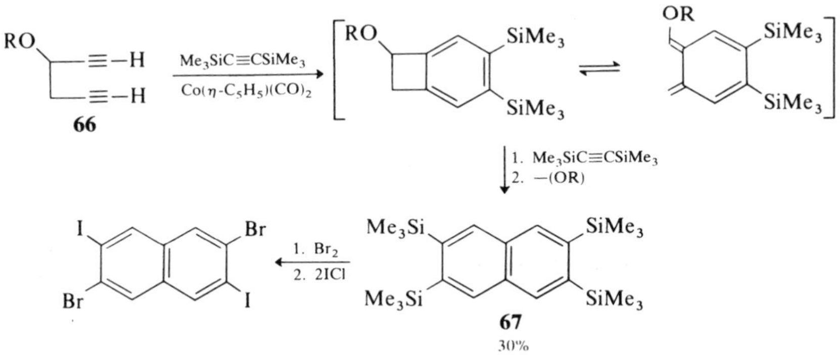

reaction [Eqs. (88), (89), (90), or (91)].[97] Interestingly, the stereochemistry of the new ring junction is almost always *trans*.

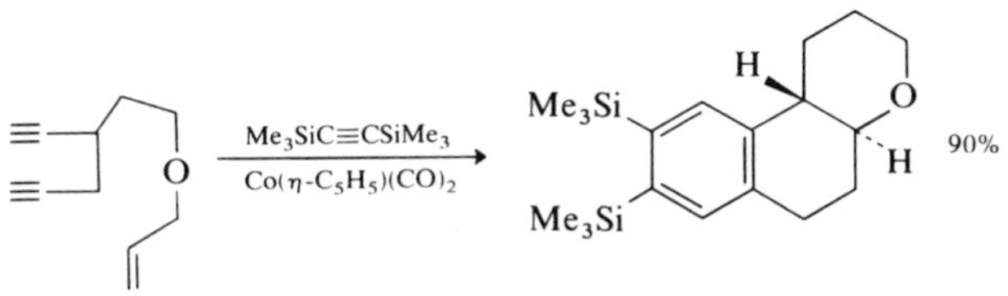

(88)

$$(89)$$

$$(90)$$

$$(91)$$

General Procedure—Synthesis of **68**, *α-Naphthopyran* [97]

The pentenyloxyhexadiyne **69** (120 mg, 0.74 mM) in degassed octane (7 ml) containing $CpCo(CO)_2$ (20 μl, 0.16 mM) is added very slowly (over 100 h using a syringe pump) to bis(trimethylsilyl)acetylene (2.1 g, 12.3 mM) in refluxing octane (7 ml) under nitrogen. The mixture is cooled and all volatiles removed by distillation (recovered diyne). The brown residue is chromatographed on two preparative T.L.C. plates (ether–petrol 1:99) to give the desired product (**68**) as an oil (147 mg, 60%).

A novel synthesis of *dl*-estrone has been reported using a related reaction [98] and further extensions of these reactions are being developed. Using a nitrile group in place of the monoalkyne, for instance, produces a route to heterocyclic compounds (Chapter 4) [Eq. (92)].

$$(92)$$

3.5.2. Formation of Nonaromatic Six-Membered Rings from Alkynes

The trimerization of alkynes has been postulated to proceed through a metallacyclopentadiene intermediate (**65**). If this intermediate can be encouraged to insert an alkene, rather than another alkyne, then a synthetic route to cyclohexadienes is possible. Attempts to perform this transformation have used activated alkenes that bind more strongly to **65** than the alkyne. In practice, only partial success has been obtained. In most instances the alkyne cyclotrimerization reaction is a competitive side reaction that limits the formation of cyclohexadiene derivatives. Also, dehydrogenation leading to aromatization appears to be a problem [Eqs. (93)–(95)].

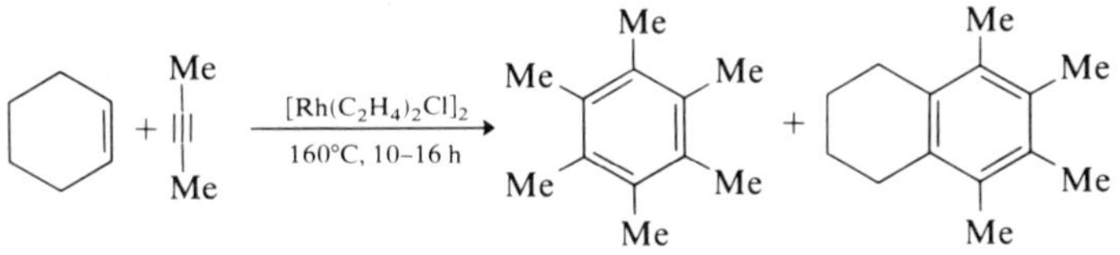

$$(93)^{99}$$

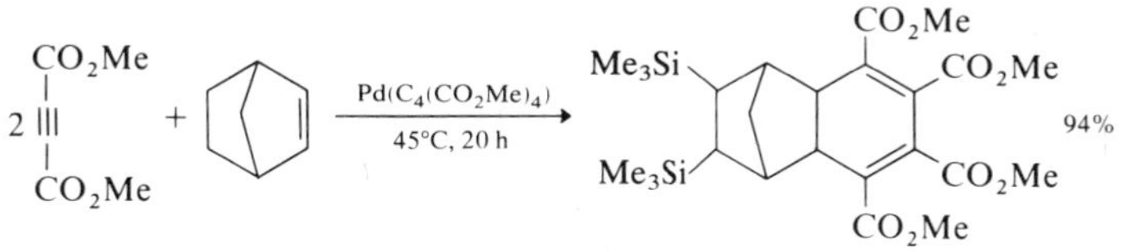

$$(94)^{100}$$

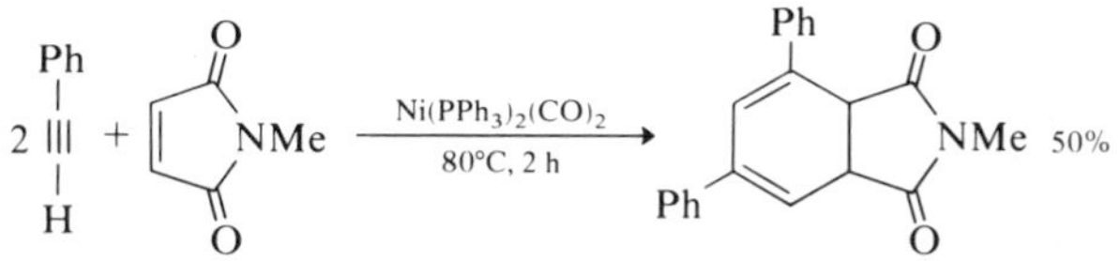

$$(95)^{101}$$

The stoichiometric reaction of an alkene with a preformed metallocycle has also been studied. The cobalt metallacycle (**70**) is prepared in reasonable yield by the stepwise reaction of two molecules of alkyne with Co(η-C_5H_5)(PPh$_3$)$_2$.[102] These air-stable complexes react with alkenes in a Diels–Alder reaction to yield, after work-up, cyclohexadiene compounds (Scheme 8). Work-up usually entails catalyst destruction by Ce(IV) oxidation to

Scheme 8

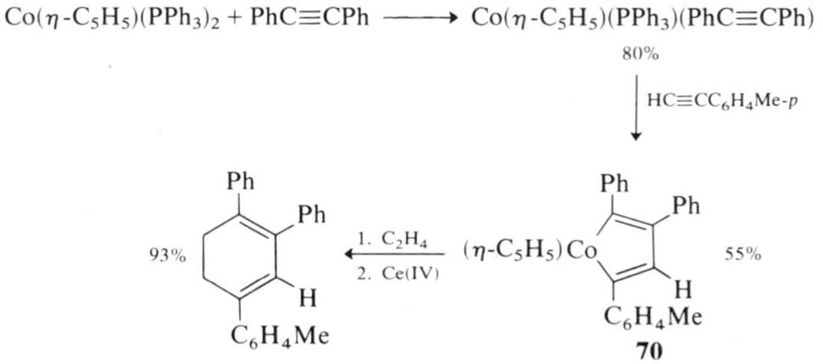

liberate the organic ring from the stable complex, although sometimes (e.g., phenyl acetylene plus acrylonitrile) the cyclohexadiene is displaced during the reaction and a catalytic reaction is possible (albeit low turnover) [Eq. (96)].

$$2PhC\equiv CH + \underset{}{\nearrow}CN \xrightarrow{Co(\eta\text{-}C_5H_5)(PPh_3)_2} \qquad \qquad (96)$$

The isolated metallacyclic intermediate can be reacted with an alkyne to produce aromatic compounds—for example, in the synthesis of quinones[103] [Eq. (97)]. The metallacycles are easily prepared from the appropriate diyne and $Rh(PPh_3)_3Cl$ in good yield. Using a tetrayne substrate more

$$\xrightarrow{R'C\equiv CR''} \qquad \qquad (97)$$

R′ = R″ = Et	R = H (15%), R = Me (74%), R = Et (68%), R = Ph (34%)
R′ = CO$_2$Me, R″ = H	R = Me (72%), R = Ph (38%)
R′ = Ph, R″ = H	R = Me (70%), R = Ph (52%)
R′ = R″ = Ph	R = Ph (50%)
R′ = R″ = CO$_2$Et	R = Me (79%), R = Ph (48%)
R′ = R″ = CH=CHPh	R = Me (64%), R = Ph (54%)

complex quinones can be prepared but in low yield [Eq. (98)]. The rhodium metallacycle can be prepared and used *in situ* but yields of product are lower (10%–20%).

$$\xrightarrow{RC\equiv CR} \qquad \qquad (98)$$

R = CO$_2$Me, 35%

3.5.3. Cyclotrimerization of Alkenes and Allenes

The cyclotrimerization of alkenes is not well documented. One of the few examples is the oligomerization of 3,3-dimethylcyclopropene by $Ni(CO)_4$, which produces a mixture of products including the cyclic compound **71** in low yield (10%).[104] The major products are linear oligomers.

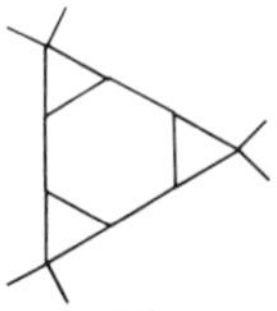

71

The reaction of vinylic halides with various alkenes in the presence of a base, triethylamine, and a palladium acetate–triphenylphosphine catalyst leads to the formation of conjugated dienes. However, in some instances, typically with less substituted reagents, the initially formed diene undergoes subsequent Diels–Alder reaction to yield a cyclohexene derivative.[105] For example, both vinyl iodide and 2-bromopropene react with excess methyl acrylate to form a cyclic product [Scheme 9 and Eq. (99)]. In the case of vinyl iodide, the presence of base (used to complex the liberated haloacid) catalyzes double-bond migration and hence the 2,3, as opposed to the 3,4, derivative is obtained (Scheme 9).

Scheme 9

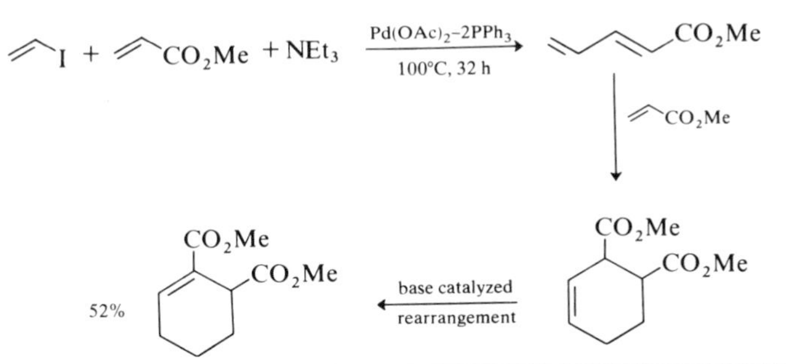

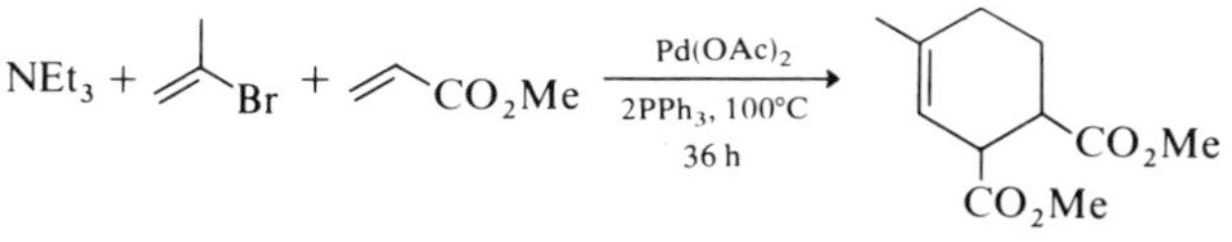

(99)

Bis(triphenylphosphine)nickel dicarbonyl catalyzes the oligomerization of allene at 80°C to produce a moderate yield of two cyclotrimers,
72 and **73**.[106] This autoclave reaction can be employed to produce large
quantities of **72** and **73** but is not amenable to the formation of other
derivatives using substituted dienes. In the presence of alkynes, however,
allene may be cooligomerized to form a mixture of substituted cyclic trienes,
74 and **75** [Eq. (101)].[107]

$$=\!\cdot\!= \xrightarrow[110°C, 6\,h]{Ni(PPh_3)_2(CO)_2} \qquad + \qquad 30\% \; (4:1) \qquad (100)$$

72   **73**

$$\equiv\!-R \; + \; =\!\cdot\!= \xrightarrow[80-85°C]{Ni(PPh_3)_2(CO)_2} \; R\!-\!\!\!\qquad + \qquad \qquad (101)$$

R = H, 43%
also R = Me, Ph, or CH = CH$_2$ **74**   **75**

3.5.4. Cyclodimerization of Dienes and Trienes

The cyclooligomerization of 1,3-butadiene has received considerable
attention and leads to the formation of 4-vinylcyclohexene (40%) together
with C$_4$ and C$_8$ compounds when Ni(COD)$_2$–P(cyclo-C$_6$H$_{11}$)$_3$ is the catalyst
[Eq. (102)].[28] This catalytic reaction proceeds at a reasonable rate (35 g
butadiene/g Ni/h) although not as fast as other butadiene cyclooligomerizations. Even so it represents a convenient route to 4-vinylcyclohexene.

$$\xrightarrow[80°C]{Ni(COD)_2-P(cyclo-C_6H_{11})_3} \qquad + \qquad \qquad (102)$$

40%   41%

Preparation of 4-vinylcyclohexene[28]

Ni(COD)$_2$ (5 mM) and P(cyclo-C$_6$H$_{11}$)$_3$ (5 mM) are slurried in benzene
(200 ml) under nitrogen. The mixture is rapidly stirred at reflux temperature
(80°C) and butadiene is passed into the solution at such a rate that it gently
bubbles through the exit bubbler. The butadiene is rapidly consumed and
after about 3 h the reaction is cooled and filtered through Celite. Solvent is
removed by distillation and the remaining liquid fractionally distilled to yield
4-vinylcyclohexene, bp 129–130°C (40%), and *cis,cis*-1,5-cyclooctadiene, bp
151–152°C (41%).

The cyclo- and especially cooligomerizations (with other dienes) of substituted butadienes produce, in the main, cyclooctadiene derivatives and not six-membered rings (Section 3.7.1). That is, in part, due to the use of $Ni(COD)_2$–$P(OC_6H_4$-o-$C_6H_5)_3$, a catalyst that favors eight-membered ring formation. However, isoprene gives a 35% yield of the six-membered rings (**76**), and 2,3-dimethylbutadiene an 86% yield of **77**.

76   **77**

Cooligomerization with butadiene, on the other hand, gives only 5.5% and 7.6%, respectively, of the cyclohexanoid compounds. One cooligomerization that does produce good yields of a cyclohexene derivative is the Diels–Alder-type reaction of butadiene with methyl sorbate [Eq. (103)].[108]

$$(103)$$

Similarly, 2,3-dimethylbutadiene and methylsorbate in the presence of the same nickel catalyst gives **78** as the major product (75%) [Eq. (104)]. It is possible that $Ni(acac)_2$–Et_3Al–PPh_3 could be replaced by a preformed nickel(0) catalyst, e.g., $Ni(COD)_2$–PPh_3, which avoids the use of highly

$$(104)$$

78

reactive aluminum alkyl. Numerous 1,3,5-trienes cyclo- and cooligomerize in the presence of nickel catalysts to produce mainly six-membered rings [Eqs. (105)–(107)].[109,110] By varying the substitution pattern on the diene and the 1,3,5-triene a wide variety of substituted cyclohexenes can be prepared. Also, altering the stereochemistry around the double bond leads to different products [cf. Eqs. (106) and (107)].

$$\text{(105)}$$

$$\text{(106)}$$

$$\text{(107)}$$

A related 1,2,4-triene cyclooligomerization has been reported.[111] 1,2,4-Pentatriene is dimerized by palladium(0)–tri(isopropyl)phosphine at 35°C to yield a mixture of six-membered rings [Eq. (108)]. Upon treatment with KOBu^t in butanol the main product **79** isomerizes to the (1-propenyl) toluene isomers **80** and **81** [Eq. (109)].

$$\text{(108)}$$

$$\text{(109)}$$

3.5.5. Cycloaddition of Dienes to Alkenes or Alkynes

The cooligomerization of butadiene with ethylene leads to the formation of 1,5-cyclodecadiene (Section 3.8.1.2) and not six-membered ring compounds. However, if high temperatures (>100°C) are employed, thermal Cope rearrangement occurs to yield *cis*-1,2-divinyl cyclohexane [Eq. (110)]. Alkynes are more readily cooligomerized with butadiene

$$(110)$$

(diene : alkyne 2 : 1) to produce initially ten-membered rings (Section 3.8.1.4); these again undergo facile Cope rearrangement to cyclohexenes [Eq. (111)].[112] The divinyl cyclohexene can be further isomerized to yield

$$(111)$$

e.g., R = Me, Bu^t, or CH$_2$OMe

the aromatic product. However, the ten-membered ring is not always isolable at room temperature. Acetylene and butadiene cooligomerize in the presence of Ni(COD)$_2$–PBu$_3^n$ to give 5-vinylcyclohexa-1,3 diene in good yield (52%) at 25°C [Eq. (112)].[113] The high yield of the cyclohexadiene as

$$(112)$$

opposed to cyclodecatriene results from the use of a trialkylphosphine, rather than an aryl-phosphine or -phosphite, and demonstrates how sensitive nickel catalysts are to the steric and electronic environment around the metal center. If phenyl acetylene is added slowly to a benzene solution of Ni(COD)$_2$–PPh$_3$ saturated with butadiene at 40°C, only diphenyl vinyl-cyclohexadiene is obtained[114] [Eq. (113)]. An alkyne to diene ratio >2 is

$$(113)$$

beneficial to formation of the cyclohexadiene compound. Butadiene reacts with alkyl- or aryl-substituted acetylenecarboxylic esters at −10°C (diene : alkyne 1 : 5) in the presence of Ni(COD)$_2$–PPh$_3$ to give substituted 5-vinylcyclohexa-1,3-dienes [Eq. (114)]. The ratio of the various isomers depends on the nature of the substituent at the alkyne; also, the conditions employed exert a considerable influence on the course of the reaction. For example, using the conditions described above (i.e., −10°C), phenyl acetylene and butadiene give only a 25% yield of cyclic product compared to

$$R—\equiv—R' + \text{(butadiene)} \xrightarrow[-10°C]{Ni(COD)_2\text{-}PPh_3}$$

(114)

R = Me, R' = CO$_2$Me (87%)
R = Bu^t, R' = CO$_2$Et (80%)
R = Bu^t, R' = CO$_2$Me (76%)
R = R' = CH$_2$OMe (26%)
R = R' = Me (20%)

78% at 40°C. 2-Butyne (R = R' = Me) and butadiene using a nickel catalyst give a poor yield of 1,2,3,4-tetramethyl-5-vinyl,1,3-cyclohexadiene; however this compound can be produced in good yield using a homogeneous TiCl$_4$–AlEt$_2$Cl catalyst.[29] These 5-vinylcyclohexa-1,3 dienes are thermally unstable and rearrange quantitatively above 80°C to tricyclo [2,2,2,0]oct-7-enes [Eq. (114a)].

(114a)

Iron catalysts have also been used to promote the reaction of butadiene with alkynes.[115] Hence, 1,2-diphenyl-1,4-cyclohexadiene (68%) and 1,2-dimethyl-1,4-cyclohexadiene (55%) can be prepared from the appropriate alkyne and butadiene using FeCl$_3$–Pr^{i}MgCl and Fe(cyclo-C$_8$H$_{12}$)$_2$, respectively. Related iron(0) catalysts have been used to codimerize butadiene and ynamines[116] [Eq. (115)]. The iron(0) catalyst is generated from FeCl$_3$ and MgPr^{i}Cl *in situ*. Hydrolysis of the 1,4-cyclohexadienamine produces either α,β- or α,δ-unsaturated cyclohexenones.

$$\text{(butadiene)} + Et_2N—\equiv—R \xrightarrow{Fe(0),\ 25°C} \text{(cyclohexadienamine)}$$

(115)

R = Me (80%), Pr^i (60%), C$_5$H$_{11}$ (50%), or Ph (40%)

3.5.6. Cyclization of Difunctional Substrates

The number of useful transition-metal-promoted cyclizations forming six-membered rings is not extensive. For example, nickel-coupling reactions can be used to prepare cyclic hexanes, but these reactions are more useful for forming larger ring systems (see Section 3.8.2.1). Titanium-induced reductive coupling of dicarbonyl compounds also provides a route to six-membered rings, but again this method is more suited to larger ring synthesis. Other cyclization methods not using transition metals, e.g., the Dieckmann condensation[117] or the acyloin cyclization,[118] are generally preferred for the synthesis of six-membered rings from linear substrates.

A few interesting and useful examples of transition-metal-catalyzed cyclizations have been reported. The formation of cyclic alcohols from unsaturated aldehydes is catalyzed by $Rh(PPh_3)_3Cl$.[119] Reaction of (+)citronellal with one equivalent of $Rh(PPh_3)_3Cl$ at room temperature for 15 h produces a mixture of two isomeric cyclic products [Eq. (116)]. This

$$(116)$$

3:1
55%

reaction is interesting because the related example, using 4-ene unsaturated aldehydes as opposed to the 6-ene analog above, produces a cyclopentanone derivative.

Combining the tricarbonyl iron diene complex **82** with excess oxalyl chloride and aluminum chloride (1:2) in dichloromethane at −78°C leads to an instantaneous reaction.[120] Aqueous work-up followed by reaction with ethanolic silver nitrate gives the cyclic compounds **83** and **84** in 35%–40% yield [Eq. (117)].

$$(117)$$

The dibromoketone plus iron carbonyl reaction for the formation of five- and seven-membered rings (see Sections 3.4.1.3 and 3.6.1, respectively) can be used to form cyclohexanoid compounds.[121] Thus the dibromoketone obtained from geraniol, and $Fe_2(CO)_9$ (1:1.2) react on heating in benzene at 100–110°C (pressure bottle) to yield a mixture of cyclic hexanes [Eq. (118)]. Similarly, the dibromoketone **85** (from E,E-farnesol) gives the cyclic terpenes **86** and **87**.

$$(118)$$

$$(119)$$

3.6. Formation of Seven-Membered Rings

Few transition-metal-catalyzed reactions are known to form seven-membered rings. This may be ascribed to the lack of suitable compounds that can serve as or generate an odd-numbered carbon unit and also to the instability of the eight-membered metallocycle intermediate generated during cyclization. A general cycloheptenone synthesis has been developed using a 3 + 4 cycloaddition reaction, but unfortunately no 2 + 2 + 2 + 1 or 2 + 2 + 3 cyclizations are known.

3.6.1. 3π-4π Cycloaddition Reactions

The reaction of an oxyallyliron(II) species, formed by the reaction of $Fe_2(CO)_9$ and a dibromoketone, with a diene provides a general synthetic route to cyclic heptenones[60] [Eq. (120)]. The reaction requires

$$(120)$$

stoichiometric quantities of expensive $Fe_2(CO)_9$, although $Fe(CO)_5$ can be used instead but is less satisfactory. Use of a preformed diene iron carbonyl complex (prepared by the photoirradiation of a mixture of diene and iron pentacarbonyl) requires an extra step but produces higher yields of cyclic product. For example, the $Fe_2(CO)_9$ promoted reaction of the dibromoketone **88** and butadiene gives only 33% yield of **89**, whereas the same reaction using butadiene iron tricarbonyl gives a 90% yield [Eq. (121)]. Reaction conditions are usually 40–80°C in benzene with reaction

$$(121)$$

times of the order of 4–80 h (shorter times of 4–20 h are required when the preformed iron diene complex is used). Both secondary [i.e., R = alkyl, R' = H, Eq. (120)] and tertiary dibromoketones (i.e., R = R' = alkyl) may be used and moderate to high yields are obtained (Table 3.7). The primary dibromoketone (R = R' = H) cannot be used in these reactions; instead, the tetrabromo analog **90** must be employed followed by bromide removal to obtain the desired product [Eq. (122)]. The reaction is not

$$(122)$$

Table 3.7. Formation of Cycloheptenones [Eqs. (120) and (123)][122]

Dibromoketone $(RR'BrC)_2C{=}O$	Diene R_a, R_b or Y	Product yield, % 1. Using $Fe_2(CO)_9$ 2. Using preformed diene iron carbonyl
R = R' = Me	$R_a = R_b$ = Me	1. 71
		2. 100
R = R' = Me	$R_a = R_b$ = H	1. 33
		2. 90
R = Me, R' = H	R_a = Me, R_b = H	2. 51
R = Me, R' = H	R_a R_b = $+CH_2+_4$	1. 80
R = Pr, R' = H	$R_a = R_b$ = H	1. 44
		2. 77
R = Me, R' = H	Y = CH_2	1. 86
R = R' = Me	Y = O	1. 89
R = R' = Ph	Y = O	1. 90
R = R' = Me	Y = NCOMe	1. 68

limited to open-chain dienes; use of cyclic dienes leads to the formation of *cis*-bicyclic systems [Eq. (123)] and Table 3.7 [where typically Y = CH$_2$,

$$\text{(123)}$$

O, NCOCH$_3$, NCO$_2$Me]. Although pyrroles containing an electron-with-drawing substituent are effective in these reactions using *N*-methylpyrrole a bicyclic adduct is not obtained. Similarly, α- and β-substituted products are formed in good yield (only α-substituted product is formed with thiophene) [Eq. (124)]. Cyclic dibromoketones have been studied but with limited success.

$$\text{(124)}$$

General Procedure—Preparation of Cyclic Heptenones

1. *Using Fe$_2$(CO)$_9$.* Diiron nonacarbonyl (2.4 mM) is placed in a thick-walled reaction ampoule (100 ml) under nitrogen. Dry benzene (10 ml), the diene (20 mM), and the α,α'-dibromoketone (2 mM) are then added and the tube sealed under nitrogen. The mixture is stirred at 60–90°C for several hours, cooled to room temperature, and the resulting precipitate filtered off through a pad of Celite. The filtrate is concentrated on a rotary evaporator to give a yellow oil, which is purified by distillation followed by preparative glc.

2. *Using Diene Iron Tricarbonyl.* Into a pressure bottle (300 ml), under nitrogen, is placed Fe(CO)$_5$ (20 mM) and the 1,3-diene (60 mM) in benzene (20 ml). The mixture is irradiated at 60°C for 6–30 h with a 200-W high-pressure mercury lamp. To the resulting solution is added the α,α'-dibromoketone (15 mM) in benzene and the mixture heated to 70–120°C under nitrogen for several hours. The mixture is cooled, filtered through Celite, and the filtrate concentrated to a pale yellow oil. The oil is dissolved in acetone (20 ml), mixed with copper chloride (10 mM), and stirred at room temperature for 15 min to decompose the remaining diene iron tricarbonyl complex. This slurry is then passed through a short Celite column and the filtrate concentrated under reduced pressure. The residue is dissolved in methylene chloride (50 ml) and washed with 5% aqueous EDTA-disodium solution (4×20 ml). The organic layer is dried (Na$_2$SO$_4$) and concentrated. Distillation gives the pure product.

These cycloaddition reactions have been used as a basis for novel organic syntheses.[123] The cycloheptenones are easily transformed into alkylated tropones by a simple bromination (pyrrolidone hydrotribromide)–dehydrobromination (LiCl/DMF) procedure in overall yield 45%–85%. Similarly, γ-tropolone derivatives are obtained in 45%–55% yield by bromination–dehydrobromination followed by acid hydrolysis. In addition, the bicyclic adducts formed from furan [Eq. (123) and Table 3.7] are converted into the analogous troponoids by catalytic hydrogenation, acid-catalyzed carbon–oxygen bond cleavage, and aromatization.

3.6.2. Cyclization of Difunctional Substrates

The bis-π-allyl nickel complex **91** formed by the reaction of Ni(COD)$_2$ and butadiene reacts with butylisocyanide at $-78°C$ to produce a mixture of cyclic C$_7$ and C$_9$ (1 : 1.6) membered rings[124] [Eq. (125)]. After hydrolysis

$$\text{Ni(COD)}_2 \;+\; \text{CH}_2\!=\!\text{CH–CH}=\!\text{CH}_2 \;+\; \text{PPh}_3 \;\longrightarrow\; \text{Ph}_3\text{P}\!-\!\text{Ni}$$

91

1. BuNC
2. H$_3$O$^+$

(125)

a mixture of conjugated and nonconjugated ketones is obtained, but the ratio of products changes with time and, after stirring with activated charcoal to remove the nickel complex, the conjugated isomers represent the major fraction. Overall yield is 30% (0.6 : 1 C$_7$: C$_9$).

Preparation of 2-Methylene Cyclohepta-5-enone

n-Butylisocyanide (97 mM) in diethyl ether (40 ml) is added slowly to a stirred solution of α,ω-octadiendiylnickel (**91**) [derived from Ni(COD)$_2$ (14.2 mM), butadiene (50 mM), and Ph$_3$P (14.2 mM)] at $-78°C$ under a nitrogen atmosphere]. The mixture is allowed to slowly warm to room temperature overnight and is then hydrolyzed with dilute acid and stirred with activated charcoal to remove nickel residues. The ketone mixture is separated from cyclooctadiene and other hydrocarbon products by column chromatography and the C$_7$ and C$_9$ ketones are purified by preparative glc.

Insertion of carbon monoxide and isocyanides into the bis-π-allyl nickel intermediate **92**, formed from Ni(COD)$_2$ and allene, has also been studied

[Eq. (126)]. A cyclic ketone was not isolated after reaction of **92** with butyl isocyanide followed by hydrolysis, but carbonylation of **92** at $-30°C$ gave 50%–80% yield of the desired ketone **93**. The lack of ketone formation in the isocyanide reaction is thought to be due to the product polymerizing under reaction conditions; indeed, some polymerization also occurs during product isolation from the carbonylation reaction.

$$Ni(COD)_2 + \text{=·=} \xrightarrow[-78°C]{} \quad \text{Ni(COD)} \xrightarrow[-30°C]{CO} \quad \text{O} \quad (126)$$

$$\textbf{92} \qquad\qquad\qquad \textbf{93}$$

3.7. Formation of Eight-Membered Rings

Two general nickel-catalyzed cycloaddition reactions have been reported for the preparation of eight-membered rings: the cyclodimerization of dienes to give cyclooctadiene derivatives and the cyclotetramerization of alkynes to yield cyclooctatetraene analogs. These two methods have been extensively studied and a number of review articles are available.[109,125]

3.7.1. Cyclodimerization of 1,3 Dienes

"Naked nickel," that is, ligand free, zero-valent nickel, catalyzes the cyclotrimerization of butadiene to 1,5,9-cyclododecatriene (see Section 3.8.1.1). The simple step of modifying the metal center by addition of a ligand (normally a phosphine or phosphite) produces a catalyst capable of cyclodimerizing 1,3 dienes. Numerous methods exist for the preparation of naked nickel but the most convenient source is nickel cyclooctadiene [Ni(COD)$_2$], which is readily prepared (Section 9.2.16) and also commercially available. Butadiene is dimerized by Ni(COD)$_2$–ligand to a mixture of products [Eq. (127)], but by careful choice of conditions it yields almost

$$\xrightarrow[\text{ligand}]{Ni(COD)_2} \quad \square + \quad \bigcirc + \quad \bigcirc \quad (127)$$

exclusively cyclooctadiene. (See Sections 3.3.1.1 and Section 3.5.4 for methods of obtaining the C$_4$ and C$_6$ compounds, respectively, in high yield.) 1,5-Cyclooctadiene is obtained in highest yield (96%) using the ligand $P(OC_6H_4\text{-}o\text{-}C_6H_5)_3$,[28] (Ni : ligand 1 : 1) at 80°C. The simplicity of the reaction (see Section 3.5.4 for experimental details) allows large quantities of cyclooctadiene to be easily prepared in a very short time (rate of reaction = 780 g butadiene/g Ni/h).

The cyclodimerization has been extended to a variety of substituted 1,3-dienes including *cis-* and *trans-* 1,3 pentadiene, isoprene, 2,3-dimethyl-butadiene, 1,3-hexadiene, and chloroprene. Isomeric mixtures of substituted cyclooctadienes are generally obtained [Eq. (128)] in yields that

$$\text{(128)}$$

55%

are dependent on the diene (Table 3.8).[125b] The rate of reaction is influenced by the substrate and falls dramatically with increased substitution.

Cocyclodimerization of a substituted 1,3-diene with butadiene proceeds readily, although a deficiency of butadiene is usually employed to avoid the preferential formation of 1,5-cyclooctadiene[125b] [Eq. (129)]. Dienes bearing electron-withdrawing substituents have also been cyclodimerized with butadiene [Eq. (130)]. Triphenylphosphine (a poorer

$$\text{(129)}$$

80%–95% (based on subst. diene)

e.g., R = Me, Et, Bu^t, OMe, or $CH_2CH{=}CHCH_3$,

$$\text{(130)}$$

45%

Table 3.8. Cyclodimerization of 1,3-Dienes^a

	1,3-Diene, % yield			
Product	Butadiene	1,3-Pentadiene	Isoprene	2,3-Dimethyl butadiene
Substituted vinyl-cyclohexene	2.3	5.3	34.8	86.3
Substituted cyclo-octadiene	97.2	90.9	55.1	6.1
Rate, g product/g Ni/h	220	31	14	0.6

^{a} Catalyst Ni(COD)$_2$–P(OC$_6$H$_4$-o-C$_6$H$_5$)$_3$ (1 : 1), 60°C.

electron acceptor) is preferred when using weaker electron-donating dienes. This helps preserve the electronic characteristics of the metal, which are of fundamental importance in these reactions.

General Procedure—Preparation of 1-Phenylcyclooctadiene[135]

Ni(COD)$_2$ (1.4 g, 5 mM), P(OC$_6$H$_4$-o-C$_6$H$_5$)$_3$ (2.7 g, 5 mM), 2-phenyl-butadiene (20 g), and toluene (75 ml) are added under nitrogen to a three-necked flask equipped with a stirrer, gas inlet, and dry-ice condenser. Butadiene is passed into the flask on demand (monitored by a gas exit bubbler connected to the top of the condenser) and the mixture heated to 80°C. After 6 h the reaction is stopped and the mixture distilled to yield 1-phenylcyclooc-tadiene, bp 155°C (14 mm Hg) 81%, based on 2-phenylbutadiene (64% consumed).

This method of cooligomerization is applicable to those substituted dienes that are not readily homocyclooligomerized. When this is a problem, e.g., with isoprene, then both butadiene and the substituted diene are added simultaneously to the flask at 80°C over the period of the reaction.

1,5-Cyclooctadiene derivatives produced by the nickel-catalyzed dimerization of dienes contain *cis* double bonds. The less stable *cis, trans* isomer has been prepared using the iron complex **94** (Scheme 10).[126] This

Scheme 10

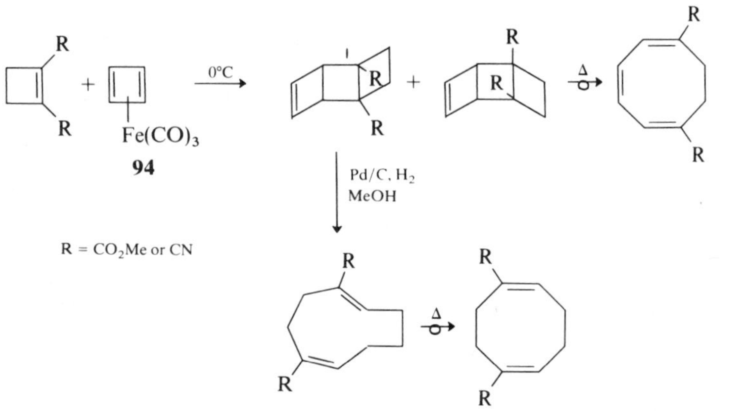

$R = CO_2Me$ or CN

strained, eight-membered ring is unstable and undergoes thermal isomeriz-ation (R = CO$_2$Me, 75°C, and R = CN, 25°C) to give the more stable *cis, cis* derivative.

3.7.2. Cyclotetramerization of Alkynes

The cyclotetramerization of acetylene to cyclooctatetraene was discovered by Reppe and Toepel in 1943. It is interesting to note that previously the preparation of cyclooctatetraene required a classical 13-stage synthesis with overall yield of only 1%–2%. The simple cyclization of acetylene employing a nickel catalyst produces yields of 70%. The preferred catalysts are labile octahedral complexes of nickel(II) containing weak-field ligands such as acetylacetonate or salicylaldehyde.[125e] The reaction is normally carried out under anhydrous conditions in benzene, THF, or dioxan at 80–120°C and 10–25 atm pressure of acetylene [Eq. (131)]. Using

$$4 \equiv \xrightarrow[\text{80–120°C, 10–25 atm}]{\text{Ni(acac)}_2} \qquad \qquad 70\% \qquad (131)$$

monosubstituted alkynes the reaction is less successful and low yields are generally obtained. However, propargyl alcohol, 2-methylbut-3-yn-2-ol,[127] and methyl or ethyl propiolate[128] have been cyclized to the cyclooctatetraene derivatives in reasonable yield under mild conditions [Eqs. (132) and (133)]. Higher yields of **95** are obtained using $Ni(CO)_4$ (~60%), but this catalyst presents many practical problems due to its toxicity. Disubstituted alkynes do not undergo this cyclization reaction.

$$\underset{Me_2C-C\equiv CH}{\overset{OH}{|}} \xrightarrow[\text{2 h, 90–120°C}]{\text{Ni(acac)}_2} \qquad \mathbf{95} \qquad R = Me_2\overset{OH}{\underset{|}{C}}- \quad (40\%) \qquad (132)$$

$$RO_2C-C\equiv CH \xrightarrow{\text{Ni(PCl}_3)_4} \qquad \begin{array}{l} R = Me, 56\% \\ R = Et, 21\% \end{array} \qquad (133)$$

Procedure

1. *Preparation of 1,2,4,6-Tetracarbomethoxycyclooctatetraene.*[128] Methyl propiolate (4.43 g, 54 mM) and $Ni(PCl_3)_4$ (0.0325 g, 0.053 mM) are stirred together in cyclohexane (60 ml) at room temperature. After a 3-min induction period an exotherm to 33°C occurs and the reaction is complete after 15 min. The reaction mixture consists of a cloudy yellow solution and an insoluble

brown viscous material. The yellow solution is decanted and distillation yields 1,2,4-tricarbomethoxybenzene. To the viscous brown residue is added ethanol and the mixture refluxed. Light crystalline flakes precipitate and are filtered, yielding 1.1 g of crude product. The solid is recrystallized from ethanol, giving pure cream platelets of 1,2,4,6 tetracarbomethoxy-cyclooctatetraene mp 182–183°C.

2. *Tetramerization of 2-Methylbut-3-yn-2-ol.*[127] Ni(acac)$_2$ (0.01 mol) is added in portions to 2-methylbut-3-yn-2-ol (1 mol) at reflux under a nitrogen atmosphere. The mixture is maintained at 90–120°C and stirred for 2 h. Work-up gives the cyclooctatetraene derivative **95** in 71% (based on reacted alkyne ~60%).

Substituted cyclooctatetraene derivatives can be prepared by the cooligomerization of acetylene with a mono- or disubstituted alkyne [Eq. (134)].[129] Yields of these mixed products are generally low (16%–25%), the major product being the more easily formed unsubstituted cyclooctatetraene.

$$\equiv \; + \; R-C\equiv C-R' \xrightarrow[\substack{80-90°C \\ 250-300\ psi}]{Ni(acac)_2} \qquad \begin{array}{l} R = H;\ R' = Me,\ CH_2OH,\ or\ Ph \\ R = R' = Me\ or\ Ph \end{array} \qquad (134)$$

Ring formation is not limited to monoalkynes. If, for example, the triple bonds of a diyne are able to closely approach each other, then intramolecular cyclization can occur. The use of diyne substrates has provided a general route for the formation of numerous synthetically useful benzene derivatives (Section 3.5.1), but only a few examples are known for the formation of C$_8$ rings [e.g., Eq. (135)].[130]

$$\xrightarrow[20°C,\ 4\ h]{Ni(CO)_4} \qquad\qquad\qquad (135)$$

3.8. Formation of Large Rings (⩾9)

Large rings are readily prepared by the organometallic-catalyzed cycloaddition of alkenes, alkynes, or dienes or by intramolecular coupling. Rings containing up to 120 carbon atoms have been detected, but the maximum ring size of an isolated compound contains around 30 carbon atoms. As noted in earlier syntheses, the preparation of odd-numbered rings is difficult and therefore most examples describe even-numbered systems.

3.8.1. Cycloaddition Reactions

3.8.1.1. Cyclooligomerization of Dienes

The cyclotrimerization of butadiene was one of the first nickel-catalyzed cycloaddition reactions to be reported. Naked nickel (zero-valent nickel containing no strongly complexing ligands) catalyzes the trimerization of butadiene to a mixture of all *trans-*, *trans,trans,cis-* and *trans,cis,cis-* 1,5,9-cyclododecatriene[131] [Eq. (136)]. The all-*cis* analog is not formed in

$$\text{\wedge\!\!\!\!/} \xrightarrow[\text{80°C}]{\text{Ni(COD)}_2} \quad \underset{85\%}{\bigcirc} \quad + \quad \underset{6.4\%}{t,t,c} \text{ and } \underset{8.6\%}{t,c,c} \text{ isomers} \qquad (136)$$

the catalytic reaction. Although naked nickel can be formed in many ways, the most convenient source is nickel bis-cyclooctadiene [Ni(COD)$_2$].

As with previously described butadiene oligomerizations, the choice of conditions is very important in obtaining high yields of the desired product. At 80°C the major product is *t,t,t*-cyclododecatriene; increasing the temperature leads to a higher proportion of the *t,t,c* isomer but also increases the amount of C$_8$ by-product. A decrease in temperature (20°C) leads to a higher proportion of the *t,t,t* analog, but the reaction rate is much slower. At lower temperatures (−90°C) a stable nickel intermediate **96** can be isolated. This compound has been used as a source of naked nickel and for the preparation of other large-ring systems (Section 3.8.1.5).

96

The cyclotrimerization of substituted 1,3-dienes has not been extensively studied. Alkyl-substituted dienes do not readily cyclotrimerize, probably because of the increased stability of the π-allyl intermediate [i.e., the analog of (**96**)].

Using the two-component catalyst system Ni(η-C$_3$H$_5$)$_2$/[Ni(η-C$_3$H$_5$)Cl]$_2$, butadiene is converted into a mixture of macrocyclic polyenes [C$_8$—>C$_{32}$ (36%)] of general structure (**97**) containing exclusively *trans* double bonds [Eq. (137)].[132] This catalyst system is interesting. Nickel *bis*-allyl alone cyclotrimerizes butadiene (i.e., a source of naked nickel),

$$(137)$$

whereas nickel allyl chloride is an active polymerization catalyst; by combining the two an intermediate effect is obtained.

3.8.1.2. Co-oligomerization of Dienes with Alkenes

The cyclodimerization or cyclotrimerization of butadiene is suppressed in the presence of alkenes or alkynes and co-oligomerization occurs. The reaction of butadiene and ethylene has received the most attention,[133] the main products being cis,trans- 1,5-cyclodecadiene (98) and 1,4,9-decatriene (99). Using a naked nickel catalyst [e.g., Ni(COD)$_2$] at ambient temperatures, selective formation of cyclodecadiene (98) (80%) occurs [Eq. (138)]. The linear triene (99), although produced in only 10%–20% yield, is difficult to separate from the cyclomer. Increasing the temperature (80°C) enhances the rate, but selectivity to the cyclic product is lost.

$$(138)$$

Procedure — Formation of Cyclodecadiene.[133]
    An autoclave is charged with butadiene, ethylene (20–30 atm, ratio $C_2H_4:C_4H_6$, 1.5:3), and the catalyst Ni(COD)$_2$ and left at 20°C for 3–4 weeks. A product mixture containing cyclodecadiene (80%) is obtained which, after destruction of the catalyst and filtration through Celite, is purified by distillation 22°C/0.3 mm Hg.

Alkyl-substituted alkenes are less reactive than ethylene and a similar reaction has only been reported for propylene. By increasing the reactivity of the double bond, e.g., by incorporating strain into the system, co-oligomerization with butadiene has been achieved. Cyclic alkenes such as 3,3-dimethylcyclobutene, dicyclopentadiene, norbornene, and methylene cyclopropane all react with butadiene [e.g., Eq. (139)].[134]
    The reactivity of the alkene can also be increased by incorporating electron-withdrawing substituents as in styrene, itaconic acid, or acrylic

$$\text{(139)}$$

esters and by using a nickel(0)–ligand catalyst (naked nickel does not perform this reaction). Although lower temperatures favor the formation of cyclic species, linear products predominate.[135]

3.8.1.3. Co-oligomerization of Dienes with Allenes

Allene and butadiene produce a mixture of two methylene cyclodecadiene compounds in the presence of a nickel(0)–ligand catalyst [Eq. (140)].[136] The reaction is dependent on allene concentration, which must

$$\text{(140)}$$

63:35
69%

remain low to obtain high yields of cyclic product. 1,1,-Dimethylallene and methoxy allene give *cis,trans*-cyclodecadiene derivatives in 40% and 45%, respectively. These ring systems are thermally unstable, and heating to 150°C leads to a Cope rearrangement [Eq. (141)].

$$\text{(141)}$$

3.8.1.4. Co-oligomerization of Dienes with Alkynes

Alkynes are readily co-oligomerized with butadiene and, unlike alkenes, have a marked tendency to form cyclic products.[138] More than one alkyne molecule is readily incorporated unless an excess of butadiene is used. Both naked nickel(0) and nickel(0)–ligand systems catalyze these reactions, although the latter is preferred. Butadiene–butyne (2 : 1) co-oligomerization has received considerable attention.[139] Using Ni(COD)$_2$–PPh$_3$ at 40°C, 4,5-dimethyl-*cis,cis,trans*- 1,4,7-cyclodecatriene is obtained in 87.5% yield [Eq. (142)]. The product (**100**) is thermally unstable, and if the temperature is not maintained below 60°C a facile Cope rearrangement occurs to give the cyclohexene derivative. The choice of ligand is important and it is

(142)

interesting to note that, whereas phosphite ligands are favored for alkene–diene co-oligomerizations, an aryl phosphine is used for the analogous alkyne reaction. This subtle change is necessary to maintain the same electron density on the nickel catalytic center, whether using an alkene or alkyne.

Procedure – Formation of 4,5-Dimethyl-c,c,t-1,4,7-cyclodecatriene.[139]

Ni(COD)$_2$ (8.5 mM) and PPh$_3$ (8.5 mM) are dissolved in a solution of butadiene in toluene (100 ml). Butyne (250 mM) is then added to give a butadiene : butyne ratio 5–10 : 1. The mixture is stirred at 20°C for 4–6 h (the reaction is monitored by observing the contraction in volume). The reaction must be terminated when all the alkyne is consumed to prevent further reaction of the product with butadiene to give higher oligomers. The catalyst is deactivated and the solution filtered through Celite. Distillation under reduced pressure, and hence low temperature, gives the pure title product 88% yield.

Alkynes containing substituents separated from the triple bond by at least two methylene units, e.g., MeO(CH$_2$)$_2$C≡C(CH$_2$)$_2$OMe, co-oligomerize to form ten-membered rings in high yield. In contrast, MeOC≡COMe and MeOCH$_2$C≡CEt form considerable amounts of a C$_{12}$ ring. Substituted alkynes containing alcohol, amine, ketone, or acid groups cannot be co-oligomerized directly because hydrogen transfer reactions cause either open-chain formation or catalyst deactivation. These functional groups must first be protected, using the 2-tetrahydropyranyl group.

Only a few substituted 1,3-dienes have been investigated, but both isoprene [Eq. (143)] and 1-methylbutadiene produce a mixture of ten- and 12-membered rings when co-oligomerized with butyne.[135]

(143)

3.8.1.5. Stoichiometric Reactions Using Nickel Dodecatrienyl

The reactions described so far are all catalytic in nickel, but it has been noted that the cyclotrimerization of butadiene by naked nickel does not occur at low temperatures (−40°C); instead nickel dodecatrienyl (**96**) can be isolated. In the absence of butadiene this complex is stable as a red crystalline material at room temperature and provides a starting point for many stoichiometric ring-forming reactions [Eqs. (144)–(148)].[140-143]

$$(144)^{140}$$

$$(145)^{140}$$

$$(146)^{141}$$

$$(147)^{142}$$

$$(148)^{143}$$

30%

Reactions on nickel dodecatrienyl (**96**) necessitate the use of inert atmospheres and hence specialized techniques. However, this reactive species can be formed *in situ* and does not require isolation and purification, which greatly simplifies the procedure.

3.8.1.6. Cyclooligomerization of Allenes, Alkenes, or Alkynes

The cyclooligomerization of allene can be controlled to produce a cyclic pentamer (50%–55%) by the reaction of allene with $Ni(COD)_2$ (0.2 mol %) at 40–70°C [Eq. (149)].[144] A related isomeric pentamer is

$$=\cdot= \xrightarrow[40-70°C]{Ni(COD)_2} \qquad 50\%-55\% \qquad (149)$$

produced quantitatively in the stoichiometric reaction of allene with $[RhCl(C_2H_4)_2]_2$, followed by liberation of the organic moiety by diphos at room temperature [Eq. (150)].[145] These allene oligomerizations require

$$=\cdot= + [Rh(C_2H_4)_2Cl]_2 \longrightarrow \qquad (150)$$
$$100\%$$

five monomer units to complete the ring and, as may be expected, reactions of this type are unusual. Alkenes and alkynes do not readily cyclooligomerize to form large rings. Propargylic acetate, for example, has been cyclotetramerized using a copper-catalyzed decomposition of the anion but only in 5% yield[146] [Eq. (151)].

$$\underset{\underset{Me}{|}}{\overset{\overset{OAc}{|}}{MeC}}-C\equiv CH \xrightarrow[\text{2. CuCl (R.T.)}]{\text{1. BuLi } (-78°C)} \qquad (151)$$

3.8.1.7. Formation of Very Large Rings from Alkenes and Dienes

A number of syntheses have been reported for the preparation of large-ring molecules. Rings up to C_{120} have been identified using alkene metathesis [Eq. (152)].[147] This reaction is not synthetically useful because reactants and products are in equilibrium and, therefore, isolation of a specific product in reasonable yield is difficult (see also section 2.2.1).

The most elegant synthesis of large-ring systems involves a three-stage reaction based on the facile cyclooligomerization of butadiene with alkynes

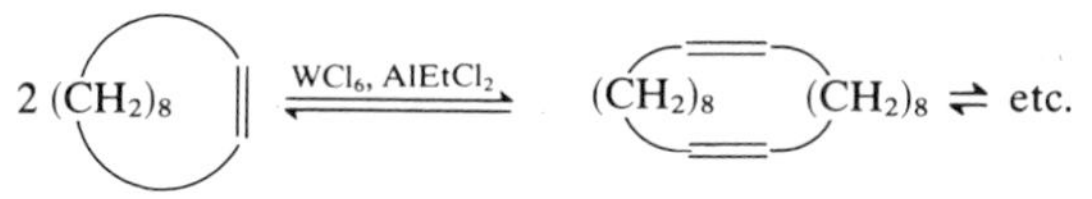

(152)

(Scheme 11).[137,148] $Ni(COD)_2$–$P(OC_6H_4$-o-$C_6H_5)_3$ catalyzes the cycloaddition of butadiene to a cyclic alkyne to produce a bicyclic system (101).

Scheme 11

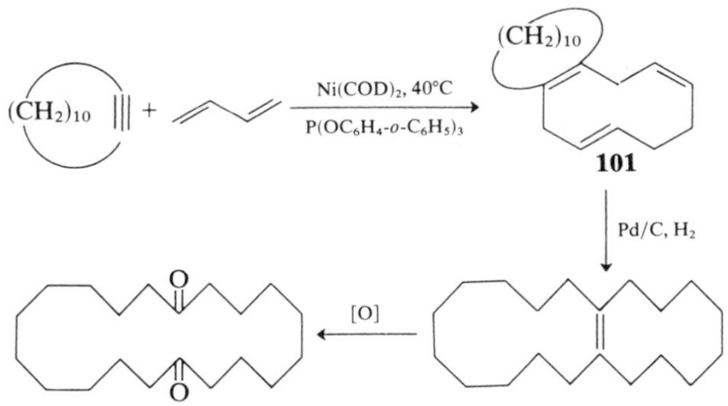

Hydrogenation [Pd/C, H_2] of 101 selectively reduces the disubstituted double bonds. Ozonolysis is then used to cleave the remaining double bond, generating the large-ring diketone in 80% yield. By using cyclic alkynes of differing ring size a variety of large rings can be prepared, but the reaction is limited to the availability of large-ring alkynes. Even so, very large rings can be synthesized using this procedure by choosing a cyclic diyne as the substrate. Respectable yields ($\sim$30%) of the C_{30} tetraketone have been obtained [Eq. (153)]. These large-ring ketones can be further reacted at the carbonyl function or reduced to give the unsaturated cyclic system.

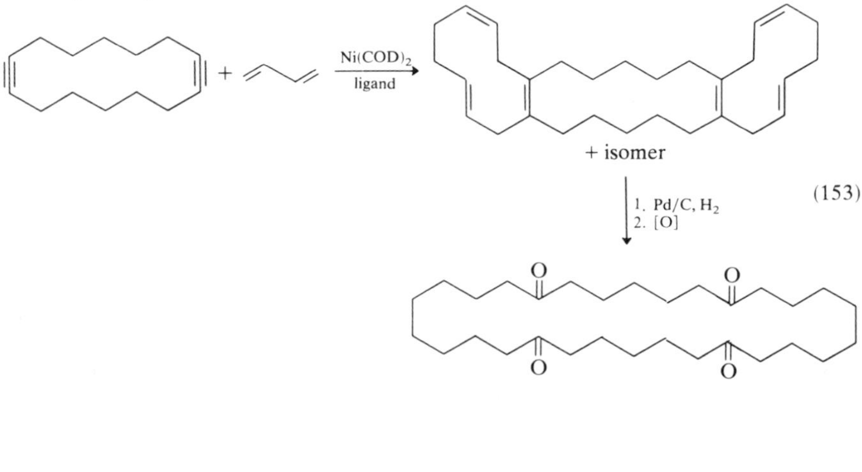

(153)

3.8.2. Intramolecular Coupling

3.8.2.1. Nickel-Promoted Coupling of Allylic Halides

The coupling of allylic halides by nickel carbonyl has been used extensively to prepare macrocyclic polyenes.[125d] In these reactions the nickel(0) reagent acts as a template upon which α,ω-diallylic dihalides can undergo coupling to produce a ring. Allylic dibromides are preferred and are prepared by alkylation of the dibromide $Br(CH_2)_n Br$ with the sodium derivative of propargyl tetrahydropyranyl ether in liquid NH_3–ether, followed by selective reduction to the ethylenic diol and bromination using PBr_3.[149] The cyclization reaction is performed by the slow addition (12 h) of the allylic dibromide in DMF or *N*-methylpyrrolidone to a solution of $Ni(CO)_4$ in DMF under an inert atmosphere at 50°C [Eq. (154)]. The slow addition maintains a high dilution and hence favors intramolecular coupling.

$$BrCH_2CH=CH(CH_2)_8CH=CHCH_2Br \xrightarrow[\text{DMF, 50°C}]{Ni(CO)_4}$$

70%–74%   (154)

The reaction is effective for 12-, 14-, and 18-membered rings in high yield ($\sim$75%). The efficiency of the cyclo-C_{18} synthesis suggests that this method should be applicable to the formation of larger rings but this has yet to be demonstrated experimentally. Eight- and ten-membered rings cannot be made this way because the formation of six-membered rings, via 1,6- or 3,8-coupling, occurs preferentially. Hence $BrCH=CH(CH_2)_4CH=CHBr$ gives 4-vinylcyclohexene (42%) and $BrCH=CH(CH_2)_6CH=CHBr$ affords 1,2-divinylcyclohexene.

The coupling reaction is sensitive to substitution at the allylic position [e.g., Eq. (155)] and to other functional groups in the molecule [e.g., Eq. (156); the third double bond is thought to hinder the cyclization reaction].

humulene
10% overall   (155)

In addition to allylic bromides, the corresponding chlorides, iodides, tosylates, and acetates are coupled effectively using $Ni(CO)_4$. (I > Br > Cl $\sim$ OTs > OAc). The geometry of the coupled product (i.e., *cis* or *trans*) is generally independent of the geometry of the starting material.

Intramolecular coupling has been utilized in, for example, the synthesis of humulene [Eq. (155)][150] and cembrene [Eq. (156)].[151]

(156)

General Procedure—Intramolecular Coupling of Allylic Dibromides Using $Ni(CO)_4$

The general method of performing nickel-promoted intramolecular coupling is described for the key step in the synthesis of cembrene [Eq. (156)]. *All operations must be performed in a good fume cupboard owing to the toxicity of* $Ni(CO)_4$ *(see p. 8).*

A three-necked flask equipped with a magnetic stirrer, dry ice condenser, stopper, and rubber serum cap is flushed with nitrogen. A slow flow of nitrogen is maintained through the apparatus, exiting through a bubbler and a conc.HNO_3 trap [for $Ni(CO)_4$]. The condenser is filled with an ice–salt mixture and N-methylpyrrolidone (20 ml) and $Ni(CO)_4$ (0.93 ml, 7.2 mM) are added. The mixture is then heated to 52°C and the allylic dibromide (685 mg, 1.43 mM) in N-methyl pyrrolidone (100 ml) added slowly over $5\frac{1}{2}$ hrs. The pale yellow $Ni(CO)_4$ solution initially turns light rust brown, finally acquiring a blue-green color after complete addition. The solution is cooled and excess $Ni(CO)_4$ removed under vacuum. The reaction mixture is then poured into HCl (50 ml, 1 M) and dried (Na_2SO_4). Volatiles are removed by rotary evaporation to yield a viscous yellow oil (300 mg), which is purified by column chromatography to give pure product (110 mg, 25%).

Nickel-promoted intramolecular coupling of allylic dibromides has been used to prepare large-ring methylene cycloalkanes[152] [Eq. (157)].

(157)

102

$n = 6, 8,$ or 12

This reaction fails completely, though, for the 12-membered analog ($n = 4$). Heterocyclic compounds have also been prepared by this method, for example, the cyclization of large-ring lactones[153] [Eq. (158)].

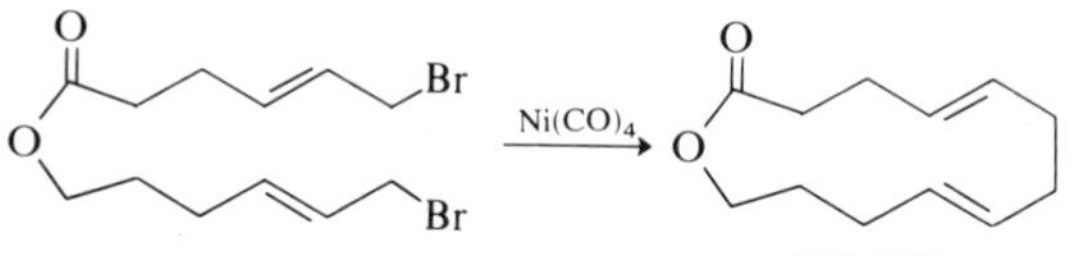
Br
Br
Ni(CO)4
O
70%–75%

(158)

Although $Ni(CO)_4$ has been used extensively for these coupling reactions, this reagent is toxic and not easily handled. An improved source of nickel(0) is $Ni(COD)_2$, but no data are available on its applicability to these reactions.

3.8.2.2. Miscellaneous Nickel-Promoted Intramolecular Coupling Reactions

Related to the above reaction is that of 1,1-dichloromethylethylene with nickel carbonyl to form 1,4,7-trimethylenecyclononane in 54% yield [Eq. (159)].[154] As noted in the introduction, routes to large odd-numbered

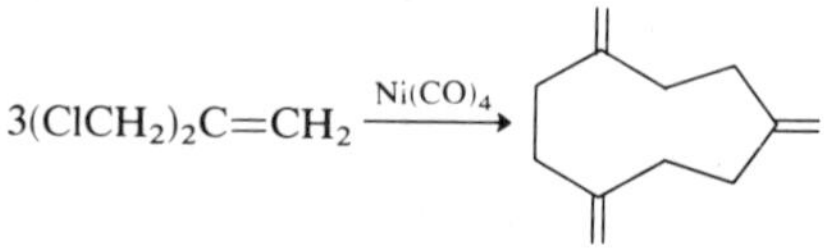
$3(ClCH_2)_2C{=}CH_2$
Ni(CO)4

103 54%

(159)

rings are few; this reaction is therefore of interest. The nine-membered ring **103** can also be prepared by reaction of 1,9-dichloro-2,5,8-trimethylenenonane with $Ni(CO)_4$ in a conventional intramolecular coupling reaction.

A nickel(0) catalyst $Ni(Ph_3P)_4$ is effective for promoting the ring closure of 1,*n*-bis(iodoaryl)alkanes to form a variety of cyclic products (i.e., bridged biphenyls) [Eq. (160)].[155] These nickel-catalyzed reactions are superior

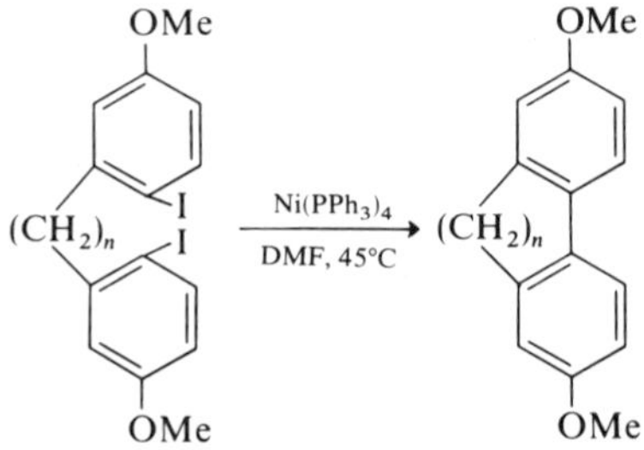
OMe
OMe
(CH2)n
I
I
Ni(PPh3)4
DMF, 45°C
(CH2)n
OMe
OMe

(160)

$n = 2$ Yield: 81%
$n = 3$ Yield: 83%
$n = 4$ Yield: 76%
$n = 5$ Yield: 85%
$n = 6$ Yield: 38%

to the analogous copper-promoted cyclocoupling of aryl halides. The choice of catalyst is important; $Ni(COD)_2$ was examined but gave low yields of the ring product owing to numerous side reactions.

3.8.2.3. Reductive Intramolecular Coupling

Difunctional compounds other than dihalides have been used in intramolecular coupling reactions. Cycloalkenes of ring size 4–16 are prepared in good yield by treatment of dicarbonyl compounds with $TiCl_3/Zn$–Cu [Eq. (161)[156] and Table 3.9]. This reaction is general for both aldehydes

$$
\begin{array}{c}
R \\
\diagdown \\
C{=}O \\
\diagup \\
C{=}O \\
\diagup \\
R'
\end{array}
\xrightarrow{TiCl_3,\ Zn(Cu)}
\begin{array}{c}
R \\
\diagdown \\
C{-}O^- \\
\diagup \\
C{-}O^- \\
\diagup \\
R'
\end{array}
\longrightarrow
\begin{array}{c}
R \\
| \\
C \\
\| \\
C \\
| \\
R'
\end{array}
\tag{161}
$$

and ketones and, although intolerant to readily reducible functional groups (e.g., esters, nitrate, nitro, sulfoxide, etc.), compares favorably with other ring-forming methods. The coupling agent is readily prepared from anhydrous $TiCl_3$ and zinc–copper couple under nitrogen in dimethoxyethane.

General Procedure—Intramolecular Coupling of Nonadecane-5,15-Dione
$TiCl_3$ (1.031 g, 6.68 mM) and Zn–Cu couple (1.01 g, 15.4 mM) are placed in a flask under nitrogen. [The couple is prepared by adding zinc dust (150 mM) to deoxygenated water (40 ml), purging the slurry with nitrogen for 15 min, and then adding $CuSO_4$ (4.7 mM, 150:4.7 Zn:Cu). The black slurry is filtered under N_2, washed with deoxygenated water, acetone, and ether, and dried *in vacuo*. The couple can be stored under N_2 indefinitely.] Anhydrous dimethoxyethane (20 ml) is added and the mixture refluxed for 1 h. Nonadecane-5,15-dione (0.182 g, 0.61 mM) in dimethoxyethane (40 ml) is added slowly (30 h) to the refluxing slurry. After an additional 14 h reflux the reaction mixture is cooled, filtered through Florisil, and concentrated on a rotary evaporator. Short-column chromatography (alumina) gives pure 1,2-di-*n*-butylcycloundecene (0.122 g, 0.46 mM, 76%).

α,ω-Bis(diazoketones) undergo intramolecular cyclization, with loss of nitrogen, in the presence of copper acetylacetonate to give cycloalk-2-ene-1,4-diones [Eq. (162)].[157] The *trans* product is the dominant isomer

$$
\begin{array}{c}
O \\
\|
\end{array}
(CH_2)_n
\begin{array}{c}
\diagup CHN_2 \\
\diagdown CHN_2 \\
\|
\\ O
\end{array}
\xrightarrow[60°C,\ 24\ h]{Cu(acac)_2}
(CH_2)_n
\begin{array}{c}
O \\
\|
\\
H \\
\\
H \\
\| \\ O
\end{array}
\ +\ (CH_2)_n
\begin{array}{c}
O \\
\| \\
{=}O
\end{array}
\tag{162}
$$

104   **105**

$n = 4$	Yield: 30%
$n = 7$	Yield: 30%
$n = 8$	Yield: 70%
$n = 9$	Yield: 25%
$n = 10$	Yield: 80%
$n = 12$	Yield: 44%
$n = 16$	Yield: 29%

Table 3.9. Intramolecular Coupling of Dicarbonyl Compounds [Eq. (161)]

Substrate	Cycloalkene	Yield, %
$BuCO(CH_2)_7COBu$		68
$BuCO(CH_2)_8COBu$		75
$BuCO(CH_2)_9COBu$		76
$OHC(CH_2)_{10}CHO$		76
$OHC(CH_2)_{11}CHO$		52
$OHC(CH_2)_{12}CHO$		71
$OHC(CH_2)_{13}COPh$		80
$OHC(CH_2)_{14}CHO$		85

except for cyclooctadienone ($n = 4$), when only the *cis* isomer is formed. No cyclic product is obtained when $n = 5$ or 6, i.e., nine- and ten-membered rings. The cyclic diketones are useful synthetic precursors to cycloalkane-1,4-diones and fused-ring cyclopentenones [Eq. (163)].

$$\textbf{104} + \textbf{105} \xrightarrow[\text{aq. EtOH}]{Na_2S_2O_4} (CH_2)_n \xrightarrow{NaOH} (CH_2)_n \qquad (163)$$

$n = 6, 8, 9,$ or 15

Chapter 4
Formation of Heterocyclic Compounds

4.1. Introduction

Whereas the formation of carbon–carbon bonds using transition metal intermediates has been relatively well studied, the synthesis of heterocyclic compounds using transition metals has attracted much less attention. Indeed in recent years only two reviews on the subject have appeared.[1] As the demand for new heterocyclic compounds increases, particularly for pharmaceutical and crop protection chemicals, then the use of transition metal species might well offer advantages over the more conventional routes in terms of milder reaction conditions, better selectivity, novel synthetic routes, or the use of more readily available starting materials. Already attractive new routes to known compounds have been discovered using transition metal catalysts. One area which could be commercially important, for example, is the cobalt-catalyzed synthesis of pyridines from acetylenes and nitriles.[2]

At first sight the lack of information on the transition-metal-catalyzed synthesis of heterocyclic compounds appears understandable since many metal catalysts are inhibited when hetero atoms such as nitrogen, oxygen, and sulfur are present in the substrate. This chapter, however, will show that a variety of transition metals have been used both as catalysts and as stoichiometric reagents for the synthesis of nitrogen-, oxygen-, and sulfur-containing heterocyclic compounds. Since the area is relatively new and few reviews are available, some of the work that will be described is of a more speculative nature than in other parts of the book, but it is hoped that it will demonstrate the enormous potential for the use of transition metal species in the synthesis of heterocyclic compounds.

To date there appears to be no particular pattern as to the type of metal or metal complex used for the synthesis of heterocyclic compounds, and the examples quoted will illustrate the wide range of transition metals that have been employed. The types of reaction that are involved are ones that appear in other parts of the book, e.g., cyclooligomerization, carbonylation and intramolecular alkylation, but to be of most use to the organic chemist the chapter is arranged in terms of the final organic product.

4.2. Nitrogen Heterocycles

4.2.1. Three-Membered Rings

N-Substituted aziridines (**1**) can be synthesized by the palladium-promoted amination of olefins using primary amines, followed by oxidation with bromine.[3] Aminopalladation of styrene with methylamine at $-50°C$ gives the σ-bonded complex **2**, which on subsequent bromination produces

the aziridine (**1**; $R^1 = Ph$, $R^2 = H$, $R^3 = Me$). Under similar conditions other olefins also react; dec-1-ene, for example, gives the corresponding N-methylaziridine in an isolated yield of 43%.

4.2.2. Four-Membered Rings

The synthesis of β-lactams, which are of interest in connection with the synthesis of analogs of such antibiotics as penicillins and cephalosporins, has been achieved using a variety of transition-metal-mediated reactions. In the presence of catalytic amounts of $Pd(OAc)_2/PPh_3$, carbon monoxide inserts into various 2-bromo-3-aminopropene (**3**) derivatives to give the corresponding α-methyl-β-lactam (**4**) in good yield.[4]

Preparation of N-benzyl-α-methylene-β-lactam (**4**; $R^1 = CH_2Ph$, $R^2 = H$)[4]
2-Bromo-3(N-benzyl)-aminopropene (**3**; $R^1 = CH_2Ph$, $R^2 = H$) (5.16 mM) and n-Bu$_3$N (6.45 mM) in HMPA (CAUTION, TOXIC) are stirred at 100°C under CO (1 atm) with Pd(OAc)$_2$ (0.1 mM) and PPh$_3$ (0.4 mM) for 5 h to afford N-benzyl-α-methylene-β-lactam in 67% yield.

Addition of the keten silyl acetal **5** to the Schiff base **6** in the presence of TiCl$_4$ gives, after hydrolysis, the β-amino-ester **7**. Treatment of the ester with base gives the β-lactam **8** in excellent yield.[5]

Preparation of 3,3-Dimethyl-1,4-diphenyl-2-azetidinone (**8**; R^1, $R^2 = Ph$, R^3, $R^4 = Me$)[5]
To a 1 M solution of TiCl$_4$ in CH$_2$Cl$_2$ (20 ml; 20 mM) is added dropwise a dichloromethane solution (20 ml) of benzylideneaniline (3.62 g; 20 mM) at room temperature with stirring. To the dark red solution is added dimethyl-ketene methyl trimethylsilyl acetal (3.48 g, 20 mM) in CH$_2$Cl$_2$ (10 ml) and the mixture stirred for 1 h. After pouring into ice water the organic layer is washed with water, dried over MgSO$_4$, and concentrated under reduced pressure. The residual crystals, after washing with pentane, give 4.79 g (85%) of the ester (**7**; R^1, $R^2 = Ph$; R^3, $R^4 = Me$). This ester is dissolved in THF (30 ml) and the solution added to a solution of lithium diisopropylamide (17 mM) in n-hexane/THF (1:1) (25 ml) at 0°C with stirring. After 10 min the reaction mixture is poured into ice water and extracted with CH$_2$Cl$_2$ to give, after the usual work-up, 3,3-dimethyl-1,4-diphenyl-2-azetidinone (4.04 g; 95%).

Nucleophilic addition of benzylamine to the complex **9** leads to the intermediate **10**, which is oxidized at −78°C with chlorine to give the β-lactam (**11**) in 34% overall yield.[6] The reaction is stereospecific; for example, *trans*-but-2-ene (**9**; R^1, R^3 = Me, R^2 = H) gives only *cis*-3,3-dimethylazetidinone (**11**; R^1, R^3 = Me, R^2 = H).

$$Fp = [\eta^5\text{-}C_5H_5Fe(CO)_2]$$

4.2.3. Five-Membered Rings

The substituted allylic acetate **12** cyclizes in the presence of catalytic amounts of $Pd(PPh_3)_4$ to produce the bicyclic compound **13** in over 50% yield.[7] Moreover, only the *cis* product is formed. Cyclization of the related compounds **14** gives the isoquinuclidine skeleton (**15**) in ~60% yield.

Isonitriles having an active α-hydrogen react with α,β-unsaturated esters or nitriles in the presence of catalytic amounts of cuprous oxide to produce 1-pyrrolines (**16**).[8]

R^1, R^2 = Ph, CO_2Me, $H_2C=CH-$, or H
X = CN or Co_2Me

Preparation of 3-Methyl-3-methoxycarbonyl-5-phenyl-1-pyrroline (**16**; $R^1 =$ *Ph, $R^2 = H$, $X = CO_2Me$*)[8]

A mixture of benzyl isocyanide (10 mM), methyl methacrylate (20 mM), and Cu_2O (0.2 mM) is heated under reflux in benzene (3 ml) for 3 h. The product is isolated by fractional distillation in 94% yield as a $1:1$ mixture of configurational isomers.

Allenic amines (**17**) in the presence of catalytic amounts of silver tetrafluoroborate cyclize readily at room temperature to give the pyrroline **18** in high yield (~90%).[9] Since allenic primary and secondary amines are readily available from the corresponding alcohols, the synthetic procedure should be valuable in the preparation of the relatively rare 3-pyrrolines (**18**).

$$R^1R^2C{=}C{=}CHCH_2NHR^3 \xrightarrow{\text{AgBF}_4}$$

17

18

Preparation of 1-Benzyl-3-pyrroline (**18**; R^1, $R^2 = H$; $R^3 = CH_2Ph$)[9]

Silver tetrafluoroborate (0.1 mM) is added to a solution of N-benzyl-2,3-butadienamine (**17**; R^1, $R^2 = H$; $R^3 = CH_2Ph$) (1 mM) in chloroform (1 ml). The mixture is stirred for 5 h at room temperature and then shaken with saturated sodium chloride solution (0.1 ml) to precipitate the silver. The mixture is diluted with ether (10 ml), dried over K_2CO_3–Na_2SO_4, and filtered. The pyrroline is precipitated from the filtrate as the oxalate in 90% yield.

The reaction between phthalaldehyde and primary amines in the presence of tetracarbonylhydridoferrate $[HFe(CO)_4]^-$ yields either 2-arylisoindoles (**19**) or 2-arylisoindolines (**20**), depending on the nature of the amine.[10] Aliphatic amines give selectively the isoindolines (**20**) in 30%–65% yield, whereas aromatic amines have a greater tendency to give

19   **20**

the isoindolines (9). By varying the condition, isoindoles can be obtained exclusively from p-toluidine, p-anisidine, and p-chloroaniline.

The palladium-catalyzed intramolecular cyclization of N-acryloyl-o-bromoanilines (21) gives the oxindole derivatives (22) in good yield.[11]

Preparation of 3-Benzylidene-2-oxindole (22; R = H)[11]

A solution of N-cinnamoyl-o-bromoaniline (10 mM), palladium acetate (0.1 mM), tri-p-tolylphosphine (0.4 mM), and triethylamine (2 ml) is heated under nitrogen at 100°C in a Pyrex tube for 18 h. The cooled, partially solid reaction mixture is transferred to a flask using methylene chloride and the volatile material removed *in vacuo*. Ether (200 ml) and water (50 ml) are added to the residue and the two layers separated. The ether extract, after drying over $MgSO_4$ and removal of the solvent *in vacuo*, gives 3-benzylidene-2-oxindole (1.28 g; 58%).

A very similar cyclization to give oxindoles takes place in hot toluene in the presence of catalytic amounts of $Ni(PPh_3)_4$, which is prepared *in situ* from $Ni(acac)_2$ and $AlEt_3$ in the presence of PPh_3.[12]

Carbonylation by means of transition metals is a useful process but often requires drastic conditions. Isoindolinones (23), however, can be synthesized in good yield by reaction of o-bromoaminoalkyl-benzenes (24) with CO at 100°C and 1 atm pressure in the presence of catalytic amounts of $Pd(OAc)_2$, PPh_3, and n-Bu_3N.[13]

Carbonylation of compounds having the generalized structure 25 provides a useful source of nitrogen heterocycles.[14] Schiff bases (25; X = CH, R = alkyl or aryl) react with carbon monoxide at 200–230°C and 100–200 atm pressure in the presence of catalytic amounts of $Co_2(CO)_8$. Benzylideneaniline, for example, gives an 85% yield of 2-phenyl isoindolinone (26; X = CH, R = Ph). With iron or rhodium catalysts lower yields are

25

X = CH or CR
R = alkyl, aryl, OH, or NHPh

26

R^1 = alkyl or aryl

obtained. The reaction goes equally well for Schiff bases containing substituted *N*-aryl groups (see page 235).

Ketoximes (**25**; X = CR; R = OH) in the presence of $Co_2(CO)_8$ and carbon monoxide also produce isoindolinones (**26**; X = CR, R' = H) rather than the *N*-hydroxy compounds, which presumably undergo hydrogenolysis. Thus benzophenone oxime gives 3-phenyl isoindolinone (**26**; X = CPh, R' = H) in 80% yield.[14] Isoindolinones are also obtained in good yield from phenylhydrazones (**25**; X = CR, R = NHPh) and appropriate semicarbazones—for example, compound **25** (X = CPh, R = NHCONH$_2$).

Tertiary-butyl isonitriles in the presence of a nickel catalyst react with alkynes to give moderate yields (~40%) of pyrroles (**27**).[15] Unsymmetrically substituted acetylenes give a mixture of the two possible pyrroles.

R^1,R^2 = H, Bu, or Ph

27

α-Dicarbonyl systems (**28**) react with vinylmagnesium bromide and acetic anhydride to give α-acetoxy-α-vinylalkanones (**29**), which, in the presence of benzylamine and Pd(PPh$_3$)$_4$, gives *N*-benzylpyrroles (**30**) with

28   **29**   **30**

substituents in the 2- and/or 3-position.[16] By selective protection of one of the carbonyl groups the reaction can be extended to unsymmetrical α-diketones. Yields for this pyrrole synthesis are variable but are generally around 50%.

A related synthesis of the *N*-substituted pyrroles (**31**) takes place on reacting 1,4-dihydroxy-*cis*-but-2-ene (**32**) with a primary amine in the presence of palladium black.[17] The reaction goes in 81% yield and is thought to proceed by a sequence of steps involving dehydrogenation, Schiff base formation, hydrogenation, and palladium-induced ring closure.

A variety of substituted pyrroles (33) have been synthesized by the nickel-catalyzed reaction of 2H-azirines (34) with activated ketones.[18] The

product is easily isolated in almost quantitative yield by precipitation with water. With sufficiently activated CH_2 groups the reaction proceeds to completion even at room temperature, and the scope of the reaction appears to be limited only by the number of 2H-azirines.

Reaction of 2-aryl-azirines (35) with either $Co_2(CO)_8$[19a] or $[RhCl(CO_2)]_2$[19b] gives the styryl indoles (36) in up to 90% yield. These products could be useful intermediates in alkaloid synthesis. From a synthetic point of view the reaction using $Co_2(CO)_8$ appears to be simpler and to give better yield.

Preparation of 2-Styrylindoles (36) from 2-Aryl-azirines (35)[19a]

A mixture of 2-phenyl-azirine and $Co_2(CO)_8$ in benzene is stirred at room temperature under nitrogen for 24 h. The solution is filtered, the filtrate concentrated to small volume and then purified by chromatography on Florisil or silica gel to give the 2-styryl indole (36) in 52%–95% yield.

2,2-Diphenyl-2H-azirines (37) are converted into indoles (38) in quantitative yield in the presence of catalytic amounts of $PdCl_2(PhCN)_2$.[20] The reaction proceeds via the complex 39, which can be isolated as a yellow precipitate.

A useful synthesis of indoles (**40**) from o-allylanilines (**41**) proceeds in the presence of either stoichiometric or catalytic amounts of palladium complexes.[21] This is a particularly neat synthesis of indoles since the starting materials for the cyclization, the o-allylaniline (**41**), can be synthesized in good yield by reaction of the appropriate π-allylnickel halide with the o-bromo-aniline (**42**).[21a]

$X = H$, Me, CO_2Et, or OMe
$R^1 = H$, Me, or COMe
$R^2 = H$ or Me

Preparation of 2-Methyl Indole (**40**; *X, $R^1 = H$; $R^2 = Me$*)[21a]

In a 250 ml flask are placed $PdCl_2(MeCN)_2$ (0.195 g; 0.75 mM), benzo-quinone (0.812 g; 7.52 mM), and LiCl (3.16 g; 75.2 mM). THF (70 ml) is then added and the mixture stirred for 5 min. 2-(2-Propenyl)aniline (**41**; X, $R^1 = H$; $R^2 = H$) (1.0 g; 7.52 mM) in THF (25 ml) is then added to the flask via a syringe and the mixture heated under reflux for 5 h. After removal of the solvent *in vacuo* the residue is stirred with ether and decolorizing charcoal for 20 min and filtered. The filtrate is washed with 1 *M* NaOH (5 × 50 ml). Removal of the solvent *in vacuo* followed by chromatography of the residue on silica (3 : 1 petrol/ether) gives 2-methylindole in 85% yield.

By conducting the cyclization of *N*-substituted o-allylanilines (**41**) in the presence of carbon monoxide and methanol, the dihydroindolacetic acid ester **43** is obtained in good yield (~70%).[21b]

A somewhat related synthesis of indoles has been studied by Japanese workers, whereby 2-chloro-*N*-methyl-*N*-allylaniline (**44**; R = Me) cyclizes to 1,3-dimethylindole (**45**; R = Me) in 46% yield in the presence of an equimolar amount of Ni(PPh$_3$)$_4$.[22] More recently an improvement in this type of cyclization has been reported using Pd(OAc)$_2$ in acetonitrile as

catalyst.[23] Apparently the catalyst is deactivated during the reaction and periodic provision of fresh catalyst gives a much improved yield. Using this technique, 3-methyl indole (**45**; R = H) was obtained from 2-iodo-*N*-allyl aniline in 87% isolated yield.

4.2.4. Six-Membered Rings

A simple synthesis of *N*-substituted piperidines (**46**) from glutaraldehyde and primary amines uses tetracarbonylhydrido-ferrate,

[HFe(CO)$_4$]$^-$.[24] The KHFe(CO)$_4$, which is generated *in situ* from Fe(CO)$_5$ and KOH, is mixed with the amine and glutaraldehyde in an atmosphere of carbon monoxide at room temperature. Yields are ~80% for a variety of aromatic and alkyl amines.

Rearrangement of the aziridine **47** with a catalytic amount of PdCl$_2$(PhCN)$_2$ gives the *N*-carboxy-nortropidine **48** in quantitative yield.[25] This is a particularly useful reaction since the product (**48**) is the backbone of the tropane alkaloids.

Bis-π-allyl complexes are active intermediates in the reaction of butadiene with alkenes, alkynes, and carbonyl compounds (see, for example, Chapter 3). All these reactions involve insertion of a multiple bond into a carbon–metal bond. The carbon–nitrogen double bonds of Schiff bases and isocyanates are also reactive toward some carbon–metal bonds. Butadiene, for example, in the presence of catalytic amounts of palladium nitrate and triphenylphosphine ($Pd:PPh_3 = 1:3$), reacts with Schiff bases in DMF at 80°C to give the substituted piperidines (**49**) in 70% yield.[26]

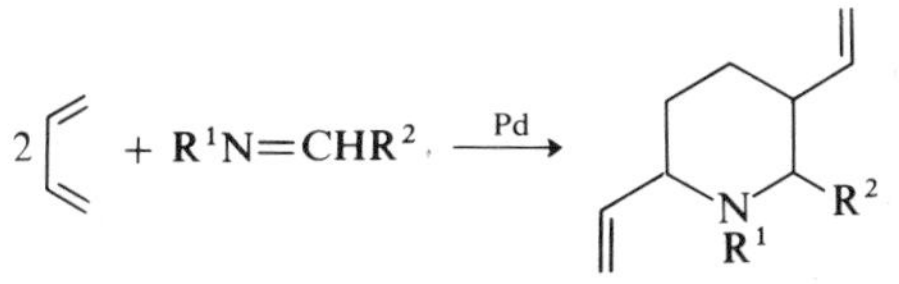

49

Similarly, cocyclization of phenyl isocyanate and isoprene in benzene at 100°C in the presence of catalytic amounts of [bis(triphenylphosphine) (maleic anhydride) palladium] produces the piperidones **50** and **51** as a 1:1 mixture in 82% yield.[27] Butadiene and phenyl isocyanate react similarly.

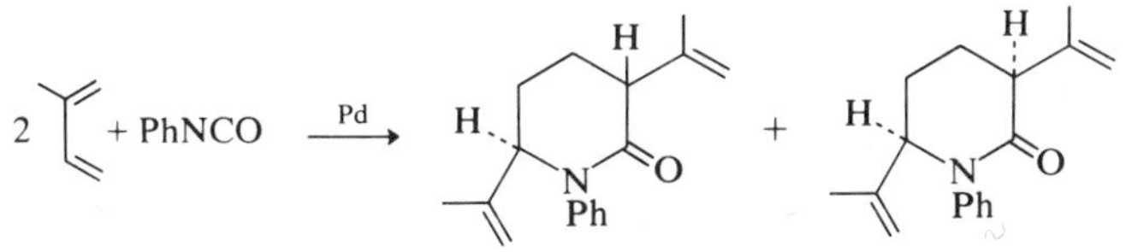

50   **51**

The $Co_2(CO)_8$-catalyzed carbonylation of the unsaturated amide **52** produces the cyclic imide **53** in 41% yield, although the reaction conditions are somewhat vigorous (300 atm CO, 250°C).[28] With acyclic α,β-unsaturated carboxamides (**54**) a mixture of five- and six-membered cyclic imides are produced, the proportions of products obtained from **54** being very dependent on the degree of alkylation of the acrylamide double bond.

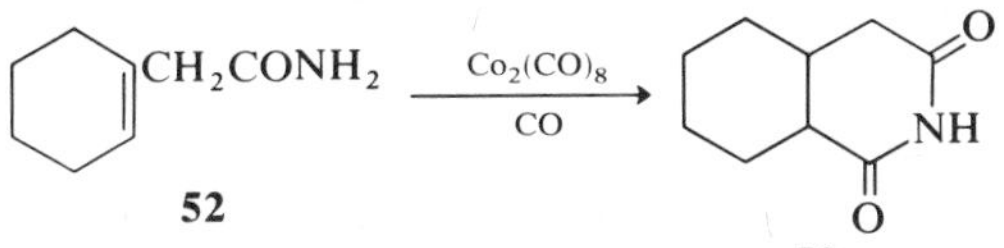

52

53

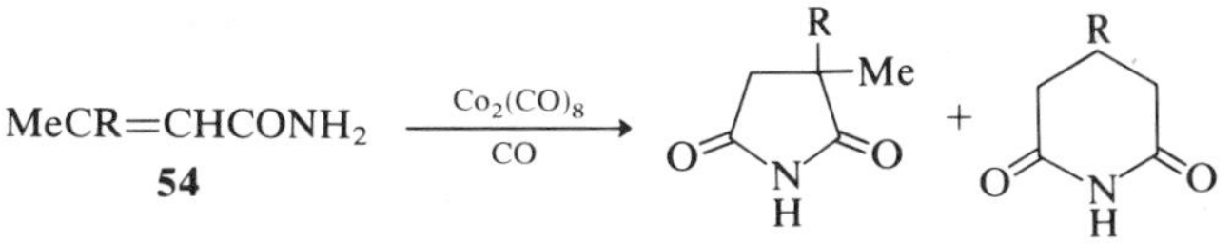

MeCR=CHCONH$_2$
54

R = H	68%	19%
R = Me	0%	67%

The cyclization of *o*-chloroallylaniline derivatives in the presence of nickel catalysts has already been discussed in relation to the synthesis of indole derivatives (Section 4.2.3). Cyclization of the related compound **55** in the presence of the Grignard reagent MeMgBr and a catalytic amount of NiCl$_2$(PPh$_3$)$_2$ gives 1-methyl-4-methylene-1,2,3,4-tetrahydroquinoline (**56**) in 91% yield.[22b]

In the presence of a catalytic amount of Pd(OAc)$_2$–PPh$_3$ and under an atmosphere of carbon monoxide, *N*-alkyl-*o*-bromophenethylamines (**57**) cyclize to give *N*-substituted 1,2,3,4-tetrahydroisoquinolin-1-ones (**58**) in good yield.[13]

Preparation of N-Benzyl-1,2,3,4-tetrahydroisoquinolin-1-one (**58**; *R = CH₂Ph*)

A mixture of *N*-benzyl-*o*-bromophenethylamine (**57**; R = CH$_2$Ph) (1 mM), *n*-Bu$_3$N (1.1 mM), Pd(OAc)$_2$ (0.02 mM), and PPh$_3$ (0.04 mM) is added to a reaction vessel that is connected to a balloon filled with CO, and heated at 100°C for 26 h. After cooling, ether is added to the solution and the ether layer washed with 10% hydrochloric acid and dried over MgSO$_4$. After removal of the solvent *in vacuo* the reaction product is purified by chromatography on silica gel (benzene/ether 1:1) to give the product (**58**; R = CH$_2$Ph) in 65% yield.

The cobaltacyclopentadiene complex **59**, which is easily obtained by reaction of η^5-cyclopentadienyl-bis-triphenylphosphinecobalt with two moles of alkyne, reacts with isocyanates to give 2-pyridones (**60**) in 70% yield.[29] With unsymmetrical complexes, for example, **59** (R^1, R^3 = CO$_2$Me,

R^2, R^4 = Ph), the reaction proceeds regiospecifically to afford one product (**60**; R^1, R^3 = CO$_2$Me, R^2, R^4 = Ph). Although this reaction might not look very attractive to the synthetic organic chemist there is the possibility of the reaction being made catalytic in that one could envisage co-trimerization of acetylenes and isocyanates to give the desired product in the presence of catalytic amounts of complex **59**.

It has already been shown that intramolecular cyclization of olefinic compounds via oxypalladation is a useful route for the synthesis of heterocyclic compounds. Another example of this type of reaction is the cyclization of 2,4-pentadienamides (**61**) with palladium salts to give the corresponding 2-pyridone (**62**).[30]

In a related reaction, *o*-vinylbenzamides (**63**) cyclize to 1-hydroxy isoquinolines (**64**) in the presence of lithium chloropalladite.[31]

Preparation of 1-Hydroxy-3-phenylisoquinoline (**64**; *R = Ph*)[31]

A solution of 2-styrylbenzamide (10 mM), Et$_3$N (20 mM), and Li$_2$PdCl$_4$ (10 mM) in dry acetonitrile (40 ml) is stirred at 60°C for 6 h. After filtering off the palladium black, the filtrate is evaporated to dryness *in vacuo* and the residue chromatographed on silica gel with benzene to give the product (**64**; R = Ph) in 62% yield.

Reaction of 2-iodoanilines (**65**) with dimethyl maleate in the presence of Pd(OAc)$_2$ gives the expected 2-amino ester intermediate **66**, which then cyclizes to the quinoline **67** in reasonable yield (30%–70%).[32] The lowest

R = H, Br, or OH

yield is from the 4-hydroxy compound, which is not unexpected in view
of the low yields obtained in a variety of similar reactions when strongly
electron-donating substituents are present in the aryl iodide.

The related palladium-catalyzed cyclization of α-substituted N-acry-
loyl-*o*-bromoanilines (**68**) does not give the expected 3-substituted but
rather the 4-substituted quinoline (**69**) in 40% yield.[11] This rearrangement
is rationalized on the basis of an initial ring closure of the organopalladium
intermediate to a five-membered ring product (**70**) containing a 3-palladio-
methyl group. In this complex there is no β hydrogen to be eliminated

with the palladium as there is when the α carbon is unsubstituted. Because
the usual decomposition reaction is not possible, elimination of the
aminocarbonyl group with palladium appears to occur and this is followed
by a reverse readdition of the aminocarbonylpalladium group to give the
adduct **71**, which can now eliminate a hydridopalladium group irreversibly
to give the observed 4-substituted-2-quinoline (**69**).

One of the most potentially useful syntheses of heterocyclic compounds
is the organocobalt-catalyzed synthesis of substituted pyridines (**72**) by
cyclotrimerization of alk-1-ynes and nitriles.[2,33] A variety of organocobalt
catalysts can be used, among the more suitable for practical applications
being 1,5-cyclooctadiene (cyclooctenyl) cobalt[2] and cobaltocene.[33] The
reaction will also proceed if the catalyst is generated *in situ*, the simplest
system being $CoCl_2 \cdot 6H_2O/NaBH_4$.[2] Using these techniques a variety of

$$R^1CN + 2R^2C\equiv CH \xrightarrow{\text{[Co]}}$$

2-substituted pyridines (**72**; R^2 = H) have been synthesized from acetylene and carbonitriles in excellent yield.

Preparation of 2-Phenylpyridine[33]
Cobaltocene (0.38 g; 2 mM) is placed in a 200-ml autoclave in an atmosphere of nitrogen. Toluene (20 mol) and benzonitrile (14.5 ml; ~140 mM) are introduced via a syringe. After flushing the vessel four times with acetylene, the acetylene is introduced up to a pressure of 9 atm while the autoclave was mechanically shaken at room temperature. The autoclave is heated to 150°C and after 2 h, when the pressure has dropped to 3 atm, more acetylene is added (13 atm at 150°C). After 7 h the reaction mixture is fractionally distilled to give 2-phenylpyridine (15.8 g; 73% yield).

Reactions of substituted alkynes with carbonitriles always produce the two isomeric pyridines **72** and **73**. The symmetrically substituted product **72** is formed as the major product, while the asymmetric type **73** is formed in about 30% relative yield.
Starting from the readily available pyridine carbonitriles **74**, reaction with terminal alkynes leads to bipyridines **75** and **76**. Use of acetylene as

the alkyne component gives the parent compound (for example, 2,2'-bipyridine from 2-cyanopyridine in 95% yield). Substituted alkynes give two positional isomers, with type **75** usually predominating.
Extensive modifications of the cyano component in the pyridine synthesis can be tolerated. Cyanamide, for example, in the presence of certain cobalt catalysts will react to give 2-aminopyridines (**77**). Alkyl thiocyanates

$$H_2N\!-\!CN + 2RC\equiv CH \xrightarrow{\text{[Co]}}$$

will also react with alkynes to give alkyl-thiopyridines (**78**), but in this case catalyst turnover numbers are low.

$$R^1SCN + 2R^2C{\equiv}CH \xrightarrow{[Co]}$$

78

A further variation on the cobalt-catalyzed synthesis of pyridines is the one-step synthesis of annulated pyridines (**79**) by the cooligomerization of diynes (**80**) with a variety of substituted nitriles.[34] The reaction goes in good yield and with remarkable selectivity using the commercially available catalyst $(C_5H_5)Co(CO)_2$.

Preparation of Ethyl-3-(5,6,7,8)-tetrahydroisoquinoline acetate (**79**; *n* = 4), R = CH₂CO₂Et)[34]

Preparation of Ethyl-3-(5,6,7,8)-tetrahydroisoquinoline acetate (**79**; $n = 4$), $R = CH_2CO_2Et$)[34]

A solution of 1,7-octadiyne (**80**; $n = 4$) (5.00 mM), ethyl cyanoacetate (5.00 nM), and $(C_5H_5)Co(CO)_2$ (0.50 mM) in xylene (15 ml) is added over 117 h (by syringe pump) to *o*-xylene (15 ml) warmed to reflux under N_2. After removal of the solvent *in vacuo*, the reaction mixture is chromatographed on silica with ether. Microdistillation of the crude product gives the product as a clear oil in 47% yield.

Although γ,δ-unsaturated ketoximes (**81**) are known to cyclize at 300°C, the reaction can be induced under much milder conditions in the presence of $PdCl_2(PhCN)_2/NaOPh$.[35] The reaction products are substituted pyridines (**82**), which are obtained in moderate yield.

The cyclopalladated complex **83**, formed in 60% yield from the appropriate acetanilide and $Pd(OAc)_2$, reacts with various α,β-unsaturated carbonyl compounds to give the intermediate **84**, which can then undergo acid-catalyzed cyclization to give compound **85**.[36] The last two steps each go in yields of 40%–80%.

$Z = H, Me, or OMe$

Primary aromatic amines react with aldehydes in the presence of a rhodium catalyst to give quinolines (**86**) in variable yields (30%–80%).[37]

*Preparation of 2-Ethyl-3-methylquinoline (**86**; R = Me, X = H)*[37]

A mixture of aniline (40 mM), propanal (88 mM), nitrobenzene (60 ml), ethanol (20 ml), and [Rh(norbornadiene)Cl]₂ (0.03 mM) is stirred under an argon atmosphere at 180°C for 4 h in an autoclave. Vacuum distillation of the reaction mixture gives 2-ethyl-3-methylquinoline in 50% yield.

4.2.5. Seven-Membered Rings

Using the same procedure as for the synthesis of the related five- and six-membered benzolactams (**24** and **58**), the benzoazepinone derivatives (**87**) have been synthesized in good yield (~50%).[13]

4.3. Oxygen Heterocycles

In general there are fewer examples of the synthesis of oxygen heterocycles than nitrogen heterocycles using transition metal reagents or catalysts. The main area of work has been the synthesis of five- and six-membered lactones by various carbonylation reactions and, as this type of reaction will be discussed in Chapter 6, and the literature reviewed up to 1973,[1] this topic will not be discussed in great detail.

4.3.1. Five-Membered Rings

In contrast to other alkenes that are oxidized to carbonyl compounds by palladium salts, allyl alcohol undergoes an oxidative cyclodimerization to 4-methylene-tetrahydrofurfuryl alcohol (**88**) and 4-methyl-2,5-dihydrofurfuryl alcohol (**89**), although the yield is poor. The other main product of the reaction is propene, which apparently results from hydrogenolysis of the allyl alcohol.[38]

$$2CH_2{=}CHCH_2OH \xrightarrow{\text{PdCl}_2} \underset{\textbf{88}}{\text{[structure]}}\,CH_2OH \;+\; \underset{\textbf{89}}{\text{[structure]}}\,CH_2OH$$

The related Pd(II)–Cu(II)-catalyzed intramolecular cyclization of γ,δ-unsaturated alcohols (**90**) gives the tetrahydrofurans **91** in moderate yield.[39a] The product is usually a mixture of the two diastereoisomers, but

$$\underset{\textbf{90}}{R^2\;R^1\text{-OH}} \xrightarrow[\text{Cu(OAc)}_2]{\text{Pd(OAc)}_2} \underset{\textbf{91}}{R^1\;R^2\,O}$$

with **91** (R^1 = Ph, R^2 = H) only one isomer, *trans*-2-vinyl-5-phenyltetrahydrofuran, is produced in 40% yield. Under similar reaction conditions 2-vinyl-2,3-dihydro-benzofuran (**92**) is prepared from 2-(2'-butenyl)phenol.[39b]

$$\underset{\text{OH}}{\text{[structure]}} \xrightarrow[\text{Cu(OAc)}_2]{\text{Pd(OAc)}_2} \underset{\textbf{92}}{\text{[structure]}}$$

Tetrahydrofuran derivatives are often formed during the hydroformylation of unsaturated alcohols. An early example was the formation of tetrahydrofurfuryl alcohol (**93**) from either 1-butene-3,4-diol or 2-butene-1,4-diol in the presence of catalytic amounts of $Co_2(CO)_8$ and under CO/H_2 pressure.[40]

$$\begin{array}{c}HOCH_2CH{=}CHCH_2OH\\ \text{or}\\ HOCH_2\underset{\text{OH}}{CH}{=}CH_2\end{array} \xrightarrow[\text{CO/H}_2]{\text{Co}_2(\text{CO})_8} \underset{\textbf{93}}{\text{[structure]}}\,CH_2OH$$

Another more recent example is the hydroformylation of coniferyl alcohol **94** to give compound **95** in 25% yield together with several other minor products.[41]

The aminoalkynes **96** undergo carbonylation in the presence of $Co_2(CO)_8$ to form the furan derivatives **97** in up to 44% yield.[42]

Dihydrofurans (**98**) are also obtained by the reaction of β-diketones or β-ketoesters with alkenes in the presence of $Mn(OAc)_3$.[43] The reaction is thought to go via a free radical mechanism in contrast to the ionic mechanism observed in the case of lead tetraacetate. The dihydrofurans produced in the $Mn(OAc)_3$-promoted reaction of acetylacetone with terminal alkenes consist of only one isomer, the 5-substituted 2-methyl-3-acetyl-4,5-dihydrofuran (**98**), in sharp contrast to those produced in the presence of thallic acetate or lead tetraacetate, where the 4-substituted isomer predominates.

R^1 = Me or OEt

Preparation of *3-Acetyl-2,5-dimethyl-5-phenyl-4,5-dihydrofuran* (**98**; R^1, $R^3 = Me, R^2 = Ph$)[43]

$Mn(OAc)_3 \cdot 2H_2O$ (0.25 mol), prepared from potassium permanganate and manganese acetate, is dissolved in glacial acetic acid (1 liter) at 45°C under hydrogen. To this solution is added a mixture of α-methylstyrene (15.3 g; 0.13 mol) and acetylacetone (75 g; 0.75 mol).

After 10 min the brown manganic color has disappeared, indicating that the reaction was complete and the dihydrofuran is isolated in quantitative yield by extraction with ether followed by distillation.

α-Bromo-ketones react with nickel carbonyl on heating in DMF to produce 2,4-disubstituted furans **99**.[44] By doing the reaction at room temperature the β-epoxy-ketone (**100**) is isolated in good yield (60%–80%), and on heating above 130°C dehydrates to give the furan (**99**).

$$RCO\cdot CH_2Br \xrightarrow{\ Ni(CO)_4\ } RCO\cdot CH_2\cdot \underset{O}{\overset{R}{C}}{-}CH_2 \xrightarrow{\ \Delta\ }$$

100   **99**

Good yields of tetrahydro-2-furanones (**101**) can be obtained by hydroformylation of α,β-unsaturated esters.[45] Typical reaction conditions involve heating the substrate at 250°C under 300 atm pressure of CO/H_2 in benzene in the presence of catalytic amounts of $Co_2(CO)_8$.

$$H_2C{=}\underset{R^1}{\overset{}{C}}{-}CO_2R^2 \xrightarrow[Co/H_2]{\ Co_2(CO)_8\ }$$

101

A simple one-step synthesis of γ-lactones (**102**) consists of the reaction of $Mn(OAc)_3$ with alkenes and carboxylic acids.[46] Other high-valent metal salts, for example, $Ce(OAc)_4$ and NH_4VO_3, also work, but the manganese salts appear to be the most readily available. High yields of lactone are obtained from both internal and terminal alkenes and, in all cases studied, the lactone obtained from terminal alkenes contains the oxygen of the carboxylic acid bonded to the more substituted 2-position of the alkene. Lactones were also obtained from both conjugated and nonconjugated dienes.

$$\underset{R^3}{\overset{R^2}{}}\!\!\underset{R^4}{\overset{R^1}{\diagup}} + \underset{R^6}{\overset{R^5}{\diagup}}CH\cdot CO_2H \xrightarrow{\ Mn(OAc)_3\ }$$

102

Preparation of γ-n-Octylbutyrolactone (102; $R^1 = C_8H_{17}$; $R^2R^3R^4R^5R^6 = H$)[46]

Mn(OAc)$_2\cdot$4H$_2$O (212 g; 0.84 mol) is dissolved in acetic acid (1200 ml) by warming to 90°C. To this solution is added KMnO$_4$ (32 g; 0.2 mol) with stirring. When the exothermic reaction has subsided and the temperature has dropped to 90°C, acetic anhydride (300 ml) and sodium acetate are added. Dec-1-ene (85 g; 0.6 mol) is added and the reaction mixture heated under reflux for 1 h. Extraction and distillation give pure γ-n-octylbutyrolactone (66.4 g; 67% yield).

The formation of α-methylene-γ-lactones (**103**) by carbonylation of the acetylenic alcohol **104** is a very attractive route to these useful compounds.[47] The original method[47a] used $Ni(CO)_4$, acetic acid, and water, but a more recent procedure[47b] uses a palladium chloride/thiourea catalyst.

In a typical reaction the acetylenic alcohol **104** in acetone is stirred overnight at 50°C under CO (50 psi) with a catalytic amount of $PdCl_2$/thiourea. With the cyclic acetylenic alcohol **104**(R^1, $R^2 = -(CH_2)_4-$) the corresponding α-methylene-γ-lactone **103** (R^1, $R^2 = -(CH_2)_4-$) is obtained in 94% yield.

In the presence of a base the homoallylic alcohols (**105**) are converted into α-methylenebutyrolactones (**106**) using nickel tetracarbonyl.[48] The

choice of base appears to be very important since in the presence of potassium acetate yields are around 60%, whereas using sodium methoxide they drop to 4%.

The anion $[Co(CO)_4]^-$, generated by vigorously stirring $Co_2(CO)_8$ in a mixture of benzene and aqueous sodium hydroxide containing the phase transfer catalyst cetyltrimethylammonium bromide, reacts with alkynes and methyl iodide in the presence of carbon monoxide to give the but-2-enolides **107**.[49] Although the reaction looks potentially very useful only a limited number of acetylenes were studied and product yields were variable.

Sodium tetracarbonyl cobaltate, $NaCo(CO)_4$, reacts with alkyl or acyl halides in the presence of CO to form the acyl cobalt tetracarbonyl **108**, which then reacts with alkynes in the presence of dicyclohexylamine to give the 2,4-pentadieno-4-lactone **109**.[50] The reaction is catalytic in cobalt and yields are around 60% for a variety of substituted alkynes and alkyl halides.

$$R^1CH_2Br + NaCo(CO)_4 \xrightarrow{CO} \underset{\textbf{108}}{R^1CH_2CO \cdot Co(CO)_4} \xrightarrow{R^2C \equiv CR^3} \textbf{109}$$

A potentially attractive route to benzo-2-furanones (**110**) is the $Co_2(CO)_8$-catalyzed reaction of *o*-methyl phenols with carbon monoxide.[51] The reaction is believed to go via the *o*-quinonemethide **111**, but the reported reaction conditions (300°C; 1000 atm) are somewhat vigorous and consequently the product is isolated in low yield (14%).

The reaction of acetylene with carbon monoxide at 100°C and 1000 atm in the presence of catalytic amounts of $Co_2(CO)_8$ produces the *trans*-bifurandione **112** in 70% yield.[52] Substituted alkynes also react to give the substituted bifurandiones in similar yield. At lower temperatures in benzene the major products are the *cis* and *trans* isomers of 2,4,6,8-decatetraene-1,4,7,10-diolide (**113**), with only minor amounts of the bifurandione.[53]

4.3.2. Six-Membered Rings

Earlier in this chapter the synthesis of piperidines (**49**) and piperidones (**50**) from butadiene and Schiff bases or isocyanates in the presence of a palladium catalyst was described. In a related reaction benzaldehyde reacts with butadiene in the presence of a catalyst generated from π-allylpalladium chloride, triphenylphosphine, and sodium phenoxide to give the pyran

114.[54] The analogous reaction of formaldehyde with butadiene produces the pyrans **115** and **116** as a 2:1 mixture.[55]

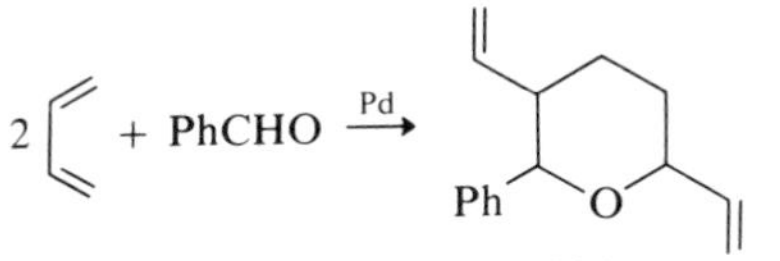

114

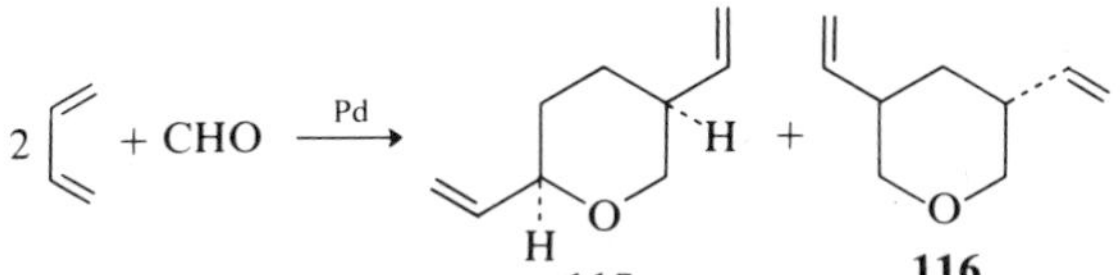

115   **116**

Monoalkenes, after complexation with palladium chloride, react with formalin to form 1,3-dioxanes.[56] 3-Methyl-1-butene, for example, gives in 70% yield a mixture of compounds **117** and **118**, with **117** predominating.

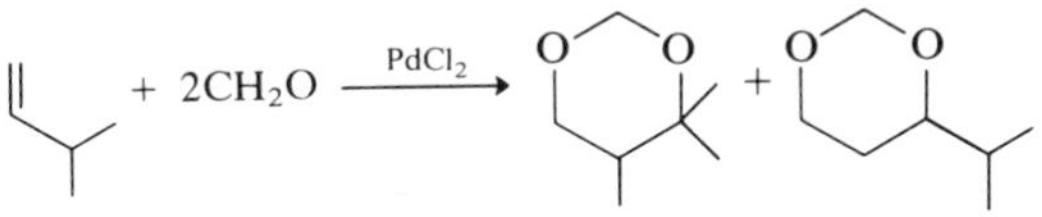

117   **118**

Hydroformylation of α,β-unsaturated esters in the presence of $Co_2(CO)_8$ produces a mixture of tetrahydro-2-furanones and tetrahydropyranones, with the latter product predominating when the double bond is able to undergo prior migration from the α,β to the β,γ position, as is illustrated by the formation of **119** and **120**.[45] This reaction can also be applied to the preparation of fused ring compounds (for example, the synthesis of **121** in 96% yield).[45]

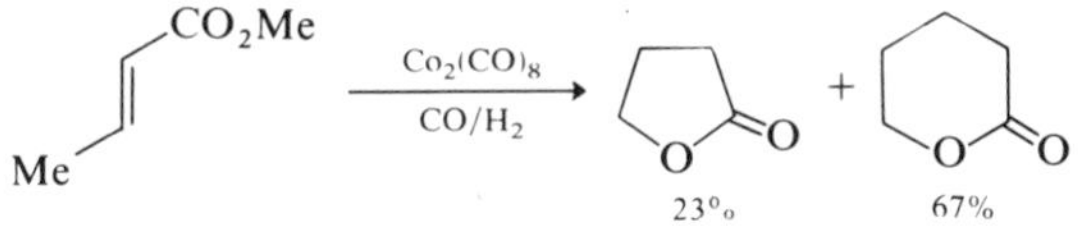

23%   67%
119   **120**

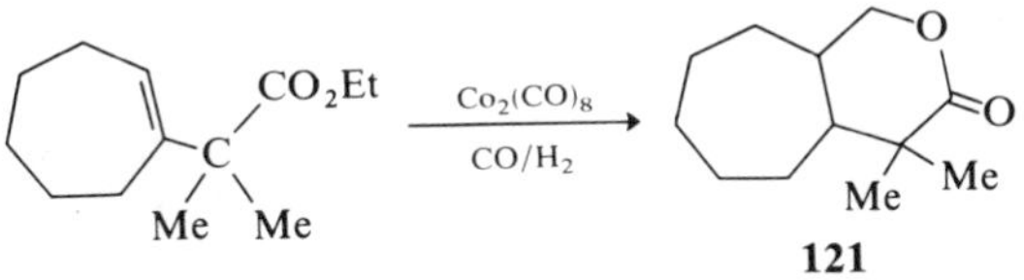

121

Carbonylation of the acetylenic alcohol **122** in the presence of aqueous acidic nickel carbonyl produces the tetrahydropyranone **123** in 20% yield.[47a]

$$HC\equiv C(CH_2)_3OH \xrightarrow{Ni(CO)_4}$$

122

123

Reaction of the epoxy-alkene **124**, prepared by epoxidation of the appropriate diene, with carbon monoxide in the presence of rhodium catalyst produced the β,γ-unsaturated lactone **125**, whereas carbonylation in the presence of iron or cobalt catalysts gives the α,β-unsaturated lactone **126**.[57] Yields are quoted as being in the range 10%–75% but experimental details are lacking.

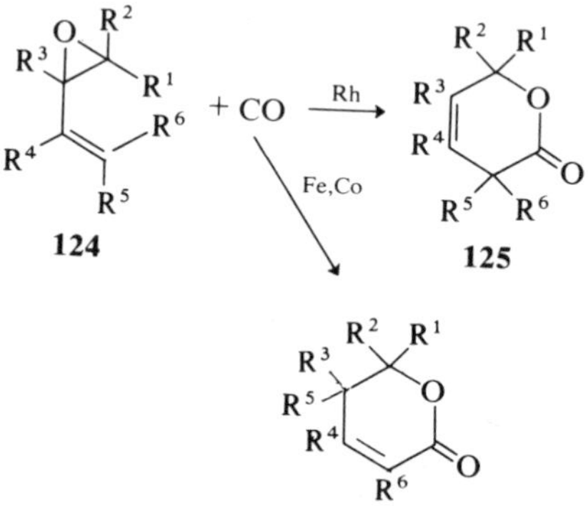

124   **125**

126

Isocoumarins (**127**) have been synthesized from 2-alkenyl benzoic acids (**128**) using a palladium-assisted cyclization reaction.[58] The reaction, which goes in high yield, is thought to proceed by initial coordination of the alkene to the palladium, attack of carboxylate on this complex to produce

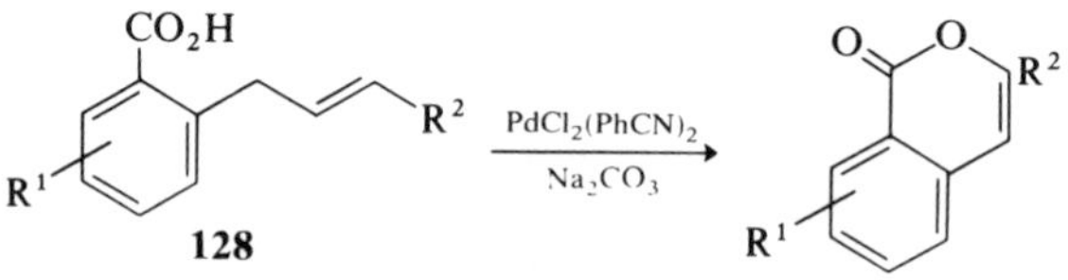

128   **127**

the cyclized *o*-alkyl palladium complex **129**, which then undergoes elimination of PdH, and rearrangement to give the product **127**. By incorporating $CuCl_2$ and oxygen into the system to reoxidize the Pd(0) the reaction can be made catalytic in palladium but the reaction is then slow. The starting materials (**128**) for this reaction are synthesized by the reaction of the appropriate 2-bromobenzoic acid with a π-allyl nickel halide complex.

129

Preparation of 3-Methylisocoumarin (**127**; $R^1 = H$; $R^2 = Me$)

To a stirred solution of 2-(2-propenyl)benzoic acid (0.2 g; 1.23 mM) and $PdCl_2(MeCN)_2$ (0.32 g; 1.24 mM) in THF (15 ml) is added Na_2CO_3 (0.19 g; 1.78 mM) and the resulting mixture stirred for 3 h at 25°C. After routine isolation and purification by preparative t.l.c. (benzene/ether 2 : 1) the product is obtained in 86% yield.

A range of di- and tri-oxabicyclo [X,2,1] systems have been prepared by a $PdCl_2$–$CuCl_2$-catalyzed oxidative intramolecular cyclization of terminal alkenes containing suitably located vicinal diols.[59] Thus the alkene diol **130** ($R^1 = H$; $R^2 = Et$), obtained via a butadiene telomerization, gives the beetle pheromone *endo*-brevicomin (**131**; $R^1 = H$; $R^2 = Et$) in 30% yield. Other related dioxabicyclo [3,2,1] and [4,2,1] systems are similarly synthesized, and using allyl ethers (**132**), the reaction can be extended to the trioxabicyclo [3,2,1] series (**133**).

130 X = CH₂
132 X = O

131 X = CH₂
133 X = O

4.4. Sulfur Heterocycles

In general this is an area that has not been well studied. To date most of the syntheses of sulfur heterocycles using transition metal species have resulted from the reaction of metallocyclopentadienes with either sulfur or carbon disulfide and they are not particularly useful reactions.[60] As an illustration, however, the cobalt complex **134** reacts with sulfur to produce the thiophene **135** in 75% yield, and with CS_2 to produce the dithio-pyranone **136** in 50% yield.

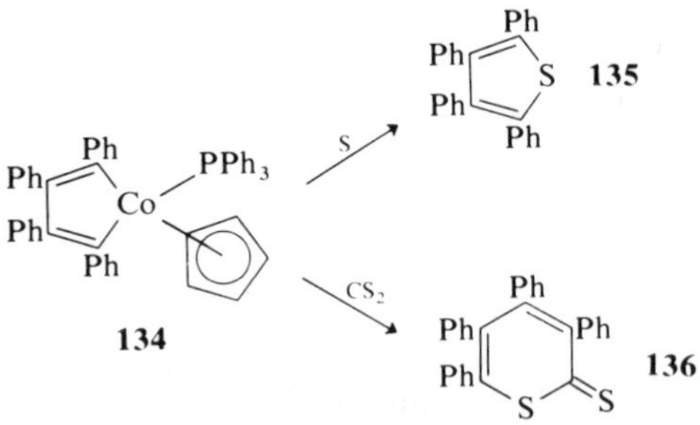

Ph
Ph
S **135**
Ph
Ph

Ph
Ph
PPh$_3$
Co
Ph
Ph

134

CS_2

Ph
Ph
Ph
S
S **136**
Ph

An alternative approach to the synthesis of sulfur heterocycles is illustrated by the reaction of thiobenzophenones (**137**) with $Fe_2(CO)_9$ to produce the *ortho*-metalated complex **138** which can then be oxidatively degraded to the thiolactone **139** in good yield.[61] With unsymmetrical thiobenzophenones one *ortho*-metalated complex **138** is usually formed predominantly.

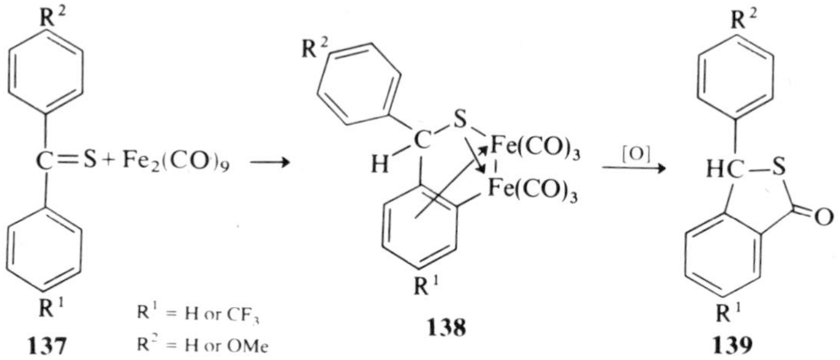

R^2

C=S + Fe$_2$(CO)$_9$ →

R^1
137

R^1 = H or CF$_3$
R^2 = H or OMe

R^2

H—C—S—Fe(CO)$_3$
Fe(CO)$_3$

R^1
138

[O] →

R^2

HC—S
O

R^1
139

4.5. Cyclic Compounds Containing Two Hetero Atoms

Azobenzenes have proved to be useful starting materials for synthesis of a variety of heterocyclic compounds using transition metal complexes. An early example is the reaction of azobenzene with carbon monoxide in the presence of $Co_2(CO)_8$.[62] Reaction at 190°C leads to incorporation of one molecule of CO to produce indazolone (**140**) in 55% yield, whereas at 230°C two molecules of CO are incorporated to give 3-phenyl-2,4-dioxo-1,2,3,4-tetrahydroquinazoline (**141**) in 80% yield. Using nickel carbonyl as the catalyst the main product is **142**.[63]

142

Reaction of azobenzene with $PdCl_2$ produces the *ortho*-metalated complex **143**, which will react with CO in water or ethanol at 100°C and 15 atm to form the indazolone **140** in 90% yield.[64] The palladium complex **143** will further react with an isocyanide to give the stable complex **144**, which thermally decomposes at 100°C to give the 3-imino-2-phenylindazoline **145** in good yield.[65]

In addition to azobenzene, palladium can form *ortho*-palladated complexes with Schiff bases, benzaldazine, acetophenone dimethylhydrazone, and 1-methyl-1-phenylhydrazones. These complexes will then undergo insertion reactions with carbon monoxide under mild conditions to produce a range of heterocyclic compounds.[66] The hydrazone–palladium acetate complex **146**, for example, reacts with CO at 100°C in xylene to produce the acetate **147** in 53% yield.

A variety of heterocycles have been synthesized by the reaction of amines with carbon tetrachloride in the presence of a metal carbonyl.[67] Benzylamine, for example, reacts with CCl_4 at 150°C in the presence of $Co_2(CO)_8$ to produce the triphenylimidazole **148** and the imidazoline **149**.[67a] At reaction temperatures below 120°C no debenzylated products

are obtained, whereas shortening the reaction time produces more of the imidazoline (**149**). In the presence of $Mo(CO)_6$ or bis(π-cyclopentadienyl-molybdenum tricarbonyl) the imidazoline (**149**) and dibenzylamine are produced.

Whereas aniline reacts with CCl_4 and carbon monoxide in the presence of a metal carbonyl to form amidine (**150**), substituted anilines react to

produce heterocyclic compounds.[67b] *m*-Chloroaniline, for example, reacts with CCl_4/CO in the presence of $Mo(CO)_6$ or $Cr(CO)_6$ to produce the quinazolinone **151** in 80% yield. *m*-Bromo- and 3,4-dichloroaniline also

$R^1, R^2 = H, Me$

152

give quinazolinones, but *m*- and *p*-toluidine give the quinazolinedione **152**. The reaction mechanisms of these CCl_4/CO reactions are not clear, but appear to go via free radical pathways.

The reaction of ethylene with carbon monoxide and concentrated aqueous ammonia in the presence of a rhodium catalyst at 150°C and 250 atm does not produce the expected propionamide as the major product, but a 52% yield of 2,4,5-triethylimidazole **153**.[68] Other terminal olefins

$$3CH_2{=}CH_2 + 3CO + NH_3 \xrightarrow{Rh} Et{\cdot}CO{\cdot}CHEt{\cdot}NH{\cdot}COEt \xrightarrow{NH_3}$$

154

153

react in a similar way. The reaction is thought to go via conversion of ethylene to 3,4-hexanedione, which then undergoes condensation with propionamide to produce **154**, which can subsequently cyclize with ammonia to give the imidazole **153**. In support of this mechanism the reaction in dilute aqueous ammonia gives the intermediate **154** in 40% yield.

Isonitriles in the presence of $PdCl_2$ react with α-amino acid esters to produce imidazolones (**155**) in about 70% yield.[69]

$$BuNC + RCH(NH_2){\cdot}Co_2Et \xrightarrow{PdCl_2}$$

155

A versatile synthesis of compounds of general type **156** involves the reaction of isonitriles with amino alcohols, with diamines, or with aminothiols in the presence of catalytic amounts of silver cyanide.[70] Yields are around 70% and the reaction can also be applied to *o*-amino phenol, *o*-phenylenediamine, and *o*-aminothiophenol to give benzoxazole (54%), benzimidazole (64%), and benzothiazole (93%), respectively.[70b] Palladium chloride is also an excellent catalyst for this reaction and can be used on substrates bearing ester groups, a reaction where the silver catalyst fails.[70a]

$$RNC + H_2NCH_2CH_2XH \xrightarrow{\text{AgCN}} \underset{\textbf{156}}{\left[N \diagup X\right]} + RNH_2$$

X = O, S, or NH

*Preparation of 5,6-Dihydro-4H-1,3-oxazine (**156**; X = O)[70a]*
A mixture of 3-amino propanol (1.5 g; 20 mM), *t*-butyl isocyanide (1.7 g; 20 mM), and silver cyanide (0.13 g; 1 mM) is heated at 90°C for 12 h with stirring in a nitrogen atmosphere. Direct distillation of the reaction mixture gives the product in 66% yield.

Isonitriles containing an acidic α-hydrogen atom undergo a cycloaddition with the carbon–oxygen double bond of carbonyl compounds in the presence of a catalytic amount of cuprous oxide to form the 2-oxazoline (**157**) in good yield.[8] The key intermediate in the reaction is thought to be the organocopper–isonitrile complex **158** formed from the isonitrile and Cu_2O.

*Preparation of 5-Methyl-4,5-diphenyl-2-oxazoline (**157**; R^1 = H, R^2, R^3 = Ph, R^4 = Me)*
A mixture of benzyl isonitrile (10 mM), acetophenone (20 mM), and Cu_2O (0.2 mM) is heated under reflux in benzene for 3 h. The product is obtained by fractional distillation in 75% yield.

An interesting (4 + 2) cycloaddition across the diheterodiene system in the copper complex **159** occurs with dimethyl acetylenedicarboxylate to give the substituted 1,4-benzoxazine **160**.[71] Yields are excellent (~90%) for a variety of substituted complexes and it is noteworthy that no reaction occurs with the uncomplexed nitroso phenols.

161

Cooligomerization of butadiene with ketazines or aldazines in the presence of a nickel phosphine or phosphite catalyst gives the 3,3,12,12-substituted 1,2-diaza-1,5,9-cyclododecatriene **161** in 50%–80% yield.[72]

Chapter 5
Isomerization of Alkenes

5.1. Introduction

A wide variety of isomerizations can be achieved using transition metal species; the more important and most extensively studied of these involve alkenes. Some are stoichiometric reactions in which it may be necessary to isolate an intermediate organometallic compound, while others employ but small amounts of transition metal in catalytic reactions. Among the more well-understood catalytic reactions are the transition-metal-induced isomerizations of strained hydrocarbons to cyclic systems or alkenes, reactions that involve carbon–carbon bond fission. However, in terms of practical synthesis the most useful transition-metal-induced isomerizations are "double-bond migration processes" involving only carbon–hydrogen bond fission. These reactions often provide short routes to high yields of a desired isomer that cannot be obtained by other means.

The interconversions of *cis/trans*-alkene isomers and positional isomers are catalyzed by several transition metal species. Reactions catalyzed by heterogeneous systems (e.g., palladium on charcoal) have been known for many years, but it is only comparatively recently that homogeneous catalysts have been investigated. Suitable homogeneous catalysts can often provide better selectivity than heterogeneous catalysts owing to the use of milder reaction conditions, which minimize side reactions.

5.2. Thermodynamic Considerations

With the more complex alkenes positional isomers generally have significantly different thermodynamic stabilities, and it is not uncommon for thermodynamically unstable isomers to be available from kinetically controlled reactions. Their subsequent isomerization to the thermodynamically stable isomer via transition metal catalysis is a useful procedure in many situations. Terminal linear monoenes are less favoured than internal isomers, but owing to there being little thermodynamic preference for one particular isomer, catalytic isomerization results in mixtures containing an equilibrium distribution of isomers. In principle it might be possible to continuously remove one isomer from an interconverting mixture of simple alkenes, but in practice differences in physical properties are not sufficiently large to permit their separation. However, in some instances separation can be achieved by selectively trapping one isomer as a "derivative," thus removing it from the equilibrium and driving the reaction to produce the "derivative" of the desired isomer. Thus, the industrial, cobalt-catalyzed hydroformylation (reaction with hydrogen and carbon monoxide) of internal alkenes affords[1] substantial yields of linear terminal aldehydes (1).

$$RCH=CHCH_3 \rightleftharpoons RCH_2CH=CH_2 \xrightarrow[\text{Co cat.}]{\text{CO/H}_2} RCH_2CH_2CH_2CHO \qquad (1)$$

Similarly, the nickel-catalyzed synthesis of adiponitrile from butadiene and hydrogen cyanide[2] involves two isomerizations in the formation of the linear product. The first hydrogen cyanide molecule adds to butadiene to form a mixture of pentenenitriles and the undesired 2-methyl-3-butenenitrile, but in the presence of the nickel catalyst this rearranges to a linear isomer. The intermediate internal alkene (2-pentenenitrile) is isomerized *in situ* to the terminal alkene, so that addition of the second molecule of hydrogen cyanide forms adiponitrile rather than other possible isomeric products as shown in Eq. (2). These industrial examples reflect

the tendency of metal alkyls to have enhanced stability when steric crowding at the metal center is minimized, and with simple alkyl chains there is least crowding at the metal center with *n*-alkyls. This effect is exploited in the hydrozirconation of alkenes,[3] which, via a series of double-bond migrations, places the metal at the least hindered position of the alkyl chain (Scheme 1). These reactions are not yet general procedures, nor has their full utility been established, but subsequent reactions of the resulting air- and moisture-sensitive zirconium *n*-alkyls permit access to a wide range of *terminal derivatives from internal linear alkenes* (see, for example, page 206).

Scheme 1

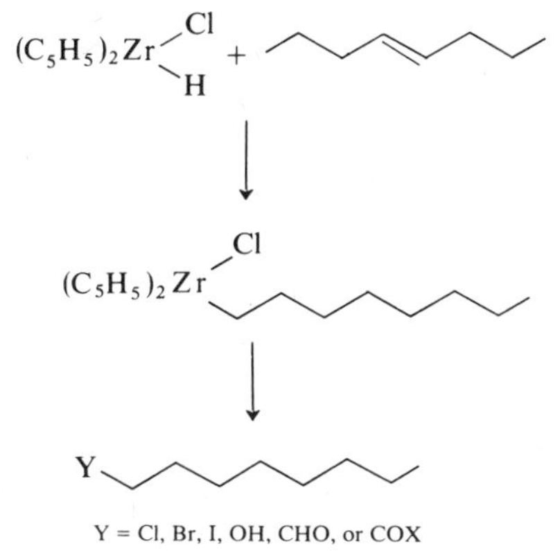

Y = Cl, Br, I, OH, CHO, or COX

5.3. *Isolation of Organometallic Intermediates*

In the previous section transition-metal-induced isomerization of an alkene is followed by stabilization of one particular isomer through selective formation of an intermediate organometallic compound, which in turn undergoes reaction to produce a derivative of that isomer. In some instances it is possible to isolate the intermediate organometallic compound, and subsequently remove the desired less stable isomer from the metal under mild conditions. Several examples of this type of induced double-bond migration have been reported. For instance, although 1,3-cyclooctadiene is the stable isomer, rhodium only forms an isolable compound with 1,5-cyclooctadiene, and since rhodium species are active in alkene isomerization the result of reacting 1,3-cyclooctadiene with rhodium trichloride is a complex of the 1,5 diene. Treatment of this with aqueous cyanide at room

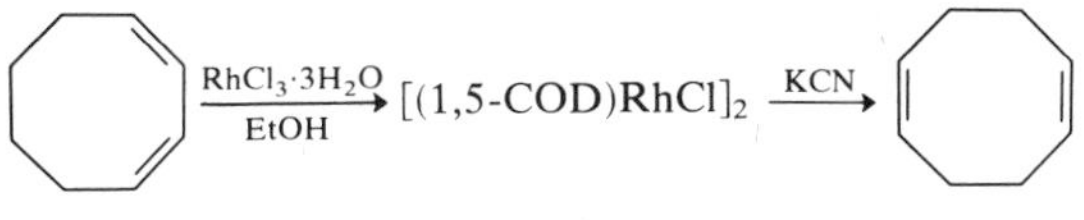

$$\xrightarrow[\text{EtOH}]{\text{RhCl}_3\cdot3\text{H}_2\text{O}} [(1,5\text{-COD})\text{RhCl}]_2 \xrightarrow{\text{KCN}}$$

(3)

temperature liberates 1,5-cyclooctadiene,[4] and so provides a means of accomplishing rearrangement to the less stable isomer.

1,5-Cyclooctadiene from 1,3-Cyclooctadiene via a Rhodium Complex[4]

A solution of 1,3-cyclooctadiene (1 ml, 8 mM) and $RhCl_3 \cdot 3H_2O$ (2 g, 8 mM) in ethanol (20 ml) is maintained overnight at 60°C under an atmosphere of nitrogen. Orange crystals of the dimeric rhodium complex are collected, washed with a little methanol, then treated with 10% aqueous potassium cyanide (~150 ml) to decompose the complex and liberate the 1,5 diene. Extraction into pentane, drying with $CaCl_2$, and evaporation of solvent affords 1,5-cyclooctadiene in about 60% yield.

The extremely strong preference of the $Fe(CO)_3$ moiety to bond with 1,3-cisoid dienes provides a means of converting heteroannular *trans*-steroidal dienes into their thermodynamically *less* stable homoannular *cis* isomers[5] (Scheme 2). The reactant transoid steroid diene is refluxed in dibutyl ether to form the iron tricarbonyl complex, which is subsequently decomposed by treatment with ferric chloride in ethanolic hydrogen chloride to liberate the cisoid isomer in good yield. Other means of decomplexing the diene include ceric ion (ceric ammonium nitrate) in ether, and trimethylamine oxide in benzene or acetone.[6] The last reagent is

Scheme 2

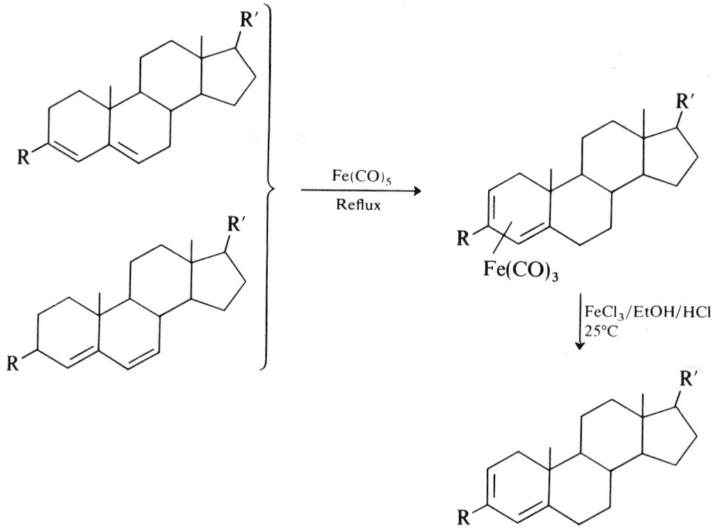

particularly suitable for oxidation-sensitive dienes. In contrast, the bonding requirements of the $Cr(CO)_4$ moiety form the basis of a highly regioselective method for isomerization of suitable *cis*-steroidal dienes to a *trans* isomer [Eq. (4)]. This reaction does not depend on the isolation of an intermediate

$$\xrightarrow[\text{36 h, 81\%}]{Cr(CO)_6}$$ (4)

organometallic compound,[7] but double-bond migration in the initially formed (*cis*-diene)$Cr(CO)_4$ complex takes place to form the *trans* isomer, which cannot meet the four-electron bonding requirements of the chromium center, and the reaction is made irreversible by subsequent loss of the chromium. Several questions about this proposed mechanism can only be answered by further work, but clearly the procedure has potential value.

The preparation of large quantities of organometallic intermediates, particularly if they are air or moisture sensitive or involve costly materials such as the platinum metals, is best avoided. In situations where there is little thermodynamic preference for one isomer this may be the only route available. However, isomers of the more complex alkenes can have significantly different thermodynamic stabilities, and catalytic methods can be used to produce the most stable isomer. This approach is often convenient, and requires the use of only small amounts of transition metal.

5.4. Mechanisms of Catalytic Alkene Isomerization

Two important mechanisms have been established for double-bond migration reactions of alkenes (i.e., hydrogen migration reactions) catalyzed by homogeneous transition metal species. The first involves addition of the alkene to a metal hydride to form a metal alkyl,[8] followed by β-elimination, reforming an alkene in which the double bond has migrated via a 1,2-hydrogen shift. This mechanism is depicted in Scheme 3 and is typified by isomerizations catalyzed by rhodium compounds. The second mechanism, although similar to the first in that addition of alkene to a metal center is followed by hydrogen transfer and elimination, involves a metal π-allyl intermediate[9] with an associated 1,3-hydrogen shift, rather than the 1,2-hydrogen shift of the metal alkyl mechanism (see Scheme 4). The metal π-allyl route requires a metal center with two available oxidation

Scheme 3. Isomerization of Terminal Alkene via Metal Alkyl Addition–Elimination Mechanism; Additional Ligands Omitted for Clarity

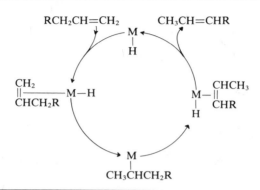

states (n and $n + 2$), since the reaction is a redox process at the metal center. Among the best examples of hydrogen migrations of this kind are those catalyzed by palladium complexes and by iron carbonyls. Deuterium labeling studies permit differentiation between the two mechanisms, and these reactions could have application in the preparation of some labeled alkenes.

In some instances several migration steps can take place while the alkene remains bound to the metal, and under favorable conditions transition-metal-catalyzed double-bond migration can be extremely rapid, even

Scheme 4. Isomerization of Terminal Alkene via Metal π-Allyl Addition–Elimination Mechanism; Additional Ligands Omitted for Clarity

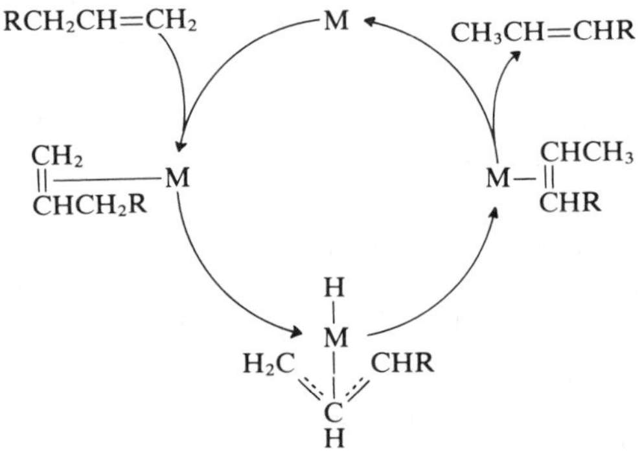

at room temperature. Because of the practical importance of catalytic alkene isomerizations the rest of this chapter is concerned with these reactions.

5.5. Formation of α,β-Unsaturated Compounds

In the absence of unfavorable steric interactions, many α,β-unsaturated compounds have enhanced stability compared with isomers having nonconjugated double bonds. A variety of transition metal species are able to catalyze double-bond migration to the conjugated isomer, usually without inducing unwanted side reactions, and in a number of reactions the α,β-unsaturated compound undergoes further rearrangement to give, for example, an aldehyde or ketone. The catalysts used in these reactions also isomerize less thermodynamically stable α,β-unsaturated compounds to the more stable isomer in situations where two such isomers exist. These reactions form the basis of a range of useful preparative procedures, which are described in the following sections.

5.5.1. Propenyl Ethers from Allyl Ethers

Primary allyl ethers are rapidly isomerized[10] to the corresponding *trans*-propenyl ethers [Eq. (5)] with very high selectivity and, in excellent yield, by the iridium complex $[Ir(solvent)_2(PMePh_2)_2]^+$ in tetrahydrofuran

$$R' \underset{R''}{\overset{}{\diagdown}} OR''' \quad \xrightarrow[25°C, 1\ h]{Ir\,(solvent)_2(PMe_2Ph)_2{}^+} \quad R' \underset{R''}{\overset{}{\diagdown}} OR''' \qquad (5)$$

R', R'' = H; R' = H or CH₃
R'' = CH₃ or H: R''' = alkyl or aryl

or dioxan at room temperature (reaction times 0.5–4 h). This very mild reaction produces exclusively *trans* products from primary allyl ethers. Secondary ethers do not react, and primary ethers carrying a methyl group on the double bond are very reluctant to react. This reactivity difference could be made use of in the selective transformation of allylic groups. The molar ratio of catalyst to substrate is about 0.001, and the catalyst is prepared *in situ* by treating a solution of $[Ir(COD)(PMePh_2)_2]^+$ with hydrogen at atmospheric pressure, which hydrogenates the diene ligand and so removes it from the metal center.

A range of other transition metal species have been investigated as catalysts for the isomerization of allyl ethers,[11–16] and the apparent order of effectiveness is $[Ir(solvent)_2(PMePh_2)_2]^+ > RhCl(PPh_3)_3 > RhCl_3 > PdCl_2 > RuCl_3 > IrCl_3$. For reasons of convenience (availability and no pretreatment) the preferred homogeneous catalysts are $RhCl(PPh_3)_3$ and $RhCl_3 \cdot 3H_2O$. The heterogeneous catalyst palladium on charcoal in refluxing benzene or toluene is convenient for isomerizing a range of allyl and 2-methylallyl ethers to mainly cis-propenyl ethers.[16]

Hydroxyl groups can be protected by converting them to simple allyl ethers. Subsequently their catalytic isomerization using, for example, $RhCl(PPh_3)_3$, followed by hydrolytic cleavage of the propenyl ether, provides an easy means of removing the protecting group.[11] An example of this procedure is given below. A similar double-bond migration catalyzed by $RhCl(PPh_3)_3$ is used in the regeneration of carbonyls protected as 5-methylene-1,3-dioxanes [Eq. (6)] via the corresponding enol ethers.[11]

(6)

Isomerization of Menthyl Allyl Ether[10]

Menthyl allyl ether (0.114 g, 0.58 mM), $RhCl(PPh_3)_3$ (0.037 g, 0.04 mM), and diazabicyclo-[2.2.2]octane† (0.037 g, 0.04 mM) in 10% aqueous ethanol is refluxed for 3 h. The 1-propenyl ether is not isolated, but hydrolyzed by pouring the reaction mixture into water extracting with ether, and washing with brine (pH ~2). After drying ($MgSO_4$) the ether extract and chromatography on silica gel, menthol is obtained in 93% yield.

α,α-Dimethylallyl benzyl ethers cannot rearrange to propenyl ethers, and at 200°C in the presence of 1 mol % $RuCl_2(PPh_3)_3$ $RuH_2(CO)(PPh_3)_3$, or $RuClH(CO)(PPh_3)_3$ they undergo clean fragmentation to the corresponding benzaldehyde and 2-methylbutene.[13] As expected,[14] when 2,5-dimethoxy-2,5-dihydrofurans are heated with $RuHCl(PPh_3)_3 \cdot PhCH_3$ or $RuHCl(CO)(PPh_3)_3$ double-bond migration takes place to form the "propenyl ether," which undergoes clean acid-catalyzed ring opening to form γ-ketoesters in good yield [Eq. (7)].

(7)

† Diazabicyclo[2.2.2]octane was added to prevent *in situ* hydrolysis of the 1-propenyl ether. The propanol so formed would react with the catalyst to form the less active $RhCl(CO)(PPh_3)_2$. This procedure is not necessary when $[Ir(solvent)_2(PMePh_2)_2]$ is used in dioxan or tetrahydrofuran.

5.5.2. γ,δ-Unsaturated Aldehydes and Ketones from Diallyl Ethers

A novel, and valuable, route for the synthesis of a range of γ,δ-unsaturated aldehydes and ketones from diallyl ethers[15] makes use of double-bond migration catalyzed by $RuCl_2(PPh_3)_3$ to form an intermediate allyl vinyl ether. This then undergoes Claisen rearrangement to the isolated carbonyl product, e.g., Eq. (8). The starting diallyl ethers are easily available

$$\text{(8)}$$

by *O*-alkylation of allyl alcohols with allyl halides. Since the double-bond migration takes place most readily with mono- and vicinally disubstituted alkenes, the rearrangement of unsymmetrical diallyl ethers is regiospecific in a predictable way, and under the reaction conditions the Claisen-rearranged products do not undergo further allylic double-bond migration.

$$\text{(9)}$$

2-Methylhex-4-enal from Allyl 1-Methylallyl Ether [Eq. (9)][15]
    Allyl 1-methylallyl ether (5 g, 45 mM) and $RuCl_2(PPh_3)_3$ (0.05 g, 0.05 mM) are sealed in a small Pyrex tube under vacuum and heated at 200°C (oil bath) for 4 h. After vacuum distillation the crude product is shaken with sodium bisulfite (6.4 g, 62 mM) in water (10 ml), and extracted with ether (4 × 2 ml) to remove starting material and a small amount of 4-methylhex-1-en-5-one. Sufficient saturated $NaHCO_3$ solution is added to the aqueous layer until no more CO_2 is evolved. The aldehyde product is obtained from the aqueous mixture by continuous ether extraction followed by distillation. Yield: 1.9 g (55%).

This approach has been successfully used to form 2-methyl-3-vinyloctanal [Eq. (10)], which was required in a synthesis of dihydrojasmone.[17]

$$\text{(10)}$$

5.5.3. Ketones from Allylic Alcohols

Rearrangements of allylic alcohols themselves are easily catalyzed at room temperature by rhodium and ruthenium complexes to give the corresponding carbonyl compounds [Eq. (11)]. This isomerization takes place particularly smoothly in the presence of small amounts of $[Rh(CO)_2Cl]_2$

$$R'CH{=}CHCHR'' \xrightarrow[\text{OH}^-,\,25°C]{[Rh(CO)_2Cl]_2} \left[\underset{|}{\overset{OH}{R'CH_2CH}}{=}CR''\right] \rightarrow R'CH_2CH_2\overset{O}{\underset{\|}{C}}R'' \quad (11)$$

in dichloromethane/aqueous sodium hydroxide, under phase transfer conditions. The phase transfer catalyst (a quaternary ammonium salt) is not essential, though the reaction is somewhat cleaner if one is used.[18] Ruthenium complexes are also efficient catalysts for the isomerization of allylic alcohols to ketones. Readily available $Ru_3O(CH_3CO_2)_7$ is one of the most active of these, and very high yields of ketone are formed after some 30 min at 85°C. Moreover, if ethylene glycol is employed as the solvent, separation of products is facilitated because of their insolubility, and the solution containing the ruthenium catalyst can be reused.[19]

5.5.4. Enamines from Allylamines

Cobalt complexes appear not to catalyze double-bond migration reactions of allyl ethers and alcohols. However, double-bond migration in allylamines (giving *trans*-enamines in high yield) is catalyzed by simple low-oxidation-state cobalt complexes.[20] For example, *N,N*-diethyl-nerylamine is isomerized to the racemic citronellal-*trans*-enamine in 85% yield by heating with 1 mol % of $CoH(N_2)(PPh_3)_3$ at 80°C for 15 h in tetrahydrofuran [Eq. (12)]. An active catalyst may be prepared *in situ* from soluble

$$\text{(structure) NEt}_2 \xrightarrow[80°C,\,15\,h,\,85\%]{HCo(N_2)(PPh_3)_3} \text{(structure) NEt}_2 \quad (12)$$

cobalt salts such as the naphthenate and aluminum alkyls (caution!) in the presence of triphenylphosphine. Moreover, when catalysts containing chiral diphosphine ligands are used in place of triphenylphosphine the reaction is enantioselective. Some 30% enantiomeric excess was obtained using a *diop*-containing catalyst.[20a] Yields are not high, and better selectivity results from the use of rhodium catalysts.[20b]

The allyl moiety is a useful protecting group for primary amines that has the advantage of being acid and base stable and also resistant to nucleophiles.[20a] *N,N*-Diallylation of primary amines is easily achieved with allyl bromide, and subsequent catalytic conversion to the propenyl isomer followed by *in situ* hydrolysis to propionaldehyde provides a convenient means of removing the protecting groups [Eq. (12a)]. The recommended catalyst is $Rh(PPh_3)_3Cl$, and it is important that the propionaldehyde is

$$RNH_2 \xrightarrow{CH_2=CHCH_2Br} RN(CH_2CH=CH_2)_2 \xrightarrow[CH_3CN/H_2O]{Rh(PPh_3)_3Cl} RNH_2 + 2CH_3CH_2CHO$$

(12a)

removed as it is produced so the less active catalyst $Rh(PPh_3)_2(CO)Cl$ is not formed via decarbonylation of the aldehyde. Continuous distillation is used to remove the propionaldehyde. This method of protecting primary amines appears to be a relatively simple, viable alternative to the formation of a phthalimide derivative.

5.5.5. Propenylamides from N-Allylamides

In contrast to the iridium-catalyzed isomerization of allyl ethers and the cobalt-catalyzed isomerization of allylamines, which afford *trans* products (see above), the transition-metal-catalyzed rearrangement of *N*-allylamides leads to predominantly *cis*-propenylamides[16] [Eq. (13)]. Rearrangement of *N*-allylamides is quite general, and iron, ruthenium, and

$$CH_2=CHCH_2NHCOCH_3 \xrightarrow[40\,h,\,110°C,\,80\%]{RuHCl(PPh_3)_3} cis\text{-}CH_3CH=CHNHCOCH_3$$

(13)

rhodium species are all active. However, as yet there is no universal catalyst, and it is necessary to match catalyst and reaction conditions with a particular *N*-allylamide. Prolonged reaction times appear to favor the formation of *trans* products.

5.5.6. α,β-Unsaturated Cyclic Ketones via Remote Double-Bond Migration

In the absence of unfavorable steric or electronic effects, the versatile catalyst rhodium trichloride/ethanol smoothly produces α,β-unsaturated ketones from unconjugated isomers via double-bond migration. With the more complex reactants, chloroform can be used as a cosolvent with ethanol to enhance the solubility of product and reactant, though ethanol alone at reflux is normally used. This isomerization is illustrated in Eq. (14), which is one of the last stages in a stereoselective synthesis of *dl*-isopetasol.[22]

(14)

Migration of quite distant exocyclic double bonds to form cyclic α,β-unsaturated ketones can be achieved with $RhCl_3 \cdot 3H_2O/EtOH$, e.g., Eq. (15), but *migration of remote external double bonds requires the use of more catalyst than does the migration of endocyclic double bonds.*[23] Typically 0.1 equivalents of $RhCl_3 \cdot 3H_2O$ are required.

$$\text{(15)}$$

This approach, of course, does not work when steric strain or electronic effects destabilize the α,β-unsaturated ketone, and the following examples illustrate situations in which α,β-unsaturated ketones are not obtained. Treatment of the reactant bicyclic ketone in Eq. (16) with $RhCl_3 \cdot 3H_2O$

$$\text{(16)}$$

(no mention of using ethanol was made; it is, however, assumed the normal $RhCl_3 \cdot 3H_2O/EtOH$ combination was used) did not give the desired α,β-unsaturated ketone,[24] but rather an unconjugated isomer as shown in Eq. (16). A similar example is shown in Eq. (17), where the unconjugated product was obtained in excellent yield.[23]

$$\text{(17)}$$

The influence of strain effects is particularly important with the smaller ring systems, and unconjugated products are common. For instance, a step in a synthesis of hirsutene involves isomerization[25] of the reactant tricyclic ketone in Eq. (18), which readily takes place on treatment with $RhCl_3 \cdot 3H_2O$

$$\text{(18)}$$

in aqueous ethanol to give the nonconjugated product. Few transition-metal-catalyzed rearrangements of unsaturated cyclobutanone derivatives have been described, although formation of the nonconjugated monoterpenoid filifolone as shown in Eq. (19) has been reported,[25] using $RhCl_3 \cdot 3H_2O$.

$$(19)$$

5.5.7. *Interconversion of α,β-Unsaturated Ketones*

Migration of the double-bond in an α,β-unsaturated ketone to form a more stable isomeric α,β-unsaturated ketone can be achieved with rhodium trichloride/ethanol, $RhCl(PPh_3)_3$, or palladium on charcoal.

5.5.7.1. *Interconversion of Five-Membered Cyclic α,β-Unsaturated Ketones*

2-Alkyl cyclopentenones are potentially valuable intermediates in the synthesis of natural products, but they are relatively inaccessible. However, the corresponding exo-isomers are readily obtained, and when treated with $RhCl_3 \cdot 3H_2O$ in ethanol they are smoothly converted[20] to the desired isomer [Eq. (20)]. Such exocyclic–endocyclic double-bond migrations are difficult to accomplish by other means.

$$(20)$$

The ease with which this rhodium-catalyzed reaction can take place is shown by the accidental formation of isodamsin as the main product during the attempted hydrogenation[28] of damsin catalyzed by $RhCl(PPh_3)_3/H_2$ as shown in Eq. (21). More recently this reaction with

$$(21)$$

the same catalyst or $RhCl_3 \cdot 3H_2O$/ethanol has been employed in a number of syntheses: for example, in the total synthesis of Rugulovasines[29] [Eq. (22)], in the final stage of a new cyclopentenone annulation[30] [Eq. (23)]

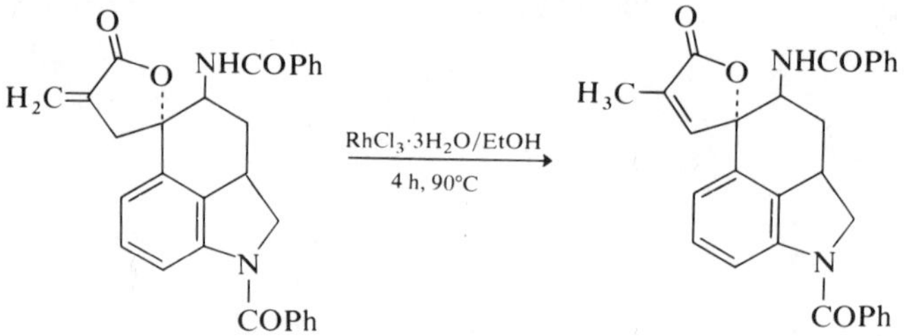

(22)

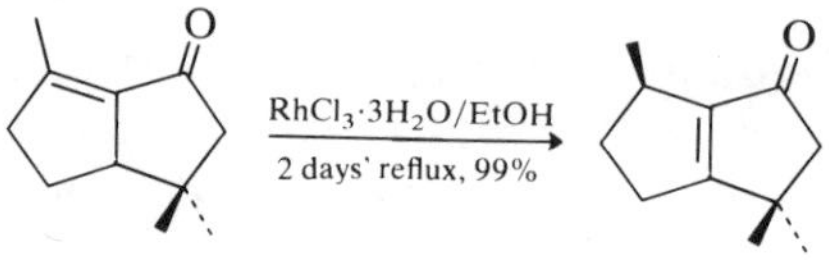

(23)

(it is not clear if the protracted reaction time was necessary), and in the preparation[32] of some modified prostaglandin precursors [Eq. (24)]. Interestingly, an analogous cyclopentene system [Eq. (25)] undergoes a similar

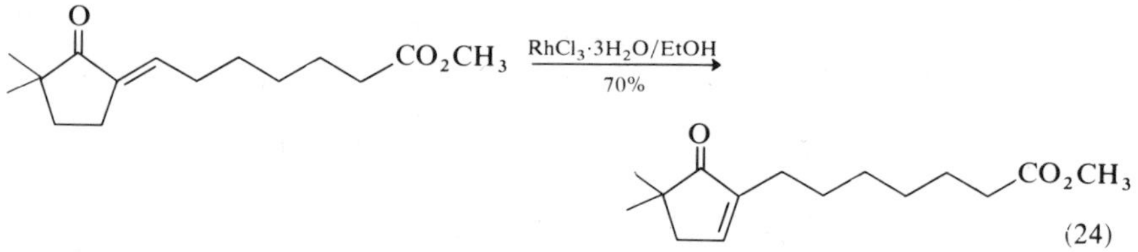

(24)

exocyclic–endocyclic reaction and this has been used in the synthesis of the [3.3.3]-propellane sesquiterpene modhephene.[31]

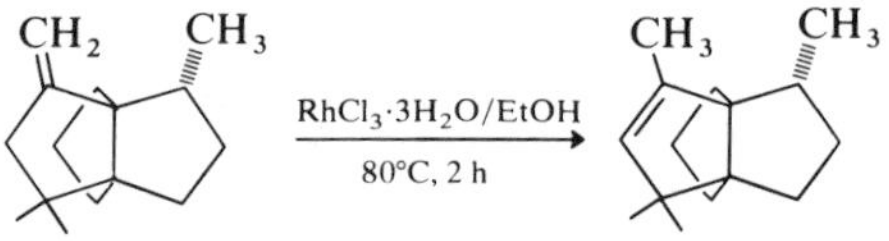

(25)

5.5.7.2. Interconversion of Six-Membered and Larger Cyclic α,β-Unsaturated Ketones

Cyclic α,β-unsaturated ketones rearrange to the more substituted alkene when treated with 2 mol % equivalents of $RhCl_3 \cdot 3H_2O$ in absolute ethanol at 100°C[23] [Eq. (26)]. Prolonged reaction times (~24 h) are not necessary, and they should be avoided since they encourage the formation of small amounts of saturated ketone via hydrogen transfer.

(26)

(27)

Isomerization of 2-Heptylcyclohex-5-enone to 2-Heptylcyclohex-2-enone [Eq. (27)][23]

2-Heptylcyclohex-5-enone (178 mg, 0.92 mM) and $RhCl_3 \cdot 3H_2O$ (4.6 mg, 0.017 mM) dissolved in ethanol (180 μl) were sealed under nitrogen in a glass tube and heated at 100°C for 3 h. Chromatography of the reaction mixture (silica gel eluted with pentane/ether) gave the 2-isomer in 94% yield.

Use of $RhCl_3 \cdot 3H_2O$/EtOH provides ready access to the otherwise difficult to obtain heterocyclic homoisoflavone skeleton by facilitating the exocyclic–endocyclic migration of the double-bond in arylmethylenechroman-4-ones [Eq. (28)]. This reaction, conducted in an ethanol–chloroform mixture at

(28)

70°C, gives a quantitative yield of homoisoflavone after 24 h. Similar results are obtained with a variety of substituted arylmethylenechroman-4-ones.[33] These conditions compare favorably with 150°C required for the thermal rearrangement, which involves a ring-opening mechanism.[34]

Isomerization of Benzylidenechromanone to Homoisoflavone [Eq. (28)][33]

Benzylidenechromanone (0.2 g, 0.85 mM) and $RhCl_3 \cdot 3H_2O$ (10 mg, 0.04 mM) with water (0.2 ml), ethanol, and chloroform (2 ml each) are stirred at 70°C for 24 h. The mixture is then poured into an excess of water and extracted with chloroform to give an almost quantitative yield of product.

Similarly, linear α,β-unsaturated ketones are easily rearranged in high yield to the more stable isomer when treated with rhodium trichloride in ethanol[35] [Eq. (29)].

(29)

5.5.7.3. Formation of Pyridones from Dibenzylidene Piperidones

Substituted 1-methyl-3,5-dibenzyl 4-pyridones can be obtained from the corresponding dibenzylidene piperidones using catalytic amounts of palladium on charcoal in the high boiling solvent ethylene glycol [Eq. (30)]. This reaction affords excellent yields of product after only 30 min.[36]

$$\begin{array}{c}\text{Pd/C} \\ \xrightarrow{\hspace{2cm}} \\ 195°C, 3\,h, 95\%\end{array} \tag{30}$$

X = H, alkyl, aryl, NO$_2$, Cl, OCH$_3$, or N(CH$_3$)$_2$

Isomerization of 1-Methyl-3,5-dibenzylidene-4-piperidone to 1-methyl-3,5-dibenzyl-4-pyridones [Eq. (30)][36]

1-Methyl-3,5-dibenzylidene-4-piperidone (0.5 g, 1.7 mM), ethylene glycol (5 ml), and 10% palladium on charcoal catalyst (0.05 g) is heated to reflux temperature (195°C) for 30 min. The hot solution is filtered to remove catalyst, which is washed with a little hot ethylene glycol. Sufficient water is added to the cooled combined filtrates to cause precipitation of the product, which is collected, washed with water, and recrystallized from aqueous ethanol. Yield is essentially quantitative.

5.6. Formation of Conjugated Dienes

Migration of isolated double bonds in linear and cyclic systems to form the more stable conjugated diene (or, in general, conjugated polyene) is brought about by the rhodium and ruthenium catalysts previously discussed. Thus, rearrangement of the cyclic diene to the conjugated isomer in Eq. (31) is achieved[37] within 5 min in the presence of RhCl(PPh$_3$)$_3$.

$$\begin{array}{c}\text{RhCl(PPh}_3\text{)}_3 \\ \xrightarrow{\hspace{2cm}} \\ 5\,\text{min}\end{array} \tag{31}$$

Related reactions involving the five-membered fused-ring system shown in Eq. (32) are catalyzed by reduced titanocene species,[38] though at much higher temperatures than needed for the related reactions described above.

$$\begin{array}{c}\text{Cp}_2\text{TiCl}_2/\text{LiAlH}_4 \\ \xrightarrow{\hspace{2cm}} \\ 165°C\end{array} \tag{32}$$

An unusual catalyst, whose scope and potential have yet to be fully explored, is nickel acetylacetonate.[39] This is particularly attractive for large-scale operation because of its cheapness. It has been reported that the yield of product in reaction (33), is almost quantitative after 2 h in the presence of 1 mol % of nickel acetylacetonate at 165°C with *N*-methyl-2-pyrollidone as solvent.

$$\xrightarrow[\text{165°C}]{\text{Ni(acac)}_2} \tag{33}$$

It is significant that in the related reaction (34) nickel acetylacetonate induces migration of an isolated double bond to form a conjugated diene, rather than migration of the double bond into conjugation with the aromatic system.[39] Nonconjugated unsaturated aldehydes and esters have been reported[40] as the products of other nickel-catalyzed isomerizations.

$$\tag{34}$$

5.7. Migration into Conjugation with Aromatic Rings

There is an appreciable thermodynamic driving force for the migration of a remote double bond into conjugation with an aromatic system. RhCl(PPh$_3$)$_3$ is an effective catalyst for this type of rearrangement. However, the most convenient catalyst is RhCl$_3$·3H$_2$O/EtOH, which, for example, rapidly converts eugenol to predominately *trans*-isoeugenol [Eq. (35)] at room temperature in an exothermic reaction.[33] The *cis/trans* ratio (90% *trans*) in this and related reactions is in contrast with the isomerization of allyl groups by strong bases at temperatures typically as high as 180°C, which affords predominantly *cis* isomers.

$$\xrightarrow[\text{20°C, 2 h, 92\%}]{\text{RhCl}_3\cdot 3\text{H}_2\text{O/EtOH}} \tag{35}$$

Isomerization of Eugenol to trans-Isoeugenol[33]

RhCl$_3$·3H$_2$O (0.1 g, 0.4 mM) is added to eugenol (20 g, 0.12 mol) in ethanol (5 ml) with external cooling to keep the temperature at 20°C, and after 2 h the mixture is poured into water. Extraction with ether followed by drying and distillation gives the product in 92% yield with at least 90% having the *trans* configuration.

5.8. Formation of Aromatic Compounds—Isoaromatization

Metal-catalyzed migration of double bonds into cyclic systems to form thermodynamically stable aromatic compounds (notably benzene derivatives) is well established. This reaction is referred to as *isoaromatization* and can be a useful synthetic route to some otherwise difficult-to-prepare aromatic compounds. Finely divided heterogeneous metal catalysts can be effective in this reaction. However, their application is not completely general because forcing conditions (200°C and higher) are usually required, which can lead to undesired side reactions such as deoxygenation, or hydrogenation via hydrogen transfer from the solvent. Nonetheless, heterogeneous catalysts have the advantages of being readily available and simple to use. Palladium is the most active heterogeneous catalyst and its tendency to catalyze isoaromatization often causes problems in hydrogenations. Raney nickel is generally inactive, even at very high temperatures, while platinum shows only slight activity. Palladium on charcoal (typically 5% or 10% by weight of palladium) is the preferred catalyst, and the amount used relative to the reactant is usually less than 0.5% by weight of palladium. Depending on reactant, amount of catalyst, and temperature, reaction times vary between 30 min and many hours. The organic substrate and catalyst may be heated (normally above 200°C) alone, or in the presence of a suitable high boiling solvent (ethylene glycol has been recommended). However, care in using solvents is necessary because of the possibility of the catalyst promoting hydrogen transfer from the solvent, leading to hydrogenated products.

During recent years a number of homogeneous catalysts have been investigated for activity in isoaromatization reactions. Some have high activity under mild conditions and, as a result, have wider application than heterogeneous catalysts, but it should be noted that not all homogeneous catalysts have high activity. For instance, IrCl(CO)(PPh$_3$)$_2$ and RhCl(PPh$_3$)$_3$ are in general less active than palladium on charcoal. One of the most active and versatile homogeneous catalysts is RhCl$_3$·3H$_2$O in the presence of ethanol [the hydride source for producing a catalytically active rhodium(I) species], and most of the isomerizations brought about by palladium on charcoal can also be achieved with this catalyst under milder conditions.

5.8.1. Formation of Substituted Phenols

Carvone is smoothly converted[41] to 2-methyl-5-isopropyl phenol at about 100°C in the presence of $RhCl_3 \cdot 3H_2O$ and ethanol [Eq. (36)], and benzylidene tetralone transforms quantitatively into 2-benzylnaphthol[24] at 70°C [Eq. (37)].

These reaction conditions compare favorably with the 200–250°C needed when reactions are catalyzed by palladium on charcoal. Although the mechanisms of such rhodium-catalyzed isomerizations have not been fully elucidated, it is likely they proceed via simple addition–elimination reactions on rhodium(I) species. However, such a mechanism does not readily explain why the presence of *some* water is important. Under strictly anhydrous conditions no reaction takes place.[42]

Isomerization of Carvone to 2-Methyl-5-isopropyl Phenol [Eq. (36)]
    (a) *With Palladium on Charcoal.*[43] Redistilled carvone (25.0 g, 0.168 mol, bp 224–226°C) and 10% palladium on charcoal (1.0 g) are heated under reflux for 1 h. The mixture is then distilled and pure 2-methyl-5-isopropylphenol collected (bp 232–234°C). Yield: 32.1 g (0.155 mol, 92%).
    (b) *With $RhCl_3 \cdot 3H_2O$ in Ethanol.*[41] A solution of carvone (100 mg, 0.66 mM) and $RhCl_3 \cdot 3H_2O$ (19 mg, 0.07 mM) in ethanol (0.175 ml) is heated in a sealed tube at 100°C for 8 h. The cooled reaction mixture is purified by chromatography on silica gel to give pure 2-methyl-5-isopropylphenol (78 mg, 78%).

Another example of the use of palladium on charcoal for isomerization of a monocyclic reactant to a phenol is the conversion of 2,6-dibenzylidene cyclohexanone (obtained from cyclohexanone and benzaldehyde) to 2,6-dibenzylphenol in high yield [Eq. (38)].

$$\text{(38)}$$

The $RhCl_3 \cdot 3H_2O$/ethanol catalyst has considerable scope in the formation of substituted phenols from cyclic ketones because the double bonds that migrate can be quite remote from the ring. Thus, a number of substituted cyclohexenones can be converted[41] to the corresponding phenols in 60%–80% yield via remote double-bond migration as illustrated in Eqs. (39) and (40).

$$\text{(39)}$$

$$\text{(40)}$$

5.8.2. Formation of Hydroxytropolones

Although the use of palladium on charcoal at high temperatures may lead to deoxygenation reactions, particularly of polycyclic compounds, this approach has been successfully employed to prepare hydroxytropolones from 3,7-dibenzylidene-1,2-cycloheptanediones[44] [Eq. (41)].

$$\text{(41)}$$

Claims[45] that 2,7-dibenzylidene cycloheptanone forms a tropolone via double-bond migration and dehydrogenation when treated with palladium on charcoal have not been confirmed.[46]

5.8.3. Formation of Substituted Anilines

The $RhCl_3 \cdot 3H_2O$/ethanol system catalyzes the isomerization of cyclohexene imines to substituted anilines [e.g., Eqs. (42) and (43)]. The reaction is carried out in the presence of potassium carbonate to remove hydrogen

$$\text{(42)}$$

$$\text{(43)}$$

chloride, which would cause hydrolysis of the imine to the enone.[41] Yields are lower (typically 50%) than those obtained for isomerization of the related cyclohexenones to phenols, but in some instances this route to substituted anilines may be more attractive than alternative syntheses.

Isomerization of Benzyl Imine of Carvone to N-Benzyl, N-2-methyl-5-isopropylaniline[41]

The benzyl imine of carvone (100 mg, 0.41 mM) and $RhCl_3 \cdot 3H_2O$ (16 mg, 0.06 mM), anhydrous K_2CO_3 (80 mg, 0.56 mM), and ethanol (0.75 ml) are heated at 100°C in a sealed tube for 30 h. Chromatography of the crude product on silica gel affords the pure aniline (47 mg, 47%).

Chapter 6

Direct Introduction and Removal of Carbonyl Groups

6.1. Preliminaries

The ability of transition metals and their compounds to coordinate and activate carbon monoxide has led to the development of many potentially valuable methods for direct introduction of carbonyl groups into organic molecules. Historically,[1] such methods have involved the use of high pressures of carbon monoxide, elevated temperatures, and the exceedingly toxic and volatile nickel tetracarbonyl [$Ni(CO)_4$] as catalyst, but in recent years the advent of high-activity catalysts, based particularly on palladium and rhodium,[2] has allowed carbonylation reactions to be achieved under very much milder, and safer, conditions.

Some idea of the scope of carbonylation as a synthetic tool may be gained from the transformations listed below. The multicomponent reactions especially indicate the unique ability of transition metal catalysts to assemble complex organic molecules with relatively high specificity, and it is clear that few such transformations could be achieved by more conventional techniques.

2-component reactions:

$$CO + RX \rightarrow RCOX \quad (X = halogen)$$
$$CO + ROH \rightarrow RCOOH$$
$$R_2O + CO \rightarrow RCOOR$$
$$RCOOR + CO \rightarrow RCOOCOR$$

3-component reactions: $CO + RX + HY \rightarrow RCOY + HX$ (X = halogen, Y = OH, OR, NR$_2$, RCO$_2$, or H)

$$R\diagdown + CO + HY \rightarrow R\overset{\text{COY}}{\diagup} + R\diagdown\diagup^{COY} \text{ (Y as above)}$$

$$RCHO + R_3SiH + CO \rightarrow R\overset{\text{OSiR}_3}{\underset{\mid}{C}}HCHO$$

4-component reactions: $\diagup\diagdown\diagup Cl + HC\equiv CH + CO + MeOH \longrightarrow$

$$\diagup\diagdown\diagup\diagdown^{CO_2Me} + HCl$$

5-component reaction: $CH_3I + HC\equiv CH + 2CO + H_2O \longrightarrow$ (structure with OH, O, O)

   Although free carbon monoxide is itself susceptible to attack by strong nucleophiles[3]

$$\text{(e.g., Na}^+\text{NH}_2^- + CO \rightarrow Na^+NH-\overset{O}{\overset{\|}{C}}-H),$$

coordination to a metal center greatly increases this susceptibility, and attack on coordinated carbon monoxide by another ligand or by an external (i.e., noncoordinated) nucleophile almost invariably forms part of any mechanistic pathway leading to carbonylated organic products. Perhaps the clearest example of this is afforded by the rhodium-catalyzed conversion of iodomethane to acetyl iodide [Eq. (1)], since the mechanism of this

$$CH_3I + CO \xrightarrow{[Rh(CO)_2I_2]^-} CH_3COI \tag{1}$$

reaction, which forms part of the catalytic cycle in the industrially important carbonylation of methanol to acetic acid,[4] has been studied in some detail.[5]

   The first step is *oxidative addition* of iodomethane to the rhodium(I) catalyst, giving a six-coordinate methylrhodium(III) anion [Eq. (2)]. *Migra-*

$$\begin{bmatrix} I & CO \\ & Rh & \\ I & CO \end{bmatrix}^- + CH_3I \rightarrow \begin{bmatrix} I & CH_3 & CO \\ & Rh & \\ I & I & CO \end{bmatrix}^- \tag{2}$$

tion of the methyl group from rhodium to carbon monoxide occurs spontaneously [Eq. (3)] and may be regarded as an internal nucleophilic addition to CO, as discussed above. (It is generally recognized, in fact, that the apparent "insertion" of an unsaturated molecule such as CO into an existing

$$\left[\begin{array}{c} I\quad CH_3\quad CO \\ Rh \\ I\quad I\quad CO \end{array}\right]^{-} \rightarrow \left[\begin{array}{c} \overset{O}{\underset{\parallel}{C}} \\ I\quad\diagdown\diagup\quad \\ Rh\qquad CH_3 \\ I\quad I\quad CO \end{array}\right]^{-} \tag{3}$$

metal–ligand bond is more realistically viewed as a 1,2 migration or internal nucleophilic attack by the ligand, but the term "insertion" provides a convenient shorthand for this process and is widely used.)

The five-coordinate acetyl–rhodium complex now coordinates a further molecule of carbon monoxide and *reductively eliminates* acetyl iodide, simultaneously regenerating the catalyst in its original form [Eq. (4)].

$$\left[\begin{array}{c} \overset{O}{\underset{\parallel}{C}} \\ I\quad\diagdown\diagup\quad \\ Rh\qquad CH_3 \\ I\quad I\quad CO \end{array}\right]^{-} + CO \rightarrow \left[\begin{array}{c} \overset{O}{\underset{\parallel}{C}} \\ I\quad CO\diagdown\diagup\quad \\ Rh\qquad CH_3 \\ I\quad I\quad CO \end{array}\right]^{-} \rightarrow \left[\begin{array}{c} I\quad\diagdown\quad CO \\ Rh \\ I\qquad CO \end{array}\right]^{-} \tag{4}$$
$$+ CH_3COI$$

The elementary steps of oxidative addition, insertion, and reductive elimination are found (or proposed) in many carbonylation reactions, but a variety of other mechanistic patterns are also observed, including nucleophilic attack by a metal carbonyl anion on an organic halide [Eq. (5)], reaction of coordinated carbon monoxide with an external nucleophile [Eq. (6)], and direct solvolytic cleavage of an acyl–metal complex [Eq. (7)].

$$CH_3I + Co(CO)_4^- \rightarrow CH_3Co(CO)_4 + I^- \tag{5}$$

$$PhLi + Fe(CO)_5 \rightarrow [Ph{-}\overset{O}{\underset{\parallel}{C}}{-}Fe(CO)_4]^-Li^+ \tag{6}$$

$$R{-}\overset{O}{\underset{\parallel}{C}}{-}Co(CO)_4 + MeO^- \rightarrow R{-}\overset{O}{\underset{\parallel}{C}}{-}OMe + Co(CO)_4^- \tag{7}$$

The latter two reactions could in fact be regarded as bimolecular analogs of insertion and reductive elimination, respectively, and in practice it is often impossible to distinguish between, say, direct alcoholysis of an acyl complex to give an ester and reductive elimination of an acyl halide followed by reaction with the solvent alcohol; indeed both pathways may be followed in the same system.

In this chapter, emphasis will generally be placed on the synthetic aspects of carbonylation chemistry rather than on mechanistic studies, but the elementary steps discussed above will often be invoked to rationalize the observed results.

6.2. Preparation of Carboxylic Acids, Esters, and Related Derivatives

Carboxylic acid derivatives may be prepared from a wide variety of starting materials (alcohols, halides, alkenes, alkynes, diazonium ions, and epoxides) by means of transition-metal-catalyzed carbonylation reactions. Such reactions generally follow the mechanistic pattern noted above, i.e. formation of a metal–carbon σ bond, insertion of carbon monoxide, and, finally, attack on the resulting acyl ligand by an internal or external nucleophile (Nu$^-$) to give the observed product and to regenerate the catalytic species (Scheme 1).

Scheme 1

$$[M-R]^{n+} \xrightarrow{CO} [M-\overset{\overset{\displaystyle O}{\|}}{C}-R]^{n+} \xrightarrow{Nu^-} R-\overset{\overset{\displaystyle O}{\|}}{C}-Nu + [M]^{(n-2)+}$$

The nature of the organic product is thus largely determined by the nucleophile; carboxylic acids, esters, amides, anhydrides, and acyl fluorides, for example, are formed when Nu = OH, OR, NHR, RCO_2, and F, respectively. A very wide range of carboxylic acid derivatives can in fact often be obtained using a single catalyst system.

If the group R itself contains the nucleophilic function —Nu, then cyclization can occur to give, for example, lactones, lactams, imides, or cyclic anhydrides, when Nu = OH, NHR, CONHR, or CO_2H, respectively (Scheme 2). The presence of a stoichiometric amount of nonnucleophilic base such as a tertiary amine is usually necessary to remove the acid formed in such reactions if high yields are to be achieved.

Scheme 2

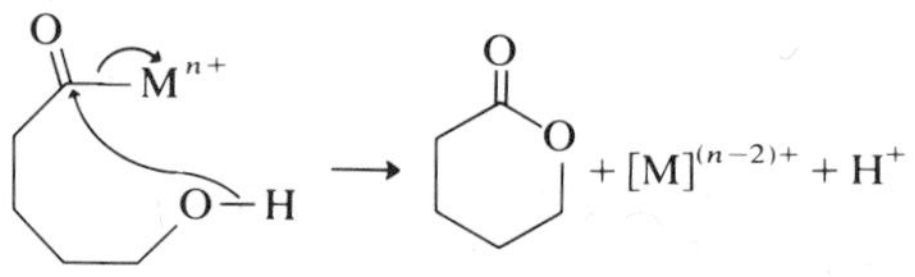

6.2.1. Carboxylic Acids

6.2.1.1. Carboxylic Acids: From Alcohols

Stoichiometrically, the conversion of an alcohol to a carboxylic acid by reaction with carbon monoxide [Eq. (8)] is one of the simplest carbonylation reactions of all. Very recently such a reaction has in fact achieved

$$ROH + CO \rightarrow RCOOH \tag{8}$$

large-scale importance, in the form of a low-pressure, rhodium-catalyzed synthesis of acetic acid from methanol.[4] This reaction has been studied mechanistically by Forster,[5] and is based on the carbonylation of iodomethane to acetyl iodide discussed in the introduction to this chapter. A low, equilibrium concentration of iodomethane is generated [Eq. (9)]

$$HI + CH_3OH \rightleftharpoons CH_3I + H_2O \tag{9}$$

by the reaction of methanol with hydrogen iodide (added as catalyst promoter). The iodomethane is catalytically converted to acetyl iodide [Eq. (10)], and finally hydrolysis yields acetic acid and hydrogen iodide [Eq. (11)], which is then available for further reaction with methanol.

$$CH_3I + CO \xrightarrow{[Rh(CO)_2I_2]^-} CH_3COI \tag{10}$$

$$CH_3COI + H_2O \rightarrow CH_3COOH + HI \tag{11}$$

It is noteworthy that the overall reaction rate is essentially independent of carbon monoxide pressure, since the *insertion* step, giving a stable, isolable acetyl complex, precedes coordination of external CO, and the rate-determining step is in fact addition of iodomethane to the metal.

The carbonylation of methanol to acetic acid can be achieved using other transition metals as catalysts (e.g., the carbonyls of iron, nickel, and cobalt), but even in the presence of hydrogen halide promoters, extreme conditions (200–700 bar; 200–300°C) are required to achieve reasonable conversions.[6]

Strongly acidic materials such as boron trifluoride, sulfuric acid, or phosphoric acid will in fact catalyze methanol carbonylation in the *absence* of transition metals, but extreme conditions (600 bar, 300°C) are again required. In this purely acid-catalyzed reaction (the Koch synthesis[7]), carbonylation occurs by the direct interaction of carbon monoxide with a carbenium ion to give an acylium ion that is subsequently hydrolyzed to the corresponding carboxylic acid [Eqs. (12), (13), and (14)]. It is therefore not surprising that the reaction proceeds under much milder conditions

$$CH_3OH + H^+ \rightleftharpoons CH_3^+ + H_2O \tag{12}$$

$$CH_3^+ + CO \rightleftharpoons CH_3CO^+ \tag{13}$$

$$CH_3CO^+ + H_2O \rightarrow CH_3COOH + H^+ \tag{14}$$

when alcohols that can yield relatively stable tertiary carbenium ions are used. Moreover, with such alcohols, the addition of a copper(I) or silver(I) salt to the reaction mixture allows the Koch synthesis to be carried out under essentially ambient conditions,[8] although the exact role of the transition metal is unclear. It may in fact simply serve to increase the effective concentration of carbon monoxide in solution by forming a loosely bound soluble carbonyl complex $[M(CO)_n]^+$ (M = Cu or Ag; n = 1, 2, 3, or 4). A particular feature of this synthesis is that, almost regardless of the structures of the starting alcohols (when C_4 or higher), the products are tertiary carboxylic acids, so that extensive isomerization of the intermediate carbenium ions must occur. Carbonylation of 1-pentanol, for example, in the presence of 98% sulfuric acid and a catalytic quantity of Cu_2O gave a 62% yield of 2,2-dimethyl butanoic acid as the only product at 30°C and 1 atm CO pressure (Scheme 3). In most cases, however, mixtures of products

Scheme 3

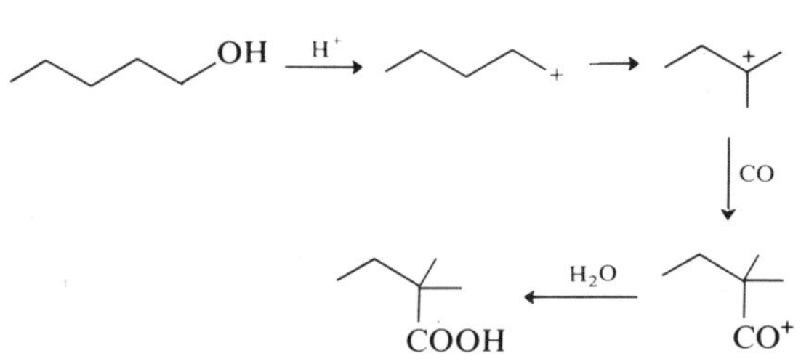

are obtained, so that carbonylation of 1-hexanol under the same conditions gives 2,2-dimethyl pentanoic acid (60%), 2-methyl-2-ethyl butanoic acid (25%), 2,2-dimethyl butanoic acid (4%), and 2,2-dimethyl propanoic acid (2%).

6.2.1.2. Carboxylic Acids: From Halides

Early studies (1930–1950) showed that, in the presence of nickel salts, chlorobenzene could be converted to benzoic acid [Eq. (15)] by reaction

$$PhCl + CO + H_2O \rightarrow PhCOOH + HCl \qquad (15)$$

with carbon monoxide and water,[9] though the conditions used were somewhat severe (typically 450 bar CO and 300°C). The yield of this reaction was found to be greatly improved by the addition of bases such as sodium acetate, but vigorous conditions were still required.

More recently it has been shown that aryl halides such as chloronaphthalenes or substituted bromobenzenes may be carbonylated in high yield under much milder conditions (100°C, 1 atm) by using a dipolar aprotic solvent such as DMSO or DMF, an insoluble base such as calcium hydroxide, and nickel tetracarbonyl as catalyst. These reactions exhibit autocatalytic behavior due to the liberation of halide ions, and a mechanism involving halogenotricarbonyl nickel anions $[NiX(CO)_3]^-$ was proposed to account for this effect.[10]

In view of the extreme toxicity and volatility of nickel tetracarbonyl, it seems unlikely that this material will ever become a common reagent in the organic laboratory, and the recent development of carbonylation techniques based on palladium, rhodium, and cobalt catalysts is therefore of great significance. Palladium, unlike nickel, forms no simple, volatile carbonyl, and it is conveniently introduced into reactions as a ligand derivative such as the readily available,[11] air-stable complex $(PPh_3)_2PdCl_2$.

A versatile procedure for the palladium-catalyzed conversion of benzyl, aryl, vinyl, or heterocyclic halides to carboxylic acids under mild conditions has been described by Cassar and coworkers.[12] It involves one of the first applications of the phase transfer technique to transition metal catalysis. The carbonylation reaction is carried out by addition of the organic halide, triphenyl phosphine, $PdCl_2(PhCN)_2$, and a phase transfer catalyst such as $Bu_4^nN^+I^-$ to a two-phase system consisting of p-xylene and 50% aqueous sodium hydroxide, followed by stirring for several hours under carbon monoxide at a pressure of 1–10 atm. A great advantage of the two-phase system is that continuous extraction of the product into the aqueous phase (as a sodium salt) leaves the catalyst and starting material in the organic layer, thus allowing easy separation and effectively heterogenizing the homogeneous catalyst. Furthermore, this technique allows selective carbonylation of polyhalogenated aromatic compounds, so that p-dibromobenzene, for example, can be converted to p-bromobenzoic acid with high selectivity and in good yield. Such selectivity is possible because transformation of the first C–X group to a C–COO⁻ function enables the product to be rapidly extracted into the aqueous phase, preventing reaction of the second C–X group with the catalyst. The probable reactions involved in this system are shown in Scheme 4.

Scheme 4

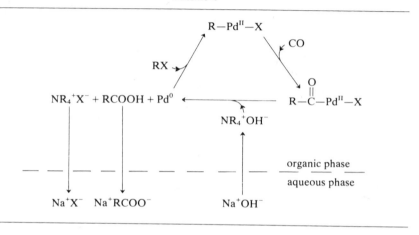

Preparation of p-Bromobenzoic Acid[12b]

Into a 100-cm^3 flask fitted with a mechanical stirrer and thermometer, under a constant 1-bar pressure of carbon monoxide, are introduced *p*-dibromobenzene (7.0 g), *p*-xylene (30 cm^3), aqueous NaOH (50% solution, 70 cm^3), PdCl$_2$(PhCN)$_2$ (0.1 g), Bu$_4^n$N$^+$I$^-$ (0.6 g), and triphenyl phosphine (1.2 g). The temperature is raised to 90°C and the mixture is stirred vigorously under carbon monoxide (1 atm) for 5 h. The phases are separated and the aqueous layer is acidified and extracted with diethyl ether. This extract is dried, filtered, and evaporated to dryness, yielding *p*-bromobenzoic acid (4.8 g, 77% based on *p*-dibromobenzene).

An analogous synthesis uses the tetracarbonylcobalt anion as catalyst, but this system is limited to substrates such as benzyl, and naphthylmethyl halides, which are susceptible to nucleophilic attack by the metal carbonyl anion. A phase transfer system can again be used, and carbonylation occurs under essentially ambient conditions.[13] The Co(CO)$_4^-$ anion is generated in the aqueous phase by the action of base on Co$_2$(CO)$_8$, and is transferred to the organic phase as its tetraalkylammonium salt. Nucleophilic attack by this anion on, say, a benzyl halide, yields benzyl tetracarbonylcobalt, PhCH$_2$Co(CO)$_4$, which inserts CO to give PhCH$_2$COCo(CO)$_4$. Hydrolysis of the latter regenerates the carbonyl anion and affords phenylacetic acid, which is extracted into the aqueous phase as its sodium salt.

Preparation of β-naphthylacetic acid[13]

In a 125-cm³ conical flask are placed $5\,M$ aqueous NaOH (25 cm³), $PhNEt_3^+Cl^-$ (1.0 mM), benzene (25 cm³), β-(bromomethyl)-naphthalene (25 mM), and $Co_2(CO)_8$ (0.5 mM). This mixture is stirred vigorously under CO (1 atm) for 10–12 h at 25°C, centrifuged, and the aqueous layer separated. Acidification with dilute sulfuric acid gives β-naphthyl acetic acid in 64% yield.

Collman has shown that disodium tetracarbonyl ferrate, $Na_2Fe(CO)_4$, may be a useful stoichiometric reagent for the synthesis of a wide range of carbonyl-containing organic compounds.[14] This commercially available salt[15] can be alkylated by reactive halides such as bromoalkanes, and by tosylate esters, to give anionic alkyltetracarbonyl iron(0) derivatives, which yield carboxylic acids on oxidation with molecular oxygen or aqueous halogens.[16] it is thought that the oxidative cleavage reaction with, say, aqueous iodine, proceeds via the four-coordinate insertion product

$$[R-\overset{\overset{\textstyle O}{\|}}{C}-Fe(CO)_3]^-,$$ which reacts as shown in Scheme 5.

Scheme 5

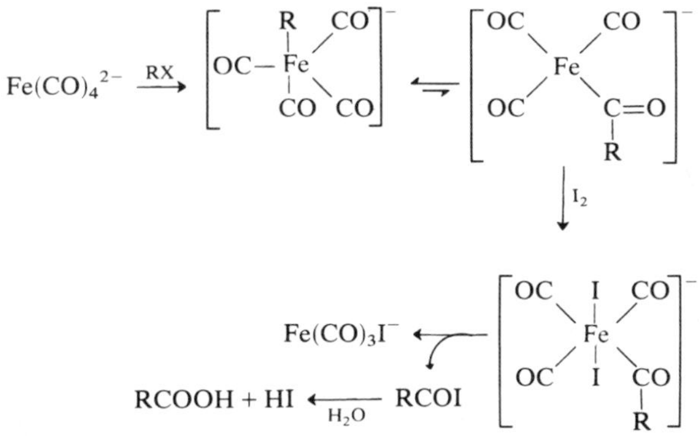

A cobalt catalyst, prepared by reaction of sodium hydride, sodium neopentoxide, and cobalt(II) acetate under carbon monoxide, allows the carbonylation of iodoarenes at 25°C and 1 atm pressure of CO, to give mixtures of benzoic acids and neopentyl benzoates in moderate yield.[17] The nature of the reaction is, however, uncertain, since the obvious product of reduction of a cobalt salt in the presence of carbon monoxide and base, i.e., the $Co(CO)_4^-$ ion, is normally unreactive toward halobenzenes.

6.2.1.3. Carboxylic Acids: From Diazonium Salts

Aryl diazonium salts react with nickel and iron carbonyls in the presence of water to give carboxylic acids in variable yield (4%–70% depending on conditions and substituents). A more reliable procedure, however, involves the direct carbonylation of diazonium salts in acetonitrile solution by carbon monoxide, with palladium acetate as catalyst.[18] The reaction also requires a molar equivalent (based on diazonium salt) of sodium acetate, and since no water is present in the system, and acids are obtained only after aqueous work-up, it seems certain that the initial product is in fact a mixed anhydride that is destroyed on hydrolysis (Scheme 6).

Scheme 6

$$ArN_2^+ + CO + CH_3CO_2^- \xrightarrow[-N_2]{Pd(OAc)_2} ArCO_2COCH_3 \xrightarrow{H_2O} ArCO_2H + CH_3CO_2H$$

The proposed reaction pathway involves reduction of the catalyst by carbon monoxide to a zero-valent palladium species that is arylated by the diazonium ion with loss of N_2. Insertion of CO into the aryl–palladium bond, followed by nucleophilic attack on the resulting acyl ligand by acetate ion, liberates the mixed anhydride with elimination of palladium in the zero-valent state (Scheme 7).

Scheme 7

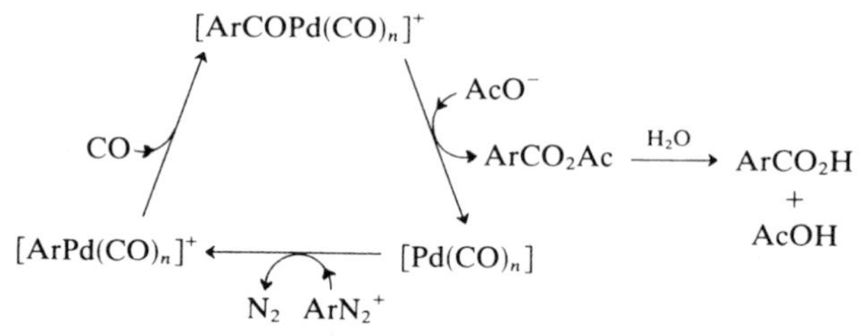

Yields are generally in the range 50%–90%, and the procedure appears substantially more reliable than the conventional method for conversion of diazonium salts to carboxylic acids, which involves the Sandmeyer reaction with cyanide ion to give nitriles that are subsequently hydrolyzed.

General Procedure for Carboxylation of Arenediazoium Salts [18]
The arenediazonium tetrafluoroborate (10 mM), sodium acetate (30 mM), palladium acetate (0.2 mM), and acetonitrile (60 cm³) are placed in a 300-cm³ autoclave cooled to 0°C under nitrogen. After replacement of nitrogen with carbon monoxide, 9 atm pressure of CO is introduced and the reaction mixture is stirred for 1 h at room temperature.

The autoclave is vented, carbon monoxide is replaced with nitrogen, and the solution is evaporated under reduced pressure. The residue is stirred for 1 h with 20 cm³ of 30% aqueous sodium hydroxide, and this mixture then treated with 40 cm³ of water and 50 cm³ of ether. The aqueous layer is separated and the organic layer washed once with aqueous sodium chloride. The combined aqueous solutions are filtered with active charcoal, and acidified to pH 1 with concentrated HCl. The acids formed are extracted three times with ether, and removal of the ether and acetic acid under reduced pressure gives pure aromatic carboxylic acid.

6.2.1.4. Carboxylic Acids: From Alkenes

The reactions of carbon monoxide with alkenes and water (hydrocarboxylations) are catalyzed by a variety of transition metal complexes including $Co_2(CO)_8$, $Ni(CO)_4$, and $H_2PtCl_6/SnCl_2$ [Eq. (16)].

$$R\diagdown + CO + H_2O \longrightarrow R\diagdown\!\!\!\!\diagdown^{CO_2H} \quad \text{and/or} \quad R\diagdown\!\!\!\diagup^{CO_2H} \tag{16}$$

In all cases, however, the actual catalytic species is thought to be a transition metal hydride, and the reaction proceeds via insertion of alkene into the metal–hydrogen bond [Eq. (17)], followed by insertion of carbon monoxide into the resulting metal–carbon bond [Eq. (18)], and finally cleavage of the acyl complex by water to give carboxylic acid and metal hydride [Eq. (19)].

$$HCo(CO)_4 + R\diagdown \longrightarrow R\diagdown\!\!\!\!\diagdown^{Co(CO)_4} \tag{17}$$

$$R\diagdown\!\!\!\!\diagdown^{Co(CO)_4} + CO \longrightarrow R\diagdown\!\!\!\!\diagdown\!\!\!\overset{O}{\diagup}^{Co(CO)_4} \tag{18}$$

$$R\diagdown\!\!\!\!\diagdown\!\!\!\overset{O}{\diagup}^{Co(CO)_4} + H_2O \longrightarrow R\diagdown\!\!\!\!\diagdown\!\!\!\overset{O}{\diagup}^{OH} + HCo(CO)_4 \tag{19}$$

The synthesis of carboxylic acids by this route, however, has not yet been developed into a generally useful procedure, since mixtures of products are often obtained due to the occurrence of both Markownikov and anti-Markownikov addition of the metal–hydride group to the alkene. For example, hydrocarboxylation of propene at 130°C/123 atm, in the presence of $HCo(CO)_4$, gave 64% butanoic acid and 20% 2-methyl propanoic acid as products.[19] Moreover, the insertion reaction leading to the metal alkyl intermediate is reversible, and this can result in alkene isomerization as shown in Scheme 8.

Scheme 8

$$RCH_2CH{=}CH_2 \underset{HCo(CO)_4}{\rightleftharpoons} RCH_2{-}\overset{\overset{\displaystyle CH_3}{|}}{CH}{-}Co(CO)_4 \rightleftharpoons RCH{=}CH{-}CH_3 + HCo(CO)_4$$

Hydrocarboxylation of the isomerized alkene could clearly give rise to further products, and indeed the cobalt-catalyzed reaction using 1-pentene at 145°C/180 atm gave a mixture of hexanoic acid (52%), 2-methyl pentanoic acid (17%), and 2-ethyl butanoic acid (5%),[20] the latter presumably arising from the carbonylation of 2-pentene.

Increased selectivity toward linear products may be obtained using a bimetallic catalyst containing chloroplatinic acid and tin(II) chloride.[21] A typical reaction is carried out in aqueous acetone at 90°C/200 atm, using a catalyst consisting of H_2PtCl_6 and $SnCl_2$ (1:5 mole ratio) and 1-dodecene as substrate. The yield of linear carboxylic acid is 68%, with a total of only 12% branched isomers.

The commercially available[22] zirconium hydride $(C_5H_5)_2Zr(H)Cl$ may be used as a stoichiometric reagent for conversion of alkenes to carboxylic acids,[23] since it is readily alkylated to give *linear* zirconium alkyls that insert CO under mild conditions (1–2 atm) forming acyl complexes. A remarkable feature of this reagent is that alkylation even by internal alkenes yields alkyls in which zirconium is bonded to a terminal carbon, presumably via the insertion–elimination–insertion sequence described above. Oxidative hydrolysis of the acyl zirconium complex gives carboxylic acids in high yield (Scheme 9), but full experimental details for the use of $(C_5H_5)_2Zr(H)Cl$ have yet to be published.

Since alkenes may be protonated to carbenium ions, the copper-promoted Koch synthesis (see above) can be applied to alkene hydrocarboxylation.[24] The conditions used are similar to those for alcohol

Scheme 9

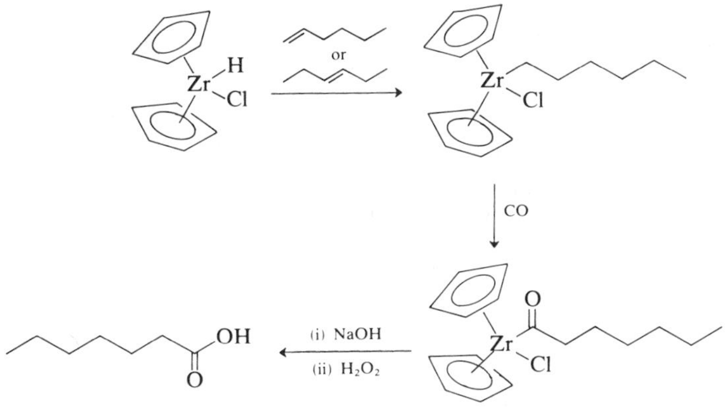

carbonylation (30°C, 1 atm CO, 98% H_2SO_4, Cu_2O catalyst), and the products again are invariably tertiary carboxylic acids due to isomerization of the intermediate carbenium ion. Cyclohexene, for example, yields only 1-methylcyclopentanecarboxylic acid [Eq. (20)].

Preparation of 1-Methylcyclopentanecarboxylic Acid[24]

In a 1-litre three-necked flask, fitted with a thermometer and a gas burette, are placed 2.86 g (0.02 mol) of Cu_2O and 105 cm³ of 98% sulfuric acid. The apparatus is evacuated and carbon monoxide introduced from the gas burette. The mixture is stirred vigorously for 40 min, and then cyclohexene (0.2 mol) added dropwise over 50 min. After 2 h the reaction mixture is poured into ice water and the products are extracted with benzene. Alkali extraction of the benzene solution, followed by acidification of the aqueous phase and reextraction into benzene gives 1-methylcyclopentanecarboxylic acid in 63% yield.

Equation (20) can be extended to include *saturated* hydrocarbons as substrates, provided these possess a tertiary carbon atom.[25] In this case,

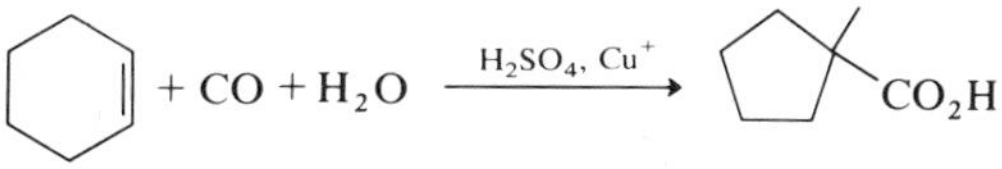

$$
\text{(cyclohexene)} + CO + H_2O \xrightarrow{H_2SO_4, \, Cu^+} \text{(1-methylcyclopentanecarboxylic acid)} CO_2H \qquad (20)
$$

addition of a mixture of 1-alkene and saturated hydrocarbon to the reaction results in formation of a secondary carbenium ion which, instead of isomerizing, abstracts hydride from the tertiary carbon atom of, say, methyl-cyclohexane, to give a tertiary carbenium ion, and it is this cation that undergoes carbonylation (Scheme 10).

Scheme 10

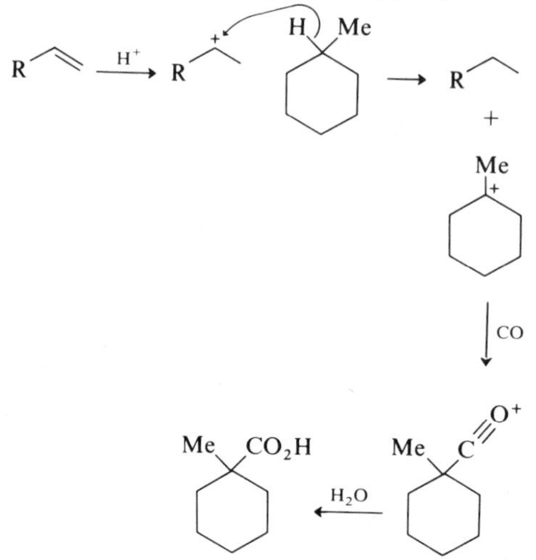

6.2.1.5. Carboxylic Acids: From Alkenes and Organic Halides

Since alkenes can insert into metal–carbon σ-bonds, it has proved possible to devise catalytic syntheses in which a carboxylic acid is assembled by the following sequence of reactions: (i) oxidative addition of an organic halide to the transition metal catalyst, (ii) insertion of an alkene into the metal–carbon σ-bond, (iii) insertion of carbon monoxide to give an acyl complex, and (iv) hydrolysis of the acyl ligand with elimination of the catalyst (Scheme 11).

Scheme 11

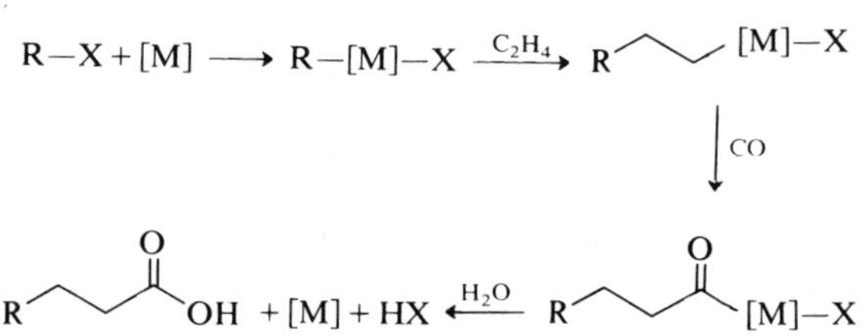

Allylic halides have been particularly studied in this respect,[26] and a typical example involves the reaction of 1-chloro-2-butene with ethylene and water in the presence of a stoichiometric amount of nickel tetracarbonyl [Eq. (21)].

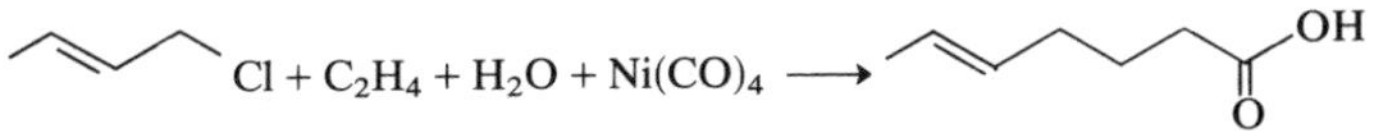

$$\text{Cl} + C_2H_4 + H_2O + Ni(CO)_4 \longrightarrow \qquad (21)$$

The yield of 5-heptenoic acid is ~40% and by-products, formed in small quantities, include 3-heptenoic acid, cyclopentanoic acid, and neutral coupled materials derived from the allylic halide. With higher alkenes, however, cyclization intervenes to give a cyclopentanone derivative that is subsequently carbonylated and hydrolyzed to a carboxylic acid (Scheme 12). Yields are, however, extremely low (3%–4%).

Scheme 12

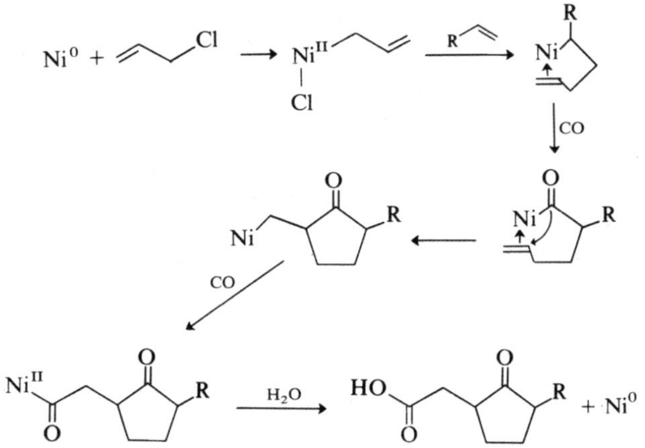

Unhindered 1,5-dienes react similarly[26] to give oxocyclopentyl-acetic acids, but yields in this case are ~35%. Moreover, since 1,5-hexadiene can be generated *in situ* by nickel-promoted coupling of allyl chloride [Eq. (22)], it is unnecessary to add the diene separately [Eq. (23)].

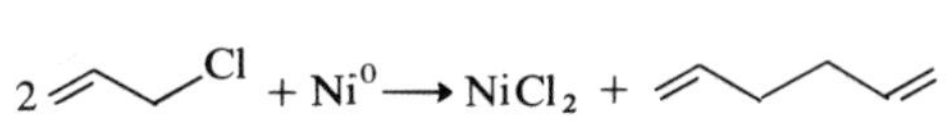

$$2 \qquad \text{Cl} + Ni^0 \longrightarrow NiCl_2 + \qquad\qquad (22)$$

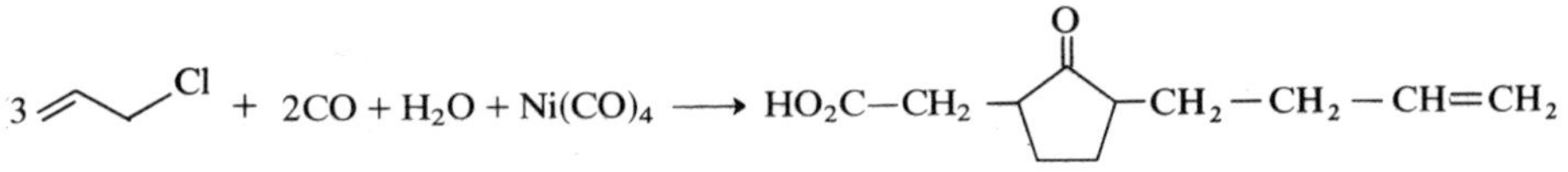

$$3 \qquad \text{Cl} + 2CO + H_2O + Ni(CO)_4 \longrightarrow HO_2C-CH_2 \qquad -CH_2-CH_2-CH=CH_2$$

$$(23)$$

6.2.1.6. Carboxylic Acids: From Alkynes

Alkynes react readily with stoichiometric amounts of nickel tetracarbonyl under essentially ambient conditions to give, in the presence of aqueous acid, good yields of *cis-α,β*-unsaturated carboxylic acids [Eq. (24)].[1,27] The acid promoter apparently serves to generate a nickel hydride species that is able to "insert" alkyne.

$$CO + RC\equiv CR' + H_2O \xrightarrow[H^+]{Ni(CO)_4} \underset{COOH}{RC=CHR'} + \underset{COOH}{RCH=CR'} \tag{24}$$

With monosubstituted alkynes the reaction rate decreases with increasing size of the substituent group, and with alkyl substituents the branched product isomer, from apparent "Markownikov" addition to the triple bond, is generally predominant.

The reaction may also be carried out catalytically in the presence of $Ni(CO)_4$ at a temperature of 180–200°C and a total pressure (alkyne + CO) of ~60 atm.[28] Other catalysts, e.g., $Fe(CO)_5$ and $RhCl_3$, are active for this synthesis, but yields and selectivities are usually lower than with nickel tetracarbonyl. The nickel-catalyzed process is used industrially for the large-scale manufacture of acrylic acid,[29] though other routes to this material, such as the autoxidation of propene, are now more competitive.

With $Co_2(CO)_8$ as catalyst, double hydrocarboxylation occurs at 80–100°C and 100–200 atm total pressure, so that succinic acid, for example, is obtained from acetylene in up to 80% yield [Eq. (25)].[30] At lower pressures acrylic acid is again formed.

$$HC\equiv CH + 2CO + 2H_2O \xrightarrow[100°C, 100\ atm]{Co_2(CO)_8,\ THF} HOOCCH_2CH_2COOH \tag{25}$$

A general synthesis of allenic acids involves the stoichiometric [$Ni(CO)_4$] carbonylation of α-chloroalkyl alkynes in the presence of aqueous sodium acetate and acetic acid (Scheme 13).[31] Although yields are often only fair,

Scheme 13

$$R'C\equiv C-CR_2Cl + CO + H_2O \xrightarrow{Ni(CO)_4} \left[\underset{COOH}{R'C=CH-CR_2Cl} \right] \xrightarrow{-HCl} \underset{COOH}{R'C=C=CR_2}$$

owing to the relative instability of the products, the selectivity achieved compares favorably with more conventional methods, which require the reaction of a propargyl Grignard reagent with carbon dioxide.

6.2.1.7. Carboxylic Acids: From Alkynes and Organic Halides

The nickel-catalyzed carbonylation of allylic halides in the presence of alkynes and water affords 2,5-dienoic acids in good yield under very mild conditions [Eq. (26)].[26,32]

$$\text{\textasciitilde}Cl + HC \equiv CH + CO + H_2O \xrightarrow{Ni(CO)_4} \text{\textasciitilde} \overset{O}{\underset{}{\underset{}{}}} OH + HCl \qquad (26)$$

As in the corresponding reaction with alkenes (see above) oxidative addition of the allyl halide to the catalyst is followed by successive insertions of alkyne and carbon monoxide, and finally by hydrolysis. The carbon–carbon double bond derived from alkyne insertion is thus conjugated with the carbonyl group, and generally has the *cis* configuration. The very small amounts of *trans* isomer that are sometimes observed are thought to arise by subsequent isomerization of the *cis* product. Other by-products include cyclopentanone derivatives, alkenes, coupling products from the allyl halide, and phenols. The latter presumably are formed by ring closure of the primary products (2,5-dienoic acids), since these are known[33] to cyclize readily in the presence of acids [Eq. (27)], and hydrogen chloride is a

$$\overset{OH}{\underset{}{}} \xrightarrow{H^+} \overset{OH}{\underset{}{}} + H_2O \qquad (27)$$

coproduct of the synthesis. This carbonylation reaction in fact provides a convenient route to certain substituted phenols that may be difficult to obtain by other means.

Alternative starting materials such as allylic alcohols, ethers, and esters may also be used in this reaction in place of allyl halides, but benzyl and alkyl derivatives fail to give unsaturated acids.[26] The intermediacy of a π-allyl nickel complex in the catalytic cycle is thus strongly implied.

6.2.1.8. Carboxylic Acids: From Aldehydes and Amides

A remarkable reaction, discovered by Wakamatsu,[34] involves the cobalt-catalyzed carbonylation of aldehydes in the presence of primary amides, which gives *N*-acyl amino acids in high yield [Eq. (28)].

$$RCHO + R'CONH_2 + CO \xrightarrow{Co_2(CO)_8/H_2} \overset{NHCOR'}{\underset{}{RCHCOOH}} \qquad (28)$$

A recent study[35] indicates that the reaction occurs rapidly at 100°C under 60 atm of CO, using ethyl acetate as solvent and a 2:1 mole ratio of acetamide to 1-butanol. Hydrogen (40 atm) should also be present to ensure the maintenance of a suitable concentration of $HCo(CO)_4$ (the active catalyst), and under these conditions the yield of N-acetyl norvaline is >90% after 20 min.

Although the mechanism of the Wakamatsu synthesis is far from certain, it seems likely that $HCo(CO)_4$ reacts with the hemi-amidal **1** (Scheme 14) to give an alkyl–cobalt complex that can insert CO. Since hydrogenolysis of the resulting acyl complex to aldehyde is not observed,

Scheme 14

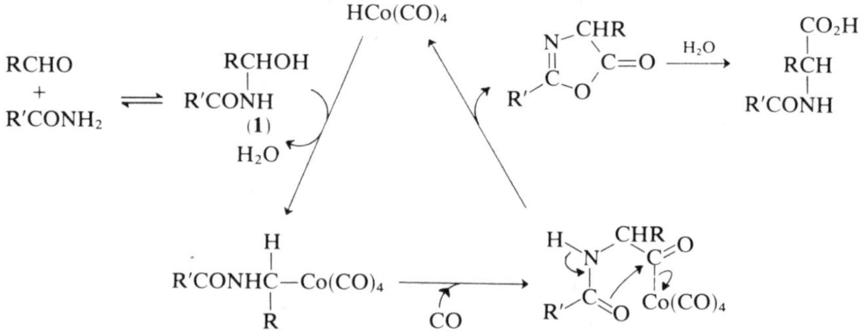

it has been suggested that rapid cyclization occurs, yielding an oxazolone that is subsequently hydrolyzed (Scheme 14).

6.2.2. Esters

6.2.2.1. Esters: From Alcohols

The carbonylation of alcohols can give rise to carboxylic acids by catalytic insertion of carbon monoxide into a carbon–oxygen bond (Section 6.2.1.1), but the alternative reaction, in which CO inserts into the oxygen–hydrogen bond giving a formate ester, also occurs under appropriate conditions. Formate esters are prepared on an industrial scale by the simple base-catalyzed reaction of carbon monoxide with alcohols in the presence of the corresponding sodium alkoxide [Eq. (29)], though transition metal

$$ROH + CO \xrightarrow{RO} HCOOR \tag{29}$$

catalysts such as $Co_2(CO)_8$ will bring about the latter reaction under neutral conditions.[36]

Carbonylation of alcohols in the presence of oxygen and a palladium catalyst affords dialkyl oxalates in moderate yield.[37a] A dehydrating agent such as an orthoformate or orthoacetate ester is necessary to remove the water formed in this reaction [Eq. (30)] and a copper cocatalyst is required

$$2ROH + 2CO + \tfrac{1}{2}O_2 \xrightarrow[\text{67 atm}]{\text{PdCl}_2/\text{CuCl}_2} (CO_2R)_2 + H_2O \tag{30}$$

to promote reoxidation of Pd(0) to Pd(II). The principal by-products are carbonate esters, carbon dioxide, and esters arising from oxidation of the solvent alcohol.

A related process from the patent literature[37b] involves the palladium-catalyzed carbonylation of alkyl nitrites to oxalate esters, with elimination of nitric oxide. The nitric oxide is then used, with oxygen, in a separate step to convert an alcohol to the corresponding nitrite, so that the overall reaction is that shown in Eq. (30).

6.2.2.2. Esters: From Ethers

Ethers may be carbonylated directly to esters [Eq. (31)] in the presence of a variety of homogeneous and heterogeneous transition metal catalysts,[38]

$$ROR + CO \rightarrow RCOOR \tag{31}$$

including $Ni(CO)_4$, Ni metal, Co/SiO_2, and NiI_2/SiO_2. The reaction is promoted by hydrogen halides, and, as in the conversion of methanol to acetic acid (Section 6.2.1.1), it is thought that the starting material is cleaved by the acid promoter to give, in this case, alkyl halide and alcohol. Catalytic carbonylation of the former gives an acyl halide or acyl–metal complex that reacts with the alcohol to give an ester [Eqs. (32)–(34)].

$$ROR + HX \rightleftharpoons RX + ROH \tag{32}$$

$$RX + CO \rightarrow RCOX \tag{33}$$

$$RCOX + ROH \rightarrow RCOOR + HX \tag{34}$$

Since relatively vigorous conditions are required (typically 200°C, 200 atm) the synthesis is generally complicated by further reaction of the product ester to give an anhydride [Eq. (35)], and a mixture of products is thus obtained.

$$RCOOR + CO \rightarrow R-\overset{O}{\overset{\|}{C}}-O-\overset{O}{\overset{\|}{C}}-R \tag{35}$$

Small-ring cyclic ethers such as trimethylene oxide and tetrahydrofuran react with carbon monoxide under similar conditions to give lactones in moderate yield [Eq. (36)], though here too further carbonylation can occur.

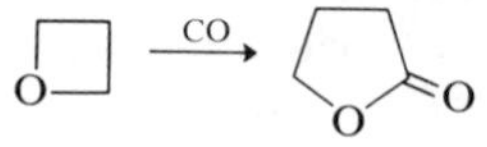

$$\text{(image)} \xrightarrow{\text{CO}} \text{(image)} \tag{36}$$

The reaction of THF with carbon monoxide has been studied[39] as a potential route to adipic acid (Scheme 15) (an intermediate in the manufacture of nylon 6.6).

Scheme 15

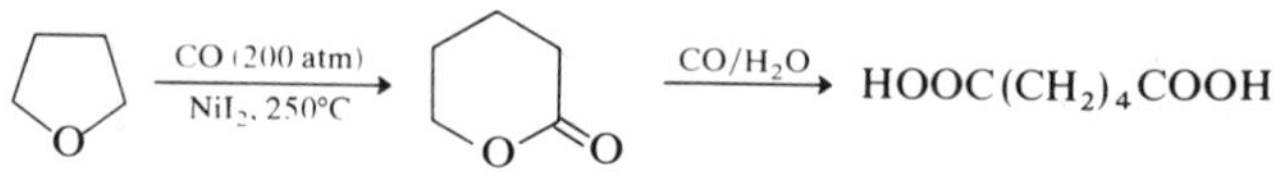

$$\text{(image)} \xrightarrow[\text{NiI}_2,\ 250°\text{C}]{\text{CO (200 atm)}} \text{(image)} \xrightarrow{\text{CO/H}_2\text{O}} \text{HOOC(CH}_2)_4\text{COOH}$$

6.2.2.3. Esters: From Epoxides

Carbonylation of ethylene oxide[40] or propylene oxide[41] in methanol solution with $Na^+Co(CO)_4^-$ as catalyst affords methyl 3-hydroxypropionate or methyl 3-hydroxybutyrate, respectively, in moderate yield. Conditions are not unduly severe (65°C, 130 atm), but the reaction [Eq. (37)] does not seem to have been generally exploited as a route to β-hydroxy esters.

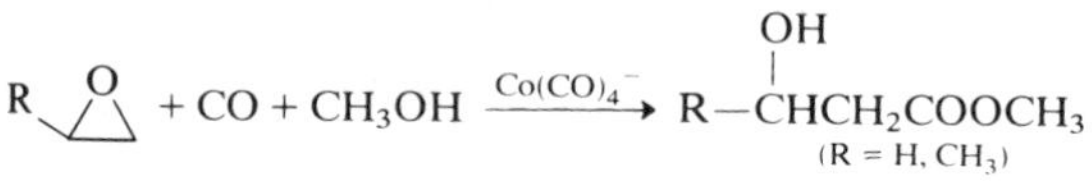

$$R\text{-(epoxide)} + CO + CH_3OH \xrightarrow{Co(CO)_4^-} \underset{(R = H, CH_3)}{R-\overset{\overset{\displaystyle OH}{|}}{C}HCH_2COOCH_3} \tag{37}$$

6.2.2.4. Synthesis of Esters: From Halides

Alkyl halides may be converted to esters by reaction with carbon monoxide and an alcohol, in the presence of base, using $Na^+Co(CO)_4^-$ as catalyst.[42] As in the corresponding carboxylic acid synthesis, reactive substrates such as iodoalkanes and benzyl halides, which are susceptible to

$$RX + CO + R'OH \rightarrow RCOOR' + HX \tag{38}$$

nucleophilic substitution by the cobalt anion, require *very mild conditions* (25°C, 1 atm CO pressure), and only the simple replacement of halogen by an alkoxycarbonyl group is observed. With less reactive halides, however, higher temperatures are necessary and isomerization of the intermediate cobalt alkyl (Scheme 16) can lead to mixtures of products.

Scheme 16

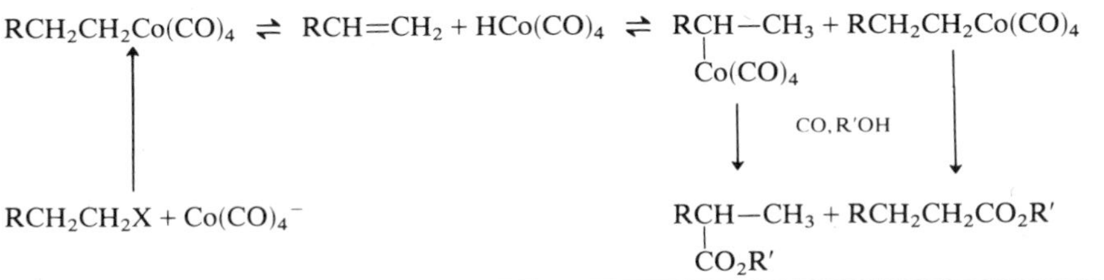
RCH2CH2Co(CO)4 ⇌ RCH=CH2 + HCo(CO)4 ⇌ RCH—CH3 + RCH2CH2Co(CO)4
Co(CO)4
CO, R'OH
RCH2CH2X + Co(CO)4−
RCH—CH3 + RCH2CH2CO2R'
CO2R'

Nickel tetracarbonyl is an efficient catalyst for the carboalkoxylation of aryl or vinyl halides,[43] which are generally inert toward nucleophilic reagents. The use of highly toxic $Ni(CO)_4$ for this type of reaction is, however, unnecessary, since it has been demonstrated by Heck and co-workers[44] that aryl and vinylic halides (particularly bromides and iodides) react with carbon monoxide and alcohols at 1 atm/60–100°C, in the presence of a stoichiometric amount of base and a palladium catalyst such as $(PPh_3)_2PdCl_2$ or $Pd(OAc)_2$, to give esters in high yield. The reaction is unaffected by the presence of a variety of functional groups (ester, ether, nitrile), and strongly electron-donating or -withdrawing substituents do not affect the yield of aromatic esters to any significant extent.

At 60–80°C, vinyl halides may be carbonylated with almost complete retention of stereochemistry at the double bond, but *cis–trans* isomerization and double-bond migration occur at higher temperatures, leading to mixtures of products. For example, at 60°C, carbonylation of E-3-iodo-3-hexene in *n*-butanol gave 74% of the corresponding E-carbobutoxylated product, with only 6% of the Z isomer, compared with yields of 69% E, 11% Z, and 19% of the nonconjugated ester when the reaction was carried out at 100°C (Scheme 17).

Scheme 17

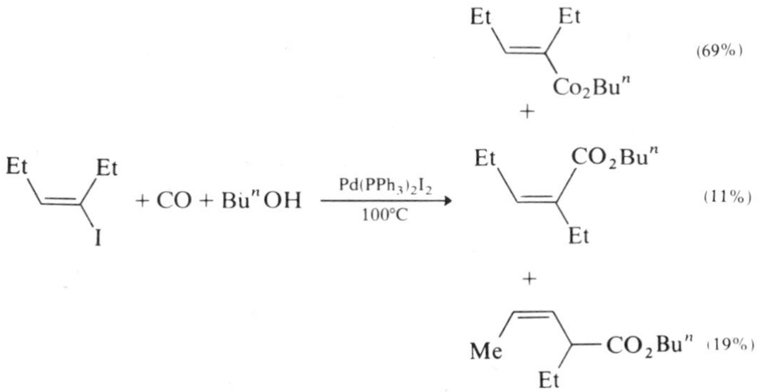
Et
Et
I
+ CO + Bu''OH
Pd(PPh3)2I2
100°C
Et
Et
Co2Bu''
(69%)
+
Et
CO2Bu''
Et
(11%)
+
Me
CO2Bu''
Et
(19%)

Table 6.1. Carbobutoxylation of Aryl and Benzyl Halides

Product [% yield bp °C]	Halide	Catalyst	Reaction time, h
$C_6H_5COOC_4H_9$ [70, 100–110 (10 mm Hg)]	C_6H_5I	$Pd(OAc)_2$	20
$4\text{-}CH_3OC_6H_4COOC_4H_9$ [63, 114–115 (0.2 mm Hg)]	$4\text{-}CH_3OC_6H_4I$	$Pd(OAc)_2$	16
$C_6H_5CH_2COOC_4H_9$ [45, 135–142 (22 mm Hg)]	$C_6H_5CH_2Cl$	$PdCl_2(PPh_3)_2$	40
$4\text{-}NCC_6H_4COOC_4H_9$ (89, mp 54–55)	$4\text{-}NCC_6H_4Br$	$PdBr_2(PPh_3)_2$	14

General Procedure for Ester Synthesis from Organic Halides[44]

Details of individual ester syntheses are given in Table 6.1. The general procedure is as follows: A Teflon cup containing 0.25 mM of the palladium catalyst is suspended in a 100-cm³ jacketed flask containing a magnetic stirring bar, and the flask then attached to a thermostatted microhydrogenation-type apparatus. The apparatus is flushed several times with carbon monoxide, and the reagents (17.2 mM of organic halide, 21.2 mM of *n*-butyl alcohol, and 19.0 mM of tri-*n*-butylamine) are injected into the flask through a side arm provided with a stopcock and rubber septum. The flask is equilibrated at the reaction temperature and 1 atm pressure of carbon monoxide, and the catalyst then added to the reaction mixture by means of a stopcock. When gas absorption ceases (420 cm³ at 25°C) the reaction mixture is cooled, dissolved in ether, and washed with several portions of 20% HCl, saturated sodium bicarbonate solution, and finally with distilled water. After drying over magnesium sulfate, the ether solution is concentrated and then distilled *in vacuo* to give the pure ester.

The palladium-catalyzed carbonylation of halides has been applied by Tsuji and coworkers to the synthesis of natural products such as Zearalenone[45] (Scheme 18) and Curvularin[46] (Scheme 19). The transformations shown were carried out at ~10-bar pressure of carbon monoxide and yields from the carbonylation steps were 70% in both cases.

This type of reaction has also been adapted for the preparation of four-, five-, and six-membered lactones [Eq. (39)]. Benzyl, allyl, aryl, and

$$\text{(39)}$$

Scheme 18

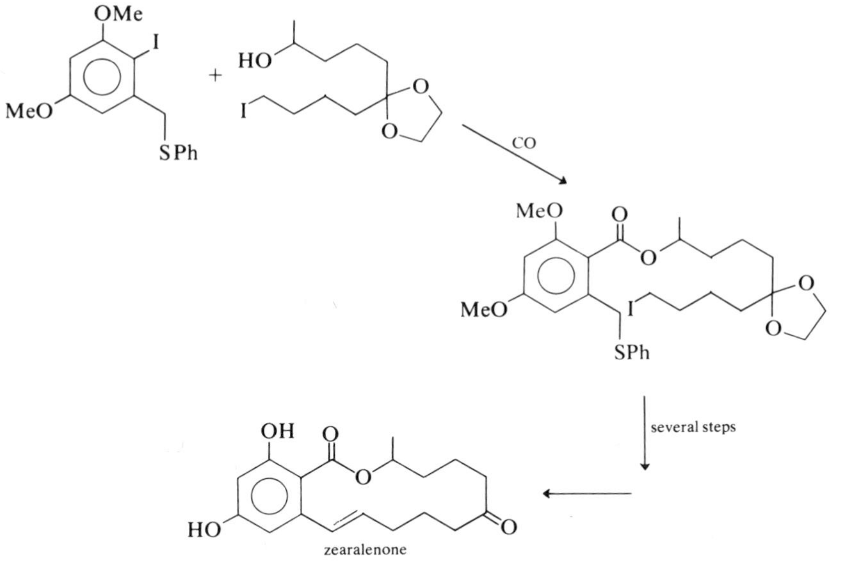

zearalenone

Scheme 19

curvularin

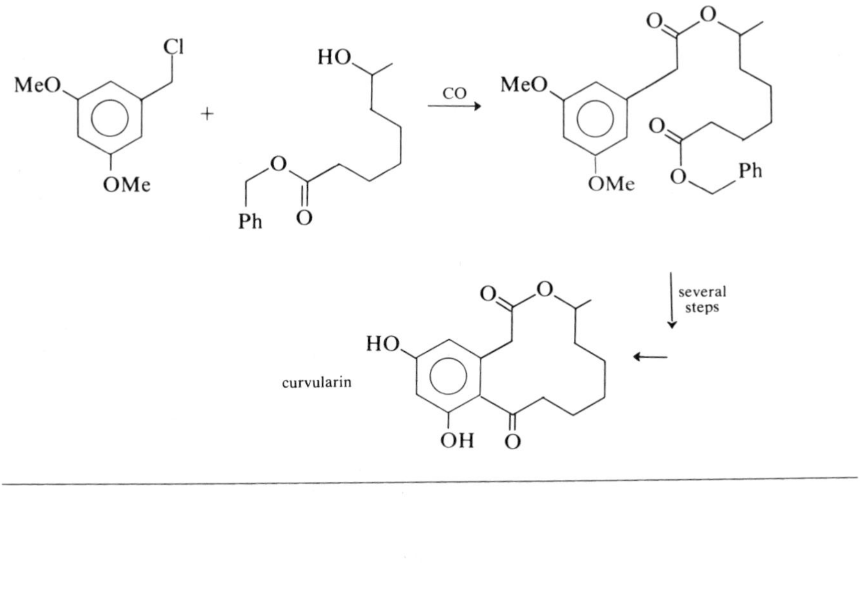

vinyl halides, containing primary, secondary, or tertiary alcohol groups, are readily carbonylated under mild conditions (1–4 atm, 25–60°C), with a high turnover of the palladium catalyst.[47] The proposed catalytic cycle is shown below (Scheme 20).

Scheme 20

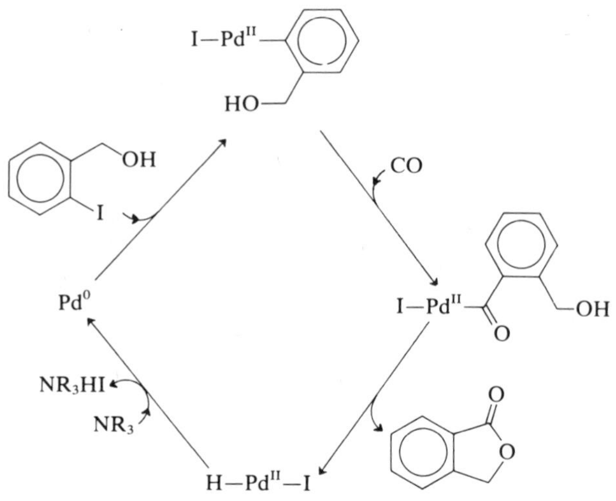

The lactone synthesis can also lead to $\Delta^{\alpha\beta}$ butenolides via carbonylation of suitable *cis*-vinyl iodides [e.g., Eq. (40)]. High yields of butenolide are

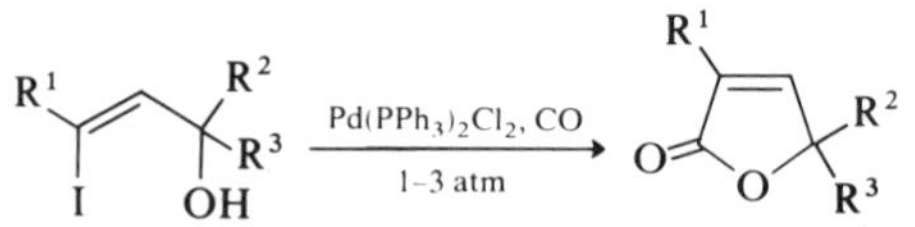

$$\text{(40)}$$

obtained when R^1 or R^2 = H or alkyl, but yields and selectivities to lactones are much lower when R^1 or R^2 = phenyl. Nevertheless, of the many reported syntheses of $\Delta^{\alpha\beta}$-butenolides, few have been applied to 3- or 3,5-substituted derivatives, and in many cases the palladium-catalyzed reaction gives improved yields with a saving in the number of synthetic steps required.

Table 6.2. Synthesis of Lactones by Carbonylation of Halo Alcohols[47]

Product, yield	Halo alcohol	Solvent	Temp., °C	Time, h
71%		THF	25	24
63%		DMF	25	24
52%		THF[a]	25	24
100%		DMF	60	72

[a] $(Ph_3P)_3Pd(CO)$ used as catalyst.

General Procedure for Synthesis of Lactones from Halo Alcohols (Details in Table 6.2)

Bis(triphenylphosphine)palladium(II)dichloride   (0.033–0.041 mM), potassium carbonate (2–12.5 mM), and the halo alcohol (2–12.5 mM) are added to a flask, which is stoppered with a serum cap, evacuated, and filled with carbon monoxide. THF or dmf (10 cm^3) is added by syringe, followed by a drop of hydrazine hydrate, and the mixture is heated to the desired temperature and stirred under a constant 1 atm of carbon monoxide for 24–72 h. The mixture is filtered, and when THF is the solvent it is removed under reduced pressure, the residue distilled, and the distillate recrystallized. When dmf is used the mixture is either fractionally distilled or else ether is added and the dmf extracted with water. The ether layer is then dried and the product either distilled or recrystallized.

Esters may be obtained from halides by using iron carbonyl derivatives as stoichiometric reagents. The carboxylic acid synthesis discussed earlier, in which an acyl halide is generated by halogenation of $Na^+[RFe(CO)_4]^-$ and hydrolyzed *in situ*, may be modified by using a solution of the halogen in an alcohol so that the acyl halide is esterified rather than hydrolyzed (Scheme 21).[16]

An alternative procedure, which allows the preparation of aromatic carboxylic esters [aryl halides do not generally react with $Na_2Fe(CO)_4$],

Scheme 21

$$RX + Na_2Fe(CO)_4 \rightarrow Na^+[RFe(CO)_4]^- \xrightarrow{I_2} RCOI \xrightarrow{R'OH} RCOOR'$$

---

involves the initial preparation of a Grignard reagent, followed by reaction with iron pentacarbonyl (Scheme 22). In this case an acyl iron intermediate is formed directly, and subsequent treatment with alcoholic halogen yields an ester precisely as in the above scheme.[48]

Scheme 22

$$ArX + Mg \rightarrow ArMgX \xrightarrow{Fe(CO)_5} Ar\overset{\overset{\displaystyle O}{\|}}{-}C-Fe(CO)_4 \xrightarrow[ROH]{I_2} ArCOOR$$

Preparation of Benzyl Benzoate[48]

    Into a THF solution (40 cm³) of phenylmagnesium bromide (22 mM), iron pentacarbonyl (3 cm³, 22 mM) is injected by syringe, and the mixture is stirred for 1 h at R.T. under N_2. A saturated solution of iodine in benzyl alcohol (10 cm³) is then added, and the solution stirred for a further hour under the same conditions. The reaction mixture is diluted with benzene, washed three times with aqueous sodium thiosulfate to remove excess iodine, dried, filtered, and the organic solvent removed *in vacuo*. The residual liquid is purified by column chromatography on silica gel and subsequent distillation. The yield of benzyl benzoate, identified by ir, nmr, ms, and gc, is ~78% based on the Grignard reagent.

6.2.2.5. Esters: From Alkenes

    The transition-metal-catalyzed reaction of carbon monoxide, an alcohol, and an alkene is often referred to as "hydroesterification," since it results in the addition of a hydrogen atom and an ester function to the double bond [Eq. (41)].

$$RCH{=}CH_2 + CO + R'OH \rightarrow RCH_2CH_2COOR' + RCH(CH_3)COOR' \tag{41}$$

    As in hydrocarboxylation, where the alcohol is replaced by water (Section 6.2.1.4), mixtures of products are generally obtained, so that, for example, the cobalt-catalyzed reaction of propene with CO and methanol typically yields methyl butanoate and methyl 1-methyl propanoate in a 3.5 : 1 mole ratio.[49] Hydroesterification is catalyzed by $Co_2(CO)_8$, palladium metal and its salts, platinum salts, and by complexes such as $(PPh_3)_2PdCl_2$ and $(AsPh_3)_2PtCl_2$. The catalytic cycle is thought to be initiated by formation of a metal hydride complex such as $(PPh_3)_2Pd(H)Cl$, followed by alkene insertion to give a metal alkyl, which itself inserts CO yielding an acyl complex. Alcoholysis of the latter then regenerates the starting hydride, with formation of free ester (Scheme 23).

Scheme 23

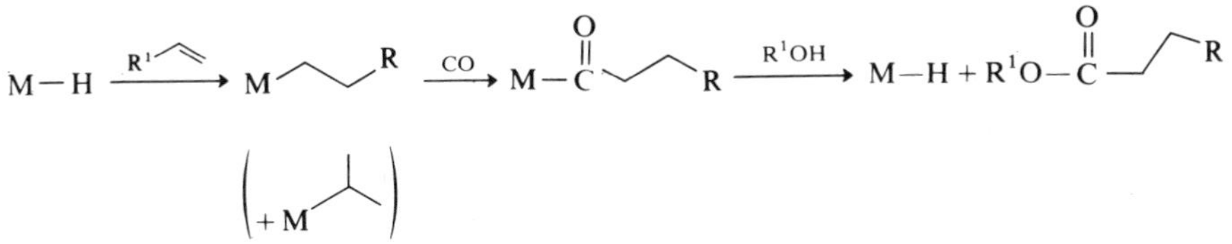

Convincing evidence for this pathway has been provided by Toniolo *et al.*, who isolated an intermediate butanoyl complex, *trans*-$(PPh_3)_2Pd$-$(COCH_2CH_2CH_3)Cl$, from the $(PPh_3)_2PdCl_2$-catalyzed reaction of propene with carbon monoxide in 1-butanol.[50] This acyl complex was characterized crystallographically and was shown to be an active hydroesterification catalyst giving yields and selectivities identical to those achieved with the original catalyst precursor.

Improved selectivity to linear ester is possible using mixed-metal catalysts. The combination of chloroplatinic acid and tin(II) chloride, mentioned earlier as a catalyst system for hydrocarboxylation (Section 6.2.1.4), may also be used for hydroesterification[21] and gives over 80% linear isomer in many cases. This approach has recently been extended by Knifton, who showed that terminal alkenes are hydroesterified at 80°C/200 atm in the presence of ligand-stabilized bimetallic catalysts containing platinum[51] or palladium,[52] and tin, and that the selectivity to linear ester can be extremely high (up to 98%). Useful platinum-based catalysts include $[(Ph_3As)_2PtCl_2/SnCl_2]$ and $[\{(PhO)_3P\}_2PtCl_2/SnCl_2]$, the highest yield of linear ester being obtained when the mole ratio Sn : Pt is around 10 : 1. A complex platinum hydride containing $SnCl_3^-$ ligands is thought to be the true catalyst, and the selectivity to linear products may be related to the high steric requirement of such ligands. For C_3–C_{20} linear 1-alkenes, the selectivity to linear ester increases with chain length, but the rate of reaction reaches a maximum at around C_7. Alkyl substituents at the double bond inhibit carbonylation, so that cyclic alkenes, for example, do not react, and the platinum-catalyzed reaction is limited essentially to unsubstituted 1-alkenes $RCH=CH_2$, though the group R may be linear or branched.

Analogous palladium-based catalysts such as $[(PPh_3)_2PdCl_2/SnCl_2]$ generally exhibit slightly lower selectivity to linear products, but they have the advantages of greater flexibility with respect to the structure of the alkene and lower cost.[52] Linear, branched, internal, and cyclic alkenes may all be carbonylated in the presence of such catalysts, and sterically hindered 1-alkenes, in which the substituent is at C_2 or C_3, provide excellent examples of regioselective carbonylation. The selectivity to linear ester in the conversion of 2-methyl-1-pentene to methyl 3-methyl hexanoate, for example, exceeds 99%, though the reaction is slow and conversions are typically only around 30%.

Synthesis of Methyl Octanoate from 1-Heptene[52]

Bis(triphenylphosphine)palladium(II)dichloride (0.5–20 mM) and tin(II) chloride dihydrate (2.5–20 mM) (optimum mole ratio 1:5) are added to an N_2-saturated mixture of methyl isobutyl ketone (75 cm³), methanol (5–15 cm³), and 1-heptene (50–200 mM). The mixture is stirred for 2–5 min to partially dissolve the solid catalyst, and the deep red liquid is transferred to an autoclave. The autoclave is sealed, deoxygenated with a purge of N_2, and heated to 80°C under 240 atm of carbon monoxide. After rocking the reactor at this temperature for 3–6 h, the apparatus is allowed to cool, and the clear reddish-brown liquid product recovered. Typical analytical data are as follows: 1-heptene conversion 95%, yield of methyl C_8-acid esters 88%, selectivity to linear methyl octanoate 88 mol %, material balance 97%.

The methyl C_8 acid esters may be recovered by fractional distillation *in vacuo*.

Conjugated dienes undergo hydroesterification in the presence of cobalt or palladium catalysts to give a number of products. With $Co_2(CO)_8$ [53] or $PdCl_2$ [54] as catalyst, the primary product of butadiene carbonylation (200 atm, 100°C) is an alkyl 3-pentenoate, but carbonylation of the second double bond may be achieved in the cobalt-catalyzed system by raising the temperature and CO pressure, to give dialkyl adipates in ~50% yield (Scheme 24). The palladium-catalyzed carbonylation of conjugated dienes.

Scheme 24

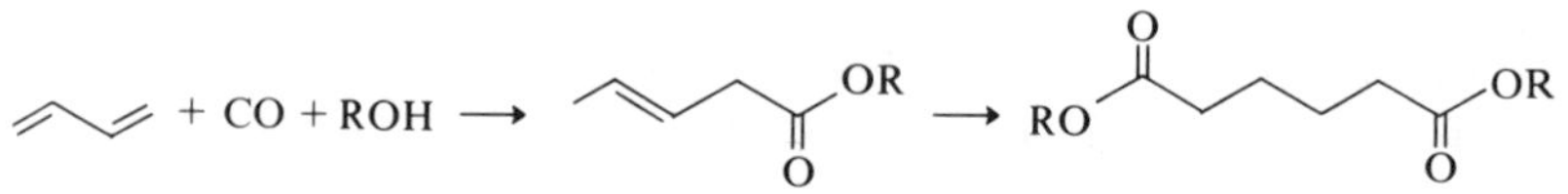

to nonconjugated unsaturated esters seems in fact to be a fairly general reaction, giving, e.g., 4-methyl-3-pentenoates from isoprene and 4-chloro-3-pentenoates from chloroprene.[54,55]

A very different type of reaction occurs, however, if *halide-free* palladium-phosphine catalysts (e.g., Pd(OAc)$_2$/PPh$_3$) are used. 3,8-Nonadienoate esters are now obtained in high yield [Eq. (42)].[56] These

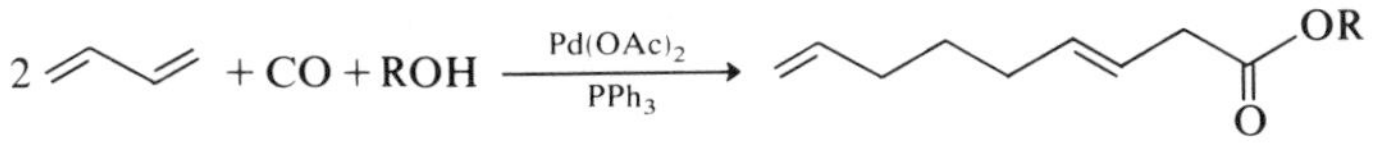

$$2 \diagup\!\!\!\diagdown\!\!\!\diagup + CO + ROH \xrightarrow[\text{PPh}_3]{\text{Pd(OAc)}_2} \diagup\!\!\!\diagdown\!\!\!\diagup\!\!\!\diagdown\!\!\!\diagup\!\!\!\diagdown\!\!\!\diagup\!\!\!\diagdown\text{OR} \qquad (42)$$

carbonylation reactions are thought to proceed via π-allylic intermediates, such as **2** and **3** for the monomeric and dimeric carbonylations, respectively.

(L = phosphine or halide bridge)

The role of halide in preventing dimeric carbonylation appears to be the blocking of a coordination site, thus preventing the binding of a second molecule of diene to the metal.

Synthesis of Isopropyl 3,8-Nonadienoate from Butadiene[56]

A mixture of palladium acetate (0.30 g), triphenylphosphine (0.70 g), and isopropanol (30 cm^3) are placed in a 200-cm^3 autoclave, and butadiene (20 g) and carbon monoxide (50 atm) are then introduced. The autoclave is shaken for 16 h at 110°C, cooled, and the product isolated by distillation (32.5 g, bp 72–76°C/1 mm Hg). The analogous ethyl ester (bp 108–112°C/18 mm Hg), methyl ester (bp 90–102°C/18 mm Hg), and *t*-butyl ester (bp 90–94°C/1 mm Hg) are similarly obtained.

Carbonylation catalysts normally isomerize unsaturated alcohols to aldehydes, but if hydrogen migration is blocked by *gem* disubstitution, then lactones can be obtained. Carbonylation of 2,2-dimethyl-3-buten-1-ol in the presence of HCo(CO)$_4$ thus yields a mixture of five- and six-membered lactones, corresponding to Markownikov and anti-Markownikov addition, respectively (Scheme 25).

Scheme 25

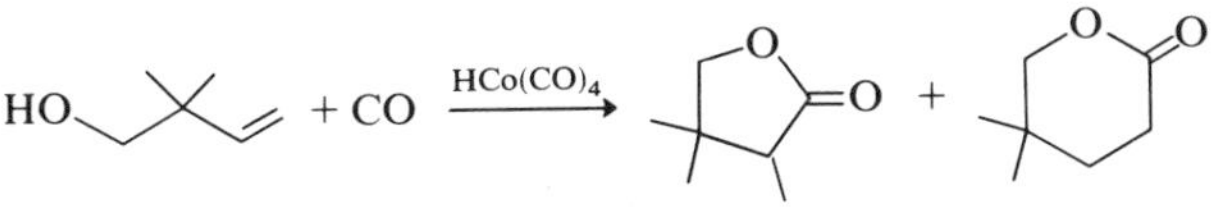

Stoichiometric hydroesterification may be carried out using the zirconium complex $(C_5H_5)_2Zr(H)Cl$, as described earlier for hydrocarboxylation (Section 6.2.1.4). In this case, however, the acyl intermediate is cleaved by a solution of bromine in methanol [Eq. (43)], giving linear ester in 50%–60% yield.[23]

$$\overset{CO}{\underset{}{\longrightarrow}}(C_5H_5)_2Zr(R)Cl \overset{alkene}{\longleftarrow} (C_5H_5)_2Zr(H)Cl$$

$$(C_5H_5)_2Zr(Cl)(COR) + \tfrac{1}{2}Br_2 + MeOH \rightarrow (C_5H_5)_2Zr(Br)Cl + RCO_2Me \qquad (43)$$

Palladium salts can bring about the *oxidative* carbonylation of alkenes, either in stoichiometric reactions or catalytically in the presence of copper(II) ions, which reoxidize metallic palladium to the divalent state.[57] Oxidative carbonylation is favored over simple hydroesterification by the presence of bases and by low temperature (25°C) and pressures (3–15 atm). The products may be α,β-unsaturated esters, dicarboxylic acid esters, or β-alkoxy esters [Eqs. (44)–(46)], which are thought to arise by initial attack of alkoxide on either coordinated carbon monoxide or on coordinated alkene.

$$R\diagup\diagup + CO + R'OH + Pd^{2+} \rightarrow R\diagup\diagdown\diagup\overset{O}{\underset{OR'}{\diagdown}} + 2H^+ + Pd \qquad (44)$$

$$R\diagup\diagup + CO + 2R'OH + Pd^{2+} \rightarrow R\diagup\overset{R'O}{\underset{}{\diagdown}}\diagdown\overset{O}{\underset{OR'}{\diagdown}} + 2H^+ + Pd \qquad (45)$$

$$R\diagup\diagup + 2CO + 2R'OH + Pd^{2+} \rightarrow R\diagdown\overset{R'O_2C}{\underset{}{\diagup}}-CO_2R' + 2H^+ + Pd \qquad (46)$$

$$(Pd + 2Cu^{2+} \rightarrow Pd^{2+} + 2Cu^+)$$

The application of this reaction to the synthesis of diesters has been studied extensively by Stille and coworkers, who have shown that, by using methanol as solvent, copper(II) chloride as reoxidant, and a sodium butyrate buffer, high yields of bis(carbomethoxylated) products may be obtained.[58] The reactions are generally carried out at 25°C and 1–6 atm and with cyclic alkenes give predominantly *cis*-disubstituted products. A range of functionally substituted alkenes (ketones, esters, alcohols) were found to give diesters by this route, but yields were somewhat variable. Conjugated dienes often gave mixtures of 1,4 addition products [Eq. (47)] in proportions that

$$CuCl_2 + \diagup\diagdown\diagup + CO + MeOH \overset{PdCl_2}{\longrightarrow} MeOCH_2CH=CHCH_2CO_2Me$$
$$+ ClCH_2CH=CHCH_2CO_2Me \qquad (47)$$
$$+ MeOCOCH_2CH=CHCH_2CO_2Me$$

depended on the reaction temperature and base concentration, but the carbonylation of 1,3-butadiene in benzyl alcohol gave dibenzyl *trans*-3-hexene-1,6-dioate exclusively in 90% yield.

The synthesis of *unsaturated* esters by oxidative carbonylation of alkenes has not been widely investigated, but it has recently been reported that methyl cinnamate can be prepared by carbonylation of styrene in methanol at 28°C, using a catalyst system containing $PdCl_2$ and $MgCl_2$, sodium acetate as buffer,[59] and a stoichiometric quantity of copper(II) chloride as reoxidant. The formation of diester is minimized by working at a partial pressure of CO of only 0.2 atm (CO/N_2 mixture) and over 100 moles of product may be obtained per mole of $PdCl_2$.

An alternative route to α,β-unsaturated esters has been described by Heck, which involves the initial preparation of a carboalkoxymercuric halide from CO, and alcohol, and a mercuric halide [Eq. (48)].

$$ROH + CO + HgCl_2 \rightarrow Cl-Hg-\overset{\overset{\text{O}}{\|}}{C}-OR + HCl \qquad (48)$$

The carbalkoxy group is then transferred to palladium in the presence of an alkene, and an insertion–elimination pathway yields the product ester and palladium metal (Scheme 26).[60]

Scheme 26

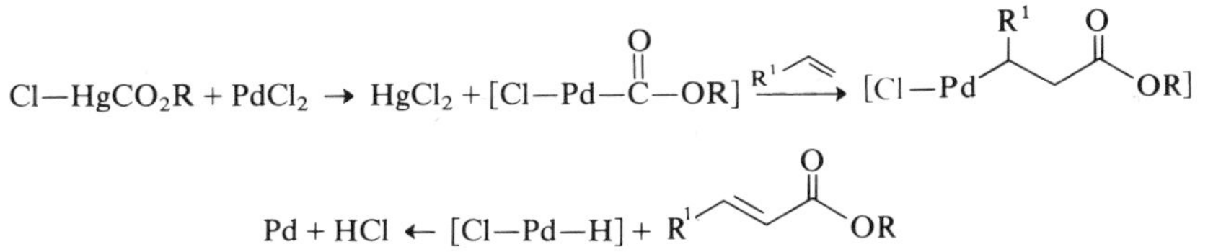

$$Cl-HgCO_2R + PdCl_2 \rightarrow HgCl_2 + [Cl-Pd-\overset{\overset{\text{O}}{\|}}{C}-OR] \xrightarrow{R^1 \diagdown} [Cl-Pd \diagup R^1 \diagdown OR]$$

$$Pd + HCl \leftarrow [Cl-Pd-H] + R^1 \diagdown OR$$

A remarkable example of oxidative carbonylation leading to a β-alkoxy ester is provided by the reaction of carbon monoxide with the norbornadiene complex of $PdCl_2$ in methanol at 25°C/1 atm.[61] The product ester, 3-*endo*-methoxycarbonyl-5-*exo*-methoxynortricyclene, is thought to arise by the reaction sequence shown in Scheme 27.

β-Lactones are sometimes obtained by oxidative carbonylation of alkenes in the presence of water, so that reaction of the cyclooctadiene complex $[(1,5-C_8H_{12})PdCl_2]$ with carbon monoxide, in aqueous sodium

Scheme 27

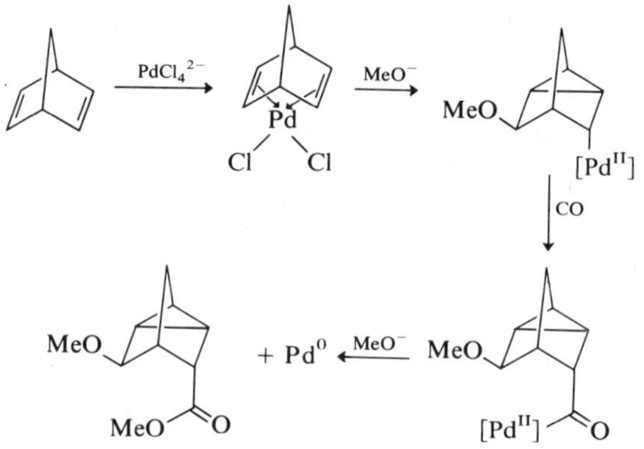

acetate, produced the lactone **4** in 79% yield [Eq. (49)].[62] Ethylene was similarly converted to β-propiolactone by carbonylation in aqueous

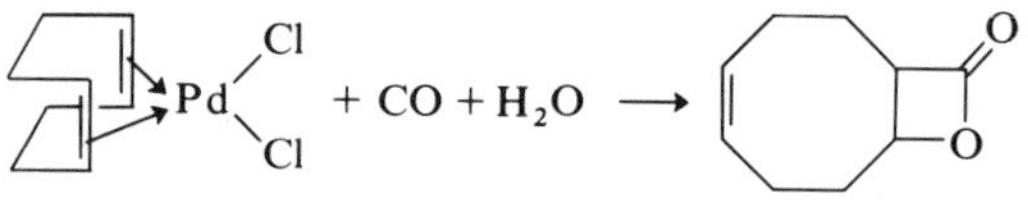

$$\text{(structure)} \ Pd \begin{array}{c} Cl \\ Cl \end{array} + CO + H_2O \longrightarrow \text{(structure with } O \text{)} \tag{49}$$

4

acetonitrile at $-20°C$, using a catalytic amount of $PdCl_2$ and a stoichiometric quantity of copper(II) chloride [Eq. (50)].[63]

$$CH_2{=}CH_2 + PdCl_2 + H_2O + CO \longrightarrow \text{(structure)} + 2HCl + Pd \tag{50}$$

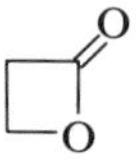

6.2.2.6. Esters: From Alkynes

Alkynes undergo simple hydroesterification under mild conditions to give α,β-unsaturated esters in the presence of nickel- or palladium-based catalysts, and this reaction has in the past been applied to the industrial synthesis of acrylate and methacrylate esters.[64] Addition of the ester function to terminal alkynes frequently occurs at the substituted carbon atom (Markownikov addition) to give branched products [Eq. (51)], but linear

$$CH_3C{\equiv}CH + CO + MeOH \xrightarrow{\text{Ni(CO)}_4/\text{HCl}} \begin{array}{c} CO_2Me \\ | \\ CH_3{-}C{=}CH_2 \end{array} \tag{51}$$

esters may be obtained with high selectivity[65] using the ligand-stabilized palladium–tin catalysts described earlier (Section 6.2.2.5) for alkene hydroesterification.

By using the anionic nickel complexes $[Ni(CO)_3X]^-$ (X = halogen) as catalysts, a dimerisation/carbonylation reaction, yielding pentadienoate esters from acetylene, may be achieved at atmospheric pressure in a mixture of dipolar aprotic solvent (e.g., dmf) and methanol [Eq. (52)]. No methyl acrylate is formed.[66]

$$2HC{\equiv}CH + CO + MeOH \xrightarrow{[Ni(CO)_3I]^-} H_2C{=}CH{-}CH{=}CH{-}CO_2Me \qquad (52)$$

The palladium-catalyzed carbonylation of 3-alkyn-1-ols gives rise to α-methylene lactones under mild conditions and in high yield [Eq. (53)],[67]

$$(53)$$

and this appears to be a useful route to bicyclic α-methylene-γ-butyrolactones, some of which possess significant pharmacological activity.

An analogous reaction with 4-pentyn-1-ol gave only traces of the corresponding six-membered lactone, but by using a substrate containing a fused-ring system in which ethynyl and hydroxy groups are fixed in the appropriate geometry, good yields of δ-lactone were obtained [Eq. (54)].[68]

$$(54)$$

Alkyne carboxylic acid esters undergo hydroesterification in the presence of $PdCl_2$ and HCl to give a range of products whose distribution depends on the concentration of hydrogen chloride and on the pressure of carbon monoxide. High concentrations and pressures favor dicarbonylation with formation of saturated esters so that, for example, the ethanetetracarboxylate ester (**4**) is a major product of the reaction shown in Eq. (55) at 100 atm and 25°C.[69]

$$RO_2CC{\equiv}CCO_2R + CO + ROH \xrightarrow{PdCl_2/HCl} (RO_2C)_2CH{-}CH(CO_2R)_2$$

4

$$+ (RO_2C)_2C{=}CH(CO_2R)$$

5

$$+ (RO_2C)_2CHCH_2CO_2R \qquad (55)$$

The unsaturated triester **5** is a product of oxidative carbonylation, and with acetylene itself the latter reaction is catalyzed cleanly in high yield by the palladium chloride–thiourea system. Dimethyl maleate, for example, is formed in >90% yield merely on passing a mixture of acetylene and carbon monoxide at 20°C into a methanol solution containing $PdCl_2$ and $(NH_2)_2CS$ [Eq. (56)].

$$HC\equiv CH + 2CO + 2MeOH + PdCl_2 \rightarrow MeO\overset{O}{\underset{}{\diagdown}}\underset{}{\diagup}\overset{O}{\underset{}{}}OMe + Pd + 2HCl \qquad (56)$$

Addition of oxygen to the system allows the reaction to be performed catalytically, even in the absence of conventional reoxidants such as copper(II) salts.[26]

Dimethyl Maleate[26]

Acetylene, carbon monoxide, and air (volume ratio 24.5:68:7.5) are passed at 20°C and atmospheric pressure into methanol (50 ml) at a rate of 6 liters/h for 40 h, while palladium(II) chloride (90 mg) and thiourea (116 mg) dissolved in methanol (80 ml) are added. The following products are obtained: dimethyl maleate (3.62 g), dimethyl fumarate (0.08 g), dimethyl *cis,cis*-muconate (0.15 g), its *cis, trans* and *trans,trans* isomers (0.13 g), and higher products (0.60 g).

The stoichiometric reaction was developed further by Heck[70] into a reasonably general route to substituted maleate esters (*cis*-products are almost invariably formed), though strongly polarizing substituents on the alkyne have a substantial deactivating effect. Heck's system requires equimolar amounts of $PdCl_2$ and $HgCl_2$ but can be made catalytic in some cases by using copper(II) chloride and oxygen to reoxidize the palladium.

An unexpected example of oxidative carbonylation, in which the product disproportionates and thus behaves as its own oxidizing agent, is provided by the carbonylation of diphenylacetylene in methanol at 100°C/100 atm, which gave diphenylcrotonolactone (60%) and dimethyl diphenylmaleate (34%) as shown in Eq. (57).[71] The oxidizing agent in this

$$PhC\equiv CPh + CO + MeOH \xrightarrow{PdCl_2/HCl} \underset{O}{\overset{Ph \quad Ph}{\diagup\diagdown}} + \underset{Ph}{\overset{Ph \quad CO_2Me}{\diagup\diagdown}}CO_2Me \qquad (57)$$

system appears to be a carbomethoxy group which is reduced to a hydroxy-methyl group, and subsequent cyclization gives the observed lactone.

Lactones are also formed in the cobalt-catalyzed reactions of carbon monoxide with alkynes in aprotic solvents [Eq. (58)], though yields are

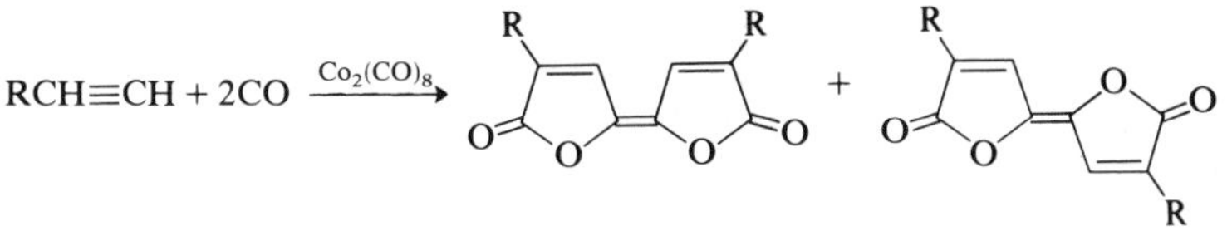

$$RCH{\equiv}CH + 2CO \xrightarrow{Co_2(CO)_8} \qquad + \qquad \tag{58}$$

variable and conditions vigorous (typically 15%–60% yield; 300 atm/100°C).[72] This reaction represents one of the few authenticated cases of homogeneous catalysis in which more than one transition metal center is involved, since two intermediate dicobalt complexes, **6** and **7**, have been isolated (Scheme 28) and characterized unambiguously by X-ray diffraction.[73,74]

Scheme 28

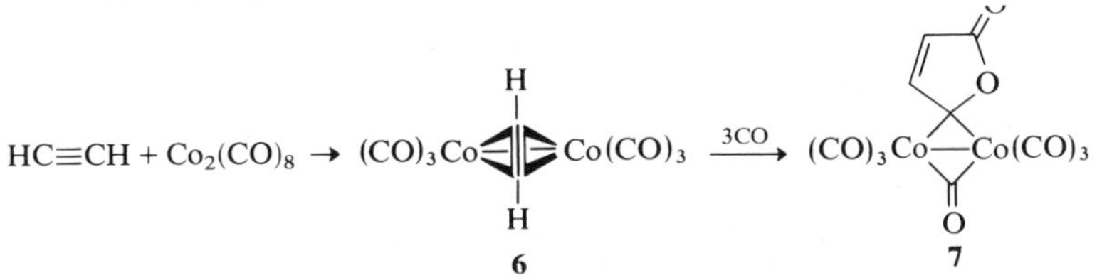

$$HC{\equiv}CH + Co_2(CO)_8 \rightarrow (CO)_3Co{\Longleftrightarrow}Co(CO)_3 \xrightarrow{3CO} (CO)_3Co\text{---}Co(CO)_3$$

6   **7**

Alkyne carboxylic acid esters can be prepared in good yield from terminal alkynes by reaction of their lithium salts with iron pentacarbonyl (Scheme 29).[75]

Scheme 29

$$RC{\equiv}CH \xrightarrow{BuLi} RC{\equiv}CLi \xrightarrow{Fe(CO)_5} [RC{\equiv}C{-}\overset{O}{\overset{\|}{C}}{-}Fe(CO)_4]^-Li^+ \xrightarrow[MeOH]{I_2} RC{\equiv}C{-}\overset{O}{\overset{\|}{C}}{-}OMe$$

6.2.2.7. Esters: From Alkynes and Organic Halides

Formation of a metal–carbon bond by reaction of a transition metal complex with an organic halide, followed by successive insertions of carbon monoxide and an alkyne, is a potentially valuable route to esters and lactones. As in the corresponding reactions leading to carboxylic acids, allyl–nickel complexes show a strong tendency to insert alkyne before CO, and the sequence of reactions shown in Scheme 30 leads to a catalytic

Scheme 30

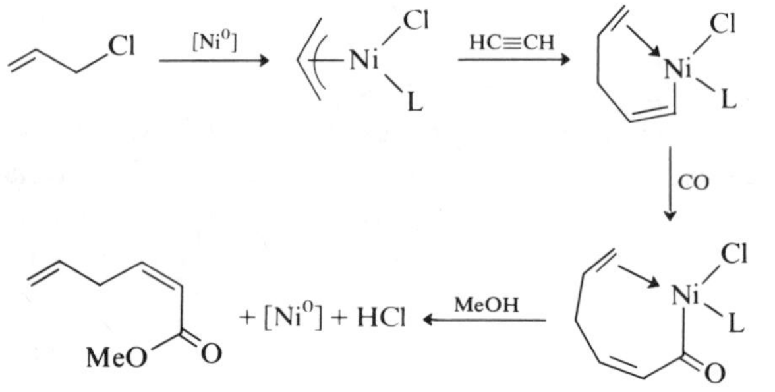

synthesis of *cis*-methyl-2,5-hexadienoate from allyl chloride, acetylene, carbon monoxide, and methanol under ambient conditions.[76] The catalyst system consists of nickel chloride, thiourea, magnesium oxide (to neutralize the HCl formed by the reaction), and a manganese–iron alloy for initial reduction of Ni(II) to Ni(0).

The reverse insertion sequence (CO *before* acetylene) is observed when a phenyl–nickel complex [from Ni(CO)$_4$ and PhI] forms the starting point for this type of synthesis. Thus reaction of iodobenzene, nickel tetracarbonyl, acetylene, methanol, and carbon monoxide at 30 atm and 130°C, affords methyl β-benzoyl propanoate in high yield,[77] most probably by the series of reactions shown in Scheme 31.

Scheme 31

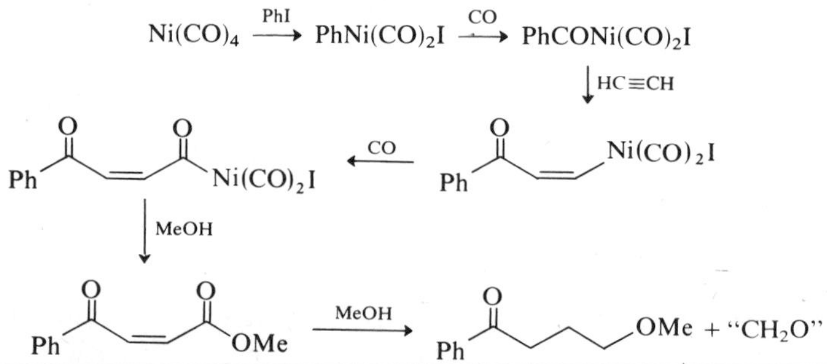

A catalytic synthesis of 2,4-pentadieno-4-lactones (α-methylene butenolides) is possible from halogenoacetate esters, alkynes, and carbon monoxide, in the presence of a stoichiometric amount of tertiary amine and a tetracarbonylcobalt salt as catalyst [Eq. (59)].[78]

$$BrCH_2CO_2Me + EtC{\equiv}CEt + CO + NR_3 \xrightarrow{Co(CO)_4^-} \cdot \quad \text{(structure)} \qquad (59)$$

The penultimate compound in the catalytic cycle leading to such lactones has been isolated and shown to be a π-allyl cobalt complex (**8**)

8

that eliminates free lactone on treatment with base. This complex is, however, stable to base in the absence of proton-activating groups such as carbomethoxy or cyano, and the reaction is thus not applicable to the synthesis of simple α-methylene butenolides.

Synthesis of 2,3-Diethyl-5-carboethoxy-2,4-pentadiene-4-lactone[78]
A 100-cm^3 reaction flask is filled with carbon monoxide at 27°C, and in it are placed 30 cm^3 of a 0.07 M solution of NaCo(CO)$_4$ in ether, 2 cm^3 of dicyclohexylethylamine, 2 cm^3 of 3-hexyne, and 2.5 cm^3 of 1M ethylbromo-acetate in ether. After stirring for $1\frac{1}{4}$ h a further 2.5 cm^3 of 1M ethylbromoace-tate in ether is added and the reaction is continued, under CO, overnight. The reaction mixture is poured into water, and the ether phase is washed with water, dilute HCl, and again with water. The solution is dried over magnesium sulfate, solvent removed *in vacuo*, and the product recrystallized from pentane at −80°C, and finally distilled (150°C/1 mm Hg) to give a pale yellow liquid, solidifying below room temperature. The uv spectrum in cyclohexane has λ_{max} at 274 nm (ε 28,500).

In a related synthesis, Alper and coworkers[79] describe the use of phase transfer techniques for a cobalt-catalyzed preparation of hydroxy-butenolides by reaction of iodomethane, water, phenylacetylene, and carbon monoxide under ambient conditions (Scheme 32).

Scheme 32

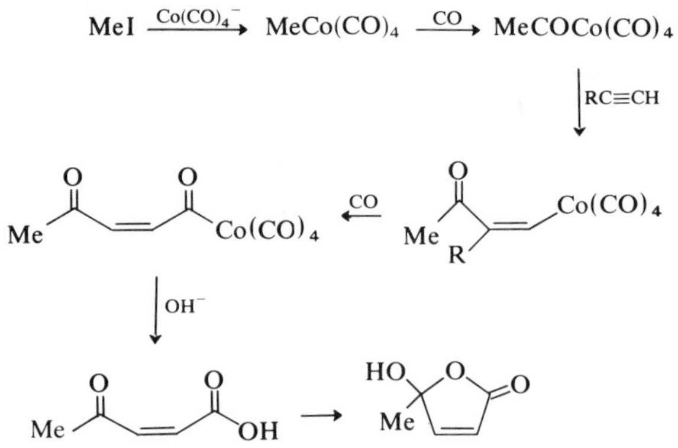

$$\text{MeI} \xrightarrow{\text{Co(CO)}_4^-} \text{MeCo(CO)}_4 \xrightarrow{\text{CO}} \text{MeCOCo(CO)}_4$$

6.2.2.8. Esters: From Organomercury Compounds

The direct carbonylation of organomercurials in the presence of alcohols is not a useful route to carboxylic esters, as extreme conditions are required and yields are generally poor.[80] This reaction can, however, be promoted by the addition of stoichiometric amounts of palladium salts to form organopalladium species that are much more readily carbonylated [Eq. (60)].

$$\text{RHgX} + \text{PdX}_2 + \text{CO} + \text{R}^1\text{OH} \rightarrow \text{RCOOR}^1 + \text{HgX}_2 + \text{Pd} + \text{HX} \qquad (60)$$

Both the transmetallation and carbonylation reactions proceed with retention of stereochemistry at carbon, but in practical terms only *vinyl* mercuric halides have been found to give good yields of ester with high specificity under mild conditions.[81] The starting materials are readily obtained by hydroboration–mercuration of the appropriate alkyne[82] and react with CO (1 atm), lithium tetrachloropalladite, and an alcohol at low temperature ($-20°C$). The choromercuri group is replaced stereospecifically by a carboalkoxy group, and α,β-unsaturated esters are obtained in near quantitative yield [Eq. (61)].

$$\text{R}\overset{}{\diagup}\diagdown\text{HgCl} + \text{Pd}^{2+} + \text{CO} + \text{R}^1\text{OH} \rightarrow \text{R}\overset{}{\diagup}\diagdown\text{CO}_2\text{R}^1 + \text{Pd} + \text{Hg}^{2+} + \text{HCl} \qquad (61)$$

This reaction is tolerant of a variety of functional groups (cyano, halogens, carboalkoxy) and can be made catalytic in palladium by addition of a stoichiometric amount of anhydrous CuCl_2 to reoxidize Pd(0) to Pd(II).

Larock and coworkers have recently extended this type of chemistry to the synthesis of butenolides[83] by direct mercuration of propargylic alcohols, followed by carbonylation as described above. The yields of mercurated intermediates are variable, but the carbonylation step is essentially quantitative (Scheme 33).

Scheme 33

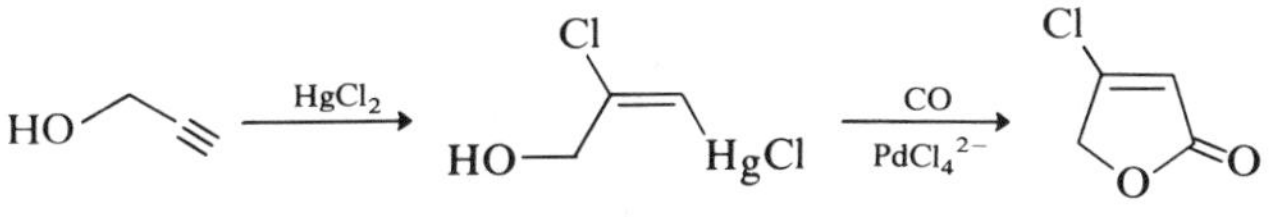

β-chloro-$\Delta^{\alpha,\beta}$-butenolide[83]

To 5 mM of palladium chloride (0.84 g), 100 mM of lithium chloride (4.25 g), and 100 mM of cupric chloride (13.45 g) in 250 cm^3 of ether at −78°C is added 50 mM of (E)-2-chloro-3-chloromercuri-2-propen-1-ol (16.4 g). The flask is flushed with carbon monoxide and a large balloon filled with carbon monoxide is connected to the top of the flask. The cold bath is removed and the reaction mixture stirred in a cold room (~5°C) for 50 h. Saturated ammonium chloride solution (10 cm^3) is added and the mixture stirred an additional hour. The resulting suspension is vacuum filtered, washed with saturated ammonium chloride, and dried over anhydrous sodium sulfate. Evaporation of the ether affords 6.18 g crude yield of product. Recrystallization from carbon tetrachloride provides a 78% isolated yield.

6.2.3. Preparation of Amides

6.2.3.1. Amides: From Amines and Carbon Monoxide

The transition-metal-catalyzed carbonylation of primary and secondary amines to formamides [Eq. (62)] has been known for many years, but

$$R_2NH + CO \rightarrow R_2N-CHO \tag{62}$$

generally requires severe conditions of temperature and pressure (200°C/300 atm).[84] Copper(I) chloride as catalyst[85] allows carbonylation of secondary amines under somewhat milder conditions (140°C/80 atm), but it has recently been found that ruthenium complexes such as $Ru_3(CO)_{12}$ catalyze this reaction at atmospheric pressure.[86] At 75°C the carbonylation typically gives 40%–50% conversion after 50 h, so that, although slow, this does seem to be a practical route to disubstituted formamides.

Ureas and oxamides have also been obtained by direct carbonylation of amines, though the product distribution varies markedly with the nature of the amine and catalyst. Aromatic amines, for example, give N,N'-diaryl ureas with $Ni(CO)_4$ as catalyst,[87] whereas aliphatic amines give almost exclusively N-aryl formamides under the same conditions. With $Mn_2(CO)_{10}$ as catalyst, however, aliphatic primary amines give dialkyl ureas with good selectivity,[88] while the corresponding palladium-catalyzed reaction yields dialkyl oxamides as the major products.[89]

6.2.3.2. Amides: From Halides

Reactions of organic halides with CO and amines, using stoichiometric quantities of $Ni(CO)_4$[43] or $Na^+Co(CO)_4^-$,[42] indicate that metal-promoted amidation is a feasible route to carboxylic amides [Eqs. (63) and (64)],

$$\text{Ph}\overset{\text{Br}}{\diagup}\diagup \;+\; \underset{\underset{N}{\overset{|}{N}}}{\bigcirc}\;+\;CO \xrightarrow{Ni(CO)_4} \;\text{Ph}\diagdown\diagup\overset{O}{\diagup}\diagdown N\bigcirc \qquad (63)$$
$$+$$
$$HBr$$

$$PhCH_2Cl + PhNH_2 + CO \xrightarrow{Co(CO)_4^-} PhCH_2CONHPh + HCl \qquad (64)$$

but a recently developed palladium-catalyzed synthesis[90] is particularly convenient and versatile. Thus, reactions of aryl, vinyl, or heterocyclic halides with primary or secondary amines and carbon monoxide (1 atm) at 60–100°C with $(PPh_3)_2PdCl_2$ as catalyst (~ 0.5 mol%) give high yields (57%–94%) of isolated, purified amides [Eq. (65)].

$$RX + CO + R^1NH_2 \xrightarrow{(PPh_3)_2PdCl_2} RCONHR^1 + HX \qquad (65)$$

A tertiary amine is generally added to neutralize the hydrogen halide formed in the reaction, though if the primary or secondary amine is a sufficiently strong base, the tertiary amine is unnecessary. Pyrrolidine, for example, does not require additional tertiary amine, whereas the more weakly basic aniline does. The function of the strong base may be to remove a proton from coordinated or uncoordinated amine, thereby increasing its nucleophilic character, but the detailed mechanism of the reaction is not well established.

The amidation reaction proceeds much more rapidly than the corresponding ester synthesis (Section 6.2.2.4), so that bromobenzene, for example, reacts with benzylamine and CO to give N-benzyl benzamide about 17 times faster than with 1-butanol and CO under the same conditions. Amidation also shows greater stereoselectivity than carboalkoxylation with

cis and *trans* vinylic halides, giving amides with essentially complete retention of configuration.

N-Phenylbenzamide[90]
    The carbonylation apparatus containing catalyst suspended in a Teflon
cup is flushed several times with carbon monoxide. A mixture of 2.7 g of
bromobenzene (17.2 mM) and 3.54 g of aniline (38.0 mM) is added to the
reaction vessel by means of a syringe. After equilibration at 100°C the catalyst,
$Pd(PPh_3)_2(Ph)(Br)$ (0.25 mM), is added by dropping the Teflon cup into
the stirred reaction mixture. 3.6 g of tri-*n*-butylamine (19.0 mM) is added
and the uptake of carbon monoxide proceeds rapidly. The mixture is cooled,
dissolved in diethyl ether (700 ml), and washed with several portions of 20%
hydrochloric acid solution to remove salts and excess amine. The extracts
are washed with several portions of distilled water. The ether layer is treated
with decolorizing carbon, dried with anhydrous magnesium sulfate, and
filtered. The solid product formed after concentration of the solution is filtered
and air-dried to give 2.89 g (94% yield) of *N*-phenylbenzamide, mp 162.5–
163°C.

    A facile synthesis of benzolactams, based on this palladium-catalyzed
amidation, involves the carbonylation of 2-bromo aminoalkyl benzenes at
100°C and 1 atm in the presence of $Pd(OAc)_2/PPh_3$. A stoichiometric
amount of tertiary amine is again included to remove the hydrogen halide
formed in the reaction [Eq. (66)], and five-, six-, and even seven-membered
lactams are obtained in good yield.[91]

$$\text{(66)}$$

    This type of cyclization has been extended to the synthesis of α-
methylene-β-lactams by carbonylation of 2-bromo-3-aminopropene
derivatives [Eq. (67)].

$$\text{(67)}$$

General Procedure for the Synthesis of Benzolactams[91]
    A mixture of *o*-halo aminoalkylbenzene (1 equiv.), *n*-Bu$_3$N (1.1 equiv.),
and $Pd(OAc)_2$ (0.04 equiv.) is added to a reaction vessel that is connected
to a balloon filled with carbon monoxide, and heated at 100°C for 26 h. After
cooling, ether is added to the solution and the ether layer is washed with
10% hydrochloric acid and dried over MgSO$_4$. The solvent is removed, and
the residue purified by chromatography or recrystallization.

6.2.3.3. Amides: From Alkenes

There has been little work on simple amide-forming analogs of the hydroesterification reaction, but a number of intramolecular ring-closure reactions of unsaturated amines with carbon monoxide have been described by Falbe and coworkers.[92] Allylamine, for example, was converted to 2-pyrrolidone in 60% yield in the presence of $Co_2(CO)_8$ under vigorous conditions (300 atm, 280°C), and a variety of N-alkyl allylamines reacted similarly [Eq. (68)].[93]

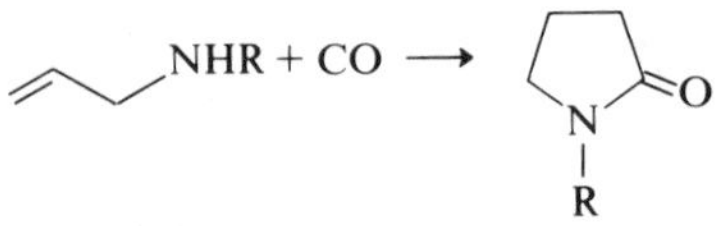

$$\text{\raisebox{0pt}{}} NHR + CO \longrightarrow \text{[2-pyrrolidone ring, N–R, C=O]} \qquad (68)$$

δ-Lactams are also accessible by this type of reaction, and indeed are sometimes formed, together with the expected γ-lactam, on carbonylation of allylic amines. Presumably initial isomerization occurs to give a γ,δ-unsaturated amine which subsequently cyclizes (Scheme 34). The cobalt-

Scheme 34

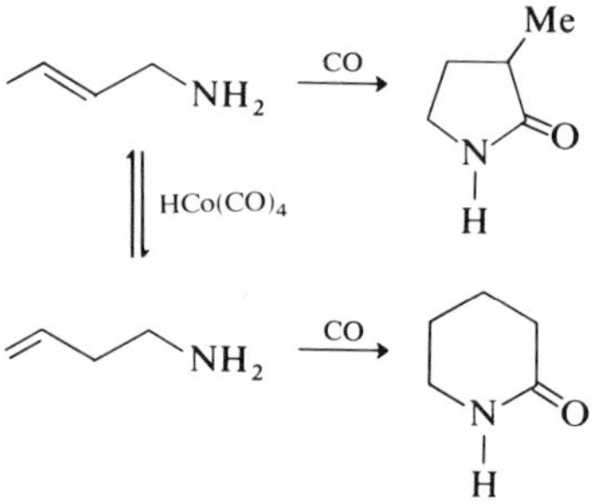

$$\text{CH}_2=\text{CH–CH}_2\text{–CH}_2\text{–NH}_2 \;\;\xrightarrow{\;CO\;}\;\; \text{[3-methyl-2-pyrrolidinone]}$$

$$\Big\updownarrow\; \text{HCo(CO)}_4$$

$$\text{CH}_2=\text{CH–CH}_2\text{–CH}_2\text{–NH}_2 \;\;\xrightarrow{\;CO\;}\;\; \text{[2-piperidinone]}$$

catalyzed carbonylation of 1-aminomethyl-3-cyclohexene thus yields a mixture of fused-ring γ- and δ-lactams [Eq. (69)].

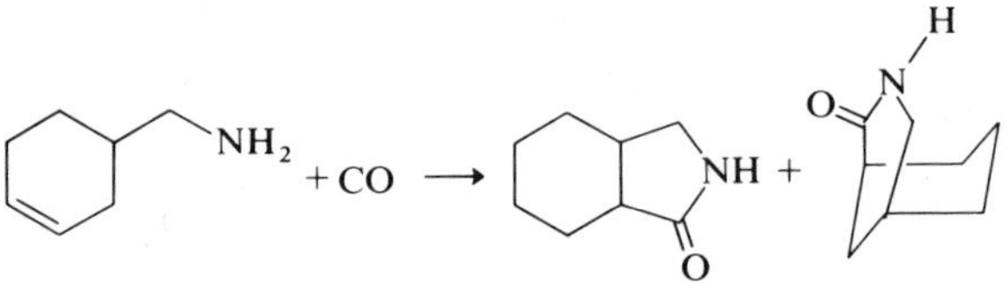

$$\text{[cyclohexene–CH}_2\text{NH}_2\text{]} + CO \longrightarrow \text{[fused bicyclic }\gamma\text{-lactam, NH, C=O]} + \text{[fused bicyclic }\delta\text{-lactam]} \qquad (69)$$

α,β-Unsaturated amides undergo a similar carbonylation–cyclization reaction to give succinimides in good yield,[94] though vigorous conditions are required (100–300 atm, 160–280°C). Isomerization again gives rise to a mixture of products in some cases, so that crotonamide, for example, yields both 3-methylsuccinimide (68%) and glutarimide (19%) (Scheme 35).

Scheme 35

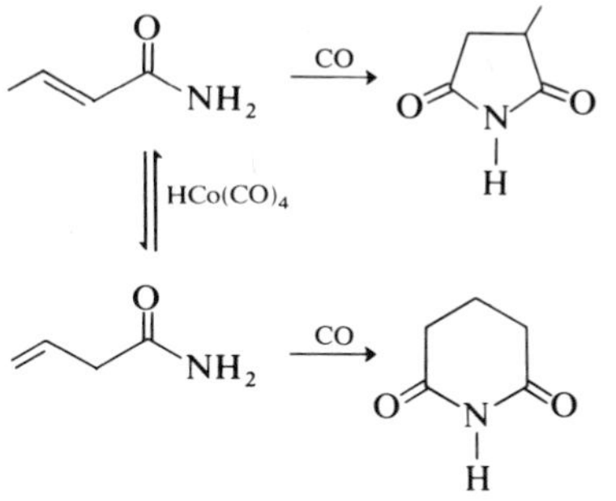

β-Lactams have been recently synthesized using $[(C_5H_5)Fe(CO)_2(\text{alkene})]^+$ complexes as stoichiometric reagents (Scheme 36).[95] The readily available propene complex **9** for example, adds benzyl-amine at $-25°C$ to give the ammonium salt (**10**) in high yield. This, on

Scheme 36

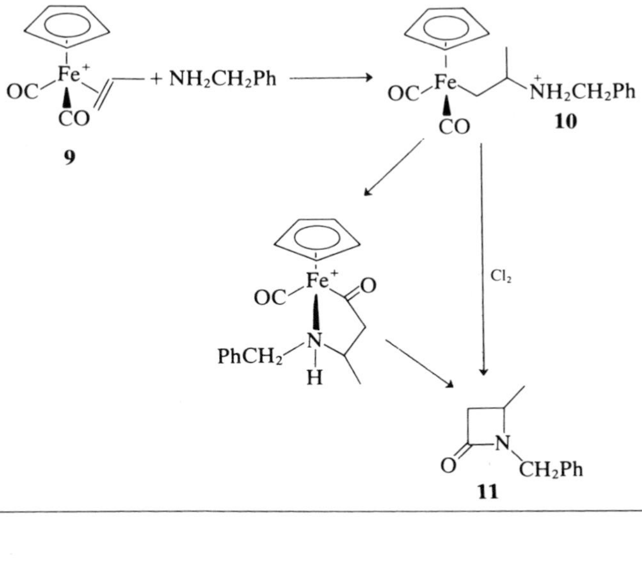

oxidation with chlorine in dichloromethane at −78°C, gives the β-lactam **11** in 47% yield. Oxidatively induced insertion of CO into the iron–alkyl bond is followed by chelation and subsequent ring closure.

Bicyclic β-lactams may also be obtained by this route, using amino-alkenes as starting materials [Eq. (70)], and this approach has been extended

$$CH_2{=}CH(CH_2)_nNH_2 + CO \rightarrow \qquad (70)$$

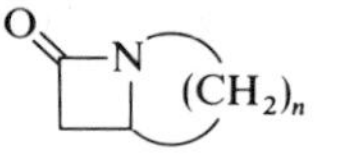

to the stereoselective synthesis of 2-methylcarbopenam, **12**, a model for β-lactam antibiotics (Scheme 37).[96]

Scheme 37

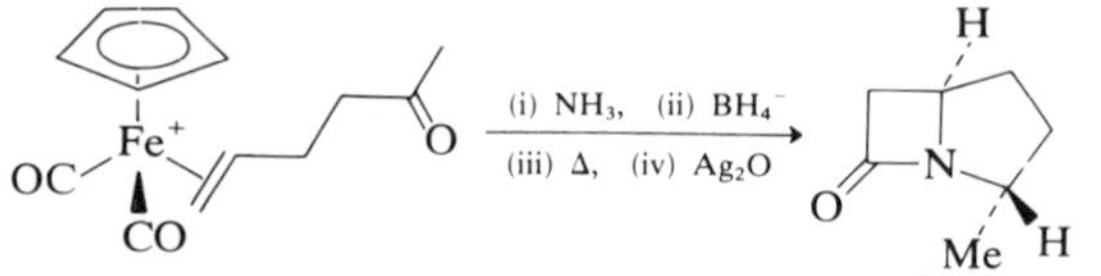

Synthesis of N-Methylsuccinimide from N-Methylacrylamide[92]

A solution of 54 g of *N*-methylacrylamide in 250 g of thiophene-free benzene is carbonylated in the presence of 7.5 g of $Co_2(CO)_8$ in a magnetically stirred autoclave at 200°C and under a CO pressure of 300 atm. After 4 h, the mixture is cooled to room temperature and the excess of carbon monoxide is vented. The reaction mixture is then refluxed for 1 h in a current of nitrogen to ensure complete decomposition of the cobalt catalyst, and the decomposition products are filtered off. Benzene is distilled off in vacuum, and the crude product is distilled over a column. Yield: 67.5 g (94%): bp 90–100°C/1–2 mm Hg; mp 68–69°C.

6.2.3.4. Amides: From Aromatic Amines, Imines, Azo Compounds, Oximes, and Hydrazones

A wide range of nitrogen-containing aromatic compounds yield amidic products on carbonylation in the presence of transition metal catalysts or reagents.[97] These reactions all depend on the phenomenon of "*ortho*-metallation," whereby an aromatic compound containing a potential donor substituent reacts with a transition metal complex or salt to give a chelated derivative containing a metal–aryl bond, *ortho* to the donor substituent.[98] Azobenzene, for example, reacts with $PdCl_2$ to give a dimeric, *ortho*-palladated complex (**13**), which is carbonylated to 2-phenyl-1-H-indazolone as shown in Scheme 38.

Scheme 38

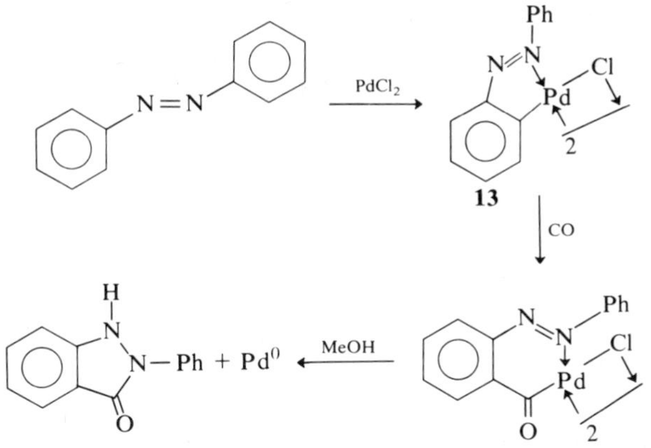

Such stoichiometric, palladium-promoted reactions have been studied in detail by Thompson and Heck,[99] who showed that a variety of cyclic compounds containing the $\left[\begin{array}{c} O \\ \parallel \\ -C-N< \end{array} \right]$ group may be obtained [Eqs. (71)–(73)], often in high yield under mild conditions (1 atm CO, 100°C).

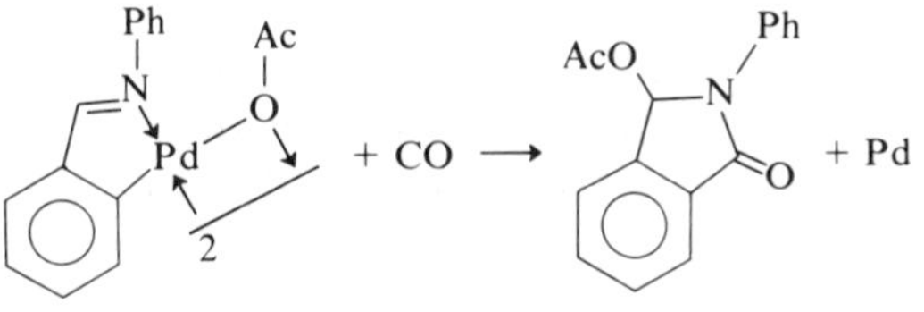

$$+ CO \longrightarrow \qquad + Pd \qquad (71)$$

$$+ CO \longrightarrow \qquad + Pd + AcOH \qquad (72)$$

$$+ CO \longrightarrow \qquad + Pd + AcOMe \qquad (73)$$

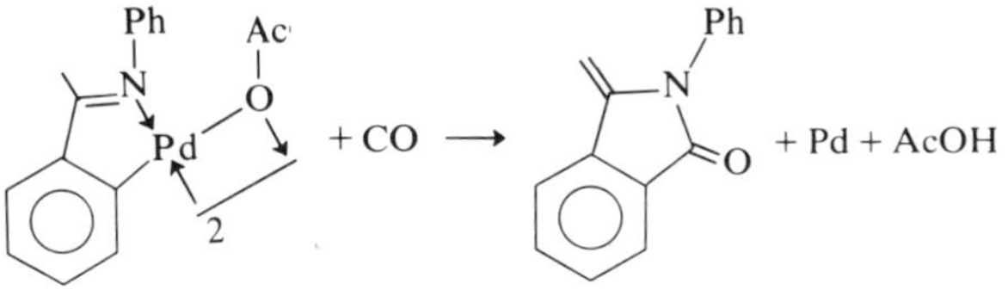

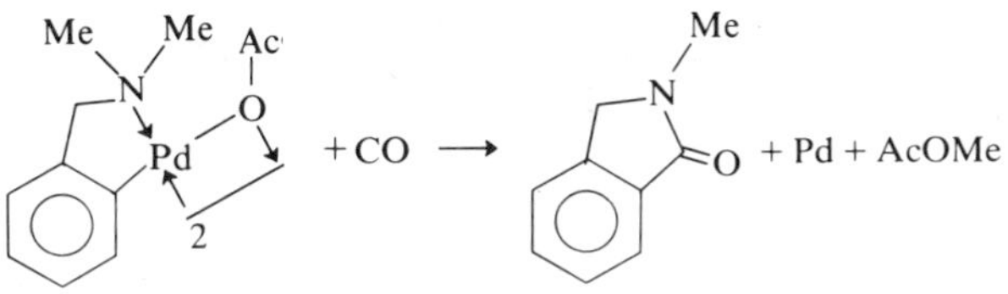

In view of the relatively high cost of palladium compounds as stoichiometric reagents, it is of interest that catalytic syntheses of this type are possible in the presence of dicobalt octacarbonyl, though much more extreme conditions are generally required. Substituted phthalimidines, for example, are prepared in >80% yield by carbonylation of aromatic imines at 200–300°C and 200 atm, with $Co_2(CO)_8$ as catalyst [Eq. (74)].[100]

$$(74)$$

Phthalimidines may in fact be obtained by carbonylation of a wide range of starting materials,[92] though a reducing agent is often necessary. This may be provided by the starting material itself, as in the carbonylation of benzaldehyde phenylhydrazone [Eq. (75)], or else small quantities of hydro-

$$(75)$$

gen (1%–2%) may be added to the carbon monoxide feed as in the reductive carbonylation of nitriles[101] [Eq. (76)] and oximes [Eq. (77)].

$$(76)$$

$$(77)$$

These cobalt-catalyzed reactions almost certainly proceed via an *ortho*-metallated intermediate such as **14**, which was isolated by Heck[102] from a ligand exchange reaction of the corresponding *ortho*-palladated complex with $Co(CO)_4{}^-$. Carbonylation of **14** in methanol leads to the ester **15** which, under normal conditions (>200°C), would cyclize to the observed indazolone product (**16**) (Scheme 39).

Scheme 39

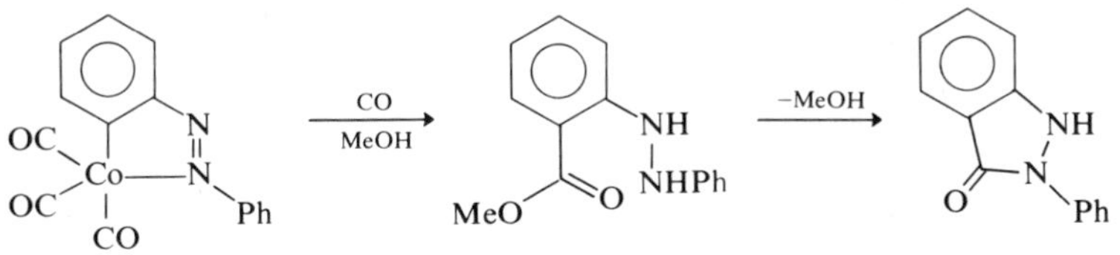

$$14 \qquad 15 \qquad 16$$

Synthesis of N-Phenylphthalimidine from Benzaldehyde Phenylhydrazone[92]

Benzaldehyde phenylhydrazone (8 g) in benzene is carbonylated in a 0.28-liter rocking autoclave for 2.2 h at 230–240°C under 250 atm of CO in the presence of $Co_2(CO)_8$. The decrease in pressure at 20°C is 6.3 atm. When the reaction is complete, 1 g of a blue-black organocobalt complex is filtered off, and the catalyst is destroyed by heating to 70°C. The solution is filtered once more, and the solvent is distilled off in vacuum. The residue is recrystallized twice from ethanol; the product is 4.2 g (50%) of *N*-phenylphthalimidine (mp 166.5–167.5°C).

Experimental details for a number of related reactions are given in the review by Falbe, which deals with ring-closing carbonylations in general.[92]

6.2.4. Preparation of Anhydrides

6.2.4.1. Anhydrides: From Esters

Carbonylation of esters to anhydrides has been noted previously (Section 6.2.2.2) as a side reaction in the nickel-catalyzed formation of esters from ethers at elevated temperatures and high pressures of carbon monoxide. Recent reports in the patent literature, however, suggest that anhydrides, and in particular acetic anhydride,[103] may be obtained under less vigorous conditions (25–80 atm and 150–200°C) by using iodine-promoted rhodium catalysts [Eq. (78)], as in the synthesis of acetic acid

$$CH_3COOCH_3 + CO \xrightarrow[\text{80 atm, 160°C}]{\text{RhCl}_3/\text{PPh}_3/\text{CH}_3\text{I}} CH_3COOCOCH_3 \qquad (78)$$

from methanol. The reaction pathway (cf. Sections 6.2.1.1 and 6.2.2.2) appears to involve cleavage of the starting material by HI, to acetic acid in this case, and iodomethane, which is catalytically carbonylated to acetyl iodide. Condensation of the latter with acetic acid then yields the anhydride and liberates HI for further reaction.

Acetic anhydride can thus be derived entirely from (coal-based) synthesis gas by the reaction sequence shown in Scheme 40, and future

Scheme 40

$$CO + 2H_2 \rightarrow CH_3OH \xrightarrow[\text{CO}]{} \overset{CH_3COOH}{\searrow} \xrightarrow[-H_2O]{} CH_3COOCH_3 \xrightarrow{CO} CH_3COOCOCH_3$$

commercial manufacture will quite probably be based on this chemistry as it avoids increasingly expensive oil-based feedstocks such as ethylene.

β-Propiolactone may be carbonylated at 150°C/100 atm, in the presence of dicobalt octacarbonyl,[104] to give succinic anhydride in 29% yield [Eq. (79)], and it has already been noted (Section 6.2.2.2) that γ-butyrolactone undergoes an analogous nickel-catalyzed reaction under more vigorous conditions to give the corresponding six-membered cyclic anhydride.

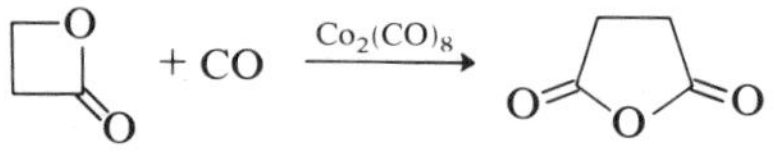

$$\text{(structure)} + CO \xrightarrow{\text{Co}_2(\text{CO})_8} \text{(structure)} \qquad (79)$$

Carbonylation of allylic esters in an aprotic solvent, with PdCl$_2$ as catalyst, yields mixed anhydrides [Eq. (80)], though high pressures of CO are generally required.[105]

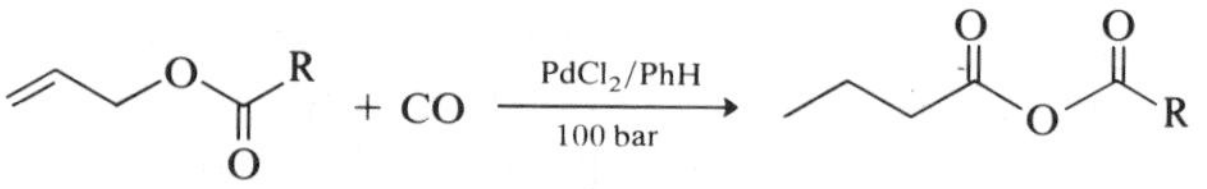

$$\text{(structure)} + CO \xrightarrow[\text{100 bar}]{\text{PdCl}_2/\text{PhH}} \text{(structure)} \qquad (80)$$

Diallyl ether reacts under similar conditions and affords 3-butenoic anhydride in ~20% yield, presumably via the allyl 3-butenoate ester (Scheme 41).

Scheme 41

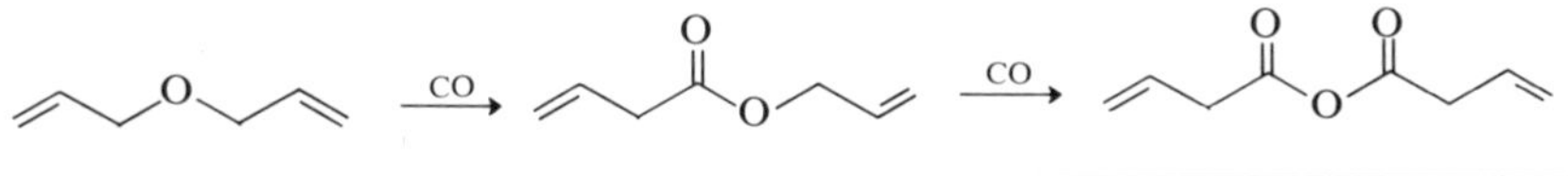

6.2.4.2. Anhydrides: From Alkynes

Substituted itaconic acid anhydrides have been prepared in fair yield (19%–45%) by the palladium-catalyzed carbonylation of propargylic alcohols at 100°C and 100 atm CO pressure [Eq. (81)].[106]

$$
HC \equiv C-CR_2OH + 2CO \xrightarrow[C_6H_6]{PdCl_2} \text{(product)} \tag{81}
$$

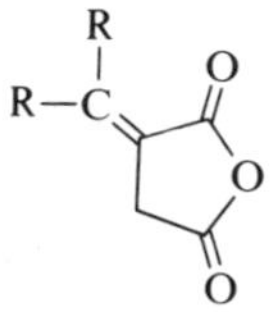

(R ≠ H)

Carbonylation of 2,5-dimethyl-3-hexyn-2,5-diol similarly yielded diisopropylidenesuccinic anhydride [49%, Eq. (82)].

$$
HOCMe_2-C \equiv C-CMe_2OH + 2CO \xrightarrow[C_6H_6]{PdCl_2} \text{(product)} \tag{82}
$$

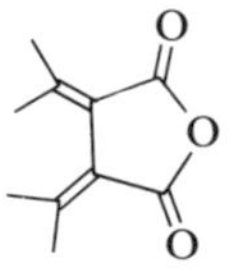

6.2.4.3. Anhydrides: From Diazonium Ions

The palladium-catalyzed carbonylation of aryl diazonium salts in the presence of carboxylate anions yields carboxylic acids on aqueous work-up (Section 6.2.1.3). It is evident, however, that anhydrides are in fact the primary reaction products,[18] and it therefore seems likely that, with further development, this could prove a useful synthesis of mixed anhydrides [Eq. (83)] under relatively mild conditions.

$$
ArN_2^+ + CO + RCO_2^- \xrightarrow[9\text{ atm, }25°C]{Pd(OAc)_2} ArCOOCOR + N_2 \tag{83}
$$

6.2.5. Preparation of Acyl Halides

6.2.5.1. Acyl Halides: From Halides

Acyl halides are only rarely isolated from carbonylation reactions, perhaps because of their relatively high reactivity, but they *have* been shown to be primary reaction products in certain cases. This rhodium-catalyzed carbonylation of iodomethane to acetyl iodide[5] is a proven step in the conversion of methanol to acetic acid, and a number of other reactions yield acyl halides by direct carbonylation of the corresponding alkyl or aryl halide.

3-Alkenoyl chlorides, for example, are formed in high yield (up to 96%) [Eq. (84)] when allylic chlorides are heated to 110°C under 500 atm

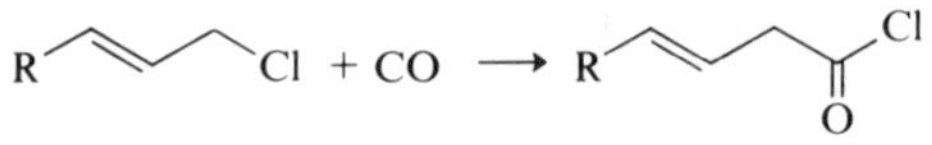

$$ R \diagdown \diagup \diagdown Cl + CO \longrightarrow R \diagdown \diagup \diagdown \diagup \overset{Cl}{\underset{O}{C}} \tag{84} $$

CO pressure, in the presence of $[(\pi\text{-allyl})PdCl]_2$,[107] and yields are almost as high when $PdCl_2$ is used as catalyst. A π-allyl palladium intermediate (**17**) is strongly implied by the observation that the same product (3-pentenoyl chloride) is obtained whether 1-chloro-2-butene or 3-chloro-1-butene is used as starting material, since both allylic chlorides would yield the same π-allyl palladium complex (**17**) (Scheme 42).

Scheme 42

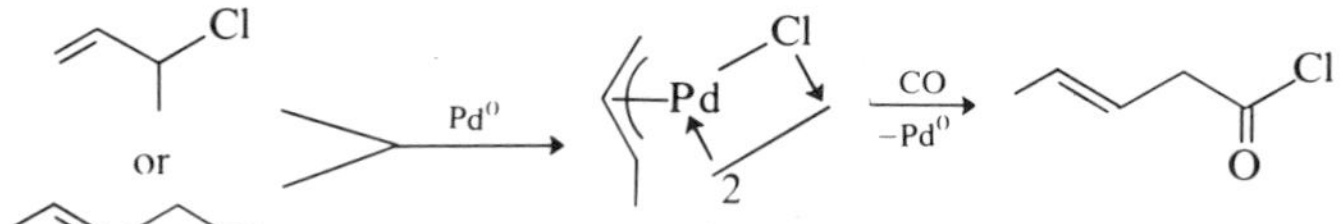

Analogous π-allyl nickel halides [from allyl halides and $Ni(CO)_4$] react with CO at 0°C/1 atm to give 3-butenoyl halides and $Ni(CO)_4$,[32b] and with a mixture of acetylene and CO under similar conditions to give 2,5-hexadienoyl halides [Eq. (85)].[108]

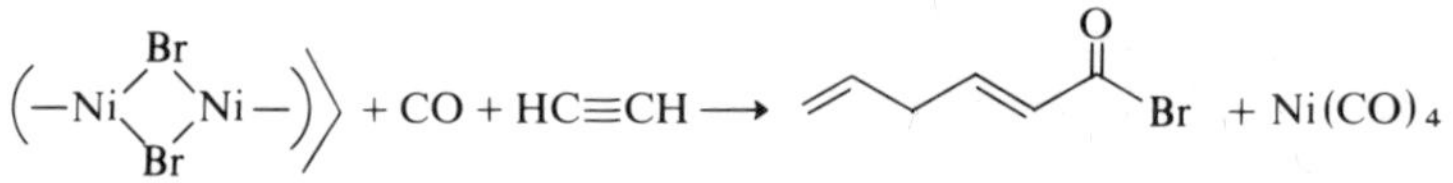

$$ \left(-Ni \overset{Br}{\underset{Br}{\diagup\diagdown}} Ni - \right) \!\!\!\Big\rangle + CO + HC\!\equiv\!CH \longrightarrow \diagup\!\!\diagdown\!\!\diagup\!\!\diagdown \overset{O}{\underset{}{C}} Br + Ni(CO)_4 \tag{85} $$

A number of patents claim direct carbonylation of aryl halides to aroyl halides at moderate temperatures and pressures. Chlorobenzene, for example, is said to be converted to benzoyl chloride in 80% yield by reaction with CO at 80 atm/160°C in the presence of $PdCl_2$,[109] and aroyl fluorides are similarly obtained[110] by carbonylation of aryl chlorides and sodium fluoride using a palladium, rhodium, or ruthenium catalyst [Eq. (86)]. Yields are, however, extremely variable.

$$ArCl + CO + NaF \xrightarrow{PdCl_2} ArCOF + NaCl \qquad (86)$$

6.2.5.2. *Acyl Halides: From Alkenes*

The addition of carbon tetrachloride and carbon monoxide to alkenes [Eq. (87)] is catalyzed by dinuclear metal carbonyls including

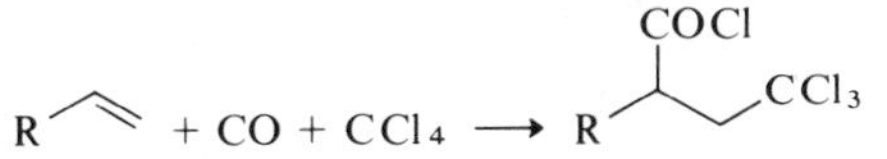
COCl
CCl3
R
+ CO + CCl4
R

(87)

$[(C_5H_5)Fe(CO)_2]_2$ and $[(C_5H_5)Mo(CO)_3]_2$, but analogous mononuclear complexes are ineffective.[111]

2-Alkyl-4,4,4-trichlorobutanoyl chlorides are obtained in fair yield at 50–160°C/200 atm, though only with ethylene is the reaction completely selective. Higher terminal alkenes give mixtures of alkyl and acyl chlorides (the latter being favored at high CO pressures), and a metal-mediated radical chain mechanism, analogous to that known for addition of CCl_4, is thought the most likely reaction pathway (Scheme 43). Curiously, however, addition of hydroquinone as a potential radical trap has little inhibitory effect.[111]

Scheme 43

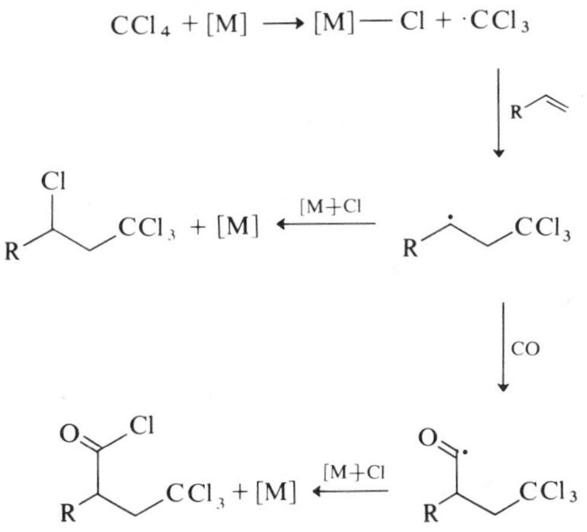
CCl4 + [M] → [M]—Cl + ·CCl3
R
Cl
CCl3 + [M]
[M]—Cl
R
CCl3
R
CO
O
Cl
CCl3 + [M]
[M]—Cl
R
O
CCl3
R

6.3. Preparation of Aldehydes

6.3.1. Aldehydes: From Halides

Alkyl bromides or iodides react readily with carbonylferrate salts such as $K^+[HFe(CO)_4]^-$ [from $Fe(CO)_5$ and ethanolic KOH], and the alkyl iron complexes so formed undergo insertion of carbon monoxide under ambient conditions. In the presence of excess CO, aldehyde is eliminated and $Fe(CO)_5$ is regenerated as shown in Scheme 44,[112] though the reaction is

Scheme 44

$$KHFe(CO)_4 \xrightarrow{RX} R(H)Fe(CO)_4 \xrightarrow{CO} RCOFe(H)(CO)_4 \xrightarrow{CO} RCHO + Fe(CO)_5$$

not catalytic and the hydridocarbonylferrate salt must be prepared in a separate step.

A similar reaction sequence can also be achieved using the commercially available salt $Na_2^+[Fe(CO)_4]^{2-}$, but since the intermediate acyl complex is now anionic, treatment with acid is necessary to liberate the aldehyde (Scheme 45).[113]

Scheme 45

$$Na_2Fe(CO)_4 \xrightarrow{RX} Na^+[RFe(CO)_4]^- \xrightarrow[L = CO \text{ or } PPh_3]{L} [RCOFe(CO)_3L]^- \xrightarrow[CO]{H^+}$$

$$RCHO + Fe(CO)_4L$$

These iron-promoted reactions are limited to alkyl halides or tosylates, which are susceptible to nucleophilic substitution by metal carbonyl anions, but aryl and heterocyclic halides may also be used if they are first converted to organolithium derivatives. Direct reaction of the latter with $Fe(CO)_5$, followed by protonolysis, yields the desired aldehydes in moderate yield (Scheme 46).[114]

Scheme 46

$$ArX \xrightarrow{Li} ArLi \xrightarrow{Fe(CO)_5} Li^+[ArCOFe(CO)_4]^- \xrightarrow{H^+} ArCHO$$

Aryl, heterocyclic, and vinylic halides are *catalytically* converted to aldehydes on treatment with synthesis gas ($CO + H_2$) at 80–100 atm/80–150°C, in the presence of $Pd(PPh_3)_2X_2$ (X = Cl or Br) and a stoichiometric amount of a tertiary amine [Eq. (88)].[115] Alkyl halides are not suitable

$$NR_3 + RX + CO + H_2 \rightarrow RCHO + NR_3H^+X^- \qquad (88)$$

substrates for this reaction, as they tend to undergo dehydrohalogenation to form alkenes, rather than carbonylation, but yields are otherwise generally good. Conditions vary according to the nature of the substrate, so that iodides, for example, are carbonylated more rapidly than chlorides or bromides, owing to the greater ease of oxidative addition of the C–I bond to the catalyst, which, as shown in Scheme 47, is probably the palladium(0) species $Pd(PPh_3)_2CO$.

Scheme 47

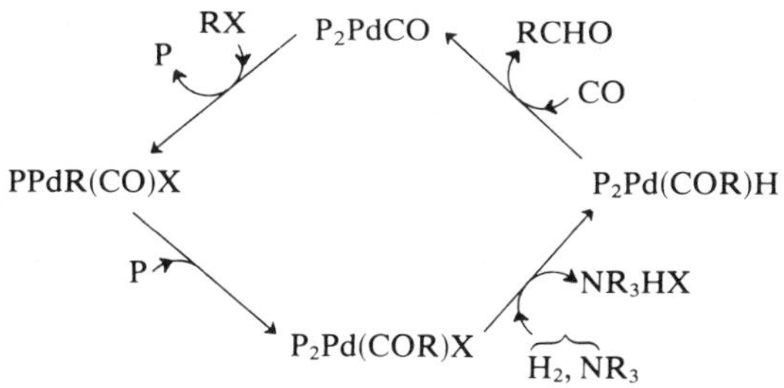

Preparation of 3-Pyridinecarboxaldehyde[115]

To a 50-cm³ autoclave are added 3-bromopyridine, [3.95 g (25 mM)], triethylamine (10 cm³), and $(PPh_3)_2PdBr_2$ (0.30 g). The vessel is flushed with argon and CO, pressurized to 80 atm with synthesis gas ($CO:H_2$ ratio 1:1), and heated to 145°C with magnetic stirring. After 24 h, gas absorption ceases and the vessel is cooled and the gases slowly vented. After addition of anhydrous diethyl ether, the reaction mixture is filtered to remove $NHEt_3^+Br^-$, evaporated at reduced pressure, and finally distilled to give 2.15 g (80%) of 3-pyridinecarboxaldehyde (bp 80°C at 12 mm Hg).

6.3.2. Aldehydes: From Alkenes

The transition-metal-catalyzed reaction of carbon monoxide and hydrogen with an alkene to give an aldehyde [Eq. (89)] is termed

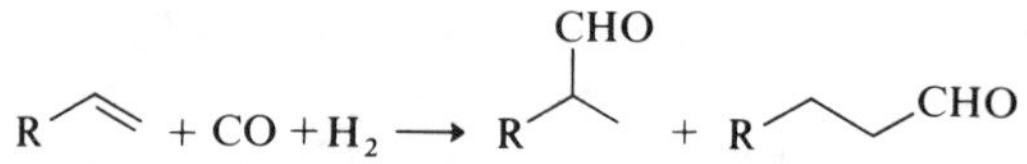

$$R\diagup\!\!\!\diagdown + CO + H_2 \longrightarrow R\diagup\!\!\!\overset{\overset{\displaystyle CHO}{|}}{\diagdown} + R\diagup\!\!\!\diagdown\!\!\!\diagdown CHO \qquad (89)$$

hydroformylation, as it results in the addition of a hydrogen atom and a formyl group to the double bond.

Since its discovery by Roelen in 1938,[116] hydroformylation has developed into an extremely important industrial process, accounting for some 4.5 million tonnes of products in 1978. Aldehydes themselves, however, rarely find direct application, and are generally hydrogenated to alcohols (*in situ* or in a separate step) for use as solvents or as intermediates in plasticiser and detergent manufacture. The two major commercial products, 1-butanol and 2-ethyl-1-hexanol, are both derived from propene via hydroformylation to butanol, followed either by direct hydrogenation or by aldol condensation and subsequent hydrogenation, respectively (Scheme 48).

Scheme 48

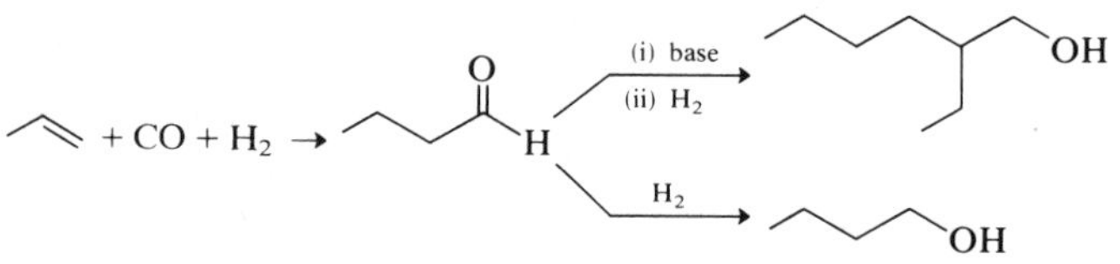

As a result of its industrial significance, a vast amount of research has been devoted to improving and understanding hydroformylation, but a detailed survey of this work would be beyond the scope of the present chapter. The reaction has, however, been recently and exhaustively reviewed by Cornils,[117] and Pruett,[118] and by Pino *et al.*,[119] so that here we discuss only briefly its development and scope, together with some recent applications in synthesis.

Until very recently, hydroformylation was carried out exclusively with cobalt catalysts, under relatively vigorous conditions (150–200°C/200–400 atm), and gave mixtures of linear and branched aldehydes in a ratio typically of around 4:1. The currently accepted mechanism[120] for the cobalt-catalyzed reaction (Scheme 49) involves generation of $HCo(CO)_4$ as the immediate catalyst precursor.

Disadvantages of the cobalt-catalyzed hydroformylation include only moderate selectivity to the desired linear isomers, low catalyst stability at

Scheme 49

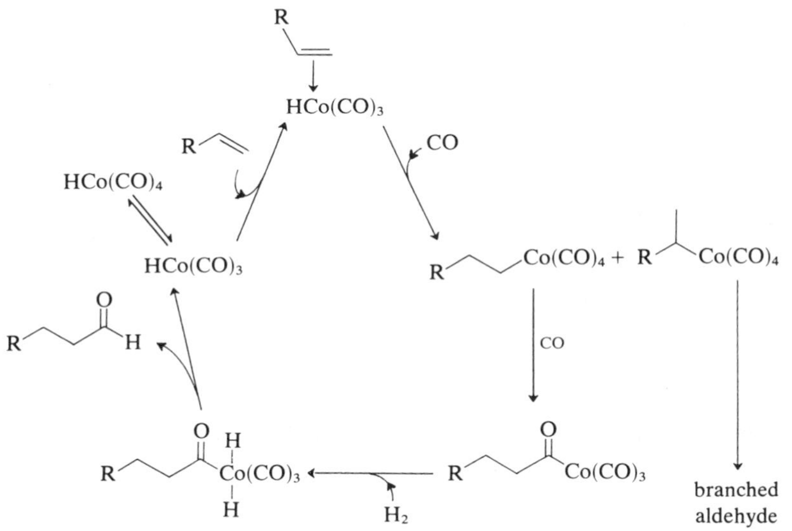

the relatively high reaction temperature (necessitating the use of high CO pressures to prevent decomposition to cobalt metal), and substantial formation of by-products including alkanes (by simple hydrogenation), ketones, and aldol polycondensation products. Some improvement may be achieved by the addition of donor ligands such as trialkyl phosphines, since these both increase catalyst stability at lower CO pressures (by coordinating to cobalt) and promote formation of linear products, though at some expense in catalyst activity. The linear:branched ratio improves typically to ~7:1. Aldehydes are no longer isolated, however, since the phosphine-substituted catalyst displays much greater hydrogenation activity, and alcohols are produced *in situ* by reduction of the first formed aldehyde.

Hydroformylation may be carried out under extremely mild conditions (25°C/1 atm) when ligand-modified rhodium compounds are used as catalysts. The activity of rhodium is in fact 10^3–10^4 times greater than cobalt for this reaction, and hydrogenation to alkane or alcohol is very much less significant. This makes possible hydroformylation of substrates such as styrene, which are merely hydrogenated by cobalt catalysts. Moreover, when the rhodium catalyst contains bulky phosphines such as PPh_3, very high selectivity to linear aldehyde may be achieved. Thus, in the presence of excess triphenylphosphine, the easily prepared[121] complex $RhH(CO)(PPh_3)_3$ catalyzes hydroformylation of 1-hexene to 1-heptanal at 1 atm/25°C with up to 94% selectivity.[112] Such selectivity is achieved only in the presence of high concentrations of PPh$_3$, and catalyst activity is much reduced under these conditions. Reasonable reaction rates can be achieved, however, by working at increased temperature and pressure, so that even in molten triphenylphosphine as solvent, the hydroformylation of 1-hexene at 30 atm/90°C [RhH(CO)(PPh$_3$)$_3$ catalyst] gives 92% conversion after 20 min.[122] Processes for hydroformylation of propene and ethylene using RhH(CO)(PPh$_3$)$_3$ in molten triphenylphosphine have in fact been operated commercially since 1976.[123]

Although cobalt- and rhodium-based catalysts have been most intensively studied, other metals including ruthenium, iron, and palladium show some activity, and it has recently been shown that phosphine-modified platinum–tin catalysts [e.g., Ph$_2$P(CH$_2$)$_4$PPh$_2$/PtCl$_2$(PhCN)$_2$/SnCl$_2$] will catalyze hydroformylation under moderate conditions with good selectivity to linear aldehyde (up to 99%).[124] By suitable choice of phosphine the activity of the platinum-based catalyst can be made to exceed that of rhodium, though by-product formation is substantially increased.

The reactivity of alkenes in hydroformylation follows a similar pattern to that observed in other carbonylation reactions, i.e., linear terminal alkenes react more rapidly than linear internal alkenes, which in turn are more reactive than branched alkenes. The inhibiting effect of alkyl substituents increases with proximity to the double bond, and an experimental "rule" due to Keulemans[125] states that "formyl groups are not produced at quaternary carbon atoms." Hydroformylation of 2-methyl propene, for example, yields 3-methyl butanal almost exclusively [Eq. (90)], with only

$$\text{=< + CO + H}_2 \xrightarrow{\text{HCo(CO)}_4} \text{O=}\hspace{-0.2em}\big/\hspace{-0.2em}\text{-H} \ (95\%) \qquad (90)$$

minor amounts of 2,2-dimethyl propanal, and if the starting alkene can *only* yield products with the formyl group at a quaternary carbon, then isomerization generally occurs before hydroformylation takes place (Scheme 50).

Scheme 50

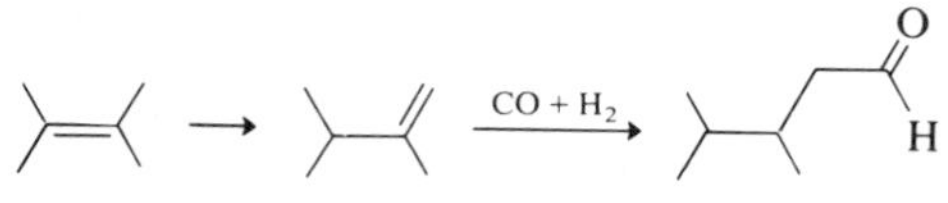

The effects of functional groups have been studied in some detail and can be broadly summarized as follows: (i) Electron-withdrawing substituents conjugated with the double bond (e.g. $>C=O$, $-C\equiv N$) promote hydrogenation rather than hydroformylation, though this effect is less marked with rhodium catalysts than with cobalt. Where hydroformylation *does* occur, the formyl group tends to add at the carbon atom β to the substituent [Eq. (91)].[126] (ii) Electron-donor substituents conjugated with

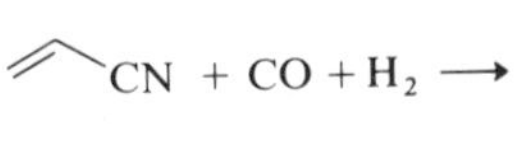

$$\text{CN} + \text{CO} + \text{H}_2 \longrightarrow \quad \underset{\text{O}}{\overset{\text{H}}{\big|}} \text{CN} \qquad (91)$$

the double bond ($-OR$, $-NR_2$, $-Cl$) do not inhibit hydroformylation, but favor addition of the formyl group at the carbon atom α to the donor group [Eq. (92)].[127] (iii) Substituents not conjugated with the double bond

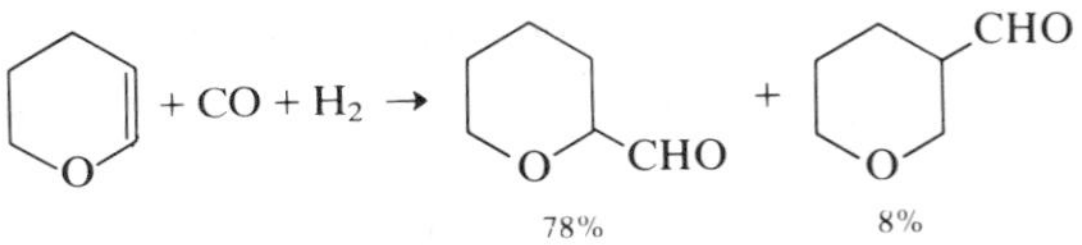

$$+ \text{CO} + \text{H}_2 \rightarrow \underset{\text{78%}}{\text{CHO}} + \underset{\text{8%}}{\text{CHO}} \qquad (92)$$

have less influence than conjugated substituents, though directional effects tend to be similar. Moreover, isomerization will often bring the substituents into conjugation with the double bond, or may even result in the disappearance of the C=C double bond altogether. Allylic alcohols, for example, isomerize to vinyl alcohol and thence to aldehydes in the presence of $HCo(CO)_4$,[128] though if hydrogen migration is blocked by chain-branching as in (**18**), then hydroformylation can occur (Scheme 51).[117] Phosphine-containing rhodium catalysts, however, allow hydroformylation of most

Scheme 51

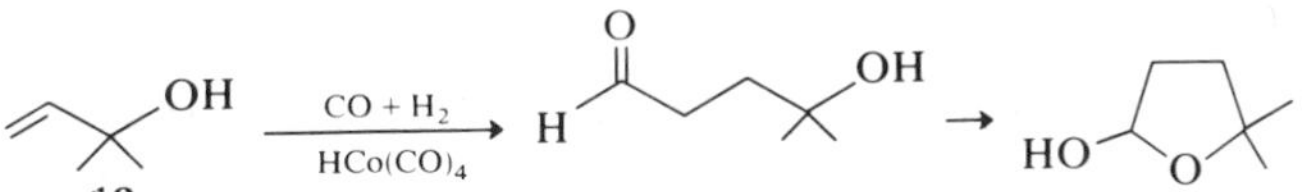

allylic alcohols since isomerization is extremely slow at high phosphine concentrations. Under such conditions, 2-buten-1-ol, for example, affords the cyclic hemi-acetal (**19**) in 85% yield (Scheme 52).[129] Hydroformylation of allyl alcohol is in fact currently being developed as an industrial route to 1,4-butanediol, a potential intermediate in polyester manufacture.

Scheme 52

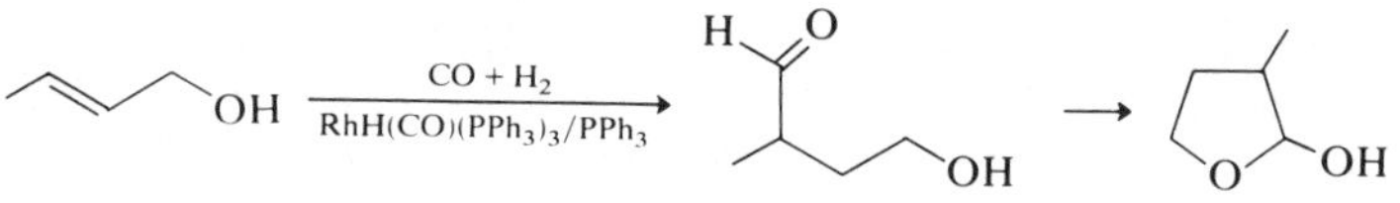

Despite its significance as an industrial process, there are relatively few descriptions of hydroformylation as applied to laboratory-scale organic synthesis, perhaps because of the high pressure and mixtures of products which were formerly inevitable. In view of the much milder conditions and greater selectivities possible with rhodium catalysts, it seems likely that hydroformylation will find increasing use as a synthetic tool, particularly as the aldehyde group is so readily converted to other functionalities by oxidation, reduction, amination, condensation, etc. Some indication that this is already happening in industrial laboratories may be found in a recent review (essentially of patent literature) by Siegel and Himmele,[130] which deals with rhodium-catalyzed hydroformylation in the synthesis of pharmacologically active compounds, heterocycles, vitamins, terpenes, and optically active natural product derivatives.

Hydroformylation of Styrene[122]

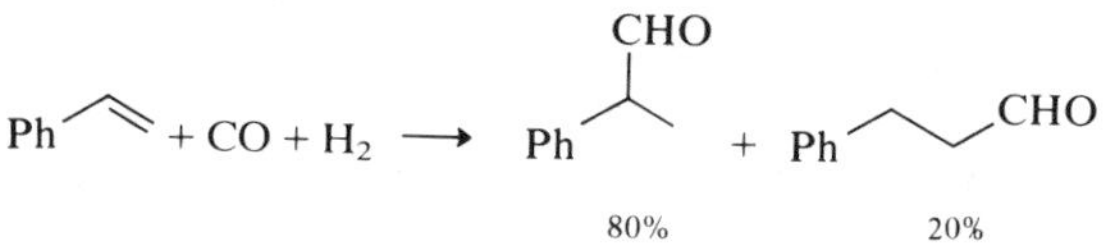

$$Ph \diagdown \diagup + CO + H_2 \longrightarrow Ph \diagdown \overset{CHO}{\underset{80\%}{|}} + Ph \diagdown \diagup ^{CHO} \qquad (93)$$

RhH(CO)(PPh$_3$)$_3$ (0.5 mM) is weighed directly into a 25 cm^3 conical flask fitted with a side arm and rubber septum cap. The side arm is connected by flexible tubing to a cylinder of synthesis gas (CO + H$_2$, 1:1) via a take-off bubbler, allowing a slight positive gas pressure to be maintained during the reaction. The flask is purged with synthesis gas for 10 min by inserting a thin steel outlet tube through the cap. After removing the outlet tube, benzene (9 g) and styrene (1.04 g, 10 mM) are successively added by syringe. The flask is shaken gently at 25°C, and the progress of the reaction [Eq. (93)] monitored by ^{1}H nmr. The product aldehydes give conveniently separated resonances: PhCH(Me)CHO at δ 3.02 (quartet) and 1.10 (doublet); PhCH$_2$CH$_2$CHO at δ 2.55 and 2.06 (triplets). The resonances may be integrated to give the linear:branched ratio ($\approx$1:8) and the integration against the highest field proton of styrene gives the conversion, which after 30 min is ~25%.

6.3.3. Aldehydes: From Aldehydes and Silanes

An aldehyde may be converted to the next highest α-siloxy aldehyde in high yield by reaction with carbon monoxide and a trialkylsilane in the presence of dicobalt octacarbonyl [Eq. (94)].[131] Addition of triphenylphosphine suppresses simple hydrosilylation of the aldehyde [Eq. (95)]. The

$$RCHO + R_3^1SiH + CO \xrightarrow[PPh_3]{Co_2(CO)_8} \overset{OSiR_3^1}{RCH-CHO} \tag{94}$$

$$RCHO + R_3^1SiH \rightarrow RCH_2OSiR_3^1 \tag{95}$$

catalytic cycle is thought to involve insertion of aldehyde into a cobalt–silicon bond, followed by insertion of CO and subsequent hydrogenolysis (Scheme 53). It has in fact been recently shown that aldehydes *can* insert

Scheme 53

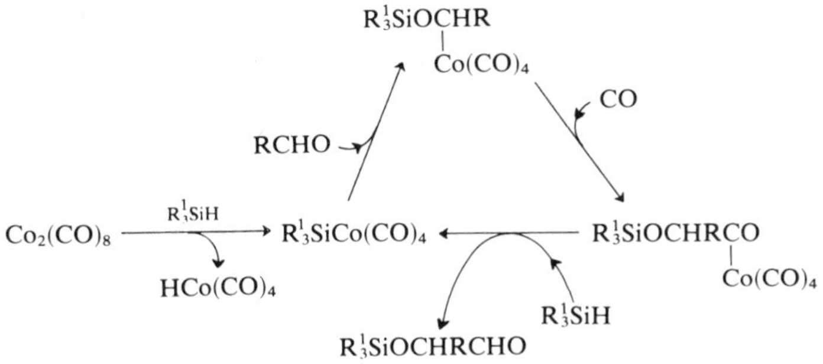

into silicon–transition metal bonds [Eq. (96)], and the siloxybenzyl manganese complex (**20**) has been isolated and characterized.[132]

$$Me_3SiMn(CO)_5 + PhCHO \rightarrow R_3SiOCH(Ph)Mn(CO)_5 \tag{96}$$
$$\mathbf{20}$$

To obtain α-siloxy aldehyde in good yield, it is necessary to use the starting aldehyde in excess, otherwise substantial amounts of 1,2-bis(siloxy)alkene are formed. The latter apparently results from hydrosilylation of the product aldehyde, followed by cobalt-catalyzed dehydrogenation.

6.4. Preparation of Ketones

6.4.1. Ketones: From Halides

As noted earlier (Section 6.3.1), alkyl halides or tosylates react with the commercially available salt $Na_2Fe(CO)_4$ to give anionic alkyl iron complexes $[RFe(CO)_4]^-$, and these can react with further alkylating agent, affording ketones in good yield (Scheme 54).

Scheme 54

$$Fe(CO)_4^{2-} + RX \rightarrow [RFe(CO)_4]^- \rightleftharpoons [RCOFe(CO)_3]^- \xrightarrow{R'Y} RCOR'$$

Primary bromides, iodides, and tosylates or secondary tosylates can be used in the first alkylation, but the second stage proceeds only with the more reactive alkylating agents such as benzylic halides or primary iodides, due to the limited nucleophilicity of the $[RFe(CO)_4]^-$ anion.[133]

Variations on this reaction include the use of activated alkenes such as acrylonitrile or ethyl acrylate in the second step (Scheme 55),[134] and

Scheme 55

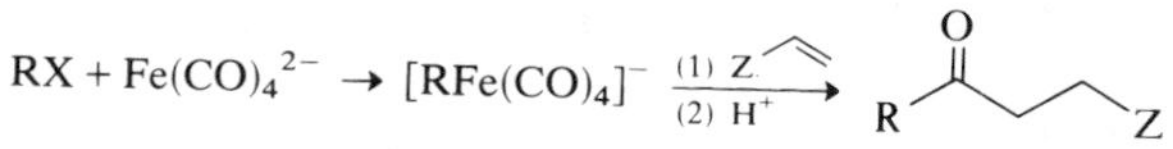

$$RX + Fe(CO)_4^{2-} \rightarrow [RFe(CO)_4]^- \xrightarrow[(2)\ H^+]{(1)\ Z\diagup\!\diagup}$$

the application of phase transfer techniques to generate $[Fe(CO)_4]^{2-}$ *in situ* from $Fe(CO)_5$ and aqueous sodium hydroxide.[135]

A particular feature of the phase transfer system is that, when reactive halides are used in excess, *two* moles of symmetrical ketone are formed per mole of $Fe(CO)_5$ consumed, so that a carbonyl anion must be regenerated at some stage.

Iron pentacarbonyl reacts with organolithium[136] or Grignard[137] reagents to give acyl iron anions $[RCOFe(CO)_4]^-$, which may be alkylated (though not by activated alkenes) to give ketones in modest yield. The benzoyl complex $[PhCOFe(CO)_4]^-$ thus reacts with benzyl bromide at $-40°C$ to give a 57% yield (after isolation and purification) of benzyl phenyl ketone. When acyl halides are used in the second step, decarbonylation occurs and monoketones are formed, though small quantities of 1,2-diketones may also be isolated (Scheme 56).[136] Aryl iodonium salts may

Scheme 56

$$[RCOFe(CO)_4]^- + R^1COX \rightarrow (RCO)Fe(COR^1)(CO)_4 \xrightarrow{-CO} (RCO)Fe(R^1)(CO)_4$$

$$\downarrow \qquad\qquad\qquad\qquad \downarrow$$

$$RCOCOR^1 \qquad\qquad\qquad\qquad RCOR^1$$

be used as arylating agents in the second step of Scheme 52 and give good yields of aromatic ketones [Eq. (97)].[138]

$$[RCOFe(CO)_4]^- + Ar_2I^+ \rightarrow RCOAr + ArI \qquad (97)$$

Synthesis of Ethyl-6-keto-octanoate[133] (Eq. (98))

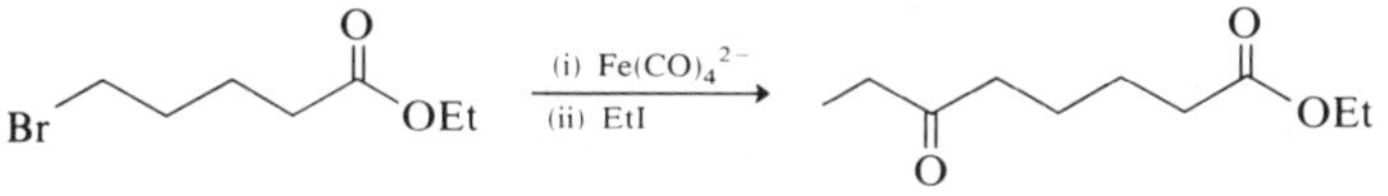

$$(98)$$

Ethyl 5-bromovalerate (2.16 g, 10.3 mM) is added under nitrogen to a solution of $Na_2Fe(CO)_4$ (11.1 mM) in 30 cm^3 of *N*-methyl-2-pyrollidone (distilled from CaH_2); the mixture is stirred for 1 h and then iodoethane (2.4 cm^3) is added. After 24 h the solution is diluted with diethyl ether, washed with brine, dried, filtered, and the ether removed on the rotary evaporator. The residue is placed on a short (130 g) silica-gel column, washed free of colored by-products with hexane, and eluted with 50% ether/hexane. Fractional distillation gives ~1.4 g of ethyl-6-keto-octanoate (74%).

6.4.2. Ketones: From Alkenes and Alkynes

Saturated ketones are sometimes obtained as hydroformylation by-products, particularly at low pressures and high concentrations of alkene. Diethyl ketones may in fact be obtained from ethylene in over 50% yield [Eq. (99)] by suitable choice of conditions, but this type of reaction is much

$$2C_2H_4 + H_2 + CO \xrightarrow{HCo(CO)_4} (C_2H_5)_2CO \qquad (99)$$

less favorable with higher alkenes.[139] It has been proposed[140] that ketones arise by condensation of alkyl and acyl cobalt complexes with elimination of $Co_2(CO)_8$, but it is perhaps more likely that the acyl cobalt intermediate (21) can insert alkene and then add hydrogen to give a ketoalkyl cobalt hydride that undergoes reductive elimination (Scheme 57).

Scheme 57

$$C_2H_5COCo(CO)_4 \xrightarrow{C_2H_4} C_2H_5COCH_2CH_2Co(CO)_4 \xrightarrow[-CO]{+H_2} C_2H_5COCH_2CH_2\overset{\displaystyle H}{\underset{\displaystyle H}{Co}}(CO)_3$$
$$\mathbf{21}$$

$$\xrightarrow{+CO} C_2H_5COC_2H_5 + HCo(CO)_4$$

Acyl dienes may be prepared in good yield by reaction of alkyl or acyl halides with conjugated dienes and carbon monoxide, in the presence of catalytic quantities of $Co(CO)_4^-$ and a stoichiometric amount of base [Eq. (100)].[141]

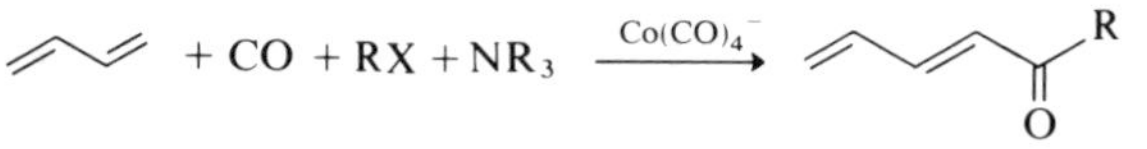

$$\diagup\!\!\diagdown\!\!\diagup + CO + RX + NR_3 \xrightarrow{Co(CO)_4^-} \diagup\!\!\diagdown\!\!\diagup\!\!\diagdown\!\!\overset{\displaystyle R}{\underset{\displaystyle O}{}} \qquad (100)$$

The reaction pathway has been established by isolation of several intermediates, including the π-allyl complex (22), and the catalytic cycle is shown in Scheme 58.

Scheme 58

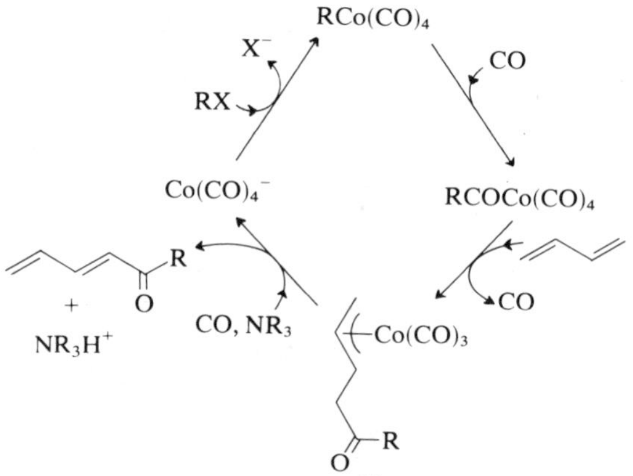

22

As in other carbonylation reactions involving anionic metal carbonyls, phase transfer techniques can be used with advantage for the acyl diene synthesis,[142] as conditions are milder (25°C/1 atm) and yields significantly better than in a single-phase system. The reaction is highly stereo- and regiospecific, since the acyl group is normally added to the least substituted carbon atom of the least substituted double bond with exclusive formation of *trans*-acyl diene.

Strained alkenes such as norbornadiene and its derivatives react photochemically with $Fe(CO)_5$ or $Fe_2(CO)_9$ to give cyclopentanones [Eq. (101)]

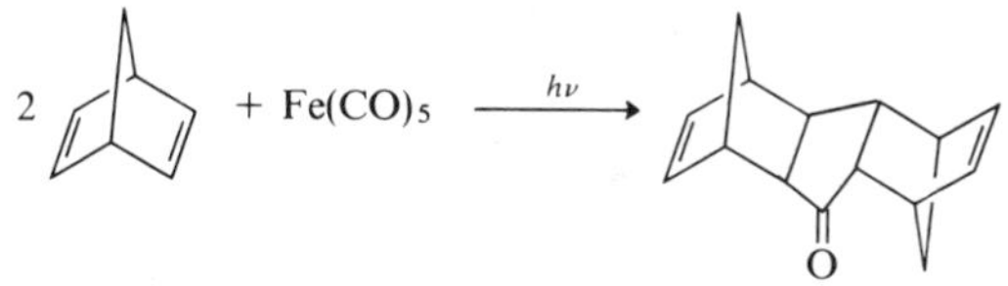

$$(101)$$

23

in up to 70% yield,[143] though much lower yields are obtained with less strained alkenes.

Product stereochemistry is invariably *exo–trans–exo* (**23**), and unsymmetrical ketones may be prepared by sequential addition of two different cycloalkenes to the system. The reaction proceeds via a bis-alkene $Fe(0)$ complex that undergoes internal coupling to a metallocyclopentane, followed by CO insertion and reductive elimination[144] (Scheme 59).

Scheme 59

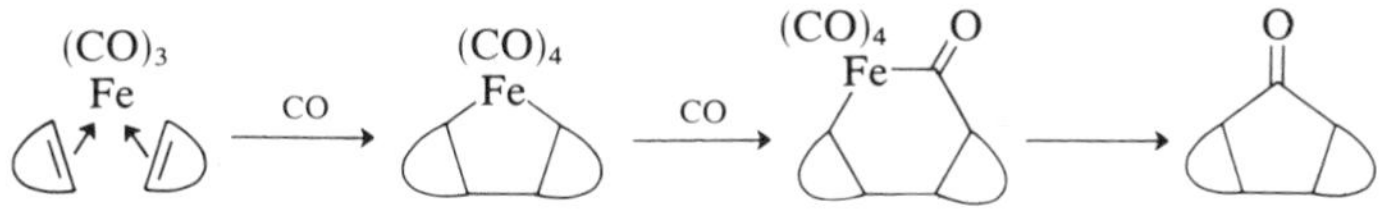

A related reaction, involving the coupling of an alkene, alkyne, and carbon monoxide to give a cyclopentenone [Eq. (102)], proceeds via an

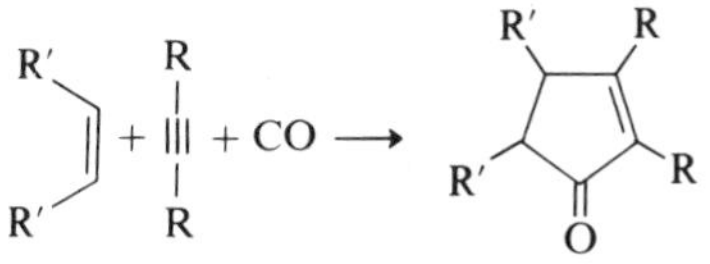

$$(102)$$

alkyne–cobalt complex (**24**), and with relatively unreactive alkenes such as cyclopentene it is preferable to synthesize the alkyne complex in a separate

$$(CO)_3 Co \Longleftarrow \| \Longrightarrow Co(CO)_3$$

with R above and R below

24

step.[145] With highly strained alkenes such as norbornadiene, however, only catalytic quantities of $Co_2(CO)_8$ need to be added to the reaction mixture.[146]

Very volatile or gaseous alkenes require the use of an autoclave, but since the reaction proceeds at 1 atm pressure of CO/acetylene, less volatile alkenes may be employed in hydrocarbon solvents. The reaction is again stereospecific, yielding *exo*-ring-fused products from norbornene and its derivatives, though the ketone (**25**) derived from norbornadiene and acetylene isomerizes [Eq. (103)] to the *endo*-isomer on prolonged contact with dicobalt octacarbonyl.[146]

$$\text{(structure)} + C_2H_2 + CO \longrightarrow \text{(structure)} \qquad (103)$$

25

Synthesis of Exo-hexahydro-4,7-methanoinden-1-one[146](**26**)

26

A solution of dicobalt octacarbonyl (1 g, 3 mM) and norbornene (3 g, 32 mM) in isooctane (100 cm³) is stirred at 60–70°C under acetylene (1 atm) and then under 1 : 1 acetylene/CO (1 atm) until gas absorption ceases (~1550 cm³). The solution is concentrated and the residue chromatographed on neutral alumina. Petroleum (bp 40–60°C)/benzene (1 : 1) elutes the acetylene complex $(C_2H_2)Co_2(CO)_6$ (70 mg), and benzene–chloroform (1 : 1) then elutes a yellow oil, which is distilled at 101–102°C (15 mm Hg) to give the product ketone (3.54 g, 74%), which slowly dimerizes when stored at room temperature (mp of dimer, 51°C).

6.4.3. Ketones: From Organomercury Compounds

Rhodium-catalyzed carbonylation of aryl- and vinyl-mercuric halides [Eq. (104)] is a potentially valuable route to the corresponding ketones,

$$2RHgX + CO \xrightarrow{[Rh(CO)_2Cl]_2} R_2CO + HgX_2 + Hg \tag{104}$$

and in at least one case (the synthesis of divinyl ketones) is by far the simplest and most convenient method.

Vinyl mercuric chlorides, which are readily prepared from alkynes via hydroboration/mercuration,[82a] are thus carbonylated at 25°C/1 atm CO in the presence of $[Rh(CO)_2Cl]_2$ to give excellent yields of divinyl ketones with complete retention of stereochemistry about the double bond.[147] The same catalyst may be used for the synthesis of diaryl ketones, but much higher temperatures and pressures are required (typically 70°C/100 atm). A somewhat analogous, though more hazardous, procedure allows the synthesis of unsymmetrical diaryl ketones by reaction of aryl iodides with aryl mercuric halides in the presence of nickel tetracarbonyl [Eq. (105)].[148]

$$ArI + Ar'HgX + Ni(CO)_4 \xrightarrow[\text{benzene}]{70°C} ArCOAr' + HgXI \tag{105}$$

Yields are generally good, and the reaction is tolerant of a number of functional groups including $-NH_2$ and $-Cl$.

Trans, trans-trideca-5,8-dien-7-one[147] *(27)*

27

Trans-1-hexenylmercuric chloride (1.60 g) and lithium chloride (0.42 g) are weighed into a 250-cm³ round-bottomed flask equipped with a serum cap, gas inlet, and magnetic stirrer bar. The flask is flushed with carbon monoxide and THF (40 cm³) is added by syringe. A solution of $[Rh(CO)_2Cl]_2$ (9.7 mg) in 10 cm³ of THF is added under CO, the flask flushed with further CO, and finally maintained under a CO atmosphere by attaching a balloon of CO to the gas inlet tube. After stirring for 24 h at room temperature, the reaction mixture is filtered to remove elemental mercury, diluted with pentane (80 cm³), and washed three times with saturated aqueous sodium chloride. The organic phase is dried over Na_2SO_4, concentrated, and bulb-to-bulb distilled (75–80°C, 0.35 mm Hg) to give 0.38 g (78%) of product as a colorless oil.

6.5. Preparation of Isocyanates

Carbonylation of aromatic nitro compounds in the presence of palladium-based catalysts such as $[PdCl_2/pyridine]$ yields isocyanates with high conversion and good selectivity [Eq. (106)].[149]

$$ArNO_2 + 3CO \rightarrow ArNCO + 2CO_2 \tag{106}$$

The conditions required are fairly vigorous (200°C/150 atm), but the simplicity of the process is attractive for industrial use[150] (isocyanates are intermediates for polyurethanes), and it seems possible that this chemistry will ultimately replace existing isocyanate processes that involve hydrogenation of nitro compounds to amines followed by treatment with phosgene.

2-Aryl azirines are carbonylated under very mild conditions (5°C, 1 atm CO) in the presence of $[Rh(CO)_2Cl]_2$ to give arylvinyl isocyanates in high yield [Eq. (107)].[151] These may be isolated as such or, more conveniently, converted to carbamates or ureas by reaction with alcohols or amines, respectively.

$$ \tag{107} $$

6.6. *Decarbonylation of Aldehydes and Acyl Halides*

Since most of the elementary steps in carbonylation reactions are reversible, it is not surprising that transition metals and their complexes promote the *decarbonylation* of organic compounds in either a stoichiometric or catalytic manner.[152] In stoichiometric reactions, carbon monoxide removed from the organic compound is retained by the metal complex [Eq. (108)], whereas for catalytic behavior this CO must be released, a reaction that often occurs only at high temperatures (>200°C).

$$ RCHO + RhCl(PPh_3)_3 \rightarrow RH + PPh_3 + RhCl(CO)(PPh_3)_2 \tag{108} $$

The rhodium(I) complex $RhCl(PPh_3)_3$ is an excellent stoichiometric reagent for decarbonylation reactions, particularly with aldehydes and acyl halides.[153] At 25–50°C, aldehydes give alkanes or arenes [Eq. (108)], whereas acyl halides containing β-hydrogens undergo both decarbonylation and dehydrohalogenation to give alkenes [Eq. (109)]. In the absence of β-hydrogens, however, alkyl or aryl halides are produced [Eq. (110)].

$$ RCH_2CH_2COX + RhCl(PPh_3)_3 \rightarrow RCH{=}CH_2 + HX + PPh_3 + RhCl(CO)(PPh_3)_2 \tag{109} $$

$$ RCOX + RhCl(PPh_3)_3 \rightarrow RX + RhCl(CO)(PPh_3)_2 + PPh_3 \tag{110} $$

In view of the extremely high cost of rhodium, it is of interest that $RhCl(PPh_3)_3$ can be regenerated from $RhCl(CO)(PPh_3)_2$ by treatment with benzyl chloride. This yields a rhodium(III) benzyl complex that reacts with triphenyl phosphine in ethanol to give $RhCl(PPh_3)_3$ in high yield (Scheme 60).[154]

Scheme 60

$RhCl(CO)(PPh_3)_2 + PhCH_2Cl \rightarrow RhCl_2PPh_3(\eta^3\text{-}CH_2Ph) \xrightarrow[EtOH]{PPh_3} RhCl(PPh_3)_3$

Under vigorous conditions (>200°C), *catalytic* decarbonylation is possible using $RhCl(PPh_3)_3$ or $RhCl(CO)(PPh_3)_2$ and is especially effective with aromatic aldehydes and aroyl halides. Aliphatic aldehydes tend to both decarbonylate and dehydrogenate, giving alkenes, and an important industrial application of this involves the decomposition of 2-methyl propanal [Eq. (111)]. The hydroformylation of propene is thus reversed, and

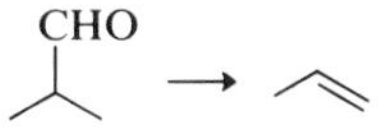

$$\overset{CHO}{\underset{}{\bigwedge}} \longrightarrow \diagdown\diagup + CO + H_2 \qquad (111)$$

the unwanted isomer is decomposed to starting materials that can be recycled.[155]

Decarbonylation of 3,4-Dichlorobenzoyl Chloride[153]

In a 35-cm³ Claisen flask, 3,4-dichlorobenzoyl chloride (10 g) and $RhCl(CO)(PPh_3)_2$ (0.3 g) are heated at 250°C for $1\frac{1}{2}$ h under a nitrogen atmosphere. During the reaction 8 g (96%) of crude 1,3,4-trichlorobenzene is collected by distillation.

Cationic rhodium(I) complexes of the type $[Rh(diphosphine)_2]^+$ [diphosphine = $Ph_2P(CH_2)_nPPh_2$ ($n = 2$ or 3)] are particularly active catalysts for decarbonylation of aldehydes, requiring substantially lower temperatures than are possible with $RhCl(PPh_3)_3$. As a result of the lower temperature, dehydrogenation of aliphatic aldehydes to alkenes is not observed, and the catalyst appears to be indefinitely stable under reaction conditions.[156]

Aldehydes and acyl halides are catalytically decarbonylated by heating to 180–200°C in the presence of palladium metal, and this technique has been used in natural product synthesis. Apopinene (**28**), for example, is prepared in a two-stage synthesis from readily available α-pinene (**26**) by oxidation to myrtenal (**27**) with SeO_2, followed by decarbonylation using palladium on barium sulfate, which gives apopinene in an overall yield of 55% (Scheme 61).[157]

Acyl halides give mixtures of isomeric alkenes on decarbonylation with palladium, so that heptanoyl bromide, for example, yields only 7.5% of 1-hexene, with 92.5% 2- and 3-hexenes.[158] Aroyl halides are not decarbonylated satisfactorily by palladium, and the homogeneous rhodium catalysts described above are preferable. Decarbonylation of aromatic *aldehydes* over palladium is fairly general, however, though even here exceptions are known. 1-Naphthaldehyde, for example, gives naphthalene cleanly and rapidly at 210°C, whereas 2-naphthaldehyde fails to react even at 250°C.[159]

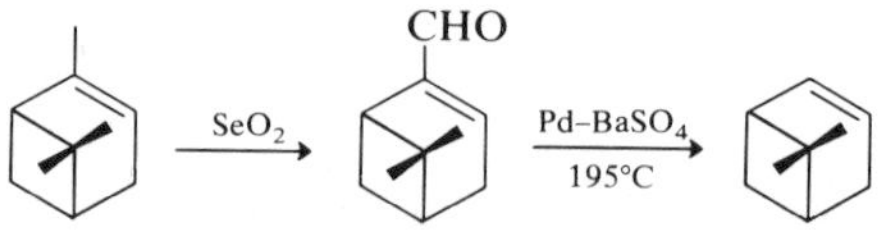

The carbonylation of amines and alcohols to formamides and formate esters, described in earlier sections (6.2.2.1 and 6.2.3.1), can be reversed by heating with palladium to ~200°C [Eq. (112)].

$$HCO_2R \rightarrow ROH + CO \qquad (112)$$

It has been proposed that decarbonylation reactions catalyzed by transition metals could involve free radical formation by analogy with known radical-initiated chain reactions (Scheme 62), but such a mechanism

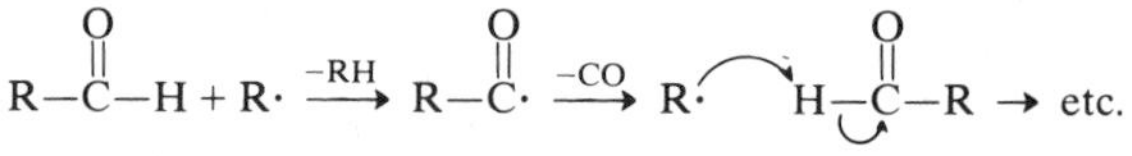

is not in keeping with the observed retention of stereochemistry of the group R, nor with deuterium labeling experiments, which indicate that, at least with rhodium catalysts, the reaction is intramolecular. The generally accepted mechanism in fact involves oxidative addition of the RCO–X bond to the transition metal, migration of R from CO to the metal center, and reductive elimination of RX (or alkene and HX) (Scheme 63).[160]

Scheme 63

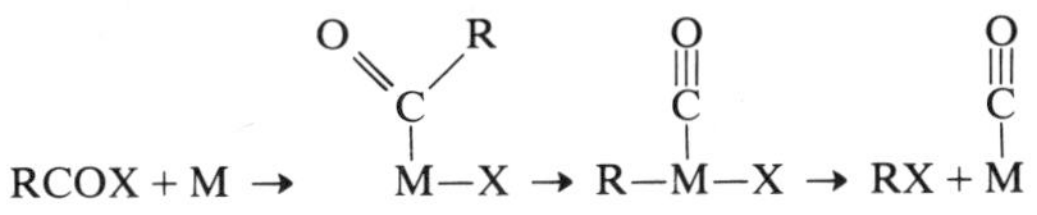

$$RCOX + M \;\rightarrow\; \underset{M-X}{\overset{\overset{\textstyle O}{\diagdown}\;\;\overset{\textstyle R}{\diagup}}{\underset{|}{C}}} \;\rightarrow\; \underset{R-M-X}{\overset{\overset{\textstyle O}{\|}}{\underset{|}{C}}} \;\rightarrow\; \underset{RX + M}{\overset{\overset{\textstyle O}{\|}}{C}}$$

Decarbonylation of Myrtenal[157]

    Myrtenal (**27**) (135 g) and a freshly prepared 5% palladium hydroxide-on-barium-sulfate catalyst (2.5 g) are placed in a 1000-cm³ flask provided with a 5 cm³ Dean–Stark distillation trap to which a reflux condenser is attached. Upon heating, the catalyst turns black and a strong evolution of carbon monoxide is observed. After 5 h of refluxing, the temperature in the flask has dropped from 195°C to 155°C and there is only a slow evolution of gas (15–20 ml/min). At this point, the distillation trap is opened and apopinene (**28**) (70 g) is collected. The remainder in the flask is refluxed for an additional 10 h to yield another 10 g of apopinene (~15 g residue). The total amount of carbon monoxide evolved amounts to 85% of the theoretical. The apopinene thus obtained is almost pure: bp 47–48°C/30 mm Hg.

Chapter 7

Reduction

7.1. Introduction

Reductions are important transformations in organic synthesis, and in these reactions transition metals can perform many roles because of their ability to coordinate the substrate. This is illustrated, for example, in functional group activation, functional group protection, and controlled and selective reduction, which can all be achieved using transition metals. Alternatively, the transition metal can activate the reducing agent, typically hydrogen, or, in transfer hydrogenation, the hydrogen donor, e.g., cyclohexene.

Three different methods of functional group reduction can be identified: (1) hydrogenation, using molecular hydrogen with heterogeneous or homogeneous catalysts; (2) transfer hydrogenation, using organic molecules to donate hydrogen; and (3) transition metal–metal hydride reductions, using combined reagents for selective reduction.

In hydrogenation, transition metals provide a template on which activated hydrogen and the substrate can combine. In transfer hydrogenation, as the name implies, transition metals facilitate the transfer of hydrogen from a donor organic molecule to the substrate. Finally, activation of a functional group by a transition metal extends the range of metal hydride reductions. The use of these methods for the reduction of a variety of unsaturated functional groups, including alkyne, alkene, carbonyl, nitro, and nitrile groups, and for the hydrogenolysis of halogeno, benzyl, and allylic functions is described in this chapter. The examples have been chosen for their practical utility or because they demonstrate a unique feature. This is especially true for the examples of hydrogenation catalysts which, for the most part, operate under mild conditions, preferably ambient, but generally <10 atm H_2 and <150°C.[8,9] Specialized inert atmosphere handling techniques are usually not required.

A brief introduction to these three reductive methods will first be given, followed by selected examples classified on the basis of the functional group being reduced.

7.1.1. Hydrogenation

The unique ability of transition metals to catalyze the hydrogenation of unsaturated organic molecules by hydrogen gas has been known for a long time,[1] and numerous industrial processes are based on this transformation (Table 7.1). Heterogeneous catalysts provide the earliest examples of hydrogenation systems and are still the most widely used. Usually heterogeneous hydrogenation catalysts are based on Group VIII transition metals, i.e., Ru, Co, Rh, Ir, Ni, Pd, or Pt, but with some notable exceptions such as chromium. Typical examples of these systems and their applications are shown in Table 7.2. Generally, heterogeneous catalysts do not favor highly selective hydrogenations; however, they are readily available and convenient to use. Side reactions including isomerization, hydrogenolysis, and hydrogen scrambling are often a problem, especially in palladium-catalyzed hydrogenations, but by careful choice of conditions and catalyst many of these problems can be avoided. The discussion on heterogeneous catalysts in this chapter enables a suitable catalyst to be selected, but often the choice of conditions and solvent play an important part in achieving high yields and these differ for each substrate. A number of excellent books are available which provide further examples and numerous experimental procedures relating to these reactions.[2,3]

Generally, when selective hydrogenation is required, homogeneous catalysts are preferred.[12,13] Group VIII transition metal complexes often produce active homogeneous catalysts, and some useful examples are shown in Table 7.3. The high selectivity observed for homogeneous catalysts results from the fact that they contain only one type of active metal center, whereas a heterogeneous surface contains a variety of active sites differing in both electronic and steric environments. The hydrogenation of 1,4-dihydroaromatics demonstrates these selectivity differences[4] [Eq. (1)]. Homogeneous catalysts are very sensitive to the nature of the unsaturated groups to be reduced. Alkynes and mono- or disubstituted alkenes are readily hydrogenated, but tri- or tetrasubstituted alkenes require more severe conditions. Similarly, carbonyl and nitro groups are not reduced by the majority of homogeneous catalysts and, as a result, functional group

Table 7.1. Hydrogenation Catalysis—Industrial Processes

Transformation	Catalyst	Conditions
$NC{+}CH_2{+}_4CN \rightarrow H_2N{+}CH_2{+}_6NH_2$ 90%–95% selective	Co–Cu	600–650 atm 100–135°C
	Raney Nickel or Pt–Li/Al$_2$O$_3$	20–40 atm 170–230°C
 >95% selective	Pd–Ca/Al$_2$O$_3$	1–2 atm 140–170°C
 98%–99% selective	Pd–C or Raney Nickel	>50 atm 100°C
 99% selective	Nickel sulfide plus Cu or Cr	300–475°C
Partial hydrogenation of fatty acids	Supported nickel	1–6 atm 120–190°C
 L-Dopa	$[Rh(diene)\{Ph(o\text{-}MeOC_6H_4)$ $P{-}CH_2CH_2P(o\text{-}MeOC_6H_4)Ph)]^{\oplus}$	3 atm 50°C

Table 7.2. Heterogeneous Hydrogenation Catalysts

Catalyst		
1. Raney Ni, Pd, Pt	Readily available catalysts	
2. PtO$_2$	Useful for the	
3. Supported Pd or Pt, e.g., Pd/C, Pd/BaSO$_4$, Pt/Al$_2$O$_3$	hydrogenation of a variety of groups, including alkynes, alkenes, carbonyl,	
4. Rh/C, or Rh/Al$_2$O$_3$	aromatic ring, nitro, and nitrile	

Selective catalysts	Transformation	Ref.
Ni(OAc)$_2$/RONa/NaH	Alkynes → alkenes	92
	Carbonyl → alcohol	
Pd/support–quinoline	Alkynes → alkenes	166
Pd–C	Partial reduction of polyaromatics	82a
Pd/BaSO$_4$	Acid chloride to aldehyde	112
PtO$_2$–FeSO$_4$–Zn(OAc)$_2$	α, β-Unsaturated carbonyl to unsaturated alcohol	106
Rh/support	Aromatic ring to saturated ring	2

selectivity presents no problems; e.g., the selective double-bond reduction of α,β-unsaturated ketones is easily achieved using [Rh(PPh$_3$)$_3$Cl].

(1)

7.1.1.1. Heterogeneous vs. Homogeneous Catalysts

A comparison of heterogeneous with homogeneous catalysts illustrates why heterogeneous systems are most widely used but demonstrates that

Table 7.3. Homogeneous Hydrogenation Catalysts

Catalyst	Transformation	Comment	Ref.
[Rh(PPh₃)₃Cl]	Alkene or alkyne to alkane	Mono- and disubstituted alkenes easily reduced, trisubstituted more difficult, and tetrasubstituted unaffected. Insensitive to other functional groups, e.g., ketone, nitro, or nitrile, but not aldehyde	33
[Ru(PPh₃)₃Cl₂]– amine	Alkene to alkane nitro to amine	As above except for NO₂	15 123b
	Acid anhydride to lactone		116 117
[Rh(diene)-(diphosphine*)]⊕	Alkene to chiral alkane	Diphosphine* ≡ chiral diphosphine; see references for examples	154
	Ketone to chiral alcohol		168
[RhCl(monoene)]₂– diphosphine*	Alkene to chiral alkane	As above	154
K₂PtCl₆–SnCl₂	Polyene to monoene	Mixed products formed because of rapid isomerization	49a
[Cr(arene)(CO)₃]	1,3 Diene to *cis* alkene	Effective but requires 30 atm pressure	53
[HCo(CN)₅]³⁻	1,3 Diene to alkene	Very sensitive to steric hindrance, solubility problems	55b
[Rh(CO)₂(PPh₃)]₂	1,3 Diene to terminal alkene	Requires 15 atm pressure	56
[CoBr(PPh₃)₃–BF₃OEt₂]	1,3 Diene to less substituted alkene	Reduces the more substituted double bond	57
[Co(π-allyl)[P(OMe)₃]₃]	Aromatic ring to saturated ring	Slow reaction giving all *cis* hydrogen addition	73
[Rh(diene)(PR₃)₂]⊕	Carbonyl to alcohol		96
[Rh(PPh₃)₃Cl] 1. HSiR₃ 2. Hydrolysis	Carbonyl to alcohol	Using chiral ligands can achieve optical induction	102a
[Rh(PPrⁱ₃)₃H]	Nitrile to amine		142

homogeneous complexes have a significant part to play and are yet to be fully exploited. The major differences are the following.

1. Heterogeneous catalysts are readily available (commercially, or are easily prepared), whereas homogeneous catalysts can require elaborate syntheses involving the use of an inert atmosphere. However, it is worth noting that some of the most active homogeneous systems are easily prepared [e.g., $Rh(PPh_3)_3Cl$ and $RuCl_2(PPh_3)_3$; see Section 9.2.1] and many can now be purchased.

2. Heterogeneous catalysts are easily separated from the reaction products by filtration, whereas homogeneous catalysts are difficult to separate and often require either catalyst destruction followed by filtration or, more usually, chromatography. To avoid such difficult separations, many workers have attempted to immobilize homogeneous catalysts on heterogeneous supports such as cross-linked polystyrene or silica.[5a] Results to date have not matched the homogeneous system alone.

3. Homogeneous catalysts offer greater selectivity compared with heterogeneous systems. Specific *cis* addition of hydrogen (deuterium), functional group selectivity, and the hydrogenation of readily isomerized or hydrogenolyzed substrates are easily achieved with homogeneous catalysts.

7.1.1.2. Mechanism of Hydrogenation

Hydrogenations using heterogeneous catalysts are difficult to study mechanistically but the reaction is thought to proceed through adjacently adsorbed species. These processes are often reversible and this accounts for the observed isomerization and scrambling reactions. Homogeneous systems are more easily studied and a number of different mechanisms have been identified.[5-7]

1. Oxidative Addition of Hydrogen. This is the dominant mechanism for many homogeneous catalysts and is typified by Wilkinson's catalyst (Scheme 1). The initial step involves the dissociation of a phosphine ligand from $[Rh(PPh_3)_3Cl]$ to give the coordinatively unsaturated Rh(I) species **1**. This 14-electron intermediate readily adds hydrogen oxidatively to form the Rh(III) dihydride **2**. Alkene coordination gives **3**, which then undergoes a migratory insertion reaction to form the hydridoalkyl rhodium complex **4**. The catalytic cycle is completed by alkane elimination to regenerate intermediate **1**.

A closely related mechanism has been proposed for asymmetric hydrogenation catalyzed by $[Rh(diene)(PR_3)_2]^{\oplus}$, and here intermediates have been identified (Section 7.7.3). Interestingly, with these cationic complexes the alkene is thought to coordinate to the metal *prior* to the addition of hydrogen.

Scheme 1

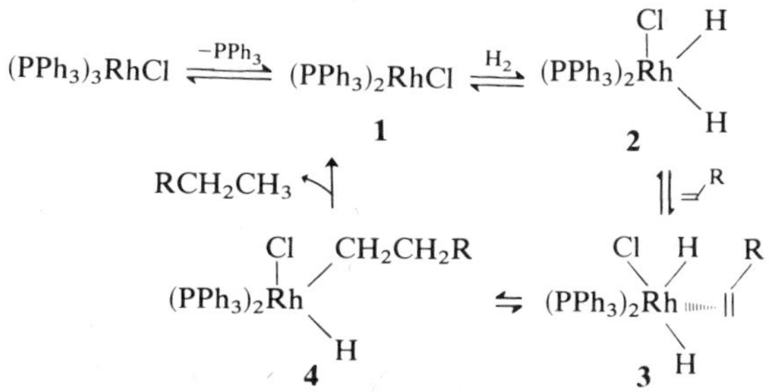

$(PPh_3)_3RhCl \underset{}{\overset{-PPh_3}{\rightleftharpoons}} (PPh_3)_2RhCl \overset{H_2}{\rightleftharpoons} (PPh_3)_2Rh\begin{smallmatrix}Cl & H \\ & \\ & H\end{smallmatrix}$

1   **2**

$RCH_2CH_3 \nwarrow$ $\parallel = R$

$(PPh_3)_2Rh\begin{smallmatrix}Cl & CH_2CH_2R \\ & \\ & H\end{smallmatrix}$ $\rightleftharpoons$ $(PPh_3)_2Rh\begin{smallmatrix}Cl & H & R \\ & & \\ & & \parallel\end{smallmatrix}$

4   **3** H

---

2. *Heterolytic Cleavage of Hydrogen.* The platinum–tin catalysts prepared by mixing H_2PtCl_6 or K_2PtCl_4 with stannous chloride react with
hydrogen by heterolytic cleavage (Scheme 2). The active species are
$[Pt(SnCl_3)_5]^{3-}$, $[PtCl_2(SnCl_3)_2]^{2-}$, and other $[PtCl_x(SnCl_3)_{4-x}]^{2-}$ complexes.

Scheme 2

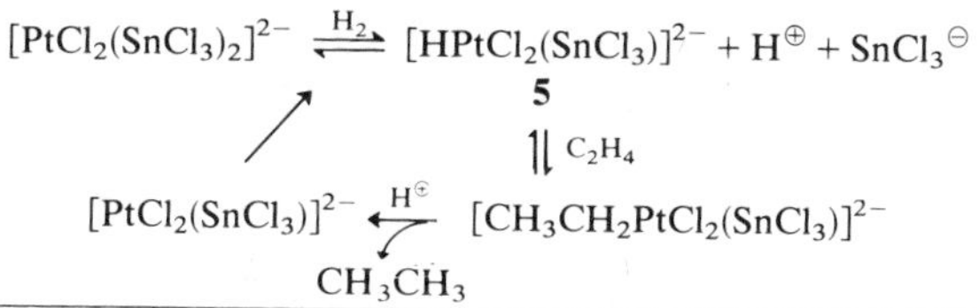

$[PtCl_2(SnCl_3)_2]^{2-} \overset{H_2}{\rightleftharpoons} [HPtCl_2(SnCl_3)]^{2-} + H^{\oplus} + SnCl_3^{\ominus}$

5

$\parallel C_2H_4$

$[PtCl_2(SnCl_3)]^{2-} \overset{H^{\ominus}}{\underset{}{\leftarrow}} [CH_3CH_2PtCl_2(SnCl_3)]^{2-}$

CH_3CH_3

The $SnCl_3^-$ ligands are labile and dissociate to produce vacant sites for
hydrogen and alkene coordination. The so-formed platinum hydride species
5 coordinates the alkene and, as with Wilkinson's catalyst, undergoes a
migratory insertion reaction. The active metal center is regenerated by
proton cleavage of the platinum–alkyl bond.

3. *Homolytic Cleavage of Hydrogen.* From kinetic studies it is proposed that metal–metal-bonded cobalt catalysts homolytically cleave hydrogen to generate the active species (Scheme 3). The reaction of the hydride

Scheme 3

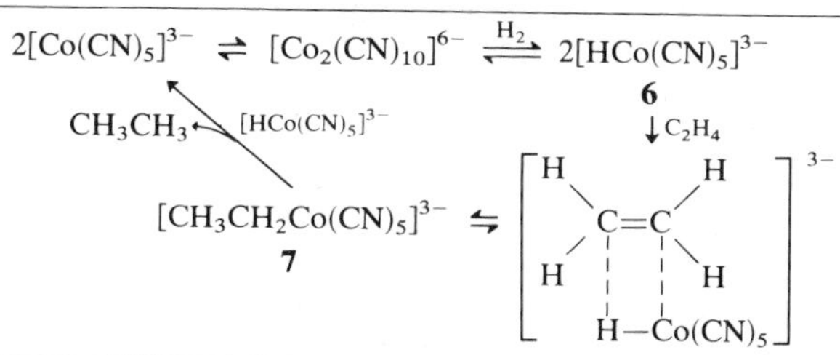

$2[Co(CN)_5]^{3-} \rightleftharpoons [Co_2(CN)_{10}]^{6-} \overset{H_2}{\rightleftharpoons} 2[HCo(CN)_5]^{3-}$

6

$CH_3CH_3 \overset{}{\underset{[HCo(CN)_5]^{3-}}{\nwarrow}}$ $\downarrow C_2H_4$

$[CH_3CH_2Co(CN)_5]^{3-} \rightleftharpoons \left[\begin{smallmatrix}H & & & H \\ & C=C & \\ H & | & | & H \\ & H-Co(CN)_5\end{smallmatrix}\right]^{3-}$

7

species **6** with the alkene is thought to occur without precoordination of the alkene in a concerted reaction. If this is correct, it is an unusual mechanism for a transition metal catalyst. The active species is regenerated by reaction of **7** with the hydride cobalt anion **6**.

7.1.2. Transfer Hydrogenation

Transfer hydrogenation offers an alternative method of reducing unsaturated substrates without the use of hydrogen gas.[10] Hydrogen is transferred from an organic donor molecule to the substrate. The donor molecule must have an oxidation potential sufficiently low that hydrogen transfer can occur under mild conditions. Typical examples are shown in Table 7.4. The common catalyst for these reactions is palladium on carbon, but Raney nickel has been used and numerous homogeneous systems including $[HIrCl_2(Me_2SO)_3]$, $[RuCl_2(PPh_3)_3]$, and $[Rh(PPh_3)_3Cl]$ have also proved effective.

The reaction requires at the most only refluxing conditions, but reaction times tend to be long, although recent work has suggested that shorter reaction periods can be achieved using an increased catalyst-to-substrate ratio. The use of palladium tends to promote hydrogenolysis reactions, which can cause serious complications. Controlled reduction is difficult to

Table 7.4. Transfer Hydrogenation Systems

System	Transformation	Ref.
H_2IrCl_6 } $P(OMe)_3-Pr^iOH$ $IrCl_4$	Cyclic ketone to axial alcohol	99a 100
$HIrCl_2(Me_2SO)_3-Pr^iOH$	α,β-Unsaturated carbonyl to unsaturated alcohol	108
$CrSO_4-H_2O$ or DMF	Internal alkyne to *trans* alkene	23
Pd/C–cyclohexene Pd/C–limonene Pd/C–tetralin	General reagent used for the reduction of alkynes, alkenes, carbonyl, nitrile, nitro, and for dehalogenation	10
Pd/C–AlCl₃–cyclohexene	Benzyl alcohols to toluene derivatives	188
$[Ru(H)Cl(\eta-C_6Me_6)PPh_3]-$ Pr^iOH	Alkene or diene to alkane	190
$Pd/C-HCO_2H$	Nitro to amine	191
$Pd/C-Et_3N-HCO_2H$	α,β-Unsaturated carbonyl to saturated carbonyl	183

achieve and in many instances total reduction of all functional groups occurs, hence selectivity is often poor. Recent developments using homogeneous transfer catalysts have produced highly selective reactions, perhaps the best known being the reduction of cyclic ketones to axial alcohols by H_2IrCl_6-2-propanol.

Transfer hydrogenation reactions are simple to perform and do not require pressure equipment. Results, certainly for simple substrates, are often good, and it is difficult to see why more use has not been made of this reductive method; perhaps this situation will change in the future.

7.1.3. Transition Metal–Metal Hydride Reductions

Metal hydride reducing agents are very useful for the reduction of a variety of functional groups.[11] In combination with transition metals the range of hydride reductions has been extended and interesting selective reactions have also been developed. Some useful examples are shown in Table 7.5. The transition metal can either activate a functional group or, in combination with the metal hydride, form a transition metal hydride, which is a more effective reducing agent than the metal hydride alone. Alternatively, the transition metal can protectively coordinate an unsaturated center, typically a carbonyl function, and allow selective reduction to be achieved. The choice of transition metal–metal hydride combination is purely empirical and the correct combination can only be obtained by experiment. This unfortunate drawback may be resolved as future work in this technique develops. Even so, novel combinations are providing some very useful synthetic procedures.

Table 7.5. Transition Metal–Metal Hydride Systems

System	Transformation	Ref.
$TiCl_4$–$LiAlH_4$	Internal alkyne → *trans* alkene	24
$CoCl_2 \cdot 6H_2O$–$NaBH_4$	Alkene → alkane; very sensitive to steric hindrance	51
$PdCl_2$–$NaBH_4$ ⎫ $Ni(OAc)_2$–$NaBH_4$ ⎭	α,β-Unsaturated aldehydes → saturated aldehydes	68
$Ni(PPh_3)_3$–$NaBH_4$	Aryl bromides → dehalogenated aryl	184
$Pd(Ph_3P)_4$–$NaBH_4$	Allyl acetates → allyl	189
CuI–$LiAlH_4$	α,β-Unsaturated ketone → saturated ketone	64
$Cu(acac)_2$–$NaBH_4$	aryl nitro → aryl amine	128
$CeCl_3 \cdot 6H_2O$–$NaBH_4$	α,β-Unsaturated ketone → saturated alcohol	105
$CeCl_3 \cdot 6H_2O$–$NaBH_4$	Ketone–aldehyde → alcohol–aldehyde	105

7.2. Reduction of Triple Bonds

7.2.1. Formation of Alkanes

Triple bonds are readily reduced by a variety of homogeneous and heterogeneous catalysts under mild conditions. They are generally the most easily reduced functional group in a molecule, being preferentially reduced over, for instance, double bonds, nitro groups, or carbonyl functions. This order can be reversed for aromatic nitro groups. Cobalt polysulfide (CoS_x) is a selective catalyst for the preferential hydrogenation of aromatic nitro groups in the presence of alkynes [Eq. (2)],[140] and, when the triple bond

$$\text{(2)}$$

is sterically crowded, ruthenium on alumina also reduces the nitro group in preference to the alkyne [Eq. (3)][134] (see Section 7.6.3.1). The ease of

$$\text{(3)}$$

reduction stems from the ability of an alkyne to strongly bond to the catalytic metal center (i.e., alkyne > alkene). If the coordination is too strong, a stable species may be formed and this will inhibit hydrogenation. This is the case for $[RuHCl(PPh_3)_3]$, which forms a stable complex with terminal alkynes, and as a result no hydrogenation occurs.[15] One of the many readily available heterogeneous catalysts, PtO_2, Pt/C, Pd/C, or Raney nickel, is the catalyst of choice for simple alkyne reduction. These catalysts reduce most simple alkynes to alkanes in high yield under mild conditions (1 atm H_2 at room temperature) and work-up is straightforward. These reactions are often exothermic and may even require cooling when a very reactive alkyne is used.

7.2.2. Formation of Alkenes

The stepwise reduction of an alkyne to an alkene is a more problematic reaction but the more useful synthetically. Of the heterogeneous catalysts, supported palladium is preferred for high selectivity, but often poisoning is required to prevent further reduction of the alkene. Commonly used supports include $BaSO_4$, $CaCO_3$, carbon, or alumina. Typical of the deactivated palladium catalysts is $Pd/BaSO_4$ inhibited with quinoline; this system reduces 6-*cis*-cyclodecenyne to 1-*cis*,6-*cis*-cyclodecadiene in good yield[16a] [Eq. (4)].

$$\text{(4)}$$

Reduction of Cyclodecenyne to Cyclodecadiene [16a]

The reaction flask is charged with 6-*cis*-cyclodecenyne (414 mg) and the hydrogenation catalyst $Pd/BaSO_4$ (50 mg) in methanol (10 ml) to which two drops of quinoline have been added to inhibit the catalyst. The flask is placed under a hydrogen atmosphere (1 atm) and maintained at room temperature with stirring. One equivalent of hydrogen is taken up in 40 min, whereupon hydrogen uptake ceases. The solution is filtered and the filtrate poured into water. The product is then extracted with petroleum ether, dried, and distilled. Yield of cyclodecadiene, 345 mg.

The method of catalyst preparation can greatly affect the success of these reactions. Detailed procedures for the preparation of one such catalyst, palladium on calcium carbonate, and its use with an inhibitor for the partial reduction of alkynes, has been described.[16b] This method of preparation of $Pd/CaCO_3$ produces a catalyst suitable for the hydrogenation of many triple bonds to double bonds. Because triple bond reduction is facile, subambient temperatures are often preferred for controlled reduction to the alkene. Similarly, addition of only one mole equivalent of hydrogen aids selectivity. In the hydrogenation of alkynes, using the heterogeneous catalyst system [NaH–RONa–Ni(OAc)$_2$] at 25°C and 1 atm of hydrogen, it was observed that little formation of alkane occurred until almost all the alkyne had been converted into alkene.[92] Therefore, by monitoring the hydrogen uptake excellent selectivity can be achieved. For example, 3-hexyne is reduced to 3-hexene, 3-*N,N*-dimethylaminoprop-1-yne to the prop-1-ene derivative, and 3-methyl,3-hydroxy pent-1-yne to the pent-1-ene analog, all in ~80% yield. The catalyst is easily prepared and can be stored for long periods with reproducible results. As well as partial hydrogenation of alkynes, it can be used for the reduction of alkenes and carbonyl compounds under ambient conditions and therefore does not require the use of high-pressure apparatus.

General Procedure—Partial Hydrogenation of 3-N,N-Dimethylamino prop-l-yne[92]

To the hydrogenation flask, under an atmosphere of hydrogen, is added methanol (50 ml) and the catalyst [NaH–RONa–Ni(OAc)$_2$] (Ni : substrate, 1 : 50 mol). [Catalyst preparation: To washed NaH (60 mM) in THF (20 ml) under nitrogen is added nickel acetate (anhydrous) (10 mM) in THF (5 ml). The mixture is heated to 45°C and *t*-amyl alcohol (20 mM) in THF (5 ml) added slowly. A black coloration develops rapidly. After 3 h at 45°C the mixture is cooled and stored. This catalyst slurry has a long lifetime stored under nitrogen.] Neutralization, by methanol, of excess NaH is immediate. 3-*N,N*-Dimethylamino prop-l-yne (20 mM) in methanol (10 ml) is then added and the mixture stirred under an atmosphere of hydrogen. Hydrogen uptake is monitored and, after 1.2 equivalents (1 equivalent for disubstituted alkynes) have been absorbed, the reaction is terminated (usually <15 min). The catalyst is filtered off and washed with pentane. The combined filtrates are then stirred under an atmosphere of dry HCl to obtain the hydrochloride adduct. The crude hydrochloride is purified by dissolving it in a minimum amount of acetone and precipitating with diethyl ether. Yield, 90%; selectivity, 90%.

The reaction proceeds through coordination of the alkyne to the metal center and results in the *cis* addition of hydrogen, in contrast to many other reduction methods, e.g., Na/NH$_3$ [17] or LiAlH$_4$/diglyme,[18] which produce the *trans*-alkene. Using a palladium catalyst isomerization, either *cis–trans* isomerization or double-bond migration, may be a problem. (Isomerization reactions are discussed in detail in Chapter 5.) When the substrate is very prone to isomerization, platinum- or nickel-based catalysts may be preferred. The catalyst system Ni(OAc)$_2$/NaBH$_4$/EtOH (sometimes with a promoter, e.g., ethylene diamine) is claimed to give highly *cis*-hydrogen addition with a *cis* : *trans* ratio as high as 200 : 1[19] [Eq. (5)].

$$\text{MeC(CH}_2)_3\text{C} \equiv \text{CEt} \xrightarrow[\text{H}_2,\ \text{EtOH}]{\text{Ni(OAc)}_2/\text{NaBH}_4} \text{MeC(CH}_2)_3\text{CH} = \text{CHEt} \qquad (5)$$

80%

Unfortunately no one heterogeneous catalyst is suitable for all alkyne reductions. Often the most appropriate catalyst system and reaction conditions are only achieved by trial and error.

A number of homogeneous catalysts have been described for the stepwise reduction to alkenes. Dicarbonylbis(cyclopentadienyl) titanium [Ti(η-C$_5$H$_5$)$_2$(CO)$_2$] is a catalyst for the selective hydrogenation of terminal alkynes to alkenes, provided the alkyne is not activated by conjugation, when total reduction occurs[20] [Eq. (6)]. Internal alkynes are not reduced by this catalyst [Eq. (7)]. Unfortunately these reactions require more

$$CH_3(CH_2)_nC\equiv CH \xrightarrow[H_2 \ (n=2 \ or \ 4)]{Ti(\eta-C_5H_5)_2(CO)_2} CH_3(CH_2)_nCH\equiv CH_2 \qquad (6)$$

$$CH_3CH_2C\equiv C(CH_2)_2CH_3 \xrightarrow[H_2]{Ti(\eta-C_5H_5)_2(CO)_2} \text{no reaction} \qquad (7)$$

vigorous conditions (50–60°C/50 atm) than with the heterogeneous systems previously discussed.

The cobalt catalyst $[CoH(CO)[P(n\text{-}Bu)_3]_3]$ hydrogenates both terminal and internal alkynes to the corresponding alkene but again requires more severe conditions (40°C/30 atm).[21] Aldehydes are also hydrogenated by this system.

$[Rh(diene)(PR_3)_{2 \ or \ 3}]^{\oplus}PF_6^{\ominus}$ allows reduction of internal alkynes to the *cis*-alkenes in virtually quantitative yield.[22a] Using [Rh(norbornadiene)(PPhMe_2)_3]^{\oplus}, $Me_2C(OH)C\equiv CMe$ is reduced quantitatively to the *cis*-alkene[22b] [Eq. (8)].

$$\underset{\displaystyle Me_2\overset{\textstyle OH}{\overset{|}{C}}-C\equiv CMe}{} \xrightarrow[\text{acetone, } H_2]{[Rh(NBD)(PPhMe_2)_3]^{\oplus}} \underset{\displaystyle Me_2\overset{\textstyle OH}{\overset{|}{C}}-CH=CHMe}{} \qquad (8)$$

The homogeneous system chromium(II) sulfate in water or dimethylformamide will reduce many alkynes stereospecifically to the *trans*-alkene at room temperature.[23] A polar group in the β position facilitates the reaction, e.g.:

$$\equiv -\!\!\!\overset{\displaystyle}{\underset{\displaystyle \ \ \ \ OH}{\diagup}}, \ \sim PhC\equiv C-CO_2H > HO\!\!-\!\!\overset{\diagup}{\ }\!\!\equiv\!\!-, \ \sim PhC\equiv CH \gg MeC\equiv CEt$$

Typical reductions are shown in Table 7.6.

Table 7.6. Partial Reduction of Alkynes Using Chromous Sulfate

Substrate	Solvent	Product	Yield, %
$HC\equiv C-CH_2OH$	H_2O	$H_2C=CH-CH_2OH$	89
$Ph-C\equiv CH$	DMF/H_2O 2:1	$PhCH=CH_2$	94
$HO_2C-C\equiv C-CO_2H$	H_2O	*trans*-$HO_2C-CH=CHCO_2H$	94
$Me-C\equiv C-CH_2-OH$	H_2O	*trans*-$MeCH=CHCH_2OH$	84
$Ph-C\equiv C-\!\!\overset{\displaystyle}{\underset{\displaystyle CO_2H}{\bigcirc}}$	H_2O	*trans*-$Ph-CH=CH-\!\!\overset{\displaystyle}{\underset{\displaystyle CO_2H}{\bigcirc}}$	85

Partial Reduction of Propargyl Alcohol[23]

To a three-neck flask (500 ml) containing aqueous $CrSO_4$ (284 mM) under nitrogen and equipped with a stirring bar, nitrogen inlet, and serum cap is added propargyl alcohol (130 mM). The blue color changes to green instantaneously. The reaction mixture is stirred for 1 h, after which time the dark green solution is saturated with ammonium sulfate and extracted with ether. The organic layer is separated, dried, and distilled. Yield of allyl alcohol, 4 g (89%).

The reagent $LiAlH_4/TiCl_4$ converts internal alkynes to the *trans*-alkenes under mild conditions[24] [Eq. (9)]. When the triple bond is terminal, further reduction to the alkane occurs.

$$CH_3CH_2CH_2C\equiv CCH_2CH_2CH_3 \xrightarrow{TiCl_4/LiAlH_4}$$

$$\begin{array}{cc} CH_3CH_2CH_2 & CH_2CH_2CH_3 \\ \diagdown & \diagup \\ C=C \\ \diagup & \diagdown \\ H & H \\ & 73\% \end{array} \qquad (9)$$

7.2.3. Selective Reductions

The selective reduction of alkynes in the presence of other functional groups generally presents no problem because the triple bond is the most reactive group toward hydrogenation. Alkyne selectivity is observed in the presence of groups such as keto, hydroxy, cyano, nitro, chloro, aldehyde, and carboxylic acid with both heterogeneous[2] and homogeneous catalysts.[13] Because selectivity is not a problem, one of the heterogeneous catalysts described earlier is preferred. Acetylenic alcohols, formed by the condensation of an alkyne and a carbonyl compound, have been extensively studied, being useful synthetic intermediates. They are readily reduced to either the unsaturated or saturated alcohol[29,30] [Eqs. (10) and (11)].

$$BuC\equiv CH \xrightarrow[2.\ MeCOEt]{1.\ EtMgBr} \begin{array}{c} Et \\ | \\ BuC\equiv CCMe \\ | \\ OH \\ 92\% \end{array} \xrightarrow{Pt,\ H_2} \begin{array}{c} Et \\ | \\ BuCH_2CH_2CMe \\ | \\ O \\ H \end{array} \qquad (10)$$

$$Bu^i\!\!\leftarrow\!\!C\equiv C\!\!\rightarrow_2 CH_2OH \xrightarrow{PtO_2,\ H_2} Bu^i(CH_2)_4CH_2OH \qquad (11)$$

The reverse selectivity, i.e., the reduction of a functional group in the presence of an alkyne, has been briefly mentioned but is more fully discussed in the appropriate functional group section.

7.2.4. Formation of Polyunsaturated Species by Partial Reduction

The selective reduction of polyalkyne compounds does not normally present a problem because of the differing reactivity of the various unsaturated groups, i.e.,

terminal triple bond	>	internal triple bond	>	terminal double bond	>	internal double bond

ease of reduction decreases →

Hence hepta-1,5-diyne is reduced to hepta-1-en-5-yne under controlled conditions and may be further reduced to the diene with more hydrogen.[25] When two internal alkynes are present the least substituted bond will be reduced first. The order described above can be affected if one triple bond is very sterically hindered; then total reduction of one alkyne group may result[26] [Eq. (12)].

$$Bu'C{\equiv}CC(Me_2)C{\equiv}CMe \rightarrow Bu'C{\equiv}CC(Me)_2CH_2CH_2Me \qquad (12)$$

The reduction of polyunsaturated derivatives provides a useful route to insect sex pheromones,[27] for example, in the preparation of 10,12-hexadeca-dien-1-ol (**8**), a pheromone of the female silkworm moth [Eq. (13)].

(13)

High selectivity becomes a problem when the substrate is a vinyl alkyne or a conjugated diene. In these systems difficulties arise because the conjugated diene once formed is very prone to 1,4-hydrogen addition[28] and results in a mixture of products being produced.

7.3. Reduction of Double Bonds (Nonaromatic)

7.3.1. Formation of Alkanes

Hydrogenation of a double bond proceeds readily under mild conditions (room temperature and 1 atm) with a variety of heterogeneous and homogeneous catalysts. Generally heterogeneous catalysts based on rhodium, nickel, palladium, or platinum are preferred for simple hydrogenations (for examples, see Table 7.2), but these systems may cause isomerization. Palladium catalysts, especially, are prone to cause double-bond migration or *cis–trans* isomerization. When these are to be avoided then nickel or platinum systems are used.

Homogeneous catalysts offer an advantage over their heterogeneous counterparts when high selectivity is required or when access to the double bond is restricted, as in an unsaturated polymer. A soluble catalyst allows transfer of the active species to the unsaturated bond in a polymer, whereas a heterogeneous catalyst requires the polymer chain to unfold to allow surface absorption in a suitable orientation.[31] High selectivity allows reactions such as specific *cis* deuteration[32] [Eq. (14)]. Whereas heterogeneous catalysts cause scrambling, the homogeneous rhodium system $[Rh(PPh_3)_3Cl]$ gives very specific D_2 addition.

$$CH_3(CH_2)_7CH{=}CH(CH_2)_7CO_2Me \begin{array}{c} \xrightarrow[D_2]{Rh(PPh_3)_3Cl} CH_3(CH_2)_7(CHD)_2(CH_2)_7CO_2Me \\ \\ \xrightarrow[PtO_2,\,D_2]{} \text{mixture of } d_2{-}d_{18} \text{ methyl sterates} \end{array} \qquad (14)$$

General Procedure—Specific Deuteration using $[Rh(PPh_3)_3Cl]$[32b]

In the reaction flask, under nitrogen, is placed the alkene (1 g) in benzene (30 ml), followed by the catalyst $[Rh(PPh_3)_3Cl]$ (0.15 g). The flask is alternately evacuated and filled with deuterium gas (four times), finally maintaining the pressure at slightly above atmospheric pressure (plus 10 mm Hg). The mixture is stirred and the deuteration monitored by withdrawing small samples and analyzing by glc. Reaction is generally complete in <3 h for mono- and disubstituted alkenes. Specific *cis* deuteration across the double bond occurs usually without any hydrogen–deuterium scrambling.

The ease of reduction varies with the degree of substitution at the double bond:

tetra- < tri- < di- < monosubstituted double bond

———————————————⟶
rate of hydrogenation increases

Problems only arise with fully substituted double bonds, which require more severe conditions, and consequently side reactions (hydrogenolysis) may occur. Homogeneous catalysts are more sensitive to substitution at the double bond than heterogeneous systems, and this may be used to advantage in selective reductions (see Section 7.3.2). One of the most

studied homogeneous systems is Wilkinson's catalyst [Rh(PPh₃)₃Cl],[33]
which hydrogenates mono- and disubstituted alkenes at rates comparable
to heterogeneous systems (*vide supra*), but tri- and tetrasubstituted
alkenes react far more slowly.[34] Hence terminal alkenes are reduced faster
than internal analogs and, interestingly, *cis* alkenes more rapidly than the
corresponding *trans* isomers. The cationic rhodium complexes [Rh(diene)-
$(PR_3)_2]^\oplus$ [35] and [RhCl(py)₂(DMF)BH₄]Cl [36] are more efficient for the
hydrogenation of hindered double bonds, whereas [RhH(CO)(PPh₃)₃] [37]
selectively reduces unhindered terminal alkenes alone.

Double bonds having electron-withdrawing substituents (i.e.,
CO_2H, CO_2R, or CN) may be reduced by aqueous chromium(II) species
in high yield[38] [Eq. (15)].

$$
\begin{array}{c}
\overset{\displaystyle CO_2Et}{\underset{\displaystyle CH}{\overset{\displaystyle CH}{\vert\vert}}} \\
\underset{\displaystyle CO_2Et}{}
\end{array}
\quad \xrightarrow[H_2O]{CrSO_4} \quad
\begin{array}{c}
CH_2CO_2Et \\
\vert \\
CH_2CO_2Et \\
95\%
\end{array}
\qquad (15)
$$

A typical reduction of a simple alkene using a heterogeneous catalyst
is exemplified by the use of platinum oxide to hydrogenate cyclododecene.
The preparation of the catalyst is important and an extremely active form is
produced by heating chloroplatinic acid in molten sodium nitrate.[39]

Hydrogenation of Cyclododecene
The reaction vessel is first purged with nitrogen and then ethyl acetate
(50 ml), acetic acid (three drops), and cyclododecene (8.4 g) are added. The
catalyst, platinum oxide (84 mg), is then carefully added (PtO₂ catalyzes the
explosive oxidation of combustible organic vapors and therefore should be
added to an oxygen-free system). The nitrogen atmosphere is replaced with
hydrogen and the hydrogenation performed at 25°C with steady agitation.
After 15 min the reaction is complete (as monitored by hydrogen uptake).
Volatiles are removed on a rotary evaporator and the residue dissolved in
hexane, filtered, and vacuum dried. Yield, 8.5 g (100%).

Numerous examples describing the use of heterogeneous catalysts for
the reduction of simple double bonds are given in *Organic Synthesis
Collected Volumes*, including, for example, Eqs. (16)–(21).

$$
C_6H_5CH{=}CHCOC_6H_5 \xrightarrow{Pt, H_2} C_6H_5CH_2CH_2COC_6H_5 \qquad (16)^{40}
$$
$$
81\%\text{–}95\%
$$

$$
C_6H_5CH{=}C\overset{\displaystyle CO_2H}{\underset{\displaystyle NHCOMe}{\big\backslash\!\!\big/}} \xrightarrow{Pt, H_2} C_6H_5CH_2C\overset{\displaystyle CO_2H}{\underset{\displaystyle NHCOMe}{\big/\!\!\big\backslash}} \qquad (17)^{41}
$$
$$
81\%
$$

$$\text{(furan)} \xrightarrow[\text{H}_2]{\text{PdO-Pd}} \text{(tetrahydrofuran)} \qquad (18)^{42}$$

90%–93%

$$\text{(dihydropyran)} \xrightarrow[\text{H}_2]{\text{Raney Ni}} \text{(tetrahydropyran)} \qquad (19)^{43}$$

100%

$$\underset{\text{84\%–88\%}}{\text{OH}} \xrightarrow{\text{Pd/C, H}_2} \text{OH} \qquad (20)^{44}$$

$$\xrightarrow{\text{PtO}_2, \text{H}_2} \qquad (21)^{45}$$

96%–98%

7.3.2. Partial Reduction of Polyunsaturated Substrates

7.3.2.1. Nonconjugated Double Bonds

The rate of reduction of an alkene decreases with increasing substitution at the double bond, although this order can be influenced by both the bulk and position of the substituents.[46] In Eq. (22) the tetrasubstituted

$$\text{Me}_2\text{C} = \underset{}{\text{Me}} \xrightarrow{\text{Pt/C, H}_2} \text{Me}_2\text{CH} - \underset{}{\text{Me}} \qquad (22)$$

double bond is reduced in preference to the trisubstituted endocyclic bond. In ring systems exocyclic double bonds are reduced before those in the ring. The order of reduction in dienes can be affected by protecting the more reactive bond prior to reduction. Cyclopentadienyl iron dicarbonyl cation $[\text{Fe}(\eta\text{-C}_5\text{H}_5)(\text{CO})_2]^{\oplus}$ complexes to the more reactive double bond in a molecule, which then becomes inert to reduction.[47] Hence the least reactive double bond is reduced $[\text{Fp}^{\oplus} = \text{Fe}(\eta\text{-C}_5\text{H}_5)(\text{CO})_2^{\oplus}]$ [Eqs (23)– (25)]. The protecting group ($\text{Fp}^{\oplus}$) is easily removed by treatment with sodium iodide in acetone.

$$\underset{\text{Fp}^{\oplus}}{} \xrightarrow{\text{Pd/C, H}_2} \underset{\text{Fp}^{\oplus}}{} \qquad (23)$$

88%

$$\text{Fp}^{\oplus}\!\!-\!\!\!\equiv\!\!\!-\!\!\!-\!\!\!\quad\xrightarrow{\text{Pd/C, H}_2}\quad\text{Fp}^{\oplus}\!\!\!-\!\!\!-\!\!\!-\!\!\!-\!\!\!-\!\!\!- \tag{24}$$

$$\xrightarrow{\text{Pd/C, H}_2}\qquad\xrightarrow{\text{NaI}}\qquad \tag{25}$$

The reduction of di- or polyenes to monoenes is not easy to control using heterogeneous catalysts, and unless steric factors influence the reaction mixed products are obtained. Homogeneous catalysts offer more control. $[RuCl_2(PPh_3)_3]$ in the presence of an alcohol or amine specifically reduces terminal alkenes, while internal double bonds are unaffected.[15] This demonstrates the response of homogeneous catalysts to steric effects. Using the related system $[RuCl_2(CO)_2(PPh_3)_2]$ (9) in the presence of excess phosphine, polyenes are effectively reduced to monoenes[48] [Eqs. (26) and (27)]. The rate of reaction decreases in the order

conjugated dienes > nonconjugated dienes > terminal alkenes > internal alkenes

$$\xrightarrow{\textbf{9}, \text{H}_2, \text{PPh}_3}\qquad + \qquad \tag{26}$$

70%   25%

overall yield, 99%

$$\xrightarrow[\text{H}_2]{\textbf{9}, \text{PPh}_3}\qquad \tag{27}$$

97%

but isomerization, migration, or transannular ring closures (in cyclic dienes) can be important side reactions. It is suggested that excess triphenylphosphine helps suppress further hydrogenation of the alkene by shifting the equilibrium in favor of formation of $[RuCl(alkyl)(CO)_2PPh_3]$ over $[RuCl(alkyl)(CO)_2]$. The former, a sterically crowded intermediate, undergoes rapid β-hydrogen elimination, regenerating the alkene, whereas the latter is able to undergo hydrogenolysis to liberate alkane. This equilibrium is unimportant with dienes since the fifth coordination site is blocked by chelation of the alkenyl ligand.

The platinum–tin mixed-metal catalyst (the active species are $[Pt(SnCl_3)_5]^{3-}$, $[PtCl_2(SnCl_3)_2]^{2-}$, and other $[PtCl_x(SnCl_3)_{4-x}]^{2-}$ complexes), prepared by reaction of chloroplatinic acid and stannous chloride, allows polyene to monoene reductions under mild conditions.[49] The relative reaction rates have been measured and demonstrate why selective reduction is possible:

alkene isomerization ≫ polyene hydrogenation ≫ monoene hydrogenation

The reactivity order highlights a major problem with this catalyst, that of isomerization. The facile isomerization reaction generally produces isomeric mixtures; rarely are clean products obtained. Conjugated double bonds are also selectively reduced to monoenes; indeed, the prime isomerization reaction involves double-bond migration to produce a conjugated system (see Chapter 5). The potential of this catalyst is not great because pure products are rarely obtained, but it is useful in some selected cases and especially when other unsaturated functional groups are present (see Section 7.3.3). It has been extensively studied for the partial reduction of vegetable oils, e.g., soya bean oil or linoleates, to remove excessive unsaturation[50] [Eq. (28)].

$$Me(CH_2)_4CH=CHCH_2CH=CH(CH_2)_7CO_2Me \xrightarrow[H_2]{[PtCl_6^{2-} + SnCl_2]} \text{monoenate} + \text{dienoate}$$

                                                                                  14%   42%

+ conjugated dienoate   (28)
43%

The complex formed by a cobalt(II) salt and sodium borohydride reduces alkenes in high yield with high steric selectivity (mono- > di- ≫ tri-, and tetrasubstituted alkenes).[51] Mono- and disubstituted alkenes are easily reduced, whereas further substitution inhibits reduction under these conditions. This is demonstrated in limonene, where only the disubstituted

(29)

exocyclic double bond is reduced and not the trisubstituted endocyclic unsaturated bond. These reactions are simple to perform (no pressure equipment) and only require a blanket of nitrogen; i.e., rigorous inert atmosphere techniques are not necessary. Reaction times are of the order 3–24 h, the longer times being required for disubstituted alkenes.

General Procedure—Reduction of 1-Dodecene[51]

To a solution of 1-dodecene (1.68 g, 10 mM) and $CoCl_2 \cdot 6H_2O$ (2.38 g, 10 mM) in ethanol (25 ml) under a nitrogen atmosphere at 0°C is added $NaBH_4$ (0.760 g, 20 mM) in portions. The solution immediately becomes dark with evolution of hydrogen. The mixture is stirred at room temperature for 3 h and then poured into a 3 *M* HCl solution. The aqueous solution is extracted with ether, washed, and dried ($MgSO_4$). Evaporation of solvent gives crude dodecane, which is then purified (distillation or crystallization). Yield, 95%.

7.3.2.2. Conjugated Double Bonds

The partial reduction of conjugated dienes or polyenes is not straightforward, and mixtures are commonly obtained due to 1,2 and 1,4 addition and further hydrogenation. With heterogeneous catalysts it is often a case of trial and error to get an adequate reaction. The use of homogeneous catalysts with these types of substrates is beneficial, although extensive studies have not been undertaken.

Homogeneous chromium-based catalysts are useful for the partial reduction of conjugated substrates. Tricarbonyl chromium methylbenzoate will selectively add hydrogen 1,4 to conjugated dienes to afford the *cis*-monoalkene.[52] The reaction is highly stereospecific toward *trans, trans* conjugated dienes or to cyclic 1,3 dienes held in the required *cisoid* configuration [Eq. (30)]. However, the reaction conditions using [Cr(arene)(CO)₃] are relatively forcing (160°C/30 atm), but the related

$$\text{ } + \text{(arene) Cr(CO)}_3 \longrightarrow \text{ Cr(CO)}_3 \qquad (30)$$

$$(31)$$

$$> 90\%$$

$$(32)$$

$$90\%$$

system $[Cr(CO)_3(CH_3CN)_3]$, in which the acetonitrile ligands are labile, operates under far milder conditions (40°C/1.5 atm).[53] Similarly, $Cr(CO)_6$ photocatalyzes 1,4-hydrogen addition to dienes at 1 atm H_2 and room temperature.[54] Nonconjugated 1,4-dienes are also reduced using [(arene)-$Cr(CO)_3$] at high temperature, probably via isomerization to the conjugated system prior to reduction. 1,5- or 1,6-dienes are not reduced owing to the inability of the chromium complex to conjugate the double bonds.

General Procedure: Partial Reduction of Conjugated Dienes

1. *Low Pressure using $Cr(CO)_3 \cdot (CH_3CN)_3$* [53] [Prepared by refluxing $Cr(CO)_6$ with excess acetonitrile under nitrogen. The yellow product is pyrophoric in air.] To neat diene (60 mM) is added either an aceto-nitrile solution of $[Cr(CO)_3(CH_3CN)_3]$ $(2–5 \times 10^{-3}$ mM) or solid $[Cr(CO)_3(CH_3CN)_3]$ under nitrogen. The mixture is pressurized with hydrogen (20 psi) and heated to 40°C. After stirring for 1–8 h work-up gives the monoene product in high yield 70%–100% (99% pure).

Note: Solid $Cr(CO)_3(CH_3CN)_3$ is only partially soluble in the neat diene. Using the isolated complex gave the higher yields but probably more catalyst was used.

2. *High Pressure using (arene)$CrCO_3$.*[52] In a 150-ml autoclave under nitrogen is placed the diene (30 mM) and methyl benzoate chromium tricar-bonyl (0.5 mM) (for preparation see Section 9.4) in pentane (50 ml). The reaction vessel is purged repeatedly with hydrogen before heating to 160°C and pressurizing to 30 atm H_2. The mixture is stirred for 6 h and then $FeCl_3$/EtOH added to destroy the catalyst. The mixture is extracted with ether and vacuum distilled to give pure product in high yield.

The pentacyano cobalt anion $[Co(CN)_5]^{3-}$ finds synthetic applications in the reduction of conjugated dienes to monoenes.[55] The catalyst is prepared by reaction of cobalt(II) chloride and potassium cyanide under a nitrogen atmosphere in either aqueous (more common) or nonaqueous (e.g., methanol) solvents. The preference for aqueous solutions has limited the applications of this reaction owing to the poor water solubility of most organic substrates. The solution requires preactivation by addition of hydrogen (1 atm) for 1–2 h at 20°C to generate the active species $[Co(CN)_5H]^{3-}$. This solution is then used immediately, because an ageing process leads to deactivation. The reduction of simple dienes, e.g., butadiene isoprene, 1-phenylbutadiene, 1,3-cyclohexadiene, and cyclopentadiene, has received the most attention and these proceed readily under mild conditions.[55] However, the catalyst is rather sensitive to steric hindrance, which inhibits hydrogenation. For example, 2,5-dimethyl-2,4-hexadiene is not reduced by this system. Generally, monoenes are not reduced by $[Co(CN)_5]^{3-}$, but activated monoalkenes, e.g., styrenes and α,β-unsaturated aldehydes and carboxylic acids, are reduced [Eq. (33)]. Interestingly, acrylic acid and acrolein are not reduced.

$$\underset{PhC=CH_2}{\overset{CO_2H}{|}} \xrightarrow[H_2]{Co(CN)_5{}^{3-}} \underset{PhCHCH_3}{\overset{CO_2H}{|}} \tag{33}$$

100%

When the substrate is a 1,3 diene, $[RhH(PPh_3)_4]$ or $[Rh(CO)_2(PPh_3)]_2$ in the presence of excess phosphine will cause reduction to the terminal alkene (50–100°C/15 atm)[56] [Eqs. (34) and (35)]. The hydrogen uptake

$$H_2 + \quad \xrightarrow[2PEt_3]{[Rh(CO)_2PPh_3]_2} \quad 83\% \tag{34}$$

$$H_2 + \quad \xrightarrow[2PEt_3]{[Rh(CO)_2PPh_3]_2} \quad + \quad + \quad \text{other products} \tag{35}$$

29% 41%

must be monitored to prevent further reduction once the concentration of the diene becomes very low. Similar to the cobalt cyanide system, substitution at the double bond inhibits hydrogenation and limits general applicability.

The novel catalyst system $[CoBr(PPh_3)_3\text{–}BF_3OEt_2]$ (**10**) is interesting because it selectively hydrogenates conjugated dienes to monoenes via 1,2-hydrogen addition to the *more* substituted double bond[57] [Eqs. (36) and (37)]. Most catalysts preferentially reduce the least substituted double

$$\xrightarrow[0°C, 1\ atm\ H_2]{\textbf{10}} \tag{36}$$

86%

$$\xrightarrow[0°C, 1\ atm\ H_2]{\textbf{10}} \tag{37}$$

61%

bond. Unfortunately hydrogenation of functionalized dienes, e.g., methyl vinyl ketone, methyl acrylate, or *n*-butyl vinyl ether, was not observed even under forcing conditions.

7.3.3. Reduction of Unsaturated Aldehydes and Ketones

The carbon–carbon double bond in a molecule is usually reduced preferentially in the presence of other functional groups, except a triple

bond or aromatic nitro group. Selective double-bond reduction is not
therefore a problem. Palladium is the preferred heterogeneous catalyst
because the reduction slows dramatically after adsorption of one mole
equivalent of hydrogen whether the carbonyl is conjugated to the double
bond or not. Problems can arise when the substrate is sensitive to hydrogenolysis or isomerization, and in these cases homogeneous catalysts may
be preferred.

$$\text{(38)}$$

[Rh(PPh$_3$)$_3$Cl] and [Ru(PPh$_3$)$_3$Cl$_2$–amine] have been used extensively
in the selective hydrogenation of 3-oxo-1,4-diene steroids where they have
been shown to be particularly effective.[59] In all cases the least sterically
hindered bond is reduced almost exclusively, although the ruthenium complex is more selective. In pregna-5,16-dien-3-ol-20-one the double bond
in the 16 position is selectively reduced[2] (Eq. 40).

$$\text{(39)}$$

$$\text{(40)}$$

General Procedure—Reduction of Ergosta-1,4,22-trien-3-one[60] [*Eq. (41)*]
    To a benzene solution (Analar, dry, 20 ml) of the ketone (A) (400 mg)
under an atmosphere of hydrogen is added [RhCl(PPh$_3$)$_3$] (100 mg). The
mixture is stirred at room temperature for 4–5 h after which hydrogen uptake

$$\text{(A)} \longrightarrow \text{(B)} \tag{41}$$

ceases (24 ml uptake). The solvent is evaporated off *in vacuo* and the residue refluxed in petroleum ether (60°C–80°C), filtered hot, and washed three times with petroleum ether (20 ml). Crystallization from methanol gives the ketone product (B) (260 mg).

In steroids the disubstituted unsaturated centers are readily hydrogenated, but more substituted double bonds, i.e., the 4,5 or 5,6 positions, are more resistant. The trisubstituted 4,5 position may be reduced by [Rh(PPh$_3$)$_3$Cl] but requires prolonged reaction times (i.e., days), and in these cases [RhCl$_3$(py)$_3$–NaBH$_4$] in dmf is preferred.[61] The 5,6 position is quite resistant to reduction by these complexes. Hydrogen addition occurs stereospecifically from the least hindered side of the molecule and is influenced by substitution at the 17 position. Good yields of product are generally obtained.

Both [Ru(PPh$_3$)$_3$Cl$_2$] and [Rh(PPh$_3$)$_3$Cl] are ineffective for the hydrogenation of unsaturated aldehydes. [Rh(PPh$_3$)$_3$Cl] fails owing to formation of inactive [RhCl(CO)(PPh$_3$)$_2$] by decarbonylation of the aldehyde function. Primary allylic alcohols behave similarly. While the decarbonylation reaction restricts hydrogenation, use may be made of the former in organic synthesis[62] [Eqs. (42)–(44) (see also Section 6.6)].

$$\text{CHO} \xrightarrow[\text{room temperature, 2 h}]{\text{Rh(PPh}_3)_3\text{Cl, H}_2} \text{H} \tag{42}$$

65%
(83% after 70 h)

$$\text{PhCHO} \xrightarrow[\text{reflux, 2 h}]{\text{Rh(PPh}_3)_3\text{Cl, H}_2} \text{PhH} \tag{43}$$

80%

$$\text{PhCOCl} \xrightarrow[\text{reflux, 20 h}]{\text{Rh(PPh}_3)_3\text{Cl, H}_2} \text{PhCl} \tag{44}$$

90% (catalytic)

When [Rh(PPh$_3$)$_3$Cl] is used as a hydrogenation catalyst in methanol with a ketone, ketal formation may occur although the reaction is relatively slow[63] [Eq. (45)].

$$\xrightarrow[\text{H}_2,\ \text{MeOH}]{\text{Rh(PPh}_3)_3\text{Cl}} \tag{45}$$

The platinum–tin catalyst [H_2PtCl_6–$SnCl_2$] described earlier (Section 7.3.2.1) reduces α,β-unsaturated ketones to the saturated derivatives[49] [Eq. (46)].

$$\text{(structure)} \xrightarrow[H_2]{H_2PtCl_6-SnCl_2} \text{(structure)} \qquad (46)$$

Whereas the reduction of α,β-unsaturated ketones with $LiAlH_4$ produces the unsaturated alcohol, pre-reaction of the hydride with CuI allows facile 1,4 reduction to give the saturated ketone[64] [Eq. (47)]. Cyclic enones

$$\underset{}{RR'C=CHCR''} \xrightarrow[\text{CuI, R.T.}]{LiAlH_4} RR'CHCH_2CR'' \qquad (47)$$

$$\begin{bmatrix} \overset{}{\underset{}{\begin{array}{c} C \\ \| \\ C \end{array}}} \begin{array}{c} C=O \\ \diagdown \\ Al \\ \diagup \\ H \end{array} \end{bmatrix}$$

11

R = R″ = Bu^t; R′ = H   (99%)
R = R′ = R″ = Me   (70%)
R = H; R′ = R″ = Me   (78%)
R = R″ = Ph; R′ = H   (100%)

(e.g., cyclohexenone) are not reduced at all by this reagent, possibly owing to lack of formation of the six-centered proposed transition state (**11**), which requires a *trans* configuration.

Cyclic enone reduction may be achieved using the closely related reagent [$LiAlH(OMe)_3$–CuBr], which effectively reduces cyclohexenone or substituted derivatives to cyclohexanone (84%) or the corresponding analogs.[65] The commercial sodium reagent Vitride, $NaAlH_2$-$(OCH_2CH_2OMe)_2$, is also effective for these reactions when pre-reacted with CuBr. The active species in all these reactions is probably a copper hydride stabilized by the aluminum moiety; cf. the use of PrC≡CCuHLi for 1,4 reductions of enones[66] and related metal hydride reductions.[67]

Unsaturated aldehydes are notoriously difficult to reduce selectively with heterogeneous catalysts because slight variations in catalyst preparation, e.g., in solvent, the choice of support, etc., result in different mixtures of products being obtained, including total reduction of the aldehyde function. Homogeneous catalysts also have their problems (see above, for decarbonylation using [$Rh(PPh_3)_3Cl$]). The selective hydrogenation of

α,β-unsaturated aldehydes may be achieved using sodium borohydride-reduced metal salts [$PdCl_2$–$NaBH_4$] (**12**) and [$Ni(OAc)_2$–$NaBH_4$] (**13**).[68] The borohydride-reduced palladium salts are of interest because they *only* hydrogenate carbon–carbon π bonds, leaving the carbonyl function unaffected [Eqs. (48)–(50)]. The reduced nickel species are effective but

$$\text{PhCHO} \begin{array}{c} \xrightarrow{\textbf{12}, \text{H}_2} \text{no reaction} \\ \\ \xrightarrow{\textbf{13}, \text{H}_2} \text{PhCH}_2\text{OH} \\ \text{(slowly over 48 h)} \end{array} \qquad (48)$$

$$\text{CH}_3\text{CH}{=}\text{CHCHO} \xrightarrow[\text{R.T., } <3\,\text{h}]{\textbf{12}, \text{H}_2 \ (30\,\text{psi})} \text{CH}_3\text{CH}_2\text{CH}_2\text{CHO} \qquad (49)$$

$$\text{PhCH}{=}\text{CHCHO} \xrightarrow{\textbf{12}, \text{H}_2} \text{PhCH}_2\text{CH}_2\text{CHO}$$
$$\downarrow {\scriptstyle \textbf{13}, \text{H}_2} \qquad\qquad (50)$$
$$\text{PhCH}_2\text{CH}_2\text{CH}_2\text{OH}$$

not so selective. Among all the available catalysts the palladium system is one of the most reproducibly selective for double-bond reduction in unsaturated aldehydes.

7.4. Reduction of Aromatic Systems

7.4.1. Formation of Cycloalkanes

Heterogeneous catalysts are very effective for the hydrogenation of aromatic compounds and they are the commonly used catalysts in organic synthesis. Despite considerable effort relatively few homogeneous systems are known that catalytically hydrogenate aromatics, and those that do are so far of only limited practical use. The resistance of the aromatic ring to hydrogenation using soluble catalysts is demonstrated by the fact that benzene is the solvent of choice when using very active homogeneous hydrogenation systems, e.g., [$Rh(PPh_3)_3Cl$] or [$Ru(PPh_3)_3Cl_2$–NEt_3].[13,69] For hydrogenation of a benzene or pyridine ring, the preferred catalyst is either rhodium on carbon (or alumina) or platinum oxide. Both of these catalysts are effective at low temperatures and pressures (50–80°C, 2–3 atm). Platinum oxide usually requires acidic conditions [Eq. (51)], which can be a disadvantage in the reduction of anilines or pyridines (i.e., formation of an insoluble quaternary salt may occur). Supported rhodium catalysts

$$\text{(51)}$$

are prone to inhibition by strong nitrogen donors, and in these cases ruthenium dioxide or supported ruthenium at higher temperatures and pressures (70–100°C, 70–100 atm) is less susceptible to poisoning and often proves effective. Palladium in acidic media hydrogenates aromatic rings slowly and is useful when partial hydrogenation is required. Unfortunately, high pressures are generally required.

General Procedure—Reduction of Mandelic Acid[71]
  Into a Parr hydrogenation apparatus,[8] previously flushed repeatedly with hydrogen, are placed *dl*-mandelic acid (7.6 g, 50 mM), rhodium (5%) on alumina (1.5 g), methanol (40 ml), and glacial acetic acid (0.5 ml). The mixture is pressurized with hydrogen to 3–4 atm and stirred or shaken for 1.5 h. The mixture is then filtered through Celite and concentrated on a rotary evaporator to yield 99% of *dl*-hexahydromandelic acid, mp 135–136°C, without recrystallization.

The major problems associated with aromatic hydrogenation are hydrogenolysis and the ready reduction of functional groups (e.g., preferential reduction of triple or double bonds or nitro groups). Particularly sensitive to cleavage are ether linkages, benzylic alcohols, and phenolic hydroxyl groups. When hydrogenolysis is to be avoided, rhodium, ruthenium, and, to a lesser extent, palladium are the preferred catalysts [Eq. (52)]. The use of higher pressures coupled with lower temperatures improves selectivity.

$$\text{(52)}$$

The earliest homogeneous catalysts for aromatic hydrogenation were the Ziegler-type catalysts prepared from a transition metal salt and aluminum alkyl. Typical transition metal derivatives include $M(acac)_n$ (M = Mn, Co, or Ni; $n = 2$, or M = Fe, or Co; $n = 3$), $M(2\text{-ethylhexanoate})_2$ (M = Co or Ni), or $[Ti(\eta\text{-}C_5H_5)_2Cl_2]$. When mixed with triethyl aluminum these systems produce dark brown or black solutions, which effectively hydrogenate aromatic ring systems (e.g., benzene, xylenes, naphthalene, pyridine, or phthalate esters), albeit at 150–190°C and 75 atm[72] [Eq. (53)].

$$\text{(structure: benzene ring with CO}_2\text{Me and CO}_2\text{Me)} \xrightarrow[\text{AlEt}_3,\,\text{H}_2]{\text{Ni(O}_2\text{CC}_7\text{H}_{15})_2} \text{(structure: cyclohexane ring with CO}_2\text{Me and CO}_2\text{Me)} \qquad (53)$$

The nature of these catalysts is ill defined but the solutions probably contain metal hydride species, stabilized by coordination to aluminum. Practically these catalysts present problems because they require the use of very air-sensitive aluminum alkyl and high pressures.

A better defined catalyst is a cobalt π-allyl complex $[\text{Co}(\eta\text{-C}_3\text{H}_5)\text{-}[\text{P(OMe)}_3]_3]$, which hydrogenates benzene to cyclohexane at room temperature and atmospheric pressure.[73] Catalytic hydrogenation has been demonstrated for benzene with substituent groups that include R, OR, CO_2R, and NR_2, but electron-withdrawing groups (e.g., halogen, NO_2, or CN) inhibit the reduction.[74a] Naphthalene and anthracene are also hydrogenated. Unfortunately the hydrogenation rate is slow and the catalyst lifetime poor. When the trimethylphosphite ligand is replaced by a more bulky phosphite [e.g., tri-iso-propylphosphite $\text{P(OPr}^i)_3$], an increase in activity is observed but at the expense of catalyst lifetime, which is now very short.

General Procedure—Hydrogenation of Benzene[74b]
The catalyst $[\text{Co}(\pi\text{-C}_3\text{H}_5)[\text{P(OMe)}_3]_3]$ (0.112 mM) is dissolved in benzene (1 ml, 11.25 mM) under nitrogen. The solution is then frozen at $-196°\text{C}$, evacuated, filled with hydrogen, the tap closed (high vacuum stopcock), and the reaction warmed to room temperature. The solution is magnetically stirred for 24 h and then volatile products are vacuum distilled. Only cyclohexane is detectable.

As with many homogeneous catalysts high selectivity is observed, hence deuteration of benzene produces all *cis*-d_6-cyclohexane 95%[74] [Eq. (54)].

$$\text{(benzene)} \xrightarrow[\text{D}_2]{\text{Co}(\eta\text{-C}_3\text{H}_5)(\text{P(OMe)}_3)_3} \text{(}cis\text{-}d_6\text{-cyclohexane structure with D, H substituents)} \qquad (54)$$

The all *cis*-deuterium addition indicates that the benzene ring is tightly bound to the cobalt and does not dissociate until complete reduction has occurred. This is reflected in the fact that no cyclohexadiene or cyclohexene is ever observed. Attempts to produce a rhodium–allyl system have met with only limited success and again catalyst lifetime is a problem.[75]

Ruthenium has been extensively studied for arene hydrogenation, especially for the partial hydrogenation of benzene. Bis-[hexamethylbenzene]ruthenium(0), $[\text{Ru}(\eta\text{-Me}_6\text{C}_6)_2]$, hydrogenates benzene to cyclohexane at 90°C/2–3 atm[76] but lacks all *cis* selectivity, indicating the potential for isolating unsaturated species that have dissociated from the metal. The related [Ru(H)Cl(η-C$_6$Me$_6$)(PPh$_3$)] is a stable long-lived catalyst for benzene hydrogenation but at higher pressure (50°C/50 atm). The phosphine-free dimeric analog [Ru$_2$(η-Me$_6$C$_6$)$_2$(μ-H)$_2$(μ-Cl)]Cl is reported to be more stable and more active than the monomeric complex.[77a] Turnover of 9000 mol benzene/mol atom Rh has been achieved in benzene to cyclohexane reductions. The catalyst is tolerant of aryl-ethers, -alcohols, -esters, and secondary amines, but halogenated benzenes are not readily reduced and some dehalogenation occurs. Hydrogenolysis of diphenyl ether is a problem and nitrobenzene is reduced only to aniline. This may be compared with the results obtained for the rhodium system [Rh(η-C$_5$Me$_5$)Cl$_2$]$_2$–NEt$_3$, which operates under similar conditions and has high stereoselectivity for all *cis*-hydrogenation but is not as active (400 mol benzene/mol atom Rh have been achieved)[77b] (Table 7.7).

Hydrogenation of a pyridine ring appears to be easier than the benzene nucleus. [RhCl$_3$(py)$_3$–NaBH$_4$] in DMF catalyzes the hydrogenation of

Table 7.7. Hydrogenation[a] of Aromatics Using [Ru$_2$(η-C$_6$Me$_6$)$_2$H$_2$Cl$_2$] and [Rh$_2$(η-C$_5$Me$_5$)$_2$Cl$_4$]–NEt$_3$

	Product, %[b]	
Substrate	[Ru$_2$(η-C$_6$Me$_6$)$_2$H$_2$Cl$_2$]	[Rh(η-C$_5$Me$_5$)Cl$_2$]$_2$–NEt$_3$
Styrene	Ethylcyclohexane, 100	Ethylbenzene, 97
Phenylacetylene	Ethylbenzene, 32	Styrene, 7
Anisole	Methoxycyclohexane, 100	Methoxycyclohexane, 6
		Methoxycyclohexene, 19
Phenol	Cyclohexanol, 62	—
Methylbenzoate	Methylcyclohexanoate, 100	Methylcyclohexanoate, 75
Acetophenone	Methylcyclohexylketone, 100	Methylcyclohexylketone, 100
Benzophenone	Dicyclohexylketone, 100	Dicyclohexylketone, 50
N,N'-Dimethylaniline	N,N'-Dimethylcyclohexyl- amine, 86	N,N'-Dimethylcyclohexyl- amine, 73
Nitrobenzene	Aniline, 38	Aniline, 96
Chlorobenzene	Cyclohexane, 8	Cyclohexane, 1
Bromobenzene	Cyclohexane, 6	Cyclohexane, 10
Fluorobenzene	Fluorocyclohexane, 15	Fluorocyclohexane, 14
	Cyclohexane, 20	Cyclohexane, 38
Diphenylether	Cyclohexane	—
	Cyclohexanol	—
	Phenol	—

[a] Conditions: 50°C, 50 atm, 36 h.
[b] Relative to benzene (100%).

pyridine to piperidine at ambient temperatures and atmospheric pressure[35] [Eq. (55)]. As the rhodium complex contains coordinated pyridine, it is easy to envisage a mechanism via an intermediate (**14**) that undergoes hydrogen transfer to the ring. Clearly with benzene this type of intermediate is impossible.

(55)

14

Overall the homogeneous catalysts are of only academic interest in this area: there is not a synthetically useful system. All the catalysts require high pressures compared to heterogeneous systems due to their low activity. High *cis* selectivity may offer an advantage in some instances.

7.4.2. Partial Reduction of Aromatic and Polyaromatic Compounds

7.4.2.1. Partial Reduction of Isolated Aromatic Systems

The partial hydrogenation of an isolated aromatic ring is very difficult unless the ring contains substituents that influence the reaction. For instance, there are no hydrogenation catalysts, which reduce benzene to cyclohexene or cyclohexadiene. This results from the fact that dienes and monoenes are more readily reduced than aromatics. The partial reduction of benzene is easier by other means, such as the Birch reduction (sodium in liquid ammonia). A typical example of where substituent effects influence the course of a hydrogenation involves phenol, which may be converted into cyclohexanone by, for instance, palladium on carbon.[78] Of the dihydridic phenols, only resorcinol forms a dicarbonyl product (using rhodium on alumina) catechol, and hydroquinone reductions stop after addition of two equivalents of hydrogen to yield 2- and 4-hydroxycyclohexanones, respectively[79] [Eqs. (56) and (57)].

(56)

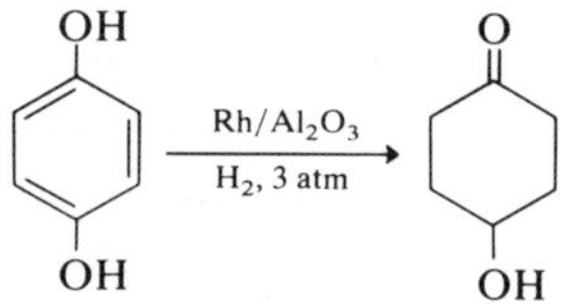

(57)

The partial reduction of pyridines can be achieved with unsaturated electron-withdrawing groups in the 3-position, e.g., acyl, formyl, or cyano.[80] The reduction of 3-acetylpyridine demonstrates the influence of different catalysts on the reaction. As a general rule, rhodium systems have a tendency to fully hydrogenate the substrate, whereas using palladium catalysts partially hydrogenated species may be obtained (Scheme 4). In

Scheme 4

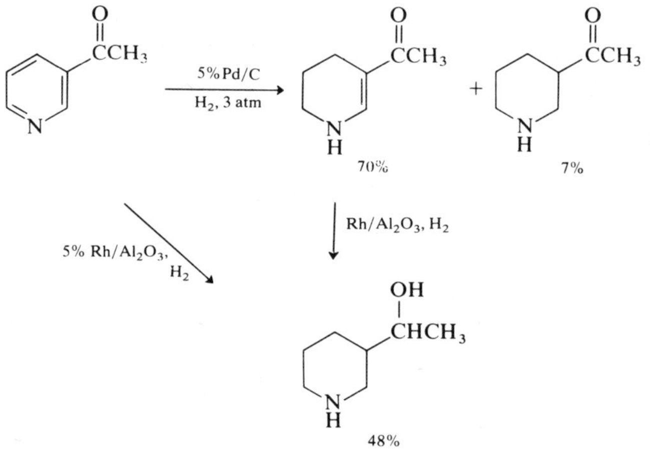

neutral or acid solution using palladium on carbon, the reaction stops after addition of two equivalents of hydrogen, and the 3-acetyl,1,4,5,6-tetrahy-dropyridine is isolated in good yield. The reaction is thought to proceed via 1,4 addition to give **15**, followed by preferential reduction of the 5,6

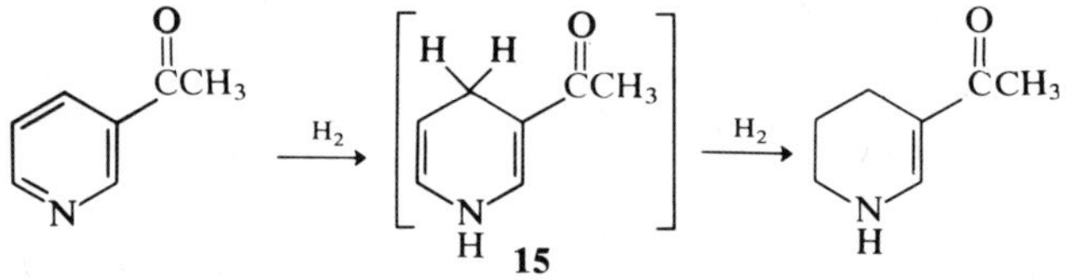

(58)

nonconjugated double bond. Using rhodium on alumina, 3-(1-hydroxyethyl)piperidine, the fully hydrogenated material, is the major product. When the pyridine ring contains a 2-amino or 2-hydroxy substituent, again only partial reduction is observed (even with rhodium on alumina) to yield 3,4,5,6-tetrahydropyridine amine or 2-piperidone, respectively.[81]

7.4.2.2. Partial Reduction of Polyaromatic Systems

A synthetically useful reaction is the selective reduction of one aromatic ring in a polyaromatic system. The highly active heterogeneous catalysts such as rhodium on alumina or carbon are prone to overhydrogenation, and poorer hydrogenation catalysts like platinum oxide in acid (low pressure) or palladium on carbon are most efficient. The hydrogenation of a series of polyaromatic compounds has been examined using platinum or palladium on charcoal, or platinum oxide as catalysts under mild conditions (25°C/20–30 psi H_2)[82a] [Eq. (59)]. Reduction over palladium affords the

(59)

internal dihydroarene regioselectively while the analogous reaction with the platinum catalysts occurs on the terminal ring to give the tetrahydroarene. These reactions complement the Birch reduction, which generally provides different hydroaromatic products.

General Procedure—Partial Reduction of Benzanthracene[82a]

Reactions are performed in a 500-ml Pyrex pressure bottle connected to a low-pressure hydrogenator. 10% Palladium on carbon (420 mg), benzanthracene (180 mg, 0.79 mM), and ethyl acetate (15 ml) are placed in the reaction vessel, which has been flushed (three times) with hydrogen before being pressurized to 20 psi. The reaction is magnetically stirred at room temperature for 10 h, after which the reaction mixture is filtered through Celite and washed several times with acetone. The filtrate is evaporated to dryness, and the residue dissolved in hexane and purified by passage through a short column of Florisil. The yield of crude 5,6-dihydrobenzanthracene is 97%. Crystallization from methanol gives pure product. The reaction is readily scaled-up, although an increase in the by-product 7,12-dihydrobenzanthracene (up to 10%) was noted. The hydrogenation over a platinum catalyst is performed in a similar manner but requires a longer time (44 h, 50 psi). Although most polyaromatic substrates studied are hydrogenated over <24 h, some require longer times, e.g., anthracene (120 h, Pd/C) and triphenylene (48 h, PtO$_2$).

When the aromatic rings are nonconjugated, high selectivity can often be achieved owing to the substituents on the ring. After the first hydrogen addition at the more susceptible ring, rapid reduction to give the perhydro ring occurs[82b] [Eq. (60)].

$$\text{(60)} \qquad 52\%$$

When the polyaromatic system contains a heteroatom ring, selectivity usually presents no problem; the heteroatom ring is preferentially reduced. Thus quinoline and isoquinoline are easily hydrogenated by palladium or platinum catalysts to give the 1,2,3,4-tetrahydro derivatives.[83] In strong hydrochloric acid, however, the hetero ring preference may be altered in favor of the benzene ring, viz., Eq. (61)].

$$\text{(61)}$$

Solvent:	Yield, %:	
MeOH	87	13
MeOH/HCl (1 M)	30	70
MeOH/HCl (4 M)	13	87

A number of homogeneous catalysts selectively or partially reduce polyaromatics. As noted earlier, hetero rings are generally reduced before the benzene system, hence [RhCl$_3$(py)$_3$–NaBH$_4$] reduces quinoline to the 1,2,3,4-tetrahydroquinoline; the benzene ring remains intact.[35]

It has long been known that, in the hydroformylation of alkenes, addition of aromatics results in the isolation of a hydrogenated ring system. Although benzene is unaffected, naphthalene is reduced to tetrahydronaphthalene by H$_2$/CO at 200°C/200 atm in the presence of Co$_2$(CO)$_8$.[84] Other polyaromatics are also partially hydrogenated, e.g., pyrene (16) to 4,5,dihydropyrene (**17**) [Eq. (62)]. The high pressures involved in this reaction

$$\text{16} \xrightarrow[\text{H}_2/\text{CO}]{\text{Co}_2(\text{CO})_8} \text{17} \quad 69\% \tag{62}$$

make this just an interesting curiosity to the synthetic organic chemist, but by using stoichiometric quantities of $[\text{HCo(CO)}_4]$ (the active species in the high pressure $\text{Co}_2(\text{CO})_8/\text{H}_2/\text{CO}$ reductions) partial polyaromatic reduction occurs at room temperature[85] under a CO atmosphere [Eq. (63)].

$$\tag{63}$$

Conditions:	Catalyst:	Yield, %:	
		cis	*trans*
150°C/3000 psi	$\text{Co}_2(\text{CO})_8/\text{H}_2/\text{CO}$	48	52
25°C/1 atm CO	HCo(CO)_4	35	65

$[\text{HCo(CO)}_4]$ can be prepared simply from $\text{Co}_2(\text{CO})_8$[86] and hence the reaction becomes more appealing for synthetic applications.

General Procedure—Partial Reduction of Anthracene using [HCo(CO)₄]
    To a toluene solution (80 ml) of anthracene (0.2 g, 1.12 mM) under carbon monoxide is added $[\text{HCo(CO)}_4]$ (6.8 mM) in toluene (20 ml). The mixture is stirred for 24 h at room temperature under a carbon monoxide atmosphere. Dimethylformamide (12 ml) is then added and the mixture stirred until the $\text{Co}_2(\text{CO})_8$ color disappears. Water is added and the lower layer removed with a syringe. The organic layer is washed (2 × 5 ml) with water and dried over Na_2SO_4. A 25% yield of 9,10-dihydroanthracene is obtained.

    The homogeneous ruthenium catalyst

$$[\overline{\text{Ru}(\text{PPh}_3)_2(\text{Ph}_2\text{PC}_6\text{H}_4)\text{H}_2}]^{\ominus}\text{K}^{\oplus}\cdot\text{C}_{10}\text{H}_8\cdot\text{Et}_2\text{O}$$

catalyzes the hydrogenation of naphthalene and anthracene to the tetrahydro derivatives with 98% selectivity.[87] The reduction of phenanthrene is very slow, and isolated aromatic rings as in benzene, toluene, tetralin, or pyridine are unaffected. Other functional groups such as ketones, esters, and nitriles are also hydrogenated. Compared to the conditions used for $Co_2(CO)_8$ catalytic reductions, this system has more potential operating at only 100°C/6 atm but its present scope is somewhat limited.

7.5. Reduction of Carbonyl Groups

7.5.1. Formation of Alcohols from Aldehydes and Ketones

Although aldehydes and ketones are readily hydrogenated to the corresponding alcohols under mild conditions, there are differences between aliphatic and aromatic derivatives. Aromatic carbonyl compounds are more reactive than aliphatic analogs, but they are also sensitive to hydrogenolysis, which produces the hydrocarbon product. Heterogeneous catalysts have been used extensively for these reductions but, as with previously described reactions, when selectivity is required homogeneous complexes are preferred.

7.5.1.1. Using Heterogeneous Catalysts

(a) *Aliphatic Substrates.* Platinum oxide is an effective catalyst for the hydrogenation of aliphatic aldehydes and ketones to the corresponding alcohol. Rapid deactivation of the catalyst often occurs so a specific amount of a promoter, typically ferric chloride or stannous chloride, is employed.[88] Interestingly, palladium is particularly ineffective for the hydrogenation of

$$C_6H_{13}CHO \xrightarrow[\text{FeCl}_3,\ \text{EtOH}]{\text{Pt, H}_2} C_6H_{13}CH_2OH \qquad (64)^{89}$$

aliphatic carbonyl compounds and hence is useful for double-bond reduction in unsaturated carbonyl systems (Section 7.3.3). Often when platinum oxide (plus a strong acid)[90] or especially palladium[91] is used, in methanol or ethanol solvent, ether formation occurs. Palladium promotes the formation of acetals and ketals during hydrogenation and these undergo hydrogenolysis to the corresponding ether. This reaction provides a neat synthetic route to ethers from aldehydes or ketones. The use of nonprimary alcohols as a solvent prevents ether formation.

$$(65)$$

Supported rhodium and ruthenium are also effective catalysts for carbonyl reduction. Ruthenium is commonly used for hydrogenations in aqueous medium (e.g., of carboxylates) because water is a unique promoter for ruthenium.

The heterogeneous nickel catalyst prepared by reaction of nickel acetate with sodium alkoxide (e.g., $EtO^{\ominus}$ or t-$C_5H_{11}O^{\ominus}$) and sodium hydride in ethanol [i.e., NaH–$NaOR$–$Ni(OAc)_2$] is a catalyst for carbonyl hydrogenations.[92] This catalyst is easily prepared and gives reproducible results (see also Section 7.2.2). For example, the reduction of 5-nonanone takes place at 25°C, 1 atm H_2, over 9 h to give 5-nonanol in virtually quantitative yield.

(b) *Aromatic Substrates.* Whereas supported palladium is ineffective for the reduction of aliphatic carbonyl compounds, palladium on carbon is the catalyst of choice for aromatic aldehyde and ketone hydrogenations. A major drawback associated with this catalyst is hydrogenolysis of the resultant alcohol to produce the hydrocarbon. Hydrogenolysis is a reaction promoted by palladium (see also Section 7.8). Indeed, a palladium-based hydrogenolysis reaction can give good yields under mild conditions and offers an alternative to the more conventional Clemmensen or Wolff–Kishner reductions [Eqs. (66) and (67)].

General Procedure—*Reduction of p-Anisoyl-propionic acid* (**27**) [93]

To a solution of p-anisoyl propionic acid (28.8 g) in acetic acid (125 ml) is added palladium (5%) on carbon (3.0 g) under a hydrogen atmosphere. The mixture is heated to 60°C and pressurized to 40 psi hydrogen. (Parr hydrogenation apparatus is used.) After 40 min hydrogen absorption is complete. The solution is filtered through Celite and the filtrate added to water (200 ml). The mixture is cooled and filtered to yield (p-anisyl)butyric acid (**28**) (14.5 g, 75%), mp 61–62°C; cf. Clemmensen reductive method, 65% yield of **28**; Wolff–Kishner reductive method, 75% yield.

Chap. 7 • Reduction

Hydrocarbon formation is favored in acidic media, using higher temperatures and excess hydrogen. Excellent yields of alcohol are achieved in palladium-catalyzed hydrogenations by monitoring hydrogen absorption to one mole equivalent and by using a neutral solvent. With palladium catalysts, ring hydrogenation is slow and the reaction can be terminated before extensive reduction of the aromatic nucleus occurs.[95] When total reduction is required (i.e., including the aromatic nucleus), supported rhodium or ruthenium is the catalyst of choice.

7.5.1.2. Using Homogeneous Catalysts

Homogeneous catalysts do not readily reduce carbonyl compounds and so they can be used for selective hydrogenation of unsaturated carbonyls. To function as a catalyst for carbonyl reduction, the metal complex should have a strong hydridic character to match the polar nature of the functional group. The most widely studied homogeneous catalytic system is the cationic rhodium complex $[Rh(diene)(PR_3)_2]^{\oplus}PF_6^{\ominus}$ (or $ClO_4^{\ominus}$) (**18**), a precursor to the active species $[RhH_2(PR_3)_2S_2]^{\oplus}$ (where S = solvent molecule), which is formed in the presence of hydrogen. Carbonyl reduction is favored by use of an alkyl-substituted phosphine (e.g., $PPhMe_2$ or PMe_3), and 1% water aids the hydrogenation[96] [Eq. (68)].

$$\underset{Me}{\overset{Et}{>}}C=O \xrightarrow[H_2, H_2O\,(1\%),\,25°C]{[RhH_2(PPhMe_2)_2(solvent)_2]PF_6} \underset{Me}{\overset{Et}{>}}\underset{H}{\overset{OH}{<}}C \qquad (68)$$

Aldehydes are also reduced by this catalyst but a rapid decrease in catalytic activity is observed (cf. Section 7.3.3). The cationic rhodium complexes **18**, but with the two monodentate phosphines replaced by a chelating asymmetric diphosphine, have been extensively studied for the asymmetric reduction of ketones (Section 7.7.5).

A neutral ruthenium complex, $[RuCl_2(CO)_2(PPh_3)_2]$, is reportedly active at 160–200°C and 15 atm H_2 for the hydrogenation of both aliphatic and aromatic aldehydes.[97] High yields of the product alcohol are reported. The iridium complex $[IrH_3(PPh_3)_3]$ in acetic acid reduces aldehydes to the corresponding alcohol under mild conditions, 50°C and 1 atm H_2, but is ineffective for ketone reduction.[98] When unsaturated aldehydes were examined the reaction was found to be nonselective, and mixtures of saturated and unsaturated alcohols and saturated aldehyde were obtained.

$$\text{CHO} \xrightarrow[H_2]{IrH_3(PPh_3)_3} \underset{73\%}{\text{OH}} \qquad (69)$$

7.5.1.3. Using Transfer Hydrogenation

Iridium complexes have been studied for carbonyl reduction via hydride transfer from a solvent. Functional group selectivity is described later (Section 7.5.2.4), but stereoselectivity can also be achieved. The system commonly used consists of chloroiridic acid (H_2IrCl_6) or iridium tetrachloride ($IrCl_4$) in aqueous isopropanol (the hydrogen source) and an equimolar amount of a phosphite (to substrate).[99] These catalysts reduce cyclic ketones to predominantly axial alcohols and have found practical utility in stereoselective steroid synthesis [Eq. (70)]. Hydrogen addition

$$\xrightarrow[\substack{P(OMe)_3, \\ Pr^iOH}]{H_2IrCl_6}$$

(70)

R = H	90% axial
R = =O	95% axial
R = COMe	78% axial

occurs from the less hindered equatorial direction. The catalyst is particularly sensitive to steric hindrance, hence carbonyl groups in the 6,7,11,12,17, or 20 positions are not reduced [Eq. (71)]. These reductions

$$\xrightarrow[\substack{P(OMe)_3, \\ Pr^iOH, reflux}]{H_2IrCl_6}$$

(71)

94% axial

do not require an inert atmosphere, but often prolonged reflux (48–72 h) is necessary to achieve high conversion. Extensive heating (>150 h) can lead to side reactions and hence mixed products.

General Procedure—Reduction of Steroidal-3-ketones by Henbest's Catalyst, *[H_2IrCl_6-$P(OMe)_3$][99]*
The steroid (10 mM), chloroiridic acid (0.5 mM), trimethylphosphite (10 mM), and 90% aqueous propanol (100 ml) are heated together under reflux; an inert atmosphere is not required. After 48–72 h little or none of the oxosteroid is detectable, and the reaction is cooled, poured into water, and the product extracted (ether–benzene). After washing with water and dilute $NaHCO_3$ the solvent is removed. Axial alcohols frequently crystallize easily from the crude reaction mixture and are further purified by chromatography. This procedure is suitable for up to 10 g of the steroid.

Simpler cyclic ketones have also been examined and demonstrate the high stereoselectivity and inhibition resulting from steric hindrance[100] [Eqs. (72)–(74)].

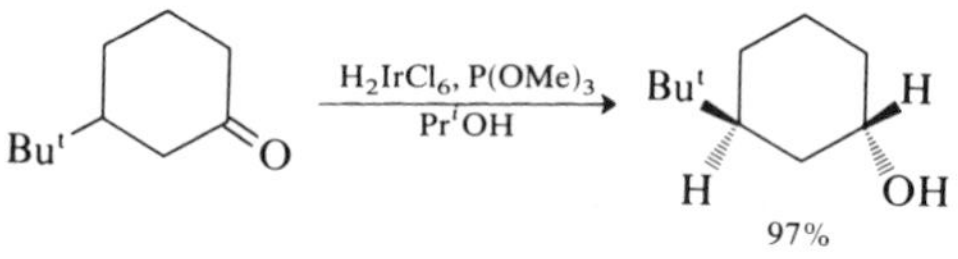

(72)

(73)

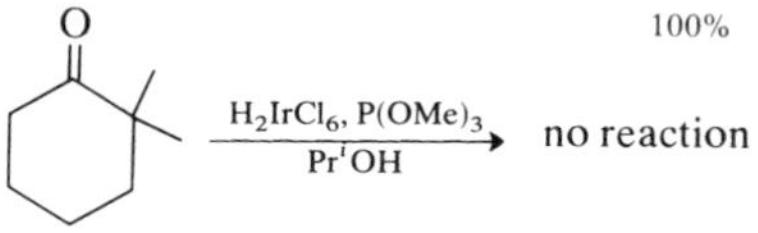

(74)

Although numerous other iridium complexes have been examined,[22094] e.g., $[IrH_3(PPh_3)_3]$ and $[HIrCl_4(Me_2SO)_2]$, only $[IrCl(C_8H_{14})_2]_2$–$P(OMe)_2H$[101] shows the high axial selectivity demonstrated by the chloroiridate–phosphite systems.

7.5.1.4. Using Hydrosilylation

The generally poor hydrogenation ability of homogeneous catalysts toward carbonyl functions is supplemented by the development of an indirect procedure via hydrosilylation. $[Rh(PPh_3)_3Cl]$ catalyzes the addition of an alkylsilane to the carbonyl group to form the silyloxy derivative, which is readily hydrolyzed by base $(MeOH/KOH/H_2O)$ to yield the alcohol[102] [Eq. (75)]. The reaction occurs rapidly under mild conditions,

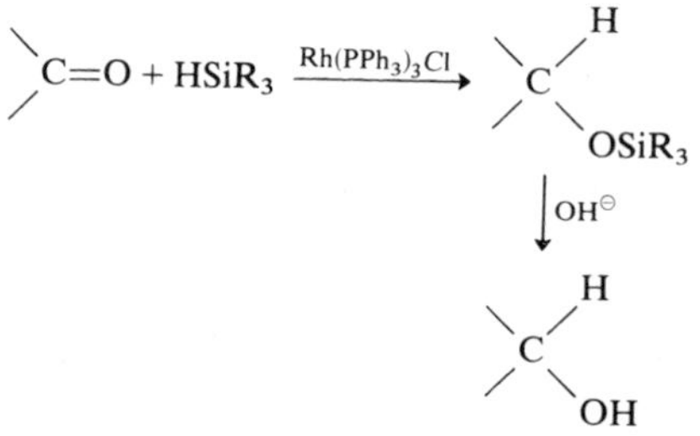

(75)

although with aromatic ketones the reaction rate is slower and heating is necessary (e.g., 15 min/60°C for acetophenone). The reaction rate increases

Table 7.8. Carbonyl Reduction via Catalytic Hydrosilylation

Reaction		Yield	Conditions
$\begin{matrix}Ph\\ \end{matrix}$ C=O $\begin{matrix}Me\end{matrix}$ 1. $EtSiH_3$, $[Rh(PPh_3)_3Cl]$ 2. $MeOH/KOH/H_2O$	$\begin{matrix}Ph\end{matrix}$ CHOH $\begin{matrix}Me\end{matrix}$	97%	60°C/15 min
$\begin{matrix}Me\end{matrix}$ C=O $\begin{matrix}Me\end{matrix}$ 1. $EtSiH_3$, $[Rh(PPh_3)_3Cl]$ 2. $MeOH/KOH/H_2O$	$\begin{matrix}Me\end{matrix}$ CHOH $\begin{matrix}Me\end{matrix}$	95%	R.T./10 min
$\begin{matrix}Bu^t\end{matrix}$ C=O $\begin{matrix}Me\end{matrix}$ 1. Ph_2SiH_2, $[Rh(PPh_3)_3Cl]$ 2. $MeOH/KOH/H_2O$	$\begin{matrix}Bu^t\end{matrix}$ CHOH $\begin{matrix}Me\end{matrix}$	97%	0°C/10 min
$Pr^n CHO$ 1. $PhMeSiH_2$, $[Rh(PPh_3)_3Cl]$ 2. $OH^\ominus$	$Pr^n CH_2OH$	98%	R.T./5 min

when a di- or trihydrosilane is used instead of a monohydro derivative (Table 7.8).

Reduction of Cyclohexanone to Cyclohexanol via Catalytic Hydrosilylation [102]
    To a mixture of cyclohexanone (2.94 g, 30 mM) and diphenylsilane (5.52 g, 30 mM) is added [Rh(PPh₃)₃Cl] (19 mg, 0.07 mol %). The mixture is stirred at 0°C for 10 min before adding hexane (100 ml). The resulting precipitate is filtered off, solvents removed, and the residual liquid distilled under reduced pressure to yield cyclohexyloxydiphenylsilane, 8.34 g (98%).
    The silyl ether is smoothly hydrolyzed by MeOH/KOH/H₂O (or alternatively p-CH₃C₆H₄SO₃H/MeOH) at ambient temperature over 1 h to give cyclohexanol in quantitative yield.

    The use of optically active diphosphine ligands (e.g., DIOP) with rhodium allows asymmetric hydrosilylation to be achieved (Section 7.7.6). Double-bond reduction in α,β-unsaturated aldehydes and ketones (via 1,4 addition) also proceeds smoothly under mild conditions. Catalytic hydrosilylation followed by hydrolysis provides a simple, effective method for the reduction of the carbonyl function to the alcohol. It does not require the use of hydrogen, and therefore special apparatus is not required.

7.5.2. Selective Reduction of Aldehydes and Ketones

The carbonyl function is not preferentially hydrogenated in the presence of triple or double bonds, aromatic nitro or imine groups, unless steric effects influence the reaction. Special catalysts can produce a reversal in reactivity and these are described below. Aromatic carbonyl compounds are easily reduced to the aromatic alcohol or alkyl (by hydrogenolysis) without affecting the ring.

7.5.2.1. Selective Reduction of a Diketone to Diol or Hydroxyketone

A diketone is readily reduced to the diol by the catalysts described earlier for monoketones, but reduction to give the hydroxyketone is more difficult and depends greatly on substrate, catalyst, and solvent.[2] Selective carbonyl reduction is achieved in the hydrogenation of steroids using hydrogen transfer from alcohol solvents in the presence of iridium catalysts. This reaction is very sensitive to the steric environment of the carbonyl function (Section 7.5.1.3).

7.5.2.2. Diketone to Monoketone

A neat reduction of 1,2 diketones to monoketones has been reported.[103] The substrate is reacted with trimethylphosphite before reduction; in this way selective hydrogenation of just one of the carbonyl groups is achieved (Scheme 5). The reactions are essentially quantitative. When both

Scheme 5

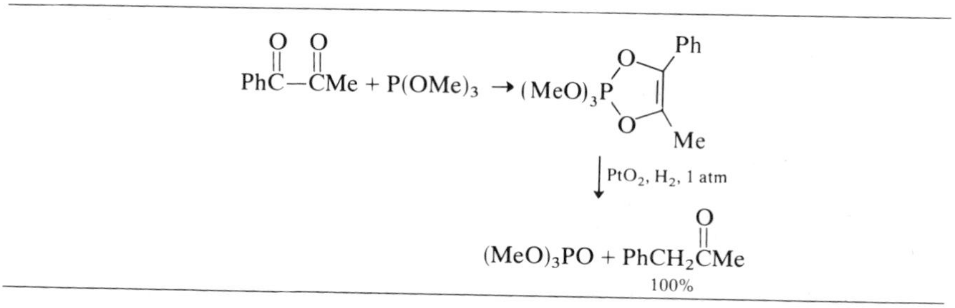

$$PhC(=O)-C(=O)Me + P(OMe)_3 \rightarrow (MeO)_3P \cdots$$

electronic and steric factors contribute to the reaction high selectivity is observed, but in the absence of electronic effects less discrimination is shown [Eq. (76)].

$$MeC(=O)-C(=O)Et + P(OMe)_3 \xrightarrow{PtO_2, H_2} MeCCH_2Et + MeCH_2CEt \qquad (76)$$

40%   60%

7.5.2.3. Preferential Ketone or Aldehyde Reduction

Aldehydes are more easily reduced than ketones, and selective reagents have been reported for the preferential reduction of aldehydes[104] (e.g., Bu$_3$SnH–silica gel). Preferential ketone reduction in the presence of an aldehyde is more difficult, but this has been achieved by using the ability of lanthanide ions to protect the aldehyde function.[105] The reduction of a ketone in the presence of an aldehyde usually necessitates a three-step process, viz., protection of the aldehyde, reduction of the ketone, and finally liberation of the aldehyde. This lengthy procedure is often poorly selective and leads to separation problems and low yields. A simple one-step route uses the reagent [CeCl$_3$·6H$_2$O–NaBH$_4$]. It is known that in aqueous solution nonconjugated aldehydes form hydrates to a greater extent than ketones, while conjugated aldehydes are not hydrated. Protection of a nonconjugated aldehyde is achieved by formation of the *gem*-diol, which is stabilized through coordination to the cerium(III) ion. The presence of cerium(III) is sufficient to ensure adequate aldehyde protection during reduction but does not hamper recovery of the aldehyde during work-up. Thus, in an equimolar mixture of hexanal and cyclohexanone, only 2% of the aldehyde is reduced compared to 100% of the ketone. When the ketone and aldehyde groups occur in the same molecule, high selectivity is again observed [Eqs. (77) and (78)].

$$\text{(77)}$$

$$\text{(78)}$$

General Procedure—Preferential Reduction of a Ketone in the Presence of an Aldehyde[105]

The substrate (C) (1 mM) and CeCl$_3$·6H$_2$O (1 mM), dissolved in aqueous ethanol (1:1.5, EtOH:H$_2$O), are cooled to −15°C. Then NaBH$_4$ (1.5 mM) is added and the mixture stirred for 10 min. Excess NaBH$_4$ is destroyed by addition of acetone and the mixture diluted with aqueous NaCl and extracted with ether. Work-up of the organic layer yields the selectively reduced product (D) in high yield (85% crude, 75% purified).

With nonconjugated aliphatic and alicyclic carbonyl compounds, ketone selectivity is excellent, but selective reduction cannot be achieved with conjugated aldehydes *vide supra* (e.g., benzaldehyde). This, however, allows a second useful selective reduction: that of a conjugated aldehyde in the presence of a nonconjugated aldehyde. For example, the reduction of a benzaldehyde–cyclohexanecarboxaldehyde mixture affords benzyl alcohol (85%) with quantitative recovery of the cyclohexanecarboxaldehyde.

7.5.2.4. Selective Carbonyl Reduction in α,β-Unsaturated Carbonyl Compounds

The selective reduction of unsaturated carbonyl compounds and especially conjugated systems has been extensively studied. Generally double-bond reduction occurs first and high selectivity to the saturated carbonyl complex is easy (Section 7.3.3). Palladium catalysts are often used to obtain saturated aldehydes because they are very effective for double-bond reduction and have poor activity for the hydrogenation of the carbonyl function.

The reduction to yield unsaturated alcohols is more difficult but numerous catalysts have been reported for this transformation. α,β-Unsaturated aliphatic aldehydes can be hydrogenated to the unsaturated alcohol by use of a platinum catalyst with added zinc and ferrous ions [e.g., PtO_2–$FeSO_4$–$Zn(OAc)_2$]. Zinc ions inhibit double-bond reduction while ferrous ions promote carbonyl saturation.[106] Unsaturated aromatic aldehydes present even more difficulties, and mixtures often result when heterogeneous catalysts are used. Platinum and ruthenium catalysts may offer some advantages but only in selected cases.

Homogeneous catalysts are more selective than heterogeneous systems but, when reducing unsaturated carbonyl complexes, only a small number of usefully selective catalysts are available. $[Rh_2Cl_2(CO)_4]$–NEt_3 is effective for the hydrogenation of aromatic substituted α,β-unsaturated aldehydes at 60°C under an atmosphere of H_2/CO (80 atm) over 1 h[107] [Eq. (79)].

$$PhCH=CHCHO \xrightarrow[NEt_3, H_2, CO]{Rh_2(CO)_4Cl_2} PhCH=CHCH_2OH \atop 85\% \qquad (79)$$

With aliphatic systems the selectivity is poorer; e.g., crotonaldehyde gives only ~50% unsaturated alcohol. It is interesting to compare this reaction using $[Rh_2Cl_2(CO)_4]$ with Wilkinson's catalyst, $[Rh(PPh_3)_3Cl]$, which under the same conditions preferentially reduces the double bond [Eq. (80)]. Addition of a phosphine to the rhodium dimer quenches all carbonyl

$$PhCH=CHCHO \xrightarrow[H_2, CO]{Rh(PPh_3)_3Cl} PhCH_2CH_2CHO \atop 100\% \qquad (80)$$

reduction. Transfer hydrogenation from a solvent using iridium catalysts has proved successful for carbonyl reduction in unsaturated systems.[108] Aqueous isopropanol in the presence of $[HIrCl_2(Me_2SO)_3]$ under reflux conditions reduces unsaturated aldehydes to the corresponding unsaturated alcohol (Table 7.9). The reasons for preferential carbonyl reduction are not clear since the same system catalyzes double-bond reduction in α,β-unsaturated ketones[109] [Eq. (81)].

$$\underset{\text{RCCH=CHPh}}{\overset{\overset{\displaystyle O}{\parallel}}{}} \xrightarrow[\text{Pr}^i\text{OH}]{\text{HIrCl}_2(\text{Me}_2\text{SO})_3} \underset{\text{RCCH}_2\text{CH}_2\text{Ph}}{\overset{\overset{\displaystyle O}{\parallel}}{}} \tag{81}$$
$$R = Ph\,(95\%)\ or\ Bu^t\,(90\%)$$

The lanthanide system $CeCl_3 \cdot 6H_2O$–$NaBH_4$ (Section 7.5.2.3) is very effective at reducing α,β-unsaturated ketones to allylic alcohols[110] [Eqs. (82)–(84)]. Without the addition of an equimolar amount of Ce(III) ion,

$$\tag{82}$$

100% (89%)^a

$$\tag{83}$$

99% (51%)^a

$$\tag{84}$$

91% (0%)^a

^{a} NaBH$_4$ alone

mixtures result (see yield in parentheses). Other functional groups (i.e., carboxylic acids, esters, amides, halides, cyano, and nitro groups) are not affected by this reagent under the mild conditions and short reaction times

Table 7.9. Selective Reduction of Unsaturated Aldehydes

		Conversion^a (90% total)		
Substrate	Time, min	Unsaturated alcohol, %	Saturated alcohol, %	Saturated aldehyde, %
Cinnamaldehyde	80	78	Trace	—
α-Methylcinnamaldehyde	250	90	—	—
Crotonaldehyde	50	85	5	Trace

^{a} 80°C under N_2, propanol:H_2O 30:1 v/v; 0.25 M substrate:0.01 M iridium catalyst.

(5 min) used in this procedure. The reaction proceeds through attack on the carbonyl function (1,2 addition) and not by 1,4 addition to the conjugated system.

7.5.3. Reduction of Carboxylic Acids and Acid Derivatives

The ease of hydrogenation of carbonyl compounds decreases in the following order[111]: acid chloride > aldehyde ~ ketone > anhydride > ester > carboxylic acid > amide. Many carboxylic acid derivatives are very difficult to hydrogenate. For example, the hydrogenation of a carboxylic acid to produce the alcohol can be achieved over heterogeneous catalysts, typically RuO_2 or Ru–C, but elevated temperatures (130–225°C) and pressures (300–600 atm) are required. The severe conditions limit any practical usefulness. Similarly, hydrogenation plays little part in the reduction chemistry of amides and esters; more accessible reduction routes (e.g., $LiAlH_4$) are preferred.

7.5.3.1. Reduction of Acid Chlorides

Acid chlorides are easily hydrogenated to produce the aldehyde selectively (Rosenmund reduction)[112] [Eq. (85)]. Palladium on barium sulfate

$$RCOCl \xrightarrow[\text{quinoline-S, xylene}]{Pd/BaSO_4, H_2} RCHO \tag{85}$$

inhibited with, typically, quinoline-S [prepared by refluxing quinoline (6 equiv) with sulfur (1 equiv)] is the catalyst of choice. The inhibitor prevents over-reduction of the aldehyde to the alcohol (or even to the hydrocarbon). The reduction is carried out by bubbling hydrogen through a hot xylene or toluene solution of the acid chloride in which the catalyst is suspended.

General Procedure [112,113]*—Preparation of 1-Acenaphthaldehyde*
To a 200-ml flask equipped with a condenser, stirrer, and gas inlet is added xylene (75 ml), 2% Pd–BaSO₄ (0.35 g), and quinoline-S (*vide supra*) (0.075 g). A slow stream of hydrogen is passed through the mixture and the solution heated to reflux. 1-Acenaphthoyl chloride is added to the hot mixture, which is stirred rapidly at reflux for 5–6 h (until not further HCl is evolved). After cooling the catalyst is filtered off and the solvent removed. The residue is distilled to yield crude 1-acenaphthaldehyde (5 g, 80%), which is then purified to give the pure product (4.5 g, 72%).

A wide variety of acid chlorides have been studied,[112] including aliphatic, aromatic, and heterocyclic analogs. Yields are generally in the range 50%–90% but are very dependent on the substrate. Further examples of halogen hydrogenolysis are described in Section 7.8.1.

An alternative route to aldehydes from acid chlorides involves the catalytic reduction by organosilanes. The reaction is catalyzed by palladium on carbon[114] or homogeneously by [Pt(PPh$_3$)$_2$Cl$_2$].[115] Using palladium on carbon, triethylsilane is the most effective reducing agent, readily reducing aliphatic acyl halides (yields 50%–75%) [Eq. (86)], but it does not perform so well with aroyl halides (yields 40%–70%) [Eq. (87)]. [Pt(PPh$_3$)$_2$Cl$_2$] is

$$Pr^iCOCl \xrightarrow[\text{5 mins}]{Pd/C, Et_3SiH} \underset{74\%}{Pr^iCHO} \qquad (86)$$

$$p\text{-MeOC}_6\text{H}_4\text{COCl} \xrightarrow{Pd/C, Et_3SiH} \underset{43\%}{p\text{-MeOC}_6\text{H}_4\text{CHO}} \qquad (87)$$

also effective as a catalyst with triethylsilane but longer reaction times are required. The related reaction between triethylsilane and some substituted benzoyl chlorides in the presence of a rhodium catalyst takes a different course, and ketones are produced[115] [Eq. (88)]. Electron-releasing substituents on the aryl group favor ketone formation with [RhCl$_3$(PBu$_2$Ph)$_3$].

$$(88)$$

63%

The mechanism of this reaction is unclear, but moderate yields of ketone (40%–65%) are obtained.

7.5.3.2. Reduction of Anhydrides

The hydrogenation of anhydrides over heterogeneous catalysts affords mixed products depending on the extent of reduction. Improved selectivity is obtained using homogeneous catalysts. Thus [RuCl$_2$(PPh$_3$)$_3$] selectively hydrogenates cyclic carboxylic acid anhydrides to γ-lactones[116] [Eq. (89)]. Succinic anhydride is converted into γ-butyrolactone (50%) and succinic

$$(89)$$

acid (50%) in the presence of $[RuCl_2(PPh_3)_3]$. The high yield of carboxylic acid is due to hydrolysis of unreacted anhydride by water formed during the reaction. The selectivity of this reaction could therefore be further improved by removal of water as it is formed. When unsymmetrical cyclic carboxylic acid anhydrides are used, the preferential reduction of the less hindered carbonyl group occurs[117] [Eq. (90)]. The reverse effect is found

$$(90)$$

Overall yield 70%
75%
72%[a]

CAT = LiAlH$_4$	19:1
Na–EtOH	11:9
$[RuCl_2(PPh_3)_3]$,	1:9
H$_2$, 20 atm	

[a] lactone acid, 1:1

when $LiAlH_4$ is used as the reducing agent; i.e., the more hindered carbonyl is reduced. This reaction has not been examined with a wide variety of substrates but may offer a general method for the conversion of cyclic anhydrides into lactones by reduction at the less hindered site.

An interesting reduction of anhydrides to aldehydes or aldehydic acids has been reported using disodium tetracarbonylferrate $[Na_2Fe(CO)_4]$[118] [Eq. (91)]. Succinic acid is converted into formylpropionic

$$(91)$$

acid (81%), and phthalic anhydride yields phthalaldehydic acid (61%). Intermolecular acid anhydrides yield a mixture of aldehydes and acids, and with mixed anhydrides mixed products are obtained [Eq. (92)]. The ratio

$$(92)$$

R = R′ = Ph 37%
R = Ph, R′ = Me 40% (RCHO:R′CHO, 64:36)

of aldehydes and acids depends upon the substituents R and R′. Treatment of the anhydride (1 equiv) with $[Na_2Fe(CO)_4]$ (Collman's reagent) (1 equiv) produces a rapid reaction (20 min) under ambient conditions, and after acid hydrolysis gives good yields of aldehydes and acids or aldehydic acid with a variety of acid anhydrides. The reaction is thought to proceed through an acylcarbonylferrate intermediate (19) [Eq. (93)].

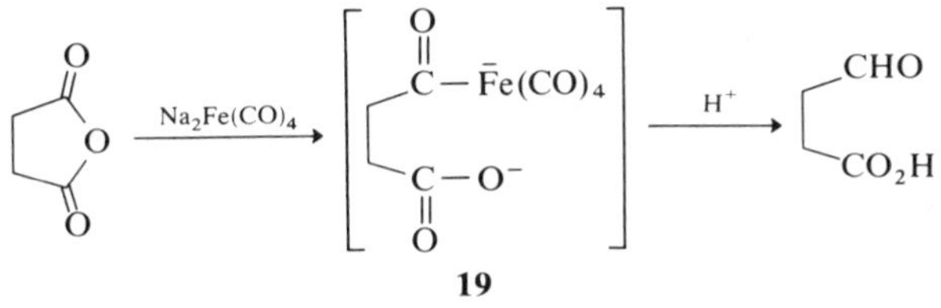

(93)

19

7.6. Reduction of Nitro, Nitrile, and Other Nitrogen-Containing Functional Groups

7.6.1. Reduction of Nitro Groups

Aromatic nitro groups are readily hydrogenated to give the amine, often with the liberation of heat. Care must therefore be taken in catalyst concentration and choice of solvent (i.e., choose a higher boiling solvent, ethanol better than methanol). Unsupported or supported Ni, Pd, or Pt heterogeneous catalysts are excellent for this reduction [Eqs. (94)–(96)].

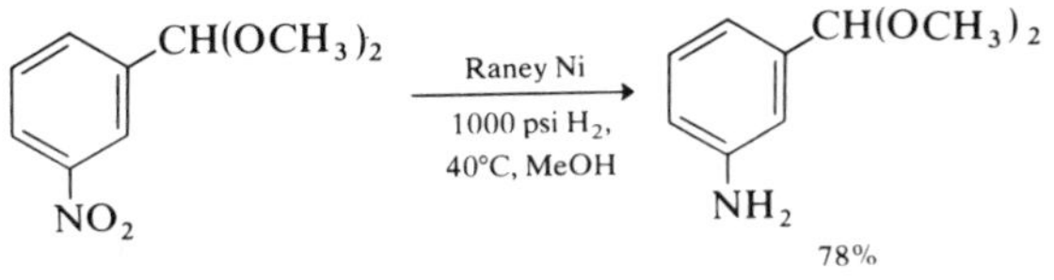

(94)[119]

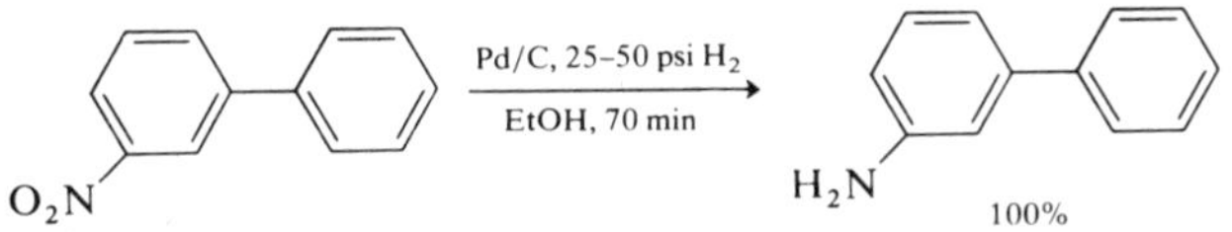

(95)[120]

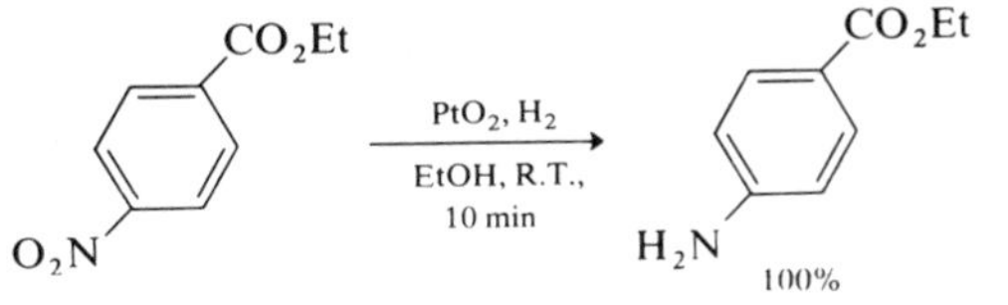

(96)[121]

Aliphatic nitro-functional groups are more difficult to reduce, possibly because the strongly basic product inhibits the catalyst. Higher catalyst concentration, longer reaction times, and elevated pressures and temperatures are often required. Palladium catalysts have received the most attention although Raney nickel has proven useful [Eq. (97)].

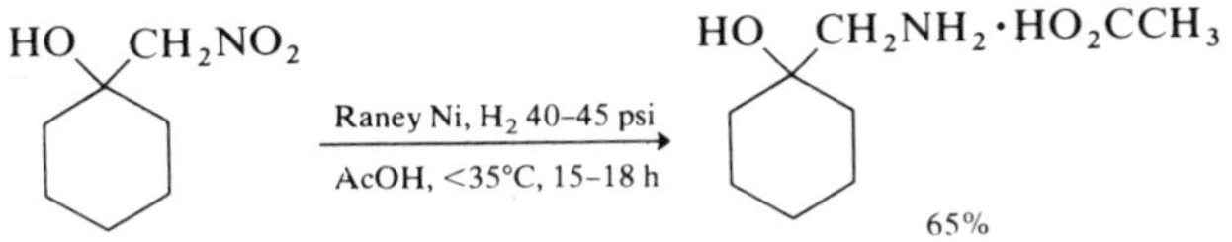

(97)[122]

Very few homogeneous catalysts have been reported for the catalytic hydrogenation of nitro compounds. Indeed, their general reluctance to reduce the nitro group allows the selective reduction of unsaturated nitro compounds to the saturated analogs [Eq. (98)]. [Ru(PPh$_3$)$_3$Cl$_2$] catalyzes

$$ \xrightarrow[\text{H}_2 \text{ 60–100 psi, 40–60°C} \atop \text{12–18 h}]{\text{Rh(PPh}_3)_3\text{Cl}} \qquad (98)^{136} $$

90%

$$ \xrightarrow[\text{KOH, H}_2 \text{ 90 atms} \atop \text{120°C, 1–6 h}]{\text{Ru(PPh}_3)_3\text{Cl}_2} $$

88%

(99)

the hydrogenation of aromatic and aliphatic nitro compounds at 50–150 atm and 90–130°C in the presence of excess alkali.[123] Aromatic nitro compounds containing substituent groups—halogen, alkoxide, ester, or nitrile—are all smoothly reduced by this catalyst without affecting the substituent. The major application of this catalyst appears to be in the selective reduction of dinitro aromatics (Section 7.6.3.2).

Catalytic transfer hydrogenation is a facile method of reducing nitro compounds without the use of hydrogen and pressure equipment. Aromatic nitro compounds containing a variety of functional groups—methoxy, hydroxy, keto, carboxyl, or nitrile, but not aldehyde, which inhibits the catalyst, or halogens, which are cleaved—are reduced to the corresponding amines by refluxing in ethanol containing cyclohexene in the presence of palladium on carbon.[124] With low catalyst-to-substrate ratios (10^{-2}:1) long reaction times are reportedly necessary, but increasing the catalyst-to-substrate ratio to 1:2 allows rapid (10 min), controlled reductions to be performed.[125] (See also Section 7.6.3.2 for further details.) RuCl$_3$·H$_2$O and RhCl$_3$·3H$_2$O using indoline as a hydrogen donor are also effective under mild conditions (80°C/4 h) for nitro group reduction.[126] Aliphatic nitro compounds are not readily reduced by transfer hydrogenation.[127]

Aromatic nitro groups are not normally reduced by sodium borohydride; however, sodium borohydride in the presence of a transition metal salt provides a simple and effective reduction method. One of the most efficient transition metal salts is copper acetylacetonate.[128] The catalyst system is not very tolerant of other functional groups, but alkoxy and, more interestingly, halogen substituents are unaffected.

$$RC_6H_4NO_2 \xrightarrow[\text{30°C, 2 h}]{\text{Cu(acac)}_2\text{-NaBH}_4} RC_6H_4NH_2 \qquad (100)$$

R = Cl, 80% (*o*), 80% (*m*), and 90% (*p*)
R = MeO, 90% (*o*), 80% (*m*), and 90% (*p*)

General Procedure—Reduction of Nitro Groups by Cu(acac)₂–NaBH₄
Into a suspension of copper acetylacetonate (2 mM) in isopropanol (20 ml) under nitrogen is dripped sodium borohydride (10 mM) in ethanol. To this mixture is added the nitro compound (10 mM) in isopropanol followed by more sodium borohydride (20 mM) in ethanol. The mixture is stirred for 2–4 h at 30°C. The reaction is then quenched with water and the solvent removed under reduced pressure. The residue is extracted with chloroform and purified. Yields, 80%–90%.

High functional group tolerance is found in nitro group reduction using stoichiometric amounts of iron carbonyl.[129] A methanolic solution of dodecacarbonyl triiron [$Fe_3(CO)_{12}$] specifically reduces the nitro group of nitroaryls to the primary amine. High product yields are obtained in the presence of sensitive groups such as double bonds, carbonyl, esters, or *N*-acetamides. The effective reagent is thought to be a hydridoiron cluster

$$RC_6H_4NO_2 + Fe_3(CO)_{12} + MeOH \xrightarrow[\substack{\text{reflux,} \\ \text{10-17 h}}]{C_6H_6} RC_6H_4NH_2 \qquad (101)$$

R = Cl, Br, Me, NH_2, NO_2, CO_2Et, OMe, OH, or $NHCOCH_3$
Yields generally >75%

generated *in situ*. The long reaction times are restrictive but this method has been improved by use of phase transfer catalysis.[130] Treatment of the aromatic nitro compound with a half molar quantity of $Fe_3(CO)_{12}$ in aqueous sodium hydroxide and benzene, using benzyl triethylammonium chloride as the phase transfer agent, gives similar or improved yields with only 1–2 h reaction times at room temperature. The phase transfer technique allows an improved preparation of the hydrido cluster catalyst $[HFe_3(CO)_{11}]^{\ominus}$ and aids its reaction in the organic phase. Although this method of reduction is interesting, practical utility is greatly inhibited by the use of stoichiometric quantities of expensive $Fe_3(CO)_{12}$. Cheaper reagents are generally preferred but do not always promote this reaction. For example, iron pentacarbonyl in refluxing *n*-butyl ether smoothly reduces substituted nitro aryls to azobenzenes in moderate yields[131] [Eq. (102)]. A serious disadvantage of this method is the toxicity of $Fe(CO)_5$,

$$2ArNO_2 \xrightarrow[\text{Bu}^n_2\text{O}]{\text{Fe(CO)}_5} ArN{=}NAr \qquad (102)$$

and, moreover, the reaction is substituent dependent: H, *o*-, *m*-, *p*-OMe, *p*-F, and *o*-, *m*-, *p*-Me produce azocompounds in >50% yield.

7.6.2. Partial Reduction of the Nitro Group

The hydrogenation of the nitro function proceeds in a stepwise fashion:

$$RNO_2 \xrightarrow{H_2} RNO \xrightarrow{H_2} RNHOH \xrightarrow{H_2} RNH_2$$

but the intermediates can rarely be isolated. Under certain conditions with selected substrates it has been possible to isolate the hydroxylamine intermediate. For example, high yields of phenylhydroxylamines have been obtained with palladium on carbon or alumina using DMSO as a promoter[132] and monitoring the hydrogen uptake [Eq. (103)]. Aliphatic

$$-\!\!\!\left\langle\!\!\!\bigcirc\!\!\!\right\rangle\!\!\!-NO_2 \xrightarrow[\substack{H_2\ 0\text{–}50\ psi,\ 25°C,\\DMSO}]{5\%\ Pd/C,\ MeOH} -\!\!\!\left\langle\!\!\!\bigcirc\!\!\!\right\rangle\!\!\!-NHOH \qquad (103)$$

hydroxylamines have also been isolated using palladium catalysts (e.g., Pd or Pd on charcoal). When the nitro group is attached to a primary or secondary carbon atom rearrangement to the oxime may occur [Eq. (104)].

$$R_2CHNO \rightleftharpoons R_2C=NOH \qquad (104)$$

The oxime has been isolated when palladium or platinum catalysts doped with lead salts [e.g., PbO, Pb(NO_3)_2 or PbSO_4] are used in the hydrogenation or by use of cuprous chloride in ethylenediamine[14] [Eq. (105)]. Copper

$$\left\langle\!\!\!\bigcirc\!\!\!\right\rangle\!\!\!-NO_2 \xrightarrow[\substack{H_2\ 35\ atm,\ 90°C}]{CuCl\text{–}EDA} \left\langle\!\!\!\bigcirc\!\!\!\right\rangle\!\!\!=NOH + H_2O \qquad (105)$$
$$\text{93\%}$$

salts in diamine solvents also promote the homogeneous reduction of nitroalkanes to oximes by carbon monoxide. With nitrocyclohexane, for example, 89% of the oxime is obtained (1 atm CO, 85°C).[132]

7.6.3. Selective Reduction of Nitro Compounds

7.6.3.1. In the Presence of Other Functional Groups

Except for triple bonds, aromatic nitro groups are preferentially hydrogenated over other functional groups. With unsaturated aromatic nitro compounds though, reduction of a double bond may occur during the latter stages of the reaction if hydrogen uptake is not monitored. Quenching the reaction after three equivalents of hydrogen have been absorbed

$(NO_2 \xrightarrow{3H_2} NH_2)$ usually leads to high yields of the unsaturated amine. Palladium or platinum catalysts are effective for these selective reductions.[133] Problems can arise in the presence of halogen substituents, which are readily cleaved by palladium; in these cases platinum on carbon or platinum oxide is the preferred catalyst.

The similar hydrogenation reactivity of alkynes and aromatic nitro groups usually leads to mixtures when these functionalities occur in the same molecule. Cobalt polysulfide at high pressure (1000 psi)[141] and supported ruthenium[134] have been reported for the selective nitro group reduction in nitroalkynes. Using ruthenium on carbon or alumina, an overall ease-of-hydrogenation sequence was observed:

$$RC{\equiv}CH > ArC{\equiv}CH \sim ArNO_2 > ArC{\equiv}CR \sim RNO_2$$

Increasing steric crowding at the triple bond allows selective reduction of the nitro group. A suitable blocking group is dimethylcarbinol, which can be easily added to and removed from a terminal alkyne [Eq. (106)]. Thus,

$$ArC{\equiv}CH + Me_2CO \xrightarrow[\text{NaOH}]{\text{liq. } NH_3,\ -50°C} ArC{\equiv}C{-}\overset{\overset{\displaystyle OH}{|}}{C}Me_2 \qquad (106)$$

the reduction of 2-methyl-4-(3-nitrophenyl)-3,butyn-2-ol occurs with 100% selectivity at the amine (50°C, 50–60 psi H_2) over ruthenium on carbon. The dimethyl carbinol blocking group is removed by a catalytic amount of sodium hydroxide in toluene (Scheme 6).

Scheme 6

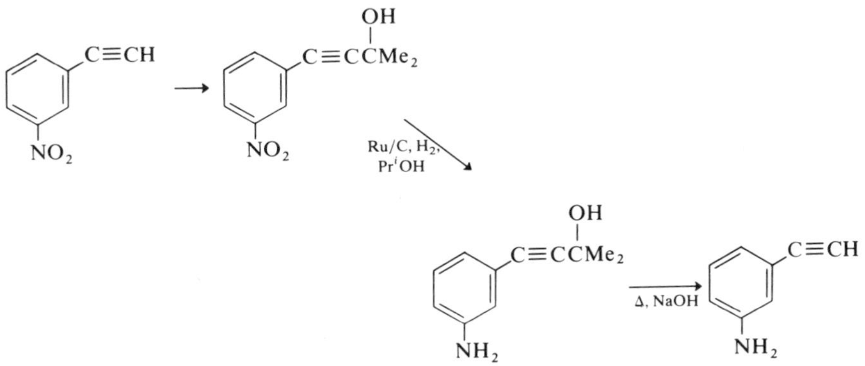

General Procedure—Selective Nitro Group Hydrogenation of a Nitroalkyne[134]

A. *Addition of Protecting Group.* A mixture of (3-nitrophenyl) acetylene (7.4 g, 50 mM) and acetone (7.9 g, 136 mM) is added over 20 min to stirred liquid ammonia (100 ml) containing crushed sodium hydroxide pellets (2 g) at −50°C. After 30 min the reaction is slowly warmed to room temperature and the ammonia allowed to evaporate off. The residue is treated with water (50 ml) and extracted with ether. After drying, evaporation of the ether, and distillation, 2-methyl-4-(3-nitrophenyl)-3-butyn-2-ol (4.1 g, 40%) is obtained.

B. *Hydrogenation.* 2-Methyl-4-(3-nitrophenyl)-3-butyn-2-ol (20.5 g) is hydrogenated in isopropanol (200 ml) over activated 5% ruthenium on alumina (1 g) in a Parr hydrogenator at 70°C and 40–60 psi hydrogen. The mixture is shaken until complete hydrogen absorption has occurred (monitored by pressure drop). The mixture is cooled, filtered, and the propanol removed on a rotary evaporator to give a tan solid (18 g). Recrystallization from toluene gives cream-colored needles (mp 114–116°C, >90%) of 3-methyl-4-(3-aminophenyl)-3-butyn-2-ol.

C. *Protecting Group Removal.* 2-Methyl-4-(3-aminophenyl)-3-butyn-2-ol (2 g) is dissolved in toluene (15 ml) containing one crushed pellet of sodium hydroxide (0.1 g). The mixture is heated to reflux for 1 h and liberated acetone collected in a Dean–Stark trap. The reaction mixture is then cooled and filtered and volatiles removed to give a quantitative yield of 3-aminophenyl-acetylene (1.4 g) 98% pure.

In unsaturated, nonconjugated aliphatic nitro compounds the situation is reversed, compared to aromatic analogs, and the double bond is preferentially hydrogenated in the presence of palladium or especially platinum catalysts.[135] Conjugated nitroalkenes are difficult to selectively reduce with heterogeneous catalysts, and mixtures generally result. Homogeneous catalysts are more selective. [Rh(PPh$_3$)$_3$Cl] readily reduces the double bond in unsaturated nitro compounds leaving the nitro group unaffected, even when long reaction times and excess hydrogen are used[136] [Eq. (107)].

$$\text{(Ar)}CH{=}CHNO_2 \xrightarrow[\text{H}_2\ 60\ \text{psi, }50°\text{C}]{\text{Rh(PPh}_3)_3\text{Cl}} \text{(Ar)}CH_2CH_2NO_2 \qquad (107)$$

7.6.3.2. Reduction of Dinitro Compounds

Heterogeneous catalysts have been used for the selective hydrogenation of dinitro aromatic compounds, but only in special cases are acceptable results achieved. For example, Raney copper has been used for the selective reduction of 2,4-dinitroalkylbenzenes to the 4-amino-2-nitro analogs.[137]

Selectivity increases with increased steric hindrance of the alkyl substituent [Eq. (108)].

$$\text{(108)}$$

Transfer hydrogenation using hydrazine as the hydrogen source has proved effective for these selective reductions. The preferred catalyst is ruthenium on carbon[138] [Eq. (109)]. Similarly, palladium on carbon with cyclohexene has proved useful [Eq. (110)] (Table 7.10). These reactions are rapid, require only mild conditions, and are insensitive to chloro, methoxy, or amine substituents.

$$\text{(109)}$$

$$\text{(110)}$$

Table 7.10. *Reduction of Nitro Compounds Using Palladium on Carbon/Cyclohexene*

Substrate	Product (yield, %)
o-Nitrophenol	*o*-Aminophenol (88)
o-Nitrotoluene	*o*-Toluidine (100)
p-Nitroanisole	*p*-Aminoanisole (83)
m-Nitrobenzaldehyde	*m*-Aminobenzaldehyde (10)
p-Nitroacetophenone	*p*-Aminoacetophenone (98)
o-Nitrobenzoic acid	*o*-Aminobenzoic acid (92)
o-Nitroacetanilide	*o*-Aminoacetanilide (100)
o-Nitroaniline	*o*-Phenylenediamine (0)

General Procedure—Selective Reduction of 4-Methoxy-2,5-Dinitroanisole[138]

4-Methoxy-2,5-dinitroanisole, 10% palladium on carbon (0.5 mol equiv), cyclohexene (5–6 mol equiv), and ethanol are refluxed vigorously for 10 min. The mixture is then filtered to remove the catalyst, which is washed and reused. The filtrate is evaporated to yield 3,6-dimethoxy-2-nitroaniline (85%).

In general, higher hydrogenation selectivity is achieved using homogeneous catalysts. For instance, [Ru(PPh$_3$)$_3$Cl$_2$] selectively hydrogenates aromatic dinitro compounds (120–135°C, 80 atm H$_2$) to give the nitroaniline derivative in high yield[123] [Eqs. (111) and (112)]. It is interesting

$$(111)$$

$$(112)$$

to note that here the hydrogenation of 2,4-dinitrotoluene occurs at the *more* hindered nitro group in contrast to heterogeneous catalysts, which reduce the less sterically hindered group.[137]

7.6.4. Reduction of the Nitrile Functional Group

The hydrogenation of nitriles to the corresponding amine over heterogeneous catalysts is complicated owing to formation of substantial amounts of secondary and even tertiary amines. Formation of the secondary amine is thought to occur by reaction of the intermediate imine **20** with the product amine (Scheme 7). The catalyst of choice is Raney nickel

Scheme 7

$$RC{\equiv}N \xrightarrow{H_2} RCH{=}NH \rightarrow RCH_2NH_2$$
$$\mathbf{20}$$

$$\downarrow {\scriptstyle RCH=NH}$$

$$(RCH_2)_2NH \xleftarrow{H_2} RCH{=}NCH_2R \xleftarrow{-NH_3} RCH(NH_2)NHCH_2R$$

---

although rhodium systems have proven useful. Various reaction modifications are employed to produce high yields of the primary amine. Scheme 7 indicates that formation of the secondary amine could be minimized by complexing the primary amine as it forms and so prevent further reaction. Hence, strong acid solutions have been employed[139] to protonate the free amine [Eq. (113)], or alternatively the reaction can be performed in the presence of chloroform[127] [Eq. (114)]. The reaction can also be run

$$C_6H_5CN \xrightarrow[\text{HCl, 2 h}]{PtO_2,\ H_2} \underset{98\%}{C_6H_5CH_2NH_2} \qquad (113)$$

$$C_6H_5CN \xrightarrow[\text{CHCl}_3,\ 90\ \text{min}]{PtO_2,\ H_2} \underset{97\%}{C_6H_5CH_2NH_2} \qquad (114)$$

in the presence of an acylating agent such as acetic anhydride.[140] Using, for example, Raney nickel with sodium acetate cocatalyst in acetic anhydride, a nitrile can be effectively reduced to the primary amine in good yield [Eqs. (115) and (116)]. The amine is trapped as the *N*-acetyl derivative, which can then be hydrolyzed to the corresponding primary amine after hydrogenation is complete.

$$C_6H_5CH_2CN \xrightarrow[\substack{\text{Ac}_2\text{O, 50°C, 1 h,}\\ \text{H}_2\ 50\ \text{psi}}]{\text{Raney Ni, NaOAc}} \underset{97\%}{C_6H_5CH_2CH_2NHAc} \qquad (115)$$

$$CH_3(CH_2)_{11}CN \xrightarrow[\substack{\text{Ac}_2\text{O, 50°C,}\\ \text{H}_2\ 50\ \text{psi}}]{\text{Raney Ni, NaOAc}} \underset{100\%}{CH_3(CH_2)_{11}CH_2NHAc} \qquad (116)$$

Performing these reductions in ammoniacal alcohol also gives high yields of the primary amine.[141] Originally a nickel catalyst was used but rhodium is effective under milder conditions. Rhodium (5%) on alumina in the presence of ammonia hydrogenates a variety of nitriles at room temperature and 2–3 atm H_2 in less than 2 h. The reaction over rhodium is particularly useful for basic nitriles and cyanoalkylethers [Eqs. (117) and (118)] but gives poor results with benzonitriles.

$$Me_2NCH_2CN \xrightarrow[\text{H}_2,\ \text{NH}_3]{Rh/Al_2O_3} \underset{68\%}{Me_2NCH_2CH_2NH_2} \qquad (117)$$

$$C_6H_{13}OCH_2CH_2CN \xrightarrow[\text{H}_2,\,\text{NH}_3]{\text{Rh/Al}_2\text{O}_3} \underset{90\%}{C_6H_{13}OCH_2CH_2CH_2NH_2} \qquad (118)$$

The most useful homogeneous catalyst for nitrile reduction is $[RhH(PPr^i_3)_3]$.[142] Under ambient conditions the rhodium hydride hydrogenates nitriles exclusively to the corresponding primary amine (Table 7.11). The catalytic activity with aromatic nitriles is not so high as for aliphatic analogs, but overall reaction times are generally short (2 h). With unsaturated nitriles hydrogenation of the double bond takes place more readily than that of the nitrile function, and mixed products are obtained. The use of this homogeneous catalyst allows high yields of the primary amine to be obtained without the necessity of added ammonia or an anhydride.

General Procedure—Reduction of Phenylacetonitrile[142]

To a solution of $[RhH(PPr^i_3)_3]$ (0.05 mM) in THF (10 ml) under a hydrogen atmosphere is added phenylacetonitrile (5 mM) at room temperature. The mixture is stirred for 2 h and after work-up gives phenethylamine in 89% yield.

7.6.5. Formation of Mixed Amines from Nitriles

The reaction of an amine with the intermediate imine (20) during hydrogenation leads to the formation of secondary or even tertiary amines (Scheme 7). This reaction is suppressed in favor of primary amine formation by addition of ammonia,[141] but, alternatively, secondary amine formation can be encouraged by increasing the amine concentration. This provides a synthesis of unsymmetrical amines via hydrogenation of a nitrile.[143] The

Table 7.11. [RhH(PPr$^i_3$)$_3$]-Catalyzed Hydrogenation of Nitriles

Substrate	Product[a]	Yield, %
$Me(CH_2)_4CN$	$Me(CH_2)_5NH_2$	100
Pr^iCN	$Pr^iCH_2NH_2$	100
Bu^tCN	$Bu^tCH_2NH_2$	67
$PhCH_2CN$	$PhCH_2CH_2NH_2$	96
$CH_2{=}CHCH_2CN$	$CH_3CH_2CH_2CH_2NH_2$	70[b,c]
$CH_3CH{=}CHCN$	$CH_3CH_2CH_2CH_2NH_2$	72[c]
$PhCN$	$PhCH_2NH_2$	45
α-Naphthyl CN	α-Naphthyl CH_2NH_2	44

[a] 20°C, 1 atm H_2, 2 h.
[b] 20 h.
[c] By-product is the unsaturated nitrile.

hydrogenation of valeronitrile in butylamine over rhodium on carbon produces butylpentyl amine in excellent yield [Eq. (119)]. These reactions

$$C_4H_9CN + C_4H_9NH_2 \xrightarrow[\text{H}_2\,50\,\text{psi}]{\text{Rh/C, 25°C}} \underset{100\%}{C_4H_9NHC_5H_{11}} \qquad (119)$$
$$\underset{(1:6.6)}{}$$

are very dependent on the catalyst, but rhodium on carbon appears to be the most effective at producing the symmetrical or unsymmetrical secondary amine.

7.6.6. Selective Reduction of the Nitrile Group

Triple or double carbon–carbon bonds and aromatic nitro groups are hydrogenated in preference to the nitrile function unless steric factors greatly inhibit double-bond reduction. Under these circumstances mixtures are generally obtained. The nitrile group is usually preferentially reduced in the presence of ketones, but complications often occur owing to condensation reactions between the carbonyl and amine. Homogeneous catalysts are very adept at removing the unsaturation in unsaturated nitriles. High double- or triple-bond selectivity is observed using, for instance, [Rh(PPh$_3$)$_3$Cl] [136] [Eq. (120)].

$$(120)$$

The partial hydrogenation of nitriles to give the imine is not well documented, but reaction of the intermediate imine has been noted. In acidic media a nitrile can be converted into an aldehyde via hydrolysis of the imine as it forms [Eq. (121)]. The imine can also be reacted on formation

$$(121)$$

R = 3-MeO (86%), 4-Me (82%),
3-Cl (82%), or 4-OH (86%)

to produce a stable species that, upon hydrolysis, produces the aldehyde[145] [Eq. (122)]. Various reagents have been used to trap the imine, notably N$_2$H$_4$, [PhNHCH$_2$]$_2$,[146] or NH$_2$NHCONH$_2$·HCl.[147] Whether these offer

$$\text{(naphthyl)}{-}CN \xrightarrow[\text{NH}_2\text{NH}_2,\ \text{EtOH}]{\text{Ni, H}_2} \text{(naphthyl)}{-}CH{=}NNH_2$$

$$\Big\downarrow \text{H}_2\text{SO}_4,\ \text{H}_2\text{O}$$

$$\text{(naphthyl)}{-}CHO \tag{122}$$

an alternative to stannous chloride–hydrochloric acid for this transformation is debatable.

7.6.7. Reduction of Oximes, Imines, and Nitroso Groups

Nitroso and oxime functions are intermediates in the hydrogenation of nitro groups; similarly, imines are the result of the partial hydrogenation of the nitrile group. These intermediates are rarely isolated but undergo further hydrogenation to give the corresponding amine. Catalysts for the reduction of nitro and nitrile groups are therefore suitable for oxime, imine, and nitroso reductions. Typically, Raney nickel, supported palladium, or platinum is used. The problems encountered with nitro and nitrile reductions manifest themselves when the partially reduced analogs are hydrogenated further. Hence imine or oxime reductions often lead to the formation of secondary amines, and precautions must be taken to obtain high yields of the primary species. As before, Raney nickel plus ammonia or acetic anhydride, or rhodium on carbon plus ammonia are suitable catalytic systems. As noted earlier, nitro group hydrogenations are often exothermic; the same applies to nitroso and oxime hydrogenations. Using Raney nickel in ammoniacal solution at elevated temperatures and pressure (70–100°C, 70–100 atm—conditions preferred for obtaining high yields), the hydrogenation of nitroso or oxime compounds may be extremely violent and extreme caution must be observed[3] (i.e., use low ratio of catalyst to compound and dilute solutions).

Homogeneous catalysts have not been extensively studied for the hydrogenation of oxime, imine, or nitroso groups, but it is to be expected that $[\text{Ru}(\text{PPh}_3)_3\text{Cl}_2]$[123] would be suitable for nitroso and oxime groups and $[\text{Rh}(\text{PPr}^i_3)_3\text{H}]$[142] for imine reduction (*vide supra*).

7.7. Asymmetric Hydrogenation

7.7.1. Heterogeneous Catalysts

It has long been the chemist's dream to emulate the high stereo and isomeric specificity of enzymic reactions, and possibly asymmetric hydrogenation is the closest yet achieved. Although heterogeneous catalysts have been examined, only modest optical yields (the excess of one enantiomer over the racemic mixture) have been achieved.[149] Hydrogenation over palladium on silk fibroin has produced optical yields of up to 70%, but these are dependent on the origin of the fibroin and its chemical treatment. Modified Raney nickel has also been moderately successful. Modification with an aqueous solution of (+) or (−) tartaric acid and sodium bromide produces a catalyst for the hydrogenation of acetylacetone to optically pure pentanediol.[150] Using R,R-tartaric acid the (−) (2R, 4R) pentanediol is obtained (optical yield 75%), whereas the S,S acid gives the (+) (2S, 4S) diol (72% optical yield) [Eq. (123)].

$$
\begin{array}{c}
\text{Me} \\ \diagdown \\ \quad\text{C}=\text{O} \\ \diagup \\
\diagdown \\ \quad \text{C}=\text{O} \\ \diagup \\
\text{Me}
\end{array}
\quad \xrightarrow[\text{tartaric acid, H}_2]{\text{Raney Ni, NaBr}} \quad
\begin{array}{c}
\text{Me} \quad \text{OH} \\ \diagdown * \diagup \\ \quad\text{C} \\ \diagdown \text{H} \\
\diagup * \diagdown \\ \quad \text{H} \\ \text{C} \\
\text{Me} \quad \text{OH}
\end{array}
\tag{123}
$$

This catalyst system has been used for the hydrogenation of 4-hydroxy-2-butanone to produce either R- or S-1,3 butanediol (70% optical yield)[151] [Eq. (124)].

$$
\underset{\displaystyle \text{MeCCH}_2\text{CH}_2\text{OH}}{\overset{\displaystyle \text{O}}{\overset{\displaystyle \|}{}}} \quad \xrightarrow[\text{NaBr, tartaric acid}]{\text{Raney Ni, H}_2} \quad \underset{*}{\overset{\displaystyle \text{H}}{\underset{}{\overset{\displaystyle \text{O}}{\overset{\displaystyle |}{\text{MeCHCH}_2\text{CH}_2\text{OH}}}}}
\tag{124}
$$

General Procedure—Asymmetric Hydrogenation of 4-Hydroxy-2-butanone[151]

4-Hydroxy-2-butanone (2 g) in THF (3 ml) is added to an autoclave containing modified Raney nickel (0.3 g) under hydrogen. [Modified Raney nickel is prepared by leaching Raney nickel (0.3 g) with 20% aqueous NaOH (20 ml) at 100°C; the catalyst is then modified with an aqueous solution (20 ml) of R,R-tartaric acid (0.2 g) and NaBr (1.6 g) at 100°C.] The hydrogenation is performed at 90 kg/cm² hydrogen pressure and 100°C for 14 h. The mixture is then cooled, filtered through Celite, and dried over K_2CO_3. The dry filtrate is distilled to give R-1,3-butanediol quantitatively, with 69% optical purity.

Modified heterogeneous catalysts have not attained a higher degree of success and it is difficult to see how identical active sites can be obtained on a catalyst surface. The variety of different sites on a surface leads to mixed products, and controlling the conformation around each active center is difficult. Indeed, the high degree of stereoselectivity achieved with this heterogeneous catalyst is remarkable.

7.7.2. Homogeneous Catalysts

Homogeneous catalysts are the obvious choice for asymmetric hydrogenation—all the metal active centers are identical and the conformation around the active center can be controlled. The hydrogenation of a prochiral substrate to produce an optically active organic compound demonstrates the features of homogeneous catalysts to maximum advantage. The environment around the metal center can be stereochemically fixed so that the unsaturated substrate preferentially coordinates in one conformation. Hydrogen addition then occurs on one side of the substrate, generating the chiral product.

It is interesting to note that most industrial hydrogenation catalysts are heterogeneous, and the only example to date of a homogeneous catalyst is in the asymmetric hydrogenation of a cinnamic acid derivative to produce L-Dopa, used to treat Parkinson's disease[152] [Eq. (125)].

$$\text{MeO} \quad \text{HO}-\text{CH}=\text{CCO}_2\text{H} \xrightarrow[\substack{1)\ \text{H}_2 \\ 2)\ \text{H}_2\text{O}}]{[\text{Rh}]^*} \text{HO}-\text{CH}_2\text{CHCO}_2\text{H} \quad (125)$$

L-Dopa

$$[\text{Rh}]^* = [\text{Rh(diene)}\text{P}_2^*]^\oplus, \quad \text{P}_2^* = \begin{array}{c} o\text{-MeOC}_6\text{H}_4 \\ \text{PCH}_2\text{CH}_2\text{P} \\ \text{Ph} \quad \text{C}_6\text{H}_4\text{OMe-}o \end{array}$$

Homogeneous rhodium catalysts have received most attention, although other systems, e.g., ruthenium and cobalt complexes, have been studied.[153] The preferred system is $[\text{Rh(diene)L}_2]^\oplus \text{BF}_4^\ominus$ (or $\text{PF}_6^\ominus$ or $\text{ClO}_4^\ominus$) or $[\text{RhCl(monoene)}_2]_2\text{-L}_2$, where diene is, e.g., norbornadiene or cyclooctadiene, monoene is, e.g., ethylene or cyclooctene, and L_2 is a chiral diphosphine, e.g., **21**, or two chiral tertiary phosphines, e.g., **22**. Diphosphines are preferred because they produce more stable catalysts and the complex is more rigid.

The investigation of asymmetric hydrogenation has revolved around the matching of the chiral phosphine ligand to the substrate to obtain

$$Me\diagdown C\diagup \genfrac{}{}{0pt}{}{O-\overset{H}{\underset{H}{C^*}}-CH_2PPh_2}{O-\overset{}{\underset{H}{C^*}}-CH_2PPh_2} \qquad *PMe(Ph)Pr^n$$

$$Me\diagup$$

$$\mathbf{22}$$

21

DIOP

maximum optical yields. Hence numerous ligands have been reported,[154] many of which are very complex, but success is often empirical and little fundamental data are available on the bonding interactions involved. Recent findings have indicated the importance of not only the chiral chelating diphosphine, but also the conformation this ligand adopts in the rhodium complex.[156b] For example, 2,4-bis(diphenylphosphino)pentane is thought to adopt a chiral conformation upon complexation, whereas 1,3-bis(diphenylphosphino)butane adopts an achiral conformation. Both produce active rhodium(I) hydrogenation catalysts, but the former generally gives high optical yields, whereas the latter is ineffective as a chiral catalyst. Successful results have been achieved with the chirality on phosphorus (**22**) or on carbon (**21**). Two ligands will be described here, for the reasons given, but an extensive review is available, with many other examples.[154]

(A) *DIOP—2,3-Isopropylidene-2,3-dihydroxy-1,4-bis(diphenylphosphino)-butane* (**21**)

This ligand is conveniently prepared from tartaric acid as the (+) or (−) isomer,[155] but is also available commercially, and because of this it has received the most attention. The asymmetry is on carbon atoms away from the active center but the ligand is still effective. One disadvantage of this ligand is that upon coordination it produces a seven-membered chelate ring that is less stable than five- or six-membered analogs.

(B) *R- or S-PROPHOS—R- or S-1,2-bis(diphenylphosphino)propane* (**23**)

$$H\diagdown \overset{Me}{\underset{Ph_2P\diagup}{C}}-CH_2PPh_2$$

23

R-prophos

Whereas the tendency has been for more complexity in ligand designs, this simple ligand has only a single methyl group at the chiral center to constrain the chirality of the chelate ring. Even so, the prophos-rhodium(I) catalyst is excellent for producing optically active amino acids from α-acylaminoacrylic acids.[156a] One interesting feature is the synthesis of this ligand. A relatively complex synthesis is required to produce the first sample

Scheme 8

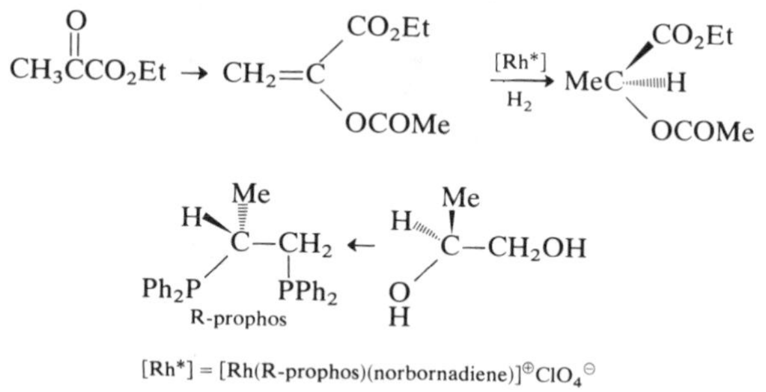

$$
CH_3\overset{\displaystyle O}{\overset{\|}{C}}CO_2Et \;\rightarrow\; CH_2=C\underset{OCOMe}{\overset{CO_2Et}{\big\langle}} \quad\xrightarrow[H_2]{[Rh^*]}\quad MeC\underset{OCOMe}{\overset{CO_2Et}{\big\langle}}{\cdots\cdots}H
$$

$$
\underset{\underset{R\text{-prophos}}{Ph_2P \qquad PPh_2}}{\overset{Me}{H{\diagdown}\overset{|}{C}-CH_2}} \;\leftarrow\; \underset{\overset{O}{\underset{H}{|}}}{\overset{Me}{H{\cdots}\overset{|}{C}-CH_2OH}}
$$

$$[Rh^*] = [Rh(R\text{-prophos})(norbornadiene)]^{\oplus}ClO_4^{\ominus}$$

of ligand, but by using this sample further ligand can be prepared by an alternative asymmetric hydrogenation procedure (Scheme 8). A similar ligand preparation cycle can be used to prepare S-prophos from a small amount of previously synthesized [Rh(S-prophos)(diene)]$^{\oplus}$. Whereas R-prophos yields products with the S configuration, S-prophos gives products with the R configuration. It may be expected naively that enantiomerically related ligands would give products differing only in the sign of their optical rotation, but this is not always the case.[176] Thus neomenthyldiphenylphosphine (NMDPP) gives a more active rhodium catalyst and higher enantiomeric excess than does menthyldiphenylphosphine (MDPP) [60% (R) using NMDPP vs. 17% (S) for MDPP].

In an attempt to eliminate the problem of separating the homogeneous catalyst from the reaction mixture, supported catalysts have been prepared.[5a] The rhodium–DIOP catalyst, for example, has been supported on cross-linked polystyrene,[160] cross-linked hydroxyethylmethacrylate/styrene polymer,[161] or silica gel.[162] Although these catalysts are active for asymmetric hydrogenation, optical induction is usually poor compared to the homogeneous analogs. Further developments are necessary in this area in order to obtain high optical yields from these heterogenized catalysts.

7.7.3. Mechanism of Asymmetric Hydrogenation

Extensive studies have been made of asymmetric hydrogenation.[157,158] Rhodium-catalyzed homogeneous hydrogenation of an alkene may proceed by either of the following routes: (1) Addition of hydrogen to the metal center followed by the alkene (hydride route); or, (2) alkene coordination to the metal followed by hydrogen addition (unsaturated route) (Scheme

Scheme 9

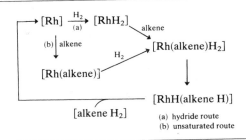

9). It was generally assumed that hydrogenation by cationic rhodium(I) complexes containing a diphosphine ligand followed the hydride route similar to [Rh(PPh$_3$)$_3$Cl], but recent evidence favors the unsaturated route.[158,159] A number of [Rh(diphosphine)alkene]$^{\oplus}$ complexes have been isolated and characterized by X-ray diffraction,[158] e.g., [Rh(S,S-chiraphos)(ethyl(Z)-α-acetamidocinnamate)]$^{\oplus}$ **(24)**. From low-temperature ^{1}H, ^{31}P, and ^{13}C nmr studies the rhodium hydrido alkyl **25**

has been identified; previously only indirect evidence was available for species of this type because they are very reactive and are not normally present in detectable concentrations.

The enantioselectivity of the reaction was thought to be determined by the preferred mode of binding of the prochiral alkene to the catalyst,[157] but the isolation of **24** has led to an alternative proposal. Addition of hydrogen to **24** would lead to the product S isomer, whereas the major reaction product is the R isomer. It has therefore been suggested that the characterized intermediate **24** corresponds to the predominant diastereoisomer in solution (confirmed by nmr), but that the other diastereoisomer is much more reactive toward hydrogen and it dominates the enantioselectivity of the reaction. That is, it is not the preferred stereochemistry of binding of the substrate to the catalyst but the difference in reactivity of the two isomeric adducts that determines the stereoisomer produced. Whether this is true for all chelating diphosphine rhodium(I) catalysts remains to be confirmed.

Understanding of the mechanism of asymmetric hydrogenation by rhodium catalysts is improving, but more data are required to establish conclusively the mechanism and ultimately allow prediction of the stereochemistry of a particular product.

7.7.4. *Asymmetric Hydrogenation of α-Acylaminoacrylic Acids*

α-Acylaminoacrylic acids have been extensively studied as substrates for asymmetric hydrogenation. Optical yields in excess of 90% have been achieved and 100% selectivity does not seem impossible. Table 7.12 indicates the variety of ligand types and the excellent results obtained.

Table 7.12. Ligands Used in Asymmetric Hydrogenation [Eq. (126)]

Ligand	Optical yield (isomer)	Ref.
$(o\text{-MeOC}_6\text{H}_4)\text{PhPCH}_2\text{CH}_2\text{PPh}(o\text{-MeOC}_6\text{H}_4)$ DIOP (21)	96% 72% (R)	163 155
S,S Chiraphos	89% (R)	164
BPPM	91% (R)	165
R–Prophos	91% (S)	156a
S,R–BPPFA	93% (S)[a]	166

[a] 20 h, 50 atm, R.T.

$$PhCH=C\begin{cases}CO_2H\\\\NHCOMe\end{cases} \xrightarrow[H_2]{[Rh]^*} PhCH_2\overset{*}{C}HCO_2H \qquad (126)$$

$$NHCOMe$$

$[Rh]^* = [Rh(diene)(diphosphine^*)]^{\oplus}BF_4^{\ominus}$, 25°C, 4 atm H_2, 4 h

General Procedure—Asymmetric Hydrogenation Using the DIOP Ligand[155]
To a benzene solution of [RhCl(cyclooctene)$_2$]$_2$ (1.5 mM) under nitrogen
is added DIOP (3 mM). The solution is stirred for 15 min and then added
to the hydrogenation flask containing the substrate (300 mM, for example)
in ethanol (benzene:ethanol, 1:2) under nitrogen, by means of a syringe.
The reaction is stirred for 1–3 h at room temperature and 1.1 atm of hydrogen.
Work-up depends on the substrate but consists of removing volatiles under
vacuum followed by dissolving the residue in either (a) water·or (b) 0.5 M
NaOH. The solution is filtered to remove catalyst residues and for (a) it is
evaporated to dryness, while for (b) it is acidified, extracted with ether, dried,
and evaporated to dryness. Chemical yields are generally >90%; the optical
yield depends on the substrate but ranges from 55%–80%.

7.7.5. Optically Active Alcohols from Asymmetric Hydrogenation of Ketones

Cationic rhodium(I) catalysts are effective for the hydrogenation of
ketones to optically active alcohols at ambient temperature and pressure,
although not as successful as with alkene substrates. Using [RhCl(hexa-1,5
diene]$_2$–DIOP, optical yields up to 50% have been achieved in the reduction
of ketones.[167] One of the best systems contains the ligand BPPFOH (**26**),

26

R, S, - BPPFOH

which is related to BPPFA described in Table 7.12. [Rh{R,S—BPPFOH}-
(norbornadiene)]$^{\oplus}$ClO$_4^{\ominus}$ effectively reduces aminomethylaryl ketones in
high optical yields (86%–90%), but elevated pressures (50 atm) and long
reaction times (2–4 days) are required [Eq. (127)].

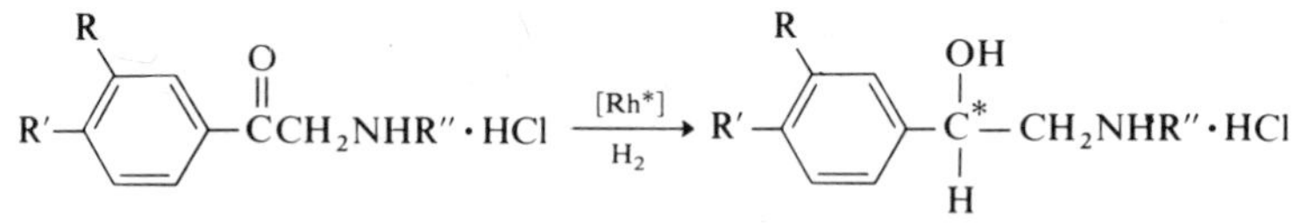

$$R'-\underset{\underset{R}{\bigcirc}}{\overset{O}{\underset{\|}{C}}}CH_2NHR'' \cdot HCl \xrightarrow[H_2]{[Rh^*]} R'-\underset{\underset{R}{\bigcirc}}{\overset{OH}{\underset{H}{\overset{|}{C}^*}}}-CH_2NHR'' \cdot HCl \quad (127)$$

R = R' = OMe, R'' = H	optical yield = 86%
R = R' = R'' = H	= 52%
R = H, R' = OH, R'' = H	= 69%
R = R' = OH, R'' = Me	= 95%

The usefulness of these reactions has yet to be fully explored, but asymmetric ketone reduction may offer novel syntheses to the pharmaceutical industry.

7.7.6. Asymmetric Hydrosilylation

The hydrogenation of ketones using homogeneous catalysts is not easy and elevated temperatures and pressures are often required (Section 7.5.1.2). Hydrosilylation offers an alternative method of reduction and the use of a chiral catalyst allows optically active products to be obtained.[169] Optical yields are not has high as those obtained with prochiral alkenes, probably as a result of the different mode of binding of C=O, compared to C=C, to the metal center (Scheme 10),[171] and they are also very dependent on the choice of silane (cf. Table 7.13). The different mode of

Scheme 10

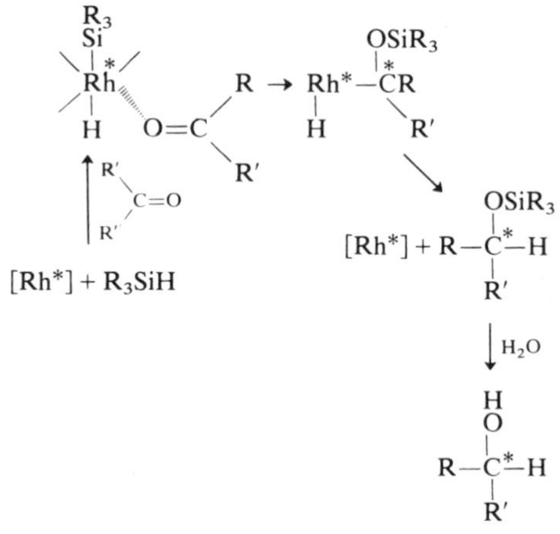

Table 7.13. Asymmetric Reduction of Ketones via Rhodium-Catalyzed Hydrosilylation

$$RCOR' + HSiX_3 \xrightarrow[\text{2. Hydrogenolysis}]{1, \text{CAT}^*, H_2} R-\underset{\underset{H}{|}}{\overset{\overset{OH}{|}}{C^*}}-R'$$

Catalyst	Ketone	Silane	Conditions, °C	Optical yield, %	Ref.
[RhCl(C$_8$H$_{14}$)$_2$]$_2$ + DIOP	MeCOPh	H$_2$Si(1-Np)Ph	20	55	173
[RhCl(C$_2$H$_4$)$_2$]$_2$ + DIOP	MeCOPh	H$_2$Si(1-Np)Ph	20	58	174
[RhCl(1,5-hexadiene)]$_2$ + R,S–MPFA	MeCOPh	H$_2$SiPh$_2$	20	49	175

$$\text{(Fe complex with } PMe_2 \text{ and } CHMeNMe_2) = MPFA$$

Catalyst	Ketone	Silane	Conditions, °C	Optical yield, %	Ref.
[Rh{R–BMPP}$_2$(diene)]$^{\oplus}$ ClO$_4$$^{\ominus}$ P(CH$_2$Ph)MePh = BMPP	MeCOPh	H$_2$SiPh$_2$	20	23	171

coordination also produces a different stereoselectivity compared to the analogous reduction of an alkene [Eqs. (128) and (129)].[170]

$$\underset{Me}{\overset{Ph}{>}}C=CH_2 + HSiCl_2Me \xrightarrow{\text{CAT}^*} \underset{Me}{\overset{Ph}{>}}C\underset{CH_2SiCl_2Me}{\overset{H}{<}} \qquad (128)$$

optical yield, 5.6% (R)

$$\underset{Me}{\overset{Ph}{>}}C=O + HSiCl_2Me \xrightarrow{\text{CAT}^*} \underset{Me}{\overset{Ph}{>}}C\underset{H}{\overset{OSiCl_2Me}{<}} \qquad (129)$$

optical yield, 7.6% (S)

CAT* = [PtCl$_2$(R$_3$P*)]$_2$, where R$_3$P* = (R), (+)-P(CH$_2$Ph)MePh

Initially, platinum catalysts were examined for asymmetric ketone hydrosilylation but poor optical yields were obtained.[170,172] Rhodium catalysts have produced better results (Table 7.13).

In asymmetric hydrogenation a careful match of ligand to substrate is required to achieve high optical yields. In asymmetric hydrosilylation a third factor, the choice of silane, also influences selectivity and adds further complications. When a correct balance of these three factors is obtained, high optical yields of chiral alcohols will be achieved, but lack of available data prevents such a rationalization at present.

7.8. Hydrogenolysis

Hydrogenolysis is usually an unwanted side reaction that occurs during a hydrogenation, but it can occasionally be a useful synthetic route. Generally, homogeneous catalysts are unlikely to produce hydrogenolysis, which is limited to heterogeneous systems. Three types of groups are particularly sensitive to hydrogenolysis: halogen, benzylic, and allylic groups. When these functions are contained in a molecule, careful choice of catalyst is required to prevent hydrogenolysis during the hydrogenation.

Palladium metal promotes hydrogenolysis reactions and is the catalyst of choice for functional group cleavage. Supported palladium is generally used, typically palladium on carbon or ·palladium on barium sulfate. For the functional groups noted above, elevated temperature and pressure are *not* required; hydrogenolysis occurs under relatively mild conditions (<100°C, <4 atm).

7.8.1. Dehalogenation

The well-known example of the hydrogenolysis of a halogen is the Rosenmund reduction, i.e., the dehalogenation of an acid chloride to produce an aldehyde (Section 7.5.3.1). Halogen hydrogenolysis has found other uses in organic synthesis.[177a] For example, the use of halogen to block reactive positions on an aromatic nucleus has been reported, the final stage of the synthesis being hydrogenolysis of the halogen[177b] (Scheme 11).

For most dehalogenation reactions, supported palladium, especially palladium on carbon, is the catalyst of choice. The acid produced during the reaction inhibits further reduction and for this reason a mole equivalent of base is usually added. Typical bases employed include tertiary amines and Group 1 or 2 hydroxides, e.g., magnesium hydroxide.

The ease of halogen removal is in the order

$$I > Br > Cl \gg F$$

and whereas iodide removal is easy, hydrogenolysis of fluoride is very

Scheme 11

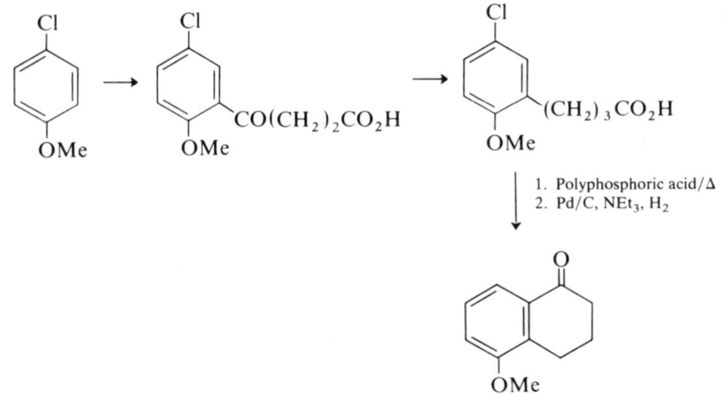

difficult and is rarely achieved. Activated halides, e.g., benzylic, allylic, or vinylic, are more readily cleaved than alkyl halides.

Heteroaromatic rings are readily dehalogenated. A useful preparation of pyrimidine involves the dechlorination of 2,4-dichloropyrimidine[178] [Eq. (130)]. The reaction is rapid (80 min) and requires only moderate pressures of hydrogen.

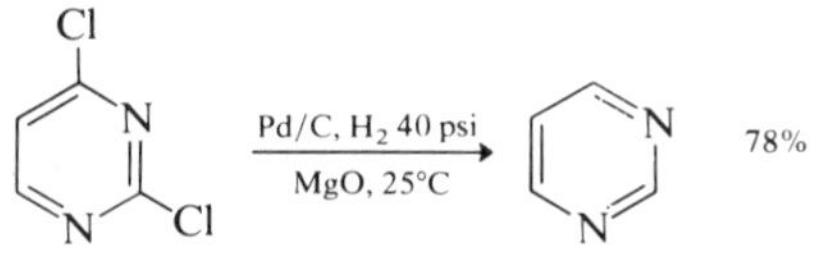

$$ \qquad\qquad (130) $$

A route to diaminopyridine from 2-aminopyridine uses halogen as a blocking group, which is readily removed in the final stage[179] (Scheme 12).

Scheme 12

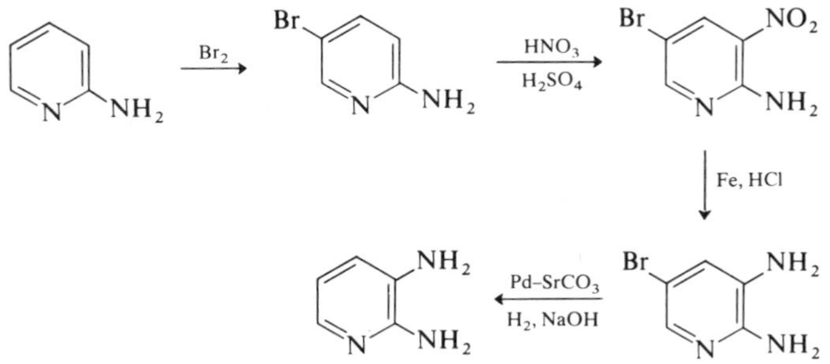

Hydrogenolysis using palladium on strontium carbonate in sodium hydroxide debrominates the substituted pyridine and gives yields of 78%– 86% under ambient conditions.

Generally dehalogenation occurs with concomitant double-bond reduction in unsaturated halides and with formation of the amine when aromatic nitro halides are hydrogenated. With aliphatic nitro compounds dehalogenation usually occurs first and the dehalogenated nitro substrate can be isolated.

Transfer hydrogenolysis (in which the hydrogen source is a donor molecule) produces similar results to direct hydrogenation.[10] Palladium on carbon is used with various hydrogen donor molecules, e.g., limonene,[180] p-menthene,[181] or cyclohexene, while palladium chloride[182] with various hydrogen-donating secondary amines, e.g., indoline or tetrahydroquinoline, has also been reported. Palladium on carbon–limonene[180] gives good results with carboxylic acids, whereas Pd/C-p-menthene[181] is used with nitriles [Eqs. (131)–(133)].

$$90\% \ (X = Cl \ or \ Br) \qquad (131)$$

$$85\%-90\% \ (o\text{- or } p\text{-Cl}) \qquad (132)$$

$$90\% \qquad (133)$$

From the examples [Eqs. (131)–(133)] it can be seen that these reactions are not very selective, that double bonds are saturated, and that cleavage of the C–N bond occurs with nitriles.

Greater selectivity in a dehalogenation reaction has been achieved using palladium on carbon with triethylammonium formate as the hydrogen source.[183] Aromatic halides readily undergo hydrogenolysis at 50–100°C without affecting nitro or nitrile groups [Eqs. (134)–(137)]. With chloro-unsaturated substrates, mixtures are obtained indicating comparable reduction rates for dechlorination and double-bond saturation. Debromination is more selective and unsaturated species can be obtained in high yield [Eq. (138)].

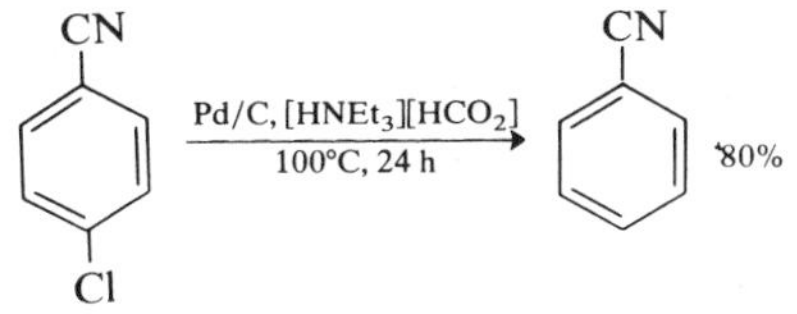

$$\text{(134)}$$

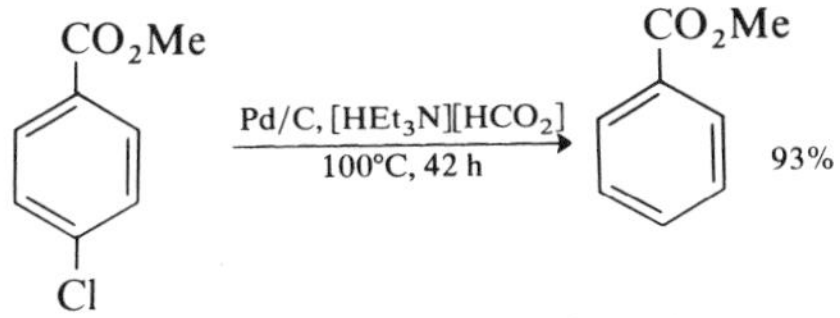

CN → (Pd/C, [HNEt$_3$][HCO$_2$], 100°C, 24 h) → CN, 80%

$$\text{(135)}$$

CO$_2$Me → (Pd/C, [HEt$_3$N][HCO$_2$], 100°C, 42 h) → CO$_2$Me, 93%

$$\text{(136)}$$

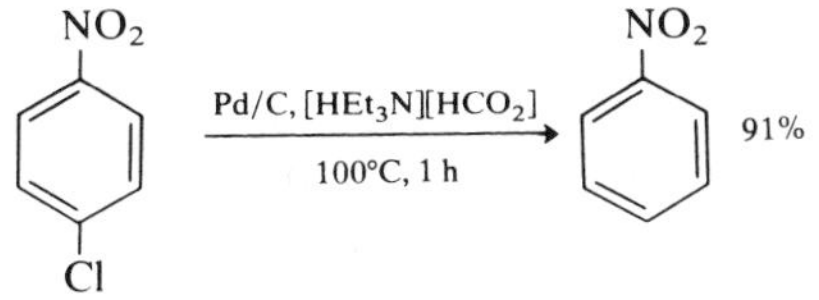

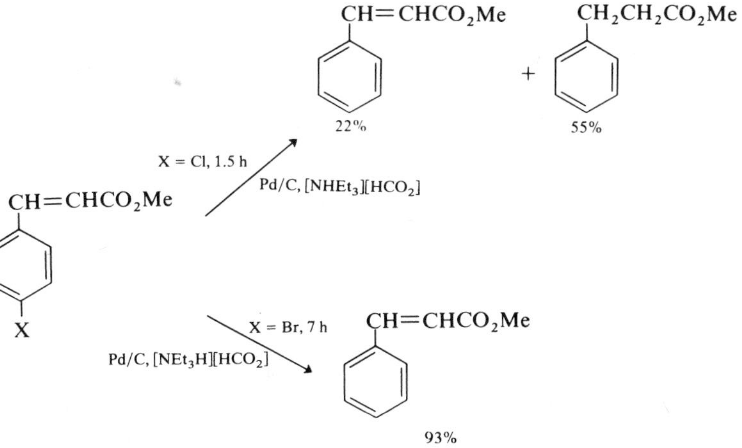

NO$_2$ → (Pd/C, [HEt$_3$N][HCO$_2$], 100°C, 1 h) → NO$_2$, 91%

$$\text{(137)}$$

CH=CHCO$_2$Me (22%) + CH$_2$CH$_2$CO$_2$Me (55%)

X = Cl, 1.5 h, Pd/C, [NHEt$_3$][HCO$_2$]

CH=CHCO$_2$Me / X

X = Br, 7 h, Pd/C, [NEt$_3$H][HCO$_2$]

CH=CHCO$_2$Me, 93%

$$\text{(138)}$$

General Procedure—Hydrogenolysis of Organic Halides[183]

Into a heavy-walled Pyrex bottle is placed the halide (20 mM) and the appropriate quantity of palladium on charcoal (0.2–1 mol %). Triethylamine (28.5 mM) is then added. The bottle is flushed with nitrogen and capped with a rubber-lined cap. Formic acid (22 mM) is syringed into the mixture through the rubber liner of the cap. The mixture is heated (50–100°C) and the progress of the reaction monitored by the increase in pressure (liberated CO$_2$) in the bottle (1–40 h). The product is isolated by addition of methylene chloride to the mixture to dissolve the unreacted lower layer of triethylammonium formate, followed by filtration and distillation.

The hydrogenolysis of aromatic bromides has been reported using nickel(0)-catalyzed sodium borohydride reduction. Generally sodium borohydride does not react with aryl halides, but the addition of Ni(PPh₃)₃ [prepared by heating (PPh₃)₂NiCl₂ in DMF with PPh₃ and excess NaBH₄ (3 : 1) under nitrogen] catalyzes halogen displacement[184] [Eq. (139)]. Aryl

$$\text{(139)}$$

A = H (90%), p-CO₂Me (100%), Cl (80%), Br (15%), or OMe (32%)

bromides are readily reduced at 70°C over 3–20 h, but, although aryl chlorides do react, the rate is very slow and of little practical significance. Functional group tolerance is poor, with both nitro and nitrile groups being readily reduced.

7.8.2. Selective Reduction of Polyhalo Compounds

The selective removal of a halogen from a polyhalo compound can be achieved, especially if the halogens are different (i.e., ease of removal I > Br > Cl ≫ F). Activated halogens are removed before nonactivated ones, and sterically hindered groups are less readily cleaved [Eq. (140)].

$$\text{(140)}^{185}$$

As noted earlier, nickel(0)-promoted sodium borohydride reductions are facile for aryl bromides but very slow with aryl chlorides. Hence 1,4-chlorobromobenzene is readily converted into chlorobenzene (80%, 70°C, 16 h) by this catalyst.[184]

7.8.3. Hydrogenolysis of Benzylic and Allylic Groups

The hydrogenolysis of benzyl alcohols, ethers, esters, or amines is relatively easy to achieve using 5% palladium on carbon. A rate order has been established[186]: amines < OH < OR < OAc. The reduction of benzyl alcohols is promoted by addition of a small amount of a strong acid, e.g., HCl, H₂SO₄, or HClO₄.[187] A general review on benzyl group hydrogenolysis is available,[187] and the range of groups readily cleaved is shown in the examples[187] [Eqs. (141)–(144)].

Transfer hydrogenation is also effective in cleaving benzylic groups. Again, palladium on carbon is the preferred catalyst with cyclohexene or tetralin as the source of hydrogen.[5] An adaptation of this method has been reported for the reduction of benzylic alcohols.[188] Reduction of the alcohol goes to completion in 36 h in refluxing cyclohexene/Pd on C/AlCl₃ [Eq. (145)].

$$ \underset{\underset{R''}{|}}{\overset{\overset{R'}{|}}{Ar\,C}}-OH \xrightarrow[\text{Cyclo–C}_6\text{H}_{10}]{\text{Pd/C, AlCl}_3} Ar-\underset{\underset{R''}{|}}{\overset{\overset{R'}{|}}{C}}-H \qquad (145) $$

R' = R'' = Ph	90%
R' = R'' = Me	84%
R' = Et, R'' = cyclo–C₃H₅	81% (3-phenylhexane)
R' = R'' = H	74%

General Method—Reduction of Benzyl Alcohols[188]

The alcohol (10 mM), 10% palladium on carbon (20% w/w), and aluminum chloride (20 mg) are mixed with cyclohexene (15 ml) and the mixture refluxed. After 36 h the reaction is complete and the product is isolated by diluting with ether, filtering, and evaporating off the solvent. The product is then purified by either distillation or recrystallization. Yields are generally high (75%–95%), but substrates with functional groups on the aromatic ring were not studied.

Allylic oxygen bonds are also hydrogenolyzed, but here competition with double-bond saturation may lead to mixed products. Palladium is again the preferred catalyst. A side reaction, which can lead to further complications, is isomerization, another reaction promoted by palladium. Overall, hydrogenolysis of allylic groups is not so straightforward as the cleavage of benzylic functions.

One interesting example of this type of reaction is the reductive displacement of allylic acetates by hydride transfer: Pd(PPh$_3$)$_4$[189] activates allylic acetates toward attack by hydride (NaBH$_4$) (Scheme 13). The

Scheme 13

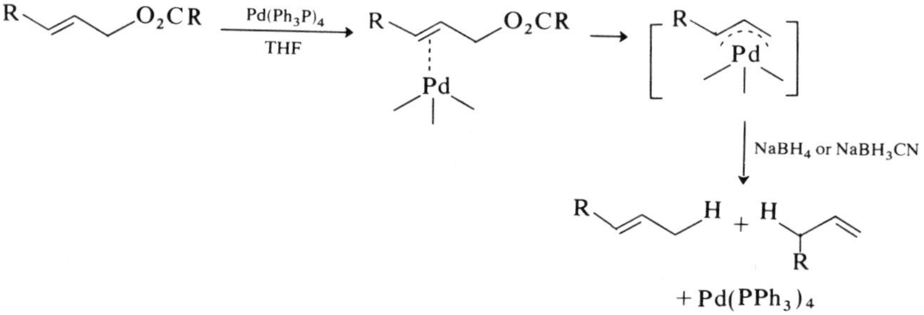

$$R\diagup\!\!\!\diagdown\!\!\!\diagup O_2CR \xrightarrow[\text{THF}]{Pd(Ph_3P)_4} R\diagup\!\!\!\diagdown\!\!\!\diagup O_2CR \longrightarrow \left[R\diagdown\!\!\diagup_{Pd} \right]$$

$$\downarrow \text{NaBH}_4 \text{ or NaBH}_3\text{CN}$$

$$R\diagup\!\!\!\diagdown\!\!\!\diagup H + \underset{R}{\overset{H}{\diagup\!\!\!\diagdown}} + Pd(PPh_3)_4$$

regioselectivity of the reaction is dependent upon the steric and electronic properties of substituents, because the reaction proceeds through a symmetrical π-allyl palladium intermediate which, controlled by the substituents, can add hydride at either end [Eqs. (146)–(148)]. These reactions proceed under mild conditions (66°C) and, although reaction times may be long (48 h), overall yields are good (60%–98%).

$$\text{PhCH=CH}\overset{\overset{\displaystyle OAc}{|}}{\text{CHPh}} \xrightarrow[\text{Pd(0)}]{\text{NaBH}_3\text{CN}} \text{PhCH=CHCH}_2\text{Ph} \qquad (146)$$
$$\underset{89\%}{}$$

$$p\text{-NO}_2\text{C}_6\text{H}_4\text{CH=CHCH}_2\text{OAc} \xrightarrow[\text{Pd(0)}]{\text{NaBH}_3\text{CN}} p\text{-NO}_2\text{C}_6\text{H}_4\text{CH=CHCH}_3 \qquad (147)$$
$$\underset{80\%}{}$$

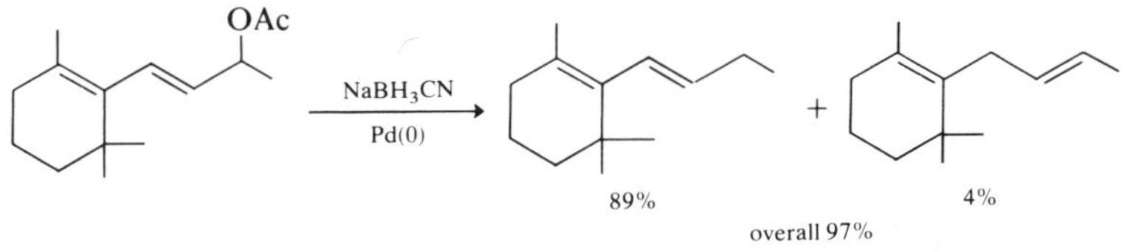

$$\begin{array}{ccc} & \xrightarrow[\text{Pd(0)}]{\text{NaBH}_3\text{CN}} & + \end{array} \tag{148}$$

89%   4%

overall 97%

General Procedure—Reduction of Allylic Acetates[189]

A solution of the acetate (3 mM), Pd(Ph₃P)₄ (0.3 mM), triphenylphos-phine (2.1 mM), and NaBH₃CN (6 mM) is placed in dry THF (30 ml) under nitrogen and stirred for 24 h at ambient temperature. The solution is diluted with saturated salt solution (2 vol) and extracted with ether (three times). The ether solution is washed with sat. NaHCO₃, dried (MgSO₄), and concentrated. Work-up then gives the product in good yield (80%–90%).

Chapter 8

Oxidation

The oxidation of organic compounds in the presence of transition metals and their complexes can not only produce a variety of oxygen-containing products such as alcohols, aldehydes, ketones, carboxylic acids, and epoxides, but it can also induce various coupling reactions. In general this chapter will be concerned with the formation of oxygen-containing products; reactions such as phenolic coupling and oxidative addition to alkenes have been discussed in Chapters 2 and 3. A short section on dehydrogenation, however, is included in this chapter.

A wide variety of transition metal species have been employed in oxidation reactions, the most widely used metals being chromium, manganese, copper, ruthenium, and osmium. Although metal complexes enhance the yields and selectivities of oxidations, the role of the metal in these reactions is often not fully understood, but the complexes are involved in one of two ways: (1) as oxidants in direct reactions, and (2) as promotors, in either catalytic or stoichiometric amounts, for other oxidants including, for example, oxygen and peroxides.

In this chapter, examples of both types of reaction will be discussed, with the emphasis being on synthetic applications, particularly reaction selectivity. Various reviews are available on the mechanistic aspects of transition-metal-mediated oxidations, the most recent of which are by Kochi.[1,2]

Industrially the oxidation of hydrocarbons to higher-value products is extremely important, and a few of the larger-scale reactions are shown in Table 8.1. More details of these and other related industrial reactions can be found in an excellent review by Parshall.[3]

Table 8.1. Industrial Application of Catalytic Oxidation

Substrate	Product	Catalyst
Cyclohexane	Cyclohexanol and cyclohexanone	Co, Mn, Cr
Butane	Acetic acid	Co
p-Xylene	Terephthalic acid	Co, Mn
Toluene	Benzoic acid	Co
Ethylene	Acetaldehyde	Pd
Ethylene/acetic acid	Vinyl acetate	Pd
Ethylene	Ethylene oxide	Ag

This short chapter can only serve as an introduction to the use of transition metals for the oxidation of organic compounds, but several reviews have been written on the topic.[4-9]

8.1. Formation of Alcohols

The selective oxidation of saturated hydrocarbons to alcohols is a highly desirable transformation, but it usually requires such vigorous conditions that the initially formed products, which are more susceptible to oxidation, are not easily isolated and further oxidation occurs. One of the best reagents for alkane oxidation is permanganate ion, but its use is often limited because of the insolubility of most alkanes in aqueous permanganate. If the alkane possesses some water solubility then the reaction will go. 2,4-Dimethyl-2-pentanol, for example, gives 2,4-dimethyl-2,4-pentanediol in up to 50% yield using a neutral aqueous solution of permanganate.[10] Cosolvents such as *t*-butanol or pyridine can be used to advantage, and the reactions are normally run at room temperature for several days to achieve the desired selectivity.

Chromium(VI) compounds have also been used for the oxidation of hydrocarbons, but usually secondary oxidation occurs. Since tertiary carbon–hydrogen bonds are more readily attacked than secondary carbon–hydrogen bonds, it is often possible to oxidize tertiary hydrogens preferentially in the presence of secondary hydrogen. *Cis*-decalin, for example, gives as the major product *cis*-9-decalol when oxidized by dichromate ion in aqueous acetic acid.[11]

In the presence of RhCl(PPh₃)₃, the tetralin **1** can be converted to either the alcohol **2** or the ketone **3**, depending on the reaction conditions.[12] Oxidation by air in refluxing benzene produces the alcohol **2** in 40% yield, whereas without solvent on a steambath the ketone is obtained in good yield. In the presence of rhodium catalysts, other alkyl benzenes undergo oxidation α to the aromatic ring to give mixtures of the secondary alcohol and ketone,[13] but oxidations using rhodium have not been widely employed.

Adipic acid, a major intermediate in the production of nylon, is made by oxidation of cyclohexanone and cyclohexanol, which are produced by aerial oxidation of cyclohexane. Oxidative attack on the C–H bonds of cyclohexane is slow and requires vigorous conditions, whereas the oxidation products, cyclohexyl hydroperoxide, cyclohexanol, and cyclohexanone, are easily oxidized. As a result the reaction is generally run at low conversions to avoid degradation of the required products. Typically the reaction is run at about 10% conversion of cyclohexane in the presence of a soluble cobalt(II) salt at 140–165°C and 10 atm pressure. The reaction mixture is continuously removed and distilled, and unchanged cyclohexane is recycled.[3]

A more subtle way of introducing an OH group into a hydrocarbon involves palladium complexation, and this is illustrated by the conversion of cholest-4-ene (**4**) into the alcohols **5** and **6**.[14] The starting material **4** is reacted with Pd(PhCN)$_2$Cl$_2$ to give the π-allyl palladium chloride dimers **7** and **8**, which are readily separated. Complex **7** undergoes bridge cleavage with pyridine, followed by oxidation with *m*-chloroperbenzoic acid, to give **5** as the only product in 66% yield. Similarly, the isomeric complex **8** gives

i, pyridine, *m*-ClC$_6$H$_4$CO$_3$H

6 as the major product in 61% yield. These results suggest that the hydroxy group becomes preferentially attached to the same diasterotopic face of the π-allyl system as that to which the palladium was attached.

In the presence of a platinum catalyst the tetracycline **9** is selectively oxidized to **10** in good yield,[15] but other examples of this type of oxidation are relatively rare.

8.2. Formation of 1,2 Diols and Amino Alcohols

The conversion of alkenes into either *cis* or *trans* glycols can be readily accomplished in the presence of transition metals. For the preparation of *cis* glycols, osmium tetroxide and alkaline permanganate are the oxidants most often used.[16] With both of these reagents dihydroxylation occurs from the least hindered side. The mechanism of the reaction involves the initial formation of a cyclic osmate **11** or permanganate ester **12**. The osmate

ester is converted to the glycol, preferably by reductive cleavage of the Os–O bond, while the glycol is obtained from the permanganate ester by hydrolysis in which the Mn–O rather than the C–O bond is cleaved. With permanganate the reaction can be carried out in a variety of polar solvents, such as aqueous ethanol, acetone, or pyridine. Typical reaction conditions involve dissolving or suspending the alkene in an alkaline medium (pH 12) and adding a slight excess of dilute potassium permanganate at room temperature or below. MnO_2, which precipitates, can be removed by filtration and the glycol extracted with an organic solvent. Under more vigorous alkaline conditions extensive carbon–carbon bond cleavage occurs, while under neutral conditions C–C bond cleavage is often the major pathway. Alkaline permanganate oxidation of bicyclo [2,2,1]-2-heptene

(13), for example, gives the diol **14** in 45% yield, but under neutral conditions the dialdehyde **15** is the major product.[17]

Although OsO_4 is poisonous and can cause damage to the eyes, it is a useful reagent for the conversion of cycloalkenes to *cis* glycols. Addition of stoichiometric quantities of OsO_4 to alkenes followed by cleavage produces *cis* glycols in excellent yield. A variety of reagents have been used to cleave the osmium complex, including acidic solutions of sodium chlorate, sodium sulfite in ethanol, alkaline solutions of mannitol or formaldehyde, and hydrogen sulfide. Since the use of OsO_4 has been extensively reviewed,[16,18] only a few developments will be discussed here.

A more convenient way of using OsO_4 to product *cis* glycols is to carry out the hydroxylation with an appropriate oxidizing agent and only catalytic amounts of OsO_4. Oxidizing agents that have been used include metal chlorates, periodate, oxygen, and hydrogen peroxide.[18] With some of the oxidants the major products may be α-ketols, aldehydes, or ketones resulting from cleavage of the carbon–carbon double bond. Metal chlorates, particularly barium or silver salts, generally give the best results. Typically the alkene (1 mol) is dissolved or suspended in water (1500 ml) containing OsO_4 (0.5 g), and barium chlorate (64 g) is gradually added with frequent shaking over a period of several hours. Any excess chlorate is reduced with SO_2 and the product isolated by extraction or crystallization. Using this procedure, 4-chlorocrotonic acid (**16**) is converted into the diol (**17**) in 80% yield.[19]

$$ClCH_2CH{=}CHCO_2H \xrightarrow{\ OsO_4/Ba(ClO_3)_2\ } \underset{\textbf{17}}{ClCH_2\overset{\overset{\displaystyle OH}{|}}{CH}{-}\overset{\overset{\displaystyle OH}{|}}{CH}CO_2H}$$
$$\textbf{16}$$

A more recent method uses tertiary amine oxides as the oxidant with catalytic amounts of OsO_4.[20] Yields are generally better than those obtained when using chlorate ions. Another oxidant that has been used is *t*-butyl hydroperoxide.[21] When used under alkaline conditions good yields are obtained, except when the substrate is sensitive to base.[21a] However, these conditions are particularly useful for tri- and tetrasubstituted alkenes, where other methods using chlorate or hydrogen peroxide give only poor yields.

With alkaline-sensitive substrates like esters, for example, *t*-butyl hydroperoxide can be used in acetone in the presence of Et_4NOAc to give good yields (70%) of the glycols, but tetrasubstituted alkenes do not react under these conditions.

Silver carboxylates and iodine combine in a $2:1$ molecular ratio to form the complex $(RCO_2)_2AgI$, which reacts with cycloalkenes under anhydrous conditions to give *trans*-diacylated glycols, which can easily be converted to the *trans*-glycols by hydrolysis.[16,22] In the presence of at least 1 molar equivalent of water, however, the reaction takes a different course to give *cis*-glycols as the final product.

Preparation of trans-Glycols[22]

Oven-dried silver benzoate (0.2 mol) and iodine (0.1 mol) are heated under reflux in benzene (300 ml) under anhydrous conditions for 30 min. The alkene (0.1 mol) is added and the reaction allowed to proceed at room temperature or under reflux for up to 50 h. Silver iodide is removed by filtration and the crude dibenzoate that is obtained after removal of the solvent *in vacuo* is either hydrolyzed directly to the corresponding glycol or it may be first purified by crystallization.

Several related procedures are available for the vicinal oxyamination of alkenes. Using stoichiometric quantities of the ethylimido–osmium complex **18**, alkenes can be converted stereospecifically into the *cis*-alkylamino-alcohols (**19**) in fair to excellent yield.[23] The complex **18** is synthesized by treatment of the amine with OsO_4 in CH_2Cl_2, but the main limitation of the reaction is the small number of stable imido complexes (**18**) that can be made.

Oxyamination of alkenes can also be carried out with chloramine-T (TsNClNa) in the presence of 1% OsO_4 to give the vicinal hydroxy-toluene-*p*-sulfonamide (**20**).[24] Although yields are generally good the reaction gives poor results with trisubstituted and unsymmetrically disubstituted alkenes. For cases where the sulfonamide protecting group is undesirable an alternative oxyamination procedure has been developed.[25] This procedure uses an osmium-catalyzed reaction of *N*-chloro-*N*-metallocarbamates with alkenes to give the *cis*-hydroxycarbamate (**21**). The *N*-chloro-*N*-metallocarbamate is generated *in situ* by reaction of the corresponding *N*-chlorosodiocarbamate with the corresponding metal salt in acetonitrile.

Of the metals tested, mercuric salts gave the most powerful oxyamination reagents, but the silver salts were also useful. Using the mercury reagent in the presence of Et$_4$NOAc, most mono-, di-, and trisubstituted alkenes can be oxyaminated, but with tetrasubstituted alkenes yields are never higher than 46%.

8.3. Acetoxylation

An indirect way of oxidizing hydrocarbons to alcohols is to introduce an acetoxy group into the substrate, followed by hydrolysis to the alcohol. This approach is often more successful than the direct introduction of an hydroxy group since the acetoxy group is not as susceptible to secondary oxidation. Typical reagents are transition metal salts, which are usually used in acetic or trifluoroacetic acid. For example:

(1) Co(OAc)$_3$ is capable of the selective oxidation of alkanes under mild conditions to give the alkyl acetate. *n*-Heptane, for example, is oxidized at 25°C with Co(OAc)$_3$ in trifluoroacetic/acetic acid mixtures under nitrogen to give the acetate **22** with 81% selectivity.[26]

(2) α-Pinene (**23**) is oxidized by *t*-butylperacetate in the presence of CuBr and acetic acid to a mixture of verbenylacetate (**24**) and pinocarveol acetate (**25**) in good yield.[27]

(3) Chromium(VI) is a versatile reagent for the oxidation of arene side chains to either carboxylic acids, aldehydes, ketones, or diacetates. Chromyl acetate, $Cr_2O_7(OAc)_2$ in acetic anhydride, for example, will oxidize *p*-nitrotoluene to the diacetate **26** in 66% isolated yield.[28]

$$O_2N\!\!-\!\!\langle\;\rangle\!\!-\!\!CH_3 \xrightarrow[5^\circ C]{Cr_2O_7(OAc)_2/Ac_2O} O_2N\!\!-\!\!\langle\;\rangle\!\!-\!\!CH(OAc)_2$$

26

(4) Methyl benzenes react readily with acetic acid in the presence of air and $Pd(OAc)_2$ to give benzyl acetate. Stannous acetate and charcoal are powerful promotors for this oxidation.[29] If excess of acetate is avoided and the reaction is done under oxygen, ring acetoxylation becomes the preferred reaction.[30]

Preparation of Benzyl Acetate [29a]

A mixture of acetic acid (8 mol), potassium acetate (1.1 mol), toluene (1.0 mol), stannous acetate (0.06 mol), palladium acetate (0.016 mol), and charcoal (33.6 g) is stirred at 100°C for 9 h while air is blown over the surface. After filtration, extraction, and distillation, benzyl acetate is obtained in 41% yield.

Nearly quantitative yields of phenyl acetate and phenol are obtained by passing benzene and acetic acid vapors in a dilute oxygen stream over palladium at 130–190°C.[31] This reaction can be made catalytic in palladium by the addition of inorganic oxidants such as $K_2S_2O_8$.[32] The reaction works for a variety of arene substituents, with, surprisingly, *meta*-substitution products predominating. If the reaction occurred by electrophilic attack by Pd^{2+}, *ortho* and *para* isomers might often be expected to be the major products.

Acetoxylation of arenes proceeds much more readily with trifluoroacetic acid than acetic acid, as well as producing a different isomer distribution. Naphthalene, for example, reacts with $Pd(O_2CF_3)_2$ in trifluoroacetic acid at room temperature to give only α-naphthyltrifluoroacetate, whereas acetic acid requires much higher temperatures and produces equal amounts of α- and β-naphthylacetates.[33]

Reaction of an alkene with a palladium(II) salt in acetic acid is a useful route to vinylic and allylic acetate. This type of chemistry is used for the industrial manufacture of vinyl acetates.[3,6,34] The original reaction was a stoichiometric oxidation of ethylene with $PdCl_2$ and sodium acetate in acetic acid. An improvement on this was a liquid phase catalytic process using a suspension of $PdCl_2$, $Cu(OAc)_2$, KCl, and KOAc in acetic acid saturated with ethylene at 40–45 atm and run at 120°C in the presence of oxygen. This process gave good yields of vinyl acetate but is hazardous owing to the mixing of ethylene and oxygen under pressure. The process was then further developed to use a heterogeneous catalyst to convert a gaseous mixture of ethylene, oxygen, and acetic acid vapor to vinyl acetate. A typical catalyst consists of porous silica impregnated with Na_2PdCl_4 and $HAuCl_4$. This is reduced to give a palladium–gold alloy, which is used at 140–170°C and about 10 atm pressure to give a 96% yield of vinyl acetate.

Most alkenes other than ethylene give allylic, rather than vinylic, acetates when oxidized in acetic acid using palladium catalysts.[3,6] Propylene, for example, is converted to allyl acetate in high yield when oxidized in the vapor phase over a heterogeneous palladium catalyst in the presence of acetic acid.

Oxidation of alkenes using chromyl chloride is notorious for producing complex mixtures of products, but in the presence of acetyl chloride the vicinal chloroacetate **27** is obtained in good yield (55%–90%).[35] Unsymmetrical alkenes show a high preference for the regioisomer in which the chlorine atom is attached to the more substituted carbon.

27

8.4. Epoxidation

The liquid phase epoxidation of alkenes with alkyl peroxides or hydrogen peroxide in the presence of transition metal catalysts is well established.[5,36,37] Soluble complexes of V, Mo, W, and Cr have been found to be the most effective catalysts for epoxidation, particularly those derived from vanadium and molybdenum.[38-43] The readily available complexes $Mo(CO)_6$, $MoO_2(acac)_2$, and $VO(acac)_2$ are the best catalysts. These complexes are probably not the active catalysts, and it appears that the active catalytic species are Mo(VI) and V(V) complexes, which are rapidly formed under the reaction conditions.

The reactions are usually conducted in hydrocarbon solvents or, where appropriate, in the alkene substrate itself. Solvent effects can be significant, alcohols and ketones tend to inhibit the reaction, most likely by competing for positions at the metal center.[40]

Although the structure of the hydroperoxide does not have as large an effect on the ease of epoxidation as the nature of the alkene or the catalyst, is has been noted that the rate of epoxidation is enhanced by electron withdrawal from the hydroperoxide.[40] In the Mo(CO)$_6$-catalyzed epoxidation of oct-2-ene, for example, the rate varies in the following order: *p*-nitrocumenyl hydroperoxide >cumene hydroperoxide > *t*-butylhydroperoxide.

The mechanism of the metal-catalyzed epoxidation of alkenes by alkyl hydroperoxides appears to be rather complex, and several mechanisms have been postulated which are discussed in a review by Lyons.[37]

A major new industrial application of homogeneous catalysis is the synthesis of propylene oxide by the metal-catalyzed epoxidation of propylene using alkyl hydroperoxide.[3] Propylene oxide is an important intermediate for propylene glycol, glycerol, and polyethers, but what makes the process attractive is the coproduct formed from the alkyl hydroperoxide. The hydroperoxides are obtained from the oxidation of ethyl benzene or isobutane, which give, after epoxidation, 1-phenylethanol and *t*-butyl alcohol, respectively. 1-Phenylethanol is dehydrated to produce styrene, and *t*-butyl alcohol is either blended with gasoline to inhibit engine knock or is oxidized to methacrylic acid.

For a large number of alkenes, vanadium- and molybdenum-catalyzed oxidations by *t*-BuO$_2$H yield epoxide and *t*-butanol as the only products. The structure of the alkene is a very significant factor in determining the selectivity of the epoxidation reaction. For simple alkenes the rate of epoxidation increases with the number of electron-donating substituents adjacent to the double bond, and, in general, cyclic alkenes react more readily.[40,41] Furthermore, both structural and geometrical isomers differ in reactivity. *Cis*-1,4-Hexadiene, for example, is epoxidized with *t*-BuO$_2$H in the presence of a molybdenum catalyst to give the internal epoxide **28** as the major product, whereas *trans*-1,4-hexadiene also gives the internal epoxide **29**, but with lower selectivity.[44]

28 92%   8%

29 85%   15%

Conjugated dienes are also epoxidized under these conditions to give mixtures of isomeric compounds.[44] Isoprene, for example, gives the two monoepoxides **30** and **31** in the ratio 4:1. Di-epoxides are rarely formed in these reactions.

The stereochemistry of the epoxidation of cyclic alkenes containing no complexing groups is determined solely by steric factors. The stereochemistry of the epoxidation of *p*-menth-1-ene (**32**), for example, is probably due to the attack of the hydroperoxide–molybdenum complex from the least hindered side.[45]

8.4.1. Epoxidation of Allylic Alcohols

When the alkene contains complexing groups, these groups usually direct the steric course of the reaction. In general, vanadium– and molybdenum–hydroperoxide reagents are more selective in the epoxidation of allylic alcohols than peracid epoxidation of the same substrate. For example, whereas geraniol (**33**) undergoes peracid epoxidation preferentially at the site furthest removed from the OH group to give **34**, the molybdenum- and vanadium-catalyzed systems show high regioselectivity for the 2,3-double bond to give **35**.[46] Similarly with the cyclooctadienol **36**, only the allylic double bond is epoxidized with *t*-BuO$_2$H in the presence of VO(acac)$_2$.[47]

Epoxidation of small- or medium-ring allylic alcohols (37) with *t*-BuO_2H/VO(acac)$_2$ produces *syn*-epoxyalcohols (38) with high selectivity and in high yield.[46–48] In all cases 91% of the *syn* isomer is formed, which is in contrast to epoxidation using peracids where there is a change from *syn* to *anti* products on increasing ring size. Epoxidation of 37 ($n = 2$) with *m*-chloroperbenzoic acid, for example, gives 95% of the *syn* isomer, whereas epoxidation of 37 ($n = 4$) gives 99.8% of the *anti* isomer.

A marked stereoselective preference is also observed in both the peracid and the metal-catalyzed epoxidations of acyclic secondary *E*-allylic alcohols.[49,50] When the alcohol 38 is treated with VO(acac)$_2$/*t*-BuO_2H in toluene, the *threo*- and *erythro*-epoxy alcohols 39 and 40 are obtained in a 2 : 1 ratio. In contrast, both peracid and Mo(CO)$_6$/*t*-BuO_2H epoxidations give the opposite specificity.

Substantial differences in behavior between the vanadium- and molybdenum-catalyzed systems are observed in substrates such as 1,5-hexadien-3-ol (41).[44] Epoxidation with *t*-BuO_2H and these catalysts highlights a

marked difference in sensitivity toward the double bond, with the vanadium-catalyzed reaction showing a marked preference for epoxidation of the allylic double bond.

The choice of a vanadium or molybdenum catalyst for the epoxidation of a particular substrate depends very much on the nature of the substrate. The presence of an alkyl group on the double bond of an unsaturated alcohol, or on the carbinol atom, or the position of the OH group relative to the double bond, can each influence the outcome of epoxidation.

General Procedure for Vanadium-Catalyzed Epoxidation of Allylic Alcohols with t-BuO$_2$H [47]

VO(acac)$_2$ (13.3 mg, 0.05 mM) in dry benzene (10 ml) is placed in a 50-ml round-bottomed flask and the air displaced by nitrogen. The allylic alcohol (10.0 mM) in benzene (5 ml) is introduced through the side arm. To this solution is added dropwise *t*-BuO$_2$H (1.17 g, 12.0 mM) in benzene (5 ml) over 15 min. The resulting solution is stirred at 40°C for 24 h until the peroxide is almost consumed. After removal of the solvent *in vacuo*, the residue is chromatographed on a Florisil column with petroleum ether to remove the metal complex. The product is isolated from the elutate by crystallization or fractional distillation.

General Procedure for Molybdenum-Catalyzed Epoxidation of Allylic Alcohols with t-BuO$_2$H [47]

To a solution of MoO$_2$(acac)$_2$ (16.3 mg, 0.05 mM) and the allylic alcohol (10 mM) in dry benzene (15 ml) at 80°C is added dropwise *t*-BuO$_2$H (1.17 g, 12.0 mM) in benzene (5 mol) over 15 min. The resulting solution is stirred for 5 h at 80°C and then worked up in a manner similar to that described above.

8.4.2. Asymmetric Epoxidation

Although chiral epoxides are useful intermediates, their preparation in good optical yield is usually difficult or tedious. The use of optically active transition metal catalysts in asymmetric hydrogenation is now well established (Chapter 7), but the field of asymmetric epoxidation using chiral transition metal catalysts has not progressed as rapidly. This is perhaps understandable in view of the lack of detailed information on the mechanism of such epoxidations.

Considering the high selectivities observed in the epoxidation of allylic alcohol under mild conditions, it is not surprising that most progress has been made in asymmetric epoxidation of such substrates. Substituted buten-1-ols are asymmetrically epoxidized with cumene hydroperoxide in the presence of the chiral molybdenum catalyst MoO$_2$(acac)-[(−)-*N*-alkylephedrinate] (**42**) to produce optical yields of up to 33%. [51]

42

Chiral *β*-diketone complexes of vanadium and molybdenum give poor results because of the rapid destruction of these ligands under the conditions of oxidation. In contrast, chiral hydroxamic acids do not suffer attack under such conditions and form moderately strong complexes with molybdenum and vanadium. The chiral molybdenum bishydroxymates, [MoO$_2$(hydroxy-mate)$_2$], produce poor optical yields (<2%), but the use of chiral vanadium–bishydroxymates gives optical yields of up to 50%.[52] The vanadium catalyst is generated *in situ* by reaction of VO(acac)$_2$ with campholylthydroxamic acid (**43**).

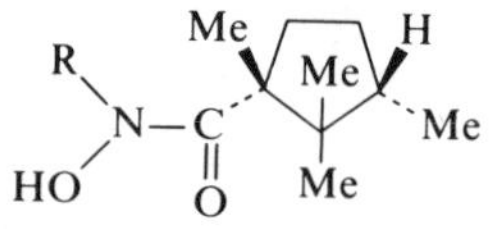

43

Stoichiometric asymmetric epoxidation of unfunctionalized alkenes has been achieved using molybdenum(VI) oxodiperoxo complexes of the type **44** with optical yields of up to 34%.[53]

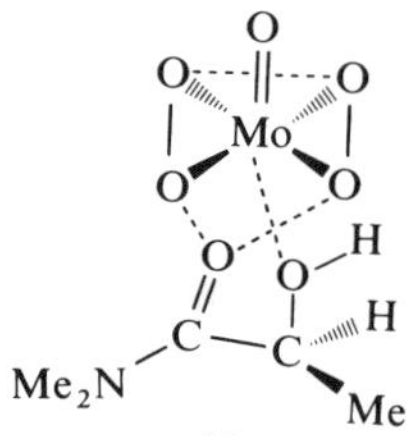

44

8.5. Formation of Aldehydes and Ketones

8.5.1. Oxidation of Hydrocarbons

The oxidation of hydrocarbons to aldehydes or ketones is best accomplished with chromium oxidants.[4,16] The oxidation of a methyl group to an aldehyde is not readily accomplished unless it is attached to an aromatic ring. Chromyl chloride, CrO$_2$Cl$_2$, for example, will oxidize *ortho*-, *meta*-, and *para*-xylenes to the corresponding tolualdehydes in yields ranging from 60%–80%.[54] Chromic acid in acetic anhydride is also an effective reagent for this transformation. *Para*-toluene methyl sulfone is oxidized with chromic acid in a mixture of acetic anhydride and acetic acid to give, after hydrolysis, the methylsulfonylbenzaldehyde (**45**) in 59% yield.[55]

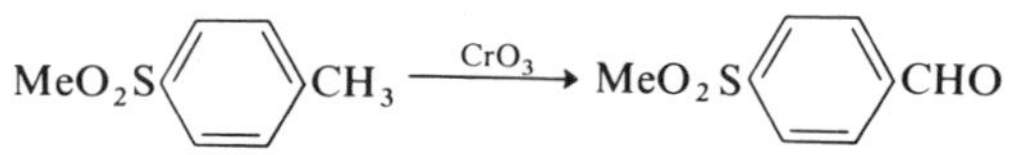

45

The oxidation of a methylene group to the corresponding ketone is much more common than the oxidation of a methyl group to an aldehyde. This is because a secondary C–H bond is more reactive than a primary C–H bond and a ketone is much less susceptible to further oxidation than an aldehyde. Again chromium(VI) reagents have been widely used, especially for activated methylene groups. An interesting oxidation occurs when 1-chlorocamphane (**46**) is treated with CrO_3 in acetic acid, the major product being 1-chloronorcamphor (**47**), which is isolated in 40% yield.[56]

The major use of chromium reagents, however, has been for allylic oxidations. Androst-5-ene-3,17-diol acetate (**48**), for example, is oxidized to the 7-keto compound **49** in 60% yield using *t*-butyl chromate.[57] Various

chromium reagents have been used for this type of oxidation, all of which can offer certain advantages. The chromium trioxide–pyridine complex in CH_2Cl_2 will oxidize a variety of allylic methylene groups to ketone at room temperature in yields of 48%–95%.[58] For the allylic oxidation of Δ^5-steroids, the 3,5-dimethylpyrazole–chromium trioxide complex (DMP·CrO_3) is very useful.[59] It is claimed that there is a rate increase of 100-fold compared with using a pyridine–chromium trioxide complex. This increased activity is thought to be partly due to the increased solubility of the complex and, more importantly, to the possibility of intramolecular acceleration due to the pyrazole nucleus.

Preparation of 7-Ketocholesteryl Benzoate[59]

CrO_3 (6.0 g; 60 mM), previously dried over P_2O_5, is suspended in CH_2Cl_2 (50 ml) at −20°C, and 3,5-dimethylpyrazole (5.76 g, 60 mM) is added in one portion. After stirring at −20°C for 15 min cholesteryl benzoate (2.44 g, 5 mM) is added and the mixture stirred for 4 h while keeping the temperature between −10° and −20°C. 5 *N* Sodium hydroxide solution

(25 ml) is added and the mixture stirred for 1 h at 0°C. The organic layer is separated, washed successively with dilute hydrochloric acid to remove the 3,5-dimethylpyrazole (which can be recovered by subsequent basification of the acid layer), water, and saturated sodium chloride. Evaporation of the methylene chloride solution gives, after crystallization from cyclohexane, 7-ketocholesteryl benzoate (1.86 g, 74%).

Oxidation of alkyl benzenes to the corresponding aromatic ketone has been accomplished using rhodium catalysts. A variety of rhodium complexes were shown to be catalytically active, including rhodium acetylacetonate and tris(triphenylphosphine) chlororhodium.[60,13] Product yields are variable and the reaction is thought to be free radical in nature.

Δ^5-Cholestenone (**50**) is oxidized to Δ^4-cholesten-3,6-dione (**51**) by aerial oxidation in the presence of cupric ions and an organic base.[61]

Polynuclear hydrocarbons can be oxidized readily using either chromium trioxide in water[62] or concentrated sulfuric acid[63] to give the appropriate quinone.

Although RuO_4 and OsO_4 have not been widely used for the oxidation of hydrocarbons, it has been reported that RuO_4 converts phenanthrene to 9,10-phenanthraquinone in 28% yield.[64]

8.5.2. Oxidation of Alkenes and Alkynes

Probably the besk-known conversion of alkenes into aldehydes or ketones is based on the chemistry of the Wacker process for the conversion of ethylene into acetaldehyde. This process is based on the stoichiometric oxidation of ethylene with palladium chloride [Eq. (1)], but is made catalytic and hence suitable for industrial application by the use of cupric salts to reoxidize the palladium [Eq. (2)]. The cuprous ion thus generated is in turn oxidized by air back to cupric ion [Eq. (3)]. This process and related

$$C_2H_4 + PdCl_2 + H_2O \rightarrow CH_3CHO + Pd^0 + HCl \tag{1}$$

$$Pd^0 + 2CuCl_2 \rightarrow PdCl_2 + Cu_2Cl_2 \tag{2}$$

$$Cu_2Cl_2 + 2HCl + 1/2O_2 \rightarrow 2CuCl_2 + H_2O \tag{3}$$

reactions are discussed more fully in various reviews.[3,6,34] The Wacker process can be used to oxidize higher alkenes. Terminal alkenes yield methyl ketones; acetone, for example, can be synthesized commercially by the palladium-catalyzed oxidation of propylene. In general, rates are slower and yields decrease with increasing chain length. Propylene is oxidized in 5 min at 20°C to give a 90% yield of acetone, whereas 1-decene requires 1 h at 70°C to give only a 34% yield of 2-decanone.[65] In an improved procedure, however, using aqueous dimethylformamide, 1-dodecene is converted into 2-dodecanone in up to 85% yield.[66] Cyclic alkenes react easily to give good yield of cyclic ketones.

Preparation of 2-Dodecanone[66]

Into a 250-ml cylindrical glass reactor that has a gas dispersion inlet tube affixed into the bottom to permit efficient passage of oxygen through the reaction solution is placed $PdCl_2$ (0.02 mol), $CuCl_2 \cdot 2H_2O$ (0.02 mol), water (7 ml), and DMF (50 ml) and the mixture stirred at 60°C. Oxygen is passed through the solution at 3.3 liters/h and 1-dodecene (0.2 mol) is added dropwise over 2.5 h via a glass delivery tube, the end of which is well below the surface of the solution. After stirring for a further 0.5 h at 60°C the solution is cooled and the upper phase separated and washed several times with water. After drying and fractional distillation the product is isolated in 85% yield.

1-Methylcyclobutene undergoes an oxidative rearrangement in the presence of palladium(II) to afford quantitative yields of cyclopropyl methyl ketone (**52**).[67] The reaction is stoichiometric but could probably be made catalytic.

The use of osmium tetroxide in conjunction with sodium periodate results in the cleavage of carbon–carbon double bond to give ketones or aldehydes.[18,68] In this procedure OsO_4 adds to the double bond to give the osmate ester **53**, followed by oxidative cleavage with sodium periodate to give the two carbonyl compounds **54** and **55**. Osmium trioxide is formed, but in the presence of excess periodate it is reconverted to OsO_4. The reaction generally proceeds in such solvents as aqueous dioxane or acetic acid to give the product in high yield. Cyclohexene, for example, gives adipaldehyde in 77% yield.[68]

Preparation of Undecanal[68]

A mixture of water (5 ml), purified dioxane (15 ml), 1-dodecene (0.77 g), and OsO_4 (11.3 mg) is stirred for 5 min, during which time the mixture becomes dark brown owing to formation of the osmate ester. To this solution at 25°C is added $NaIO_4$ (2.06 g) in portions over 30 min. The tan-colored slurry is stirred for an additional 1.5 h and then extracted thoroughly with ether (4 × 50 ml), and the combined organic layers are filtered through anhydrous sodium sulfate. Undecanal is isolated as the 2,4-dinitrophenyl-hydrazone in 68% yield by treating the ether solution with a solution of 2,4-dinitrophenylhydrazine (1.0 g) in H_2O (7.5 ml), conc. H_2SO_4 (5 ml), and 95% EtOH (35 ml). The two-phase mixture is stirred for 1 h and then evaporated to about 50 ml when the 2,4-dinitrophenylhydrazone crystallizes out of solution.

Ruthenium tetroxide–periodate can also be used to oxidize alkenes, but this tends to be a more vigorous reagent and initially formed aldehydes are usually oxidized further to the acid.[18] Typically, oxidations are carried out using 4 mol of $NaIO_4$ and 2 mol % RuO_4 in aqueous acetone.

Aldehyde enamines are selectively cleaved to the corresponding ketone in the presence of catalytic amounts of a cuprous salt under very mild conditions.[69] The mild conditions permit the rapid and selective cleavage of the enamine double bond in the presence of other sites of unsaturation in the substrate. Oxygenation of **56**, for example, occurs at 0°C in the presence of cuprous chloride to give a quantitative yield of progesterone (**57**).

56   **57**

Alkynes possessing some water solubility are converted to the corresponding diketone by treatment with neutral aqueous $KMnO_4$.[70] Sodium stearolate, for example, is converted into the diketone **58** in 95% yield at 25°C. When the reaction is carried out under either acidic (pH 1) or basic

$$Me(CH_2)_7C\equiv C(CH_2)_7CO_2Na \xrightarrow[pH\,7]{KMnO_4} Me(CH_2)_7CO\cdot CO(CH_2)_7CO_2Na$$
58

(pH 12) conditions, only cleavage products are isolated. Similar transformations can be performed using RuO_4 in a carbon tetrachloride/water, two-phase system to give α-diketones in good yield.[71]

8.5.3. Oxidation of Alcohols

The oxidation of alcohols can be effected by a variety of metal oxidants; among the more commonly used are chromium(VI) compounds, potassium permanganate, and manganese dioxide. Other metals that have been used include copper, platinum, silver, ruthenium, and vanadium.

The oxidation of alcohols by chromium(VI) compounds is a complex process.[72] With primary alcohols carboxylic acids are usually obtained in aqueous solutions, whereas aldehyde formation is favored in nonaqueous solvents. Secondary alcohols are readily oxidized to the corresponding ketone using 2.67 M chromic acid (Jones reagent),[73] which is prepared by dissolving CrO_3 (26.72 g) in conc. H_2SO_4 (23 ml) and then diluting the mixture to 100 ml with water. One milliliter of this reagent oxidizes 4 mM of a monohydric alcohol according to the equation

$$3R_2CHOH + 2H_2CrO_4 \rightarrow 3R_2CO + 2Cr(OH)_3 + 2H_2O \,\rightharpoondown$$

Various solvents can be used including acetone and acetic acid.[74-76]

When chromium trioxide is dissolved in pyridine it forms a complex that is capable of oxidizing alcohols.[77] Although oxidations using this reagent are slow, requiring up to 24 h at room temperature, it is a very useful reagent because it selectively reacts with alcohols in the presence of such easily oxidizable groups as double bonds and thioethers. Care, however, must be exercised in preparing this reagent or it may ignite. The solution must always be prepared by cautious addition of CrO_3 to pyridine, which has been carefully purified by distillation from $KMnO_4$; the reverse order of addition may cause the mixture to inflame. A useful modification of the procedure, developed by Collins,[78] involves isolating the chromium trioxide–pyridine complex in a solid form, which can then be stored indefinitely at 0°C. The reagent is usually used in dichloromethane, a solvent in which it is soluble to the extent of 12.5 g/100 ml. Even though the reagent is used in excess, excellent yields of aldehydes and ketones are obtained. Using 6 molar equivalents of the complex, the unsaturated alcohol **59** is oxidized to the aldehyde **60** in 94% yield after only 15 min.[79]

59 → **60**

A two-phase ether–water system has also been used in the oxidation of secondary alcohols with CrO_3 to give ketones of high purity and in excellent yield.[80] 1-Menthol, for example, gives menthone (**61**) in 97% yield.

61

A variety of other chromium(VI) reagents have been developed, particularly for the mild oxidation of primary alcohols to aldehydes. In addition to the pyridine–chromium trioxide complexes, pyridinium chlorochromate[81] and pyridinium dichromate[82] are particularly useful.

Preparation of Heptanal Using Pyridinium Chlorochromate[81]

To 6 *M* hydrochloric acid (184 ml, 1.1 mol) is rapidly added CrO_3 (100 g, 1 mol) by stirring. After 5 min the homogeneous solution is cooled to 0°C, and pyridine (79.1 g, 1.0 mol) is added carefully over 10 min. Cooling to 0°C gives a yellow-orange solid that is collected by filtration and dried *in vacuo* to give the pyridinium chlorochromate (181 g, 84% yield).

In a 500-ml round-bottomed flask fitted with a reflux condenser is suspended pyridinium chlorochromate (32.3 g, 150 mM) in anhydrous CH_2Cl_2 (200 ml). 1-Heptanol (11.6 g, 100 mM) in $CHCl_2$ (20 ml) is added in one portion to the stirred solution. After 1.5 h 200 ml of dry ether is added and the supernatant decantered from the black gum. The insoluble residue is washed thoroughly with dry ether (3 × 50 ml), and the combined organic solution is passed through a short pad of Florisil. Removal of the solvent *in vacuo* gives a residual oil, which is distilled through a short Vigreux column to give 8.87 g (78%) of heptanal.

A recent development in the use of chromium(VI) reagents is the use of such compounds on an inert support. This not only makes the reaction products easier to isolate but the supported reagent is often more selective. Inorganic supports such as silica, alumina, or Celite have been used to absorb chromyl chloride,[83] pyridinium chlorochromate,[84] chromic acid,[85] and chromium trioxide.[86] Organic polymer reagents have been prepared using polyvinyl pyridinium chlorochromate[87] and polyvinyl ammonium chlorochromate.[88] All of these reagents are very selective for the oxidation of primary or secondary alcohols to the corresponding aldehyde or ketone.

Preparation of Carvone (62) Using Pyridinium Chlorochromate Absorbed on Alumina[84]

62

To a solution of CrO_3 (6 g) in 6 N hydrochloric acid (11 ml) is slowly added pyridine (4.75 g) at 40°C. The mixture is then kept at 10°C until a yellow-orange solid forms. The mixture is reheated to 40°C and alumina (50 g) is added with stirring. After evaporation of the solvent *in vacuo*, the orange solid is dried *in vacuo* for 2 h at room temperature. The capacity of the reagent is ~1 mM/g of alumina.

This reagent (7.5 g) is added to a flask containing a solution of carveol (0.6 g, 3.8 mM) in *n*-hexane (10 ml). After stirring for 2 h, the solution is filtered and washed with ether (3 × 10 ml). The combined filtrates are evaporated and vacuum distilled to give carvone in 93% yield.

The reaction between secondary or tertiary allylic alcohols and pyridinium chlorochromate (PCC) results in oxidative rearrangement leading to α,β-unsaturated aldehydes or ketones (63) in very good yield.[89]

63

PCC = pyridinium chlorochromate

Since the substrate can be readily made by reaction of vinyl–lithium reagents with ketones, this process provides a simple and efficient method of carrying out mixed aldol condensations.

Potassium permanganate readily oxidizes primary and secondary alcohols to carboxylic acids or ketones in either basic or acidic solutions, the rate being much higher in basic solutions.[89] Since permanganate is a fairly vigorous and nonselective oxidant, it has not found wide use in organic solvents, although it can be used in acetic acid. Using the solid reagent $KMnO_4/CuSO_4 \cdot 5H_2O$, secondary alcohols are oxidized to ketones at room temperature, but primary alcohols and alkenes do not react effectively.[90]

The reagent is useful in that it can be filtered off easily at the end of the reaction, but it is used in relatively large amounts.

The readily available and stable barium manganate is a mild oxidizing agent for the formation of carbonyl compounds from primary and secondary alcohols.[91] It has similar activity to MnO_2 but is claimed to be better for the preparation of certain aldehydes. The furan dialdehyde **64**, for example,

is produced in 80% yield—whereas with MnO_2 the yield is less than 20%—but in general it offers no advantage over the more readily available MnO_2.

Manganese dioxide is a particularly useful reagent for the selective oxidation of allylic and benzylic alcohols.[92–94] It is usually used in large excess as a suspension in a variety of organic solvents at room temperature. Active MnO_2 is prepared by the dropwise addition of concentrated aqueous solutions of $KMnO_4$ to aqueous manganese sulfate at 90°C until the pink permanganate color persists. After stirring for a few minutes at 90°C the product is collected by filtration, washed with water, methanol, and ether, and dried at 120°C to constant weight.[95]

In the presence of platinum, primary and secondary alcohols are readily oxidized by oxygen.[96] Primary alcohols are usually oxidized more readily than secondary alcohols, and with polyhydroxy compounds such as carbohydrates, oxidation of only the primary alcohols can be achieved. Under alkaline conditions primary alcohols are converted in high yield to the corresponding carboxylic acid, but under neutral conditions in an organic solvent the products are aldehydes.[97] Steroids containing two or more hydroxy groups, with one being in the 3-position, undergo selective oxidation of the 3-position. The ketone **65**, for example, is obtained in 70% yield from the triol **66**.[98] The platinum-catalyzed oxidations are usually carried out under conditions similar to those used for catalytic hydrogenation, a useful catalyst being 5% or 10% platinum on charcoal.

Ruthenium tetroxide in the presence of sodium periodate is a useful reagent for the oxidation of secondary alcohols to ketones. The steroid alcohol **67** is converted to the corresponding ketone **68** in 93% yield,[99] and the hydroxylactone **69** is oxidized to the ketolactone **70** in 80% yield.[100]

The reactions are usually carried out in a two-phase system by adding a 10% aqueous solution of sodium periodate with vigorous stirring to the substrate and RuO_4 in carbon tetrachloride.

Potassium ruthenate, $KRuO_4$, can be used in catalytic amounts in the presence of persulfate to oxidize primary and secondary alcohols to the corresponding carboxylic acid or ketone.[101] There is no significant reaction with tertiary alcohols, alkenes, or alkynes. The reaction proceeds at room temperature in high yield (>80%) and, although catalyst turnover numbers have not been fully determined, initial results suggest they will be reasonably high.

Ruthenium complexes, such as $RuCl_2(PPh_3)_3$ or $Ru_3(CO)_{12}$, for example, catalyze the amine *N*-oxide oxidation of primary and secondary alcohols to the corresponding aldehyde or ketone.[102] The reaction proceeds readily at room temperature and, although a variety of *N*-oxides can be used, the best yields are obtained with *N*-methylmorpholine *N*-oxide. Yields are generally in the range 80%–100% for a variety of primary, secondary, and allylic alcohols, but some homoallylic alcohols—cholesterol, for example—are not oxidized, probably because of the formation of stable ruthenium alkoxyalkene complexes.

The oxidation of sterically unhindered primary or secondary alcohols can be achieved in high yield using silver carbonate on Celite.[103] This is a particularly useful reagent since the metallic silver formed in the reaction is readily removed by filtration. The reagent is prepared by treating a mixture of silver nitrate (34 g) and purified Celite (30 g) in distilled water with an aqueous solution of $Na_2CO_3 \cdot 10H_2O$ (30 g). The product, after filtration, water washing, and drying, contains ~1 mM of silver carbonate per 0.57 g of dried reagent. The reactions are usually carried out in dry benzene and the order of reactivity is benzylic or allylic > secondary > primary alcohol. By using acetone or methanol as the solvent instead of benzene, it is possible to oxidize an allylic alcohol selectively in the presence of a secondary alcohol. Thus testosterone (71) is obtained in 95% yield from the diol 72.[103b]

72   71

Cupric ion in aqueous acetic or pyridine solution converts α-ketols into the corresponding diketones. Sebacoin (73), for example, is oxidized to sebacil (74) in 89% yield in only 1 min on heating under reflux in acetic acid in the presence of cupric acetate.[104]

73   74

8.5.4. Oxidation of Amines

Argentic salts have been used to oxidize amines to aldehydes and ketones in yields of between 30% and 60%.[105] With primary amines the silver reagent is best prepared by addition of sodium persulfate to catalytic amounts of silver nitrate, whereas for the oxidation of secondary amines it is preferable to first isolate the argentic ion as the picolinate. From primary amines the initial isolable products are imines, which are probably formed by reaction of the carbonyl products with the parent amine. Aldimines are stable in basic media and usually appear as a separate liquid phase when water is used as the solvent, but if the solution is acidic hydrolysis occurs and the product is the aldehyde (75). Primary amines of the type RR^1CHNH_2 produce ketones directly.

$$RCH_2NH_2 \xrightarrow[Na_2S_2O_3]{AgNO_3} RCH=NCH_2R \xrightarrow{H^+} RCHO$$

75

Primary aliphatic amines are also converted to the corresponding aldehyde when heated under reflux in an aqueous suspension containing a fivefold excess of active MnO_2.[106]

Secondary or tertiary amines can be oxidized by MnO_2 at room temperature, but the reaction often gives a mixture of products. *N*-Methylaniline, however, affords formanilide (**76**; R = H) in over 80% yield, and *N,N*-dimethylaniline gives *N*-methylformanilide (**76**; R = Me) in similar

$$PhN\overset{R}{\underset{Me}{<}} \xrightarrow{MnO_2} PhN\overset{R}{\underset{CHO}{<}}$$

76

yield.[107b] Electron-donating substituents in the aromatic ring promote the reaction, whereas strong electron-withdrawing groups (e.g., *para*-nitro) completely inhibit the reaction at room temperature. *N,N*-dimethylcyclohexylamine is oxidized to cyclohexanone and *N,N*-dimethylbenzylamine is oxidized to benzaldehyde using MnO_2, both reactions going in good yield.

The oxidation of steroidal amines can be carried out using chromium trioxide in pyridine.[108] Acetylrehine (**77**), for example, is oxidized to the *N*-formyl derivative (**78**) in 96% yield. In this and other examples oxidation

results in the preferential cleavage of the primary C–H bond, the normally more reactive tertiary bond not being attacked at all. This is explained on the basis that formation of the carbinolamine involving the tertiary C–H bond is sterically hindered. For example, in acetylrehine (**77**) attack on the C-20 hydrogen is obstructed by the methyl group at C-13. The mechanism of the oxidation of a tertiary amine with CrO_3/pyridine is postulated to involve the initial formation of an amine–chromium trioxide complex **79**, which subsequently abstracts an active proton, usually from the α-carbon, to give the carbenium ion (**80**). This is then converted to the carbinolamine **81**, which can be oxidized further to give the product **82**.

$$-\overset{\overset{\displaystyle H}{|}}{\underset{\underset{\displaystyle H}{|}}{C}}-N + CrO_3 \rightarrow -\overset{\overset{\displaystyle H}{|}}{\underset{\underset{\displaystyle H}{|}}{C}}-\overset{+}{N}- \rightarrow -C{=}\overset{+}{N}-$$

79

$$\underset{\mathbf{82}}{-\overset{\overset{\displaystyle O}{\parallel}}{C}-N-} \xleftarrow{[O]} \underset{\mathbf{81}}{-\overset{\overset{\displaystyle OH}{|}}{\underset{\underset{\displaystyle H}{|}}{C}}-N-} \leftarrow \underset{\mathbf{80}}{-\overset{+}{\underset{\underset{\displaystyle H}{|}}{C}}-N-}$$

8.6. Formation of Acids

8.6.1. Oxidation of Alkanes and Alkenes

Although future processes for the manufacture of acetic acid will probably be based on the carbonylation of methanol, most synthetic acetic acid is now made by oxidation processes, the largest-scale process being the oxidation of butane and other aliphatics in the presence of cobalt(II) catalysts.[3] The reaction is typically carried out at 100–200°C and 60–80 atm pressure to give a mixture of acetic, propionic, and butyric acids, together with some 2-butanone. The products are separated by distillation, the acetic acid yield being ~45% at 30% conversion.

The largest-scale industrial homogeneous catalytic oxidation is the formation of terephthalic acid from p-xylene. Other large-scale processes are the oxidation of toluene to benzoic acid and m-xylene to isophthalic acid.[3] The oxidation of methylbenzenes is generally much easier than other methyl derivatives since benzylic C–H bonds are more susceptible to free radical attack than alkyl C–H bonds. The oxidations are usually carried out industrially using aerial oxidation in the presence of mixed cobalt and manganese salts as the catalyst; yields are very high. Typically the reaction is carried out at 200°C and 15 atm pressure of air in the presence of a mixture of cobalt acetate, manganese acetate, and sodium bromide.

On a laboratory scale the more common oxidants for converting aromatic alkyl groups to carboxylic acids are chromium trioxide, sodium dichromate, and potassium permanganate. Thus toluene is oxidized with aqueous sodium dichromate at 200°C in an autoclave to give benzoic acid in 90% yield.[109] Oxidation of ethyl benzene under similar conditions gives a mixture of acetophenone and benzoic acid.[109b]

Potassium permanganate is a vigorous oxidant for arenes under either acidic or basic conditions. It readily converts alkyl benzenes to the corresponding benzoic acids,[110] and polycyclic aromatic hydrocarbons suffer destruction of one or more rings when oxidized by basic permanganate. Oxidation of naphthalene gives phthalonic acid (**83**) in moderate yield.[111]

Ruthenium tetroxide used in catalytic amounts with sodium periodate will also attack aromatic systems vigorously, but will not affect alkanes. *Para-tert*-butylphenol, for example, is oxidized to pivalic acid, and phenylcyclohexane to cyclohexane carboxylic acid using this system.[112] Controlled oxidation of 5-methoxy-1-naphthol with this reagent produces 3-methoxyphthalic acid (**84**) in 50% yield.[113]

The RuO_4/$NaIO_4$ system has been successfully used in the oxidation of steroids. Oxidation of estrone (**85**), for example, yields the diacid **86**, with degradation of ring A terminating at carbons 5 and 10.[114]

Using a catalyst prepared from cuprous chloride and pyridine, *o*-benzoquinone, catechols, and even phenols are oxidatively cleaved in the presence of methanol to the corresponding monomethyl ester of muconic acid (**87**).[115] The product is obtained in 85% yield from catechol and in 50%–60% yield from phenol.

*Preparation of cis,cis-Muconic Acid Monomethyl Ester (**87**; R = H)*[115]

A three-necked 500-ml flask equipped with a mechanical stirrer, a dropping funnel, and an inlet attached to a gas buret is charged with purified CuCl (5.93 g; 60 mM), MeOH (2.4 g, 75 mM), and pyridine (60 ml) under a blanket of nitrogen. After flushing with oxygen without stirring, the system is exposed to the oxygen in the buret and the mechanical stirrer started.

Oxygen uptake ceases after one equivalent is consumed. From the dropping funnel is added a solution of catechol (1.1 g, 10 mM) in pyridine (20 ml) and methanol (0.5 ml) dropwise over 1 h. The reaction mixture is evaporated to dryness and the yellow-brown solid residue hydrolyzed with dilute hydrochloric acid in chloroform under N_2. Drying and evaporation of the chloroform solution gives the product **87** (R = H) in 85% yield. Addition of ammonia to the catalytic system produces a new copper reagent that reacts with the same substrates in the presence of oxygen to give the corresponding mononitrile of muconic acid (**88**) in good yield.[116] The mechanism of these reactions is unclear, as is the exact nature of the active catalyst.

Alkenes can be oxidized to acids or ketones using $RuO_4/NaIO_4$, the product depending on the structure of the alkene. The alkene **89**, for example, is oxidized to the acid **90** in excellent yield at room temperature in acetone.[117]

8.6.2. Oxidation of Alcohols

Primary alcohols are oxidized to aldehydes, which are very susceptible to further oxidation to the corresponding carboxylic acid. Hence several reagents that convert an alcohol to an aldehyde can also be used to oxidize the alcohol to an acid. The most commonly used reagents are chromium(VI) compounds and potassium permanganate, but platinum-catalyzed oxidation of carbohydrates is well established.

The preferred chromium compound is chromic acid; other chromium(VI) compounds, especially when used in nonaqueous solvents, can give good yields of the aldehyde (Section 8.5.3). The oxidation of a cyclic tertiary alcohol to give a long-chain keto acid (**91**) is achieved in excellent yield using chromium trioxide in anhydrous acetic acid (Fieser's reagent), provided there is no other oxidizable group present.[118]

91

Primary alcohols are readily oxidized to carboxylic acids using potassium permanganate in basic solutions.[89] Under neutral or acid conditions the rate of reaction is generally very slow. In general, good yields are obtained using alkaline permanganate, except when the initial product is an enolizable aldehyde or ketone. In such cases oxidation of the enolic carbon–carbon double bond may be a serious side reaction.

Catalytic oxidation of carbohydrates over platinum catalysts is usually very selective.[96] Selective oxidations of mono- and oligosaccharides has led to the efficient preparation of uronic acids, aminouronic acids, uronosides, aldonic acids, and ascorbic acid intermediates. In general, primary hydroxyls are preferentially oxidized in the presence of secondary hydroxyls.

8.6.3. Oxidation of Aldehydes and Ketones

Aldehydes are, in general, oxidized very easily, and therefore many reagents that oxidize primary alcohols to aldehydes cause further oxidation to the carboxylic acid when used in excess. The reagents most commonly used are chromic acid, permanganate, and silver oxide.

Potassium dichromate in aqueous sulfuric acid oxidizes furfural to furoic acid (**92**) in 75% yield at 100°C.[119] In contrast to alcohol oxidation,

92

the rate of reaction is not greatly increased by using acetic acid as the solvent instead of water in chromium(VI) oxidations of aldehydes. Thus the oxidation of primary alcohols to aldehydes is favored in acetic acid solutions containing very little water.

Acidic, neutral, or alkaline solutions of permanganate can be used to oxidize both aromatic and aliphatic aldehydes. In spite of the fact that both acid and alkaline conditions tend to promote the cleavage of enolizable aliphatic aldehydes, good yields of the acids can be obtained. Heptanal, for example, is oxidized with a mixture of potassium permanganate and concentrated sulfuric acid to give heptanoic acid in 76% yield.[120]

For sensitive aldehydes, such as those with double bonds and other easily oxidizable groups in the molecule, silver oxide, freshly prepared from silver nitrate and sodium hydroxide, is particularly useful.[121] The reaction is generally carried out in an aqueous medium, but organic solvents can be used. 3-Thiophenecarboxylic acid (**93**), for example, is produced in 96% yield from the corresponding aldehyde.[121a] The silver is easily isolated from the reaction mixture and can be readily reconverted to silver oxide.

93

The oxidation of ketones to acids can be catalyzed by $Rh_6(CO)_{16}$, $IrCl(CO)(PPh_3)_2$, or $Pt(PPh_3)_3$.[122] Using $Rh_6(CO)_{16}$, cyclohexanone is oxidized to adipic acid at 100°C and under a pressure of 34 atm of a 3:1 mixture of oxygen and carbon monoxide. One thousand millimoles of adipic acid are produced per millimole of $Rh_6(CO)_{16}$. The carbon monoxide is present to prevent excessive decomposition of the rhodium catalyst. In the presence of methanol the reaction product is dimethyl adipate.

8.7. Formation of Esters

Ethers can be oxidized under very mild conditions to give esters in high yield using either ruthenium tetroxide or chromium(VI) reagents.

Using RuO_4 in carbon tetrachloride, a variety of ethers have been oxidized to the corresponding ester in virtually quantitative yield. Tetrahydrofuran is oxidized to γ-butyrolactone in quantitative yield and di-n-butyl ether is converted to n-butyl-n-butyrate in similar yield.

Anhydrous chromium trioxide in acetic acid has been used to convert methyl ethers into the corresponding formate (**94**) in yields of ~50%.[124] Since the formate can be easily hydrolyzed to the corresponding alcohol,

this reaction has application in the removal of methyl groups that have been introduced to protect aliphatic hydroxyl groups. Oxidation of secondary alkyl ethers leads to cleavage products. Diisopropyl ether, for example, when oxidized by chromic acid in 45% aqueous sulfuric acid, gives a 97% yield of acetone.[125]

A convenient method for the conversion of enol ethers to esters (**95**) is achieved using pyridinium chlorochromate.[126] Cyclic ethers give the corresponding lactone. In a typical experiment the enol ether (5 mM) in CH_2Cl_2 (10 ml) is added at room temperature to a suspension of pyridinium chlorochromate (10 mM) in CH_2Cl_2 (10 ml). After 1 h the solution is diluted with ether, filtered, and evaporated to give the crude product, which is purified by chromatography on silica to afford the pure ester or lactone in excellent yield. The reaction mechanism is thought to involve initial electrophilic attack by pyridinium chlorochromate on the alkene to give the unstable intermediate **96**, which, after heterolytic cleavage of the Cr–O bond and a 1,2-hydride shift, gives the product **95**.

Cyclic α-hydroxyethers are oxidized by MnO_2 to the corresponding lactones by stirring at room temperature in chloroform.[127] Thus, 2-hydroxytetrahydrofuran is converted to the lactone (**97**).

Oxidation of 1,4 diols by chromium(VI) reagents in both aqueous and nonaqueous media gives γ-lactones in good yield when no stereochemical barrier to cyclization is present.[128] The *cis* diol **98** is readily converted to the corresponding lactone (**99**) in 60% yield by oxidation with sodium dichromate in aqueous sulfuric acid, whereas the corresponding *trans* diol (**100**) gives only the *trans* dialdehyde (**101**). γ-Lactones are also obtained from 1,4 diols by oxidative cyclization with neutral potassium permanganate in similar yield.[128]

8.8. Dehydrogenation

By definition dehydrogenation involves removal of one or more pairs of hydrogen atoms to provide an unsaturated bond or bonds. Catalytic dehydrogenation of hydroaromatic compounds is one of the classical methods of synthesis of polycyclic aromatic compounds.[129] The most generally useful catalysts are platinum or palladium used either as the finely divided metal or supported on activated charcoal. Nickel and copper catalysts have also been used, but generally require higher reaction temperatures. Reaction conditions usually consist of heating at temperatures up to 350°C or refluxing in a solvent of boiling point around 200°C.

Whereas dehydrogenation of the enol lactone **102** with reagents such as selenium dioxide, mercuric acetate, or bromine failed, reaction with 10% palladium on charcoal in boiling p-cymene produced the pyrone **103** in 75% yield.[130]

Manganese dioxide in refluxing chloroform or benzene can be used to convert α,β- and β,γ-unsaturated sterols to conjugated dienones.[131] Thus the β,γ-unsaturated alcohol **104** is converted to the dienone **105** via the ketone **106**. The corresponding α,β-unsaturated alcohol **107** undergoes the same transformation but in lower yield, since the initially formed α,β-unsaturated ketone **108** is less prone to undergo further reaction than the β,γ-unsaturated ketone **106**.

A simple dehydrogenation of the substituted cyclopentanone **109** with palladium chloride and chloranil produces the cyclopentenone **110** in 70% yield.[132] This one-step conversion is particularly useful for small-scale preparations.

Related to dehydrogenation is the dehydrosilylation of the silyl enol ethers (**111**) to give α,β-unsaturated carbonyl compounds (**112**) in the presence of palladium acetate.[133] Using stoichiometric amounts of

$$R^1CH_2CH=CR^2 \xrightarrow{\ Pd(OAc)_2\ } R^1CH=CHCR^2$$

with OSiMe$_3$ on **111** and O on **112**.

Pd(OAc)$_2$ yields are quantitative, the reaction taking place at room temperature. The regioselectivity of the reaction is illustrated by the reaction of 2-methyl-1-trimethylsilyloxycyclohexene (**113**) to give only **114** in 96%

yield, and the dihydrosilylation of an (E) and (Z) mixture of 1-trimethyl-silyloxycyclodecene to produce selectively (E)-cyclodecen-2-one.

Chapter 9
Preparing and Handling Transition Metal Catalysts

9.1. Introduction

Most of the transition metal species discussed in this book are employed in small quantities as catalysts rather than as stoichiometric reagents, and this chapter is concerned with some of the practical problems associated with transition metal catalysts. Some transition metal catalysts are relatively commonplace compounds [e.g., $RhCl_3 \cdot 3H_2O$, $Pd(CH_3CO_2)_2$, $Co_2(CO)_8$], while others are less familiar. However, it is not always necessary to prepare these compounds in the laboratory because many of them are now available from either the large general chemical supply companies, or specialist suppliers of inorganic and organometallic compounds (for names and addresses see Table 9.1). Nonetheless, the commercial price of some catalysts makes their preparation in the laboratory attractive. Moreover, this is often straightforward, and not particularly time-consuming. The preparation of stoichiometric reagents is given in the chapters in which their use is described, and this chapter contains preparative details for a range of the more important homogeneous and heterogeneous transition metal catalysts. In each case sufficient practical detail is given for the reader to undertake the preparations, but it is recommended that additional information be obtained from the cited references.

A number of catalysts (and some reagents) contain precious metals whose recovery, after isolation of the organic product, can be attractive. Methods for doing this are discussed in this chapter.

Table 9.1. Names and Addresses of Specialist Suppliers of Inorganic and Organometallic Reagents and Catalysts

---

Strem Chemicals, Inc.
7 Mulliken Way
Dexter Industrial Park
P.O. Box 108
Newburyport, Massachusetts 01950

Available in Britain through:
Fluorochem Ltd.
Dinting Vale Trading Estate
Dinting Lane
Glossop
Derbyshire SK13 9NU, United Kingdom
Available in Japan through:
Iwai Kagaku Yakuhin Co Ltd.
7, 3-Chome, Nihonbashi-Honcho
Chuo-Ku
Tokyo, Japan
and:
Nakarai Chemicals Ltd.
Nijyo Karasuma, Nakagyo-Ku
Kyoto, Japan
Available in West Germany through:
Ventron GMBH
D-7500 Karlsruhe 1
Zeppelin Strasse 7
Postfach 6540, West Germany

Alpha Division
Ventron Corporation
P.O. Box 299
152 Andover Street
Danvers, Massachusetts 01923

Available in West Germany through:
Ventron GMBH, as above

Available in Britain through:
Lancaster Synthesis Ltd.
St. Leonard's House
St. Leonardgate
Lancaster LA1 1NB, United Kingdom
Available in Japan through:
Nippon Alfa Chemical Co. Ltd.
Ishikura Building
1-1 Nihonbashi-Odenmacho
Chuo-Ku
Tokyo (103), Japan

Research Organic/Inorganic Chemical
Corporation
11686 Sheldon Street
Sun Valley, California 91352

PCR Inc.
Research Chemicals Division
P.O. Box 1466
Gainesville, Florida 32601

Grilyt
Emser Werker AG
Market Development Department
CH-8039 Zurich, Switzerland

Table 9.1. (cont.)

Pressure Chemical Company
3419 Smallman Street
Pittsburgh, Pennsylvania 15201

Precious metal refiners—compounds and catalysts:
Johnson Matthey Chemicals Ltd.
Orchard Road
Royston
Hertfordshire SG8 5HE, United Kingdom

Johnson Matthey Inc.
Malvern, Pennsylvania 19355

Engelhard Sales Ltd.
Chemical Group
Valley Road
Cinderford
Gloucester GL14 2PB, United Kingdom

Engelhard Industries Division
529 Delancy Street
Newark, New Jersey 07105

The use of air-stable materials, rather than air-sensitive ones, is emphasized throughout this book. However, in circumstances where no alternative is available it may be necessary to use an inert atmosphere, and in this chapter a section is devoted to simple, convenient means of working in the absence of air.

9.2. Preparation of Catalysts

Attention to detail is important when preparing catalysts. With heterogeneous catalysts, care is needed to ensure that they are not accidentally poisoned by even minute traces of heavy elements (particularly by sulfur, arsenic, phosphorus compounds, and nitrogen bases). For example, fume cupboards used for sulfur compounds should be avoided. In some instances halide ions are catalyst poisons, and distilled or deionized water, rather than tap water, should always be used when preparing catalysts. Heterogeneous catalysts are usually considerably more susceptible to poisoning than homogeneous catalysts.

Many so-called catalysts are in fact catalyst precursors that are converted to the active species during use. For example, PtO_2 (Adam's catalyst) is converted to platinum metal by reaction with hydrogen, and $RhCl(PPh_3)_3$

(Wilkinson's catalyst) undergoes phosphine dissociation in solution to form the active coordinatively unsaturated species. Usually, though not always, catalyst precursors are reasonably stable materials. However, care is often needed when handling active catalysts. For instance, activated Adam's catalyst will ignite (explosively in some cases) flammable solvent vapor–air mixtures, and solutions of Wilkinson's catalyst absorb oxygen, which can cause deactivation. It is therefore prudent to store catalysts under an atmosphere of nitrogen and to exercise care when handling them.

9.2.1. *Chlorotris(triphenylphosphine)rhodium(I)—$RhCl(PPh_3)_3$ (Wilkinson's Catalyst)*

Wilkinson's catalyst was the first effective homogeneous catalyst for the hydrogenation of alkenes at room temperature and atmospheric pressure.[1] Only unhindered double bonds undergo reaction, so polyenes may be selectively hydrogenated. In the solid, $RhCl(PPh_3)_3$ exists in red and orange forms, and both are soluble in benzene and toluene—solvents in which it is commonly used. Erratic results are sometimes obtained with old catalyst (e.g., long induction period).

Procedure[2]

$$RhCl_3 \cdot 3H_2O + 4PPh_3 \rightarrow RhCl(PPh_3)_3 + Ph_3PO$$

A solution of triphenylphosphine (12 g, 46 mM) in hot ethanol (350 ml) is added to a solution of $RhCl_3 \cdot 3H_2O$ (2 g, 8 mM) in hot ethanol (70 ml). The hot solution is refluxed for 30 min under nitrogen, and red crystals of product separate. These are quickly filtered off,† washed with ether (50 ml), and dried *in vacuo*. Yield, 6.3 g, 86% (mp 157–158°C).

9.2.2. *trans-Chlorocarbonylbis(triphenylphosphine)rhodium(I)— trans-$RhCl(CO)(PPh_3)_2$*

This complex is a particularly active hydroformylation catalyst. The efficient single-stage preparation described here has the advantage of using formaldehyde as the source of carbon monoxide. *Trans*-$RhCl(CO)(PPh_3)_2$ is a yellow, air-stable, crystalline material (mp 195–197°C) with moderate solubility in benzene.

---

† Addition of water to the filtrate precipitates excess triphenylphosphine. This can be collected, recrystallized from ethanol, and reused.

Procedure[3]

$$RhCl_3 + PPh_3 + H_2CO \rightarrow \textit{trans}\text{-}RhCl(CO)(PPh_3)_2$$

A solution of $RhCl_3 \cdot 3H_2O$ (2.0 g, 8 mM) in ethanol (70 ml) is slowly added to a refluxing solution of triphenylphosphine (7.2 g, 28 mM) in ethanol (300 ml). When the resulting solution becomes clear, sufficient formaldehyde solution (10–20 ml of 40%) is added to make the red solution yellow in about a minute. Yellow crystals of the product are formed. When cool these are filtered off and washed with ethanol and ether (2 × 50 ml). Yield, 4.5 g (85%).

9.2.3. Hydridocarbonyltris(triphenylphosphine)rhodium(I)— RhH(CO)(PPh₃)₃

Bright yellow $RhH(CO)(PPh_3)_3$ is moderately soluble in aromatic solvents, but insoluble in water, ethanol, and saturated hydrocarbons. In air it melts with decomposition at ~120°C. It can be obtained by the reaction of *trans*-$RhCl(CO)(PPh_3)_2$ with sodium borohydride in the presence of excess triphenylphosphine. It is more conveniently prepared from $RhCl_3 \cdot 3H_2O$, triphenylphosphine, and formaldehyde in refluxing ethanolic potassium hydroxide.

Procedure[4]

$$RhCl_3 \cdot 3H_2O + PPh_3 \xrightarrow[\text{EtOH}]{\text{KOH, }H_2CO} RhH(CO)(PPh_3)_3$$

To a refluxing solution of triphenylphosphine (2.64 g, 10 mM) in ethanol (100 ml) is added $RhCl_3 \cdot 3H_2O$ (0.26 g, 1.0 mM) dissolved in ethanol (20 ml); then almost immediately (~15-sec delay), aqueous formaldehyde (10 ml of 40%) and hot ethanolic potassium hydroxide (0.8 g, 14 mM in 20 ml) are added. After refluxing for 10–20 min the mixture is allowed to cool and the product crystallizes. Yield, 0.8–0.9 g (~95% based on rhodium).

9.2.4. Dichlorotetrakis(ethylene)dirhodium(I)—[RhCl(C₂H₄)₂]₂

Reaction of $RhCl_3 \cdot 3H_2O$ in aqueous methanol with ethylene at atmospheric pressure and room temperature forms $[RhCl(C_2H_4)_2]_2$, which precipitates in good yield over 8 h. $[RhCl(C_2H_4)_2]_2$ is a useful intermediate in the preparation of other rhodium(I) compounds; for instance, Wilkinson's catalyst can be prepared *in situ* by reaction with PPh_3.

Procedure[5]

$$2RhCl_3 \cdot 3H_2O + 6C_2H_4 \rightarrow [RhCl(C_2H_4)_2]_2 + 2CH_3CHO + 4HCl + 4H_2O$$

A solution of $RhCl_3 \cdot 3H_2O$ (10 g, 37 mM) in water (15 ml) is added to methanol (250 ml), and a stream of ethylene bubbled through the solution (fume cupboard). The product precipitates over 8 h and is filtered off, taking care to minimize exposure to air. The red-brown product is washed once with methanol (30 ml) and dried *in vacuo*. Yield, 5.0 g (65%). This material is best stored under ethylene at ~0°C.

9.2.5. Dichlorotetracarbonyldirhodium(I)—[RhCl(CO)₂]₂ (Rhodium Carbonylchloride Dimer)

This useful catalyst is soluble in most organic solvents and can be obtained by reaction of carbon monoxide with $RhCl_3 \cdot 3H_2O$ at 100°C.[6] However, this reaction does not always work well, and $[RhCl(CO)_2]_2$ is best obtained by displacement of ethylene from $[RhCl(C_2H_4)_2]_2$ by carbon monoxide.

Procedure[6]

$$[RhCl(C_2H_4)_2]_2 + 4CO \rightarrow [RhCl(CO)_2]_2 + 4C_2H_4$$

Caution: Carbon monoxide requires the use of an efficient fume cupboard.
Carbon monoxide is bubbled through a stirred mixture of $[RhCl(C_2H_4)_2]_2$ (1 g, 2.6 mM) in diethyl ether (30 ml) at room temperature for about an hour. The mixture is then filtered, the filtrate concentrated to about 10 ml (rotary evaporator), and the deep-red product filtered from the ice-cold liquor. Yield, 0.6 g (60%). Storage under carbon monoxide in a refrigerator is recommended.

9.2.6. cis-Dichlorobis(benzonitrile)palladium(II)—PdCl₂(PhCN)₂

The catalyst $PdCl_2(PhCN)_2$ is an air-stable, light-yellow solid soluble in aromatic solvents, but insoluble in saturated hydrocarbons. It is readily prepared from anhydrous $PdCl_2$.

Procedure[7]

$$PdCl_2 + 2PhCN \rightarrow PdCl_2(PhCN)_2$$

A suspension of $PdCl_2$ (2.0 g, 11 mM) in benzonitrile (50 ml) is heated to 100°C to form a red solution (20 min). Undissolved material is removed by filtering the hot solution. The filtrate is poured into low-boiling petroleum (300 ml) to precipitate the product. This is filtered off and washed with low-boiling petroleum. Yield, 4.0 g (93%).

9.2.7. Dichlorobis(triphenylphosphine)palladium(II)—PdCl₂(PPh₃)₂

Pale yellow, air-stable $PdCl_2(PPh_3)_2$ is a useful carbonylation catalyst, and, in conjunction with stannous chloride (which enhances its mild hydrogenation activity), polyenes may be selectively hydrogenated to monoenes. Other catalytic applications of $PdCl_2(PPh_3)_2$ include alkene isomerization and hydrosilyation of alkenes. It can also be used in various carbon–carbon bond-forming reactions. It is soluble in chloroform, moderately soluble in aromatic solvents and alcohols, but insoluble in water, saturated hydrocarbons, ether, and carbon tetrachloride. It can be prepared by dissolving palladium chloride in molten triphenylphosphine, but sometimes this method gives poor yields.[8a] A better procedure uses aqueous ethanol as solvent.

Procedure[8]

$$PdCl_4^{2-} + 2PPh_3 \rightarrow PdCl_2(PPh_3)_2 + 2Cl^-$$

A solution of palladium chloride (3.0 g, 17 mM) in dilute hydrochloric acid (0.5 ml concentrated hydrochloric acid in 150 ml water) is slowly added to a stirred warm solution of triphenylphosphine (9.0 g, 34.5 mM) in ethanol (300 ml). The stirred mixture is maintained at ~60°C for 3 h before the yellow product is collected. After washing with 100 ml portions of warm water, ethanol, and ether, the crude product (11.5 g) is sufficiently pure for most purposes. It can be purified by precipitation from chloroform by hexane.

9.2.8. Tetrakis(triphenylphosphine)palladium(O)—Pd(PPh₃)₄

Many procedures are available for the preparation of this highly versatile catalyst. The procedure given here is convenient because it uses readily available reagents and is known to work well in different laboratories. $Pd(PPh_3)_4$ is a yellow solid (mp ~ 115°C) that is soluble in most organic solvents except saturated hydrocarbons. In the solid it is air stable, but in solution it is sensitive to oxygen, forming $[Pd(PPh_3)_2O_2]$. It is recommended that the solid be stored under nitrogen.

Procedure[9]

$$PdCl_2 + 4PPh_3 + NH_2NH_2 \cdot H_2O \rightarrow Pd(PPh_3)_4 + N_2\uparrow$$

PdCl$_2$ (17.7 g, 0.1 mol), PPh$_3$ (131 g, 0.5 mol), and dimethylsulfoxide (1.2 liters) are heated with stirring to ~140°C under nitrogen. Once a clear solution is obtained, hydrazine hydrate (19 ml, 0.4 mol) is added in one portion (with exclusion of air). Nitrogen is evolved in the *vigorous* ensuing reaction. The reaction flask is then cooled immediately (water bath) to reduce the temperature of the mixture to ~100°C. After crystallizing overnight, the product is filtered off under nitrogen on a glass sinter, and washed (under nitrogen) with ethanol and ether (2 × 50 ml portions). Yield, 103–113 g (90%–98%).

9.2.9. *Palladium(II) Acetate Trimer—[Pd(CH$_3$CO$_2$)$_2$]$_3$*

Palladium is one of the most commonly used transition metals in organic synthesis. The acetate has a variety of applications, particularly in carbon–carbon bond-forming reactions. It has a trimeric structure with each palladium atom bridged by two acetate groups. The air-stable brown needles are soluble in some organic solvents (e.g., chloroform, dichloromethane, acetone, and diethylether), but insoluble in water and alcohols. In glacial acetic acid it is monomeric, and in benzene a trimer. Palladium acetate is commercially available at a reasonable price, and in general its preparation is not therefore recommended. However, the following procedure for obtaining the acetate from the metal can be used in the recovery of useful compounds from palladium residues.

Procedure[10]

Palladium sponge (10 g), glacial acetic acid (150 ml), and concentrated nitric acid (6 ml) are boiled under reflux until no brown fumes are produced.† The boiling solution is filtered from excess metal and the filtrate cooled overnight. The orange-brown crystals are collected and washed with a little acetic acid and water before air drying. Yield, 80%–95%. A small amount of additional product can be obtained by concentrating the pale-brown filtrate. Alternatively, the filtrate can be reused in subsequent preparations or added to palladium residues for later recovery.

---

† A small amount of metal should remain undissolved; if none remains, some must be added, and refluxing must be continued until no brown fumes are formed. This avoids contamination of the product with Pd(NO$_2$)(CH$_3$CO$_2$).

9.2.10. Palladium on Charcoal

Palladium (5% or 10% by weight) on charcoal is available commercially. Its preparation, however, is straightforward and financially attractive. Like many other heterogeneous catalysts, it can be poisoned by impurities adsorbed during preparation or use. Once prepared, catalyst should be dried only at room temperature, since *at high temperatures it may ignite.* Care should be taken to keep heterogeneous catalysts of this type out of contact with combustible vapors. In use, it is advisable that reaction vessels first be blanketed with nitrogen, and the organic solvent (particularly when using low-boiling alcohols) added in large portions to the catalyst, because once wet the catalyst is less likely to cause ignition. In some situations it is convenient to use carbon impregnated with a palladium salt (e.g., $PdCl_2$/HCl) and to activate this by reduction *in situ* before reaction, rather than handle the more active prereduced catalyst. Reducing agents used to convert the supported chloride to the metal include hydrazine, formates, formaldehyde, hydrogen, and sodium borohydride. The procedure given here uses formaldehyde, which is economical and practically convenient.

Procedure [11]

A solution of palladium chloride (8.2 g) in hydrochloric acid (20 ml concentrated acid and 50 ml water) is obtained by warming (about 2 h). This is added to a stirred, hot (80°C) suspension of nitric acid-washed charcoal† (93 g) in water (1.2 liters). Formaldehyde (8 ml of 37% solution) is then added, followed by sufficient sodium hydroxide solution (30%) to make the suspension strongly alkaline. After 10 min the catalyst is filtered off, washed with water (10 × 250 ml), and dried *in vacuo* over calcium chloride, before storing in a tightly closed bottle. Yield, 93–98 g (5% palladium on charcoal).

9.2.11. Lead-Conditioned Palladium on Calcium Carbonate—Selective Hydrogenation Catalyst (Lindlar Catalyst)

Palladium conditioned with lead is a selective heterogeneous catalyst for hydrogenation of alkynes to alkenes in the presence of a small amount of amine (usually quinoline). Suitable solvents for selective hydrogenations include saturated hydrocarbons, toluene, and acetone. Unlike hydrogenations using more vigorous platinum catalysts, alcohols are usually unsatisfactory solvents. Lead-conditioned palladium on calcium carbonate is grey in appearance, and it is stable over a long period.

† Almost any high-surface-area carbon can be used after treating with nitric acid (10%) for 2–3 h, followed by washing free of acid with water and drying at 100°C.

Procedure[12]

Distilled water (45 ml) is added to a solution of palladium chloride (1.48 g, 8.3 mM) in warm concentrated hydrochloric acid (2.6 ml), and sodium hydroxide solution (3 N) is added until the pH is 4.0–4.5. After diluting to 100 ml with distilled water, "precipitated" (not "powdered") calcium carbonate (18 g) is added to the liquid, and the mixture stirred while it is heated to 75–85°C. After 20 min the liquid becomes colorless, and sodium formate solution (6 ml, 0.7 N) is added. The foaming mixture is stirred, and maintained at 75–85°C for an hour before filtering off the catalyst, which is washed with distilled water (8 × 65 ml). The moist, black catalyst is reslurried with distilled water (60 ml), and a solution of lead acetate (1.6 g in 18 ml water) is added. The slurry is stirred for another 45 min at 75–85°C before filtering and washing with distilled water (4 × 50 ml). After drying at 60–70°C the yield is 19 g.

9.2.12. Tetrakis- and Tris(triphenylphosphine)platinum(0)— $Pt(PPh_3)_4$ and $Pt(PPh_3)_3$

Pale yellow $Pt(PPh_3)_4$ melts in air, with decomposition at ~120°C. Being moderately air sensitive it must be kept under nitrogen. It dissolves in benzene with dissociation. Reaction takes place with chlorinated solvents; with CCl_4 good yields of *cis*-$PtCl_2(PPh_3)_2$ can be obtained. Boiling an ethanolic suspension (under nitrogen) affords $Pt(PPh_3)_3$ in acceptable yield. The properties of this yellow complex are similar to those of $Pt(PPh_3)_4$.

Procedure[13]

$$K_2PtCl_4 + 4PPh_3 + 2KOH + C_2H_5OH \rightarrow Pt(PPh_3)_4 + 4KCl + CH_3CHO + 2H_2O$$

Potassium hydroxide (1.4 g, 25 mM) in aqueous ethanol (32 ml ethanol and 8 ml H_2O) is added to a hot (65°C) stirred solution of PPh_3 (1.5 g, 60 mM) in ethanol (200 ml), followed by addition (over 20 min) of a solution of K_2PtCl_4 (5.2 g, 12.5 mM) in water (50 ml). During this period the temperature of the stirred solution is maintained at 65°C, and the yellow product begins to form shortly after the first addition of K_2PtCl_4. The product is separated from the cooled solution and rapidly washed with warm (35°C) ethanol (100 ml), cold water (50 ml), and cold ethanol (50 ml), before drying *in vacuo* and storing under nitrogen. Yield of $Pt(PPh_3)_4$, 12.4 g (79%).

$$Pt(PPh_3)_4 \xrightarrow{\text{EtOH}} Pt(PPh_3)_3 + PPh_3$$

A stirred suspension of $Pt(PPh_3)_4$ (5.8 g, 4.7 mM) in ethanol (250 ml) is refluxed under nitrogen for 2 h. The hot solution is quickly filtered and the yellow product washed with cold ethanol (30 ml), before drying *in vacuo* and storing under nitrogen. Yield of $Pt(PPh_3)_3$, 3.0 g (66%).

9.2.13. Platinum Oxide Catalyst (Adam's Catalyst)

On contact with hydrogen some forms of platinum oxide ($PtO_2 \cdot H_2O$) produce very finely divided platinum metal, which is a highly active hydrogenation catalyst. In this active form it needs to be handled with care, since in its presence inflammable vapors are readily ignited, and explosions can result from hydrogen–air and solvent vapor–air mixtures. The danger of this happening is minimized by excluding air and keeping the active catalyst moist. Platinum oxide presents no storage or handling problems. Brown and black forms of platinum oxide have been described, with the former producing higher-activity catalyst. The procedure given below is a relatively easy means of obtaining the brown oxide from chloroplatinic acid. Since this can be recovered from platinum residues, this procedure represents a step in recycling platinum.

Procedure[14]

Sodium nitrate (9 g) is dissolved in hot chloroplatinic acid solution (10 ml, 10%) and the mixture evaporated to dryness using a rotary evaporator. The dry powder is carefully added in one portion to molten sodium nitrate (100 g, 500–530°C) contained in a Pyrex beaker. A brown precipitate forms, and nitrogen oxides are evolved. The mixture is then poured into a metal container and allowed to cool (if left in the beaker, the beaker will crack when the nitrate solidifies). The cold sodium salts are dissolved in distilled water (2 liters), and the product is collected on a No. 3 glass sinter and carefully washed with distilled water (1 liter). The product must not become dry until washing is complete. (This prevents the product from becoming colloidal. If the product does become colloidal, washing should be terminated; traces of sodium nitrate do not affect catalytic performance.) The product is dried *in vacuo* over calcium chloride. Yield almost quantitative.

9.2.14. Dicyclopentadienylcobalt(II)—Co(C₅H₅)₂ (Cobaltocene)

Cobaltocene is air sensitive, and whenever possible it should be handled *in vacuo* or under an atmosphere of nitrogen. Large crystals may, however, be exposed to air for short periods (e.g., transfer to reaction flask). Like ferrocene and its other analogs, cobaltocene is soluble in a variety of organic solvents. The procedure outlined below makes use of the reaction between anhydrous cobalt chloride and sodium cyclopentadienide, and a number of other metal cyclopentadienyl compounds can be obtained in this way by using the appropriate anhydrous metal chloride.[15]

Procedure[15]

$$2C_5H_5Na + CoCl_2 \rightarrow Co(C_5H_5)_2 + 2NaCl$$

Under an atmosphere of nitrogen, anhydrous cobalt chloride† (65.0 g, 0.5 mol) is added in small quantities to a stirred tetrahydrofuran solution (500 ml) of sodium cyclopentadienide, previously prepared from a dispersion of sodium (23.0 g, 1.0 mol) and freshly distilled (and hence monomeric) cyclopentadiene (80 g, 1.2 mol). The resulting warm mixture is refluxed (under nitrogen) before removing solvent under reduced pressure. The residue is transferred under nitrogen to a sublimation apparatus, and pure cobaltocene is obtained as dark violet crystals by sublimation at 70–200°C/0.1 mm Hg. Yield, 70–80 g of crystals, which must be stored under nitrogen.

9.2.15. Nickel Dichlorobis(phosphine) Complexes—NiCl₂(PPh₃)₂, NiCl₂(Ph₂PCH₂CH₂PPh₂), and NiCl₂(DIOP)

Nickel(II) phosphine complexes activate aryl halides in reactions with Grignard reagents and lithium alkyls. Asymmetric cross-coupling reactions of secondary Grignard reagents with organic halides are also possible in the presence of nickel complexes containing an optically active phosphine (e.g., DIOP). The dichlorobis(phosphine)nickel(II) complexes are not very soluble in common solvents, which facilitates their preparation, but they are sufficiently soluble for catalytic applications. Thus, the complex $NiCl_2(Ph_2PCH_2CH_2PPh_2)_2$ is obtained as a dull red precipitate by mixing acetone or ethanol solutions of hydrated nickel chloride and the phosphine. The red-violet complex containing the optically active phosphine DIOP is similarly obtained, as is the blue complex containing triphenylphosphine.

Procedure[16]

$$NiCl_2 \cdot 6H_2O + Ph_2PCH_2CH_2PPh_2 \rightarrow NiCl_2(Ph_2PCH_2CH_2PPh_2)$$

A solution of 1,2-bis(diphenylphosphino)ethane (4.0 g, 10 mM) in hot ethanol (400 ml) is added to a solution of NiCl₂·6H₂O (2.4 g, 10 mM) in ethanol (20 ml). The mixture is allowed to stand overnight, and the red precipitate is collected, washed with a small amount of ethanol, and dried *in vacuo*. Yield, almost quantitative.

---

† Obtained by dehydrating CoCl₂·6H₂O at ~160°C/0.1 mm Hg.

9.2.16. *Bis(1,5-cyclooctadiene)nickel(O)—Ni(C₈H₁₂)₂*

Bis(1,5-cyclooctadiene)nickel catalyzes oligomerization of alkenes. It is a low-melting (~60°C), yellow, crystalline solid that is very air sensitive. It should be prepared, handled, and stored under nitrogen.

Procedure[17]
    Under nitrogen, Ni(pyridine)₄Cl₂† (89.2 g, 0.2 mol) and 1,5-cyclooctadiene (64.8 g, 0.6 mol) in tetrahydrofuran (120 ml) are treated with small pieces of sodium metal (9.2 g, 0.4 mol) over 4 h at room temperature. The brown solution is then concentrated under reduced pressure, and methanol is added to induce crystallization of the product, which is collected, washed well with methanol, and dried *in vacuo*. Yield, 30 g (50%).

9.2.17. *trans-Chlorocarbonylbis(triphenylphosphine)iridium(I)— trans-IrCl(CO)(PPh₃)₂*

Bright yellow *trans*-IrCl(CO)(PPh₃)₂ is air stable, though in solution it reacts with oxygen to form an adduct that is converted to the pure compound by heating *in vacuo* at 100°C. It is soluble in benzene and chloroform, but insoluble in alcohols. *trans*-IrCl(CO)(PPh₃)₂ is known as "Vaska's compound," and it is easily obtained by refluxing IrCl₃·3H₂O and PPh₃ in dimethylformamide.

Procedure[18]

$$IrCl_3 \cdot 3H_2O + PPh_3 + HCONMe_2 \rightarrow trans\text{-}IrCl(CO)(PPh_3)_2$$

A mixture of IrCl₃·3H₂O (3.52 g, 10 mM), PPh₃ (13.1 g, 50 mM), and dimethylformamide (150 ml) is vigorously boiled under reflux overnight. The hot red-brown solution is filtered and warm methanol (300 ml) is added to the filtrate, which is then cooled in an ice bath. The crystalline product is collected and washed with cold methanol (50 ml) before drying *in vacuo*. Yield, 6.8–7.0 g (87%–90%).

9.2.18. *Dichlorotris(triphenylphosphine)ruthenium(II)— RuCl₂(PPh₃)₃*

Under appropriate conditions RuCl₂(PPh₃)₃ reacts with hydrogen to form Ru(H)Cl(PPh₃)₃, which is a useful specific hydrogenation catalyst.

---

† Ni(pyridine)₄Cl₂ is obtained by crystallizing hydrated nickel chloride from pyridine. The analogous iron compound is similarly prepared and is useful for its solubility in organic solvents.

Black crystalline $RuCl_2(PPh_3)_3$ is air sensitive. It dissolves in warm chloroform, acetone, and aromatic solvents to form brown solutions.

Procedure[19]

$$RuCl_3 \cdot 3H_2O + PPh_3 \rightarrow RuCl_2(PPh_3)_3$$

An atmosphere of nitrogen is maintained throughout this preparation. A solution of $RuCl_3 \cdot 3H_2O$ (1.0 g, 3.8 mM) in methanol (250 ml) is refluxed for 5 min and cooled. Maintaining the inert atmosphere, triphenylphosphine (6.0 g, 22.9 mM) is added to the solution, and the mixture is refluxed for another 3 h. After cooling, the black crystalline product is filtered off under nitrogen, washed well with degassed diethyl ether, and dried *in vacuo*. Yield, 2.5 g (70%).

9.3. Metal Acetylacetonates

Acetylacetone (2,4-pentanedione, Hacac) and similar β-diketones readily form enolate anions, as well as stable metal derivatives (salts) containing six-membered chelate rings in which the oxygen atoms are coordinated to the metal. Many monomeric metal β-diketonates have properties typical of organic compounds, and their solubility in common solvents is exploited in applications such as solvent extraction of metal ions from aqueous solutions, the use of paramagnetic chromium acetylacetonate as a nmr relaxation agent, the use of lanthanide β-diketonates as nmr shift reagents, and in the preparation of organometallic compounds. Several transition metal acetylacetonates have catalytic properties. For instance, $VO(acac)_2$ catalyzes formation of epoxides from alkenes and hydrogen peroxide, and $Ni(acac)_2$ is a catalyst for the isomerization of alkenes. Several reviews are available that deal with the preparation of metal acetylacetonates.[20] Most of the common transition metal derivatives can be prepared by adding sodium carbonate to a solution of acetylacetone and a metal salt.

Procedure[21]

$$V_2O_5 + 2H_2SO_4 + C_2H_5OH \rightarrow 2VOSO_4 + CH_3CHO + 3H_2O$$
$$VOSO_4 + 2C_5H_7O_2^- \rightarrow VO(C_5H_7O_2) + SO_4^{2-}$$

Vanadium pentoxide (20 g, 0.11 mol), distilled water (50 ml), concentrated sulfuric acid (30 ml), and ethanol (100 ml) are stirred at the boiling point for 30 min. The dark blue solution† is filtered, and freshly distilled acetylacetone (50 ml, 0.49 mM) is added to the filtrate, followed by slow addition of sodium carbonate solution (80 g Na$_2$CO$_3$, 500 ml H$_2$O). The green product is filtered off, washed with water, and air-dried. Yield, ~50 g. Recrystallization from chloroform affords green crystals.

9.4. Arene Metal Tricarbonyls—ArM(CO)$_3$ (M = Cr, Mo, or W)

The arene tricarbonyl derivatives of chromium, molybdenum, and tungsten are mildly air-sensitive solids. In solution they are slightly air sensitive when cold, and more so when hot. Refluxing solutions should be protected by an inert atmosphere. As stoichiometric intermediates they provide a means of facilitating nucleophilic attack on arenes. As catalysts they have activity in hydrogenation, alkene metathesis, and Friedel–Crafts reactions. They are conveniently prepared by refluxing the parent hexacarbonyl with the arene under an inert atmosphere. Owing to the volatility of the group VI hexacarbonyls, they sublime from the reaction flask into the reflux condenser. Use of an air condenser, rather than one cooled with water, allows the arene–solvent vapor to rise high into the condenser and wash the hexacarbonyl back into the reaction flask. Addition of a nucleophile (e.g., diglyme, tetrahydrofuran, or pyridine) can accelerate the reaction by forming a partially substituted labile intermediate. This provides a convenient means of preparing Cr(C$_6$H$_6$)(CO)$_3$ via the *in situ* generation of Cr(CO)$_5$ (donor), which, being involatile and rapidly formed, removes the problem of Cr(CO)$_6$ subliming from the reaction flask. This approach cannot, however, be extended to the preparation of the molybdenum and tungsten compounds because their monosubstituted intermediates are stable and fail to react further.

Procedure[22]

$$Cr(CO)_6 + C_6H_6 \xrightarrow{\text{2-picoline}} Cr(C_6H_6)(CO)_3 + 3CO$$

Under an atmosphere of nitrogen, a mixture of benzene (100 ml, 1.0 mol), 2-picoline (100 ml, 1.2 mol), and chromium hexacarbonyl (8.8 g, 0.04 mol) is refluxed for 96 h. Excess benzene, 2-picoline, and chromium hexacarbonyl are removed under reduced pressure on a rotary evaporator. The yellow-green residue is extracted with portions of hot ether until the

---

† Alternatively, commercial vanadyl sulfate in dilute sulfuric acid may be used.

extract is colorless. The combined extracts are evaporated to dryness (rotary evaporator), and the residue is washed with pentane to remove traces of 2-picoline. Yield, 7.7 g (90%) of $Cr(C_6H_6)(CO)_3$, mp 160–161°C. Sublimation (85°C, 1×10^{-3} mm Hg) or recrystallization from ether affords material with a slightly higher melting point.

9.5. Transition Metal Carbonyls

The binary transition metal carbonyls have extensive chemistries, and many have useful catalytic properties or stoichiometric applications in organic chemistry (see Table 9.2). Thus, $Co_2(CO)_8$ is a catalyst for the hydroformylation of alkenes, $Fe(CO)_5$ catalyzes isomerization of alkenes, and $Ni(CO)_4$ is a good catalyst in several carbon–carbon bond-forming reactions. Iron pentacarbonyl is easily converted to iron tetracarbonyl anions [e.g., Collman's reagent $Na_2Fe(CO)_4 \cdot (dioxane)_{1.5}$] and these provide a means of achieving a wide range of transformations, typified by the formation of higher aldehydes from alkyl halides.

Most of the more common metal carbonyls are reasonably stable and can be stored over prolonged periods, but if heated in air they become pyrophoric. Light sensitivity is not uncommon, and metal carbonyls are best stored under oxygen-free conditions in a refrigerator. Solubility in hydrocarbons and other solvents is slight to moderate, and these solutions tend to be oxygen sensitive, but manipulating metal carbonyls is not difficult. They do, however, have varying degrees of toxicity and should be handled in a well-ventilated fume cupboard. Owing to their volatility, the flammable, poisonous liquids iron pentacarbonyl (bp, 103°C) and nickel tetracarbonyl (bp, 43°C) are the most hazardous. *Nickel tetracarbonyl is extremely toxic* (**TLV, 0.001 ppm**) **and its general use in organic laboratories is not** recommended.

Preparations of transition metal carbonyls are inconvenient because they almost invariably involve carbon monoxide at high pressure and high temperature. Fortunately, many are available commercially at moderate prices, with iron pentacarbonyl being the least expensive, and in some instances it is more economical to convert this to $Fe_2(CO)_9$ or $Fe_3(CO)_{12}$ [23] rather than purchasing them. However, in general the preparation of transition metal carbonyls is not recommended.

9.6. Recovery of Precious Metals

Although by definition precious metal salts and their complexes are costly, when they are used catalytically only small amounts are required

Table 9.2. Properties of the More Common Transition Metal Carbonyls (arranged according to their position in the periodic table)

$Cr(CO)_6$ White microcrystals MW 220.0 readily sublimes	$Mn_2(CO)_{10}$ Yellow crystals MW 390.0 mp 154°C	$Fe(CO)_5$ Amber liquid MW 195.9 mp −21°C bp 103°C	$Co_2(CO)_8$ Orange crystals MW 342.0 mp 51°C	$Ni(CO)_4$ Colorless liquid MW 170.8 mp −25°C bp 43°C
		$Fe_2(CO)_9$ Golden brown plates MW 363.8 mp 100°C (dec)	$Co_4(CO)_{12}$ Black solid MW 571.9	
		$Fe_3(CO)_{12}$ Black crystals MW 503.7 mp 140°C (dec)		
$Mo(CO)_6$ White microcrystals MW 264.0 readily sublimes		$Ru_3(CO)_{12}$ Orange crystals MW 639.3 mp 154°C	$Rh_4(CO)_{12}$ Dark red crystals MW 747.8 mp 150°C (dec)	
$W(CO)_6$ White microcrystals MW 351.9 readily sublimes	$Re_2(CO)_{10}$ White crystals MW 625.5 mp 170°C	$Os_3(CO)_{12}$ Pale yellow crystals MW 906.7 mp 224°C (dec)	$Ir_4(CO)_{12}$ Yellow solid MW 1104.9	

because of their high activity and they can be very cost effective. Moreover, in some instances heterogeneous catalysts can be recovered (e.g., by filtration) and may be reused without serious loss of activity. This is generally more difficult to do with homogeneous catalysts because they are normally destroyed during work-up of the product. When precious metal compounds are used as stoichiometric reagents, considerable amounts of metal can be involved, and it is essential to recover this material from the reaction mixture after removal of the organic product.

If precious metals are used frequently, residues containing even small amounts of metal should be retained for subsequent work-up. It is important that residues containing different metals be kept separate; recovery of a single metal can be straightforward, but the time-consuming separation of metals from mixed residues by selective precipitation, or liquid–liquid extraction procedures,[24] is not practical in the organic laboratory. If arrangements can be made for returning precious metal residues to a refiner for credit, it is recommended that this be done, since it is the most convenient means of recycling. However, occasionally it is very useful to be able to obtain specific compounds from precious metal residues in the laboratory. Although there is relatively little work published in this area, only simple inorganic reactions[25] are needed to recover some metals. The procedure usually involves evaporation of the residue (if liquid is present) to dryness, followed by ignition at red heat in air in order to remove organic materials by oxidation and leave the metal behind. This is then treated with a suitable reagent to produce a useful pure salt.

CAUTION

Before heating to dryness, a very small quantity of the residue should be tested before working on a larger scale. Mixtures containing organic material and oxidizing agents (e.g., NO_3^-, NO_2^-, ClO_4^-) can explode violently when heated! Ignitions must always be carried out in an efficient fume cupboard behind a suitable shield.

The difficulties in recovering precious metals stem from the inertness of the metals and problems associated with converting them into salts. Their reactivities are detailed in Table 9.3. Palladium metal dissolves in hot concentrated nitric acid, though a more useful reagent is hot acetic acid containing about 5% concentrated nitric acid, which provides a particularly convenient[26] means of obtaining palladium acetate (see Section 9.2.9). Similarly, silver metal is readily converted to the nitrate by nitric acid. All of the other precious metals are considerably more inert to attack by mineral acids.

Platinum and gold will, however, dissolve in hot aqua regia (3 volumes of concentrated hydrochloric acid to 1 volume of concentrated nitric acid).

Table 9.3. Chemical Reactivity of Precious Metals and Procedures for Their Recovery from Ignited Residues

Metal (atomic weight)	Reactivity	Conversion of metal to useful salt
Palladium (106.4)	Dissolves in concentrated nitric acid, and even in aerated hydrochloric acid.	Refluxing in acetic–nitric acid affords palladium acetate (see Ref. 26) and the chloride via treatment with aqua regia.
Platinum (159.1)	Almost completely inert to mineral acids other than aqua regia. Reacts with chlorine at red heat, and rapidly with fused alkali.	Dissolving in aqua regia, followed by repeated evaporation with hydrochloric acid, gives chloroplatinic acid (see Ref. 28), which is easily converted to platinum dichloride (see Ref. 32).
Rhodium (102.9)	Reacts at red heat with chlorine to form $RhCl_3$, and in the presence of potassium chloride, K_3RhCl_6. Dissolves in fused alkali containing oxidizing agents. Insoluble in mineral acids, but when very finely divided will dissolve slowly in aqua regia.	Chlorination of metal mixed with potassium chloride at 575°C produces K_3RhCl_6, which can be converted to $RhCl_3 \cdot 3H_2O$ (see Ref. 27).
Iridium (192.2)	Very inert. Insoluble in aqua regia. Dissolves in fused alkali nitrates. At red heat reacts with chlorine to form $IrCl_3$.	Chlorination of metal mixed with sodium chloride at 625°C and suitable work-up gives $(NH_4)_2IrCl_6$ (see Ref. 29).
Ruthenium (101.1)	Ruthenium dioxide is slowly formed when heated in air. Attacked by chlorine, and dissolves in fused alkali. Insoluble in mineral acids including aqua regia, but can be dissolved in aqua regia containing a little $KClO_3$ (caution!).	Fusion with sodium peroxide gives Na_2RuO_4, which in aqueous solution is converted to toxic RuO_4 by chlorine. Not a procedure recommended for general use.
Osmium (190.2)	When heated in oxygen, slowly forms OsO_4. Inert to mineral acids, including aqua regia, but reacts with fused sodium peroxide/hydroxide.	Similar to ruthenium. Fusion with sodium peroxide and work-up to obtain toxic OsO_4. Not a procedure for general use.
Silver (107.9)	Easily dissolved in moderately strong nitric acid. Not attacked by fused alkali.	Dissolution in nitric acid to afford silver nitrate.
Gold (197.0)	Attacked by chlorine at red heat. Insoluble in all mineral acids except aqua regia.	When a solution of gold in aqua regia is evaporated, yellow $HAuCl_4 \cdot 4H_2O$ is obtained.

Repeated evaporation of the platinum solution with hydrochloric acid removes the nitric acid and affords chloroplatinic acid,[28] which crystallizes as brownish-red, very deliquescent prisms ($H_2PtCl_6 \cdot 6H_2O$) but is more conveniently used as a solution. Solutions of gold in aqua regia, when evaporated, afford yellow crystals of chloroauric acid ($HAuCl_4 \cdot 4H_2O$). Treatment of palladium with aqua regia gives aqueous H_2PdCl_4, and the useful hydrated chloride $PdCl_2 \cdot 2H_2O$ is obtained by repeatedly evaporating to near dryness with a little hydrochloric acid.

Recovery of Platinum from Residues[28]

Residues are dried and strongly ignited in air to form finely divided platinum. (*Caution*: A small sample must first be examined for possible explosive behavior.) Soluble impurities are removed by washing with water, before dissolving the metal in aqua regia. After destroying any remaining nitro complexes by repeated evaporation to dryness with hydrochloric acid, the resulting solution of chloroplatinic acid can be used as such, or converted to the sparingly soluble ammonium salt. This is done by adding ammonium chloride, followed by the addition of alcohol, to precipitate yellow $(NH_4)_2PtCl_6$, which is collected and washed with portions of ice-cold ammonium chloride solution, ethanol, and finally ether.

The remaining precious metals (rhodium, iridium, ruthenium, and osmium) are more inert than platinum and very reluctant to dissolve in any acid. As a result of their inertness these metals are difficult to recycle in most laboratories. It is possible to convert them to their chlorides by heating strongly in chlorine. This is not an attractive proposition, though it is the recommended procedure for recycling rhodium[27] and iridium.[29]

Even the most inert of the precious metals can (with some difficulty) be dissolved in fused potassium hydroxide containing potassium nitrate to form oxide species, which can subsequently be dissolved in acid.[30] This works well for ruthenium and osmium, but the technique is not recommended for routine work with these metals because subsequent purification steps invariably involve distillation of the highly toxic tetroxides.[31]

CAUTION: RuO_4, OsO_4

Ruthenium and osmium form extremely toxic, volatile tetroxides (RuO_4, bp 100°C; OsO_4, bp 131°C) that are injurious to both eyes and lungs. Appropriate precautions (use of well-ventilated fume cupboard, etc.) must always be taken whenever there is the possibility of forming them. Similarly, care is needed when osmium tetroxide is used as an oxidizing agent, and the catalytic procedures that use only small amounts of ruthenium or osmium are preferred (see, for example, Sections 8.2 and 8.5.2).

Recommendations. It is not difficult to recover palladium, platinum, silver, and gold from residues in which they are the only metal. On the other hand, rhodium, iridium, ruthenium, and osmium can be very difficult to recycle without special facilities, and it is recommended that residues containing these metals be returned to refiners for metal recovery.

9.7. Handling Air-Sensitive Compounds

9.7.1. Reactions under Nitrogen

Although many organometallic compounds are sensitive to either moisture, oxygen, heat, or, in some instances, light, most of the transition metal species discussed in this book are stable under ambient conditions, and only a few of them are oxygen sensitive, thus requiring storage and handling under nitrogen. A number of reactions do have to be performed under an inert atmosphere, but the familiar techniques used in organic chemistry, in which glassware with ground joints is used, can be very satisfactory. Thus, in most situations suitable oxygen-free reaction conditions are obtained by bubbling pure nitrogen through the reaction mixture under a reflux condenser equipped with a bubbler. An alternative procedure favored by some is to purge the reaction system with nitrogen and maintain a positive pressure of inert gas by attaching a nitrogen-filled balloon to the sealed system, which, if heated, must be equipped with an efficient reflux condenser.

9.7.2. Handling Solids

Over a number of years a variety of relatively simple techniques have been developed for handling solids with the exclusion of air. These are described in several books,[33] and that by Shriver[34] is particularly recommended as a valuable source of detailed information.

Slightly air- or moisture-sensitive compounds can be handled by crude methods, such as under an inverted funnel conveying a flow of nitrogen. "Glove boxes" suitably equipped with nitrogen-purging facilities provide a convenient means of handling air-sensitive materials—for transferring them into reaction flasks, for example, as well as for carrying out small-scale reactions. However, glove boxes are relatively expensive, and a cheaper alternative is the polythene glove-bag, illustrated in Figure 9.1. Sensitive materials are easily transferred under nitrogen once air has been purged from the bag by repeatedly filling with nitrogen and collapsing the bag, or by a continuous flow of nitrogen. However, only straightforward reactions,

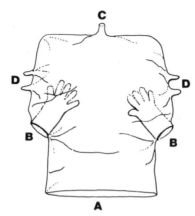

Figure 9.1. The polythene glove bag enables an entire apparatus to be manipulated in an inert atmosphere. Materials are inserted through flap **A**, which is sealed by rolling it up and applying clips. Inert gas enters via an aperture cut in one of the inlets (**C** and **D**), and inward pointing gloves (**B**) facilitate handling equipment within the bag.

such as those that take place at room temperature, are easily carried out within the confines of a glove-bag.

9.7.3. Schlenk-Type Glassware

By far the most common techniques used by organometallic chemists for carrying out the complete range of manipulations under an inert atmosphere involve glassware referred to as Schlenk apparatus. This type of equipment provides a convenient method for handling the most air-sensitive compounds, and the necessary operations are simple to perform, requiring only a minimum of specialized apparatus. A glass manifold is usually used, consisting of a series of two-way taps that connect to an inert gas line (typically nitrogen or argon) and to a simple vacuum pump (see Figure 9.2). Even a water pump with a silica-gel-filled column between the pump and the manifold to prevent back-diffusion of water will suffice, although a small rotary oil pump is preferred. The inert gas line is vented to the atmosphere via a bubbler. There are several kinds of bubblers that prevent ingress of air into the manifold and maintain a slight positive pressure in the system. In practice, two Dreschel bottles in series, one containing paraffin oil, are quite satisfactory when a bubbler is not available. The reaction apparatus is connected to the manifold by thick-walled rubber or PVC tubing.

All reaction flasks require a side arm to allow connection to the manifold, and conventional two- or three-necked, round-bottomed flasks

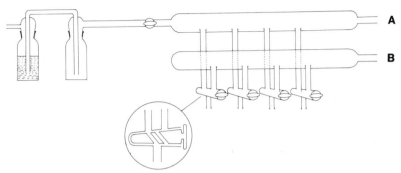

Figure 9.2. The two-way taps of a manifold connect an inert gas (**A**) or vacuum line (**B**) via tubing to a piece of equipment. A slight positive pressure of inert gas is maintained by a bubbler. The improvised bubbler illustrated consists of two Dreschel bottles, one of which contains paraffin oil.

can be used with a tap adapter in one of the necks. Alternatively, Schlenk flasks (now commercially available) can be used. All ground-glass joints must be lubricated with vacuum grease to obtain good seals and to prevent them from sticking under vacuum.

A typical reaction setup using only one connection to a manifold is illustrated in Figure 9.3. With this equipment one solution can be added

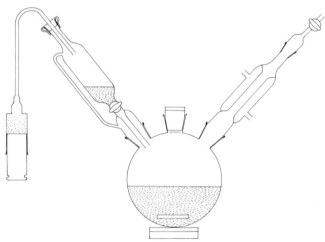

Figure 9.3. Typical equipment for conducting reaction under reflux in an inert atmosphere. Connection to the manifold via a tap at the top of the reflux condenser enables removal of air by applying vacuum and its replacement with inert gas. Liquid is transferred to the pressure-equalized dropping funnel through a self-sealing rubber septum from a long stainless steel hypodermic needle attached to a syringe.

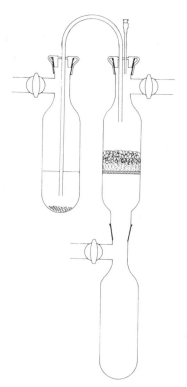

Figure 9.4. Filtration under inert atmosphere using Schlenk equipment. Application of a slight positive pressure of inert gas to the left Schlenk tube transfers liquid and solid through stainless steel or plastic tubing onto a bed of "Celite" filter-aid covered with a pad of glass wool.

to another under reflux in an inert atmosphere. The basic initial procedure is as follows:

(i) The apparatus is set-up and connected to the manifold.

(ii) The two-way tap is opened to vacuum and the apparatus evacuated.

(iii) When fully evacuated the two-way tap is opened carefully to the inert gas line, which contains a rapid flow of gas. *CAUTION*: If this is done too rapidly air may enter the manifold through the bubbler.

(iv) Operations (ii) and (iii) are repeated sequentially three or four times.

The apparatus, now under an inert atmosphere and maintained at a slight positive pressure via the manifold and oil level in the bubbler, permits removal of a stopper for addition of reagents while maintaining a steady flow of gas through the flask. Preweighed solids can be added to the reaction flask by using a transfer bend (see below), and liquids by using a syringe or stainless steel tube. Solvents should normally be dried over sodium or a molecular sieve, and degassed by bubbling the inert gas through them for half an hour or so (care must be taken with inflammable solvents). Air-sensitive liquids and solids must always be stored in suitable flasks under an inert atmosphere.

Transfers under an inert atmosphere are not difficult. As mentioned earlier, liquids are readily transferred using a syringe that has been preflushed with inert gas. A second method uses a length of flexible thin stainless steel tubing (2–3 mm diameter). This is passed through a rubber septum in the neck of a nitrogen-filled flask attached to the manifold. The stainless steel tube is flushed with nitrogen before passing its free end through a rubber septum in the neck of an empty flask, and in order to prevent buildup of pressure in the receiver a syringe needle is inserted through its septum cap. Inserting the tube into the liquid in the first flask and increasing the pressure of nitrogen in that flask causes liquid to flow through the tube into the empty flask. This technique is also very useful for filtering and washing a solid under an inert atmosphere see Figure 9.4. When only the filtrate is required, it is often expedient to allow the solid to settle before transferring the liquid and to filter it through a short bed of "Celite" (diatomaceous earth) supported on a plug of glass wool. Solids are readily transferred using a "transfer bend"—a bent short length of glass tube with ground joints at each end, which is first attached to the empty flask and whose open end is closed by a small cap. The air in the flask is then replaced by nitrogen as described above, and with a small flow of nitrogen the two flasks are connected. Solid can then be tipped through the bend from the first flask to the second.

With a little practice the techniques described briefly above, and more fully elsewhere,[33,34] are easily carried out and permit routine use of very oxygen-sensitive materials. Moreover, reactive gases can be handled equally well in this way—for instance, carbon monoxide for carbonylations, or hydrogen for hydrogenations at atmospheric pressure.

References

References for Chapter 1

1. G. H. Posner, *Organic Reactions* **19**, 1 (1972); **22**, 253 (1975).
2. J. F. Normant, *J. Organomet. Chem. Library* **1**, 219 (1976); *Synthesis*, 63 (1972).
3. A. E. Jukes, *Adv. Organomet. Chem.* **12**, 215 (1974).
4. H. O. House, *Acc. Chem. Res.* **9**, 59 (1976).
5. G. H. Posner, *An Introduction to Synthesis Using Organocopper Reagents*, John Wiley and Sons, New York (1980).
6. G. W. Parshall, *Homogeneous Catalysis*, J. Wiley and Sons, New York (1980).
7. A. Nakamura and M. Tsutsui, *Principles and Application of Homogeneous Catalysis*, J. Wiley and Sons, New York (1980).
8. G. W. Parshall, *J. Mol. Catal.* **4**, 243 (1978).
9. D. Forster, *Adv. Organomet. Chem.* **17**, 255 (1979).
10. R. Fowler, H. Connor, and R. A. Baehe, *Chemtech*, 772 (1976); *Hydrocarbon Process.* 247 (September 1979); *Chem. Eng.*, 110 (December 1977).
11. R. Jira, Manufacture of acetaldeyhyde directly from ethylene, in *Ethylene and Its Industrial Derivatives*, ed. S. A. Miller, Ernest Benn Ltd., London (1969).
12. E. S. Brown, Addition of hydrogen cyanide to olefins, in *Organic Syntheses via Metal Carbonyls*, Vol. 2, ed. I. Wender and P. Pino, Wiley–Interscience, New York (1977).
13. G. W. Parshall, *Homogeneous Catalysis*, John Wiley and Sons, New York (1980), Chap. 4, and references therein.
14. G. W. Parshall, *Homogeneous Catalysis*, John Wiley and Sons, New York (1980), Chap. 3.
15. G. W. Parshall, *Homogeneous Catalysis*, John Wiley and Sons, New York (1980), pp. 63–65.
16. A. L. Waddans, *Chemicals from Petroleum*, 4th ed., John Murray, London, England (1978). Chap. 13.
17. E. G. Hancock, ed., *Propylene and Its Industrial Derivatives*, Ernest Benn Ltd., London (1973).

References for Chapter 2

1. T. Alderson, E. L. Jenner, and R. V. Lindsey, Jr., *J. Am. Chem. Soc.* **87**, 5638 (1965).
2. C. G. P. Dixon, E. W. Duck, and D. K. Jenkins, *Organomet. Chem. Synth.* **1**, 77 (1970).

3. P. W. Jolly and G. Wilke, *The Organic Chemistry of Nickel*, Vols. 1 and 2, Academic Press, New York (1974, 1975).
4. R. Cramer, *J. Am. Chem. Soc.* **87**, 4717 (1965).
5. P. M. Maitlis, *The Organic Chemistry of Palladium*, Vol. 2, Academic Press, New York (1971).
6. S. M. Neumann and J. K. Kochi, *J. Org. Chem.* **40**, 599 (1975).
7. M. Kumada, in *Prospects in Organotransition Metal Chemistry*, ed. Y. Ishii and M. Tsutsui, Plenum Press, New York (1975).
8. K. Tamao, K. Sumitani, and M. Kumada, *J. Am. Chem. Soc.* **94**, 4374 (1972).
9. Y. Kiso, K. Tamao, N. Miyake, K. Yamamoto, and M. Kumada, *Tetrahedron Lett.*, 3 (1974).
10. M. Kumada, T. Hayashi, M. Tajika, and K. Tamao, *J. Am. Chem. Soc.* **98**, 3718 (1976).
11. T. Hayashi, Y. Katsuro, and M. Kumada, *Tetrahedron Lett.*, 3915 (1980).
12. (a) M. F. Semmelhack, *Org. React.* **19**, 115 (1972); (b) R. Baker, *Chem. Rev.* **73**, 487 (1973).
13. E. J. Corey, M. F. Semmelhack, and L. S. Hegedus, *J. Am. Chem. Soc.* **90**, 2416 (1968).
14. K. Sato, S. Inoue, S. Ota, and Y. Fujita, *J. Org. Chem.* **37**, 462 (1972).
15. E. Negishi and S. Baba, *J. Chem. Soc. Chem. Commun.*, 596 (1976).
16. J. Schwartz, M. J. Loots, and H. Kosugi, *J. Am. Chem. Soc.* **102**, 1333 (1980).
17. D. W. Hart, T. F. Blackburn, and J. Schwartz, *J. Am. Chem. Soc.* **97**, 679 (1975).
18. E. Negishi and D. E. Van Horn, *J. Am. Chem. Soc.* **99**, 3168 (1977).
19. (a) D. E. Van Horn and E. Negishi, *J. Am. Chem. Soc.* **100**, 2252 (1978); (b) N. Okukado and E. Negishi, *Tetrahedron Lett.*, 2357 (1978).
20. E. Negishi, N. Okukado, A. O. King, D. E. Van Horn, and B. I. Spiegel, *J. Am. Chem. Soc.* **100**, 2254 (1978).
21. H. Yatagai, Y. Yamamoto, and K. Maruyama, *J. Chem. Soc. Chem. Commun.*, 702 (1978).
22. Y. Yamamoto, H. Yatagai, A. Sonoda, and S. I. Murahashi, *J. Chem. Soc. Chem. Commun.*, 452 (1976).
23. S. B. Bowlus and J. A. Katzenellenbogen, *Tetrahedron Lett.*, 1277 (1973).
24. A. Alexakis, G. Cahiez, and J. F. Normant, *Synthesis*, 826 (1979).
25. P. Boontanonda and R. Grigg, *J. Chem. Soc. Chem. Commun.*, 583 (1977).
26. R. C. Larock and J. P. Burkhart, *Synth. Commun.* **9**, 659 (1979).
27. B. M. Trost and P. J. Metzner, *J. Am. Chem. Soc.* **102**, 3572 (1980).
28. B. M. Trost, *Acc. Chem. Res.* **13**, 385 (1980).
29. I. T. Harrison, E. Kimura, E. Bohme, and J. H. Fried, *Tetrahedron Lett.*, 1589, (1969).
30. B. M. Trost and T. J. Fullerton, *J. Am. Chem. Soc.* **95**, 292 (1973).
31. B. M. Trost and T. R. Verhoeven, *J. Am. Chem. Soc.* **100**, 3435 (1978).
32. B. M. Trost and T. R. Verhoeven, *J. Org. Chem.* **41**, 3215 (1976).
33. K. Takahashi, A. Miyaki, and G. Hata, *Bull. Chem. Soc. Jpn.* **45**, 230 (1972).
34. B. M. Trost and P. E. Strege, *J. Am. Chem. Soc.* **99**, 1649 (1977).
35. S. Takahashi, T. Shibano, and N. Hagihara, *Tetrahedron Lett.*, 2451 (1967).
36. H. Tom Dieck and A. Kinzel, *Angew. Chem. Int. Ed.* **18**, 324 (1979).
37. D. Medema and R. Van Helden, *Recl. Trav. Chim. Pays-Bas* **90**, 324 (1971).
38. (a) R. Baker, *Chem. Ind.*, 816 (1980); (b) J. Tsuji, *Adv. Organomet. Chem.* **17**, 141 (1979); (c) J. Tsuji, *Top. Curr. Chem.* **91**, 29 (1980).
39. E. J. Smutny, *J. Am. Chem. Soc.* **89**, 6793 (1967).
40. W. E. Walker, R. M. Manyik, K. E. Atkins, and M. L. Farmer, *Tetrahedron Lett.*, 3817 (1970).
41. K. Takahashi, A. Miyake, and G. Hata, *Bull. Chem. Soc. Jpn.* **45**, 1183 (1972).

42. T. Mitsuyasu and J. Tsuji, *Tetrahedron* **30**, 831 (1974).

43. H. P. Dang and G. Linstrumelle, *Tetrahedron Lett.*, 191 (1978).

44. I. D. Webb and G. T. Bocherdt, *J. Am. Chem. Soc.* **73**, 2654 (1951).

45. E. J. Corey, M. F. Semmelhack, and L. S. Hegedus, *J. Am. Chem. Soc.* **90**, 2416 (1968).

46. F. Guerrieri and G. P. Chiusoli, *Chim. Ind. (Milan)* **51**, 1252 (1969).

47. K. Sato, S. Inoue, S. Ota, and Y. Fujita, *J. Org. Chem.* **37**, 462 (1972).

48. G. P. Chiusoli, *Proc. Int. Congr. Pure. Appl. Chem.* **6**, 169 (1971).

49. M. F. Semmelhack, Ph.D. thesis, Harvard University (1967).

50. E. I. Negishi and S. Baba, *J. Am. Chem. Soc.* **98**, 6729 (1976).

51. N. Okukado, D. E. Van Horn, W. L. Klima, and E. Negishi, *Tetrahedron Lett.*, 1027 (1978).

52. J. Yoshida, K. Tamao, M. Takahashi, and M. Kumada, *Tetrahedron Lett.*, 2161 (1978).

53. K. Kaneda, T. Uchiyama, Y. Fujiwara, T. Imanaka, and S. Teranishi, *J. Org. Chem.* **44**, 55 (1979).

54. A. O. King, N. Okukado, and E. I. Negishi, *J. Chem. Soc. Chem. Commun.*, 683 (1977).

55. H. D. Hodes and K. M. Nicholas, *Tetrahedron Lett.*, 4349 (1978).

55a. S. Padmanabhan and K. M. Nicholas, *Synth. Commun.* **10**, 503 (1980).

56. A. O. King, E. Negishi, F. J. Villani, Jr., and A. Silveira, Jr., *J. Org. Chem.* **43**, 358 (1978).

57. G. Giacomelli and L. Lardicci, *Tetrahedron Lett.*, 2831 (1978).

58. J. Schwartz, D. B. Carr, R. T. Hansen, and F. M. Dayrit, *J. Org. Chem.* **45**, 3053 (1980).

59. D. W. Clack and W. Smith, *Inorg. Chim. Acta* **20**, 93 (1976).

60. G. Jaouen, in *Transition Metal Organometallics in Organic Synthesis*, Vol. 2, ed. H. Alper, Academic Press, New York (1978), p. 66.

61. C. A. L. Mahaffy and P. L. Pauson, *Inorg. Synth.* **19**, 154 (1979).

62. A. Ceccon and G. Catelani, *J. Organomet. Chem.* **72**, 179 (1974); A. Ceccon, *J. Organomet. Chem.* **72**, 189 (1974).

63. R. E. Davis, H. D. Simpson, N. Grice, and R. Pettit, *J. Am. Chem. Soc.* **93**, 6688 (1971).

64. A. Meyer and G. Jaouen, *J. Chem. Soc. Chem. Commun.*, 787 (1974).

65. G. Carganico, P. Del Buttero, S. Maiorana, and G. Riccardi, *J. Chem. Soc. Chem. Commun.*, 989 (1978).

66. C. A. L. Mahaffy and P. L. Pauson, *J. Chem. Res.* (S)126, (M)1752 (1979).

67. (a) J. F. Helling and D. M. Braitsch, *J. Am. Chem. Soc.* **92**, 7207 (1970); (b) A. N. Nesmeyanov, N. A. Vol'kenau, and I. N. Bolesova, *Proc. Acad. Sci. USSR* **175**, 661 (1967).

68. P. L. Pauson and J. A. Segal, *J. Chem. Soc. Dalton Trans.*, 1677 (1975).

69. M. F. Semmelhack and H. T. Hall, *J. Am. Chem. Soc.* **96**, 7091 (1974).

70. M. F. Semmelhack and H. T. Hall, *J. Am. Chem. Soc.* **96**, 7092 (1974).

71. M. F. Semmelhack, H. T. Hall, M. Yoshifuji, and G. Clark, *J. Am. Chem. Soc.* **97**, 1247 (1975).

72. M. D. Rausch, *J. Org. Chem.* **39**, 1787 (1974).

73. M. F. Semmelhack, *Ann. N.Y. Acad. Sci.*, 295 (1977).

74. M. F. Semmelhack, H. T. Hall, and M. Yoshifuji, *J. Am. Chem. Soc.* **98**, 6387 (1976).

75. A. J. Birch, P. E. Cross, J. Lewis, and D. A. White, *Chem. Ind. (London)*, 838 (1964).

76. A. J. Pearson, *Chem. Ind. (London)*, 741 (1982), and references therein.

77. G. Simonneaux and G. Jaouen, *Tetrahedron* **35**, 2249 (1979).

78. G. Jaouen and G. Simonneaux, *Inorg. Synth.* **19**, 197 (1979).

79. G. Jaouen and A. Meyer, *Tetrahedron Lett.*, 3547 (1976).

80. J. F. Fauvarque and A. Jutand, *J. Organomet. Chem.* **132**, C17 (1977).

81. A. Sekiya and N. Ishikawa, *J. Organomet. Chem.* **125**, 281 (1977).

82. E. Negishi, A. O. King, and N. Okukado, *J. Org. Chem.* **42**, 1821 (1977).

83. L. Cassar, *J. Organomet. Chem.* **54**, C57 (1971).
84. K. Tagaki, T. Okamoto, Y. Sakakibara, A. Ohno, S. Oka, and N. Hayama, *Bull. Chem. Soc. Jpn.* **49**, 3177 (1976).
85. H. A. Dieck and R. F. Heck, *J. Organomet. Chem.* **93**, 259 (1975).
86. K. Sonogashira, Y. Tohda, and N. Hagihara, *Tetrahedron Lett.*, 4467 (1975).
87. C. E. Castro, R. Havlin, V. K. Honwad, A. Malte, and S. Mojé, *J. Am. Chem. Soc.* **91**, 6464 (1969).
88. A. M. Malte and C. E. Castro, *J. Am. Chem. Soc.* **89**, 6770 (1967).
89. G. H. Posner, *Org. React.* **22**, 253 (1975).
90. G. M. Whitesides, W. F. Fischer, J. San Filippo, R. W. Bashe, and H. O. House, *J. Am. Chem. Soc.* **91**, 4871 (1969).
91. C. R. Johnson and G. A. Dutra, *J. Am. Chem. Soc.* **95**, 7777, 7783 (1973).
92. C. Jallabert, N. T. Luong Thi, and H. Riviere, *Bull. Soc. Chim. Fr.*, 797 (1970).
93. G. H. Posner, *Org. React.* **19**, 1 (1972).
94. S. Murahashi, Y. Tamba, M. Yamamura, and N. Yoshimura, *J. Org. Chem.* **43**, 4099 (1978).
95. (a) R. F. Heck, *Ann. N.Y. Acad. Sci.*, 201 (1977); (b) K. Mori, T. Mizoroki, and A. Ozaki, *Bull. Chem. Soc. Jpn.* **46**, 1505 (1973).
96. R. F. Heck and J. P. Nolley, Jr., *J. Org. Chem.* **37**, 2320 (1972).
97. H. A. Dieck and R. F. Heck, *J. Org. Chem.* **40**, 1083 (1975).
98. N. A. Cortese, C. B. Ziegler, B. J. Hrnjez, and R. F. Heck, *J. Org. Chem.* **43**, 2952 (1978).
99. B. A. Patel, C. B. Ziegler, N. A. Cortese, J. E. Plevyak, T. C. Zebovitz, M. Terpko, and R. F. Heck, *J. Org. Chem.* **42**, 3903 (1977).
100. C. B. Ziegler and R. F. Heck, *J. Org. Chem.* **43**, 2941 (1978).
101. (a) A. J. Chalk and S. A. Magennis, *J. Org. Chem.* **41**, 1206 (1976); (b) J. B. Melpolder and R. F. Heck, *J. Org. Chem.* **41**, 265 (1976).
102. K. Kikukawa and T. Matsuda, *Chem. Lett.* 159 (1977).
103. F. Akiyama, S. Teranishi, Y. Fujiwara, and H. Taniguchi, *J. Organomet. Chem.* **140**, C7 (1977).
104. Y. Fujiwara, I. Moritani, S. Danno, R. Asano, and S. Teranishi, *J. Am. Chem. Soc.* **91**, 7166 (1969).
105. A. J. Bingham, L. K. Dyall, R. O. C. Norman, and C. B. Thomas, *J. Chem. Soc. C*, 1879 (1970).
106. R. S. Shue, *J. Catal.* **26**, 112 (1972).
107. A. Sekiya and N. Ishikawa, *J. Organomet. Chem.* **125**, 281 (1977).
108. A. Sekiya and N. Ishikawa, *J. Organomet. Chem.* **118**, 349 (1976).
109. E. Negishi, A. O. King, and N. Okukado, *J. Org. Chem.* **42**, 1821 (1977).
110. H. Iataaki and H. Yoshimoto, *J. Org. Chem.* **38**, 76 (1973).
110a. R. Selke and W. Thiele, *J. Prakt. Chem.* **313**, 875 (1971).
111. M. F. Semmelhack, P. M. Helmquist, and L. D. Jones, *J. Am. Chem. Soc.* **93**, 5908 (1971).
112. M. F. Semmelhack and L. S. Ryono, *J. Am. Chem. Soc.* **97**, 3873 (1975).
113. P. E. Fanta, *Chem. Rev.* **64**, 613 (1964); *Synthesis*, 9 (1974).
114. A. Cairncross and W. A. Sheppard, *J. Am. Chem. Soc.* **93**, 247 (1971).
115. H. O. House, D. G. Koepsell, and W. J. Campbell, *J. Org. Chem.* **37**, 1003 (1972).
116. T. Kauffmann, *Angew. Chem. Int. Ed. Eng.* **13**, 291 (1974).
117. G. H. Posner, *Org. React.* **22**, 253 (1975).
118. J. F. Normant, *Synthesis*, 63 (1972).
119. M. S. Karasch and P. O. Tawney, *J. Am. Chem. Soc.* **63**, 2308 (1941).
120. H. O. House, W. L. Respess, and G. M. Whitesides, *J. Org. Chem.* **31**, 3128 (1966).
121. W. E. Parham and L. J. Czuba, *J. Org. Chem.* **34**, 1899 (1969).

122. (a) G. Van Koten and J. G. Noltes, *J. Chem. Soc. Chem. Commun.*, 940 (1972); *J. Organomet. Chem.* **84**, 129 (1973); (b) J. A. J. Jarvis, B. T. Kilbourn, R. Pearce, and M. F. Lappert, *J. Chem. Soc. Chem. Commun.*, 475 (1973).

123. H. O. House, *Acc. Chem. Res.* **9**, 59 (1976).

124. H. O. House and M. J. Uman, *J. Org. Chem.* **38**, 3893 (1973).

125. H. O. House and W. F. Fischer, *J. Org. Chem.* **33**, 949 (1968).

126. J. B. Siddall, M. Biskup, and J. H. Fried, *J. Am. Chem. Soc.* **91**, 1853 (1969).

127. H. O. House and W. F. Fischer, *J. Org. Chem.* **34**, 3615 (1969).

128. D. B. Ledlie and G. Miller, *J. Org. Chem.* **44**, 1006 (1979).

129. G. H. Posner, J. J. Sterling, C. E. Whitten, C. M. Lentz, and D. J. Brunelle, *J. Am. Chem. Soc.* **97**, 107 (1975).

130. F. Näf, R. Decorzant, and W. Thommen, *Helv. Chim. Acta* **58**, 1808 (1975).

131. R. K. Boeckman and K. J. Bruza, *J. Org. Chem.* **44**, 4781 (1979).

132. Y. Yamamoto and K. Maruyama, *J. Am. Chem. Soc.* **100**, 3240 (1978).

133. Y. Yamamoto, H. Yatagai, and K. Maruyama, *J. Org. Chem.* **44**, 1744 (1979).

134. R. Noyori, in *Transition Metal Organometallics in Organic Synthesis*, Vol. 1, ed. H. Alper, Academic Press, New York (1976), p. 84.

135. V. Schneider and P. K. Frohlich, *Ind. Eng. Chem.* **23**, 1405 (1931).

136. T. J. Katz, *Adv. Organomet. Chem.* **16**, 283 (1977); J. J. Rooney and A. Stewart, in *Catalysis*, Vol. 1, Specialist Periodical Reports, The Chemical Society, London (1977); N. Calderon, E. A. Ofstea, and W. A. Judy, *Angew. Chem. Int. Ed.* **15**, 401 (1976); R. J. Haines and G. J. Leigh, *Chem. Soc. Rev.* **4**, 155 (1974); R. H. Grubbs, in *New Applications of Organometallic Reagents in Organic Synthesis*, ed. D. Seyferth, Elsevier, New York (1976); W. B. Hughes, *Ann. N.Y. Acad. Sci.* **295**, 271 (1977); R. Streck, *Chem. Z.* **99**, 397 (1975).

137. K. Maruyama, K. Terada, and Y. Yamamoto, *J. Org. Chem.* **45**, 737 (1980).

138. F. W. Kupper and R. Streck, *Z. Naturforsch* **31b**, 1256 (1976).

139. F. W. Kupper and R. Streck, *Chem. Z.* **99**, 646 (1975).

140. F. W. Kupper and R. Streck, German Pat., 2,512,741 [*Chem. Abs.* **85**, 176806m (1976)]; German Pat., 2,531,959 [*Chem. Abs.* **87**, 22390a (1977)]; German Pat., 2,533,247 [*Chem. Abs.* **86**, 155187r (1977)].

141. R. Nakamura and E. Echigoya, *Chem. Lett.* 1227 (1977).

142. R. Baker and M. J. Crimmin, *Tetrahedron Lett.*, 441 (1977); E. Verkuijlen, F. Kapteijn, J. C. Mol, and C. Boelhouwer, *J. Chem. Soc. Chem. Commun.*, 198 (1977).

143. P. Chevalier, D. Sinou, G. Descotes, R. Mutin, and J. M. Basset, *J. Organomet. Chem.* **113**, 1 (1976).

144. J. P. Laval, A. Lattes, R. Mutin, and J. M. Basset, *J. Chem. Soc. Chem. Commun.*, 502 (1977).

145. W. Ast, G. Rheinwald, and R. Kerber, *Makromol. Chem.* **177**, 1341, 1349 (1976).

146. L. G. Wideman, *J. Org. Chem.* **33**, 4541 (1968).

147. G. A. Olah and G. K. S. Prakash, *Synthesis*, 607 (1976).

148. T. Mukaiyama, T. Sato, and J. Hanna, *Chem. Lett.* 1041 (1973).

149. J. E. McMurry and M. P. Fleming, *J. Am. Chem. Soc.* **96**, 4708 (1974); J. E. McMurry and D. D. Miller, *J. Am. Chem. Soc.* **105**, 1660 (1983).

150. L. Castedo, J. M. Saa, R. Suau, and G. Tojo, *J. Org. Chem.* **46**, 4292 (1981).

151. J. E. McMurry and M. P. Fleming, *J. Org. Chem.* **41**, 896 (1976).

152. J. E. McMurry, M. P. Fleming, K. L. Kees, and L. R. Krepski, *J. Org. Chem.* **43**, 3255 (1978).

153. J. E. McMurry and K. L. Kees, *J. Org. Chem.* **42**, 2655 (1977).

154. J. E. McMurry and L. R. Krepski, *J. Org. Chem.* **41**, 3929 (1976).

155. S. Nishida and F. Kataoka, *J. Org. Chem.* **43**, 1612 (1978).

156. K. B. Sharpless, *J. Chem. Soc. Chem. Commun.*, 1450 (1970).

157. F. Bertini, P. Grasselli, G. Zubiani, and G. Cainelli, *J. Chem. Soc. Chem. Commun.*, 144 (1970).

158. T. Fujisawa, K. Sugimoto, and H. Ohta, *Chem. Lett.*, 883 (1974).

159. J. K. Kochi, D. M. Singleton, and L. T. Andrews, *Tetrahedron* **24**, 3503 (1968).

160. J. E. McMurry and M. P. Fleming, *J. Org. Chem.* **40**, 2555 (1975).

161. J. E. McMurry, M. G. Silvestri, M. P. Fleming, T. Hoz, and M. W. Grayston, *J. Org. Chem.* **43**, 3249 (1978).

162. W. P. Giering, M. Rosenblum, and J. Tancrede, *J. Am. Chem. Soc.* **94**, 7170 (1972).

163. M. Rosenblum, M. R. Saidi, and M. Madhavarao, *Tetrahedron Lett.*, 4009 (1975).

164. P. Dowd and K. Kang, *J. Chem. Soc. Chem. Commun.*, 384 (1974).

165. T. Imamoto and Y. Yukawa, *Chem. Lett.*, 165 (1974).

166. J. E. McMurry and T. Hoz, *J. Org. Chem.* **40**, 3797 (1975).

References for Chapter 3

1. P. J. Garratt, in *Comprehensive Organic Chemistry*, Vol. 1, ed. J. F. Stoddart, Pergamon Press, Oxford (1979), p. 361; also T. Clark and M. A. McKervey, p. 96; G. H. Whitham, pp. 141, 158.

2. J. P. Schaefer and J. J. Bloomfield, *Org. React.* **15**, 1 (1967).

3. L. Ruzicka, M. Stoll, and H. Schinz, *Helv. Chim. Acta.* **9**, 249 (1926); L. Ruzicka, W. Brugger, M. Pfeiffer, H. Schinz, and M. Stoll, *Helv. Chim. Acta* **9**, 499 (1926).

4. J. J. Bloomfield, D. C. Owsley, and J. M. Nelke, *Org. React.* **23**, 259 (1976).

5. K. Stockel and F. Sondheimer, *Org. Synth.* **54**, 1 (1974).

6. H. Buchholz, P. Heimbach, H. Hey, H. Selbeck, and W. Wiese, *Coord. Chem. Rev.* **8**, 129 (1972).

7. F. J. Brown, *Prog. Inorg. Chem.* **27**, 1 (1980).

8. D. J. Cardin, B. Cetinkaya, M. J. Doyle, and M. F. Lappert, *Chem. Soc. Rev.* **2**, 99 (1973).

9. H. E. Simmons, T. L. Cairns, S. A. Vladuchick, and C. M. Hoiness, *Org. React.* **20**, 1 (1973).

10. N. Kawabata, M. Naka, and S. Yamashita, *J. Am. Chem. Soc.* **98**, 2676 (1976).

11. V. Dave and E. W. Warnhoff, *Org. React.* **18**, 217 (1970).

12. D. S. Wulfman *et al.*, *Tetrahedron* **32**, 1231, 1241, 1251, 1257 (1976).

13. U. Mende, R. Raduchel, W. Skuballa, and H. Vorbruggen, *Tetrahedron Lett.*, 629, (1975).

14. M. P. Doyle and J. G. Davidson, *J. Org. Chem.* **45**, 1538 (1980).

15. A. J. Hubert, A. F. Noels, A. J. Anciaux, and Ph. Teyssie, *Synthesis*, 600 (1976).

16. D. Arlt, M. Jautelat, and R. Lantzsch, *Angew. Chem. Int. Ed. Engl.* **20**, 703 (1981).

17. T. Aratani, Y. Yoneyoshi, and T. Nagase, *Tetrahedron Lett.*, 2599, (1977); 1707, (1975).

18. A. Nakamura, *Pure Appl. Chem.* **50**, 37 (1978).

19. A. Nakamura, A. Konishi, Y. Tatsuno, and S. Otsuka, *J. Am. Chem. Soc.* **100**, 3443, 3449 (1978).

20. A. Nakamura, T. Koyama, and S. Otsuka, *Bull. Chem. Soc. Jpn.* **51**, 593 (1978).

21. N. Petiniot, A. J. Anciaux, A. F. Noels, A. J. Hubert, and Ph. Teyssie, *Tetrahedron Lett.*, 1239 (1978).

22. H. Sakuria, T. Imai, and A. Hosomi, *Tetrahedron Lett.*, 4045 (1977).

23. A. L. Baumstark, C. J. McCloskey, T. J. Tolson, and G. T. Syriopoulos, *Tetrahedron Lett.*, 3003 (1977).
24. A. Bury, M. D. Johnson, and M. J. Stewart, *J. Chem. Soc. Chem. Commun.*, 622 (1980).
25. A. Efraty, *Chem. Rev.* **77**, 691 (1977).
26. R. B. Woodward and R. Hoffmann, *Angew. Chem. Int. Ed.* **8**, 781 (1969).
27. P. Heimbach and W. Brenner, *Angew. Chem. Int. Ed.* **6**, 800 (1967).
28. W. Brenner, P. Heimbach, H. Hey, E. W. Muller, and G. Wilke, *Annalen*, **727**, 161 (1969).
29. P. Heimbach, *Angew. Chem. Int. Ed.* **12**, 975 (1973).
30. W. E. Billups, J. H. Cross, and C. V. Smith, *J. Am. Chem. Soc.* **95**, 3438 (1973).
31. E. G. Chepaikin and M. L. Khidekel, *Bull. Acad. Sci. USSR* **20**, 1052 (1971).
32. L. G. Cannell, *J. Am. Chem. Soc.* **94**, 6867 (1972).
33. G. N. Schrauzer, *Adv. Catal.* **18**, 373 (1968).
34. A. Greco, A. Carbonaro, and G. Dall 'Asta, *J. Org. Chem.* **35**, 271 (1970).
35. J. Kiji, S. Yoshikawa, E. Sasakawa, S. Nishimura, and J. Furukawa, *J. Organomet. Chem.* **80**, 267 (1974).
36. R. Noyori, T. Ishigami, N. Hayashi, and H. Takaya, *J. Am. Chem. Soc.* **95**, 1674 (1973).
37. P. Binger, *Angew. Chem. Int. Ed.* **11**, 309 (1972).
38. M. Catellani, E. Dradi, G. P. Chiusoli, and G. Salerno, *J. Organomet. Chem.* **177**, C29 (1979).
39. R. G. Salomon and A. Sinha, *Tetrahedron Lett.*, 1367 (1978).
40. R. G. Salomon, K. Folting, W. E. Streib, and J. K. Kochi, *J. Am. Chem. Soc.* **96**, 1145 (1974).
41. K. Mizuno, J. Ogawa, M. Kamura, and Y. Otsuji, *Chem. Lett.*, 731 (1979).
42. F. J. Weigert, R. L. Baird, and J. R. Shapley, *J. Am. Chem. Soc.* **92**, 6630 (1970).
43. P. Binger, G. Schroth, and J. McMeeking, *Angew. Chem. Int. Ed.* **13**, 465 (1974).
44. P. Binger and M. J. Doyle, *J. Organomet. Chem.* **162**, 195 (1978).
45. F. W. Hoover and R. V. Lindsey, *J. Org. Chem.* **34**, 3051 (1969).
46. D. R. Coulson, *J. Org. Chem.* **37**, 1253 (1972).
47. T. Mitsudo, K. Kokuryo, and Y. Takegami, *J. Chem. Soc. Chem. Commun.*, 722 (1976).
48. T. Mitsudo, K. Kokuryo, T. Shinsugi, Y. Nakagawa, Y. Watanabe, and Y. Takegami, *J. Org. Chem.* **44**, 4492 (1979).
49. Y. Yamamoto and H. Yamazaki, *J. Organomet. Chem.* **137**, C31 (1977).
50. K. P. C. Vollhardt, *Acc. Chem. Res.* **10**, 1 (1977).
51. W. G. L. Aalbersberg, R. L. Funk, A. J. Barkovich, R. L. Millard, and K. P. C. Vollhardt, *J. Am. Chem. Soc.* **97**, 5600 (1975).
52. R. L. Funk and K. P. C. Vollhardt, *J. Chem. Soc. Chem. Commun.*, 833 (1976).
53. R. L. Hillard and K. P. C. Vollhardt, *J. Am. Chem. Soc.* **99**, 4058 (1977).
54. T. Kiji, K. Masui, and J. Furukawa, *Bull. Chem. Soc. Japan* **44**, 1956 (1971).
55. H. Felkin, L. D. Kwart, G. Swierczewski, and J. D. Umpleby, *J. Chem. Soc. Chem. Commun.*, 242 (1975).
56. H. Minematsu, S. Takahashi, and N. Hagihara, *J. Chem. Soc. Chem. Commun.*, 466 (1975).
57. M. Rosenblum, *Acc. Chem. Res.* **7**, 122 (1974); A. Cutter, D. Ehnholt, W. P. Giering, P. Lennon, S. Raghu, A. Rosan, M. Rosenblum, J. Tancrede, and D. Wells, *J. Am. Chem. Soc.* **98**, 3495 (1976).
58. T. S. Abram and R. Baker, *J. Chem. Soc. Chem. Commun.*, 267 (1979).
59. J. P. Williams and A. Wojcicki, *Inorg. Chem.* **16**, 2506 (1977).
60. R. Noyori, *Acc. Chem. Res.* **12**, 61 (1979).
61. Y. Hayakawa, K. Yokoyama, and R. Noyori, *J. Am. Chem. Soc.* **100**, 1791 (1978).

62. Y. Hayakawa, K. Yokoyama, and R. Noyori, *J. Am. Chem. Soc.* **100**, 1799 (1978).
63. Y. Hayakawa, F. Shimizu, and R. Noyori, *Tetrahedron Lett.*, 993 (1978).
64. B. M. Trost and D. M. T. Chan, *J. Am. Chem. Soc.* **101**, 6429 (1979).
65. H. Takaya, N. Hayashi, T. Ishigami, and R. Noyori, *Chem. Lett.*, 813 (1973).
66. P. Binger, *Synthesis*, 427 (1973).
67. R. Noyori, Y. Kumagai, I. Umeda, and H. Takaya, *J. Am. Chem. Soc.* **94**, 4018 (1972).
68. R. Noyori, T. Odagi, and H. Takaya, *J. Am. Chem. Soc.* **92**, 5780 (1970).
69. A. Baba, Y. Ohshiro, and T. Agawa, *J. Organomet. Chem.* **110**, 121 (1976).
70. R. Noyori, I. Umeda, H. Kawauchi, and H. Takaya, *J. Am. Chem. Soc.* **97**, 812 (1975).
71. J. E. Lyons, H. K. Myers, and A. Schneider, *J. Chem. Soc. Chem. Commun.*, 636 (1978).
72. I. U. Khaud and P. L. Pauson, *J. Chem. Res. (S)*, 8, (1977); 9, (1977).
73. W. Best, B. Fell, and G. Schmitt, *Chem. Ber.* **109**, 2914 (1976).
74. B. F. G. Johnson, J. Lewis, and D. J. Thompson, *Tetrahedron Lett.*, 3789 (1974).
75. H. Yamazaki, K. Aoki, Y. Yamamoto, and Y. Wakatsuki, *J. Am. Chem. Soc.* **97**, 3546 (1975).
76. N. Adachi, K. Kikukawa, M. Takagi, and T. Matsuda, *Bull. Chem. Soc. Jpn.* **48**, 521 (1975).
77. S. Takahashi, Y. Suzuki, K. Sonogashira, and N. Hagihara, *J. Chem. Soc. Chem. Commun.*, 839 (1976).
78. J. Y. Merour, J. L. Roustan, C. Charrier, and J. Collin, *J. Organomet. Chem.* **51**, C24 (1973); A. Guinot, P. Cadiot, and J. L. Roustan, *J. Organomet. Chem.* **128**, C35 (1977).
79. J. K. Crandall and W. J. Michaely, *J. Organomet. Chem.* **51**, 375 (1973).
80. J. K. Crandall, P. Battioni, R. Bindra, and J. T. Wehlacz, *J. Am. Chem. Soc.* **97**, 7171 (1975).
81. W. C. Kossa, T. C. Rees, and H. C. Richey, *Tetrahedron Lett.*, 3455 (1971).
82. A. Z. Rubezhov, *Tetrahedron Lett.*, 2189 (1977).
83. W. Muzenmaier and H. Straub, *Annalen*, 313 (1977).
84. E. J. Corey, R. L. Danheiser, and S. Chandrasekaran, *J. Org. Chem.* **41**, 260 (1976).
85. W. Munzenmaier and H. Straub, *Synthesis*, 49 (1976).
86. R. Grigg, T. R. B. Mitchell, and A. Ramasubba, *J. Chem. Soc. Chem. Commun.*, 27 (1980).
87. R. Grigg, T. R. B. Mitchell, and A. Ramasubba, *J. Chem. Soc. Chem. Commun.*, 669 (1979).
88. Y. Ito, H. Aoyama, T. Hirao, A. Mochizuki, and T. Saegusa, *J. Am. Chem. Soc.* **101**, 494 (1979).
89. J. Furukawa, J. Kiji, T. Tojo, and K. Yamamoto, *Tetrahedron* **29**, 3149 (1973).
90. B. M. Trost, T. A. Runge, and L. N. Jungheim, *J. Am. Chem. Soc.* **102**, 2840 (1980).
91. Y. Ito, K. Nakayama, K. Yonezawa, and T. Saegusa, *J. Org. Chem.* **39**, 3273 (1974).
92. (a) C. W. Bird, *Transition Metal Intermediates in Organic Synthesis*, Logos Press, London (1967); (b) P. N. Rylander, *Organic Synthesis with Nobel Metal Catalysts* Academic Press, New York (1973); (c) P. M. Maitlis, *Pure Appl. Chem.* **33**, 489 (1973); (d) F. L. Bowden and A. B. P. Lever, *Organomet. Chem. Rev. Sect. A*, **3**, 227 (1968); (e) W. Reppe, N. Kutepow, and A. Magin, *Agew. Chem. Int. Ed.* **8**, 727 (1969).
93. L. S. Meriwether, E. C. Colthup, M. F. Leto, and G. W. Kennerly, *J. Org. Chem.* **27**, 3930 (1962).
94. K. Moseley and P. M. Maitlis, *J. Chem. Soc. Chem. Commun.*, 1604 (1971).
95. R. L. Hillard and K. P. C. Vollhardt, *Angew. Chem. Int. Ed.* **14**, 712 (1975).
96. R. L. Funk and K. P. C. Vollhardt, *J. Am. Chem. Soc.* **102**, 5245 (1980).
97. R. L. Funk and K. P. C. Vollhardt, *J. Am. Chem. Soc.* **98**, 6755 (1976).
98. R. L. Funk and K. P. C. Vollhardt, *J. Am. Chem. Soc.* **102**, 5253 (1980).

99. D. M. Singleton, *Tetrahedron Lett.*, 1245 (1973).
100. H. Suzuki, K. Itoh, Y. Ishii, K. Simon, and J. A. Ibers, *J. Am. Chem. Soc.* **98**, 8494 (1976).
101. A. J. Chalk, *J. Am. Chem. Soc.* **94**, 5928 (1972).
102. Y. Wakatsuki, T. Kurumitsu, and H. Yamazaki, *Tetrahedron Lett.*, 4549 (1974).
103. J. Hambrecht and E. Muller, *Annalen*, 387 (1977); J. Hambrecht and E. Muller, *Z. Naturforsch.* **32B**, 68 (1977); E. Muller and G. Odenigbo, *Annalen*, 1435 (1975).
104. P. Binger and A. Brinkmann, *Chem. Ber.* **111**, 2689 (1978).
105. H. A. Dieck and R. F. Heck, *J. Org. Chem.* **40**, 1083 (1975).
106. R. E. Benson and R. V. Lindsey, *J. Am. Chem. Soc.* **81**, 4247 (1959).
107. R. E. Benson and R. V. Lindsey, *J. Am. Chem. Soc.* **81**, 4250 (1959).
108. P. J. Garratt and M. Wyatt, *J. Chem. Soc. Chem. Commun.*, 251 (1974).
109. P. W. Jolly and G. Wilke, *The Organic Chemistry of Nickel*, Vol. 2, Academic Press, New York (1975).
110. H. Buchholz, P. Heimbach, H. Hey, H. Selbeck, and Wiese, *Coord. Chem. Rev.* **8**, 129 (1972).
111. H. Siegel, H. Hopf, A. Germer, and P. Binger, *Chem. Ber.* **111**, 3112 (1978).
112. W. Brenner, P. Heimbach, K. Ploner, and F. Thomel, *Annalen*, 1882 (1973).
113. D. R. Fahey, *J. Org. Chem.* **37**, 4471 (1972).
114. P. Heimbach, K. Ploner, and F. Thomel, *Angew. Chem. Int. Ed.* **10**, 276 (1971).
115. A. Carbonaro, A. Greco, and G. Dall'Asta, *J. Org. Chem.* **33**, 3948 (1968).
116. J. P. Genet and J. Ficini, *Tetrahedron Lett.*, 1499 (1979).
117. J. P. Schaefer and J. J. Bloomfield, *Org. React.* **15**, 1 (1967).
118. J. J. Bloomfield, D. C. Owsley, and J. M. Nelke, *Org. React.* **23**, 259 (1976).
119. K. Sakai and O. Oda, *Tetrahedron Lett.*, 4375 (1972).
120. A. J. Birch and A. J. Pearson, *J. Chem. Soc. Chem. Commun.*, 601 (1976).
121. R. Noyori, M. Nishizawa, F. Shimizu, Y. Hagakawa, K. Maruoka, S. Hashimoto, H. Yamamoto, and H. Nozoki, *J. Am. Chem. Soc.* **101**, 220 (1979).
122. H. Takaya, S. Makino, Y. Hagakawa, and R. Noyori, *J. Am. Chem. Soc.* **100**, 1765 (1978).
123. H. Takaya, Y. Hayakawa, S. Makino, and R. Noyori, *J. Am. Chem. Soc.* **100**, 1778 (1978); Y. Hagakawa, Y. Baba, S. Makino, and R. Noyori, *J. Am. Chem. Soc.* **100**, 1786 (1979).
124. R. Baker and A. H. Copeland, *Tetrahedron Lett.*, 4535 (1976).
125. (a) G. N. Schrauzer, *Adv. Organomet. Chem.* **2**, 1 (1964); (b) P. Heimbach, P. W. Jolly, and G. Wilke, *Adv. Organomet. Chem.* **8**, 29 (1970); (c) R. Baker, *Chem. Rev.* **73**, 487 (1973); (d) M. F. Semmelhack, *Org. React.* **19**, 115 (1972).
126. H. D. Martin, M. Hekman, G. Rist, H. Sauter, and D. Bellus, *Angew. Chem. Int. Ed.* **16**, 406 (1977).
127. P. Chini, N. Palladino, and A. Santambrogio, *J. Chem. Soc. C*, 836 (1967).
128. J. R. Leto and M. F. Leto, *J. Am. Chem. Soc.* **83**, 2944 (1961).
129. A. C. Cope and D. S. Smith, *J. Am. Chem. Soc.* **74**, 5136 (1952); A. C. Cope and H. C. Campbell, **73**, 3536 (1951); A. C. Cope and D. F. Rugen, **75**, 3215 (1953).
130. F. Wagner and H. Meier, *Tetrahedron* **30**, 773 (1974).
131. B. Bogdanovic, P. Heimbach, M. Kroner, and G. Wilke, *Annalen*, **727**, 143 (1969).
132. A. Miyake, H. Kondo, and M. Nishino, *Angew. Chem. Int. Ed.* **10**, 802 (1971).
133. P. Heimbach and G. Wilke, *Annalen*, **727**, 183 (1969).
134. P. Heimbach, R. V. Meyer, and G. Wilke, *Annalen*, 743 (1975).
135. G. Wilke and P. Heimbach, Ger. Patent 1,493,221 (1969); see *Chem. Abst.* **67**, 73242 (1957).
136. P. Heimbach, H. Selbeck, and E. Troxler, *Angew. Chem. Int. Ed.* **10**, 659 (1971).

137. W. Brenner and P. Heimbach, *Annalen*, 660 (1975).
138. K. J. Ploner and P. Heimbach, *Annalen*, 54 (1976).
139. W. Brenner, P. Heimbach, K. J. Ploner, and F. Thomel, *Angew. Chem. Int. Ed.* **8**, 753 (1969).
140. H. Breil and G. Wilke, *Angew. Chem. Int. Ed.* **9**, 367 (1970).
141. R. Baker, P. Bevan, and R. C. Cookson, *J. Chem. Soc. Chem. Commun.*, 752, (1975); see also R. Baker, P. C. Bevan, R. C. Cookson, A. H. Copeland, and A. D. Gribble, *J. Chem. Soc. Perkin Trans. 1,* 480 (1978).
142. R. Baker, B. N. Blackett, and R. C. Cookson, *J. Chem. Soc. Chem. Commun.*, 802 (1972); R. Baker, R. C. Cookson, and J. R. Vinson, *J. Chem. Soc. Chem. Commun.*, 515 (1974).
143. R. Baker and M. G. Kelly, *J. Chem. Soc. Chem. Commun.*, 307, (1980).
144. S. Otsuka, A. Nakamura, K. Teni, and S. Ueda, *Tetrahedron Lett.*, 297 (1969).
145. F. N. Jones and R. V. Lindsey, *J. Org. Chem.* **33**, 3838 (1968).
146. L. T. Scott and G. J. DeCicco, *Tetrahedron Lett.*, 2663 (1976).
147. E. Wasserman, D. A. BenEfraim, and R. Wolovsky, *J. Am. Chem. Soc.* **90**, 3286 (1968).
148. P. Heimbach and W. Brenner, *Angew. Chem. Int. Ed.* **5**, 961 (1966).
149. E. J. Corey and E. K. W. Wat, *J. Am. Chem. Soc.* **89**, 2757 (1967).
150. E. J. Corey and E. Hamanaka, *J. Am. Chem. Soc.* **89**, 2758 (1967).
151. W. G. Dauben, G. H. Beasley, M. D. Broadhurst, B. Muller, D. J. Peppard, P. Pesnelle, and C. Suter, *J. Am. Chem. Soc.* **97**, 4973 (1975).
152. E. J. Corey and P. Helquist, *Tetrahedron Lett.*, 4091 (1975).
153. E. J. Corey and H. A. Kirst, *J. Am. Chem. Soc.* **94**, 667 (1972).
154. E. J. Corey and M. F. Semmelhack, *Tetrahedron Lett.*, 6237 (1966).
155. M. F. Semmelhack and L. S. Ryono, *J. Am. Chem. Soc.* **97**, 3873 (1975).
156. J. E. McMurry and K. L. Kees, *J. Org. Chem.* **42**, 2655 (1977).
157. S. Kulkowit and M. A. McKervey, *J. Chem. Soc. Chem. Commun.*, 1069 (1978).

References for Chapter 4

1. C. W. Bird, *J. Organomet. Chem.* **47**, 281 (1973); J. L. Davidson and P. H. Preston, *Adv. Heterocyclic Chem.*, **30**, 321 (1982).
2. H. Bonnemann, *Angew. Chem. Int. Ed.* **17**, 505 (1978).
3. J. E. Backvall, *J. Chem. Soc. Chem. Commun.*, 413 (1977).
4. M. Mori, K. Chiba, M. Okita, and Y. Ban, *J. Chem. Soc. Chem. Commun.*, 698 (1979).
5. I. Ojima, S. Inaba, and Y. Yoshida, *Tetrahedron Lett.*, 3643 (1977).
6. P. K. Wong, M. Madhavarao, D. F. Marten, and M. Rosenblum, *J. Am. Chem. Soc.* **99**, 2823 (1977).
7. B. M. Trost and J. P. Genet, *J. Am. Chem. Soc.* **98**, 8516 (1976).
8. T. Saegusa, Y. Ito, H. Kinoshita, and S. Tomita, *J. Org. Chem.* **36**, 3316 (1971).
9. A. Claesson, C. Sahlberg, and K. Luthman, *Acta. Chem. Scand. Ser. B* **33**, 309 (1979).
10. Y. Watanabe, S. C. Shim, H. Uchida, T. Mitsudo, and Y. Takegami, *Tetrahedron* **35**, 1433 (1979).
11. M. O. Terpko and R. F. Heck, *J. Am. Chem. Soc.* **101**, 5281 (1979).
12. (a) M. Mori and Y. Ban, *Tetrahedron Lett.*, 1807 (1976); (b) M. Mori and Y. Ban, *Tetrahedron Lett.*, 1133 (1979).
13. M. Mori, K. Chiba, and Y. Ban, *J. Org. Chem.* **43**, 1684 (1978).
14. C. W. Bird, *J. Organomet. Chem.* **47**, 296 (1973).
15. M. Jautelat and K. Ley, *Synthesis*, 593 (1970).

16. B. M. Trost and E. Kienan, *J. Org. Chem.* **45**, 2741 (1980).
17. S. Murahashi, T. Shimamura, and I. Moritani, *J. Chem. Soc. Chem. Commun.*, 931 (1974).
18. P. F. dos Santos Filho and U. Schuchardt, *Angew. Chem. Int. Ed.* **16**, 647 (1977).
19. (a) H. Alper and J. E. Prickett, *Tetrahedron Lett.*, 2589 (1976); (b) H. Alper and J. E. Prickett, *J. Chem. Soc. Chem. Commun.*, 483 (1976).
20. K. Isomura, K. Uto, and H. Taniguchi, *J. Chem. Soc. Chem. Commun.*, 664 (1977).
21. (a) L. S. Hegedus, G. F. Allen, J. J. Bozell, and E. L. Waterman, *J. Am. Chem. Soc.* **100**, 5800 (1978); (b) L. S. Hegedus, G. F. Allen, and D. S. Olsen, *J. Am. Chem. Soc.* **102**, 3583 (1980).
22. (a) M. Mori and Y. Ban, *Tetrahedron Lett.*, 1803 (1976); (b) M. Mori, S. Kudo, and Y. Ban, *J. Chem. Soc. Perkin Trans. 1*, 771 (1979).
23. R. Odle, B. Blevins, M. Ratcliff, and L. S. Hegedus, *J. Org. Chem.* **45**, 2709 (1980).
24. Y. Watanabe, S. C. Shim, T. Mitsudo, M. Yamashita, and Y. Takegami, *Bull. Chem. Soc. Jpn.* **49**, 2302 (1976).
25. G. R. Wiger and M. F. Rettig, *J. Am. Chem. Soc.* **98**, 4168 (1976).
26. J. Kiji, K. Yamamoto, H. Tomita, and J. Furukawa, *J. Chem. Soc. Chem. Commun.*, 506 (1974).
27. K. Ohno and J. Tsuji, *J. Chem. Soc. Chem. Commun.*, 247 (1971).
28. J. Falbe and F. Korte, *Chem. Ber.* **95**, 2680 (1962).
29. P. Hong and H. Yamazaki, *Synthesis*, 50 (1977).
30. A. Kasahara and T. Saito, *Chem. Ind.*, 745 (1975).
31. A. Kasahara, T. Izumi, and O. Saito, *Chem. Ind.*, 666 (1980).
32. N. A. Cortese, C. B. Ziegler, B. J. Hrnjez, and R. F. Heck, *J. Org. Chem.* **43**, 2952 (1978).
33. Y. Wakatsuki and H. Yamazaki, *Synthesis*, 26 (1976).
34. A. Naiman and K. P. C. Vollhardt, *Angew. Chem. Int. Ed.* **16**, 708 (1977).
35. T. Hosokawa, N. Shimo, K. Maeda, A. Sonoda, and S. I. Murahashi, *Tetrahedron Lett.*, 383 (1976).
36. H. Horino and N. Inoue, *Tetrahedron Lett.*, 2403 (1979).
37. Y. Watanabe, M. Yamamoto, S. C. Shim, T. Mitsudo, and Y. Takegami, *Chem. Lett.*, 1025 (1979).
38. W. Hafner, H. Prigge, and J. Smidt, *Justus Liebigs Ann. Chem.* **693**, 109 (1966).
39. (a) T. Hosokawa, M. Hirata, S. I. Murahashi, and A. Sonoda, *Tetrahedron Lett.*, 1821 (1976); (b) T. Hosokawa, H. Ohkata, and I. Moritani, *Bull. Chem. Soc. Jpn.* **48**, 1533 (1975).
40. L. E. Craig, R. M. Elofson, and I. J. Ressa, *J. Am. Chem. Soc.* **72**, 3277 (1950).
41. L. S. Nahum, *J. Org. Chem.* **33**, 3601 (1968).
42. J. C. Sauer, B. W. Howk, and R. T. Stiehl, *J. Am. Chem. Soc.* **81**, 693 (1959).
43. E. I. Heiba and R. M. Dessau, *J. Org. Chem.* **39**, 3456 (1974).
44. E. Yoshisato and S. Tsutsumi, *J. Am. Chem. Soc.* **90**, 4488 (1968).
45. J. Falbe, N. Huppes, and F. Korte, *Chem. Ber.* **97**, 863 (1964).
46. E. I. Heiba, R. M. Dessau, and P. G. Rodewald, *J. Am. Chem. Soc.* **96**, 7977 (1974).
47. (a) E. R. H. Jones, T. Y. Shen, and M. C. Whiting, *J. Chem. Soc.*, 230 (1950); (b) J. R. Norton and K. E. Shenton, *Tetrahedron Lett.*, 51 (1975).
48. I. Matsuda, *Chem. Lett.*, 773 (1978).
49. H. Alper, J. K. Currie, and H. Des Abbayes, *J. Chem. Soc. Chem. Commun.*, 311 (1978).
50. R. F. Heck, *J. Am. Chem. Soc.* **86**, 2819 (1964).
51. H. E. Holmquist, *J. Org. Chem.* **34**, 4164 (1969).
52. J. C. Sauer, R. D. Cramer, V. A. Englehardt, T. A. Ford, H. E. Holmquist, and B. W. Howk, *J. Am. Chem. Soc.* **81**, 3677 (1959).
53. C. W. Bird, *J. Organomet. Chem.* **47**, 295 (1973).

54. K. Ohno, T. Mitsuyasu, and J. Tsuji, *Tetrahedron Lett.*, 67 (1971).
55. P. Haynes, *Tetrahedron Lett.*, 3687 (1970).
56. S. Sakai, Y. Kawashima, Y. Takahashi, and Y. Ishii, *J. Chem. Soc. Chem. Commun.*, 1973 (1967); *Chem. Abstr.* **72**, 21646 (1970).
57. R. Aumann and H. Ring, *Ang. Chem. Int. Ed.* **16**, 50 (1977).
58. D. E. Korte, L. S. Hegedus, and R. K. Wirth, *J. Org. Chem.* **42**, 1329 (1977).
59. N. T. Byrom, R. Grigg, and B. Kongkathip, *J. Chem. Soc. Chem. Commun.*, 216 (1976).
60. A. P. Kozikowski and H. F. Wetter, *Synthesis*, 586 (1976).
61. H. Alper and A. S. K. Chan, *J. Am. Chem. Soc.* **95**, 4905 (1973).
62. S. Murahashi and S. Horiie, *J. Am. Chem. Soc.* **78**, 4816 (1956).
63. W. W. Pritchard, U.S. Pat. 2,769,003 [*Chem. Abs.* **51**, 7412 (1967)].
64. H. Takahashi and J. Tsuji, *J. Organomet. Chem.* **10**, 511 (1967).
65. Y. Yamamoto and H. Yamazaki, *Synthesis*, 750 (1976).
66. J. M. Thompson and R. F. Heck, *J. Org. Chem.* **40**, 2667 (1975).
67. (a) Y. Mori and J. Tsuji, *Tetrahedron* **27**, 4039 (1971); (b) Y. Mori and J. Tsuji, *Tetrahedron* **27**, 3811 (1971).
68. Y. Iwashita and M. Sakuraba, *J. Org. Chem.* **36**, 3927 (1971).
69. Y. Ito, T. Hirao, and T. Saegusa, *J. Organomet. Chem.* **131**, 121 (1977).
70. (a) T. Saegusa and Y. Ito, *Synthesis*, 292 (1975); (b) Y. Ito, Y. Iubushi, M. Zenbayashi, S. Tomita, and T. Saegusa, *J. Am. Chem. Soc.* **95**, 4447 (1973).
71. A. McKillop and T. S. B. Sayer, *J. Org. Chem.* **41**, 1079 (1976).
72. P. Heimbach, B. Hugelin, H. Peter, A. Roloff, and E. Troxler, *Ang. Chem. Int. Ed.* **15**, 49 (1976).

References for Chapter 5

1. F. Piacenti, P. Pino, R. Lazzaroni, and M. Bianchi, *J. Chem. Soc. C*, 488 (1966).
2. E. S. Brown, Addition of hydrogen cyanide to olefins, in *Organic Synthesis via Metal Carbonyls*, Vol. 2, ed. I. Wender and P. Pino, John Wiley and Sons, New York, (1977).
3. J. Schwartz and J. A. Labinger, *Angew. Chem. Int. Ed.* **15**, 333 (1976).
4. R. E. Rinehart and J. S. Lasky, *J. Am. Chem. Soc.* **86**, 2516 (1964).
5. H. Alper and J. T. Edward, *J. Organomet. Chem.* **14**, 411 (1968).
6. Y. Shvo and E. Hazun, *J. Chem. Soc. Chem. Commun.*, 336 (1974).
7. D. H. R. Barton, S. G. Davies, and W. B. Motherwell, *Synthesis*, 265 (1979).
8. D. Bingham, D. E. Webster, and P. B. Wells, *J. Chem. Soc. Dalton*, 1514, 1519 (1974); M. Orchin, *Adv. Catal.* **16**, 1 (1966); M. Turner, J. V. Jouanne, H-D. Brauer, and H. Kelm, *J. Mol. Catal.* **5**, 425, 433, 447 (1979); R. Cramer, *J. Am. Chem. Soc.* **88**, 2272 (1966); R. Cramer and R. W. Lindsey, *J. Am. Chem. Soc.* **88**, 3534 (1966); D. Evans, J. A. Osborn, and G. Wilkinson, *J. Chem. Soc. A*, 3133 (1968).
9. C. P. Casey and C. R. Cyr, *J. Am. Chem. Soc.* **95**, 2248 (1973); M. A. Schroeder and M. S. Wrighton, *J. Am. Chem. Soc.* **98**, 551 (1976); J. F. Harrod and A. J. Chalk, *J. Am. Chem. Soc.* **88**, 3491 (1966); P. M. Maitlis, *The Organic Chemistry of Palladium*, Vol. 2, Academic Press, New York (1971), pp. 128–142.
10. D. Baudry, M. Ephritikhine, and H. Felkin, *J. Chem. Soc. Chem. Commun.*, 694 (1978).
11. E. J. Corey and J. W. Suggs, *J. Org. Chem.* **38**, 3224 (1973).
12. E. J. Corey and J. W. Suggs, *Tetrahedron Lett.*, 3775 (1975).
13. R. G. Salomon and J. M. Reuter, *J. Am. Chem. Soc.* **99**, 4372 (1977).
14. K. Hirai, H. Suzuki, H. Kashiwagi, Y. Moro-Oka, and T. Ikawa, *Chem. Lett.*, 23 (1982).
15. J. M. Reuter and R. G. Salomon, *J. Org. Chem.* **42**, 3360 (1977).
16. H. A. J. Carless and D. J. Haywood, *J. Chem. Soc. Chem. Commun.*, 980 (1980).

17. J. Tsuji, Y. Kobayashi, and I. Smihizu, *Tetrahedron Lett.*, 39 (1979).
18. H. Alper and K. Hachem, *J. Org. Chem.* **45**, 2269 (1980); D. Baudry, M. Ephritikhine, and H. Felkin, *Nouv. J. Chim.* **2**, 355 (1977).
19. Y. Sasson and G. L. Rempel, *Tetrahedron Lett.*, 4133 (1974).
20. H. Kumobayashi, S. Akutagawa, and S. Otsuka, *J. Am. Chem. Soc.* **100**, 3949 (1978).
20a. B. C. Laguzza and B. Ganem, *Tetrahedron Lett.*, 1483 (1981).
20b. K. Tani, T. Yamagata, S. Otsuka, S. Akutagawa, H. Kumobayashi, T. Taketomi, H. Takaya, A. Miyashita, and R. Noyori, *J. Chem. Soc. Chem. Comm.* 600 (1982).
21. J. K. Stille and Y. Becker, *J. Org. Chem.* **45**, 2139 (1980).
22. S. Torii, T. Inokuchi, and K. Kawai, *Bull. Chem. Soc. Jpn.* **52**, 861 (1979).
23. P. A. Grieco, M. Nishizawa, N. Marinovic, and W. J. Ehmann, *J. Am. Chem. Soc.* **98**, 7102 (1976).
24. S. R. Wilson and R. N. Misra, *J. Org. Chem.* **45**, 5079 (1980).
25. T. Hudlickly and T. Kutchan, *Tetrahedron Lett.* **21**, 691 (1980).
26. T. Hudlicky, T. M. Kutchan, S. R. Wilson, and D. R. Mao, *J. Am. Chem. Soc.* **102**, 6351 (1980).
27. J. Andrieux, D. H. R. Barton, and H. Patin, *J. Chem. Soc. Perkin Trans. 1*, 359 (1977).
28. J. F. Biellmann and M. J. Jung, *J. Am. Chem. Soc.* **90**, 1673 (1968).
29. J. Rebek and Y. K. Shue, *J. Am. Chem. Soc.* **102**, 5426 (1980).
30. L. A. Paquette, W. E. Fristad, D. S. Dime, and T. R. Bailey, *J. Org. Chem.* **45**, 3017 (1980); W. E. Fristad, D. S. Dime, T. R. Bailey, and L. A. Paquette, *Tetrahedron Lett.*, 1999 (1979).
31. M. Karpf and A. S. Dreiding, *Helv. Chim. Acta.* **64**, 1123 (1981); *Tetrahedron Lett.* **21**, 4569 (1980).
32. R. D. Little and L. Brown, *Tetrahedron Lett.* **21**, 2203 (1980).
33. J. Andrieux, D. H. R. Barton, and H. Patin, *J. Chem. Soc. Perkin Trans. 1*, 359 (1977); see also A. J. Chalk, Ger. Pat. 2,508,347 (1975) [*Chem. Abs.* **83**, 205922n (1975)].
34. D. Mulvagh, M. J. Meegan, and D. Donnelly, *J. Chem. Res. (S)*, 137 (1979).
35. D. F. Taber and B. P. Gunn, *J. Org. Chem.* **44**, 450 (1979).
36. N. J. Leonard and D. M. Locke, *J. Am. Chem. Soc.* **77**, 1852 (1955).
37. A. J. Birch and G. S. R. Subba Rao, *Tetrahedron Lett.*, 3797 (1968).
38. F. Turecek, H. Antropiusova, K. Mach, V. Hanus, and P. Sedmera, *Tetrahedron Lett.*, 637 (1980).
39. T. Onishi, Y. Fujita, and T. Nishida, *Chem. Lett.*, 765 (1979).
40. C. F. Lochow and R. G. Miller, *J. Org. Chem.* **41**, 3020 (1976).
41. P. A. Grieco and N. Marinovic, *Tetrahedron Lett.*, 2545 (1978).
42. D. G. Parker, Ph.D. thesis, University of Newcastle, 1975.
43. E. C. Harnig, *J. Org. Chem.* **10**, 263 (1945); see also R. P. Linstead, K. O. A. Michellis, and S. L. S. Thomas, *J. Chem. Soc.*, 1139 (1940).
44. N. J. Leonard and J. W. Berry, *J. Am. Chem. Soc.* **75**, 4989 (1953).
45. N. J. Leonard, L. A. Miller, and J. W. Berry, *J. Am. Chem. Soc.* **79**, 1482 (1957).
46. Z. Aizenshtat, M. Hausmann, Y. Pickholtz, D. Tal, and J. Blum, *J. Org. Chem.* **42**, 2386 (1977); Y. Pickholtz, Y. Sasson, and J. Blum, *Tetrahedron Lett.*, 1263 (1974).

References for Chapter 6

1. C. W. Bird, *Chem. Rev.* **62**, 283 (1962).
2. (a) J. Falbe (ed.), *New Syntheses with Carbon Monoxide*, Springer-Verlag, Berlin (1980); (b) I. Wender and P. Pino (eds.), *Organic Syntheses via Metal Carbonyls*, Vol. 2, Wiley, New York (1977).

3. R. Nest and P. Dilly, *Angew. Chem. Int. Ed. Engl.* **6**, 357 (1967).
4. J. F. Roth, J. H. Craddock, A. Hershman, and F. E. Paulik, *Chem. Technol.*, 600 (1971).
5. D. Forster, *J. Am. Chem. Soc.* **98**, 846 (1976); *Adv. Organomet. Chem.* **17**, 255 (1979).
6. W. Reppe, *Acetylene Chemistry*, C. A. Meyer and Co., Boston (1949).
7. H. Bahman, Reference 2a, Chap. 5, p. 372.
8. Y. Souma and H. Sano, *Bull. Chem. Soc. Jpn.* **46**, 3237 (1973).
9. Y. Yamamoto and K. Sato, *Bull. Chem. Soc. Jpn.* **27**, 389 (1954).
10. L. Cassar and M. Foà, *J. Organomet. Chem.* **51**, 381 (1973).
11. J. Tsuji, *Organic Synthesis with Palladium Compounds*, Springer-Verlag, Berlin (1980), p. 3.
12. (a) L. Cassar, M. Foà, and A. Gardano, *J. Organomet. Chem.* **121**, C55 (1976); (b) Montedison, U.S. Pat. 4,034,044 (1977).
13. H. Alper and H. des Abbayes, *J. Organomet. Chem.* **134**, C11 (1977); L. Cassar and M. Foà, *J. Organomet. Chem.* **134**, C15 (1977).
14. J. P. Collman, *Acc. Chem. Res.* **8**, 342, (1975).
15. Sold by Alfa Ventron Inc. as Collman's reagent.
16. J. P. Collman, S. R. Winter, and R. G. Komoto, *J. Am. Chem. Soc.* **95**, 249 (1973).
17. J. J. Brunet, C. Sidot, B. Loubinoux, and P. Caubere, *J. Org. Chem.* **44**, 2199 (1979).
18. K. Nagira, K. Kikukawa, F. Wada, and T. Matsuda, *J. Org. Chem.* **45**, 2365 (1980).
19. Reference 2b, p. 245.
20. F. Piacenti, M. Bianchi, and R. Lazzaroni, *Chim. Ind. (Milan)* **50**, 318 (1968).
21. L. J. Kehoe and R. A. Schell, *J. Org. Chem.* **35**, 2846 (1970).
22. Sold by Alfa Ventron Inc. as Schwartz's reagent.
23. C. A. Bertelo and J. Schwartz, *J. Am. Chem. Soc.* **97**, 228 (1975).
24. Y. Souma, H. Sano, and J. Iyoda, *J. Org. Chem.* **38**, 2016 (1973).
25. Y. Souma and H. Sano, *J. Org. Chem.* **38**, 3633 (1973).
26. L. Cassar, G. P. Chiusoli, and F. Guerrieri, *Synthesis*, 509 (1973).
27. W. Reppe, *Justus Liebigs Ann. Chem.* **582**, 1 (1953).
28. R. Toepel, *Chim. Ind. (Paris)* **91**, 139 (1964).
29. *Hydrocarbon Processing*, 120 (November 1971).
30. P. Pino and G. Braca, Reference 2b, p. 419.
31. E. R. H. Jones, G. H. Whitham, and M. C. Whiting, *J. Chem. Soc.* 4628 (1957).
32. (a) G. P. Chiusoli and L. Cassar, Reference 2b, p. 297; (b) *Angew. Chem. Int. Ed. Engl.* **6**, 124, (1967).
33. G. P. Chiusoli and G. Agnès, *Z. Naturforsch., Teil B.* **17**, 852 (1962).
34. H. Wakamatsu, J. Uda, and N. Yamakami, *J. Chem. Soc. Chem. Commun.*, 1540 (1971).
35. J. J. Parnaud, G. Campari, and P. Pino, *J. Mol. Catal.* **6**, 341 (1979).
36. L. Marko and P. Szabo, *Chem. Tech. (Leipzig)* **13**, 482 (1961).
37. (a) D. M. Fenton and P. J. Steinwand, *J. Org. Chem.* **39**, 701 (1974); (b) Ube Industries Ltd., Belg. Pat. 870,268 (1979).
38. F. Piacenti and M. Bianchi, Reference 2b, p. 18; S. K. Bhattacharyya and S. K. Palit, *J. Appl. Chem.* **12**, 174 (1962).
39. W. Reppe, H. Kroper, H. J. Pistor, and O. Weissbarth, *Justus Liebigs Ann. Chem.* **582**, 87 (1953).
40. R. F. Heck, *J. Am. Chem. Soc.* **85**, 1460 (1963).
41. J. L. Eisenmann, R. L. Yamartino, and J. F. Howard, Jr., *J. Org. Chem.* **26**, 2102 (1961).
42. R. F. Heck and D. S. Breslow, *J. Am. Chem. Soc.* **85**, 2779 (1963).
43. E. J. Corey and L. S. Hegedus, *J. Am. Chem. Soc.* **91**, 1233 (1969).
44. A. Schoenberg, I. Bartoletti, and R. F. Heck, *J. Org. Chem.* **39**, 3318 (1974).
45. T. Takahashi, T. Nagashima, and J. Tsuji, *Chem. Lett.*, 369 (1980).

46. T. Takahashi, H. Ikeda, and J. Tsuji, *Tetrahedron Lett.* **21**, 3885 (1980).

47. A. Cowell and J. K. Stille, *J. Am. Chem. Soc.* **102**, 4193 (1980).

48. M. Yamashita and R. Suemitsu, *Tetrahedron Lett.*, 1477 (1978).

49. P. Pino, F. Piacenti, and M. Bianchi, Reference 2b, p. 233.

50. R. Bardi, A. del Pra, A. M. Piazzesi, and L. Toniolo, *Inorg. Chim. Acta.* **35**, L345 (1979).

51. J. F. Knifton, *J. Org. Chem.* **41**, 793 (1976).

52. J. F. Knifton, *J. Org. Chem.* **41**, 2885 (1976).

53. A. Matsuda, *Bull. Chem. Soc. Jpn.* **46**, 524 (1973).

54. S. Hosaka and J. Tsuji, *Tetrahedron* **27**, 3821 (1971).

55. J. Tsuji, S. Hosaka, J. Kiji, and T. Susuki, *Bull. Chem. Soc. Jpn.* **39**, 141 (1966).

56. J. Tsuji, Y. Mori, and M. Hara, *Tetrahedron* **28**, 3721 (1972).

57. J. K. Stille and D. E. James, in *Transition Metal Catalysed Carbonylation of Olefins. The Chemistry of Functional Groups, Supplement A. Double-bonded Functional Groups*, ed. S. Patai, Wiley, London (1976), p. 1099.

58. J. K. Stille and R. Divakaruni, *J. Org. Chem.* **44**, 3474 (1979).

59. G. Cometti and G. P. Chiusoli, *J. Organomet. Chem.* **181**, C14 (1979).

60. R. F. Heck, *J. Am. Chem. Soc.* **90**, 5518 (1968); R. F. Heck, *J. Am. Chem. Soc.* **91**, 6707 (1969).

61. J. K. Stille and L. F. Hines, *J. Am. Chem. Soc.* **92**, 1798 (1970).

62. J. K. Stille and D. E. James, *J. Am. Chem. Soc.* **97**, 674 (1975); *J. Organomet. Chem.* **108**, 401 (1976).

63. J. K. Stille and R. Divakaruni, *J. Am. Chem. Soc.* **100**, 1303 (1978).

64. A. Mullen, Reference 2a, Chap. 3, p. 298.

65. J. F. Knifton, *J. Mol. Catal.* **2**, 293 (1977).

66. M. Foà and L. Cassar, *Gazz. Chim. Ital.* **102**, 85 (1972).

67. J. R. Norton, K. E. Shenton, and J. Schwartz, *Tetrahedron Lett.*, 51 (1975).

68. T. F. Murray, V. Varma, and J. R. Norton, *J. Org. Chem.* **43**, 353 (1978).

69. J. Tsuji and T. Nogi, *J. Org. Chem.* **31**, 2641 (1966).

70. R. F. Heck, *J. Am. Chem. Soc.* **94**, 2712 (1972).

71. J. Tsuji and T. Nogi, *J. Am. Chem. Soc.* **88**, 1289 (1966).

72. J. C. Sauer, R. D. Cramer, V. A. Engelhardt, T. A. Ford, H. E. Holmquist, and B. W. Howk, *J. Am. Chem. Soc.* **81**, 3677 (1959).

73. H. W. Sternberg, J. G. Shukys, C. D. Donne, R. Markby, R. A. Friedel, and I. Wender, *J. Am. Chem. Soc.* **81**, 2339 (1959).

74. O. S. Mills and G. Robinson, *Inorg. Chim. Acta.* **1**, 61 (1967).

75. I. Rhee, M. Ryang, and S. Tsutsumi, *Tetrahedron Lett.*, 4593 (1969).

76. G. P. Chiusoli, M. Dubini, M. Ferraris, F. Guerrieri, S. Merzoni, and G. Mondelli, *J. Chem. Soc. C*, 2889 (1968).

77. G. P. Chiusoli, S. Merzoni, and G. Mondelli, *Tetrahedron Lett.*, 2777 (1964).

78. R. F. Heck, *J. Am. Chem. Soc.* **86**, 2819 (1964).

79. H. Alper, J. K. Currie, and H. Des Abbayes, *J. Chem. Soc. Chem. Commun.*, 311 (1978).

80. J. M. Davidson, *J. Chem. Soc. A*, 193 (1969).

81. R. C. Larock, *J. Org. Chem.* **40**, 3237 (1975).

82. (a) R. C. Larock and H. C. Brown, *J. Organomet. Chem.* **36**, 1 (1972); (b) R. C. Larock, S. K. Gupta, and H. C. Brown, *J. Am. Chem. Soc.* **94**, 4371 (1972).

83. R. C. Larock, B. Riefling, and C. A. Fellows, *J. Org. Chem.* **43**, 131 (1978).

84. D. Durand and C. Lassau, *Tetrahedron Lett.*, 2329 (1969).

85. T. Saegusa, S. Kobayashi, K. Hirota, and Y. Ito, *Bull. Chem. Soc. Jpn.* **42**, 2610 (1969).

86. J. J. Byerley, G. L. Rempel, N. Takebe, and B. R. James, *J. Chem. Soc. Chem. Commun.*, 1482 (1971).

87. N. S. Imyanitov and D. M. Rudkovskii, *Kinet. Catal.* (*Engl. transl.*) **9**, 859 (1968).
88. F. Calderazzo, *Inorg. Chem.* **4**, 293 (1965).
89. J. Tsuji and N. Iwamoto, *J. Chem. Soc. Chem. Commun.*, 380 (1966).
90. A. Schoenberg and R. F. Heck, *J. Org. Chem.* **39**, 3327 (1974).
91. M. Mori, K. Chiba, and Y. Ban, *J. Org. Chem.* **43**, 1684 (1978).
92. J. Falbe, in *Newer Methods of Preparative Organic Chemistry*, Vol. 6, ed. W. Foerst, Academic Press, New York (1971), p. 193.
93. J. Falbe and F. Korte, *Chem. Ber.* **98**, 1928 (1965).
94. J. Falbe and F. Korte, *Angew. Chem. Int. Ed. Engl.* **1**, 266 (1962).
95. P. K. Wong, M. Madhavarao, D. F. Marten, and M. Rosenblum, *J. Am. Chem. Soc.* **99**, 2823 (1977).
96. S. R. Berryhill and M. Rosenblum, *J. Org. Chem.* **45**, 1984 (1980).
97. A. Rosenthal and I. Wender, in *Organic Syntheses via Metal Carbonyls*, Vol. 1, ed. I. Wender and P. Pino, Wiley, New York (1968), p. 405.
98. M. I. Bruce, *Angew. Chem. Int. Ed. Engl.* **16**, 73 (1977).
99. J. M. Thompson and R. F. Heck, *J. Org. Chem.* **40**, 2667 (1975).
100. S. Horiie and S. Murahashi, *Bull. Chem. Soc. Jpn.* **33**, 247 (1960).
101. A. Rosenthal and J. Gervay, *Can. J. Chem.* **42**, 1490 (1964).
102. R. F. Heck, *J. Am. Chem. Soc.* **90**, 313 (1968).
103. N. Rizkalla (Halcon International Inc.), Ger. Pat. 2,610,036 (1976).
104. Y. Mori and J. Tsuji, *Bull. Chem. Soc. Jpn* **42**, 777 (1969).
105. J. Tsuji, J. Kiji, S. Imamura, and M. Morikawa, *J. Am. Chem. Soc.* **86**, 4350 (1964).
106. T. Nogi and J. Tsuji, *Tetrahedron* **25**, 4099 (1969).
107. W. T. Dent, R. Long, and G. H. Whitfield, *J. Chem. Soc.*, 1588 (1964).
108. R. F. Heck, *J. Am. Chem. Soc.* **85**, 2013 (1963).
109. National Distillers and Chemical Corp., Neth. Pat. Appl. 6,614,185 (1967).
110. W. W. Prichard (E.I. DuPont de Nemours and Co.), U.S. Pat. 3,632,643 (1972).
111. T. Susuki and J. Tsuji, *J. Org. Chem.* **35**, 2982 (1970).
112. Y. Takegami, Y. Watanabe, H. Masada, and I. Kanaya, *Bull. Chem. Soc. Jpn.* **40**, 1456 (1967).
113. M. P. Cooke, Jr., *J. Am. Chem. Soc.* **92**, 6080 (1970).
114. M. Ryang, I. Rhee, and S. Tsutsumi, *Bull. Chem. Soc. Jpn.* **37**, 341 (1964).
115. A. Schoenberg and R. F. Heck, *J. Am. Chem. Soc.* **96**, 7761 (1974).
116. O. Roelen (Ruhrchemie A.G.), Ger. Pat. 849,548 (1938).
117. B. Cornils, Reference 2a, Chap. 1, p. 1.
118. R. L. Pruett, *Adv. Organomet. Chem.* **17**, 1 (1979).
119. P. Pino, F. Piacenti, and M. Bianchi, Reference 2b, p. 43.
120. R. F. Heck and D. S. Breslow, *J. Am. Chem. Soc.* **83**, 4023 (1961).
121. N. Ahmad, J. J. Levison, S. D. Robinson, and M. F. Uttley, *Inorg. Synth.* **15**, 59 (1974).
122. C. K. Brown and G. Wilkinson, *J. Chem. Soc. A*, 2753 (1970).
123. B. Cornils, Reference 2a, p. 172.
124. Y. Kawabata, T. Hayashi, and I. Ogata, *J. Chem. Soc. Chem. Commun.*, 462 (1979).
125. A. J. M. Keulemans, A. Kwantes, and T. van Bavel, *Recl. Trav. Chim. Pays-Bas* **67**, 298 (1948).
126. Y. Ono, S. Sato, M. Takesada, and H. Wakamatsu, *J. Chem. Soc. Chem. Commun.*, 1255 (1970).
127. J. Falbe and F. Korte, *Chem. Ber.* **97**, 1104 (1964).
128. J. Falbe, H. J. Schulze-Steinen, and F. Korte, *Chem. Ber.* **98**, 886, (1965).
129. B. Fell and M. Barl, *Chem. Z.* **101**, 343 (1977).
130. H. Siegel and W. Himmele, *Angew. Chem. Int. Ed. Engl.* **19**, 178 (1980).

131. S. Murai and N. Sonoda, *Angew. Chem. Int. Ed. Engl.* **18**, 837 (1979).
132. D. L. Johnson and J. A. Gladysz, *J. Am. Chem. Soc.* **101**, 6433 (1979).
133. J. P. Collman, S. R. Winter, and D. R. Clark, *J. Am. Chem. Soc.* **94**, 1788 (1972).
134. M. P. Cooke, Jr., and R. M. Parlman, *J. Am. Chem. Soc.* **99**, 5222 (1977).
135. Y. Kimura, Y. Tomita, S. Nakanishi, and Y. Otsuji, *Chem. Lett.*, 321 (1979).
136. Y. Sawa, M. Ryang, and S. Tsutsumi, *Tetrahedron Lett.*, 5189 (1969).
137. M. Yamashita and R. Suemitsu, *Tetrahedron Lett.*, 761 (1978).
138. R. C. Cookson and G. Farquharson, *Tetrahedron Lett.*, 1255 (1979).
139. E. A. Naragon, A. J. Millendorf, and J. H. Vergilio, U.S. Pat., 2,699,453 (1955).
140. J. A. Bertrand, C. L. Aldridge, S. Husebye, and H. B. Jonassen, *J. Org. Chem.* **29**, 790 (1964).
141. R. F. Heck, *J. Am. Chem. Soc.* **85**, 3381, 3383 (1963).
142. H. Alper and J. K. Currie, *Tetrahedron Lett.*, 2665 (1979).
143a. C. W. Bird, R. C. Cookson, and J. Hudec, *Chem. Ind.* (*London*), 20 (1960).
143b. J. Mantzaris and E. Weissberger, *J. Am. Chem. Soc.* **96**, 1873 (1974).
144. J. Mantzaris and E. Weissberger, *J. Am. Chem. Soc.* **96**, 1880 (1974).
145. I. U. Khand and P. L. Pauson, *J. Chem. Res.* (*S*) **9**, (M), 0168 (1977).
146. I. U. Khand, G. R. Knox, P. L. Pauson, W. E. Watts, and M. I. Foreman, *J. Chem. Soc. Perkin Trans. 1*, 977 (1973).
147. R. C. Larock and S. S. Hershberger, *J. Org. Chem.* **45**, 3840 (1980).
148. I. Rhee, M. Ryang, T. Watanabe, H. Omura, S. Murai, and N. Sonoda, *Synthesis*, 776 (1977).
149. F. J. Weigert, *J. Org. Chem.* **38**, 1316 (1973).
150. K. Weissermel and H. J. Arpe, *Industrial Organic Chemistry*, Verlag-Chemie, Weinheim (1978).
151. T. Sakakibara and H. Alper, *J. Chem. Soc. Chem. Commun.*, 458 (1979).
152. J. Tsuji and K. Ohno, *Synthesis*, 157 (1969).
153. K. Ohno and J. Tsuji, *J. Am. Chem. Soc.* **90**, 99 (1968).
154. R. W. Fries and J. K. Stille, *Synth. React. Inorg. Metal-Org. Chem.* **1**, 295 (1971).
155. J. Falbe, H. Tummes, and H. Hahn, *Angew. Chem. Int. Ed. Engl.* **9**, 169 (1970).
156. D. H. Doughty and L. H. Pignolet, *J. Am. Chem. Soc.* **100**, 7083 (1978).
157. H. E. Eschinazi and H. Pines, *J. Org. Chem.* **24**, 1369 (1959).
158. J. Tsuji and K. Ohno, *J. Am. Chem. Soc.* **90**, 94 (1968).
159. J. O. Hawthorne and M. H. Wilt, *J. Org. Chem.* **25**, 2215 (1960).
160. J. Tsuji, Reference 2b, p. 595.

References for Chapter 7

1. C. Ellis, *Hydrogenation of Organic Substances*, D. Van Nostrand Co., New York (1930).
2. P. N. Rylander, *Catalytic Hydrogenation in Organic Syntheses*, Academic Press, New York (1979).
3. M. Freifelder, *Catalytic Hydrogenation in Organic Synthesis, Procedures and Commentary*, John Wiley and Sons, New York (1978).
4. J. J. Sims, V. K. Honwad, and L. H. Selman, *Tetrahedron Lett.*, 87 (1969).
5. (a) F. R. Hartley and P. N. Vezey, *Adv. Organomet. Chem.* **15**, 189 (1977); (b) B. R. James, *Homogeneous Hydrogenation*, John Wiley and Sons, New York (1973).
6. B. R. James, *Adv. Organomet. Chem.* **17**, 319 (1979).
7. G. Dolcetti and N. W. Hoffman, *Inorg. Chim. Acta.* **9**, 269 (1974).

8. H. O. House, *Modern Synthetic Reactions*, W. A. Benjamin, New York (1965).
9. R. Adams and V. Voorhees, in *Organic Syntheses, Coll. Vol. I*, ed. H. Gilman, John Wiley and Sons, New York (1951), p. 61.
10. G. Brieger and T. J. Nestrick, *Chem. Rev.* **74**, 567 (1974).
11. T. L. Ho, *Synthesis*, 1 (1979).
12. F. J. McQuillin, *Homogeneous Hydrogenation in Organic Chemistry*, Vol. 1, Reidel, Dordrecht (1976).
13. A. J. Birch and D. H. Williamson, *Org. Reac.* **24**, 1 (1976).
14. A. Onopchenko, E. T. Sabourin, and C. M. Selwitz, *J. Org. Chem.* **44**, 3671 (1979).
15. P. S. Hallman, B. R. McGarvey, and G. Wilkinson, *J. Chem. Soc. A*, 3143 (1968).
16. (a) D. J. Cram and N. L. Allinger, *J. Am. Chem. Soc.* **78**, 2518 (1956); (b) H. Lindlar and R. Dubuis, in *Organic Syntheses, Coll. Vol. V*, ed. E. Baumgarten, John Wiley and Sons, New York (1973), p. 880.
17. A. T. Blomquist, L. H. Liu, and J. C. Bohrer, *J. Am. Chem. Soc.* **74**, 3643 (1952).
18. E. F. Magoon and L. H. Slaugh, *Tetrahedron* **23**, 4509 (1967).
19. (a) T. E. Bellas, R. G. Brownlee, and R. M. Silverstein, *Tetrahedron* **25**, 5149 (1969); (b) C. A. Brown and V. K. Ahuja, *J. Chem. Soc. Chem. Commun.*, 553 (1973).
20. K. Sonogashira and N. Hagihara, *Bull. Chem. Soc. Jpn.* **39**, 1178 (1966).
21. (a) G. F. Pregaglia, A. Andreetta, G. F. Ferrari, and R. Ugo, *J. Organomet. Chem.* **30**, 387 (1971); (b) M. Hidai, T. Kuse, T. Hikita, Y. Uchida, and A. Misono, *Tetrahedron Lett.*, 1715 (1970).
22. (a) R. R. Schrock and J. A. Osborn, *J. Am. Chem. Soc.* **93**, 3089 (1971); (b) R. R. Schrock and J. A. Osborn, *J. Am. Chem. Soc.* **98**, 2143 (1976).
23. C. E. Castro and R. D. Stephens, *J. Am. Chem. Soc.* **86**, 4358 (1964).
24. P. W. Chum and S. E. Wilson, *Tetrahedron Lett.* 15 (1976).
25. A. A. Petrov and M. P. Frost, *J. Gen. Chem. USSR* **34**, 3331 (1964).
26. A. I. Zakharova and G. D. l'lina, *J. Gen. Chem. USSR* **34**, 1391 (1964).
27. (a) A. Butenandt and E. Hecker, *Angew. Chem.* **73**, 349 (1961); (b) A. Butenandt, E. Hecker, M. Hopp, and W. Koch, *Annalen* **658**, 39 (1962); (c) A. Henrick, *Tetrahedron* **33**, 1845 (1977).
28. S. G. Morris, S. F. Herb, P. Magidman, and F. E. Luddy, *J. Am. Oil Chem. Soc.* **49**, 92 (1972).
29. M. L. Capmau, W. Chodkiewicz, and P. Cadiot, *Tetrahedron Lett.*, 1619 (1965).
30. J. I. DeGraw and J. O. Rodin, *J. Org. Chem.* **36**, 2902 (1971).
31. A. J. Birch and K. A. M. Walker, *Aust. J. Chem.* **24**, 513 (1971).
32. (a) A. J. Birch and K. A. M. Walker, *J. Chem. Soc. C*, 1894 (1966); (b) J. R. Morandi and H. B. Jensen, *J. Org. Chem.* **34**, 1889 (1969).
33. F. H. Jardine, *Prog. Inorg. Chem.* **28**, 63 (1981).
34. (a) F. H. Jardine, J. A. Osborn, and G. Wilkinson, *J. Chem. Soc. A*, 1574 (1967); (b) W. Strohmeier and R. Endres, *Z. Naturforsch* **25B**, 1068 (1970).
35. P. Abley, I. Jardine, and F. J. McQuillin, *J. Chem. Soc. C*, 840 (1971).
36. R. R. Schrock and J. A. Osborn, *J. Am. Chem. Soc.* **98**, 2134, 2143, 4450 (1976).
37. C. O'Connor and G. Wilkinson, *J. Chem. Soc. A*, 2665 (1968).
38. C. E. Castro, R. D. Stephens, and S. Moje, *J. Am. Chem. Soc.* **88**, 4964 (1966).
39. V. L. Frampton, J. D. Edwards, and H. R. Henze, *J. Am. Chem. Soc.* **73**, 4432 (1951).
40. R. Adams, J. W. Kern, and R. L. Shriner, *Organic Syntheses, Coll. Vol. I*, ed. H. Gilman, John Wiley and Sons, New York (1951), p. 101.
41. R. M. Herbest and D. Shemin, in *Organic Syntheses, Coll. Vol. II*, ed. A. H. Blatt, John Wiley and Sons, New York (1955), p. 491.

176. A. M. Aguiar, C. J. Morrow, J. D. Morrison, R. E. Burnett, W. F. Masler, and W. S. Bhacca, *J. Org. Chem.* **41**, 1545 (1976).
177a. A. R. Pinder, *Synthesis*, 425 (1980); 177b. J. W. Huffman, *J. Org. Chem.* **24**, 1759 (1959).
178. N. Whittaker, *J. Chem. Soc.*, 1646 (1953).
179. B. A. Fox and T. L. Threlfall, in *Organic Syntheses, Coll. Vol. V*, ed. H. E. Baumgarten, John Wiley and Sons, New York (1973), p. 346.
180. K. Kindler and K. Luhrs, *Justus Liebigs Ann. Chem.* **685**, 36 (1965).
181. K. Kindler and K. Luhrs, *Chem. Ber.* **99**, 227 (1966).
182. H. Imai, T. Nishiguchi, M. Tanaka, and K. Fukuzumi, *J. Org. Chem.* **42**, 2309 (1977).
183. N. A. Cortese and R. F. Heck, *J. Org. Chem.* **42**, 3491 (1977).
184. S. T. Lin and J. A. Roth, *J. Org. Chem.* **44**, 309 (1979).
185. D. P. Brust, S. Tarbell, S. M. Hecht, E. C. Hayward, and L. D. Colebrook, *J. Org. Chem.* **31**, 2192 (1966).
186. A. M. Khan, F. J. McQuillin, and I. Jardine, *J. Chem. Soc. C*, 136 (1967).
187. W. H. Hartung and R. Simonoff, *Org. React.* **7**, 263 (1953).
188. G. A. Olah and G. K. Surya Prakash, *Synthesis*, 397, (1978).
189. R. O. Hutchins, K. Learn, and R. P. Fulton, *Tetrahedron Lett.*, 27 (1980).
190. M. A. Bennett, T. Huang, A. K. Smith, and T. W. Turney, *J. Chem. Soc. Chem. Commun.*, 582 (1978).
191. I. D. Entwistle, A. E. Jackson, R. A. W. Johnston, and R. P. Telford, *J. Chem. Soc. Perkin Trans. 1*, 443 (1977).

References for Chapter 8

1. J. K. Kochi, *Organometallic Mechanism and Catalysis*, Academic Press, New York (1978).
2. R. A. Sheldon and J. K. Kochi, *Adv. Catal.* **25**, 272 (1976).
3. G. W. Parshall, *Homogeneous Catalysis*, Wiley–Interscience, New York (1980).
4. J. L. Chinn, *Selection of Oxidants in Synthesis*, Marcel Dekker, New York (1971).
5. R. L. Augustine (ed.), *Oxidation*, Vols. 1 and 2, Marcel Dekker, New York (1969, 1971).
6. P. N. Rylander, *Organic Synthesis with Noble Metal Catalysts*, Academic Press, New York (1973).
7. P. M. Henry, *Palladium-Catalyzed oxidation of Hydrocarbons*, Reidel, Dordrecht (1980).
8. K. B. Wiberg (ed.), *Oxidation in Organic Chemistry*, Part A, Academic Press, New York (1965).
9. W. S. Trahanovsky (ed.), *Oxidation in Organic Chemistry*, Part B, Academic Press, New York (1973).
10. R. H. Eastman and R. A. Quinn, *J. Am. Chem. Soc.* **82**, 4249 (1960).
11. K. B. Wiberg and G. Foster, *J. Am. Chem. Soc.* **83**, 423 (1961).
12. A. J. Birch and G. S. R. S. Rao, *Tetrahedron Lett.*, 2917 (1968).
13. V. P. Kurkov, J. Z. Pasky, and J. B. Lavigne, *J. Am. Chem. Soc.* **90**, 4743 (1968).
14. D. N. Jones and S. D. Knox, *J. Chem. Soc. Chem. Commun.*, 166 (1975).
15. H. Muxfeldt and W. Rogalski, *J. Am. Chem. Soc.* **87**, 933 (1965).
16. D. G. Lee, in *Oxidation*, Vol. 1, ed. R. L. Augustine, Marcel Dekker, New York (1969), p. 5.
17. K. B. Wiberg and K. A. Saegebarth, *J. Am. Chem. Soc.* **79**, 2822 (1957).
18. Reference 6, Chap. 4, p. 121.

19. G. Braun, *J. Am. Chem. Soc.* **52**, 3176 (1930).
20. V. Vanrheenan, R. C. Kelly, and D. Y. Cha, *Tetrahedron Lett.*, 1973 (1976).
21. (a) K. B. Sharpless and K. Akashi, *J. Am. Chem. Soc.* **98**, 1986, (1976); (b) K. Akashi, R. E. Palermo, and K. B. Sharpless, *J. Org. Chem.* **43**, 2063 (1978).
22. H. Wittcoff and S. E. Miller, *J. Am. Chem. Soc.* **69**, 3138 (1947).
23. (a) D. W. Patrick, L. K. Truesdale, S. A. Biller, and K. B. Sharpless, *J. Org. Chem.* **43**, 2628, (1978); (b) S. G. Hentges and K. B. Sharpless, *J. Org. Chem.* **45**, 2257 (1980).
24. E. Herranz and K. B. Sharpless, *J. Org. Chem.* **43**, 2544 (1978).
25. E. Herranz and K. B. Sharpless, *J. Org. Chem.* **45**, 2710 (1980).
26. J. Hanotier, Ph. Camerman, M. Hanotier-Bridoux, and P. de Radzitzky, *J. Chem. Soc. Perkin Trans. 2*, 2247 (1972).
27. A. R. Doumaux, in *Oxidation*, Vol. 2, ed. R. L. Augustine and D. J. Trecker, Marcel Dekker, New York (1971), p. 170.
28. T. Nishimura, in *Organic Syntheses, Coll. Vol. IV*, ed. N. Rabjohn, John Wiley and Sons, New York (1963), p. 713.
29. (a) D. R. Bryant, J. E. McKeon, and B. C. Ream, *J. Org. Chem.* **33**, 4123 (1968); (b) D. R. Bryant, J. E. McKeon, and B. C. Ream, *J. Org. Chem.* **34**, 1106 (1969).
30. L. Eberson and L. Gomez-Gonzales, *J. Chem. Soc. Chem. Commun.*, 263 (1971).
31. L. Hornig and T. Quadflieg, U.S. Pat. 3,642,873 (1972).
32. L. Eberson and L. Jonsson, *Acta. Chem. Scand. Ser. B* **30**, 361 (1976).
33. G. G. Arzoumanidis, F. C. Rauch, and G. Blank, *Abstr. 163rd Meet. Amer. Chem. Soc. Inorg. Sec.*, paper 73 (1972).
34. C. W. Bird, *Transition Metal Intermediates in Organic Synthesis*, Logos Press, London (1967), p. 88.
35. J. E. Backvall, M. W. Young, and K. B. Sharpless, *Tetrahedron Lett.*, 3523 (1977).
36. M. Pralus, J. C. Lecod, and J. P. Schirmann, in *Fundamental Research in Homogeneous Catalysis*, Vol. 3, ed. M. Tsutsui, Plenum Press, New York (1979), p. 327.
37. J. Lyons, in *Fundamental Research in Homogeneous Catalysis*, Vol. 1, ed. M. Tsutsui and R. Ugo, Plenum Press, New York (1977), p. 28.
38. N. Indicator and W. F. Brill, *J. Org. Chem.* **30**, 2074 (1966).
39. E. S. Gould, R. R. Hiatt, and K. C. Irwin, *J. Am. Chem. Soc.* **90**, 4573 (1968).
40. M. N. Sheng and J. G. Zajacek, *Adv. Chem. Ser.* **76**, 418 (1968).
41. R. A. Sheldon, *Rec. Trav. Chim.* **92**, 253 (1973).
42. M. N. Sheng, *Synthesis*, 1974 (1972).
43. G. Descotes and P. Legrand, *Bull. Soc. Chim. France*, 2942 (1972).
44. M. N. Sheng and J. G. Zajacek, *J. Org. Chem.* **35**, 1839 (1970).
45. V. P. Yur'ev, I. A. Gailyunas, L. V. Spirikhin, and G. A. Tolstikov, *J. Gen. Chem. USSR* **45**, 2269 (1975).
46. K. B. Sharpless and R. C. Michaelson, *J. Am. Chem. Soc.* **95**, 6136 (1973).
47. T. Itoh, K. Jitsukawa, K. Kaneda, and S. Teranishi, *J. Am. Chem. Soc.* **101**, 159 (1979).
48. R. E. Dehnel and G. H. Whitman, *J. Chem. Soc. Perkin Trans. 1*, 953 (1979).
49. E. D. Mihelich, *Tetrahedron Lett.*, 4729 (1979).
50. B. E. Rossiter, T. R. Verhoeven, and K. B. Sharpless, *Tetrahedron Lett.*, 4733 (1979).
51. S. Yamada, T. Mashiko, and S. Terashima, *J. Am. Chem. Soc.* **99**, 1988 (1977).
52. R. C. Michaelson, R. E. Palermo, and K. B. Sharpless, *J. Am. Chem. Soc.* **99**, 1990 (1977).
53. H. B. Kagan, H. Mimoun, C. Mark, and V. Schurig, *Ang. Chem. Int. Ed. Engl.* **18**, 485 (1979).
54. Reference 4, p. 122.
55. B. Eistert, W. Schade, and H. Selzer, *Chem. Ber.* **97**, 1471 (1964).

56. K. B. Wiberg, B. R. Lowry, and T. H. Colby, *J. Am. Chem. Soc.* **83**, 3998 (1961).
57. K. Heusler and A. Wettstein, *Helv. Chim. Acta* **35**, 284 (1952).
58. W. G. Dauben, M. Lorber, and D. S. Fullerton, *J. Org. Chem.* **34**, 3587 (1969).
59. W. G. Salmond, M. A. Barta, and J. L. Havens, *J. Org. Chem.* **43**, 2057 (1978).
60. J. Blum, H. Rosenman, and E. D. Bergmann, *Tetrahedron Lett.*, 3665 (1967).
61. H. C. Volger, W. Brackman, and J. W. F. M. Lemmers, *Rec. Trav. Chim.* **84**, 1203 (1965).
62. L. F. Fieser, W. P. Campbell, E. M. Fry, and M. D. Gates, Jr., *J. Am. Chem. Soc.* **61**, 3216 (1939).
63. R. Wendland and J. Lalonde, in *Organic Syntheses, Coll. Vol. IV*, ed. N. Rabjohn, John Wiley and Sons, New York (1963), p. 757.
64. C. Djerassi and R. R. Engle, *J. Am. Chem. Soc.* **75**, 3838 (1953).
65. J. Smidt, W. Hafner, R. Jira, J. Sedlemeier, R. Sieber, R. Ruttinger, and H. Kojer, *Ang. Chem.* **71**, 1976 (1959).
66. W. H. Clement and C. M. Selwitz, *J. Org. Chem.* **29**, 241 (1964), and U.S. Pat. 3,370,073 (1968).
67. J. E. Byrd, L. Cassar, P. E. Eaton, and J. Halpren, *J. Chem. Soc. Chem. Commun.*, 40 (1971).
68. R. Pappo, D. S. Allen, Jr., R. U. Lemieux, and W. S. Johnson, *J. Org. Chem.* **21**, 478 (1956).
69. (a) V. van Rheenan, *J. Chem. Soc. Chem. Commun.*, 314 (1969); (b) T. Ho, *Synth. Commun.* **4**, 135 (1974).
70. N. A. Khan and M. S. Newman, *J. Org. Chem.* **17**, 1063 (1952).
71. H. Gopal and A. J. Gordon, *Tetrahedron Lett.*, 2941 (1971).
72. Reference 1, p. 109.
73. K. Bowden, I. M. Heilbron, E. R. H. Jones, and B. C. L. Weedon, *J. Chem. Soc.*, 39 (1946).
74. Reference 4, p. 42.
75. Reference 16, p. 58.
76. H. O. House, *Modern Synthetic Reaction*, 2nd ed., W. A. Benjamin, New York (1972), pp. 257–285.
77. (a) G. I. Poos, G. E. Arth, R. E. Beyler, and L. H. Sarett, *J. Am. Chem. Soc.* **75**, 422 (1953); (b) J. R. Holum, *J. Org. Chem.* **26**, 4814 (1961).
78. J. C. Collins, W. W. Hess, and F. J. Frank, *Tetrahedron Lett.*, 3363 (1968).
79. W. S. Johnson, C. A. Harbert, and R. D. Stipanovic, *J. Am. Chem. Soc.* **90**, 5279 (1968).
80. H. C. Brown and C. P. Garg, *J. Am. Chem. Soc.* **83**, 2952 (1961).
81. E. J. Corey and J. W. Suggs, *Tetrahedron Lett.*, 2647 (1975).
82. E. J. Corey and G. Schmidt, *Tetrahedron Lett.*, 399 (1979).
83. J. San Filippo, Jr. and C. I. Chern, *J. Org. Chem.* **42**, 2182 (1972).
84. Y. S. Cheng, W. L. Liu, and S. Chen, *Synthesis*, 223 (1980).
85. E. Santaniello, F. Ponti, and A. Manzocchi, *Synthesis*, 534 (1978).
86. S. J. Flatt, G. W. J. Fleet, and B. J. Taylor, *Synthesis*, 815 (1979).
87. J. M. J. Frechet, J. Warnock, and M. J. Farrall, *J. Org. Chem.* **43**, 2618 (1978).
88. G. Cainelli, G. Cardillo, M. Orena, and S. Sandri, *J. Am. Chem. Soc.* **98**, 6737 (1976).
89. Reference 16, p. 64.
90. F. M. Menger and C. Lee, *J. Org. Chem.* **44**, 3446 (1979).
91. H. Firouzabadi and E. Ghaderi, *Tetrahedron Lett.*, 839 (1978).
92. Reference 16, p. 66.
93. Reference 4, p. 54.
94. A. J. Fatiadi, *Synthesis*, 65, 133 (1976).

95. O. Mancera, G. Rosenkranz, and F. Sondheimer, *J. Chem. Soc.*, 2189 (1953).
96. Reference 6, p. 99.
97. K. Heyns and L. Blazejewicz, *Tetrahedron* **9**, 67 (1960).
98. R. P. A. Sneeden and R. B. Turner, *J. Am. Chem. Soc.* **77**, 130, 190 (1955).
99. H. Nakata, *Tetrahedron* **19**, 1959 (1963).
100. R. M. Moriarty, H. Gopal, and T. Adams, *Tetrahedron Lett.*, 4003 (1970).
101. M. Schroder and W. P. Griffith, *J. Chem. Soc. Chem. Commun.*, 58 (1979).
102. K. B. Sharpless, K. Akashi, and K. Oshima, *Tetrahedron Lett.*, 2503 (1976).
103. (a) M. Fetizon, M. Golfier, and J. M. Louis, *J. Chem. Soc. Chem. Commun.*, 1102 (1969); (b) M. Fetizon and M. Golfier, *C. R. Acad. Sci. Paris* **267**, 900 (1968).
104. A. T. Blomquist and A. Goldstein, in *Organic Syntheses, Coll. Vol. IV*, ed. N. Rabjohn, John Wiley and Sons, New York (1963), p. 838.
105. (a) R. G. R. Bacon and W. J. W. Hanna, *J. Chem. Soc.*, 4962 (1965); (b) R. G. R. Bacon and D. Stewart, *J. Chem. Soc. C*, 1384 (1966); (c) R. G. R. Bacon, W. J. W. Hanna, and D. Stewart, *J. Chem. Soc. C*, 1388 (1966).
106. M. Z. Barakat, M. F. Abdel-Wahab, and M. M. El-Sadr, *J. Chem. Soc.*, 4685 (1956).
107. (a) H. B. Henbest and M. J. W. Stratford, *J. Chem. Soc. C*, 995 (1966); (b) H. B. Henbest and A. Thomas, *J. Chem. Soc.*, 3032 (1957).
108. A. Cave, C. Kan-Fan, P. Potier, J. LeMen, and M. M. Janot, *Tetrahedron* **23**, 4691 (1967).
109. (a) R. H. Reitsema and N. L. Allphin, *J. Org. Chem.* **27**, 27 (1962); (b) L. Friedman, D. L. Fishel, and H. Schechter, *J. Org. Chem.* **30**, 1453 (1965).
110. C. F. Cullis and J. W. Ladbury, *J. Chem. Soc.*, 4186 (1955).
111. J. H. Gardner and C. A. Naylor, in *Organic Syntheses, Coll. Vol. II*, ed. A. H. Blatt, John Wiley and Sons, New York (1955), p. 523.
112. J. A. Caputo and R. Fuchs, *Tetrahedron Lett.*, 4729 (1967).
113. D. C. Ayres and A. M. M. Hossain, *J. Chem. Soc. Chem. Commun.*, 428 (1972).
114. D. M. Piatak, G. Herbst, J. Wicha, and E. Caspi, *J. Org. Chem.* **34**, 116 (1969).
115. M. M. Rogic and T. R. Demmin, *J. Am. Chem. Soc.* **100**, 5472 (1978).
116. T. R. Demmin and M. M. Rogic, *J. Org. Chem.* **45**, 2737 (1980).
117. G. Stork, A. Meisels, and J. E. Davies, *J. Am. Chem. Soc.* **85**, 3419 (1963).
118. L. F. Fieser and J. Szmuszkovicz, *J. Am. Chem. Soc.* **70**, 3352 (1948).
119. C. D. Hurd, J. W. Garrett, and E. N. Osborne, *J. Am. Chem. Soc.* **55**, 1082 (1933).
120. J. R. Ruhoff, in *Organic Syntheses, Coll. Vol. II*, ed. A. H. Blatt, John Wiley and Sons, New York (1955), p. 315.
121. (a) E. Campaigne and W. M. Leseuer, in *Organic Syntheses, Coll. Vol. IV*, ed. N. Rabjohn, John Wiley and Sons, New York (1963), p. 919; (b) I. A. Pearl, in *Organic Syntheses, Coll. Vol. IV*, ed. N. Rabjohn, John Wiley and Sons, New York (1963), p. 972; (c) H. E. Zimmerman and P. S. Mariano, *J. Am. Chem. Soc.* **91**, 1718 (1969).
122. G. D. Mercer, W. B. Beaulieu, and D. M. Roundhill, *J. Am. Chem. Soc.* **99**, 6551 (1977).
123. L. M. Berkowitz and P. N. Rylander, *J. Am. Chem. Soc.* **80**, 6682 (1958).
124. I. T. Harrison and S. Harrison, *J. Chem. Soc. Chem. Commun.*, 752 (1966).
125. R. Brownell, A. Leo, Y. W. Chang, and F. H. Westheimer, *J. Am. Chem. Soc.* **82**, 406 (1960).
126. G. Piancatelli, A. Scettri, and M. D'Auria, *Tetrahedron Lett.*, 3483 (1977).
127. R. J. Highet and W. C. Wildman, *J. Am. Chem. Soc.* **77**, 4399 (1955).
128. V. I. Stenberg and R. J. Perkins, *J. Org. Chem.* **28**, 323 (1963).
129. P. P. Fu and R. G. Harvey, *Chem. Rev.* **78**, 317 (1978).
130. D. Rosenthal, P. Grabowich, E. F. Sabo, and J. Fried, *J. Am. Chem. Soc.* **85**, 3971 (1963).
131. F. Sondheimer, C. Amendolla, and G. Rosenkranz, *J. Am. Chem. Soc.* **75**, 5932 (1953).

132. S. Wolff and W. C. Agosta, *Synthesis*, 240 (1976).
133. Y. Ito, T. Hirao, and T. Saegusa, *J. Org. Chem.* **43**, 101 (1978).

References for Chapter 9

1. F. H. Jardine, *Prog. Inorg. Chem.* **28**, 63 (1981).
2. J. A. Osborn, F. H. Jardine, J. F. Young, and G. Wilkinson, *J. Chem. Soc. A*, 1711 (1966).
3. D. Evans, J. A. Osborn, and G. Wilkinson, *Inorg. Synth.* **11**, 99 (1968).
4. N. Ahmad, J. J. Levison, S. D. Robinson, and M. F. Uttley, *Inorg. Synth.* **15**, 59 (1974).
5. R. Cramer, *Inorg. Synth.* **15**, 14 (1974).
6. J. A. McCleverty and G. Wilkinson, *Inorg. Synth.* **8**, 211 (1966).
7. J. R. Doyle, P. E. Slade, and H. B. Jonassen, *Inorg. Synth.* **6**, 216 (1960).
8. (a) H. A. Tayim, A. Bouldoukian, and F. Awad, *J. Inorg. Nucl. Chem.* **32**, 3799 (1970); H. Colquhoun, unpublished results; (b) H. Itatani and J. C. Bailar, *J. Am. Oil Chem. Soc.* **44**, 147 (1967).
9. D. R. Coulson, *Inorg. Synth.* **13**, 121 (1972).
10. T. A. Stephenson, S. M. Morehouse, A. R. Powell, J. P. Heffer, and G. Wilkinson, *J. Chem. Soc.*, 3632 (1965).
11. R. Mozingo, in *Organic Syntheses, Coll. Vol. III*, ed. E. C. Horning, John Wiley and Sons, New York (1955), p. 685; A. I. Vogel, *A Text Book of Practical Organic Chemistry*, 3rd ed., Longmans, New York (1964), pp. 948–951.
12. H. Lindlar and R. Dubuis, *Org. Synth.* **46**, 89 (1966).
13. R. Ugo, F. Cariati, and G. la Monica, *Inorg. Synth.* **11**, 105 (1968).
14. V. L. Frampton, J. D. Edwards, and H. R. Henze, *J. Am. Chem. Soc.* **73**, 4432 (1951); A. I. Vogel, *A Text Book on Practical Organic Chemistry*, 3rd ed., Longmans, New York (1964), pp. 470–471.
15. R. B. King, *Organometallic Syntheses*, Academic Press, New York (1965), pp. 64–81.
16. G. Booth and J. Chatt, *J. Chem. Soc.*, 3238 (1965); L. M. Venanzi, *J. Chem. Soc.*, 719 (1958); M. J. Hudson, R. S. Nyholm, and M. H. B. Stiddard, *J. Chem. Soc. A*, 40 (1968); Y. Kiso, K. Tamao, N. Miyake, K. Yamamoto, and M. Kumada, *Tetrahedron Lett.*, 3 (1974).
17. N. Yamazaki and T. Ohta, *Polym. J.* **4**, 616 (1973); S. Otsuka and M. Rossi, *J. Chem. Soc. A*, 2630 (1968).
18. J. P. Collman, C. T. Sears, and M. Kubota, *Inorg. Synth.* **11**, 101 (1968).
19. P. S. Hallman, T. A. Stephenson, and G. Wilkinson, *Inorg. Synth.* **12**, 237 (1970).
20. J. P. Fackler, *Prog. Inorg. Chem.* **7**, 361 (1966); W. C. Fernelius and B. E. Bryant, *Inorg. Synth.* **5**, 105 (1957).
21. R. A. Rowe and M. M. Jones, *Inorg. Synth.* **5**, 114 (1957).
22. M. D. Rausch, *J. Org. Chem.* **39**, 1787 (1974).
23. R. B. King, *Organometallic Syntheses*, Academic Press, New York (1965), pp. 82–104.
24. P. Charlesworth, *Platinum Met. Rev.* **25**, 106 (1981).
25. S. E. Livingstone, *The Platinum Metals*, in *Comprehensive Inorganic Chemistry*, Vol. 3, Pergamon Press, New York (1973), pp. 1163–1370; W. P. Griffiths, *The Chemistry of the Rarer Platinum Metals*, Interscience, New York (1967); N. V. Sidgwick, *The Chemical Elements and Their Compounds*, Vol. 2, Oxford University Press, New York (1950), pp. 1454–1628.
26. T. A. Stephenson, S. M. Morehouse, A. R. Powell, J. P. Heffer, and G. Wilkinson, *J. Chem. Soc.*, 3632 (1965).

27. S. N. Anderson and F. Basolo, *Inorg. Synth.* **7**, 214 (1963).
28. G. B. Kauffman and L. A. Teter, *Inorg. Synth.* **7**, 232 (1963).
29. G. B. Kauffman and R. D. Myers, *Inorg. Synth.* **18**, 131 (1978).
30. G. L. Silver, *J. Less-Common Met.* **40**, 265 (1975); **45**, 335 (1976).
31. F. E. Beamish and J. C. Van Loon, *Analysis of Noble Metals*, Academic Press, New York (1977), Chap. 7; G. A. Stein, H. C. Vogel, and R. G. Valerio, U.S. Pat. 2,610,907 (1952); J. Harkema, U.S. Pat. 3,582,270 (1971).
32. W. E. Cooley and D. H. Busch, *Inorg. Synth.* **5**, 208 (1957).
33. S. Herzog, J. Dehnert, and K. Lühder, in *Techniques of Inorganic Chemistry*, Vol. 7, ed. H. B. Jonassen and A. Weiddberger, Interscience, New York (1968); W. L. Jolly, *The Synthesis and Characterization of Inorganic Compounds*, Prentice-Hall, Englewood, Cliffs, New Jersey (1970).
34. D. F. Shriver, *The Manipulation of Air-Sensitive Compounds*, McGraw-Hill, New York (1969).

Subject Index

Compound Index

Page numbers in italic type refer to compounds mentioned in experimental procedures.

The Woman of Substance

THE LIFE AND WORKS OF
BARBARA TAYLOR BRADFORD

Piers Dudgeon

HarperCollins*Publishers*

HarperCollins*Publishers*
77–85 Fulham Palace Road,
Hammersmith, London W6 8JB

www.harpercollins.co.uk

Published by HarperCollins*Publishers* 2005
1

Copyright © Piers Dudgeon 2005

Piers Dudgeon asserts the moral right to
be identified as the author of this work

A catalogue record for this book
is available from the British Library

ISBN 0 00 716568 4

Set in PostScript Linotype Sabon by
Rowland Phototypesetting Ltd, Bury St Edmunds, Suffolk
Printed and bound in Great Britain by
Clays Ltd, St Ives plc

All rights reserved. No part of this publication may be
reproduced, stored in a retrieval system, or transmitted,
in any form or by any means, electronic, mechanical,
photocopying, recording or otherwise, without the prior
permission of the publishers.

CONTENTS

LIST OF ILLUSTRATIONS

12. Freda and daughter Barbara – 'the kind of little girl who always looked ironed from top to toe'. (Bradford Photo Archive)
13. Barbara aged three. (Bradford Photo Archive)
14. Christ Church Armley, where Barbara was baptised and received her first Communion.
15. Barbara as a fairy in a Sunday School pantomime. (Bradford Photo Archive)
16. Barbara with bucket and spade, aged five on holiday at Bridlington. (Bradford Photo Archive)
17 and 18. Leeds Market, where Marks & Spencer began and the food halls in Emma Harte's flag-ship store in *A Woman of Substance* were inspired. (*Yorkshire Post* and Leeds Library)
19. Top Withens, the setting for *Wuthering Heights*. (Yorkshire Tourist Board)
20. 'The roofless halls and ghostly chambers' of Middleham Castle, North Yorkshire. (Skyscan Balloon Photography, English Heritage)

Section 2
21. 1909 Map of Ripon, when Barbara's mother, Freda, was five.
22. Ripon Minster. (Ripon Library)
23. Ripon Market Place as Freda and her mother, Edith, knew it. (Ripon Library)
24. The Wakeman Hornblower, who still announces the watch each night at 9 p.m. (Ripon Library)
25. Water Skellgate in 1904, the year that Edith Walker gave birth there to Freda. (Ripon Library)
26. The stepping stones on the Skell where Freda fell. (Ripon Library)
27. One of Ripon's ancient courts, like the one where Freda was born.

PREFACE

Exploring one of the world's most successful writers through the looking glass of her fiction is an idea particularly well suited in the case of Barbara Taylor Bradford, whose fictional heroines draw on their creator's character and chart the emotional contours of her own experience, and whose own history so often emerges from the shadowland between fact and fiction.

She turned out to be unstintingly generous with her time, advising me about real-life places, episodes and events in the novels, despite a hectic round of her own, which included the writing of two novels, the launch of her nineteenth novel, *Emma's Secret* (2003), a high-profile legal action in India against a TV film company suspected of purloining her books and films, a grand party celebrating a quarter of a century with publishers HarperCollins, and a schedule of charity events, which film producer, business manager and husband Robert Bradford arranged for her – oh, and a week or so's holiday.

Barbara's first novel, *A Woman of Substance*, is, according to *Publishers Weekly*, the eighth biggest-selling novel ever to be published. It has sold more than twenty-five million copies worldwide. In it, so reviewers will tell you, we have the classic

Cinderella story. Emma Harte rises from maid to matriarch; the impoverished Edwardian kitchen maid comes, through her own efforts, to rule over a business empire that stretches from Yorkshire to America and Australia.

What it took to escape the constraints of the Edwardian and later twentieth-century English class system is at the heart of Barbara's family's own story too. Her rise to bestselling novelist and icon for emancipated womanhood, currently valued at some $170 million, from a two-up two-down in Leeds is by any standard extraordinary. Her elevation coincided with the post-war drift from an Edwardian upstairs/downstairs class system (into which Barbara's mother Freda was born), reconstructed by socialism in the period of Barbara's own childhood, to one ultimately sensible to merit, a transformation which finds symbolic incidence in the year 1979, in which *A Woman of Substance* was first published and that champion of meritocracy, Margaret Thatcher, who had risen from the lower middle classes to become Britain's first female Prime Minister, arrived at No. 10 Downing Street.

Barbara's novels, which encourage women to believe they can conquer the world, whatever their class or background and despite the fact that they are operating in a man's domain, tapped into the aspirational energy of this era and served to expedite social change. Indeed, it might be said that Barbara Taylor Bradford would have invented Margaret Thatcher if she had not already existed. When they met, there was a memorable double take of where ambition had led them. 'I was invited to a reception at Number Ten,' Barbara recalls. 'I saw a picture of Churchill in the hall outside the reception room and slipped out to look at it. Mrs Thatcher followed me out and asked if I was all right. I just said: "I never thought a girl from Yorkshire like me would be standing here at the invitation of the Prime Minister looking at a portrait of Churchill inside ten Downing Street," and she whispered: "I know what you mean."'

More intriguingly, in the process of writing fiction, ideas arise which owe their genesis not to the culture of an era, but to the author's inner experience, and here, as any editor knows, lie the most compelling parts of a writer's work. Barbara is the first to agree: 'It's very hard when you've just finished a novel to define what you've really written about other than what seems to be the verity on paper. There's something else there underlying it subconsciously in the writer's mind, and that I might be able to give you later.' She promised this to journalist Billie Figg in the early 1980s, but never delivered, though the prospect is especially enticing, given that she can also say: 'My typewriter is my psychiatrist.'

There is no pearl without first there being grit in the oyster. The grit may lie in childhood experience, possibly only partly understood or deliberately blanked out, buried and unresolved by the defence mechanisms of the conscious mind. Unawares, the subconscious generates the ideas that claw at a writer's inner self and drive his or her best fiction.

Barbara shies away from such talk, denouncing inspiration – it is, she says, something that she has never 'had'. She admits that on occasion she finds herself strangely moved by a place and gets feelings of *déjà vu*, even of having been part of something that happened before she was born, but mostly she sees herself as a storyteller, a creator of stories, happy to draw on her own life, all of it perfectly conscious and practical. Later we will see how she works up a novel out of her characters, which is indeed a conscious process. But subconscious influences are by their very nature not known to the conscious mind, and we will also see that echoes of a past unknown do indeed inhabit her writing.

Barbara will tell you that it was her mother, Freda, who

made her who she is today. Mother and daughter were so close, and Freda so determined an influence, that their relationship reads almost like a conspiracy in Barbara's future success. Freda was on a mission, 'a crusade', but it wasn't quite the selfless mission that Barbara supposed, for Freda had an agenda: she was driven to realise ambition in her daughter by a need to resolve the disastrous experiences of her own childhood, about which Barbara, and possibly even Freda's husband, knew nothing.

In fact, Barbara's story turns out to be an inextricable part of the story of not two but three women – herself, her mother Freda, and Freda's mother, Edith, whose own dream of rising in the world turned horribly sour when the man she loved failed her and reduced her family to penury. It left her eldest daughter Freda with enormous problems to resolve, and a sense of loss which, on account of her extraordinarily close relationship with Barbara, found its way into the novels. One might even say that in resolving it in the lives of her characters Barbara appeased Freda and, in her material success, actually realised Edith's dream.

Barbara told me very little about her mother and grandmother before I began writing this book, and as my own research progressed I had to assume that she did not know their incredible story. If she had known, surely she would have told me. If she had not wanted me to know, why set me loose on research geared to finding out? At the end, when I showed her the manuscript, her shock confirmed that she had not known, and she found it very difficult to accept that she had written about these things of which she had no conscious knowledge.

That the novels own something of which their creator is not master, to my mind adds to their magic, the subconscious process signifying Freda's power over Barbara, which Barbara would not deny. She may not have been aware of all that happened to Freda, but she was 'joined at the hip', as she put it, to one who was aware.

No one was closer to Barbara at the most impressionable moments of her life than Freda. They were inseparable, and the child's subconscious could not have failed to imbibe a sense of what assailed her mother. Even if Freda withheld the detail, or Barbara blanked it out, the adult Barbara allows an impression of 'a great sense of loss' in Freda, which is indeed the very feeling that so many of her fictional characters must overcome.

Freda's legacy of deprivation – her loss – became the grit around which the pearls of Barbara's success were layered. So closely woven were the threads of these three women's lives that Barbara's part in the plot – the end game, the novels – seems in a real sense predestined. In reality, everything in Freda's history, and that of Edith, required Barbara to happen. In the garden of her fiction, they are one seed. None is strictly one character or another, although Barbara regards Freda as most closely Audra Kenton in *Act of Will*. Nevertheless, their preoccupations are what the novels are about.

Of secrets, she writes in *Everything to Gain*, 'there were so many in our family . . . I never wanted to face those secrets from my childhood. Better to forget them; better still to pretend they did not exist. But they did. My childhood was constructed on secrets layered one on top of the other.' In *Her Own Rules*, Meredith Stratton quantifies the challenge of the project: 'For years she had lived with half-truths, had hidden so much, that it was difficult to unearth it all now.'

Secrets produce rumour, which challenges the literary detective to run down the truth at the heart of the most fantastic storylines. Known facts act like an adhesive for the calcium layers of fictional storyline which make up the 'pearl' that is the author's work. These secrets empower Barbara's best fiction. They find expression in the desire for revenge, which Edith will have felt and which drives Emma Harte's rise in *A Woman of Substance*. They find expression in the unnatural force of Audra Kenton's determination in *Act of Will* that her daughter Christina will live a better life than she, and in the driving ambition of dispossessed Maximilian West in *The Women in His Life.*

My project is not, therefore, as simple as matching Barbara's character with the woman of substance she created in the novels, but I believe it takes us a lot closer to explaining why *A Woman of Substance* is the eighth biggest-selling novel in the history of the world than merely observing such a match or studying the marketing plan. That is not to belittle the marketing of this novel, which set new records in the industry; nor is it to underestimate the significance of the timing of the venture; nor to underrate the talents of the author in helping fashion Eighties' *Zeitgeist*.

It is just that these 'secrets', which come to us across a gulf of one hundred years, have exerted an impressive power in the lives of these women, and although we live in a time when the market rules, deep down we still reserve our highest regard for works not planned for the market but which come from just such an elemental drive of the author, especially when, as in the case of *A Woman of Substance*, that drive finds so universal a significance among its readers in the most pressing business of our times – that of rising in the world.

PART ONE

CHAPTER ONE
The Party

'The ambience in the dining room was decidedly roman-
tic, had an almost fairytale quality ... The flickering
candlelight, the women beautiful in their elegant gowns
and glittering jewels, the men handsome in their dinner
jackets, the conversation brisk, sparkling, entertaining ...'
Voice of the Heart

The Bradfords' elegant fourteen-room apartment occupies
the sixth floor of a 1930s landmark building overlooking
Manhattan's East River. The approach is via a grand ground-
floor lobby, classical in style, replete with red-silk chaise long-
ues, massive wall-recessed urns, and busy uniformed porters
skating around black marbled floors.

A mahogany-lined lift delivers visitors to the front door,
which, on the evening of the party, lay open, leaving arrivals
naked to the all-at-once gaze of the already gathered. Fortunately I had been warned about the possibility of this and
had balanced the rather outré effect of my gift – a jar of
Yorkshire moorland honey (my bees, Barbara's moor) – by
cutting what I hoped would be a rather sophisticated, shadowy, Jack-the-Ripper dash with a high-collared leather coat.
If I was successful, no one was impolite enough to mention it.

One is met at the door by Mohammed, aptly named spiriter away of material effects – coats, hats, even, to my chagrin, gifts. Barbara arrives and we move swiftly from reception area, which I would later see spills into a bar, to the drawing room, positioned centrally between dining room and library, and occupying the riverside frontage of an apartment which must measure all of five thousand square feet.

The immediate impression is of classical splendour – spacious rooms, picture windows, high ceilings and crystal chandeliers. These three main rooms, an enfilade and open-doored to one another that night, arise from oak-wood floors bestrewn with antique carpets, elegant ground for silk-upholstered walls hung with Venetian mirrors, and, as readers of her novels would expect, a European mix of Biedermeier and Art Deco furniture, Impressionist paintings and silk-upholstered chairs.

This is not, as it happens, the apartment that she draws on in her fiction. The Bradfords have been here for ten years only. Between 1983 and 1995 they lived a few blocks away, many storeys higher up, with views of the East River and exclusive Sutton Place from almost every room. But it was here that Allison Pearson came to interview Barbara in 1999, and, swept up in the glamour, took the tack that from this similarly privileged vantage point it is 'easy to forget that there is a world down there, a world full of pain and ugliness', while at the same time wanting some of it: 'Any journalist going to see Barbara Taylor Bradford in New York,' she wrote, 'will find herself asking the question I asked myself as I stood in exclusive Sutton Place, craning my neck and staring up at the north face of the author's mighty apartment building. What has this one-time cub reporter on the *Yorkshire Evening Post* got that I haven't?' It was a good starting point, but Allison's answer: 'Well, about $600 million,' kept the burden of her question at bay.

Before long the river draws my gaze, a pleasure boat all lit up, a full moon and the clear night sky, and even if Queens is not exactly the Houses of Parliament there is great breadth that the Thames cannot match, and a touch of mystery from an illuminated ruin, a hospital or sometime asylum marooned on an island directly opposite. It is indeed a privileged view.

Champagne and cocktails are available. I opt for the former and remember my daughter's advice to drink no more than the top quarter of a glass. She, an American resident whose childhood slumbers were disturbed by rather more louche, deep-into-the-night London dinner parties, had been so afeared that I would disgrace myself that she had earlier sent me a copy of Toby Young's *How To Lose Friends & Alienate People*.

I find no need for it here. People know one another and are immediately, but not at all overbearingly, welcoming. In among it all, Barbara doesn't just Europeanise the scene, she colloquialises it. For me that night she had the timbre of home and the enduring excitement of the little girl barely out of her teens who had not only the guts but the *joie de vivre* to get up and discover the world when that was rarely done. She is fun. I would have thought so then, and do so now, and at once see that no one has any reason for being here except to enjoy this in her too.

It is a fluid scene. People swim in and out of view, and finding myself close to the library I slip away and find a woman alone on the far side of the room looking out across the street through a side window. Hers is the first name I will remember, though by then half a dozen have been put past me. I ask the lady what can possibly be absorbing her. I see only another apartment block, more severe, brick built, stark even. 'I used to live there,' she says. 'My neighbour was Greta Garbo . . . until she died.' This, then, was where the greatest of all screen

5

goddesses found it possible finally to be alone, or might have done had it not been for my interlocutor.

'Where do you live now?' I venture.

She looks at me quizzically, as if I should know. 'In Switzerland and the South of France. New York only for the winter months.'

Then I make the faux pas of the evening, thankful that only she and I will have heard it: 'What on earth do you *do*?'

Barbara swoops to rescue me (or the lady) with an introduction. Garbo's friend is Rex Harrison's widow, Mercia. She does not *do*. Suddenly it seems that I have opened up the library; people are following Barbara in. I find myself being introduced to comedienne Joan Rivers and fashion designer Arnold Scaasi, whose history Barbara peppers with names such as Liz Taylor, Natalie Wood, Joan Crawford, Candice Bergen, Barbra Streisand, Joan Rivers of course, and, as of now, all the President's women. Barbara and movie-producer husband Bob are regular visitors to the White House.

It is November 2002 and talk turns naturally to Bob Woodward's *Bush at War*, which I am told will help establish GB as the greatest president of all time. I am asked my opinion and once again my daughter's voice comes like a distant echo – her second rule: no politics (she knows me too well). Barbara has already told me that she and Bob know the Bush family and I just caught sight of a photograph of them with the President on the campaign trail. They are Republicans. When I limit myself to saying that I can empathise with the shock and hurt of September 11, that I had a friend who died in the disaster, but war seems old-fashioned, so primitive a solution, Barbara takes my daughter's line and confesses that she herself makes it a rule not to talk politics with close friend Diahn McGrath, a lawyer and staunch Democrat, to whom she at once directs me.

Barbara is the perfect hostess, this pre-dinner hour the complete introduction that will allow me to relax at table, even to contribute a little. There's a former publishing executive, one Parker Ladd, with the demeanour of a Somerset Maugham, or possibly a Noel Coward (Barbara's champagne is good), who tells me he is a friend of Ralph Fields, the first person to give me rein in publishing, and who turns out – to my amazement – still to be alive.

So, even in the midst of this Manhattan scene I find myself comfortable anchorage not only in contact through Barbara with my home county of Yorkshire, but in fond memories of the publishing scene. It was not at all what I had expected to find. I am led in to dinner by a woman introduced to me as Edwina Kaplan, a sculptor and painter whose husband is an architect, but who talks heatedly (and at the time quite inexplicably) about tapes she has discovered of Winston Churchill's war-time speeches. Would people be interested? she wonders. Later I would see a couple of her works on Barbara's walls, but for some reason nobody thought it pertinent until the following day to explain that Edwina was Sir Winston's granddaughter, Edwina Sandys. Churchill, of course, is one of Barbara's heroes; in her childhood she contributed to his wife Clementine's Aid to Russia fund and still has one of her letters, now framed in the library.

The table, set for fourteen, is exquisite, its furniture dancing to the light of a generously decked antique crystal chandelier. The theme is red, from the walls to the central floral display through floral napkin rings to what seems to be a china zoo occupying the few spaces left by the flower-bowls, crystal tableware and place settings. Beautifully crafted porcelain elephants and giraffes peek out from between silver water goblets and crystal glasses of every conceivable size and design.

There are named place-cards and I begin the hunt for my

seat. I am last to find mine, and as soon as I sit, Barbara erupts with annoyance. She had specially chosen a white rose for my napkin ring – the White Rose of Yorkshire – which is lying on the plate of my neighbour to the right. Someone has switched my placement card! Immediately I wonder whether the culprit has made the switch to be near me or to get away, but as I settle, and the white rose is restored, my neighbour to the left leaves me in no doubt that I am particularly welcome. She tells me that she is a divorce lawyer, a role of no small importance in the marital chess games of the Manhattan wealthy. How many around this table might she have served? Was switching my place-card the first step in a strategy aimed at my own marriage?

I reach for the neat vodka in the smallest stem of my glass cluster and steady my nerves, turning our conversation to Bob and Barbara and soon realising that here, around this table, among their friends, are the answers to so many questions I have for my subject's Manhattan years. I set to work, both on my left and to my right, where I find Nancy Evans, Barbara's former publisher at the mighty Doubleday in the mid-1980s.

By this time we have progressed from the caviar and smoked salmon on to the couscous and lamb, and Barbara deems it time to widen our perspectives. It would be the first of two calls to order, on this occasion to introduce everyone to everyone, a party game rather than a necessity, I think, except in my case. Thumbnail sketches of each participant, edged devilishly with in-group barb, courted ripostes and laughter, but it was only when she came around the table and settled on me, mentioning the words 'guest of honour' that I realised for the first time that I was to be the star turn. I needn't have worried; there was at least one other special guest in Joan Rivers and she more than made up for my sadly unimaginative response.

Barbara tells me that Joan is very 'in' with Prince Charles:

8

'She greeted me with, "I've just come back from a painting trip . . . with Prince Charles." A friend of hers, Robert Higdon, runs the Prince's Trust. Joan is very involved with that, giving them money. So she was there with Charles and Camilla and the Forbses at some château somewhere. She always says, "Prince Charles likes me a lot; he always laughs at my jokes." But Joan is actually a very ladylike creature when she is off the stage, where she can be a bit edgy sometimes. In real life she is very sweet and she loves me and Bob.'

Joan is deftly egging Arnold Scaasi on as he heaps compliments on Barbara from the far end of the table. At the very height of his paean of praise, the comedienne rejoins that Arnold's regard for his hostess is clearly so great that he will no doubt wish to make one of his new creations a *gift* to her. The designer's face is a picture as he realises he has walked straight into a game at his expense, in which the very ethos of celebrity Manhattan is at stake. The table applauds his generosity, while Arnold begins an interminable descent into get-out: Alas, he does not have the multifarious talents of Joan to allow such generosity, he cannot *eat* publicity, etc., etc.

I felt I was being drawn in to Bob and Barbara's private world. When we first met I had said to Barbara that I would need to be so, and she had begun the process that night. After cheese and dessert, coffee and liqueurs were served in the drawing room and one after another of her guests offered themselves for interview.

The evening reminded me of the glittering birthday party in the Bavarian ski resort of Konigsee in *Voice of the Heart*. The table setting was remarkably similar – the candlelight and bowls of flowers that 'march down the centre of the table', interspersed with 'Meissen porcelain birds in the most radiant of colours', the table itself 'set with the finest china, crystal

and silver . . . The flickering candlelight, the women beautiful in their elegant gowns and glittering jewels, the men handsome in their dinner jackets . . .', and the conversation 'brisk,
sparkling, entertaining . . .'

In *Act of Will* the Manhattan apartment is added to the
mix. As guests of Christina Newman and her husband Alex in
their Sutton Place apartment, we, like Christina's mother,
Audra Crowther (née Kenton, and the fictional counterpart of
Barbara's mother, Freda Taylor), are stung by the beauty of
'the priceless art on the walls, two Cézannes, a Gauguin . . .
the English antiques with their dark glossy woods . . . bronze
sculpture by Arp . . . the profusion of flowers in tall crystal
vases . . . all illuminated by silk-shaded lamps of rare and
ancient Chinese porcelains'.

Barbara's passion for antique furniture and modern Impres-
sionist paintings was born of her own upbringing in Upper
Armley, Leeds. 'My mother used to take me to stately homes
because she loved furniture, she loved the patinas of wood.
She often took me to Temple Newsam, just outside Leeds
[where the gardens also found a way into the fiction – Emma
Harte's rhododendron walk in *A Woman of Substance* is
Temple Newsam's], and also to Harewood House, home of
the Lascelles family . . . to Ripley Castle [Langley Castle in
Voice of the Heart], and to Fountains Abbey and Fountains
Hall at Studley Royal.' These were Barbara's childhood
haunts, and it was Freda Taylor who first tuned her in to
beautiful artefacts and styles of design, as if preparing her for
the day when such things might be her daughter's: 'I always
remember she used to say to me, "Barbara, keep your eyes
open and then you will see all the beautiful things in the
world."'

In *A Woman of Substance* the optimum architecture is
Georgian, and Emma Harte's soul mate Blackie O'Neill's

dream is to have a house with Robert Adam fireplaces, Sheraton and Hepplewhite furniture, 'and maybe a little Chippendale'.

In *Angel*, Johnny dwells on the paintings and antiques in his living room – a Sisley landscape, a Rouault, a Cézanne, a couple of early Van Goghs, 'an antique Chinese coffee table of carved mahogany, French bergères from the Louis XV period, upholstered in striped cream silk . . . antique occasional tables . . . a long sofa table holding a small sculpture by Brancusi and a black basalt urn . . .' Costume designer Rose Madigan's attention is caught by a pair of dessert stands, 'each one composed of two puttis standing on a raised base on either side of a leopard, their plump young arms upstretched to support a silver bowl with a crystal liner', the silver made by master silversmith Paul Storr. There are George III candlesticks also by Storr dated 1815.

In *Everything to Gain*, Mallory Keswick feasts her eyes on a pair of elegant eighteenth-century French, bronze doré candlesticks, and her mother-in-law Diana buys antiques from the great houses of Europe, specialising in eighteenth- and nineteenth-century French furniture, decorative objects, porcelain and paintings.

For a dozen or so years leading up to publication of her first novel, Barbara wrote a nationwide syndicated column in America three times a week, about design and interior decor. She also wrote a number of books on interior design, furniture and art for American publishers Doubleday, Simon & Schuster and Meredith, long before the first commissioned *A Woman of Substance*. So, this design thing is, if not bred in the bone, part and parcel of her being.

But these mother and daughter trips out into the countryside had a more fundamental effect: they introduced Barbara to the landscape and spirit of Yorkshire, in which her fiction is

11

rooted. In *A Woman of Substance*, Barbara sets Fairley village, where teenager Emma Harte lives with her parents and brother, Frank, in the lee of the moors which rise above the River Aire as it finds its way down into Leeds. 'It was an isolated spot,' she wrote, 'desolate and uninviting, and only the pale lights that gleamed in some of the cottage windows gave credence to the idea that it was inhabited.'

Today she will say: 'Fairley village is Haworth, but not exactly; it is the Haworth of my imagination.' It could be anywhere in the area of the Brontës' Haworth, Keighley or Rombalds moors. Barbara knows the area well. It lies within the regular expeditionary curtilage of her childhood home in Leeds.

> *The hills that rise up in an undulating sweep to dominate Fairley village and the stretch of the Aire Valley below it are always dark and brooding in the most clement of weather. But when the winter sets in for its long and deadly siege the landscape is brushstroked in grisaille beneath ashen clouds and the moors take on a savage desolateness, the stark fells and bare hillsides drained of all colour and bereft of life. The rain and snow drive down endlessly and the wind that blows in from the North Sea is fierce and raw. These gritstone hills, infinitely more sombre than the green moors of the nearby limestone dale country, sweep through vast silences broken only by the mournful wailing of the wind, for even the numerous little becks, those tumbling, dappled streams that relieve the monotony in spring and summer, are frozen and stilled.*
>
> *This great plateau of moorland stretches across countless untenanted miles towards Shipley and the vigorous industrial city of Leeds beyond. It is amazingly feature-*

*less, except for the occasional soaring crags, a few black-
ened trees, shrivelled thorns, and abandoned ruined
cottages that barely punctuate its cold and empty spaces.
Perpetual mists, pervasive and thick, float over the rugged
landscape, obscuring the highest peaks and demolishing
the foothills, so that land and sky merge in an endless
mass of grey that is dank and enveloping, and everything
is diffused, without motion, wrapped in unearthly soli-
tude. There is little evidence here of humanity, little to
invite man into this inhospitable land at this time of year,
and few venture out into its stark and lonely reaches.*

Near here, at Ramsden Ghyll (Brimham Rocks in the film), 'a
dell between two hills . . . an eerie place, filled with grotesque
rock formations and blasted tree stumps', Lord of the Manor
Adam Fairley seduces Emma's mother, Elizabeth. There, years
later, Adam's son Edwin Fairley makes love to teenage virgin
Emma Harte, the Fairley Hall kitchen maid who conceives
their illegitimate child, Edwina, this episode the impetus
behind a succession of events that will realise Emma's destiny.

*The heather and bracken brushed against her feet, the
wind caught at her long skirts so that they billowed out
like puffy clouds, and her hair was a stream of russet-
brown silk ribbons flying behind her as she ran. The sky
was as blue as speedwells and the larks wheeled and
turned against the face of the sun. She could see Edwin
quite clearly now, standing by the huge rocks just under
the shadow of the Crags above Ramsden Ghyll. When he
saw her he waved, and began to climb upwards towards
the ledge where they always sat protected from the wind,
surveying the world far below. He did not look back, but
went on climbing.*

13

'Edwin! Edwin! Wait for me,' she called, but her voice was blown away by the wind and he did not hear. When she reached Ramsden Crags she was out of breath and her usually pale face was flushed from exertion.

'I ran so hard I thought I would die,' she gasped as he helped her up on the ledge.

He smiled at her. 'You will never die, Emma. We are both going to live for ever and ever at the Top of the World.'

When Edwin abandons Emma she wreaks vengeance on the Fairleys, at length razing Fairley Hall to the ground. Meanwhile, the geography moves some miles to the north. Emma's centre in Yorkshire becomes Pennistone Royal, with its 'Renaissance and Jacobean architecture ... crenellated towers ... mullioned leaded windows' and 'clipped green lawns that rolled down to the lily pond far below the long flagged terrace'. The model is Fountains Hall on the Studley Royal Estate, Ripon, gateway to the Yorkshire Dales and another of Barbara's childhood haunts, while Pennistone Royal village is neighbouring Studley Roger.

Why should an author who left North Yorkshire as soon as she could, found success and glamour in London as a journalist on Fleet Street, married a Hollywood film producer and moved lock, stock and barrel to a swish apartment in New York City, return to her homeland for the setting of her first novel, a novel that featured a character whose spirit seems at first sight more closely in tune with the go-getting ethos of Manhattan than the dour North Yorkshire moors? The answer to that is, broadly, the text of this book.

Barbara's novels are novels principally of character. The dominant traits are the emotional light and shade of the landscape of her birth. When she came to write the novels, she had no hesitation in anchoring them there, even though she was, by then, cast miles away in her Manhattan eyrie.

The county is blessed with large tracts of wide-open spaces – breathtaking views of varied character – so that even if you are brought up in one of the great industrial cities of the county, as Barbara was, you are but a walk away from natural beauty. There is a longing in her for the Yorkshire Dales which living in Manhattan keeps constantly on the boil. Like Mallory Keswick in *Everything to Gain*, 'I had grown to love this beautiful, sprawling county, the largest in England, with its bucolic green dales, vast empty moors, soaring fells, ancient cathedrals and dramatic ruins of mediaeval abbeys . . . Wensleydale and the Valley of the Ure was the area I knew best.'

The author's sense that landscape is more than topography may first have been awakened when Freda introduced Barbara to the wild workshop out of which Emily Brontë's Heathcliff was hewn. 'My mother took me to the Brontë parsonage at Haworth, and over the moors to Top Withens, the old ruined farm that was supposed to be the setting for *Wuthering Heights*. I loved the fact that this great work of literature was set right there. I loved the landscape: those endless, empty, windswept moors where the trees all bend one way. I loved Heathcliff.'

There are many allusions to *Wuthering Heights* in Barbara's novels. For example, *Voice of the Heart* tells of the making of a film of it. Shot in the late 1950s, the film stars heart-throb Terence Ogden as Heathcliff and dark-haired, volatile, manipulative Katharine Tempest as Catherine Earnshaw. *The Triumph of Katie Byrne* is about an actress whose first big break is to play Emily Brontë in a play-within-the-novel about life in Haworth parsonage. In *A Woman of Substance* the principal love story between Emma and Edwin Fairley, though Edwin is no Heathcliff, draws on Brontë's idea of Cathy's sublimation of her self in Heathcliff and in the spirit of the moor: 'My love for Heathcliff resembles the eternal rocks beneath . . .' Brontë wrote. 'Nelly, I AM Heathcliff.' When Emma makes love with Edwin literally within 'the eternal rocks beneath' the moor – in a cave at Ramsden Ghyll – 'Emma thought she was slowly dissolving under Edwin, becoming part of him. Becoming him. They were one person now. She *was* Edwin.'

There is scarcely any landscape description as such in *Wuthering Heights*, but Emily Brontë (1818–48) was the greatest of all geniuses when it comes to evocation of place. Charlotte, her sister, worried what primitive forces Emily had released from the bleak moorland around Haworth, 'Whether it be right or advisable to create things like Heathcliff, I do not know,' she wrote, 'I scarcely think it is.' She compared her sister's genius to a genius for statuary, Heathcliff hewn out of 'a granite block on a solitary moor', his head, 'savage, swart, sinister', elicited from the crag, 'a form moulded with at least one element of grandeur . . . power'. The mark of genius was the writer working an involuntary act – 'The writer who possesses the creative gift owns something of which he is not always master – something that at times strangely wills and works for itself . . . With time and labour, the crag took human shape; and there it stands colossal, dark, and frowning . . . terrible and goblin-like . . .'

Readers of *A Woman of Substance* will know just how central this 'element of grandeur . . . power' is to the character of the woman of substance. Are we to understand that it is hewn from the same granite.crag whence *Wuthering Heights* came? The natural assumption is that Barbara takes from the imagery of that 'nursling of the moors' and transports it to the boardrooms and salons of Manhattan, London and Paris. Certainly, wherever the settings of Barbara's novels take us, her values are Yorkshire based, but hers is a moral focus on the *history* of place, and the spirit of Yorkshire speaks to her through its history as much as through Nature's demeanour.

She owes to her mother Freda's expeditions the sense of drama she shares with mediaeval historian Paul Murray Kendall from 'this region of wild spaces and fierce loyalties and baronial "menies" of fighting men, with craggy castles and great abbeys scattered over the lonely moors . . . a breeding ground of violence and civil strife'. Freda saw to that; she took her to castles – Middleham and Ripley – and to ruined abbeys – Kirkstall Abbey in Leeds and Fountains Abbey on the Studley Royal Estate in Ripon.

Centuries before Emily trod the Brontë 'heath, with its blooming bells and balmy fragrance', and created Heathcliff out of its darker aspects, a real-life personification of power came forth in Wensleydale, the most pastoral, gentle and green of all the Yorkshire Dales, and appealed to Barbara's imaginative sense that the spirit of place is the spirit of the past. For her, Yorkshire is a living ideological and architectural archive of the past, a palimpsest or manuscript on which each successive culture has written its own indelible, enduring text.

Wensleydale lies less than a half-hour's drive from Ripon, the tiny city north of Leeds where Freda was brought up. The dale has two centres of power, Middleham and Bolton castles, and it is the former that commanded her attentions. Middleham was the fifteenth-century stronghold of the Earl of Warwick, one of the most dynamic figures in English history. 'The castle at Middleham is all blown-out walls and windows that no longer exist,' Barbara told me, 'but Richard Neville, Earl of Warwick, who was raised there and lived there, was devastating as a young man, devastating in the sense that he was very driven and ambitious . . . and a great warrior.' Within 'the roofless halls and ghostly chambers' of Middleham Castle,

Freda introduced her daughter to the story of the Earl of Warwick, the 'reach' of his ambitions and many of the traits that would define her woman of substance. 'She told me all about Richard Neville, the Kingmaker . . . He put Edward IV on the throne of England, and he was one of the last great magnates. He held a fascination for my mother.'

Warwick's tireless constitution was rooted in the hard-bitten culture of the North. When Richard was a boy he lined up next to his father to repel attempts to wrest their lands away from them. At eighteen he won his spurs and was hardened further by action in skirmishes to avenge rustling and looting of villages within family territories. He was instinctively the Yorkshire man, but he was also someone who, like the woman of substance herself, was not bound to his home culture. The vitality of his character awakened him to recognise and seize his moment in the wider world when it occurred.

It was in the Wars of the Roses (1455–85), the struggle between the houses of York and Lancaster for the throne of England, that he really came to the fore. His role in changing the English monarchy in the fifteenth century affected England for two centuries afterwards, but his relevance is for all times, as his biographer Paul Murray Kendall records: 'The pilgrim-age of mankind is, at bottom, a story of human energy, how it has been used and the ends it has sought to encompass . . . Warwick's prime meaning is the *reach* of human nature he exemplifies and – type of all human struggle – the combat he waged with the shape of things in his time.'

For Barbara the spirit is all, and in Warwick, as in Middle-ham Castle itself, it is powerfully northern. Born on 28th November 1428 to Richard, Earl of Salisbury, and his wife Alice, 'on his father's side he was sprung from a hardy northern tribe who had been rooted in their land for centuries . . . The North was in Richard's blood, and it nourished his first experiences with the turbulent society of his day,' Kendall writes in *Warwick the Kingmaker*. And yet Richard would hold sway over lands so far distant – more than fifty estates from South Wales across some twenty counties of England – that he, like Barbara, could never be said to have been anchored down by the northern culture in which he was raised.

Neither Kendall nor Barbara go along with the Warwick that Shakespeare gives us in the three parts of *Henry VI* – a 'bellicose baron of a turbulent time'. Kendall's Warwick is 'an amalgam of legend and deeds', a figure whose character and actions attracted heroic levels of adulation and gave him mythic status throughout the land, as he rode in triumph through his vast estates; a figure who, like Barbara herself and her charismatic heroines, seems to have been marked with a strong sense of destiny from the start. Warwick, writes Kendall, never doubted for one moment that he could achieve what he set out to do: 'He refused to admit there were disadvantages he could not overcome and defeats from which he could not recover, and he had the courage, and vanity, to press his game to the end. In other words, he is a Western European man, and in him lies concentrated the reason why that small corner of the earth, in the four centuries after his death, came to dominate all the rest.'

From an early age he gave the impression of a man awaiting his moment, of a 'depth of will' as yet untapped but equal to any challenge that truly merited his time. And when the moment came, when the dream promised to become the man, he recognised it, gave up his subordinate role without second thought, seized it and won it, not with sleight of hand, subterfuge or trickery, but with valour, the occasion the defeat of the King's troops in the city of St Albans in 1455.

His role had been as back-up to the dukes of York and Salisbury against forces raised by Somerset from a full quarter of the nobility of England. They had approached the city making clear their intention to rescue the King from the clutches of Margaret of Anjou, beautiful and feisty niece of Charles VII of France and now wife of King Henry and *the* divisive force in the land. When battle commenced in the narrow lanes that led up to Holywell, York and Salisbury found themselves in serious difficulties and it was then that Warwick took it upon himself to lead his men forward on the run, dashing across domestic gardens and through private houses to attack Somerset's men from the rear. From the moment his archers burst into St Peter's Street shouting 'A Warwick! A Warwick!' his reputation flew. With 'Somerset's host broken,' as Kendall describes, 'Warwick, York and Salisbury approached the peaked King, standing alone and bewildered in the doorway of a house, his neck bleeding from an arrow graze. Down on their knees they went, beseeching Henry the Sixth for his grace and swearing they never meant to harm him. Helplessly, he nodded his head. The battle was over.'

There is in Kendall's Warwick the same unifying robustness to which the nation rises when the England rugby team presses its game to the end, seizing the Webb Ellis trophy against a background of fans clad in the livery of St George. What Kendall is identifying is what attracts Barbara to Winston Churchill and Maggie Thatcher: the character that won us an Empire and coloured what is understood to be our very Englishness.

It is a spirit often given to excess, bigotry, even fanaticism, so that Barbara can say defensively and with evident contradiction: 'There was no bigotry in our family. The only thing my father said was, "Nobody listens to Enoch Powell."' But there is no hint of fanaticism in Barbara's ideals. It is not in her character to support it, and through husband Bob, a German Jew, dispossessed by the Nazis as a boy, Barbara is alert to

20

the danger more than most. She would probably avoid politics altogether if she could, and draws any political sting in the novels by introducing a crucial element of compassion in her heroic notion of power.

In the young Warwick, Barbara found the epitome of the person of substance for whom integrity is all. In her novels, power is 'the most potent of weapons', and it only corrupts 'when those with power will do anything to hold on to that power. Sometimes,' she tells us in full agreement with the Warwick legend, 'it can even be ennobling.'

The character of Warwick that got through to Barbara encompassed more than soldier values. The fierce loyalties of those times were, in young Warwick's case, not forged in greed, nor were they all about holding on to, or wresting, power from an opponent for its own sake. Long before he fell out with his protégé Edward and, embittered, took sides against him; long before he 'sold what he was for what he thought he ought to be', as Kendall put it, his purpose really was to defend the values which true Englishmen held as good.

Freda made sure that Barbara picked up on this heroic aspect. As a child, her mother 'instilled in her a sense of honour, duty and purpose', the need for 'integrity in the face of incredible pressure and opposition' and 'not only an honesty with those people who occupied her life, but with herself'. These noble values arise in *Act of Will* and *A Woman of Substance*, but they first found impetus in Freda's expeditions into Wensleydale; they are what Barbara always understood to be the values of the landscape of her birth. The seed took root when Freda led her by the hand up the hill through Middleham into the old castle keep, even if she was unable to articulate and bring it to flower until she sat down many years later to write *A Woman of Substance*.

In the novel, Paul McGill recognises the woman of substance

in Emma with reference to Henry VI – 'O tiger's heart wrapp'd in a woman's hide'. The heroic values Barbara garnered in her childhood as a result of Freda's influence – the sense of honour, duty and purpose – ensures a strong moral code. 'Emma has such a lot of inner strength,' as Barbara says, 'physical and mental strength, but also an understanding heart. She is tough, but tough is not hard,' an allusion that brings us from Shakespeare to Ernest Hemingway, who once said, 'I love tough broads but I can't stand hard dames.'

Emma is tireless, obsessive, ruthlessly determined and dispassionate. She has a 'contained and regal' posture, there is an imperiousness about her, but she is also 'fastidious, honest, and quietly reserved'. She wears a characteristically inscrutable expression and cannot abide timidity where it indicates fear of failing, which she says has 'stopped more people achieving their goals than I care to think about.' She is physically strong and has a large capacity for hard work. 'Moderation is a vastly overrated virtue,' she believes, 'particularly when applied to work.' Emma is 'tough and resilient, an indomitable woman', with 'strength of will' and 'nerves of steel'. To her PA, Gaye Sloane, she is 'as indestructible as the coldest steel'.

To Blackie's wife, sweet Laura Spencer, with whom Emma lodges, 'there was something frightening about her', the feeling that 'she might turn out to be ruthless and expedient, if that was necessary. And yet, in spite of their intrinsic difference, they shared several common traits – integrity, courage, and compassion.' While 'understanding of problems on a personal level, [she] was hard-headed and without sentiment when it came to business. Joe [Lowther, her husband] had once accused her of having ice water in her veins.' But granddaughter Paula admires Emma's 'integrity in the face of incredible pressure and opposition', and while she can be 'austere and somewhat stern of eye' and there is a 'canny Yorkshire wariness' about her, when her guard is down it is 'a vulnerable face, open and fine and full of wisdom.'

References to Middleham are legion in the novels. In *Angel*, research for a film takes us there. In *Where You Belong* Barbara chooses it as the site for the restaurant, Pig on the Roof, and there's a lovely Yorkshire Christmas there. In *Voice of the Heart*, Francesca Cunningham guides Jerry Massingham and his assistant Ginny to the castle in search of film locations. Key scenes in the film of *A Woman of Substance* were shot in the village, and when you climb up the main street towards the castle you will see to your right the iron-work canopied shop, which, though placed elsewhere in Barbara's imagination, became the film location for Harte's Emporium (Emma's first shop in her empire).

When I visited Middleham with Barbara, an army of horses clattered down the road from the castle to meet us, descending from the gallops and tipping me straightaway into the pages of *Emma's Secret* and *Hold the Dream*, where Allington Hall is one of the greatest riding stables in all England. Barbara, however, was back in her childhood with Freda: 'We'd get the bus to Ripon and then my mother had various cousins who drove us from Ripon to Middleham . . .'

In *Hold the Dream*, past and present find a kind of poetic resolution in this place. Shane O'Neill believes that he is linked to its history through an ancestor on his mother's side. It is 'the one spot on earth where he felt he truly belonged', and at the end he and Emma's granddaughter Paula come together there. This sense of belonging plays an important role in the author's own imaginative life: 'I have very strange feelings there. I must have been about eight or nine when we first went. I thought, I know this place, as if I had lived there. I want to come back'

No matter whether it is Middleham Castle, Studley Royal

or Temple Newsam, Barbara readily enters into an empathic relationship with Freda's favourite places, feeling herself into their history, and it is a strangely intense and markedly subjective relationship. Talking to me about Temple Newsam in Leeds, she said, 'I can't really explain this to you – how attracted I was to the place, my mother and I used to go a lot. It was a tram ride, you'd go on the tram to town and then take another tram . . . or was it a bus? I loved it there, I always loved to go and I felt very much *at home*, like I'd been there before. Yes, déjà vu. Completely.'

'Can you think why that was?' I asked her.

'No. I have no idea.'

'Did you say anything about it to your mother at the time?'

'No, she just knew I loved to go.'

When I drove Barbara to Middleham Castle, we had a similar conversation while exploring what remains of the massive two-storey twelfth-century Keep with Great Chamber and Great Hall above, 'the chief public space in the castle', I read from a sign. 'The Nevilles held court here. Walls were colourful with hangings and perhaps paintings. Clothes were colourful and included heraldic designs . . .'

Barbara interrupted me: 'I have always been attracted to Middleham and I have always had an eerie feeling that I was here in another life, hundreds of years ago. I know it; why do I know it all? *How* do I know it all? Was I here? I know this *place*, and it is not known because I came in my childhood.'

From outside came the sound of children playing. We made our way gingerly up steps nearly one thousand years old, the blue sky our roof now, held in place by tall, howling, windowless walls that supported scattered clumps of epiphytic lichen and wild flowers. Barbara stood still in the Great Hall, taking it all in with almost religious reverence. Then, inevitably, the larking children burst in. She turned, her look silencing them before even she opened her mouth: 'Now look, you've got to stop making a lot of noise. You're disturbing other people. This is not a place for you to play!' It was as if they had desecrated a church. We descended to areas which were once kitchens and inspected huge fireplaces at one time used as roasting hearths, and discovered two wells and a couple of circular stone pits, which a signpost guide suggested may have been fish tanks.

'It was much taller than this, it has lost a lot,' she sighed, and then asked, 'Would it have been crenellated?'

I said I thought that likely, adding, 'It is gothic, dark,' before my eyes returned to the wild flowers in search of a lighter tone. 'Look at the harebells,' I said, but Barbara was not to be deterred. She had come from New York to be there, she wanted me to grasp a point.

'I don't understand why I have this feeling. I don't understand why it is so meaningful to me.'

'There *is* a very strong sense of place here,' I agreed.

'For *me* there is.'

I felt a compulsion to test the subjectivity of Barbara's vision. 'I think *anyone* would find that there is a strong sense of place here,' I said.

She leapt back at me immediately: 'No, no, I *know* this, I have been here, not in this life.' Then, as suddenly, the spell was broken: 'And then you see, you can go down here . . . I had the feeling as a child, I thought I knew it. I had this really strong pull, and I don't know why. I feel I was here in that time, in the Wars of the Roses. I feel that I lived here in the time of Warwick.'

An ability to empathise with the spirit of place is a characteristic of all writers grouped together in the nineteenth-century Romantic movement, not least William Wordsworth, whose poem, 'I wandered lonely as a cloud . . .' was one of Freda's favourites and crops up time and again in Barbara's novels. The verses tell of an empathic moment in the woods beyond Gowbarrow Park, near Ullswater in the Lake District, where the poet and his sister, Dorothy, come upon the most beautiful daffodils they have ever seen: 'Some rested their heads upon these stones as on a pillow for weariness and the rest tossed and peeled and danced and seemed as if they verily laughed with the wind that blew them over the lake . . . ever glancing, ever changing,' Dorothy recorded in her diary. But Barbara's déjà vu experiences are different in an important respect from those of the Romantics. For her, sympathetic identification with Middleham Castle or Temple Newsam or Studley Royal always carries with it a conviction not only that the past is contained in the present, but of *herself* as part of it. The Romantic notion of empathy is absolutely the opposite of this: it is the disappearance of self. Empathy between Keats and the nightingale was contingent on the poet *becoming* the immortal spirit of the bird. Barbara's feeling that she has been to a place before, in another life perhaps, comes from somewhere else. The 'experience' carries a sense of *belonging*. She seems on the verge of finding out more about herself by being there. It has something to do with identity.

Also inherent in what she terms déjà vu (literally 'already seen') is a feeling of *disassociation* with what is felt to have been experienced before; a sense of loss, a sense that there is a past which was hers and has been lost to her. Such a sense of loss can be a powerful inspiration for an author. For instance, Thomas Hardy's novels were inspired by the loss he felt deeply of the land-based, deep-truth culture into which he had been born at Bockhampton in Dorset in the nineteenth century, and we will see that only after Barbara made her return in imagination to the landscape of her birth, and drew on the values that she associated with it, could she write the novels that made her famous.

But unlike Hardy, Barbara was *not* born into the culture or spirit of the times that inspired these values, and there was nothing that she could give me about her past to suggest that something in her identity had been lost to the passing of the times of which Middleham, Temple Newsam or Studley Royal belonged. I was, however, strongly aware that these experiences occurred and had been repeated on many occasions in the company of her mother. The image came to mind of Freda standing hand-in-hand with her daughter in the Keep at Middleham. Everything seemed to lead back to Freda. Why had Freda thought it so important to take Barbara to these places? Was it a committed mother's desire to share their history, or can we see in the intensity of feeling that the trips engendered something more?

Interestingly, Wordsworth's 'Daffodils' poem is used in Barbara's novel *Her Own Rules* to demonstrate that Meredith Stratton has a problem of identity – a terrible feeling of loss, of being robbed, of being incomplete, which is resolved in the novel when she discovers who her mother is. Meredith hears the poem and thinks she has heard it before – but not here, not in this life. It is the first of many so-called déjà vu experiences linked to Meri's true identity, her secret past. '*Her Own Rules* is about a woman who doesn't know who she really is,' as Barbara confirmed.

Was this how it was for Freda? Was she, like Meredith Stratton, drawing something from the spirit of the place that answered questions about her own identity? Was she sublimating the sense of loss, which her daughter noticed in her but could never explain, in the noble spirit of places like Middleham, Temple Newsam, Fountains Abbey and Studley Royal? And did the intensity of the experience encourage her daughter Barbara, with whom she was 'joined at the hip', to identify with their history and experience this déjà vu?

Freda's very being was redolent of the sense of loss which permeates not only the narrative but also some of the best imagery of Barbara's novels, as when the winter sets in 'for its long and deadly siege' and the landscape is 'brush-stroked in grisaille' – a technique to which Barbara alludes not only in *A Woman of Substance* but also in *The Women in His Life* and *Act of Will* invariably to describe a beauty pained by loss.

Barbara, who knew no more about Freda's problems than I did at the time of our trip to Middleham, allowed only that her mother did definitely want her to have a fascination for the history of the places they visited. But she herself had connected these déjà vu experiences with Meredith Stratton's search for her roots of existence, and, as I mulled over our trip to Middleham, I remembered her appraisal that the fundamental theme of all her novels – including *A Woman of Substance* – is one of identity: 'to know who you are and what you are'.

It would be some time, however, before the burden of the theme could be laid at Freda's door.

CHAPTER TWO

Beginnings

'I was the kind of little girl who always looked ironed from top to toe, in ankle socks, patent leather shoes and starched dresses. My parents were well dressed, too.'

Barbara was born on 10th May 1933 to Freda and Winston Taylor of 38 Tower Lane, Upper Armley, on the west side of Leeds. 'Tower Lane was my first home,' Barbara agreed, 'but I was born in St Mary's Hospital in the area called Hill Top. My mother, being a nurse, probably thought it was safer.'

Hill Top crests the main road a short walk from Tower Lane. St Mary's Hospital is set back from the road and today more or less hidden behind trees within its own large site. A map dated the year of Barbara's birth still carries the hospital's original name, 'Bramley Union Workhouse' (Bramley is the next 'village' to the west of Armley). The Local Government Act of 1929 had empowered all local authorities to convert workhouse infirmaries to general hospitals, and by the time Barbara was born, it was probably already admitting patients from all social classes.

As the crow flies, Armley is little more than a mile and a half west of the centre of Leeds, which is the capital of the North of England, second only to London in finance, the law, and for theatre – the Yorkshire Playhouse being known as the National Theatre of the North. More than 50,000 students of its two universities and arts colleges also ensure that it is today one of the great nights out in the British Isles. Straddling the River Aire, which, with the Aire & Calder Navigation (the Leeds canal), helped sustain its once great manufacturing past, Leeds is positioned at the north end of the M1, Britain's first motorway, almost equidistant between London and Edinburgh.

Armley, now a western suburb of the city, sits between the A647 Stanningley Road, which connects Leeds to Bradford, and Tong Road a mile to the south, where the father of playwright Alan Bennett, a contemporary of Barbara's at school, had his butcher's shop.

Armley's name holds the secret of its beginnings, its second syllable meaning 'open place in a wood' and indicating that once it was but a clearing in forest land. Barbara will appreciate this. Oft heard celebrating the 'bucolic' nature of the Armley of old (it is one of her favourite adjectives both in the novels and in life), she recalls: 'In the 1930s this was the edge of Leeds. There were a lot of open spaces . . . little moors – so called – fields, playing fields for football, as well as parks, such as Gott's Park and Armley Park.' There is still a fair today on Armley Moor, close to where Barbara first went to school: 'Every September the fair or "feast" came, with carousels, stalls, candy floss, etc. We all went there when we were children.'

The Manor of Armley and, on the south side of Tong Road, that of Wortley, appear in the *Domesday Book* of 1086 as Ermelai and Ristone respectively. Together they were valued at ten shillings, which was half what they had been worth before the Normans had devastated the North in 1069. In King William's great survey of England, Armley is described as comprising six carucates of taxable land for ploughing, six acres of meadow and a wood roughly one mile by three-quarters of a mile in area.

Not until the eighteenth century did the village come into its own, thanks to one Benjamin Gott, who was *the* outstanding figure among the Leeds woollen manufacturers of the industrial revolution. He was born in Woodall, near Calverley, a few miles west of the town, in 1762. At eighteen, he was apprenticed to the leading Leeds cloth merchant, Wormald and Fountaine. By 1800 the Fountaines had bowed out and in 1816 the Wormald family sold up too. Just how far all this was down to manoeuvring on the part of the acquisitive Gott does not come down to us. What is clear is that long before the firm was renamed Benjamin Gott and Sons it was his energy that made it the most successful woollen firm in England.

Gott's mills – Bean Ing on the bank of the Aire, and a second one in Armley – brought railway terminals, factories and rows of terraced houses for workers, so that Armley was already part of Leeds by the mid-nineteenth century and the whole area was covered in a pall of smoke. So bad was the pollution that as early as 1823 Gott was taken to court. At his trial, the judge concluded that 'in such a place as Leeds, which flourishes in consequence of these nuisances, some inconveniences are to be expected.'

Such attitudes made Gott a rich man. He bought Armley lock, stock and smoky barrel, built himself a big house there and hung it with his European art collection. Like many Victorian entrepreneurs, he was a philanthropist – he built a school and almshouses, organised worker pensions and gave to the Church's pastoral work in the area. After he died in 1840, two sons carried on the business, made some improvements to the mill, but refused to compromise the quality of their high-grade cloths and take advantage of the ready-made clothing industry, which burgeoned after 1850, preferring to exercise their main interest as art and rare-book collectors. Inevitably their markets shrank. When one of the next generation went into the Church, parts of Bean Ing were let out, and by 1897 one tenant had a lease on the entire building.

William Ewart Gott, the third-generation son who stayed in Armley, is lambasted by David Kallinski in *A Woman of Substance* for having built statues and fountains rather than helping the poor, although in fact he provided the land for the foundation – in 1872 – of Christ Church, Armley, where Barbara was christened, received her first Communion and attended service every Sunday, going to Sunday School there as well. He gave towards the building of it and appointed its first vicar, the Reverend J. Thompson, who served a longer term (thirteen years) than any vicar since.

Barbara likes to say that she was 'born in 1933 to ordinary parents in an ordinary part of Leeds and had a similarly ordinary childhood,' but there was nothing unexceptional in the times into which she was born. The industrial revolution had finally ground to a halt. Two years before she was born, in the General Election of 1931, the Conservatives had romped home with some twelve million votes, the party having been elected to stem the economic crisis. It was the last year they would enjoy anything like that tally for some time to come.

The steps leading up to economic crisis and the Tory majority in 1931 led also to Adolf Hitler becoming German Chancellor two years later, and, seemingly inexorably, to war six years after that. Those who lived through it will tell you that the slump started in 1928 in the North of England, but it became world news in October 1929 with the Wall Street crash. Between 1930 and 1933, following President Hoover's decision to raise tariff barriers, world trade fell by two-thirds.

Unemployment in America rose to twelve million (it had been but two million in 1920); in 1931 nearly six million were out of work in Germany. In Britain, in the January of the year of Barbara's birth – 1933 – the same year that Walter Greenwood's classic novel of life in a northern town during the slump, *Love on the Dole*, was published – it reached an all-time peak of 2,979,000. The Depression was on. Barbara was born at the height of it.

There is no doubt that there was great suffering in areas of the northwest and northeast of England. Figures of the unemployed seeking 'relief' in the workhouses in these regions confirm it, but for some there was a less drastic and emotive story. Barbara's family seem not to have suffered too badly, even though her father was unemployed 'for most of my childhood', and was once reduced to shovelling snow, getting paid sixpence for his work and later telling his daughter: 'At that time, Barbara, there was a blight on the land.' The memory went into *Act of Will*, the 'blight on the land' line causing Barbara some grief when her American editor cut it out.

So how did the Taylors make ends meet? 'My mother worked. She worked at nursing and she did all sorts of things. She was a housekeeper for a woman for a while. Do you remember that part of *Act of Will* when Christina gives her mother Audra a party? I remember that party, and I remember having those strawberries. My editor in England, Patricia Parkin, said nothing in the book summed up the Depression better than the strawberries. I cried when I wrote it because I remembered it so clearly – when I say, "their eyes shone and they smiled at each other . . ." I mean, I still choke up now!'

'It's time for the strawberries, Mam, I'll serve,' Christina cried, jumping down off her chair. 'And you get to get the most, 'cos it's your birthday.'

'Don't be so silly,' Audra demurred, 'we'll all have exactly the same amount, it's share and share alike in this family.'

'No, you have to have the most,' Christina insisted as she carefully spooned the fruit into the small glass dishes she had brought from the sideboard. They had not had strawberries for a long time because they were so expensive and such a special treat. And so none of them spoke as they ate them slowly, savouring every bite, but their eyes shone and they smiled at each other with their eyes. And when they had finished they all three agreed that these were the best strawberries they had ever eaten . . .

The dole, or unemployment benefit, was £1 a week in 1930, thirty shillings for man, woman and child. Barbara's father may also have received some sort of disability allowance, for he had lost a leg. A day's work might bring Freda in five shillings, say eighteen shillings a week, cash in hand. That's only 90p in the British decimalised economy, an old shilling being the current 5p piece, but its value was many times greater. In 1930, best butter cost a shilling a pound, bacon threepence for flank, fourpence-halfpenny for side, fivepence or sixpence for ham, two dozen eggs (small) were a shilling, margarine was fourpence a pound, and one pound of steak and rabbit was a shilling. It was quite possible to live on the Taylor income.

Indeed, there was money over to maintain Barbara's 'ironed look from top to toe, in ankle socks, patent leather shoes and starched dresses'. Her parents, she tells us, were well dressed, too. Of course, it was easier to be well dressed in those days. Men wore suits to work whether they were working class or middle class, and few changes of clothing were actually required; women for their part were adept at making do. There is no doubt also that there was the usual Yorkshire care with money in the Taylor household, which Barbara will tell you she has to this day. Certainly, when she was a child there was money left over for her father's beer, a flutter on the horses and even for summer holidays, taken at the east coast resort of Bridlington, the seaside holiday being a pastime whose popularity was on the increase, while foreign travel remained an elite pursuit for the very rich.

Another apparent anachronism of the depressed 1930s is that it was also the decade of the mass communication and leisure revolution, which facilitated industries that would be Barbara's playground as an adult. British cinema began as a working-class pastime, films offering escapism, excitement and a new focus for hero worship more palatable than the aristocracy, as well as a warm, dark haven for courting couples. The first talkie arrived in Britain in 1929. By 1934 there was an average weekly cinema attendance of 18.5 million (more than a third of the population), and more than 20 million people had a radio in the home.

Sales of newspapers also burgeoned, with door-to-door salesmen offering free gifts for those who registered as readers – it was rumoured that a family could be clothed from head to foot for the price of reading the *Daily Express* for eight weeks. In 1937 the typical popular daily employed five times as many canvassers as editorial staff. It is interesting that Barbara is wont to say in interview, 'I'd read the whole of Dickens by the time I was twelve,' because complete sets of Dickens were a typical 'attraction' offered to prospective readers – perhaps to *Daily Mirror* readers in particular, for 'When I was a child,' Barbara once said, 'we had the *Mirror* in our house and I have always been fond of it.'

Literacy increased throughout the country at this time, partly due to the expansion of the popular press, the sterling work of libraries and the coming of the paperback book. Allen Lane founded the Penguin Press in 1936 and Victor Gollancz set up the Left Book Club in the same year. Literature, as well as books not classifiable as such, was now available to the masses: 85.7 million books were loaned by libraries in 1924, but in 1939 the figure had risen to 247.3 million.

Freda took full advantage. 'She was a great reader and force-fed books to me. I went to the library as a child. My mother used to take me and plonk me down somewhere while she got her books.'

Armley Library in Town Street was purpose-built in 1902 at a cost of £5,121.14s. It is five minutes' walk from the family's first house in Tower Lane and even less from Greenock Terrace, to which the Taylors repaired during the war. Libraries in the North of England are often supreme examples of Victorian architecture, like other corporation buildings an excuse to shout about the industrial wealth of a city. Though Armley's is relatively small, there is something celebratory about its trim, and the steps leading up to the original entrance give, in miniature, the feeling of grandeur you find in Leeds or Manchester libraries, for example. What's more, the architect, one Percy F. Robinson, incorporated a patented water-cooled air-conditioning system in the design. 'It was a beautiful building,' Barbara agreed when I told her that I was having difficulty getting access to local archives because it was now closed. 'Don't tell me they are destroying it!' she exclaimed in alarm. It was closed in fact for renovation, and today there is a pricey-looking plaque commemorating Barbara's reopening of it in November 2003.

'My mother exposed me to a lot of things,' Barbara continued. 'She once said, "I want you to have a better life than I've had." She showed me – she *taught* me to look, she taught me to read when I was very young. She felt education was very important. She would take me to the Theatre Royal in Leeds to see, yes, the pantomime, but also anything she thought might be suitable. For instance, I remember her taking me to see Sadlers Wells when it came to the Grand Theatre in Leeds. I remember it very well because Svetlana Beriosova was the dancer and I was a young girl, fifteen maybe. I loved the theatre and I would have probably been an actress if not a writer. I remember all the plays I was in, the Sunday School plays: I was a fairy – I have a photograph of myself! – and a witch! And then I was in the Leeds Amateur Dramatics Society, but only ever as a walk-on maid. We did a lot of open-air plays at Temple Newsam, mostly Shakespeare. I have a picture somewhere of me in an Elizabethan gown as one of the maids of honour.' The involvement of Barbara and some of her schoolfriends in these plays was organised by Arthur Cox, a head teacher in the Leeds education system, whose wife was a teacher at Northcote School, which Barbara attended from 1945. A friend at the time, June Exelby, remembers: 'We used to go and be extras in things like *Midsummer Night's Dream* – as fairies and things like that. Barbara used to particularly enjoy it. I can't remember whether she was any good at it.'

Affluence in Armley seemed to rise and fall with the topography of the place. Going west from Town Street at Wingate Junction, which was where the Leeds tram turned around in Barbara's day, up Hill Top Road and over the other side to St Mary's Hospital and St Bede's Church, where Barbara went to dances as a girl, the houses were bigger and owner-occupied by the wealthier professional classes: 'It was considered to be the posh end of Armley,' she recalled.

Tower Lane, where Barbara lived with Freda and Winston, is a pretty, leafy little enclave of modest but characterful, indigenous-stone cottages. It is set below Hill Top but hidden away from the redbrick industrial terraces off Town Street to the east, in which most of the working-class community lived. It must have seemed a magical resort to Barbara in the first ten years of her life, and certainly she remembered Armley with a fairytale glow when she came to write about it in *A Woman of Substance* and *Act of Will*, Emma Harte and Audra Kenton both coming upon it first in the snow.

> *Audra saw at once that the village of Upper Armley was picturesque and that it had a quaint Victorian charm. And despite the darkly-mottled sky, sombre and presaging snow, and a landscape bereft of greenery, it was easy to see how pretty it must be in the summer weather.*

In *A Woman of Substance*, it is 'especially pretty in summer when the trees and flowers are blooming,' and in winter the snow-laden houses remind Emma explicitly of a scene from a fairytale:

> *Magically, the snow and ice had turned the mundane little dwellings into quaint gingerbread houses. The fences and the gates and the bare black trees were also encrusted with frozen snowflakes that, to Emma, resembled the silvery decorations on top of a magnificent Christmas cake. Paraffin lamps and firelight glowed through the windows and eddying whiffs of smoke drifted out of the chimneys, but these were the only signs of life on Town Street.*

It is a little girl's dream. Although the description is unrecognisable of Armley today, and its 'mundanity' is again deliberately discarded when Barbara selects Town Street as the spot where Emma Harte leases a shop and learns the art of retail, setting herself on the road to making millions, we accept it because it was plausible to the imaginative little girl who lived and grew up there: 'There are a number of good shops in Town Street catering to the Quality trade,' Blackie tells Emma when she first arrives:

They passed the fishmonger's, the haberdasher's, the chemist's, and the grand ladies' dress establishment, and Emma recognised that this was indeed a fine shopping area. She was enormously intrigued and an idea was germinating. It will be easier to get a shop here. Rents will be cheaper than in Leeds, she reasoned logically. Maybe I can open my first shop in Armley, after the baby comes. And it would be a start. She was so enthusiastic about this idea that by the time they reached the street where Laura Spencer lived she already had the shop and was envisioning its diverse merchandise.

Today, beyond Town Street's maze of subsidiary terraces, where Barbara's father Winston's family once lived, stand Sixties tower blocks and back-to-back housing with more transient tenants not featured in Barbara's fiction. And at the end of the line stands Armley Prison, its architectural purpose clearly to strike terror into the would-be inmates. This does register in *A Woman of Substance* – future architect Blackie O'Neill calls it a 'horrible dungeon of a place'. Nearly a century later, multiple murderer Peter Sutcliffe – 'the Yorkshire Ripper' – added to its reputation.

Now, twenty-five per cent of Armley's inhabitants are from ethnic minorities where English is a second language. The great change began as Barbara left for London in the 1950s. As a result, the culture of Armley village today is unlike anything she remembers, even though, according to local headmistress Judy Blanchland, inhabitants still feel part of a tradition with sturdy roots in the past, and have pride in the place. Certainly there is continuity in generations of the same families attending Christ Church School. The school, and the church opposite, remain very much the heart of the local community, with around 100 attending church on Sunday, seventy adults and some thirty children. There always has been a lot of to-ing and fro-ing between the two, even if changing the name of Armley National School, as it was in Barbara's day, to Christ Church School did cause something of a stir.

Barbara enrolled there on 31st August 1937, along with eleven other infants. Her school number was 364 until she was elevated to Junior status in 1941, when it became 891. Alan Bennett, born on 9th May 1934, one year after Barbara, joined on 5th September 1938, from his home at 12 Halliday Place. The families didn't know one another. 'My mother used to send me miles to a butcher that she decided she liked better [than Bennett's shop on Tong Road]. It was all the way down the hill, almost on Stanningley Road.' After leaving the school, the two forgot they had known one another until the day, fifty years later, when they were both honoured by Leeds University with a Doctor of Letters *Honoris Causa* degree.

Bennett became a household name in England from the moment in 1960 that he starred in and co-authored the satirical review *Beyond the Fringe* with Dudley Moore, Peter Cooke and Jonathan Miller at the world-famous Edinburgh Arts Festival. Later the show played to packed audiences in London's West End and New York. He was on a fast-track even at Christ Church School, passing out a year early, bound for West Leeds High School, according to the school log. From there the butcher's son won a place at Oxford University.

Barbara and I walked the area together in the summer of

2003, mourning the fact that generally little seems to have been done to retain the nineteenth-century stone buildings of her birthplace. Even many of the brick-built worker terraces, which have their own period-appeal, have been daubed with red masonry paint in a makeshift attempt to maintain them. There was, however, enough left to remind Barbara of her childhood there.

We drove up Town Street towards Tower Lane, where she lived until she was about ten. At Town Street's west end, you can filter right into Tower Lane or left into Whingate, site of the old tram terminal and the West Leeds High School, now an apartment block. (See 1933 map in the first picture section.) The small triangular green between Town Street and Whingate which appears at this point must have been a talking point for Barbara and her mother from earliest times, if only on account of its name – Charley Cake Park – mentioned in both *A Woman of Substance* and *Act of Will*. 'Laura told me that years ago a man called Charley hawked cakes there,' says Blackie. Emma believes him, but only because no-one could invent such a name for an otherwise totally insignificant strip of grass.

As we wind our way towards Barbara's first home, she has a mental picture of 'me at the age of three, sitting under a parasol outside 38 Tower Lane, near a rose bush. It *is* a lane, you know,' she emphasises, 'and it was a tiny little cottage where we lived. Do you think it is still there? We got off the tram here . . . Whingate Junction . . . then we walked across the road and up Tower Lane, and there was a very tall wall, and behind that wall were . . . sort of mansions; they were called The Towers.'

At the mouth of the lane she points to a cluster of streets called the Moorfields: 'That used to be where the doctor I went to practised – Doctor Stalker was his name. One of those

41

streets went down to the shop where I got the vinegar. Did I tell you about the vinegar? Boyes, a corner shop, that's where I used to get it. I wonder if that's still there?'

The vinegar turns my mind to Barbara's penchant for fish and chips. I had heard that when she comes to Yorkshire she likes nothing better than to go for a slap-up meal of fish and chips, mushy peas, and lashings of vinegar. That very night I would find myself eating fish and chips with her in Harrogate. Nothing odd, you might say, about a Yorkshire woman eating a traditional Yorkshire meal, only Barbara has her posh cosmopolitan heroes and heroines do it in the novels too, and has herself been known to request, and get, a bottle of Sarson's served at table in the Dorchester Grill.

What is Emma Harte, the woman of substance's favourite dish? Fish and chips, preceded by a bowl of vegetable soup, served in Royal Worcester china of course, the only concession to Emma's transformation. Again there is this feeling of fairy-tale about it all, except that one knows that the writer has herself made the same journey as Emma Harte, and that she does in fact order fish and chips too. The desire seems to pass down the generations, so that Emma's grandson, the immensely wealthy Philip McGill Amory, insists on eating fish and chips with his wife Madelena – what matter if she is wearing a Pauline Trigère evening gown?

'My mother used to send me to get the household vinegar from Mr Boyes,' Barbara continues, 'and she sent me with a bottle because it was distilled from a keg, and when I returned with it she'd always look at the bottle and say, "Look at this, he's cheating me!" Until one day she went in herself with the bottle and she said to Mr Boyes, "You're cheating me. I *never* get a full bottle," and apparently Mr Boyes replied, "Eeh, ah knows. Tha' Barbara's drinking it." He was very broad Yorkshire, and it's true, I did drink a bit of it on the way home.

42

Even today I like vinegar on many things, but especially on cabbage . . .

'There's Gisburne's Garage!' I slow down as we pass the garage on our right at the mouth of the lane, and she points out an old house, pebble-dashed since she was a girl. 'This was where Mrs Gisburne lived and it had a beautiful garden in the back. But where this is green there used to be a pavement, surely . . . but maybe it wasn't, perhaps I am seeing . . .'

This is the first time that Barbara has set foot in the place for fifty years. What will turn out to be real of her childhood memories? What part of imagination? Childhood memories play tricks on us. She looks for the 'tall wall' that she remembers should be on our left, containing the mansions known as The Towers. There is a wall, but it is not tall, nor have the original blackened stones been touched since the four- or five-foot construction was built all those years ago.

'That wall used to seem so *high* when I was a child,' she says in amazement. 'Anyway, these are called The Towers and this is where Emma had a house and they were considered to be very posh. It was all trees here.'

The Towers stretched many floors above us and must have seemed to a child's eye to reach into the sky. Their castellated construction of blackened West Riding stone gives them a powerful, gothic feel, and it was the majesty of the site that captured Barbara's wonder when she was growing up here. Her eyes must have fallen upon the building virtually every day during her most impressionable years, whenever she emerged from the garden of her house opposite:

The Towers stood in a private and secluded little park in Upper Armley that was surrounded by high walls and fronted by great iron gates. A circular driveway led up to the eight fine mansions situated within the park's precincts, each one self-contained, encircled by low walls and boasting a lavish garden. The moment Emma had walked into the house on that cold December day she had wanted it, marvelling at its grandness and delighted with its charming outlook over the garden and the park itself.

A Woman of Substance

But where was No. 38 Tower Lane, supposedly opposite the tall wall of The Towers? There is No. 42 and 44 ... but no label indicating No. 38. 'We *were* thirty-eight,' Barbara insists as she alights from the car to get a better look. We move through a gate into a front garden, and, set back from the lane, we see what might have once been a row of three tiny, terraced stone cottages, all that was left of the courtyard where they lived, 'the small cul-de-sac of cottages,' as she wrote in *A Woman of Substance*, describing the neighbourhood of Emma's childhood home.

'This seems very narrow,' she says as we make our way gingerly down the flagged path like trespassers in time. 'They've knocked it all down, I think, and turned it into this. All right, well, I'll find it! There was a house across the bottom,' she muses for her own benefit. 'This, the first of the line [of cottages] was number thirty-eight. You went down some steps. Here it was a sort of garden bit, and where the trees are ... There were three cottages along here and then a house at the bottom, which has gone. Wait a minute, are there three cottages or only two? Have they torn our house down? Well, this is the site of it anyway.' Her voice breaks as she says this. 'There *were* three houses there. There was our cottage, the people in the corner and the lady at the bottom. There *were* three houses.' She then shows me the site of their air-raid shelter, where she and Freda would sit when the sirens sounded during the early years of the war. In the whole of the war only a handful of bombs actually fell on Leeds, but the preparations were thorough, the windows of trams and shops covered with netting to prevent glass shattering all over the place from bomb blast, entrances to precincts and markets sandbagged against explosions.

'I went to school with a gas mask, I remember,' said Barbara. 'We all had them in a canvas bag on our shoulders and there used to be a funny picture of me with these thin little legs – I've got thin legs even today – thin little legs with the stockings twisted and a coat and the gas mask and a fringe. My mother was cutting her rose bushes and I was playing with my dolls' pram that day in 1940 when a doodlebug, a flying bomb, came over, and she just dropped everything and dragged me into the air-raid shelter. I vaguely remember her saying to my father later – he was out somewhere – "Oh, I never thought I'd see that happen over England."'

No. 38 Tower Lane had two rooms downstairs and two bedrooms on the upper floor. That is all: a sweet, flat-fronted cottage; a tiny, humble abode. The house at the bottom of the garden is long gone. Its absence offers by way of recompense a spectacular view across the top of Leeds, although Barbara's interest, as we walk the area, is only in how things were, and how they are no more.

Being an only child had various repercussions. Her parents will have been able to feed and clothe Barbara to a better standard than most working-class children, which we know to have been the case. But it would have set Barbara apart for other reasons, too – single-child families were unusual in those days before family planning, and in the single-child home the emphasis was on child-parent relationships rather than sibling friendships and rivalries, which can affect a child's ability to relate to other boys and girls at school; although when things are going well between child and parents it can make the relationship extra-special. 'There were plenty of times,' she says, 'when I just knew that we were special, the three of us. I always thought that we were special and they were special. I think when you are an only child you are a unit more. I always adored them. Yes, rather like Christina does in *Act of Will.*'

The closeness and reliance of Barbara on the family unit was never more clearly shown than in the only time she spent away from home during her early childhood, as an evacuee. The school log reads: '*1st September 1939, the school was evacuated to Lincoln this morning. Time of assembly 8.30, departure from school, 9, to Wortley Station, departure of train, 9.43.*'

The school stayed closed until 15th January 1940: '*Re-opened this morning, three temporary teachers have been appointed to replace my staff, which are still scattered in the evacuation areas. Miss Laithwaite is at Sawbey, Miss Maitland at Ripon, Miss Musgrave at Lincoln and Miss Bolton is assist-ing at Meanwood Road. The cellars have been converted into air-raid shelters for the Infants. Accommodation in the shelters, 100. Only children over 6 can be admitted for the present.*'

'I went to Lincoln,' remembers Barbara, 'but I only stayed three weeks. It was so stupid to send us to Lincolnshire. I remember having a label on me, a luggage label, and my mother weeping as the school put us on a train. I was little. I wasn't very happy, that I know, I missed my parents terribly. I was very spoiled, I was a very adored child. My mother sent me some Wellington boots, so it must have been in winter. She'd managed to get some oranges and she'd put them in a boot with some other things, but the woman had never looked inside. So, when my father came to get me the oranges were still there and had gone bad. My mother was furious about that.

'Daddy came to get me. He'd gone to the house and they said, "She'll be coming home from school any moment." He said, "Which way is it? I'll go to meet her." And I saw him coming down the road and I was with the little girl who was at the house also. I remember it very well because I started to run – he was there on the road with his stick, walking towards me . . . and I'm screaming, "Daddy, Daddy, Daddy!" He said, "Come on, our Barbara, we're going home." We stood all the way on the train to Leeds. I was so happy because I missed my parents so much it was terrible. I was crying all the time – not all the time, but I cried a lot, I didn't like it. I didn't like being away from them. I loved them so much.'

I lead us back out of the gate and we return to the present with more sadness than joy. Walking further up the lane, which has a dogleg that leads eventually out onto Hill Top Road, we explore a steep track down the hill to the right, which Barbara calls 'the ginnel'. Later I discover from Doreen Armitage, who also grew up in the area and let us in to the church, that this is an ancient weaver's track: 'They would bring up the wool to the looms from the barges on the canal there.' And, sure enough, I see on today's map that it is marked as a quarter-mile cut-through to the canal across Stanningley Road. Barbara was lost once more in her own memories – the fun she had as a little girl skipping down the ginnel – before again being arrested by the intrusive present: the gardens behind Gisburne's Garage, once so lovely, have been built upon and obscured.

Despondently we make our way back up the ginnel towards the moor where she would often play after school. Past the main gate of The Towers we emerge from a tunnel of trees into a wide-open space, flanked on our left by an estate of modern houses, which has replaced the 'lovely old stone houses' of her youth. Off to the right, we come to what was always referred to as 'the moor', but is no more than half an acre of open ground, where now a few strongly built carthorses are feeding. On the far side of it is a wall and some trees. Barbara at once exclaims: 'That's the wall! When you climbed over *that* wall, you were in something called the Baptist Field – I don't know why it was called that, but . . . we used to play in that field, some other children and I, we used to make little villages, little fairylands in the roots of the trees, which were all gnarled, with bits of moss and stones and bits of broken glass, garnered from that field, and flowers.'

Memories of the Baptist Field had been magical enough to earn it, too, a place in *A Woman of Substance* all of forty years later. In the novel, the field promises entry to Ramsden Crags and the Top of the World, symbol of the spirit of Yorkshire.

Barbara's fictional recipe may involve real places, but it is the feelings recalled from her youth that are especially true in the novels, and overlooking the Baptist's Field I felt in at the very source of a little girl's teeming imagination. As with Emma Harte, the years peeled away on her feelings as a child and 'she had a sudden longing to go up to the moors, to climb that familiar path through the Baptist Field that led to Ramsden Crags and the Top of the World, where the air was cool and bracing and filled with pale lavender tints and misty pinks and greys . . . Innumerable memories assailed her, dragging her back into the past.' (*A Woman of Substance*)

'My father used to walk up here sometimes and go for a drink at the Traveller's Rest,' Barbara says, breaking the silence. I had already noted the pub on Hill Top Road. Walking as far as we could up Tower Lane would bring us round to it. Later, Doreen Armitage would recall sitting up there 'with me father and me Uncle Fred; we used to sit there on a Sunday

Barbara's mother, Freda Walker, aged about 18 when working at Ripon Fever Hospital.

Freda in her early twenties with baby Tony Ellwood.

Winton's sister, Laura, Barbara's favourite aunt who died of lung cancer during the Second World War. She was the model for Laura Spencer in *A Woman Of Substance*.

Winston Taylor, Barbara's father, in the Royal Navy, aged about 16.

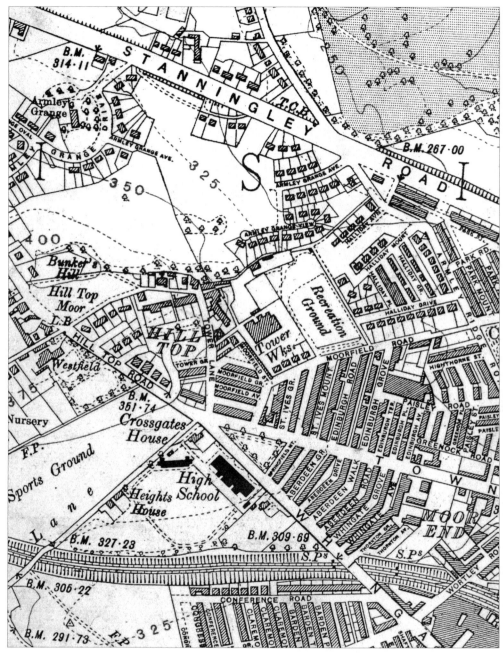

The map, dated 1933 (the year Barbara was born) shows Town Street, Armley, leading north-west past Charley Cake Park (the triangular green that figures in *A Woman Of Substance* and *Act Of Will*) into Tower Lane – Barbara's birthplace. Thence, past The Towers (Emma Harte's residence in the first novel) onto Hill Top Moor and the Baptist Field (another childhood haunt present in the fiction, shown here as Bunker's Hill). Opposite the moor, on Hill Top Road, still stands The Traveller's Rest, which Winston frequented, the road leading off the map to Bramley Union Workhouse (later St Mary's Hospital, where Barbara was born). To the north of Town Street can be seen St Ives Mount, where Winston's parents lived; and eastwards lie the Greenocks, to which Barbara's family moved during the war. Greenock Terrace, not named, is the second street at right angles to Greenock Road; Barbara lived at No. 5. The church and school she attended are off the map, a few streets further north.

Barbara's parents, Winston and Freda Taylor in 1929, the year they were married.

Barbara, a chubby two-year-old, walking in Gott's Park, Armley.

Left: Tower Lane, a surprise piece of rural bliss on the edge of dense industrial terraces. Barbara's first home was off to the right, The Towers are beyond the 'tall wall' on the left.

All that survives of the courtyard in which was set 'the tiny little cottage', No. 38, bucolic sanctuary of Barbara's earliest years.

Above: The Towers, the gothic, castellated construction of blackened West Riding stone which so commanded Barbara's attention as a child, reaching high above her home, that she included the house in her first novel.

Below: Armley National School (now Christ Church School), which Barbara attended from 31st August, 1937. Alan Bennett, son of a butcher on Tong Road, joined her a year later.

Freda in fur jacket and cloche hat, Barbara in her Whitsuntide clothes – leghorn straw bonnet, pale green coat and cream gloves – 'the kind of little girl who always looked ironed from top to toe'.

Barbara aged three. She remembers always wearing gloves.

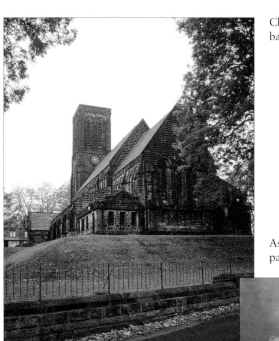

Christ Church Armley, were Barbara was
baptised and received her first Communion.

As a fairy in a Sunday School
pantomime. Freda made the costume.

Barbara with bucket and spade, aged five
– on holiday at Bridlington in Yorkshire.

Above: Leeds Market. 'Mummy and I used to go on Saturday mornings,' Barbara recalls. 'The first Mr Marks [of Marks & Spencer] had a stall there, which had the sign: "Don't ask the price it's a penny". So, I had Emma go into that market.' The food halls in Emma Harte's flag-ship store in *A Woman of Substance* owe much to the weekly trips. *Below:* 'Game Row' as it was then.

Barbara's mother took her over the moors to Top Withens, the setting for *Wuthering Heights*: 'I loved the fact that this great work of literature was set right there. I loved the landscape: those endless empty, windswept moors where the trees all bend one way.' Yorkshire awakened in her the notion that landscape can hold the the spirit of the past.

Within 'the roofless halls and ghostly chambers' of Middleham Castle, north of her own birthplace, Ripon, Freda introduced her to the story of Richard Neville, Earl of Warwick, and many of the traits that would define the character of her woman of substance.

morning.' Barbara said she used to sit there with Winston: 'We were probably waiting for the pub to open, I should think! I used to come here with my father, and we would sit out-side and my mother would have a shandy. He liked to go out for his pint, you know, and have his bet ... Ripon, York, Doncaster races. He bred in me a love of horses and racing.'

Barbara's father, Winston Taylor, was born on 13th June 1900 at 6 Wilton Place, Armley, the first child of Alfred and Esther Taylor. 'Daddy was called after Winston Churchill, who had just escaped from the Boer War.' On Barbara's admission, Winston is Vincent Crowther in *Act of Will*, the firstborn of Alfred and Eliza Crowther. He is also Emma Harte's brother, Winston, in *A Woman of Substance*. 'The fictional Winston Harte looked like him, thought like him, and had many of his characteristics,' Barbara told me.

I also learn that she based Emma Harte's father, 'Big Jack Harte', who, as a Seaforth Highlander, fought the Boers in 1900 and 'could kill a man with one blow from his massive fist', on Winston's father, big Alfred Taylor, 'because a certain ingredient of physical strength was required in the character of Emma', the woman of substance she was concocting.

'I loved my grandfather. He was in my mind when I created Jack Harte. He had a moustache, all lovely and furry white, and white hair. Big man, big moustache. He used to hold me on his knee, give me peppermints and tell me stories about when he was a Sergeant Major in the Seaforth Highlanders. He loved me and I loved him.'

Grandpa Alfred Crowther in *Act of Will* also serves as a Sergeant Major in the Seaforth Highlanders, and in *A Woman of Substance* Emma's first husband, Joe Lowther, and Blackie O'Neill join the regiment in the First World War.

The real-life Alfred Taylor, who occasioned these many ref-erences in Barbara's fiction, is described as a forgeman on his son Winston's birth certificate and as a cartman on Winston and Freda's marriage certificate. Barbara remembers him as the latter, as the Co-op drayman – the man who looked after the horses and drove the cart carrying stores for the Armley link in Britain's first supermarket chain.

The senior Taylors lived five minutes away from Tower Lane in Edinburgh Grove, off Town Street. 'The house was a Victorian terraced house with a series of front steps, but also a set of side steps leading down to the cellar kitchen. It was a through-house, not a back-to-back, but I can't remember if there were doors on both sides. They then moved to 5 St Ives Mount, two streets closer. This was also a Victorian terraced house with front steps down to the cellar kitchen, where you'd always see my grandmother baking. Both were tall houses as I recall.

'I also liked my grandmother, Esther Taylor. Her maiden name was Spence. She was very sweet and not quite Alfred Crowther's wife in *Act of Will*. Eliza Crowther is always the voice of doom, saying, "Happiness, that's for them that can afford it." My real grandmother was full of other sayings like, "A stitch in time saves nine", "You'd better watch your p's and q's", though I do remember her baking like Mrs Crowther – bacon-and-egg pies (we'd call them quiche Lorraine today), also apple pie, and sheep's-head broth and lamb stews. On Saturday morning I used to go for her to the Co-op, but we didn't call it the Co-op, it was called the *Cworp* – Armley dialect, I suppose. You could buy everything at the Cworp. It had a meat department, vegetables, groceries, cleaning products . . . She loved me. I think I was her favourite.'

All was as Barbara remembered of her grandparents' houses, and she went on to tell me about the rest of the Taylor brood. 'Winston had a sister called Laura, who lived in Farsley, was married and had a child. She was my favourite aunt. She died of lung cancer during the war, when I was about seven; she died at home and it was a horrible death. I remember going to see her. She was a very heavy smoker. After Laura died, her husband went to live with his family with their little boy, but died not long after. Laura Spencer in *A Woman of Substance* was based on my Aunt Laura, except our family were not Catholic. She was sweet and gentle and so good.

'Then there was Olive, married to Harry Ogle. He had a motorbike and sidecar. He was in the RAF during the war. They never had children. And Aunt Margery, always called Madge . . . she was beautiful.'

Aunt Margery, I was to learn, had the 'uncommon widow's peak above the proud brow' that was Emma Harte's and granddaughter Paula's in *A Woman of Substance*, and Vincent's in *Act of Will*. It did not belong to Barbara, as I had imagined.

'Madge lived in Lower Wortley. She was married, but didn't have children either. I remember I was her bridesmaid when I was six. Olive was another favourite aunt. Everybody loved Olive. She managed a confectioner's shop, Jowett's at Hyde Park Corner – in Leeds, not London! – and I used to go and see her there. She used to let me serve a customer now and then. She lived in this house at twenty-one Cecil Grove, just the other side of the Stanningley Road, by Armley Park.

'We would go on picnics together. She always took her knitting. On one occasion, so the story goes, I nearly drowned. Olive suddenly looked up and said to Uncle Harry, "Where's Barbara?" They couldn't see me. Then, far out in the river, they saw my dress caught on a dead branch, they saw me actually bob up and go under the water again and I was flailing around because I couldn't swim. But the branch had hooked into my dress and was holding me up. Today I have a terrible fear of the sea where I can't put my feet on the bottom. I panic, which I think must go back to that. They got me out in the

51

end, soaked and crying, and dried the dress in the sun, but you know the English sun! They took me home to their house and Auntie Olive ironed the dress and got me back home looking like the starched child that I was.

'So, Daddy had three sisters, and they spoiled me to death, of course, because two of them didn't have children. Their brothers were Winston – my father – Jack, Bill, and Don, the youngest, who sadly died just recently. He and his wife, Jean, my one surviving aunt, had a daughter, Vivienne, my only cousin.

'Jack and Bill Taylor lived at home with my grandmother and never married. I didn't know them very well. Jack was in the army during the war, I think, and Bill was in minesweepers off the coast of Russia and places like that. Afterwards they came home and lived with my grandma, and then when she died they continued to live in that house – two bachelors, very straight, but very *dour*. I used to go and see my grandmother, and they would be sitting in their chairs and nobody spoke. I hated it. I wasn't scared of them, I paid not a blind bit of notice because my focus was somewhere else.

Barbara says she owes her good looks to her father, and that 'he was very good-looking, dark-haired, green-eyed. My father in particular was always very well dressed and very well groomed, and even today I can't stand ungroomed men. Any man that I ever went out with before I knew my husband was always good-looking, always well dressed and well groomed. I loved my father. He would have charmed you.' In *Act of Will*, Vincent, Winston's fictional persona, boasts an almost feminine beauty, though 'there never was any question about his virility'. Barbara describes him in fact and fiction as a natural star, charismatic, one who drew others (especially women) to him by force of personality, dashing looks and more than his fair share of beguiling charm ... plus he had 'the gift of the gab'.

Family legend boosted his reputation further with the adventurous story that at fourteen he ran away to sea and signed up in the Royal Navy, forging his father's signature on the application form. Barbara gives the story to Emma Harte's brother Winston in *A Woman of Substance*. It is not too far from the truth. The real Winston did join the Navy, but not until shortly before his sixteenth birthday. His naval record shows that he signed on as Boy (Class II) on 20th May 1916 in Leeds, that he had previously worked as a factory lad, and that he was at this stage quite small – height 5 ft 1 in, chest 31.5 ins. His first attachment was to *HMS Ganges*, a second-rate, 2284 ton, 84-gun ship with a ship's company of 800, built of teak in 1816 at the Bombay Dockyard under master shipbuilder Jamsetjee Bomanjee Wadia. By the time Winston joined it, *Ganges* was a shore-bound training establishment for boys at Shotley Gate. He remained in the Navy until 16th July 1924, when he was invalided out with a serious condition, which could have led to septicaemia and may have been the reason why he subsequently had one leg amputated.

They took him to what was then a naval hospital, Chapel Allerton in Leeds. Today there is still a specialist limb unit there. Barbara recalls that when he was lying in hospital, 'He said to his mother, "I don't want to have it off because I won't be able to dance." His mother said, "Winston, forget dancing. If you don't have the leg off, you won't live." Gangrene was travelling so rapidly that it had got to the knee. That was why it was taken off very high up. And so he couldn't dance, but he did go swimming. You can swim with one leg and two arms. He went swimming in Armley Baths.'

One can only speculate on the impact the loss would have had on one so sociable, though Barbara marks him out as pragmatic: 'I think I take after my father in that way. He didn't really give a damn what people thought, it was take it or leave me.' Yet, however brave Winston was, losing a leg would have been a huge thing; it would have taken tremendous determination to lead a normal life. Barbara agreed: 'He had to prove himself constantly, I think. I didn't know that when I was growing up.'

The artificial leg he was given was the best available and made of holed aluminium, which according to the specialist at Chapel Allerton would have been light and possibly easy enough to manoeuvre to allow Winston to engage in his favourite pastime, though perhaps not the jitterbug, which was sweeping the country in the years leading up to the war. Some years before rock 'n' roll, the jitterbug involved the disgraceful practice of leaving hold of your partner and improvising fairly frantic steps on your own. In *Act of Will*, when Audra meets Vincent at a dance, it is the less demanding waltz that brings them together:

> *Audra had almost given up hope that he would make an appearance again when he came barrelling through the door, looking slightly flushed and out of breath, and stood at the far side of the hall, glancing about. At the exact moment that the band leader announced the last waltz he spotted her. His eyes lit up, and he walked directly across the floor to her and, with a faint smile, he asked her if she would care to dance.*
>
> *Gripped by a sudden internal shaking, unable to speak, Audra nodded and rose.*
>
> *He was taller than she had realised, at least five feet nine, perhaps six feet, with long legs; lean and slenderly built though he was, he had broad shoulders. There was an easy, natural way about him that communicated itself to her instantly, and he moved with great confidence and panache. He led her on to the floor, took her in his arms*

masterfully, and swept her away as the band struck up 'The Blue Danube'.

During the course of the dance he made several casual remarks, but Audra, tongue-tied, remained mute, knowing she was unable to respond coherently. He said, at one moment, 'What's up then, cat got your tongue?'

She managed to whisper, 'No.'

Barbara's parents married on 14th August 1929. Winston was living at 26 Webster Row, Wortley, at the time, and described himself on their marriage certificate as a general labourer, while Freda, of 1 Winker Green, Armley, described herself as a domestic servant even though she had been working as a nurse. 'I don't actually know where they met,' admitted Barbara. 'Probably at a dance. If my father couldn't really dance any more because of the leg, perhaps he went just to listen to the music. I actually do think he met her at a dance; they used to have church dances and church-hall dances.'

Although Barbara adored her father, when she was a child there was little openly expressed of the love they shared. 'He didn't verbalise it perhaps in the way that Mummy did,' but the depth of it was expressed in a touching scene one day, which had to do with his artificial leg, symbol of the man's vulnerability.

It had been snowing and Barbara was walking with her father in Tower Lane when he fell and couldn't get up because of the slippery snow. 'He was down on his back, and there was nobody around, and he told me what to do. He said, "Go and find some stones and pile them up in the snow, near my foot." He was able to wedge his artificial leg against the stones in order to lever himself up. I got him sitting up, I couldn't lift him. I was six or seven years old. But he managed to heave himself to his feet eventually.'

But at least she had helped him, as she had always longed to do, and now he realised what a practical, efficient *doer* of a little girl he had fathered. The leg brought him close to her again when he died in 1981. 'My mother said, "Your father wanted his leg taken back to the hospital." So my Uncle Don drove me there with it, and three spare legs, and when I handed them over I just broke down in floods of tears. It was like giving away part of him and myself. I was very close to Mummy, but I was close to my father in a different way.'

After Barbara was born in 1933, however, relations between Winston and Freda were not all they might be. I knew that Barbara's mother Freda didn't always see eye to eye with Grandma Taylor. In *Act of Will* tension between wife and mother-in-law is created by Vincent being the favourite son – Grandma Crowther is forever undermining her daughter-in-law's position. In reality as in the fiction, Winston, I learned, was Esther Taylor's favourite, and Barbara recalls her father going 'maybe every day to see his mother. Why do I think my mother always used to say, "I know where you've been, you've been to . . ."?'

'Did it used to annoy Freda, his going to see his mother every day?' I asked.

'Probably. I should imagine it would. Don't you think it would?'

I thought it would be perfectly natural, particularly given the proximity of the houses. Extended families were supposed to be the great boon of working-class life, and Freda with her small child would surely have warmed to such support. Wortley, Farsley and Armley – Freda was surrounded by the entire Taylor family and she had been otherwise alone, having been brought up in Ripon.

Barbara's mother was not by nature big on company how-ever, so perhaps she didn't find it easy to 'fit in' to this extended family scene, which at first must have seemed quite overpowering. She was 'a very sweet, rather retiring, quiet woman' in Barbara's words, 'rather reserved, shy, but with an iron will'. Freda told Barbara little about her past, only that her parents had brought her up in Ripon and that her father had died. 'Her mother Edith then married a man called Simpson. There were three daughters, Freda, Edith and Mary, and two sons, Frederick and Norman. I don't know, to tell you the honest truth, but as far as I remember there was only one Simpson, Norman Simpson.'

I would discover that Freda always had a particularly strong desire to return to the tiny city where she was brought up: 'My mother *always* wanted to be in Ripon,' Barbara recalled. 'All the time.'

'You mean, some time after you were born, when you were eight, nine or ten?' I asked.

'No, much younger, I went back as a baby.' Freda took Barbara back to Ripon from when she was a baby and constantly throughout her childhood. Was this a kind of escape? I wondered. Was marriage into the Taylor family so difficult an adjustment? It seemed unlikely, given that Barbara made Winston out to be such a catch and Freda as a woman who, in spite of her reserve, could fight her own corner with Esther.

In the novels we are given a number of reasons for a marriage hitting hard times. In *Act of Will* Audra is criticised by her mother-in-law for failing to see that her desire to go out to work is flouting the working man's code, which says that if a wife works, her husband loses his dignity. This is odd, as the mills of Armley and Leeds were filled with working wives, and pretty soon we discover that the real problem is that Audra comes from a better class background. She has fallen on hard times, but – she the lady, he the working man – it can never work, the mother-in-law says.

There may have been similar strife for Freda and Winston. There is speculation that Freda had come, or may have had the impression that she had come, from better stock than Winston. We know that mother-in-law Esther Taylor sounded warnings to Freda that giving Barbara ideas above her station would lead only to trouble. Her counterpart in *Act of Will* does the same. In the North of England there was a nasty word for girls with big ideas – 'upstart'. Barbara notes in the novel, 'the lower classes are just as bad as the aristocracy when it comes to that sort of thing. Snobs, too, in their own way.'

In *Everything to Gain*, on the other hand, the cause of strife is laid firmly at the husband's door. Mallory Keswick's father is 'very much a woman's man, not a man's man . . . He adores women, admires them, respects them . . . he has that knack, that ability to make a woman feel her best – attractive, feminine and desirable. [He] can make a woman believe she's special, *wanted*, when he's around her, even if he's not particularly interested in her for himself.'

In *Act of Will* we sense the same about Vincent. There are arguments not only about Audra's highfalutin ideas, but about Vincent's drinking and a 'fancy woman', and fears that he will leave home and never come back. ' "Go back to your fancy woman," I heard that,' said Barbara. 'That happened when I was little, I remembered it and I did ask Grandma, "What's a fancy woman?" And I know that my mother rejected Winston constantly. No, she didn't talk about it, but I knew about it somehow when I was in my teens. How do children know things?'

Recently, as guest on BBC Radio Four's *Desert Island Discs*, Barbara admitted that she hadn't been able to write *Act of Will* until after her parents had died, 'because I really wrote in a sense about this very tumultuous marriage that they had. They were either in each other's arms or at each other's throats. He was very good-looking and he had an eye for the girls at times.' Later, she said to me, 'But that's life. I think Winston had a bit of an eye for the ladies, but that doesn't mean that he did much about it. Listen, the more I write, the more I read of other people's lives, the more I realise how terribly flawed we all are.'

In the novels this is the message about fathers in general. In *Everything to Gain*, Mallory Keswick says: 'He was a human being after all, not a God, even if he had seemed like one to me when I was growing up. He had been all golden and shining and beautiful, the most handsome, the most dashing, the most brilliant man in the world. And the most perfect . . . Yes, he had been all those things to me as a child.'

In *Act of Will*, although Vincent's family all dote on him, Grandpa Alfred has 'no illusions about him'. Vincent has 'temperament, stubbornness and a good measure of vanity', and is easily sidetracked from his purpose. His daughter gets her strength of purpose from elsewhere, from her mother's 'iron will', like Barbara.

Despite what Barbara said, that 'when you are an only child you are a unit more', one can imagine that this difference in character between mother and father was fertile ground for disagreement, and it is not uncommon for Barbara's fictional heroines to recall a childhood trauma of expecting the father to up and leave the family home. In *Everything to Gain*, Mallory is suddenly shaken one day 'not only by the memory but by the sudden knowledge that all the years I was growing up I had been terrified my father would leave us for ever, my mother and I, terrified that one day he would never come back.'

Mal and her mother discuss Edward, her archaeologist father, in this vein. Mal cannot understand 'why Dad was always away when I was a child growing up. Or why we didn't go with him.' Her mother talks about his not wanting either of them along 'on his digs'. Mallory is no fool, however. She remembers 'that fourth of July weekend so long ago, when I had been a little girl of five . . . that awful scene in the kitchen . . . their terrible quarrel [which] had stayed with me all these years.'

In the electoral records of Upper Armley, there is a period when Winston is not included as an inhabitant of the family home. Freda is the sole occupant of electoral age when they are living at Greenock Terrace in 1945, the first record available after the war years (when none were kept), and Barbara considers that 'the trauma [of expecting the father to leave] must spring from the war years. [In Tower Lane] we had an air-raid shelter at the end of the garden. My mother and I would go in with a torch and I'd worry about where my father was.'

'Your father was very often not there?'

'No, he was out having a drink. That was Daddy.'

Besides the Traveller's Rest on a Sunday, his favourite watering holes were 'the White Horse, and the other was the Commercial in Town Street. He'd have to walk home, and during the war I thought he was going to be killed,' said Barbara.

'So the picture I have is of you and your mother sitting in the shelter. Were you there alone or were the neighbours in there too?'

'No, it was ours. There were three in a row, but they were awful. There were seats to sit on, but no radio because you couldn't plug it in, could you? Yes, you put bottles of water and some things in there and a Thermos flask Mummy would fill. A woman wrote a very chastising letter about six months ago saying, "I don't know who did your research for the Anderson shelters, but they weren't like you made it out in *The Women in His Life*. I belong to the Society of Anderson

Shelters People," or whatever . . . she will have been all of eighty!'

'It must have been strange to be in the dark with nothing to look at or do, in a makeshift shelter and in the knowledge that bombs could rain down on you at any time.'

'Well,' Barbara remembered, 'we had candles and my mother always took a book, because she was a reader. She didn't knit like my Auntie Olive.'

'Did you take a book as well?'

'I can't remember, but I know that I listened for that unique, very particular step. It was like a missed step – because of the artificial leg, his was not an even step. There was a *lot* of worry about my father. I used to worry about my father, it's funny, isn't it.'

'Children do worry about their parents,' I say.

'Why? A fear of losing them?'

'Sometimes.'

'I used to worry about him being out when the sirens began to shrill. He always went out, not every night, but some nights a week he'd go down to the local for a pint. Usually he was down at the pub, locked in during the raid, and then later we'd hear his step down the garden. And I'd be so relieved I thought I would cry.'

What we have here is the classic 'only child' situation, touching in the extreme. You want to reach back in time and wipe the worry from the busy mind of this girl who took the responsibility for family relations upon herself. In reality, as in *Act of Will*, the only child was doing a balancing act between mother and father, which can't have been easy. Barbara would have had to stand alone under the burden of any unhappiness in her parents' marriage, and it was not done to complain about this. She would have had nobody to understand her worries and grief, and, clearly,

61

the situation between Freda and Winston did become sadly polarised.

One day Barbara said to me about Freda, 'She neglected my father,' and, later, that Freda 'shut the bedroom door' on Winston. In *Everything to Gain*, Mal's parents, like Vincent and Audra in *Act of Will*, sleep in separate bedrooms, and this is deemed by Mal 'with a sinking feeling' to have been the reason why eventually her father must have romanced other women. And it comes out that, yes, her father was having affairs. The hindsight conclusion reached is that the Keswicks were 'a dysfunctional family', and the uncertainty of her parents' marriage made Mal want 'to have the perfect family when I got married. I wanted to be the perfect wife to Andrew, the perfect mother to Jamie and Lissa. I wanted it all to be . . . to be . . . *right* . . .'

We see this in Barbara also, when, after her marriage to Robert Bradford in 1963, she wrote a trio of manuals for the American publisher Simon & Schuster about being the perfect wife – *How to be the Perfect Wife: Etiquette to Please Him*, *Entertaining to Please Him* and *Fashions that Please Him*. When a journalist discovered these in the 1980s, the cry went up: Can this be the same woman who created Emma Harte? No journalist had an inkling of the fear out of which Barbara's desire for an even-keel marriage came.

However, it is also clear in both *Act of Will* and in reality that Winston, even if he was a bit flighty, was not the crucial factor. What led to separate bedrooms was the mother switching her attention away from the husband to the child – to Barbara – and that happened for reasons that went deeper than sex. One event that conspired to sharpen Freda's focus on Barbara at the expense of Winston was the tragic death of their firstborn, a boy called Vivian. 'He died from meningitis six months after he was born and some time before I was born,' Barbara told me. In a confusing and quite extraordinary and upsetting coincidence, Alfred and Esther Taylor also had a late son called Vivian, who died. 'She [Esther] would never lock the door at night and her youngest son, Don, who was probably still living at home in those days, said, "Mam, you've got to lock your door at night; it's not safe." And she'd say, "No, I can't in case Vivian wants to get in." He was her last child, I think, and he died as a baby, as a little boy. Then my parents had a son before me, who is Alfie in *Act of Will*, which is why I had Alfie also die of meningitis. Our Vivian Taylor was six or eight months old when he died, not even a year. It certainly affected my mother's relationship to me, because she focused every bit of love and attention on me. If there was a purpose in my mother's life it was *me*. That's rather sad actually.'

But there was more to it than Vivian. 'My mother didn't want any more children because she wasn't going to let anything stop her from giving me a better life than she had had,' Barbara told me. In *Act of Will*, Audra is fired by her need to *redeem her own lost opportunity*. No sacrifice is too great to enable her daughter, Christina, to realise the opportunities that were denied her in childhood. There is an obsessive quality about it from the moment Christina is born and her mother announces, 'I am going to give her the world.'

Barbara remembers well how this was expressed for real in her relationship with Freda: 'We were very close. I was very close to Mummy. She totally and completely believed in me. There wasn't a day of her life that if she spoke to me, even after I'd gone to live in London and then America, when she didn't say, "I love you." There wasn't a time when she didn't tell me that I was the most beautiful and the cleverest and the most talented and the most charming and the most wonderful person and of course that's not true, we all know that we have

63

faults. But what it did ... it gave me tremendous self-confidence and a self-assurance that I had even when I was fifteen and sixteen. And she instilled in me a desire to excel. Her message was: "There's nothing you can't have if you try hard enough, work hard enough and strive towards a goal. And never, never limit yourself."'

Barbara took away from their relationship an absolute conviction that she was capable of anything to which she set her mind. Inadequacy was not a concept ever entertained. Her friend Billie Figg noted this as her defining characteristic in her early twenties: 'What she had was enormously high expectations of herself and a lot of assurance.'

In *Act of Will*, Vincent fears that Audra's motivation to do the same for Christina is tinged with obsession. He notes a possessiveness about his wife's relationship with their daughter, which seems to exclude him, and comes to frighten him. 'There was a cold implacability in the set of the mouth and the thrust of the jaw, a terrible relentlessness in those extraordinary cornflower-blue eyes...' And Vincent fears, 'She's going to make it a crusade.'

Audra announces her intention to give her daughter the world in the hospital, shortly after she is born. Both Audra's husband, Vincent, and his doctor friend, Mike Lesley, bridle at her naked aggression, not seen before.

When it is all over in the novel, and Audra's daughter, Christina, is the success she has made her, the girl says: 'I'll never be able to thank you enough, or repay you for everything you've done for me, Mummy. You've been the best, the most wonderful mother in the world.' But, as Emma's brother Winston says of his sister towards the end of *A Woman of Substance*, her success is attributable to '*Abnormal* ambition. *Abnormal* drive.'

In *Act of Will*, Vincent, convinced that his wife is victim to irrational forces, shows his mettle in his response to her. He is tender and loving. He masterminds a surprise birthday party for her. There is no hint of violence towards his wife, even when she brings him to his wits' end with her obsession. Moreover, he gives his wife one hundred per cent support over her sense of loss of status. He could have gained the whip hand in the class turmoil of their relationship – that she always believed she came from a better class than he – but nowhere does he use it as a weapon against her.

Knowing now what happened to Freda in her childhood, and the loss she suffered, knowing what it was that made her so determined that Barbara should have the opportunities that were denied herself, it is safe to say that Winston's response to Freda (if it is reflected in Vincent's) was the very best that could have been made. There was in Freda something running deeper even than the loss of her first child, something which possibly no project of success – not even Barbara – could ever quite resolve. Maybe Freda's mother-in-law, Esther, sensed it was never going to be resolved by her son Winston either – however good he was to her. Reason enough for her unsettled relationship with Freda.

The novels first tipped my research in Freda's direction, and it was the novels that gave me a sense of the deepest roots of dysfunction I would find. Turning again to *Everything to Gain*, Mal searches for the reason for her mother's unhappiness: 'Perhaps [she] had experienced humiliation and despair and more heartache than I ever realised. But I would never get the real truth from her. She never talked about the past, never confided in me. It was as if she wanted to bury those years, forget them, perhaps even pretend they never happened.'

In *Act of Will*, Audra's in-laws are all around her. She loves Vincent, but there is something getting in the way, a feeling of apartness certainly. Is it class, as in the novel? Is the belligerent 'outsider' in her really being outed by her better birth? Or is it, as in *Everything to Gain*, something in her childhood, some loss she suffered?

We never get to the heart of the matter in the fiction (because Barbara didn't know), but, like Mallory Keswick, we cannot but suspect there is something we are not being told; indeed we only accept Audra's strangely aggressive love for her daughter in *Act of Will* because we entertain such a suspicion.

In reality, I was to discover, there was every reason for Freda to behave so. Her story provides the crucial dysfunctional and motivational forces that led to her unique relationship with her daughter and Barbara's extraordinary will to succeed. Much of it remained hidden during Freda's lifetime, for Barbara's childhood 'was constructed on secrets layered one on top of the other,' as she wrote in *Everything to Gain*.

These secrets provided Barbara with many of the narrative possibilities of her best novels, and one reason why they have been so successful is that Barbara is not simply writing good ideas, but ideas that are her inheritance. The novels are the means by which she shares in the experience of her past, her mother's past, and that of her mother's own mother. More strangely still, she does so without knowing anything about Freda's history or that of her maternal grandmother, the extraordinary and beguiling Edith Walker.

CHAPTER THREE

Edith

'I think we ought to go to Ripon. We've quite a lot of things to review, and to discuss . . .'
Meredith Stratton in *Her Own Rules*

Edith Walker, Freda's mother, was the daughter of John and Mary Walker (née Scaife). John Walker was a slater's apprentice when Edith was born on 4th September 1880, the youngest of six children. The occupation seems odd in that John was thirty-seven at the time, which is late to consider being an apprentice in anything. The family lived at Primrose Hill, Skipton Road (the Ripon side of Harrogate), close to what is now a roundabout by a pub called The Little Wonder.

It seems that Edith never knew her mother, for John Walker was registered in the 1881 census as a widower. Perhaps Mary Walker died in childbirth. Living in the Primrose Hill house with John and Edith in that year were his other children: Thomas (sixteen), Elizabeth Ann (twelve), John William (ten), Minnie (five), and Joseph (three).

One cannot help but wonder how her arrival on the scene was received by the rest of the family. Was she regarded as the final drain on the meagre resources available to a family of eight, in which the breadwinner was on apprenticeship wages?

Was she rejected for being the cause of their mother's death? Or was her advent greeted with great love and pity for the poor little mite, and was she spoiled and fussed over by her elder sister, Elizabeth Ann, and, indeed, doted on by her father, John, who may even have caught glimpses of his lost wife in her?

Being motherless, even with the care of others in the family, Edith may have suffered from maternal deprivation, a condition believed to be as harmful as poor nourishment. It can have a child yearning throughout its life for the kind of unconditional love that only a mother can give, and which no substitute can hope to assuage. Edith may also have suffered physically, for feeding infants was mainly by mother's breast in those days.

The family does seem to have been a close one. It was largely still intact ten years later, and some members of it remained close for a considerable time into the future. It is indeed tempting to surmise that Edith was the apple of her father's eye. If so, it is all the more tragic that he was unable or, for some reason beyond resolution, unwilling to save Edith from the abyss into which she was to fall.

Typically for a working man of the period, John Walker stayed in one area for his entire lifetime. He was born in 1844 in Burton Leonard, a tiny village midway between Harrogate and Ripon, his parents having moved there from Pateley Bridge, ten miles to the west, where his elder sister Mary was born in 1839. After marrying, he went with his wife to live in Scotton, a village three miles to the south of Burton Leonard. Their first son, Thomas, was born there in 1865. The family then moved to Ripon in or around 1869, where siblings Elizabeth Ann, John William and Minnie were born. Then, around 1877, came the move to the outskirts of Harrogate, where first Joseph and then Edith (Barbara's grandmother) were born. By 1891 the family had moved back to Ripon.

The city, twenty miles or so north of Leeds, lies just west of the Vale of York, which runs north–south between the North York Moors to the east and the softer Dales to the west. To anyone who doesn't know Yorkshire, the names of these two great land masses may be misleading, because there are plenty of dales in the North York Moors and plenty of moors in the Dales. Dales are valleys; moors are high, largely unsettled tracts of wilderness, naturally enough separated by dales. Indeed, as we have seen, perhaps the most famous moor in all England – Haworth Moor – is not in the North York Moors at all, but way to the southwest, near Leeds.

In the immediate vicinity of Ripon there is some of the most beautiful countryside in all England. It was built at the confluence of three rivers. From Middleham to the north, a sense of Warwick's struggles washes down east of the city through Wensleydale on the Ure, the river's Celtic name – Isura – meaning physical and spiritual power. There it is met by the Skell, which is itself met southwest of the city by the River Laver.

In *Act of Will*, Audra refers to Ripon as 'a sleepy old backwater' compared to 'a great big metropolis like Leeds', but the description belies the tiny city's unique and many-faceted appeal. Its population was a mere 7500 at the time of the Walker family's incursion. Today it is little more than double that size, and retains the feel of a small, busy, rural community with tremendous reserves of history at its fingertips. There is a twelfth-century cathedral or minster, a seventeenth-century House of Correction, a nineteenth-century workhouse and debtor's prison, old inns bent and worn by time, and, nearby, twelfth-century monastic ruins – Fountains Abbey at Studley Royal.

Bronze-age earthworks and henges to the northeast suggest human habitation thousands of years BC, but Ripon itself can be said to have first drawn breath in 634 AD with the birth of the city's patron saint, Wilfrid, in Allhallowgate, Ripon's oldest street, just north of what was then a newly established Celtic Christian monastery.

The monks sent Wilfrid, an unusually able boy, to be educated in Lindisfarne (Holy Island) off the Northumberland coast, where St Aidan had founded a monastery in 635. Later, Wilfrid championed Catholicism over Celtic Christianity as the faith of the Church in England, and was appointed Abbot of the Ripon monastery, then Archbishop of York. The church he built in Ripon became a 'matrix' church of the See of York, and his work inspired the building of Ripon Minster in 1175 and the foundation of various hospital chapels, which established the city's definitive role in providing food and shelter for the poor and sick.

St Mary Magdalen's hospital and chapel were built in the twelfth century on the approach to the city from the north, and had a special brief to care for lepers and blind priests. St John the Baptist's chapel, mission room and almshouses lie at the southeast Bondgate and New Bridge approach to the city over the River Skell, which also offers a view of seventeenth-century Thorpe Prebend House, situated off High St Agnesgate in a peaceful precinct with another hospital chapel, St Anne's, already a ruin in Edith's time, but founded by the Nevilles of Middleham Castle. Thorpe Prebend was one of seven houses used by the prebendaries (canons) of the minster, and a house with a special place in Barbara's memory.

When Freda brought Barbara to Ripon as baby and child, they would invariably stay at Thorpe Prebend with a family by name of Wray, who were caretakers of the house. Joe Wray was married to Freda's cousin, Lillie, and Barbara became firm friends with Joe's niece, Margery Clarke (née Knowles). 'I spent a lot of weekends with Margery when I was a young girl,' Barbara recalls, and later as teenagers they would go to dances together at the Lawrence Café in the Market Place. The Lawrence had a first-floor ballroom famous because the dance floor was sprung: 'We used to go, she and I, to the Saturday night dance there at fifteen, sixteen . . . When I reminded her and I recalled that we stood waiting to be asked to dance, her swift retort was, "But not for long!" Apparently we were very popular.'

Margery still lives in Ripon today and took me to the spot beside Thorpe Prebend House where Freda and Margery's father, who went to school together, used to play on stepping-stones across the Skell. Later I learned from Barbara of the celebrated occasion when, according to Freda, he pushed her in. 'She was wearing a dress her mother Edith had dyed duck-egg blue and she dripped duck-egg blue all the way home on her white pinafore!'

Jim Gott, no obvious relation of the acquisitive Gott of Armley, but a working-class lad born seven years before Freda at 43 Allhallowgate, recalled in his memoir that playing on these stones was a popular pastime. What comes across in Jim Gott's book, *Bits & Blots of T'Owd Spot*, is the fun he had as a child and the empathy he and his contemporaries enjoyed with the spirit of Ripon – past and present fusing in its ancient architecture and traditions, and the daily round. These were the riches of a life that in other ways was hard, and Freda, despite the particular difficulties attached to her childhood, would have shared in them too, and, as an adult, looked back with similar wonder.

The spirit of the place, matured over time, was celebrated on a weekly basis at the twelfth-century Market Place, with its covered stalls, self-styled entertainers and livestock pens. It features in Barbara's novel, *Voice of the Heart*. Every Thursday, long before the market bell sounded at 11 a.m. to declare trade open, folk poured in from the surrounding moors and dales, and Ripon awoke to the clatter of vehicles laden with fresh produce and squawking hens, and the drovers' fretting sheep and lowing cattle as flocks and herds made their way through the city's narrow lanes to the colourful square on the final leg of what was often a two- or three-day journey. All came to an end at 9 p.m., as Gott recorded, when the Wakeman Hornblower announced the night watch, a tradition that survives in Ripon to this day.

Home for Edith and her family in 1891 was a small stone cottage just two minutes' walk south of the Market Place at No. 8 Water Skellgate, a whisper from where Barbara and Margery would dance the night away half a century later. (The 1909 map in the second picture section charts the area clearly.)

By this time Edith's father, John, was forty-seven and had married again, his second wife, Elizabeth, being four years his junior. Edith herself was ten, going on eleven, and very likely a pupil at the nearby Minster Girls Primary School. The only other alternative would have been the Industrial School, reserved for the very poor and substitute for the Workhouse School, which had closed its doors for the last time a few years earlier.

Siblings Elizabeth Ann, John William and Joseph (twenty-two, twenty and thirteen respectively) were still at home, but there is no sign of eldest brother Thomas or of Minnie, who, if she was alive, would have been fifteen, old enough to be living out as a maid. Also sharing the house is a lodger, John Judson, and a 'grandson' (possibly Elizabeth's or even Minnie's child) by name of Gabriel Barker. It must have been a tight squeeze.

John had quite possibly made the move back to Ripon to avail himself of the wealth of job opportunities. The city's position on the western fringe of the Vale of York – good sheep-rearing country – had made it the centre of the mediaeval wool trade, its three rivers diverted into a millrace or water course, which flowed from High Cleugh, southwest of the city, through Water Skellgate, to Bondgate Green in the southeast, serving three mills in the process. Ripon's pre-eminence in this had continued into the late fifteenth century, but then steadily declined and various other industries had come to the fore. In the seventeenth century, Ripon rowels (spurs) had gained a worldwide reputation, so that Ben Jonson could write in his play *The Staple of News* (1626):

> *Why, there's an angel, if my spurs*
> *Be not right Rippon.*

The city was also renowned for its saddletrees, frames of saddles worked from local ash, elm and beech – horse racing has been part of Ripon's life since 1664, and fourteen days of flat racing are still staged between April and August at Yorkshire's Garden Racecourse. Button-making made another important economic contribution and, a century later, out of the French Revolution came a gift of an industry from a band of Ripon-bound French refugees, who were befriended by a French-speaking Yorkshireman called Daniel Williamson. In return for his welcome the immigrants handed Williamson the secret details of a varnish-making process, and the lucky man set up the first varnish-making business in England, becoming so successful that today Ripon manufacturers are still doing business in every major country of the world.

Besides these, an iron foundry made a significant contribution from the nineteenth century, and the city also supported a hive of little industries, such as bone- and rope-millers, tanners, fell-mongers, coachbuilders and chandlers. Finally, as many as nineteen per cent of those in employment in the city at this time were able to find work as unskilled labourers.

Upon arrival, Edith's brother, Joseph, found work as 'a rope twister', while his father, John, took up as a lamplighter. 'One of my earliest memories is of kneeling on a chair by a window waiting for the lamplighter,' wrote one J. Hilsdon, a Ripon citizen of those days. 'He was a tall figure, who carried a pole with a light on the end. The street lamps were gaslights and he tipped the arm to release the gas and lit the mantle. Presumably he made a second round to extinguish them.'

When one soaks up the history of Ripon in late Victorian and Edwardian times, the chief impression is of the huge divergence of lifestyle between rich and poor, who nevertheless lived on top of one another in so small a place as this. When Beryl Thompson arrived in 1956 she was spellbound by the accessibility of the city's history, and has spent her life looking into the plight of the poorest citizens ever since. 'I have done this over a period of about thirty or forty years,' she told me, 'and some of the places that children lived in Ripon at this time – if you read the medical inspectors' reports – were not fit to be in. They were no better than pigsties down St Marygate or Priest Lane [this is the oldest part of the city, site of the original Celtic monastery]. If you look at the early maps you will find a lot of courts all over the city – there's Foxton's Court and Thomson's Court and Florentine's Court and so on. I think there were a lot of hovels in these courts and they probably shared lavatories or there'd be a cesspool outside or a midden. They were trying to clean it up at the turn of the century, but the reports still comment on this.'

These courts were a throwback to mediaeval times and a common sight in the nineteenth and early twentieth centuries, not only in Ripon. The imagination of Charles Dickens, as a boy alone in London, his parents incarcerated in the

Marshalsea debtors' prison, was set alight by 'wild visions' as he lurked around their entries and looked down into their inky depths. In Ripon, the city centre retained its mediaeval street plan right into the twentieth century, and the labouring poor were consigned to cottages in just such dark and dismal courts or yards, set behind the narrow streets. Very often at the top of them you'd find workshops – a blacksmith, bakery or slaughterhouse.

There was tremendous movement by the poor within them. In the first decade of the twentieth century, Edith's brother Joseph (Barbara's great uncle) moved at least four times in the space of four years through one maze of them. In 1907 he could be found at 2 York Yard, Skellgarths (a continuation to the east of Water Skellgate), having just moved there from 3 Millgate Yard. By 1910 he had moved to 1 York Yard, then to 4 York Yard. In 1911 he took up residence at 3 Johnson's Court, between 14 and 16 Low Skellgate (continuation to the west of the street), settling there and getting wed. Sadly, he was then almost certainly killed in the First War, for when records resumed in 1918 the sole occupant of 3 Johnson's Court was his widow, Ruth Matilda Walker. Within a year she remarried, her second husband a man called James Draper. And life went on.

These yards were private, often close-knit communities, safe from outsiders because few dared venture into them unless they had business to perform, but Beryl was right that conditions were grim. In 1902, a Ripon sanitary inspector reported: 'The really antiquated and disgusting sanitary arrangements now existing in the poorer quarters – where it is not unusual to find numbers of homes crowded round a common midden, which is constructed to hold several months' deposit of closet and vegetable matter, ashes and every sort of household refuse imaginable, and which remains in a festering and decaying condition, poisoning what pure air may reach the narrow courts . . .'

A Ripon Council report, dated seven years before Freda was born, records one domicile 'where ten human beings have been herded together in a space scarcely adequate for a self-respecting litter of pigs,' and it is recorded in *A Ripon Record 1887–1986* that in 1906, when Freda was two, increasing numbers of children were turning up at school without shoes.

The awful conditions in which unskilled working-class people lived in Edwardian England was so widely appreciated that in the hard winter of 1903, the New York *Independent* could find no more deserving case than England to cover: 'The workhouses have no space left in which to pack the starving crowds who are craving every day and night at their doors for food and shelter. All the charitable institutions have exhausted their means in trying to raise supplies of food for the famishing residents of the garrets and cellars of London lanes and alleys.'

As in London, so in Ripon. These horrors were to be found 'within a stone's throw of our cathedrals and palaces', as William Booth, founder of the Salvation Army, noted in his book, *In Darkest England and the Way Out* (1890). George R. Sims, a respected journalist writing a year earlier, had conjured up a picture that would not have seemed out of place in Dickens's *A Tale of Two Cities*, of 'underground cellars where the vilest outcasts hide from the light of day . . . [where] it is dangerous to breathe for some hours at a stretch an atmosphere charged with infection and poisoned with indescribable effluvia.'

In Ripon, at precisely the time that William Booth published his book about the slums of England being 'within a stone's throw of our cathedrals and palaces', life for the Walkers in Water Skellgate, a stone's throw from Ripon Minster, will have been hard by our standards today, but it would also have given Edith a keen sense of what life was like on the other side of the wealth divide, and an opportunity perhaps to dream of how to make the transition from one to the other.

When John Walker and his first wife first lived in the city in 1869 there would still have been water running openly through Water Skellgate – a section of the millrace already described – and one can imagine that the speedy brook provided a useful if smelly flush in the days before water sewerage systems. In 1875, however, the foetid stream was covered over, perhaps after pressure from the more well-to-do who had begun to frequent the area. For, by the time Edith was living there in the 1890s, Water Skellgate was masquerading as downtown Ripon, with its pubs and a musical hall, theatre and dance hall which attracted the nobs – all cheek-by-jowl with an intriguing mix of small businesses, dismal courts and low-rent accommodation in easy reach of a pawnshop in High Skellgate, and various charitable organisations, including the Jepson Hospital and Bluecoat School, the Girls' Friendly Society Lodge (run by a Miss Taylor) and a Wesleyan Mission Chapel to which Edith attached her allegiance whenever required to state her religion officially.

The leisure facilities in the vicinity included the Palace Theatre at the lower end of adjacent Kirkgate, and in Water Skellgate itself the Victoria Hall, a music hall and theatre, which later became the Opera House. Next door was the Constitutional Club, and then the Crown Vaults, a pub owned by a Mr Lavin, who was also a horse dealer. At one stage, in a yard set back from the main street, between Nos 8 and 16, there was even a covered casino, a skating rink, where young skaters would 'glide away to the strains of the Volunteer Band', and the delights of a permanent fun fair.

It was the Victoria Hall in particular that would have given young Edith an opportunity to see how the other half lived. In late Victorian and Edwardian England such places were accessible to rich and poor alike, and full to overflowing. Jim Gott recalled the Victoria Hall as a place that drew the nobility in droves:

> *See the carriages sway and whirl to the main entrance. In the long-necked brass candle lamps, the flames flicker and dance as the carriages trundle away.*
>
> *Intermingling shadows appear and disappear into the Hall, amid the flickering yellow lights. Flashing gems highlight the colourful satins and silks swished by my ladies, echoed by the sheen of their escorts' black velvet toppers. This is their hour – the local nobility – as they vanish within the gaslit auditorium. Can we forget those Primrose League, Conservative, and Volunteer balls – sparkling and colourful? From a distance we watched these grand spectacles, with the Beckwith orchestra playing music, soothing, simple and haunting.*

Edith was living in the thick of it all, a few doors up in Water Skellgate, and no doubt wanting it all too. Did she have her first thrilling taste of theatre at the Victoria Hall, and later did her daughter Freda come bustling along in the wake of her mother and perhaps some male escort, the little girl thrilling to the brief suspension of daily care? 'Freda loved opera and she loved all the musical things like Gilbert and Sullivan and all of the things that I mention [that Audra likes] in *Act of Will* . . .' Barbara told me. Had her mother's love of it been sparked at the Victoria Hall?

The presence of local nobility on such occasions meant, in Ripon's case, the presence of the Studley Royal party. The Studley Royal Estate is set in a secluded valley just four miles southwest of the Market Place obelisk, a monument to one William Aislabie designed by the celebrated London architect Nicholas Hawksmoor in 1781.

Studley Royal, which Aislabie then owned, has been called 'the Wonder of the North'. It is a World Heritage site and contains over 900 years of history with a twelfth-century abbey, a mediaeval deer park, a Jacobean Hall, an eighteenth-century water garden, and a Victorian church.

There are other stately homes close to the city. Newby Hall, to the southeast, is an eighteenth-century Robert Adam house with a magnificent tapestry room, library, statue gallery and some of Thomas Chippendale's finest furniture, all of which leave readers of Barbara's novels in no doubt as to the part played by the Hall in her education at Freda's hands. Less than three miles to the north lies Norton Conyers, claimed as the original of Thornfield Hall in *Jane Eyre*, although so too is Rydings in Birstall, five miles southwest of Leeds, the crenellated family home of Ellen Nussey, one of Charlotte Brontë's two best childhood friends. Charlotte did visit Norton Conyers, however, and family pictures, furniture and costumes are on display.

But neither was as influential in the lives of Edith and Freda as Studley Royal, the Estate the most pervasive reminder of the influence of power, money and politics on all Ripon's citizens since 1180. Studley Royal reached into their lives as one of the city's biggest employers, and owned many of the properties in which they dwelt, particularly in the area where the Walkers lived. For example, in February 1889 Lord Ripon put up six houses for sale in Water Skellgate – Nos. 18 to 23 – No. 20 including in its lot a blacksmith's shop, a joiner's, a brewhouse, stable buildings and yard. These must have been leasehold sales because Studley Royal tenants are still recorded as paying rent on them through 1991, when records stop.

Studley Royal's influence was so ubiquitous in the city that on High Skellgate there traded a cycle-maker, whose top-end product was 'the famous Studley Royal model'. Almost certainly he was supplier by appointment to the Estate, where visitors – amounting to some 40,000 annually before the First World War – could hire bicycles, 'motors', or a bath chair. Even the front stage cloth at the Victoria Hall carried on it a painting of Fountains Abbey (part of Studley Royal since the eighteenth century), and Jim Gott writes about this with great reverence. The Estate was the unseen hand in everything, it impinged on life in Ripon at every level, as employer, as property owner, as political presence, and the male line of it no doubt also impinged on more than a few young girls' dreams, the Estate's very name seeming to recommend it in the matter of breeding.

The Aislabie family had succeeded to Studley Royal in 1664, when one George Aislabie, a merchant from York, married into the ancient, landed Mallory family. A long history led up to this succession. The estate is called Royal because the manor was originally held directly from Henry II, king from 1154 to 1189, its earliest owner being one Richard le Aleman. Succession was often carried by means of daughters from one family to another, the Mallory family getting it through marriage into the Tempest family in 1444.

The Tempests had themselves come into the Estate early in the fourteenth century when Sir Richard Tempest married a member of the Le Gras family who succeeded from the Le Alemans in similar fashion. The Tempests died out in the male line, as did the Mallorys in the seventeenth century, but it is perfectly in line with Barbara's interest in Ripon's history that names of these powerful, knightly Ripon families should live on in some of her fictional heroines. One thinks of Mallory Keswick in *Everything to Gain* and Katharine Tempest in

Voice of the Heart, for example. The film of *A Woman of Substance* actually began shooting at Broughton Hall, near Skipton, which was once the home of the Tempest family.

Ten years after George Aislabie inherited Studley Royal, he was killed in a duel and the Estate passed to his son, John, who, twenty years later, became Member of Parliament for Ripon and rose to Chancellor of the Exchequer, only to be run out of office in 1710 following the most celebrated financial scandal in English history, the South Sea Bubble, a disastrous scheme for paying off the National Debt in which Aislabie appeared personally to profit. Aislabie was arrested and imprisoned in the Tower of London, and although at his trial he was unable to explain his great wealth, he somehow managed to keep Studley Royal, dedicating himself thereafter to improving its house and gardens.

The Tudor house, Studley Hall, had been partly destroyed by fire in 1716. Aislabie's eighteenth-century rebuild was never finished and much of it was said to have rather spoiled what was left of the original architecture, that it had been reduced by Aislabie to a grand architectural mess, a terrible muddle of styles. This and the fact that it had been fired for the second time when Barbara was a child, instantly recommended it to me as the model for Fairley Hall, where Emma works as a maid and which she later takes great delight in razing to the ground to avenge her treatment at the hand of Edwin Fairley, father of her illegitimate child. In particular, the passage came to mind when Blackie O'Neill, with his eye for perspective and line, surprises us with his condemnation of it as 'the most grotesque house he had ever seen . . . it had no redeeming features at all . . . an ugly, hodgepodge of styles.'

In essence, Fairley Hall was a hodgepodge of diverse periods that competed with each other to create a façade that was without proportion, symmetry, or beauty. The house was large, solid, and rich, a veritable mansion, in fact, but its architectural inconsistencies made it hideous. Blackie sighed. He loved Georgian houses in Ireland, with their fluid lines and classical proportions that gave them such perfect balance. He had not expected to find such a house on these rough Yorkshire moors, but not unaware of the standing and importance of the Fairley family, and their great wealth, he had anticipated a structure that had more taste and refinement than this.

The four-hundred-acre deer park and garden were another matter. In 1726, one hundred men were at work making canals and waterworks at Studley Royal, and it is worth noting that while the formal geometric design and extraordinary vistas were the brainchild of Aislabie, the works were carried out by local labour under the direction of a man called John Simpson, almost certainly a member of a Ripon family that would one day play a significant role in Edith's story.

In 1768, William Aislabie took over the estate and purchased the ruined Fountains Abbey and adjacent Fountains Hall, unifying them with Studley Hall by landscaping the valley that stands between them. The terraced landscaping of the gardens, sweeping vistas and incorporation of the abbey ruins, make Studley Royal today one of the most romantic places in all England.

Barbara denies the original Studley Hall as model for Fairley Hall, but, as already mentioned, she did use the front façade of Fountains Hall as that of Pennistone Royal in the six-novel Emma Harte series which leaves no doubt as to the call the Estate's very Englishness makes on her.

By 1901 Edith Walker was living and working a few miles away from the Hall as a domestic servant at No. 10 Skell

Bank, which leads off the west end of Water Skellgate. Her father, John, had moved from 8 Water Skellgate to 9 Bedern Bank, a short walk to the east in the shadow of the minster, and a property owned by the Studley Royal Estate. He would live there until 1910, but the house would be leased to the family for years afterwards.

Here, with John and wife Elizabeth in 1901, are son Joseph (who has not yet set up on his own), grandson Gabriel Barker (aged ten) and various lodgers, including a builder called Leo Walker, a boy of nine called Chris (surname illegible) and a girl, Rebecca, who is thirteen. Next door at No. 8, John's eldest son Thomas has reappeared and has followed in his father's footsteps as a slater. Thomas's wife, whose name is difficult to read but could be Frances, is with him, along with their son, Jamie, who is fifteen, and three daughters, Lillie, Elizabeth and Minnie, who are five, four and two respectively. Lillie – Freda's cousin – is the future wife of Joe Wray, with whom Barbara and Freda would later lodge at Thorpe Prebend House.

This ancient street of Bedern Bank, once called Betherom Bank, runs down the west side of the minster precinct, by the old cathedral graveyard, leading to the minster's south boundary at High St Agnesgate. The street was part of a localised area for census in Freda's time, which included thirty-two houses employing fifty-one servants – high density relative to the city as a whole, where only twenty-three per cent employed any servants at all. Nevertheless, Bedern Bank appears from a map of 1909 to have been made up of quite tiny cottages, except for an early Georgian brick house known as The Hall at the top of the hill by the minster. Very likely, Bedern Bank was where some of the servants of the area lived.

At the top of the Bank were '2 *cottages with yards and outbuildings behind*', described as such in a leasehold sale by the Studley Royal Estate held at the Unicorn Hotel on 29th November 1887. At that time these cottages – actually 8 and 9 Bedern Bank – were 'in the occupation of William Gott Senior and William Gott Junior', none other than grand-fatherly relations (great-great and great respectively) of our witness of these times, Jim Gott. Thanks to him we have the following description of two dwellings inhabited by members of Barbara's family from the start of the twentieth century until (with one short gap) 1935, cottages that Edith and Freda knew particularly well:

> On top of Bedern Bank, two low, yellow-washed cottages used to stand. One in particular was of great significant [sic] to me, being the residence of a great, great grand-father of mine – William Gott – who was a jeweller, clock and watchmaker ... The style of the establishment was unique itself, especially the front with its low window. The pathway continued the sill, and the gradient of the hill made it seem lower still. In fact, one had to bend down to see into the window. Steep steps led down to the inner doorway. From inside, one got a grand view through the window of feet and legs.

So, in 1901, for the first time in her life, Edith was living independently of her father. Her employer, thirty-seven-year-old Richard Guy, hailed from Waterford in Ireland and lived in Skell Bank with his wife Annie and their four children. Richard Guy was a tailor; he had his own business and shop. His younger brother George also lived in the house and worked in the business with him.

You did not have to be rich to afford a servant in those days – they cost little to hire – but as early as 1897 Richard Guy was prosperous enough to pay for an advertisement in the

West Riding edition of *Kelly's Trade Directory*, and by 1908 he had moved from Skell Bank to No. 6 North Street, to the north of the Market Place, a better address though nothing special. Jim Gott recalled Richard Guy's shop in North Street, and offers a rather unsettling glimpse of it, its contents seeming to carry the dust of ages: 'Day after day, year after year, I have gazed into the window of the clothier's shop [Mr Guy's] where time virtually stands still,' he wrote. 'The same old wax models in the same old place with their sweaty, sickly complexions tanned by the grime and dust of age, fixedly staring with death-like ghostly impressions. There they stand bedecked in the very latest – a la Fauntleroy; or the choicest design – the Norfolk pattern; or the *pièce de résistance* – the ever popular children's sailor outfit complete with wooden whistle. Surrounding these . . . are the well-known chicken chokers, dicky fronts, straw boaters and Beau Brummel bows.' Richard Guy was still at 6 North Street in 1936. By 1923 he carried the letter 'J' beside his name, indicating that he was respected enough to serve as a juror.

When I discovered the Guy family it occurred to me what a coincidence it was that servant-girl Edith will have been surrounded by the same kind of Irish cheer that servant-girl Emma Harte experiences in the close company of Blackie O'Neill at precisely the same period of her life, when she is kitchen maid at Fairley Hall. The Guys' home county even supplies the origins of Blackie's 'spectacular Waterford crystal chandelier', which hangs from the 'soaring ceiling' of his Harrogate mansion after he and Emma become rich. The Irish influence in the novels, particularly in *A Woman of Substance*, is completely in tune with the Irish influence in Ripon at this time. Nearly a third of the unskilled workforce (which, you will recall, represented nineteen per cent of all workers in the city) were born in Ireland. So, there were a lot of Irish about.

In the novel, Blackie is Emma's succour. His liveliness and gaiety, his distinctive character, an 'unquenchable spirit and a soul that was joyous and without rancour', transform her. From the start he calls her 'mavourneen', later telling her that it's the Irish equivalent of 'luv' in Yorkshire, an Irish term of endearment, which turns up time and again in the novels.

Blackie O'Neill was, in fact, an exceptionally handsome young man. But it was his manner and his attitude that were most intriguing and which, in many ways, set Blackie apart from other men. He exuded liveliness and gaiety. His face was full of vivacity, and it had great mobility and not a little wit. An easy, careless charm was second nature to him, and he was buoyant of spirit, as if he accepted life for what it was, and was constantly entertained by it. There was a light-hearted self-confidence inherent in him, and to Emma, observing him, he seemed untouched by the weariness and the fear and the hopelessness that haunted the men of the village, bowing them down and ageing them prematurely.

Blackie and his Irishness are central to Emma's ambition, for it is of course he who opens Emma's eyes to the very idea that she can make a fortune. 'It was the magical word "fortune" that had made the most profound impact on her . . . Emma's heart was pounding so hard she thought her chest would burst . . . "Can a girl like me make a fortune in Leeds?" she asked, breathless in her anticipation of his answer.' It's the word 'fortune' that holds Emma spellbound. When she savours it, Blackie sees in her face 'ambition, raw and inexorable'.

It is not wholly fanciful to suppose that Richard Guy was similarly encouraging of Edith, that she, too, first considered she might make herself a fortune and lift herself out of the

86

poverty into which she was born, buoyed up by an Irishman's blarney. Certainly Richard was on the 'up' and would have been full of his own ambitions, and for a girl like Edith (as for Emma Harte), dreams of success would have been measured in material terms, in terms of *making a fortune*.

In 1902, when Jack London came to England from America to conduct an investigation into the living conditions of the poor, he commented that they were as materialistic, indeed perhaps more so than the rich, and this is so important in charting the lives of Edith, Freda and, later, Barbara – the one life, the seed in the garden of Barbara's fiction, where success means money. In *A Woman of Substance* Emma Harte's purpose is money, 'Vast amounts of it. For money was power. She would become so rich and powerful she would be invulnerable to the world.'

Given the conditions of life for Emma, as for Edith and Freda in the real world, and other of Barbara's fictional heroines whose money 'arms them against the world', one can see why.

> *Emma knew that without money you were nothing, just a powerless and oppressed victim of the ruling class, a yoked and shackled beast of burden destined to a life of mindless drudgery, and an existence so wretched and so without hope, so filled with terror and despair that it was hardly worth the contemplation let alone the living. Without money you were susceptible to all the capricious whims and moods and fancies of the careless, thoughtless rich, to all the vicissitudes of life. Without money you were vulnerable to the world.*

Barbara never knew the want and vulnerability she described, but, like Emma Harte, her grandmother Edith did. In the novel, as Emma becomes richer her hatred of the Fairley family, who have exploited and brought her down, diminishes, so that by the end of the book she can forgive and forget, and allow Jim Fairley, the last of the Fairley line, to become engaged to her favourite granddaughter, Paula. At that point, money no longer has the same status: 'Money is only important when you're truly poor, when you need it for a roof over your head, for food and clothes,' writes Barbara. 'Once you have these essentials taken care of and go beyond them, money is simply a unit, a tool to work with.' But at the start, it is making a fortune that motivates her.

Barbara gave me a picture of Edith in her mid to late twenties wearing her hair in the Edwardian style and a lace-trimmed dress. It is impossible to date the photograph accurately, but it was almost certainly taken after she had ceased working for Richard Guy and was enjoying a period of uncharacteristic prosperity, to which we will come. It is, however, quite possible that a perk of working for the Guys was that Edith had access to far better clothes than she would otherwise have been able to afford. Mistresses often gave their maids cast-offs, so how much better to work in a tailor's. There are references to this practice in *A Woman of Substance*, where Emma discovers from Adele Fairley, 'the smell of expensive perfumes, the touch of good linens and supple silks, and the sparkle of brilliant jewels . . .' Such things fan the flickering flames of ambition in Emma to become 'a grand lady . . .' Having access to decent hand-made clothes would have made Edith feel better about herself too, showing her that there was more to life than the hopelessness she saw all around her, and perhaps also giving her confidence to dream of making something of her life.

For maidservants the dream was usually as far as it went, their only realistic hope being to find a man from a higher class, a prospect fraught with the risk of exploitation and one that invariably led to complete disaster. Not so for Emma Harte of course, whose exploitation by Edwin Fairley ironically gives her the guts, the fury, to claw herself up by her own efforts. The picture of Edith Walker in her mid to late twenties (see second photo section), with her hand set purposefully on hip, suggests that she might have had a similarly forceful character. She would soon have occasion to put it to the test, for towards the end of 1903 Edith became pregnant, and on 3rd June 1904 she gave birth out of wedlock to a baby girl called Freda.

Being an unmarried domestic servant, the birth could not take place at the Guys' house, and for some reason it did not take place at the house that Edith's father was occupying at 9 Bedern Bank. Freda was born at 9A Water Skellgate in that yard of dwellings between Nos. 8 and 16 already described.

In the 1901 census, the one most recently available to public scrutiny, there is a note that 9A Water Skellgate was leased to the Irish League. Perhaps the Guys had co-opted it for Edith out of kindness, though the census return suggests that the address was something of a secret hideaway. Normally an occupier's name is given for each address, but in the case of 9A Water Skellgate there is no name. Instead the words 'in occupation' are appended. This is unusual even for somewhere leased by an organisation, where occupancy might change frequently. There was frequent occupancy drift elsewhere in Ripon's yards and the census records names in those cases. It is not the purpose of census to establish permanent occupancy. Again, if the house was known to be occupied, an occupant's name would be available. Nevertheless, the name of the occupant has been withheld. It is an interesting anomaly or fudge, which would have needed explanation, or sanctioning by someone high up.

There are so many unanswered questions attending Freda's

birth. Why did Edith not have the child at the family home in Bedern Bank? Was there simply no room? Or had Edith fallen out with her father, John Walker, over the fact of this illegitimate birth? Perhaps having it elsewhere was a mark of her independent will. Or was it at the insistence of Freda's father that Edith took up occupancy of 9A Water Skellgate, because it was in his gift?

Then there are the more interesting questions posed by Freda's birth certificate. The name of the father has been withheld. Edith retains her maiden name, Walker, and lines are drawn through the boxes headed 'Name and Surname of Father' and 'Occupation of Father'. Again, this is most unusual, and indicates a definite decision by Edith or the father or both to keep paternity a secret. In an official Ripon illegitimacy return dated 1857, where eight out of nine single mothers are granted five shillings per week for the first six weeks and one shilling and sixpence until the child is thirteen, there is *in every case* appended the name of a father, whether or not authentic. It was done even if the mother wasn't sure who the father was, because in those days illegitimacy was regarded as a terrible stigma, and a mother would do all in her power to alleviate the burden on the illegitimate child, inventing a name if necessary. In the case of the Tyneside novelist Catherine Cookson (born illegitimately only two years after Freda), the name of a man – Alexander Davies – appeared on her birth certificate as father even though he (if indeed the name is genuinely that of her father) had known her mother for a very short time and had long left her in the lurch by the time Catherine was born. My contact at the Harrogate area registry confirmed that it is unusual for a name not to have been conjured up from somewhere, especially on three consecutive birth certificates, as was the case for Freda. For, as we shall see, Edith gave birth to three illegitimate children between

90

1904 and 1910, and in each case no father was named on the certificate.

We may safely conclude that a definite decision was taken not to name the father and that there was good reason not to name him. Very likely he was a married man with a position to uphold, a man whose reputation would suffer were his 'indiscretions' made known. Either the father insisted on it or Edith wanted to maximise her chances of keeping him by making paternity a secret.

Either way, not naming the father on three certificates in 1904, 1907 and 1910 suggests continuity, that the father was the same man in each case, that each birth was not a one-time fling. If Edith had been leading a profligate life with different men, there would be no reason not to name the children's fathers, or to invent names, as was the way in such cases. The three children are bound to one father by the very fact of his not being named on all three certificates, and most convincingly, as we shall see, by the choice of their Christian names.

Moreover, if one man fathers three children by the same woman out of wedlock over a period of seven years it seems likely that love was involved in the relationship. Perforce money will also have been involved. Being pregnant so often would have made it impossible for Edith to hold down a job as a maid. No one would have hired her. And I have it from Barbara that Edith didn't work because – so Freda told her – she didn't need to. Edith's support would appear to have been bound up in the deal.

It is of course logically possible that the father was a working man who had Edith in thrall, a tradesman perhaps, like Richard Guy, her erstwhile employer. But this is, I think, unlikely. In the case of Guy, was he not busy enough building his own family, with four children under ten? Would he undertake a parallel family of three with a former maid? Would such a man – respectable but hardly wealthy – have been able to persuade Edith he was worth waiting for through three births and the passage of seven years or more, from at least 1903 to 1910? And would he have been able to afford to keep her during this time out of that dusty shop in North Street? How would any ordinary, working, married man have kept two families in this way?

If support was part of the arrangement, there would indeed have been a reputation to protect, and the very fact that Edith chose not to register a false name may even suggest a certain pride in who the father was, that *at all costs* the name had to be withheld from the record but that that also made it vital, on another less expedient plane, not to compromise the truth with a false name. Perhaps subconsciously this was her way of drawing attention to the secret, to what it was she was hiding. Perhaps, after all, Edith saw Freda's birth as anything but a calamity, rather as the first step to realising a dream of great fortune ... If she could persuade the fellow to marry her.

All this takes us into the emotional epicentre of *A Woman of Substance* and *Act of Will*, where problems attached to illegitimacy figure strongly, and as we pursue the shadows of the past it is increasingly difficult to be sure where fact ends and fiction begins.

Barbara would register deep shock when she learned from my researches that her mother Freda was illegitimate, yet Audra Kenton in *Act of Will*, who Barbara admits is the fictional equivalent of Freda and has a mother called Edith, is herself born illegitimately. And when we look for who Audra's father was, he turns out to have been the fictional Edith's 'benefactor and protector'. Again, Audra is set apart from her working-class husband, Vincent, by her better birth. Writes Barbara in the novel (her italics): '*If the circumstances of her life had been different she would never have been permitted to marry him* [viz working-class Vincent].' The point, and it is a crucial one, is that what makes Audra different is her high birth of which she has been dispossessed. Is that what Freda believed made her different? Did she believe that her father had been a member of the nobility? If so, it would likely have infuriated Winston's mother, just as it did Vincent's mother in the novel.

So, was Freda, like Audra, dispossessed of some great inheritance by the illegitimacy of her birth? 'Who leaves what to whom' is an enduring preoccupation of the Emma Harte novels. They are bound up with what is lost when a character is dispossessed, cheated out of the rights attached to their birth. In *Act of Will*, Audra has to remind herself who she really is and where she really comes from by going to her 'memory place', from which she can look out at the long, low, eighteenth-century manor house called High Cleugh, where she was born and brought up, and mourn what might have been.

High Cleugh attracts Audra 'like a magnet'. Barbara, for some reason unexplained, positions High Cleugh northwest of Ripon, instead of southwest, where in fact it lies. For High Cleugh, I discovered, is a real place. In reality, it is not a house at all, but a delightful, tree-lined green space on the bank of the River Laver, just before it melds with the Skell, as you exit the city by means of Mallorie Park Drive and Studley Road.

High Cleugh is so close to the Studley Royal Estate that when Lord Wemyss set fire to a butt on the Estate one hundred years ago during a shoot, and the flames took hold of the dry turf and spread out of control, a workforce was raised in Ripon and the flames were extinguished by means of buckets passed hand-to-hand from the river there. High Cleugh is, in effect, the *perfect* 'memory place' for Freda to sit and consider what might have been hers if she, like Audra, had been connected to someone there.

But Barbara doesn't know why she used High Cleugh as a memory place. She was not trying to suggest that Audra/Freda was linked to Studley Royal by birth. She had no reason to do so. So, on the face of it, there can be no significance in High Cleugh . . . except that there is.

Three years after Freda was born, on 21st July 1907, Edith had a second illegitimate child, whom she named Fred. Again, the father is unnamed on the official certificate. Transparently, the names of Edith's first two illegitimate children are diminutives of 'Frederick'. I am told by the Register of Births that it was common practice for a single mother to include the father's name as a Christian name of the illegitimate child. There is a desire to acknowledge the father's identity which the circumstances of the birth must keep hidden, and, perhaps for the child's sake, to establish its identity however clandestinely. In Edith's case it may even have been the compromise to which she had the unnamed father agree.

In *A Woman of Substance* there is a similar desire for secrecy over the identity of the father of Emma Harte's illegitimate daughter Edwina, the son of the lord of the manor, Edwin Fairley. Yet the girl is named after him in that case too. Emma wants no one to know who Edwina's father is. She imparts only half-truths about him to the man she later marries, Joe Lowther, and for brothers Winston and Frank she invents 'a nebulous gentleman of doubtful background . . . whom she had met in Leeds'. Yet, as was often done, she uses the derivative of 'Edwin' to seal Edwina's identity, leaving the key to the secret for posterity as it were, or perhaps for her daughter to find when she is older.

In the context of common practice the names of Edith's first two illegitimate children being diminutives of 'Frederick' cannot be regarded as chance. In our search for Freda's father we are surely looking for a man in a prominent, wealthy, local family whose name is Frederick and with whom Edith could conceivably have come into contact.

Asking a local studies archivist, steeped in Ripon's history, whether she could nominate a prominent Ripon citizen called Frederick in Edwardian times turned out not to be too searching a request. Given the tidy nature of the city (no more heavily populated than some so-called villages), the fantastic amount of research available, the archivist's own knowledge and the fact that one prominent family is well-known even today for the sheer number of Fredericks in it meant that she did not pause for one moment before replying.

Frederick has been the family name of the Robinsons of Ripon since at least 1746, when one Frederick Robinson first drew breath. In 1782 a Frederick John was added to the family, in 1816 a Hobart Frederick and in 1830 a Frederick William. In 1832, Lady Mary Gertrude Robinson married into the Vyners of Gautby, Lincolnshire, and soon Fredericks began to litter their family records too – in 1836 Henry Frederick Clare Vyner was born and Frederick Grantham Vyner in 1847. Meanwhile, in 1827, Frederick John Robinson had a son named George Frederick Samuel. And in 1852 George Frederick's son by (his cousin) Henrietta Ann Theodosia Vyner, was named Frederick Oliver Robinson. Both George Frederick Samuel and his son Frederick Oliver were contemporaries of Edith in Ripon. There couldn't have been a more visible or, as it turns out, more accessible family of Fredericks available to her.

The Robinsons' history takes us back to 1522, the year in which one William Robinson was born in the city of York. William went on to make a fortune in trade with Germany, became Lord Mayor of the city in 1581 and was subsequently its Member of Parliament. Like many a successful Tudor merchant, he invested in land, leaving estates both in Yorkshire and Lincolnshire, an inheritance which caught its first whiff of noblesse in 1660 when William's Royalist grandson picked up a baronetcy at the Restoration.

It was not, however, until 1761 that the transformation of the Robinsons from trade to aristocracy was made official, when a Thomas Robinson was rewarded for diplomatic services to the Crown with the Barony of Grantham. Subsequent Robinsons added to the family coffers, but now found themselves with a new opportunity to enhance the value of Robinson bloodstock by means of marriage into truer, blue-blooded veins of nobility, and it was on account of this that the family eventually came to Ripon.

In 1780, the second Baron Grantham took as his spouse one Mary Gemima Grey Yorke, daughter of the second Earl of Hardwicke, which brought within the Robinson orbit the de Grey earldom and, through a connection in Mary's line with one George Aislabie, a royal estate. In 1845, Miss Elizabeth Lawrence, the last of the Aislabies, died, and Studley Royal passed into the hands of Thomas Philip Robinson, elder brother of Frederick John Robinson, by virtue of his descent from a daughter of George Aislabie. By 1859, the way lay clear for George Frederick Samuel Robinson (nephew of Thomas Philip, son of Frederick John, who died that year) to inherit Studley Royal. George Frederick was a highly successful politician and would be rewarded with the title Marquess of Ripon by Gladstone for drawing up the Treaty of Washington. With his death in 1909, George Frederick's son, Frederick Oliver Robinson, inherited the title of Marquess and took up the reins at the estate to which Freda remembered Edith dragging her constantly when she was a child and on which Freda would have gazed, mournful of what might have been, if High

Cleugh had been her memory place, as indeed it was in the fiction.

I have kept until now the name of Edith's third illegitimate child, a girl born on 18th February 1910. Again, no father is named. Edith called the child Mary. Frederick Oliver Robinson, Second Marquess of Ripon, had only one sibling, a dear sister who died in infancy. Her name was Mary.

Frederick Oliver was born on 29th January 1852, and when he fell dead in 1923 while out grouse shooting, he was, like Jim Fairley of Fairley Hall in *A Woman of Substance*, the last of the family line, the last legitimate Robinson who could lay claim to Studley Royal.

Like his father he stood for Parliament. The Robinsons had quite a track record in Westminster. Frederick John, Frederick Oliver's grandfather, had even made it to 10 Downing Street as Prime Minister, succeeding George Canning in the summer of 1827, but earning the distinction of holding office for so short a time – the autumn and winter of 1827–8 – that he never once addressed the House as PM. His son, George Frederick Samuel, was a good deal more successful, entering the House of Commons as member for Hull in 1852, the House of Lords seven years later, and serving in every Liberal government for the next half-century, rising to Lord Privy Seal.

Frederick Oliver was a rather different kettle of fish. He took little interest in the affairs of the House of Commons and spent most of his time shooting on the Studley Royal Estate. A fantastic shot he made shooting his career; indeed, he had a reputation as one of the two best shots in Britain (the other was his friend Lord Walsingham). There are countless legendary stories of his skills, for example the speed with which he could change guns is supposed to have led to his accounting for seven birds dead in the air before any hit the ground. In

1905 he took 306 at Studley Royal, and that was small compared to the 576 out of 2745 partridges at a shoot in Austria.
On Dallowgill Moor the best score ever made was 1216 grouse
to four guns in 1915. Frederick Oliver personally took 588 of
them. He expired while picking up his grouse after a drive on
Dallowgill Moor, on 22nd September 1923. A month earlier
he had killed 249 birds in one day. He was seventy-three.

Frederick Oliver will have called frequently at Water Skellgate, which at one time or another he or his father largely
owned, and not only at the places of entertainment which
were points of contact between rich and poor in the Ripon
community, but to attend the Masonic Lodge there that bore
his titles and to pursue his particular social concerns as patron
of the Ripon Home For Girls close by. His father and mother
had done a huge amount of social and philanthropic work in
Ripon and were held in esteem by its inhabitants – Frederick
Oliver was again very much in the shadow of his father in this,
but he made the Ripon Home For Girls his special thing.

The de Grey & Ripon Lodge (No. 837) met first of all in
the Town Hall on the city's Market Place. But in 1902, a
Masonic Temple, a Studley Estate property built on the site of
a stables they had also owned, opened on the corner of High
Skellgate and Water Skellgate, opposite the Victoria Hall, the
subsequent festive board being held in the nearby Unicorn
Hotel, another property of Studley Royal, on the southeast
side of the Market Place. The lodge still meets there today, in
every month of the year except August.

The First Marquess had been a leading light in the Masonic
Craft, holding the position of Provincial Grand Master for
thirteen years, and then Grand Master for four years (that's
top dog nationally, a most unusual appointment outside
London), before resigning in 1874 after a sudden conversion
to the Roman Catholic Church, which he believed mistakenly

to be incompatible due to some outstanding Papal Bulls against the Craft. He was received into the Church at the London Oratory on 4 September 1874. I have it from the secretary of the de Grey & Ripon Masonic Lodge that he did so 'after marrying a Catholic', in other words at Henrietta Ann Theodosia Vyner's behest. News of his conversion was received with astonishment. *The Times* launched a vicious attack, accusing Ripon of renouncing 'his mental and moral freedom. A statesman who becomes a convert to Roman Catholicism forfeits at once the confidence of the English people.' Such was the anti-Catholic feeling of the time, following the assertion of the Vatican Council of 1870 of papal infallibility.

The Masonic Hall on Water Skellgate still has the Marquess's Provincial Grand Master regalia on display. Ripon gave it to his gardener to burn, but the gardener, considering it was of value, disobeyed his instructions and the costume found its way to the local lodge.

Frederick Oliver would not himself have been deterred by his father's resignation, and there even seems to have been second thoughts on the First Marquess's part about quitting Freemasonry after he discovered that the Pope's position was not as intractable as he had thought. He helped bring the Prince of Wales (afterwards King Edward VII) to the Craft in 1870, and subsequently the other Royal Princes, the Duke of Connaught and his younger brother, the Duke of Albany. It was a period of unprecedented support for Freemasonry by the Royal Family. In the Marquess of Ripon's circle, 'to be a Freemason was to be the quintessential Englishman, a part of the establishment at the height of the British Empire.'

One crisp winter afternoon I walked in the freezing sunlight from Skellgarths through Water Skellgate, via High Cleugh, across the fields through Studley Roger onto the Marquess's estate. It was as crystal clear to me as the air I breathed that

High Cleugh really is a window onto the past, a past that was, for Barbara's mother, the reality in which she truly dwelt. There are other points of circumstance, to which we will come, that recommend a liaison between Edith and Frederick, and in the final chapter I bring all of them together for the reader to judge. But even if one prefers to agree with Vincent in *Act of Will*, who says, 'Only Edith knows the truth,' in a sense it doesn't matter. For, if Freda mistakenly thought of herself as high-born – if she was given that impression deceitfully by Edith – her fantasy turned out to be just as significant as if it had been true. For the belief empowered Freda, as it did Audra in her relationship with her daughter in *Act of Will*, to lift Barbara out of her situation and enable her to succeed far beyond Edith's wildest dreams.

As biographer of the late Catherine Cookson, I am aware how important a belief in a genetic connection with the aristocracy could be in this business of rising in the world from humble origins in the first decade of the twentieth century. Catherine's conviction that she was the illegitimate daughter of a gentleman saw her through the entire process of her rise. Enormous self-belief was born of it, and even after she came to doubt the truth of it, the comfort the fantasy had provided stayed with her, because it shored up her sense of identity. Perhaps Studley Royal gave Freda an identity, just as the house, High Cleugh, gave Audra hers in *Act of Will*.

My walk there, which retraces one in *Her Own Rules* from the fictional Skell Garth Hotel to Fountains Abbey on the Estate, took me less than an hour, no distance at all in the days when a pair of legs was the most likely mode of transport. There is a public footpath between High Cleugh and Studley Royal to ease one's passage. Before you know it, you are walking through the gates of the Estate and passing along the '*stately avenue of lime trees*', the vista which leads up to

'Studley church', as Barbara writes in the novel. It is in fact the church of St Mary the Virgin, where all the Robinsons I have described lie buried, and which was built by Frederick Oliver's mother, Henrietta, from 1871–8, at a cost of £15,000. There is a monument to the Marchioness, a big white marble effigy on a tomb chest dated 1908.

It was she, according to my informant within Frederick's Water Skellgate Masonic Lodge, who persuaded Frederick Oliver's father to convert to Catholicism. In *A Woman of Substance* and elsewhere (for example in *Everything to Gain*) the Catholic attitude to divorce has a crucial bearing on Barbara's plot. In the first novel, wealthy, dashing Australian Paul McGill, with whom Emma falls deeply in love after Joe Lowther dies in France, cannot break free of his Catholic wife, Constance. It is as a direct result of this that Paul sires Daisy, Emma's second illegitimate child. A Catholic woman as strong as Henrietta clearly was, strong enough to have her husband George Frederick become a Catholic when it was politically a very unwise thing to do, would have been the best excuse a son of hers could find for not divorcing his own wife and marrying Edith Walker, or at least the kindest way of disabusing Edith of the notion that he did not love her enough to marry her.

A hundred head of deer crossed my path as I walked the route Edith, Freda and Barbara trod so many times. They came thundering past me in the crisp light as I gazed in awe at the sublime marriage of picturesque landscape art and wild nature that is Studley Royal. There are ancient, broken-down trees on which Edith's eyes would have rested when they were saplings. At the top of the rise, like Meri in *Her Own Rules*, I 'glanced over at Studley Church, so picturesque in the snow, and at the obelisk nearby, and then directed my gaze to the lake below, glittering in the sunlight. The River Skell flowed beyond it, and there, just a short distance upstream, was the abbey . . .'

Could this indeed have been the trip made so often by Edith to see the man she believed might one day leave his wife and marry her? Time and again the *grit* of the novels comes out of the speculative possibilities that will have haunted Freda as Edith's real-life illegitimate daughter, and which might solve the problem of her identity – possibilities which perhaps Barbara blanked out as a child, sensing that they were already a source of disquiet in the tight little unit in Tower Lane, Armley, and which were only to emerge years later when she was writing her novels.

The coincidence of fictional plot and real-life fact is too great to allow any other conclusion. Here was I caught up in a web of imaginative truth spun by three women at the centre of which was an incomplete birth-certificate dated 3 June 1904, exactly a century before I was writing.

One is bound to look for similarities between Freda and Barbara and the two Fredericks, father and son. George Frederick was described by M. Bence-Jones in his book *Viceroys of India* (1982) as 'a rather timid' fellow, 'serious-minded, a voracious reader of books . . . noted for his sincerity and moral earnestness . . . prosaic, endowed with neither grace nor sparkle . . . *Persistence* was Ripon's outstanding quality.' The character fits Freda to a T and introduces the key element of the woman of substance, of which Freda's 'iron will' was not short – *persistence*.

In photographs, his son, the Second Marquess – Frederick Oliver – looks a dapper fellow, fine looking and very well groomed. He is supposed to have inherited his father's serious, 'prosaic' qualities. In *The Big Shots* Jonathan Garnier Riffler writes that 'he could never have been described as good company.' But this needs deeper consideration, for clearly he was a man who, unlike his father, pursued his pleasures rather than his duties.

Studley Royal was the principal shoot in England at this time and in the Prince of Wales's Marlborough House set, Frederick Oliver was a popular figure. He may not have been much of a party animal, but that is not to say that he was dull. There was, for example, something of the poet in him. He had a real sense of the spirit of the Yorkshire moors, and loved to talk about the beauty of the birds of the moors in flight (albeit oblivious to the fact that he killed more of them than anybody else), once writing: 'Maybe a generation will spring up to whom all these things [the beauty of the moors, the colour and characteristic sounds of the birds of the shoot he knew so well] will be a closed book; but when that day comes England will lose her most attractive and distinctive feature, and one of her most cherished traditions. For the England of whom the poets have sung will have ceased to exist.'

Also, there was something in Frederick Oliver that attracted, and was attracted by, women of great vitality and vigour – in that sense women that were the opposite of him. On 7 May 1885 he married Constance Gladys Lonsdale, widow of the Fourth Earl of Lonsdale. In hindsight, commentators have wondered how this prosaic character had attracted her, and concluded that she must have grabbed him for the promise of a marquessate and 24,000 acres, to which, at that time, he was heir.

This may well have been the case, but it is unlikely to have been the whole story, for Frederick Oliver will have been in no doubt when he married Gladys, as she was known, as to what she promised him, and very often opposite poles attract.

Gladys Ripon was a bit of a Lillie Langtry figure. She knew how to have a good time, and had a reputation for living in the fast lane. Edward VII once referred to her as 'a professional beauty'. Her first husband, Lonsdale, himself once an ardent admirer of Lillie Langtry, the actress mistress of Edward VII, died in a brothel in 1882.

Gladys was away from the Ripon estate a lot, leaving Frederick to his own devices. She owned the lease of a house – Coombe Court, Kingston Hill in Surrey, on the edge of London, and when she died she breathed her last not at Studley Royal but at 13 Bryanston Square in London.

'Lady Ripon was the uncrowned queen of a smart, artistic, bohemian, and frivolous social set,' Donald Taylor reports in an article entitled, 'The Bohemian Mistress of Studley Royal'. The bohemian and artistic elements of her reputation were deserved. Oscar Wilde dedicated the printed edition of his play *A Woman of No Importance* (1893) to her (whether this was a wry comment on Gladys or a genuine compliment goes unrecorded). She became a close friend of the Russian dancer Nijinsky, attracted to him because of his poverty-stricken childhood, and managed him socially during the Russian ballet's first season in Covent Garden in 1911 – booking him into the Savoy and advising him on where to go and which invitations to accept. Later she would galvanise support for him when he was detained as a prisoner of war in Hungary.

The party atmosphere Gladys engendered was inspired by a reverence for art, which was spirited, idealistic and yet racy: 'When one begins to sacrifice Art to personal matters,' she once said, 'it takes away every wish to have anything to do with it.'

Art apparently justified anything, and as tireless fundraising patron of the arts she once had the puritanical Raymond Asquith, son of the Prime Minister, view 'with a jaundiced eye' her seduction of a millionaire neighbour of his at dinner: 'By the end of dinner she had got £35,000 out of him, and probably had more before she went to bed – or after,' Asquith was reported to have said.

When Gladys was around, you could expect a party, and often a noisy one. Her party-piece was to arrange a surprise smashing of quantities of crockery at dinner, an event which in fact guests looked forward to, and which was, according to the Duchess of Marlborough, originally precipitated by a real incident: 'Once a footman had dropped a tray,' she recounted, 'producing the amusement that the misfortune of others usually creates; since then the incident was repeated, with the china specially bought for making a noise.'

Gladys was as much at home in the company of royals as that of artists. Queen Alexandra, Edward VII's wife, became a close friend, and arranged for her to be on the house-party guest list of all the royal estates, another reason why she was frequently absent from Studley Royal. There was, however, no danger of Gladys sidelining her husband. As a result of her contacts, Edward VII became a shooting companion of Frederick Oliver's, and she brought her party-set lifestyle often enough to the estate for the gardeners to know never to mow the lawns until late in the day when she was in residence, owing to her 'delicate nature'. (Delicate she may have been the morning after, but not so by nature.)

It is, therefore, to Gladys Ripon that we can ascribe an exciting, *laissez faire* attitude within Studley Royal, which servants and employees clearly enjoyed and which was in line with her dropping of her first name, which spelled steadfast qualities she abhorred. She and Frederick were a childless couple and one cannot help but wonder whether during her many absences from the Estate, the sudden removal of her vivacity left her husband with expectations which he then pursued on his own account. Did Gladys create in Frederick Oliver a thirst for a good time that did not disappear when she withdrew? Or did she perhaps make him feel rather inadequate in company and in need of relationships in which he could exercise control? If he did, and Gladys knew about them, she was sufficiently a woman of the world not to object, having earlier come to terms with her first husband meeting his maker in a brothel possibly for similar reasons.

It is likely, as I have suggested, that a love of glamour and a natural vitality were defining elements in Edith's character, too, until the world finally buried her. In *Act of Will* Audra's mother is 'the beautiful Edith Kenton. That was how they always spoke about her hereabouts.' It was also how people spoke of Edith Walker. 'She was called "the beautiful Edith" by members of the family,' Barbara said to Billie Figg in the mid-1980s. It is possible that this glittering something, which Edith had (and which Barbara probably inherited from her), and which Gladys Ripon clearly had in spades (and which attracted Frederick Oliver to her), was also what attracted the otherwise prosaic Frederick Oliver to Edith.

We have a possible match, access, and on both sides a need. It is of course possible that Edith came at length to be in the Studley Royal employ. Many from Ripon were. Sadly there are no records either to substantiate this or to rule it out.

What we do know is that the Estate had a special allure for her, which she passed on to her daughter Freda, and which Freda impressed upon Barbara in her childhood sufficiently for Studley Royal to play a central role in the novels, as in *Her Own Rules* when Meredith, who has come halfway across the world to discover her true identity, experiences the sense Barbara had there of déjà vu:

It was just turning nine o'clock on Tuesday morning. Meredith was bundled up in boots and a sheepskin coat, walking through Studley Park. The stately avenue of lime

*trees down which she hurried led to Studley Church, just
visible on top of the hill at the end of the avenue . . .*

*I'm almost there, Meredith told herself, as she finally
reached the top of the hill at the end of the avenue of
limes. She glanced over at Studley Church, so picturesque
in the snow, and at the obelisk nearby. She then directed
her gaze to the lake below, glittering in the sunlight.
The River Skell flowed beyond it, and there, just a short
distance upstream, was the abbey . . .*

*A strange sensation came over her. She stood very still,
all of her senses alert. Instantly, she knew what it was . . .
a curious feeling that she had been here before, that she
had stood on this very spot, on this very hill, gazing down
at those mediaeval ruins . . . She shivered again. Déjà vu,
the French call it, already seen, she reminded herself. But
she had not been here before; she had never even been to
Yorkshire . . .*

*As she looked around, absorbing everything, her heart
clenched, and she felt a strange sense of loss. So acute, so
strong, so overwhelming was this feeling, that tears came
into her eyes. Her throat closed with such a rush of emotion she was further startled at herself.*

*Something was taken from me here . . . something of
immense value to me. I have been here before. I know
this ancient place . . . somehow it is a part of me. What
was it I lost here? . . . She closed her eyes, not understanding what was happening to her; it was as though her heart
was breaking. Something had been taken from her . . .
The only thing she really knew at this moment was that
she was experiencing an immense sense of deprivation . . .*

*Pain, she thought. Why do I feel pain and hurt and
despair?*

CHAPTER FOUR

The Abyss

'Who are you, Meredith Stratton. And why are you so troubled? Where does that deep well of sadness spring from? Who was it that hurt you so badly they've scarred your soul?'

Her Own Rules

If Edith meant to protect the father of her children by withholding his name from the authorities, it was a favour which, for one reason or another, was not – in the last resort – returned by the man in question, who, after an affair of around eight or nine years, appears to have deserted Edith and consigned her to a fate that was bound to scar their children for life.

Having an illegitimate child, with no partner to bring in the money, was next to suicide in Edwardian England. As I have said, Edith would not have been able to go out to work. There was no crèche or child support mechanism available, no State financial allowance that made any difference. That she managed to keep her head above water financially between 1904 and 1910 must indicate that she was receiving support from the father of her children. Her own father, John Walker, was in no position to help a family of four. Since his days as

108

lamplighter, John (by 1910 a man of sixty-six years) had become a builder, working latterly for the Ripon Corporation.

It is interesting that the tenancy of 8 and 9 Bedern Bank, known to have been owned by the Studley Royal Estate, remained with the Walkers into the mid-1930s (except for one short but pertinent period). As the Walkers could not have afforded to buy them, could a lease have been in the gift of Frederick as part of his support for Edith?

There are two crucial dates in Edith and her children's lives, which coincide with crucial moments in those of the Robinsons of Studley Royal, and further support a connection between them. The first is the death on 9th July 1909, of the First Marquess, George Frederick, which heralded a long and terrible stint in the Ripon Workhouse for Edith, Freda, Fred and Mary.

As aforementioned, the birth of Edith's third illegitimate child, Mary, took place on 18th February 1910, which suggests a late spring/early summer conception in 1909, a few months prior to George Frederick's death. Did the First Marquess's death precipitate their incarceration, spelling the end of their protection? Barbara chooses the same year for the death of Adrian Kenton in *Act of Will*, which precipitates the fall of Audra – Freda's fictional self – apparently quite coincidentally.

Did Frederick foreclose on Edith in 1909–10? The First Marquess's wife, Henrietta, had died two years earlier. With the removal of the heavy Catholic influence of his parents, Frederick Oliver's 'good reason' for not divorcing Gladys and marrying Edith had disappeared. Did Edith pile pressure on him to do the right thing by her, to the extent that Frederick decided to rid himself of the bind of his second family by withdrawing support? Or was he wiping the slate clean before taking on the onerous responsibilities of the Estate (which doubtless did not appeal to him)? Or had Edith's protection been in the gift of the First Marquess, and simply ended with his death?

Interestingly, the electoral rolls show that no person eligible to vote lived at 9 Bedern Bank from 1910. John seems to have disappeared, yet we know he was still alive until at least 1912. Does his sudden absence and Edith's simultaneous fall in 1910 suggest that the lease on 9 Bedern Bank had been withdrawn from the Walkers during the period of Edith's despair in the workhouse?

Ripon Workhouse records, handwritten at the time, are contained within half-year periods. They show that in the first half-year of 1910, Edith, Freda, Fred and Mary spent continuously 52, 49, 49 and 41 consecutive days in the workhouse respectively, that the four of them then spent the entire 184 days of the next half-year there, and fifteen days into the first half-year of 1911.

Edith arrived at the workhouse gates on her own in May 1910, three days before Freda and Fred, who were six and three years old respectively, and eleven days before Mary, who was only three months old, the children's late coming suggesting that an arrangement to look after them had fallen through. What this means is that Barbara's mother was not only illegitimate, but was delivered into the care of the workhouse when she was a child of six.

It is difficult to imagine truly what this meant to Edith, Freda and her siblings. The picture we have of the workhouse generally comes to us from Dickens, and the illustrations of Cruikshank or Phiz. Today, it is all too easy to admire the art with which Dickens writes almost at the expense of the agonies of the workhouse he depicts. The ironic tone he adopted as Oliver Twist faced the Workhouse Board for the first time may suggest that he greatly exaggerated the situation, though he did not.

110

'What are you crying for?' inquired the gentleman in the white waistcoat. And to be sure it was very extraordinary. What could the boy be crying for?

'I hope you say your prayers every night,' said another gentleman in a gruff voice, 'and pray for the people who feed you, and take care of you, like a Christian.'

'Yes, sir,' stammered the boy . . .

'Well! You have come here to be educated, and taught a useful trade,' said the red-faced gentleman in the high chair.

'So you'll begin to pick oakum tomorrow morning at six o'clock,' added the surly one in the white waistcoat.

For the combination of both these blessings in the one simple process of picking oakum, Oliver bowed low by the direction of the beadle, and was then hurried away to a large ward, where, on a rough, hard bed he sobbed himself to sleep. What a noble illustration of the tender laws of this favoured country! They let the paupers go to sleep! . . .

The members of this board were very sage, deep, philosophical men, and when they came to turn their attention to the workhouse, they found out at once what ordinary folk would never have discovered – the poor liked it! It was a regular place of public entertainment for the poorer classes; a tavern where there was nothing to pay; a public breakfast, dinner, tea, and supper all the year round; a brick and mortar Elysium, where it was all play and no work. 'Oho!' said the board, looking very knowing: 'we are the fellows to set this to rights; we'll stop it all, in no time.' So, they established the rule, that all poor people should have the alternative (for they would compel nobody, not they) of being starved by a gradual process in the house, or by a quick one out of it. With this view,

*they contracted with the water-works to lay on an unlimi-
ted supply of water; and with a corn-factory to supply
periodically small quantities of oatmeal; and issued three
meals of thin gruel a day, with an onion twice a week,
and half a roll on Sundays. They made a great many other
wise and humane regulations having reference to the
ladies, which it is not necessary to repeat; kindly undertook to divorce poor married people, in consequence of
the great expense of a suit in Doctors' Commons; and,
instead of compelling a man to support his family, as they
had theretofore done, took his family away from him and
made him a bachelor!'*

The Ripon Union Workhouse that Edith, Freda and her siblings knew was built in 1854 to the north of the city at 75
Allhallowgate. Twenty years earlier the Poor Law Amendment
Act had brought an end to 'outdoor relief' (top-up wages when
earnings fell below a certain level). Henceforth, only asylum
within the workhouse, in exchange for work, was on offer and
all claimants had to prove themselves paupers. If a vagrant
possessed even a shilling he would be turned away. The pauper
test, along with the conditions of the new residential workhouses – made awful to deter claimants – were the demeaning
aspects of the law that caused such a furore.

The declared purpose of the Act was to encourage unemployed, able-bodied men to get a job and all employers to
pay at least a subsistence wage; while the asylum aspect would,
it was felt, prevent paupers from breeding because once in the
workhouse a man would be separated from his wife. Also, as
shown in a report on the West Riding of Yorkshire, it was
expected that 'the separation of husband and wife would have
a beneficial tendency in rousing the indolent to exert themselves.' Records depict firm discipline and terrible sorrow. In

The map, dated 1909, shows Ripon when Barbara's mother, Freda, was five. To the south, Water Skellgate, where Freda was born to Edith Walker on June 3, 1904. A few minutes walk east, in the shadow of the Minster, Bedern Bank, where members of Edith's family lived in two low-lying, yellow-washed cottages until the mid-1930s. To the north of the Market Place, Allhallowgate and Ripon Workhouse.

Above: Ripon Minster at the north end of Bedern Bank, was founded in 1175. Barbara's grandmother, Edith Walker, is first recorded as resident in the city in 1891 with her father, John, a lamplighter, and her five brothers and sisters.

Below: Ripon Market Place as Edith knew it, the huge obelisk testimony to the commanding presence of the Ripons of Studley Royal. It carries an inscription by Frederick Oliver Robinson, Marquess from 1910–1923. Note the Lawrence Hotel to its right, where Barbara and her friend Margery Knowles would dance in the 1940s.

Ripon, redolent of times past. The Wakeman Hornblower *(right)* still announces the watch each night at 9.

Water Skellgate in 1904, the year that Edith Walker gave birth there to Freda, Barbara's mother – in poverty, but with great expectations.

The stepping stones on the Skell where Freda fell and dripped duck-egg blue all the way home.

Below left: one of Ripon's ancient courts off Water Skellgate, not unlike the one where Freda was born in 1904. *Below right:* Freda dressed in her best at 14, a decade after her workhouse experience and the year she took her life in her own hands and began working at the Fever Hospital.

Studley Royal Hall, home of the Marquesses of Ripon. The hall was mysteriously burned down in 1946.

Fountains Hall was made part of the Studley Royal Estate in the 18th century. Barbara used it as Pennistone Royal in the Emma Harte novels. It is the one place in which Emma finds peace.

Edith Walker, Barbara's grandmother, an Edwardian through and through, with the fashionable hairstyle of the day, lace trimmings to her collar and cuffs, a diamond brooch blazing at her neck, and no small amount of determination in her hand-on-hip stance. But Edith's appearance belies her situation, a domestic servant who fell into the same trap as Emma Harte in *A Woman Of Substance* and gave birth to an illegitimate child. This was no one-night stand, however, her comfortable position suggesting a long-term lover of distinction. Two more children would follow, their birth certificates and pattern of their fortunes pointing the finger ever more obviously at the father.

The one clear favourite *(right)*: Frederick Oliver Robinson of Studley Royal, 2nd Marquess of Ripon, the estate to which Freda remembered Edith dragging her constantly when she was a child and to which Freda took Barbara repeatedly from as early as she can remember.

The Grubber, Ripon Union Workhouse, to which Edith and her children were reduced after the death of the First Marquess of Ripon, which left them destitute. Edith arrived first, in May 1910, three days before her children Freda and Fred, six and three years old respectively, and eleven days before three-month-old Mary was delivered to its doors. The stigma of being in the workhouse was burned into a child's very soul.

reality, the agonies of the receiving wards, where husbands were separated from wives, and children from parents, were devastating.

What had also been overlooked was the effect upon the *needy*, that the unemployed and vagrants were not all shirkers, that unmarried mothers were not all prostitutes, that orphans were not a lower order, and that if you lived into your sixties you could easily become destitute because there were no state pensions in those days. All had effectively been criminalised by the Act. Adding to their burden, workhouses were made to take in 'lunatics', mentally defective children and the physically sick. Guardians displayed their patronising attitude by lumping their charges – 'idiots and lunatics, bastards, venereals, the idle and dissolute' – together. Yet segregation was by gender and age only.

Today, visitors to Ripon Workhouse – part of it is open to the public – will approach the same nineteenth-century portico through which Barbara's mother passed in 1910. The entrance to the Grubber, as it was known, is a low, one- and two-storey, brick-built construct with a stone gateway and imposing twelve-foot doors, behind which lie a village-sized collection of buildings within very much higher walls. There were adult wards (into which sane and insane intermingled) and receiving wards, where the separations – male/female, child/adult – took place, and bodies were stripped, disinfected and clothed in workhouse attire before being taken to segregated dormitories, exercise yards or casual ward cells barely big enough for a bed.

If inmates wanted to eat they must work: breaking stone, chopping wood (there was a chopping shed built in 1903), weeding, lifting potatoes (Ripon Workhouse had its own garden), or – as in Dickens – picking oakum, the loose fibre obtained by unravelling old rope. Respite was only given after the task was complete. Inmates would dine in gender-segregated areas, sitting on benches at long tables, all facing the same way and eating in silence. Edith's daily diet consisted of less than half a pint of milk, just under a pound of flour (bread was baked on the premises until 1916), a few potatoes (equivalent to four small new potatoes), a little sugar, butter and cheese (about 3 oz a week), a quarter of a pound of meat, and a cup full of oatmeal. Coffee was not introduced until 1917. Bedtime was '8.30 p.m., rising in summer at 6.15 a.m. and in winter at 6.45 a.m.'

There were also children's nurseries and at one time a schoolroom, and a hospital block (male and female wards, and a special maternity ward) and mortuary. The maternity or lying-in ward was available to unmarried mothers and others who, in the days before National Health Service hospitals, could not afford hospitalisation. Edith, perhaps because she had had a difficult birth with Freda, availed herself of the facility in 1907 and 1910 for the births of Fred and Mary. She would have been in and out as quickly as possible, for the ward was notably without frills: 'Traditionally the noisiest and worst-behaved part of the workhouse was the "lying-in" ward,' writes Beryl Thompson. Mostly the women in the ward were regarded as of 'low moral character'. Some were youngsters in their teens expecting their first baby, generally frightened and apprehensive, but most were 'regulars' with a lifetime of prostitution behind them. Quarrelling and fighting often broke out and the workhouse staff found these inmates resentful and uncooperative.

Jack London's famous book, *The People of the Abyss*, gives us a fascinating glimpse of conditions in a workhouse at around the time Barbara's mother and grandmother were suffering the one in Ripon, its relevance accentuated by Barbara's own choice of 'The Abyss' as title of Part Two of *A Woman of Substance*, where she introduces us to the poverty in which Emma Harte's own mother lived in Fairley Village. Here, on the title page of the latter, a quote from John Milton leaves us in no doubt that clawing herself out of it was no mean achievement:

> *Long is the way*
> *And hard, that out of hell leads up to light.*
>   Paradise Lost

Jack London actually lived among the homeless of the East End of London. This he did over seven weeks in the summer of 1902, repairing regularly to a safe house to recover. Immediately in his book there is a sense of division between two worlds, of Light and Dark, of Hope and Despair, and of the one-way passage between them. For him, the workhouse casual ward (for vagrants) was the pit. Interestingly, the inmates there were of the opinion that the metropolis workhouses were better than those in the countryside, for less food was offered in rural parts. Here he describes his first night:

> *Many hours passed before I won to sleep. It was only seven in the evening, and the voices of children, in shrill outcry, playing in the street, continued till nearly midnight. The smell was frightful and sickening, while my imagination broke loose, and my skin crept and crawled till I was nearly frantic. Grunting, groaning, and snoring arose like the sounds emitted by some sea monster, and several times, afflicted by nightmare, one or another, by his shrieks and yells, aroused the lot of us. Towards morning I was awakened by a rat or some similar animal on my breast. In the quick transition from sleep to waking, before I was completely myself, I raised a shout to wake*

115

*the dead. At any rate, I woke the living, and they cursed
me for my lack of manners.*

In Edith and Freda's ears the children playing in the street
might well have included Jim Gott, who was a thirteen-year-
old lad living in Allhallowgate when the Walkers passed
through the workhouse gates for the first time. In his memoir
he gives a sense of how children on the outside would be lured
towards the workhouse walls in the hope of the odd glimpse
of an inmate. How the imagination would take flight when so
rewarded – as when Tommy Jackson suddenly bobbed his
head above the side gates: 'It was like a nightmare in the
daylight. He used to stare with eyes like lemonade glass alleys
set in the circular rings of raw liver; and when he laughed
it was like the opening of the Mersey Tunnel on a wet day.'
Then there was Tommy Coates with his 'one upper molar',
and 'Bill Sykes cap . . . perched on his thin, worn head . . . He
always reminded me, when he walked, of those jointed
wooden dolls, his every movement stiff and mechanical as
though he had to be wound up like a clockwork model.'

One who had not actually witnessed life within the Ripon
Workhouse walls assured me that, 'This was a small, rural
workhouse. Forget Charles Dickens. There would be gramo-
phone evenings from people in the city, special treats at Christmas, they even went to the theatre.'

It is true, as specialist Anthony Chadwick points out in an
excellent booklet on the subject, that 'some Yorkshire workhouses were reprimanded by London for allowing such comforts as tobacco or seaside trips,' and Ripon's own records
point to a Magic Lantern Entertainment for aged inmates in
1884, a mayoral tea and concert eight years later, a pint of
beer and an ounce of tobacco on the occasion of the marriage
of the Duke of York in 1893, and outings to the Victoria Hall

(I have a note that in November 1912, inmates were treated to a production of Gilbert and Sullivan's *Pirates of Penzance* there), Studley Royal and Fountains Abbey – what would Edith have made of these! But they were very occasional and anyone who has braved the Ripon Workhouse museum, housed in the old male casual wards, on a chilly autumnal morning, as I did, will beg to differ. The tiny vagrant cells, one complete with restraining chair for rowdies or lunatics, were as cold as stone. The ignominy, the uncertainty and fear, the absence of warmth of any kind, love and comfort, the loneliness, the sheer despair are missing from any equation that delivers a pleasure verdict to a child on such a place as this. And unaccounted for is the authoritarian and patronising attitude of such a system in respect of the poor, and the stigma and humiliating aspect of being in receipt of charity. As Jack London wrote: 'It is the way of the world that when one man feeds another he is that man's master . . .'

> *Poverty, misery, and fear of the workhouse, are the principal causes of suicide among the working classes. 'I'll drown myself before I go into the workhouse,' said Ellen Hughes Hunt, aged fifty-two. Last Wednesday they held an inquest on her body at Shoreditch. Her husband came from the Islington Workhouse to testify. He had been a cheesemonger, but failure in business and poverty had driven him into the workhouse, whither his wife had refused to accompany him.*

As for the view that Ripon Workhouse was a place of entertainment, Dickens's 'brick and mortar Elysium', let Jim Gott give us the benefit of actual experience, for he was once an entertainer there:

Years ago, when I was throng in the entertainment world, I appeared at No. 75 Allhallowgate ... and really the memory of the night will always be with me. I remember I sang one of my compositions, 'Down at Fishergreen'. I've often wondered since what were their thoughts as I reminded them of their early days.

Many old familiar faces were before me – some that I had thought had passed on. How strange to see them there like a long-forgotten community cut off from the rest of the world. Not that I am decrying the administration of those in charge, but through my eyes, the inmates seemed so lonely with a pitiful hunger for the warmth of their own kith and kin ...

The stigma and the fear lived on. The stigma of being a workhouse child was burned into the child's very soul. However solicitous some may have been in lightening the burden, 75 Allhallowgate was the gateway to hell, as the endless lists of those who died there bears terrible testimony. Ripon Workhouse was no more independent of the system that Dickens was set against than any other.

If Edith had been let down by Frederick it would have been very difficult for her to make enough of a rumpus to put a stain on his reputation. It would have taken time to make accusations stick or raise much credibility from inside the workhouse. Yet this would have been her only hope.

In Barbara's novel *Three Weeks in Paris*, we have a similar scenario to the Fairley family's cruel disassociation of Emma Harte in *A Woman of Substance*, but wound on a few years and with the male suitor as more obviously predatory from the start, and ending in the same desperate situation for heroine Kay Andrews as for Emma, and as, in reality, for Edith: pregnancy, with no man to support her.

Kay Andrews goes with her mother to run the house of a titled member of the aristocracy in magnificent grounds just outside Edinburgh, on the Firth of Forth. Kay is molested by 'his lordship', their relationship beginning at so young an age that she does not protest until such time as she becomes pregnant, when she tells her mother what's been going on. The squire arranges for an abortion for Kay and three weeks' severance for Kay's mother, who, far from accepting the situation, sits down and considers her options.

> *Suddenly she understood she was holding all the cards. His lordship sat in the House of Lords, he was a businessman, and very well known socially. He moved in all of the top social circles, and so did her ladyship. It struck Mam very forcibly that the last thing he wanted was a scandal . . .*

So she demands money, and, 'frightened out of his wits', his lordship pays up. 'It was blackmail,' confesses Kay to her aged school headmistress Anya Sedgwick. 'My mother saw an opportunity to help me, not only then, but in the future. And so she blackmailed him.'

> *There was a silence.*
> *Everything was very still in the garden. Not a leaf stirred, nor a blade of grass. Nothing moved at all. Even the birds were quiet.*
> *But Anya's head buzzed with all that she had heard.*

Echoes of the past inhabit Barbara's fiction in an almost uncanny way. Secrets abound around every corner, the only question is which flight of fancy to follow in pursuit of them. The facts in Edith's case are that on 15 October 1910, she was taken from the workhouse to Ripon Minster and married to one John Thomas Simpson, a labourer six years her junior, and then she was returned to 75 Allhallowgate to resume her 'sentence'. Who was this John Thomas Simpson? He describes himself on their marriage certificate as a labourer, and gives his address as 8 Belle Vue Yard, Somerset Place, where we know his father, Thomas, lived (though the son is not included on the return).

John Thomas Simpson was born in 1886 and had three brothers: James, Arthur and Alfred. He was only twenty-four when he was married. Edith's sister-in-law, Ruth Matilda Walker (wife of Edith's brother Joseph), was one of two witnesses to the wedding.

What seems so strange is that Edith returned afterwards to the workhouse. What kind of marriage was this? Had Simpson stepped forward to rescue her from the abyss? It seems unlikely, in that a twenty-four-year-old labourer offered little financial security. Had she known him before entering the workhouse and fallen in love with him perhaps? Again, unlikely, given that Edith returned to the workhouse rather than joining her new husband.

Edith finally left the workhouse in January 1911, three months after the marriage. On 3 February 1912 she produced a second son, Norman. On the birth certificate, Simpson is named as the child's father, but gives his residence as 9 Bondgate, Ripon, while Edith gives hers as 9 Bedern Bank. Why the two addresses if she and Simpson were husband and wife in the normal way? Furthermore, from the electoral roll we also know that from 1913, Freda's cousin Gabriel Barker was also living at Bedern Bank. What sort of circumstances could have suddenly made the Bedern Bank address available once more to the Walker family?

Increasingly, the whole thing feels like a put-up job, a

marriage of convenience and not one that necessarily involved any emotional input on Simpson's part. Could he have been suggested as a safe way forward? Perhaps he already worked on the Studley Royal Estate. Consider the ramifications. Simpson would be guaranteed employment. Edith would have her respectability, and as soon as the keys of Bedern Bank could be returned to her, she could quit the workhouse and repair to her own home. As far as Frederick was concerned, he was off the hook. So long as there was a husband registered on the birth certificate, his name – the Robinson name – was in the clear. Edith was at last a married woman.

It was a solution that appealed to everyone for their own reasons, and it is quite possible for it to have been cooked up in the workhouse. Ripon Workhouse records for the period show various Simpsons as inmates. One Elizabeth Ann Simpson had an almost permanent residence there – could this be Edith's elder sister who had also married a Simpson? And there's a Thomas Simpson (born 1845), who is in and out of the Grubber constantly – could this be John Thomas's father (he would be the right age)? Also a Joseph Simpson crops up time and again, as does a labourer called James Simpson – could this be John Thomas's brother, James?

Furthermore, just as Edith had contact with the Simpson family through the workhouse, so there is a distinct possibility that Frederick had contact with the family too. At the time that Edith married John Thomas Simpson, a branch of the Simpson family had a gardening business at Kirby Road in Ripon and, among other duties, looked after the cemetery gardens. Edith's husband described himself on the marriage certificate as a labourer, but later at the birth of Edith's first child after their marriage, he was more specific, calling himself a 'grocer's waggoner'. This commercial gardening connection cries out for us to make the link between the time-honoured gardening Simpsons of Ripon and the Studley Royal Estate, one John Simpson heading the gardening workforce there when the main works were done. Ripon was a very small city, people did not move around like they do today. The Simpsons of Ripon that Edith knew may well have been that same family, further down the line.

Other facts also suggest a connection between Simpson and Studley Royal. Records show that on 3rd July 1914, Edith had her fifth child, called Frances, a child who, quite clearly, Barbara knew nothing about. Nor was it easy to find out much about her. The only Frances Simpson born in 1914, the year recorded as the year of her birth in a later workhouse record, to which I will come, was indeed registered as born to John Thomas Simpson and his wife, Edith Simpson (née Walker), but baby Frances was born in Millar's Lane, Morpeth, in Northumberland, not in Ripon at all.

Had Simpson taken his wife on holiday? Not very likely. Given paucity of funds and general inclination, holidays eighty miles away to the northeast are an unlikely part of the agenda. Could it be that work had taken him up there, possibly at the behest of the Studley Royal Estate? I rang the local library at Morpeth and asked a helpful young lady whether she knew of any connection between the Ripons of Studley Royal and any estates in the area. 'Funnily enough,' she said, 'there is a connection. In my village of Swarland, just north of Morpeth, Commander Clare Vyner [who would take up the reins of Studley Royal after Frederick Oliver Robinson died] had an estate there, where workers could relocate, live in their own house and work a piece of land.'

In fact, the Swarland Settlement itself, as the Vyner project became known, didn't happen until 1934, but the land owner-ship is a fact and it evinces the particular character of the Ripons of Studley Royal: land ownership and social conscience is what engaged them, which is evident at other times and which is clearly consistent with the notion of Edith being 'looked after' there and kept out of the way.

The Studley Royal Estate had land holdings in the area at least as early as 1575 and the specific connection between Morpeth and Ripon goes as far back in time as the twelfth century when Ranulph de Merlay, Second Baron of Morpeth, financed the monks of Fountains to build a replica of their abbey, calling it Newminster Abbey, in a similar position southwest of the town. In Victorian, Edwardian and more modern times there were ever-closer collaborations. Frederick Oliver, being one of the two best shots in England, would have enjoyed fraternisation with the other big grouse and pheasant shooting areas, of which Northumberland is key. And Commander Clare Vyner would seal the connection by marrying into the Northumberland aristocracy. In 1923 he married Lady Doris Gordon Lennox, who was the niece of Helen, Duchess of Northumberland.

This has strange echoes in *Act Of Will*. Lady Doris and Helen, Duchess of Northumberland were granddaughter and daughter of the Duke of Richmond and Gordon, and their maiden names were both Gordon Lennox. In *Act of Will*, Matron Lennox is a women's libber of the old school – 'she was of a new breed of woman, very modern in her way of thinking, some said even radical. She was well known in the North of England for her passionate *espousal of reforms in woman and child welfare*, and for her dedication to the advancement of women's rights in general.' Matron Lennox is willing and able to pull strings for Audra at a time when work was becoming difficult to find. Was Edith's Northumberland sojourn perhaps a sign that she was a beneficiary of the Lennox family social conscience?

Was Frederick Oliver Robinson's reputed interest in the

social welfare of females – as patron of the Ripon Home For Girls – Edith's ticket to this otherwise unlikely destination? Did the Ripon Masons smooth the path to the Edith Simpson 'solution' perhaps? The question is relevant because in Morpeth, at Winton House, was the headquarters of the Freemasons in the North. The Robinsons of Studley Royal were leading lights in the Craft, not only in Ripon but nationally, and will have been hand-in-glove with the Morpeth Masons. That Winton House had previously been the home of the famous suffragette, Emily Davison, may be deemed coincidence even in the light of the Lennox/Robinson interest in women's welfare, but it is perfectly consistent with practice elsewhere at this time that the Masons would have been involved in welfare projects of this sort.

The Morpeth Masons would first and foremost have included in their membership the local landed gentry, the most visible of whom in the town was one J. R. Blackett-Ord. Just as Frederick Oliver Robinson owned Fountains, so Blackett-Ord had the Newminster Abbey Estate among other significant holdings in Morpeth. He employed an estate manager called John Sadler who looked after Newminster and owned various properties in the town on his own account. We know from an assessment undertaken in 1910 following the Finance Act, who owned and occupied every property in England and Wales. For a few years afterwards, in localities up and down the country, copies of the assessment were updated as land or properties changed hands and new occupiers moved in. The 1910 names were simply crossed out and new names added. Imagine my surprise to find the name J. Simpson replacing that of the 1910 occupier of a small house in a yard in the same block as Winton House. The humble abode was owned by Blackett-Ord's manager, John Sadler. Was Simpson working for Sadler/Blackett-Ord? It seems likely.

If so, this was surely an ongoing protection strategy of Edith of an extraordinary kind, and one that, for Frederick, happily distanced his little problem geographically from Ripon. But whatever the reason for their going there, the Northumbrian experiment came to an end around 1916, and the couple returned to live not in Bedern Bank, which Edith had probably sublet, but at Yorkshire Hussar Yard. The Yorkshire Hussars was an inn, marked on the 1909 map on the east side of the Market Place, by what is now the archway entrance to the main city car park. There, on 14th August 1917, Edith gave birth to her sixth and last child, a girl, also called Edith.

It is odd that Simpson had not been called up. At the start of the First World War, married men were not required to serve in the forces, but by 1916 the killing fields of France were not so choosy and men avoiding the draft would be shamed with white feathers. From that time it would have been very difficult for Simpson to remain in Morpeth, which was headquarters of C Squadron of the Northumberland Yeomanry and of A Company of the 7th Battalion of the Northumberland Fusiliers. Perhaps that is why they returned. Ripon was, by Freda's account, also busy with soldiers, but the situation may have been more manageable there under the Marquess's personal protection.

In 1918, Freda, who had experienced life without Edith during her mother's Northumberland sojourn, struck out on her own. Perhaps she had begun to form her own opinions about the helter-skelter ride which had been her life to date and decided to take it into her own hands. What we know is that Freda began working at the Ripon Fever Hospital in Stonebridgegate, off Allhallowgate – the workhouse and fever hospital were in fact neighbouring buildings. Erected in 1878 and enlarged in 1916, it provided for twenty-five patients.

Although it was to one side of the Ripon Union Workhouse,

the Fever Hospital was not part of it, and the workhouse had its own hospital. It features, of course, in *Act of Will*, where Audra, like Freda, is beset by 'all the gruelling work . . . the endless scrubbing and cleaning and washing and ironing, the terrible drudgery . . . [but] there was something very special in Audra Kenton, call it stubbornness, that made her stick it out until she could graduate to a nurse's training . . . she had unusual stamina, as well as mental energy and toughness of mind . . . inner resources which she was able to draw on for inner strength. And so she had valiantly continued to scrub and clean and polish endlessly . . .' It is, now, easy enough to believe it. One might even call it *persistence*.

No. 9 Bedern Bank had by this time been let to a couple called John and Lily Taylor. With the First World War coming to an end, and with the Taylors suddenly entering upon the scene, it is as if one generation is handing over to the next. Could John and Lily be related to the Taylors of Armley, the family into which Barbara would be born? When Freda eventually found her way to Leeds looking for work, did she travel with an introduction to John and Lily's big-city relations? Is this how Freda met Winston? It is possible, although when Freda first left Ripon she did not go immediately to Leeds, as we shall see.

During this period, when Edith is listed as living at Yorkshire Hussar Yard, Freda was almost certainly living at No. 8 Bedern Bank, next door to the Taylors. Women were eligible to vote from 1918, but Freda was not yet of voting age and is not listed as resident there on the electoral roll, although another member of the Walker family is . . . a woman called Frances Walker, who might well have been looking after Freda. Who was Frances Walker? Could she have been Edith's elder brother Thomas Walker's widow, whose name was a scrawl in the earlier records, but just could have been read as 'Frances'?

126

Frances Walker is, I discovered, yet another name to emerge from the shadows between fact and fiction, where truth so often lies in our story. In *Act of Will*, a certain Frances is the mother of wicked Aunt Alicia, who has Audra's brother and sister transported to Australia after their mother Edith's death. In real life, Freda's brothers and sisters were similarly transported, as was the way in those days, not only in the case of orphans but in any case where proper care was deemed inadequate.

While unaware of much of Edith and Freda's story, Barbara did know that her uncles and aunts had been so dealt with, believing the event to have been precipitated by Edith's death. Indeed, she made the whole heartless practice of child deportation the wellspring of *Her Own Rules*. Early on, Barbara said to me: 'Freda had, I think, two brothers, Frederick and Norman, and two sisters, Mary and Edith [she was unaware of Edith's fourth-born child, Frances]. One brother and two sisters were sent to Australia. One brother to Canada. I think only Norman went to Canada. Australian emigration could have had something to do with the scheme, the emigration of thousands of orphaned children in the 1920s. There was a Granada TV film about it – *The Lost Children of the Empire*.'

Later she would also tell me that there had indeed been an Aunt Alicia figure in real life – different name and not as heartless, but someone who ultimately handed over Freda's brothers and sisters to the authorities, and that Freda had been given the choice whether to go or stay, and that being older and with a job at the fever hospital, had opted to stay.

It was a wrenching moment for the three young Kentons when Frederick and William took leave of their sister on that bitter winter morning. Before setting out for London and the boat to Sydney, they had huddled together in

127

the front hall, saying their goodbyes, fighting back their tears.

Act of Will

In the novel, the children's 'fiercest protestations and anguished pleadings to stay together had made no impression. They were helpless in the face of their aunt's determination.' And so, against their wishes, and those of Alicia Drummond's mother, 'Great-Aunt Frances', they had been forced to do as Alicia had said. Now, I was wondering, could that fictional Great-Aunt Frances be one and the same woman as Frances Walker, listed as living at 8 Bedern Bank in 1918?

Barbara gave me the answer before I knew even to ask for it, in a note on a letter I faxed to her at the start of my research when I queried what her childhood friend Margery Clarke had told me about Edith – Freda's mother – being still alive and living in Bedern Bank when Barbara was a small girl. Barbara appended a note to my letter and faxed it back: 'NO,' she wrote in capitals. 'She's thinking of Great-Aunt Frances.'

Searching for confirmation of the transportation of Freda's siblings, I came up against all kinds of problems. Records of a practice that sometimes amounted to child deportation are jealously guarded, no one denying that many were taken without parental consent if their mothers were sick or otherwise deemed incapable.

The official justification for the scheme had been that it would relieve Britain's overcrowded cities of orphans and children at risk, while the receiving nation would benefit from the cheapness of labour these children promised. The Barnardo migrations to Canada began around 1870. For seventy years, over fifty British childcare organisations sent 100,000 children aged between four and fifteen to work as indentured farm workers and domestic servants. Later there was also a scheme to Australia and New Zealand, the scheme persisting until the mid-1960s.

Allegations of abuse and exploitation have been rife. In the case of Barnardo's, one of the main vehicles for the emigration of child victims of poverty, illegitimacy or broken homes to Canada, there was an employer agreement with the receiving farmer, which covered the period of 'adoption', stating that the child would be provided with sufficient and proper board, that he or she would attend church and school, and a wage was agreed, which was remitted to Barnardo's in trust. In addition a fee was paid by the Canadian government to Barnardo's for each child that emigrated to Canada. Each child had a trust account with the Bank of Commerce in Canada, but the passbook was sent to Barnardo's office in Toronto and no withdrawal could be made by a child without Barnardo's permission. At the ages of eighteen and twenty-one respectively, Barnardo's boys and girls could withdraw money from their account.

There was also a system of inspection to protect a child from abuse, but 'rewards for endurance' were given to those who remained on Canadian farms for long periods of time, and very few if any children were ever removed for reasons of abuse or neglect. Meanwhile, many reports reached the ears of the public, and Barnardo's employees admitted that there were employers 'who are far from desirable guardians or associates of young children.'

Newspaper articles of the period proclaim only a positive message about the opportunities the colonies provided: 'Open Door to Canada – free loans to nominated immigrants . . . No Exploitation . . . cooperative policies of Great Britain and Canada.' In Ripon, the *Gazette* urged locals to rally to an 'Australia Calls' slogan, which heralded the arrival of a certain Mr Stabler of the Australian Immigration Office. At the

YMCA Lecture Hall, Stabler waxed lyrical about 'the splendid opportunities afforded to young men of this country desiring to emigrate to the far-distant colony of Australia'. On 12th February 1925, a party of Barnardo's boys were promoting the organisation in the YMCA Lecture Hall in Ripon, as the *Gazette* reported: 'A very delightful entertainment was given by the boys, who exhibited remarkable skill in the manipulation of the different instruments, the playing of the handbells being especially good.'

Just one year earlier, however, Dr Barnardo's bells tolled a more mournful note at 'the verdict of a coronor's jury, which heard evidence [in Ontario, Canada] in connection with the suicide of a Barnardo boy, John Payne, on the farm of Mr Charles Fee . . . The evidence showed that Payne had read in the newspapers about the suicide of another boy from the Marchmont Home, and the jury found that this, combined with an alleged threat of Mrs Fee that he would be sent back to the Barnardo Home if he did not mend his ways, was the only cause that they could find from the evidence . . . On the day he killed himself, Mrs Fee had given him a slap in the face for calling her a liar.' Such suicides were not infrequent.

On 13th June 2002 a British Home Children Class Action Law Suit was commenced against Barnardo's on behalf of the class of persons whom Barnardo's emigrated to Canada. It was moved by one Harold Warneford Vennell, who was left destitute as a boy with his mother and older brother after his father left them. Around 1923 he contracted the pauper's disease – rickets. Taken in by Barnardo's, he claims that records show he was denied visits from his mother, and that without her permission, and despite the provisions of a contract she had signed, Barnardo's sent the fourteen-year-old boy to Toronto, and thence to a farm in Pakenham, Ontario, where he undertook hard farm labour eighteen hours a day (with no schooling or church). Harold tells of meagre food, poor clothing, neglect and mistreatment, even assaults by the farmer and his wife.

This, then, was the fate to which the real-life equivalent of *Act of Will*'s wicked Aunt Alicia – the daughter of Great-Aunt Frances of No. 8 Bedern Bank – had consigned Freda's brothers and sisters. For Freda herself it must have been a terrible blow. In the novel, Audra tells Vincent of 'the endless, endless miles of ocean stretching from here to Australia. And quite suddenly I did miss William and Frederick so, it was like a tight pain in my chest, a terrible constriction . . .' In reality, Barbara told me that 'Norman Simpson did come back and see us during the Second War,' but by then the past must have seemed to Freda like another world. At the time, the pathos of their departure cannot have failed to touch her and perhaps hardened her will further to get out and make something of her life, which we see Audra attempting to do in *Act of Will* before she finds her purpose in her daughter.

I got nowhere with Barnardo's on a search for Freda's siblings – they will only deal with family members and Barbara did not want to jump through the hoops. But my search was rewarded by chance with the saddest of all entries appertaining to Edith's family in the records of the Ripon Workhouse: on 22nd December 1924, just three days before Christmas, Norman, Frances and Edith Simpson (twelve, ten and seven respectively) were admitted *on their own* to 75 Allhallowgate by order of the Children's Act 1908. They were held there until 10th February 1925, when they were discharged into the care of the NSPCC *'for removal to Dr Barnardo's Homes'*, and onward transportation to the colonies. It is to be assumed that Fred and Mary went with them.

Whatever the level of treatment, the terrible psychological effect of enforced removal from one's roots and country emerges from the Granada TV film aforementioned by Barbara, *The Lost Children of the Empire*, as a lasting problem. Leslie Shaw was sent to Halifax, Nova Scotia on 23rd March 1927, and a lifetime later asserts that the problem this set in motion was one of identity, still needing solution: 'It is important to know who I am,' she said, even at this late stage seeing the TV team as an opportunity too good to lose. An eighty-year-old grandmother says that in all the years she has been in her adopted country, 'my mother was never out of my mind.' The migration scheme was Freda's own sorrow writ large. Accusations of sexual abuse fly in the film, as terrible memories lift the veil on indignity, degradation and suffering. But beneath it all there is this sense of dispossession, of loss, lost identity, which can never be made good. 'I love my children,' says one woman. 'But even they don't completely fill that gap. I feel as if I was robbed ... Even today I am very insecure, deep down. I feel a nobody.'

In 1925, at the very moment that the Barnardo boys were being applauded at Ripon's YMCA, all three Simpson children were rubbed out of the city for ever, and, as far as Barbara is aware, so were Fred and Mary too. Only Freda escaped, she being twenty in 1924 and making her way on her own.

Barbara was always told that the children were sent abroad because Edith died, but if this terrible act was due to their mother's death I can find no record of it. If Edith was still alive at this stage, she would only have been forty-four.

It is possible that the children were taken from Edith without her permission, for this was a common-enough occurrence, especially if, as is the case in *Her Own Rules*, the sole charge of the children – the mother – was ill when the 'care-takers' swung into action.

Edith, as ever, remains a mystery. I searched for records of her demise, either as Edith Simpson or Walker, from 1917 (the year in which she gave birth to her last child) until 1942, and came up with a list of possibles but no clear claims. The only Edith Simpson to die during the period in Ripon did so in 1934, but she is a farmer's wife, quite distinct and aged forty-three, whereas Barbara's grandmother would have been fifty-four. There are other Edith Walkers and Edith Simpsons listed who are the same age as our Edith, but they are in other places: Dewsbury (1917), Spilsby, Lincoln (1922), Halifax (1927), Barnsley (1929), Birkenhead (1935), Marylebone (1937), Leeds (1939), and none matches on other grounds.

Could it be that, like Meredith's mother in *Her Own Rules*, Edith had not in fact died when her daughter supposed? Could it be that she married again and moved on, perhaps even abroad after her children, or even changed her name after the devastation of losing them? In 1924 when the last three children were put in care, Edith was only in her mid-forties, which is young to die, though of course TB was an early killer of many.

There is one compelling possibility other than Edith's death in explaining how Freda's brothers and sister were brought to this pass. It is the second date that points to a connection between Studley Royal and Edith's fortunes, for in 1923, one year before the children were taken into care, the Second Marquess of Ripon died.

If Edith and the children had been living under his protection, then for sure, now, it had come to an end, and who else could afford to look after so many children? Not John Thomas Simpson, the putative father of three of them, that's for sure. Edith's husband was still alive in 1924, but when he did go, he perished in penury in Ripon Workhouse from cardiac arrest after a bout of pneumonia and lies buried in Ripon cemetery. He was only forty-four.

Secrets, there were so many in our family ... I never wanted to face those secrets from my childhood. Better to forget them; better still to pretend they did not exist. But they did. My childhood was constructed on secrets layered one on top of the other.

As I write, I am left with no other possibility as to Edith's demise, unless the death records are incomplete (which, I am told, can be the case). It sometimes seems that she simply disappeared.

At Bedern Bank, following the children's enforced emigration, Frances Walker was joined for a while by an Elizabeth Walker – could this be her daughter? (She and Thomas had three – Lillie, who married Joe Wray, Elizabeth, and Minnie.) Could this be the real-life Alicia Drummond? Then, in 1933, the year that Barbara was born, there is no sign of Frances, and in 1935 no sign of 8 Bedern Bank itself, which is appropriate, for by then the focus of the action, both in fiction and in reality, had switched from Ripon to Leeds.

And yet the chapter is not quite ended, because there is a question mark that hangs over Freda, who seems to me so private a person. She must have shown tremendous inner strength. She had known real loss, real deprivation, real destitution. Barbara, though quite unaware of the enormity, recognised as much: 'I believe that my mother always had a great sense of loss, in fact I know that she did.' Fate – her age and independent streak – had saved her from the Migration Scheme, she was able to choose not to go, but Freda knew what it was to feel robbed, insecure, and – fatherless – uncertain deep down of who she was. Yet, I have this feeling about Freda, however private, however difficult she may have been with people, that incredibly she was always in control. It was an essential element of her character, which Barbara, when I put it to her, recognised too.

In *A Woman of Substance* revenge is a major theme. The gentry get their comeuppance. Emma Harte ruins Edwin Fairley's family and razes Fairley Hall to the ground – 'I do not want one rosebud, one single leaf left growing,' Emma orders. Emma has been hurt so deeply that she cannot bear even to smell the perfume of roses (symbolic of romance and Freda's favourite flower), which filled the air when Edwin dumped her. Strange as it may seem, in reality the Robinsons' house – Studley Hall – was also razed, first in 1716 for reasons sadly lost to us, though arson was suspected and the finger pointed at a woman, a local, one Anne Gill – 'so divilish a woman that there is no mischief she could invent'. Could this act have been born of vengeance too? One could be forgiven for speculating. Two hundred and thirty years later, on unlucky Saturday, 13th April 1946, when Barbara was in her first year at Northcote Private School and her mother Freda was, according to her account, working her fingers to the bone to send her daughter there, Studley Royal was set on fire again . . .

> *Studley Royal, the home of Commander Clare Vyner and Lady Doris Vyner, was until a fortnight ago being used by the Queen Ethelburga School of Harrogate. Commander and Lady Vyner were expected to return from Ross-shire at the weekend to arrange for the mansion's redecoration before taking up residence.*
>
> *Within half an hour of the outbreak the whole length of the two-storey building was ablaze, and the light of the fire could be seen thirty to forty miles away. Firemen from eight areas in the North Riding were called out. Water had to be pumped about three-quarters of a mile from a lake in the grounds and an adjoining reservoir. An eye-witness stated that valuable paintings, tapestries, and furniture were saved by villagers, estate workers, and*

police, who formed themselves into a chain and passed the articles from hand to hand.

The Times, Monday, 15th April 1946

In the middle of May, Emma made a second trip to Fairley Hall. She walked along the terrace, which still remained intact, and regarded the great tract of rough, bare ground where the house and stables had formerly stood. Not one brick was left and the rose garden, too, had disappeared. Emma felt an enormous surge of relief and an unexpected sense of liberation. Fairley Hall, that house where she had suffered such humiliation and heartache, might never have existed. It could no longer hurt her with the painful memories it evoked. She had exorcised all the ghosts of her childhood. She was free at last . . .

A Woman of Substance

PART TWO

CHAPTER ONE

A New Start?

'Yesterday was now. The past was immutable.'
A Woman of Substance

Precisely when Freda Walker upped and left Ripon is not clear. Barbara says that like Audra in *Act of Will*, 'Mummy wanted to be a nurse. After working as a ward maid [at the Ripon Fever Hospital] she became a trainee student nurse, then she went to Bradford to be a nanny.'

On her marriage certificate in 1929, Freda would describe herself as a domestic servant, which allows service as a nanny. In Bradford she worked for a short time for a family of mill owners, looking after their son, before moving to Leeds where she was employed in the posher part of Upper Armley (in the direction of Hill Top), as is Audra in *Act of Will*.

In the novel, Audra's Leeds employer is a Mrs Irene Bell, a suffragette and modern thinker and an important influence on Audra. Irene's son (Audra's charge) is called Theophilus, actually the name of the boy that Freda looked after in Bradford and one that fascinated Barbara after she discovered it meant 'beloved of God'.

In reality, Freda's Leeds employer was a Mrs Ellwood and quite unlike Irene Bell, though her son, Tony, always believed he was the model for Theo in *Act of Will* and surprised Barbara one day in the mid-Eighties by stepping out of the shadows while she was over from America signing copies of the novel.

'I was in Leeds at one of the bookstores. They had this enormous queue of people,' she told Billie Figg in 1986. 'It being my home town, all sorts of people turned up, girls I was at school with ... Then this man suddenly appeared and he looked at me, and I looked at him, and he said, "You don't know who it is, do you?" And I did immediately, I said "Yes", and said his name, and then he said, "These are my grand-children and I *had* to come and buy your book." It was the child my mother had brought up! He said, "Freda would be so proud of you." I was so touched my eyes actually filled with tears.'

Barbara and Tony saw each other again: 'He always said he had the best nanny in the world, and his daughter said, "That was your mother." After that, Tony Ellwood and Janice, his daughter, always came to my book signings and he always managed to make me cry.'

When Freda emerged from Ripon and went to Bradford she would have needed to be at least in her early twenties in order to be offered these responsible child-caring positions. In which case she probably left Ripon around 1926–7, aged twenty-two or twenty-three, just two or three years before she met and married Winston Taylor. As Barbara confirmed, 'She was actually working in Upper Armley for the Ellwoods when she met my father.'

It is difficult to imagine Freda's frame of mind. She had lost everything, her father, her mother, her brothers and sisters. She was alone, stigmatised in her own mind by the calamity of life that had destroyed her family. Yet she did not hold her mother Edith responsible for the mess she'd got the family into. Barbara insisted: 'Freda missed her mother a lot. She loved her mother a great deal and spoke about her a lot. She was *very* close to her.' This can only have exacerbated the loneliness Freda now felt.

In her stints of nannying in Bradford and Leeds, Freda poured into her boys the love she needed so much to feel. Tony Ellwood recognised this in speaking so lovingly of Freda. It was why Barbara nearly cried whenever they met, though of course neither Tony nor Barbara knew the half of it.

In *Act of Will*, Irene Bell, with her modern politics and feminist principles, gives Audra tremendous confidence to stand up for herself and move forward, and it may be that there was someone like this in Freda's life, possibly her matron at the Fever Hospital, with whom she will have remained in touch if only to supply references to her new employers. Freda would have appreciated the support. In the end, of course, she met Winston Taylor, who swept her off her feet, just as Audra meets Vincent in the novel, and, like Vincent, Winston proved his value.

Freda found the love she needed in Winston, but, as we have seen, she couldn't accept it. She neglected him and turned instead to their daughter, pouring her whole life into Barbara. I was at first surprised to learn that Freda had not blamed Edith for what had become of her family. But if she had, she would very likely not have turned to Barbara in the way she did. On the contrary, she may even have spurned and neglected her daughter rather than her husband. Instead, Freda re-created in her relationship with Barbara the strong bond she had had with her own mother. Freda was Edith's first child, they would have been always together, there being no father around. They too would have been 'joined at the hip' in the early years. Now, with Barbara, she would revivify that bond and relive the period of her childhood when Edith was living her dream and Freda had enjoyed the most security.

As a child Barbara was so bound up in Freda's will that she only allowed herself to see the positive aspects, her father's anguish coming to her only occasionally as a missed step in the night when she worried what had become of him.

Others saw a different woman. So withdrawn and shy did Freda become as a result of her childhood experience that Billie Figg found it 'very hard to associate her with the input to Barbara's success with which Barbara credits her', even harbouring a suspicion that Barbara might have been adopted: 'She [Freda] was a very quiet woman, not one that you would expect to be behind such a powerfully driven woman like Barbara,' Billie explained.

Adult Barbara became aware of their different personalities, once saying to me, 'My mother might have irritated you because she was very self-effacing, shy.' Knowing what it must have meant to have been fatherless, to have suffered the indignities of the workhouse and other terrible uncertainties from six years of age, to have lost all her brothers and sisters as a teenager, and then to have lost a son, I should have expected nothing from Freda, but I have a feeling I would not have left empty-handed. For, like Audra in *Act of Will*, Freda could say, 'There was a time in my life when I had absolutely nothing and no-one.' Yet she, a mother, had made something out of her nothing, and given it to her daughter.

Certainly, Freda was not the driving personality that people who feel they know Barbara might expect of her tutor in life, and yet Barbara herself knows she has achieved everything as a result of her mother. What went on between them went on below the line – no one else was a party to it. Their union ran very deep and was in a real sense a continuation of the best part of Freda's earlier experience.

However different, Freda and Barbara were one, and that was Freda's doing, her loss the vital ingredient in Barbara's success. When I said, 'She may have been shy, but she was also single-minded,' Barbara was already ahead of my thoughts: 'Single-minded about her daughter, about me ... A sweet woman.'

'She was never going to be happy, though, was she?'

'My father used to say, "Freda, you cross your bridge before you get to it. You are always worrying." My mother was the biggest worrier.'

The thought flashed through my mind that Freda had kept her past from Winston, that even the love of Freda's life might not have known the depth of her suffering. Later Barbara told me that she was sure Winston didn't know. If so, one can only marvel at the fortitude of a woman who could hold so much to herself, and not be surprised that the effect of such extreme repression should be felt in the obsessional nature of her relationship with Barbara, which had Freda telling her every day of her life how much she loved her and that she was the 'most beautiful and the cleverest and the most talented and the most charming and the most wonderful person in the world', and which was entered into at the expense of her own marriage. Years later Freda confessed to her daughter that she had sidelined Winston because, '"I couldn't have given you everything if I had had another child. And *I wanted you to have everything* . . ." My mother was even glad I wasn't a boy because she felt that a boy would go off with the father, you know, but that I was hers. I am conscious of her always saying to me, "I prayed for a girl and my prayers came true, my prayers were answered."'

The loss of Freda's first child, a psychological catastrophe in the making for any mother, must surely have finally triggered this extreme reaction. The boy who contracted meningitis is called Adrian Alfred in *Act Of Will*, Alfie for short, and everybody loves him. As for his mother, 'she had never known such happiness as this tiny person gave her.' Then one day the fever is upon him, delirium, coma, and he is gone. Billie asked Barbara whether Freda ever got over it. Barbara replied, 'Yes, I suppose she did. She didn't discuss it much.'

In the novel Audra suffers depression after Alfie's death, and this was probably the state in which Freda found herself leading up to Barbara's birth. Says Vincent: 'Audra was so odd, she was like a stranger to me.' His friend, Mike Lesley, concedes she is in a bad way, 'but a lot of women react as she did when they lose a child, especially when it's the first baby. They're demented for a while. And her loss was a terrible one, in that she's had so many other losses in her life already.'

Winston Taylor left the Navy on 16th July 1924, and married Freda five years later. He was working as a general labourer at the time. Barbara says that when they married, Freda ceased working, as was often the way, but she went back to work before Barbara was born in 1933, which is when we get the proud outburst from Vincent in the novel: 'No wife of mine is going out to work and that's that!' Freda went to work anyway, of course, because Winston, along with many others, had lost his job.

She was at work all Barbara's childhood. So what was life like as the child of a working mum? 'Lovely, because she was so special, and later when I was older I used to come home from school and always turned on all the gas rings, you know, [for] the pots and pans. And then she'd get home for tea, she was working in nursing . . . and Daddy took me to my grandma's every day for lunch. Yes, he'd have lunch too, and try to get work. So I never felt deprived because my mother worked. Oh no, she taught me everything . . .'

In selecting for her daughter's education every aristocratic seat and museum of antiques and *objets d'art* in the neighbourhood to which Barbara might aspire – from Ripley Castle to Middleham, from Temple Newsam to Harewood House, from Newby Hall to Norton Conyers, and in particular Studley Royal, where her own identity lay buried – Freda was of course holding up a mirror to the life of which Edith, and by extension Freda herself and Barbara, had been dispossessed. By earthing her daughter's adolescent emotions and natural ambition in values inherent in the literature and history of Yorkshire and fusing them in the landscape – noble values that would, in the end, characterise the novels – Freda also ensured a strong sense within Barbara of her own destiny.

Freda's focus may have seemed to be wholly an educational strategy – the trips to the theatre, the ballet, the reading of books – and at one level it was. But as important was her introduction of Barbara to fine things, beautiful and, above all, expensive artefacts, and to the aristocratic lifestyle that spoke of her loss. Manners were 'a very important part' of Freda's strategy – table manners and social etiquette. Though apparently there was little money being earned, Barbara's clothes, other than hats and coats and shoes, were all professionally handmade. This was most unusual in working-class Armley, as was calling her parents Mummy and Daddy rather than Mam and Dad. She was given shoe-trees as a child and hangers for her clothes, which she noticed others didn't have. All of it spoke of Freda's need to realise a purpose in the rubble of her childhood, to resurrect the dream that Edith had shared with her when she was a child.

In Daphne du Maurier's *Rebecca*, the ruined Manderley is the sepulchre of the unnamed heroine's dreams. They lie buried in the burned-out shell, and will remain so for ever. But in Freda's case there was a determination that *there should be a resurrection* ... She and her daughter would roll away the stone on her past, on Edith and what might have been. Freda's project was, as it turned out, a success, for Barbara went on to tell it as it might have been, and to live the life that Edith saw and so nearly attained, but which she lost to the damnation of her six children.

This is why Freda drew Barbara back to Ripon time and again. Appropriately, resurrection was the theme of Barbara's remembrance of it: 'Easter Sunday. My mother used to *love* to go to Ripon Cathedral on Easter Sunday and she used to take me there.' Good Friday was spent in Armley, and Barbara would busy herself with comforting food. 'Good Friday I always think of hot cross buns, at home with mother and father, for tea. I used to go and get them and I see this long-legged skinny little girl of ten running down to the local baker my mother liked in Town Street.' Then it was off to celebrate Easter, the Feast of the Resurrection, in the very sepulchre of Edith's dreams – Studley Royal: 'I loved to be outdoors all the time and especially when we went to Ripon,' Barbara said. 'We (mother and I) would go for long walks picking flowers in the hedgerows. We would especially go to Studley Royal. But Studley Royal House burned down during the war years, some accident . . . it always fascinated me.'

They would take picnics, home-made sandwiches, and there is a 'Manderley moment' in Chapter Two of *Act of Will*, where Audra comes upon High Cleugh, not unlike the way du Maurier's heroine comes upon the house of her dreams in *Rebecca* – 'Last night I dreamt I went to Manderley again . . .' Audra sits down upon the grass and looks at High Cleugh from a distance. The house 'appeared to slumber in the brilliant sunshine as if it were not inhabited.'

A peacefulness lay over the motionless gardens. Not a blade of grass, not a single leaf stirred . . . Audra's gaze became more intense than ever. She saw beyond the

146

exterior walls to the inner core of the house. She closed her eyes, let herself sink down into her imagination, remembering, remembering . . .

As in *Rebecca*, Barbara's heroine imagines herself inside the house: 'She closed her eyes, let herself sink down into her imagination, remembering, remembering . . .' and there SHE is. Edith standing before her. Freda, as Audra, is a little girl again, mindful of . . .

> *. . . her mother's laughter, the swish of her silk gown, as she joined her by the fire. The Beautiful Edith Kenton. That was how they always spoke of her hereabouts. Sapphires blazed at her throat, on her cool white arms. Blue fire against that translucent skin. Hair the colour of new pennies, an aureole of burnished copper light around the pale heart-shaped face. Warm and loving lips were pressed down to her young cheek. The smell of gardenias and Coty powder enveloped her. A slender, elegant hand took hold of hers, guided her out of the room . . .*

This was Edith's style. Barbara was born to it, which is why she acquiesced in Freda's strategy so readily. In *Act of Will*, Gwen sees the hallmarks of it in Audra. This style – if not aristo blue blood itself – ran in Audra's veins, even if she (like Freda) can only avail herself of it in the limited environs of the little place she shares with Vincent:

> *'The place does you proud, lovey.'*
> *Audra beamed at her. 'Thanks, Gwen, and I am glad you like it. The room's small, of course, but that makes it cosy and comfortable, don't you think?'*
> *Gwen nodded, then let her eyes roam around. 'And*

what did you say the name of this funny green on the wall is?'

Audra laughed. 'Eau-de-Nil.'

'What a strange name.' Gwen made a face.

'It means water of the Nile in French, and it's a very popular colour at the moment . . . in fashion, I mean.'

'Oh is it. Fancy that. Well, you always did keep up with the latest trends in clothes, lovey, didn't you? I keep telling Mum that you're the expert on fashion and styles and fabrics and all that kind of thing. I hope you realise how much I value your advice. Yes, you've got the best taste of anybody I know.'

Gwen accepted the drink from Audra. 'Thanks, lovey,' she said. The two women clinked glasses and Audra stepped up to the fireside, sat down in one of the Chippendale-style chairs which had belonged to her mother. 'I'm glad you think so – that I have taste, I mean.'

The novelist Muriel Spark, who, like Barbara, creates characters 'dizzy with power', once said: 'You can't separate style from the person. You bring that package into the world with you.' Barbara's novels explore this as a theme. Time and again we see the mother figure nurture what is inherent in the daughter. For example, *To Be the Best*, the third novel in the Emma Harte series, asks how Paula's PA, Madelana O'Shea, a poor Irish-American Catholic girl from the South, has so many things in common with 'an aristocratic Englishwoman, heiress to one of the world's great fortunes and a noted international business tycoon'. The answer is that she has been born with a certain inherited style, which amounts to a potential to succeed, but she has been beset by loss: 'her parents are dead, her brothers have been killed in Vietnam, her little sister died when she was four from complications following a tonsillectomy.

Loss, loss, loss . . .' But then a new mother figure emerges, Sister Bronagh of the Sisters of Divine Providence at the Catholic hostel on West 24th Street in New York, where Madelana lives. The nun tells her that the courage was 'with you, already part of you then . . . If I did anything at all, it was to show you that it was there.' And Madelana is defining Freda's role in Barbara's life when she says to Sister Bronagh: 'Your belief in me has been so important . . . it has mirrored the belief my mother had in me. She encouraged me the way you have. I'll try never to let you down.'

Also significant in terms of Barbara's inherited style, it is the grandchild who inherits more completely than the daughter, as if the qualities actually skip a generation. In *A Woman of Substance* it is granddaughter Paula who inherits Emma Harte's style, and not Daisy, Emma's second illegitimate daughter and Paula's mother.

Edith's vanity and vivacious style seem to lie dormant in Freda (or have been knocked out of her), but are evident in her granddaughter, so that Freda plays nurse to what she sees and recognises in Barbara: 'My mother saw in me something she felt had to be cultivated.' And cultivate it she did – in effect nursing Edith's ambition in her with all the persistence that allied her, perhaps, to her natural father.

I once asked Barbara why she thought her mother took her to these historic houses with their things of expensive beauty. She said: 'She loved those places, she loved to see *furniture* – it was seeing beautiful things. She used to say to me, "Barbara, you must always use your eyes, you must always look."'

'She was saying this to you because she wanted you to be a writer, or to draw or to be a painter or –?'

'No. She wanted me to *reach* . . .'

'Were any of your friends taken to these places?'

'Not with us.'

149

'No, but by their parents?'

'I doubt it. I don't think Margery was ever taken . . .'

'So this was most unusual for a girl from Armley or Ripon.'

'My mother exposed me to a lot of things. She would say to me, "I want you to have a better life than I've had."'

The style of the places to which Freda introduces her daughter *becomes* Barbara because the style is endemic, and what encouraged Freda, I believe, is that she saw other qualities in Barbara that were endemic, too, and would enable her daughter to succeed where Edith had not; qualities of persistence but also of self-discipline. 'I know a lot of talented people who fail because they don't have that kind of focus or determination or discipline . . .' said Barbara when discussing the legacy Freda gave her to excel. Freda knew that Barbara would have to work hard for success, but that the inclination to do so was also in her.

In another child, without the genetic inheritance that I believe Barbara had, Freda's ambitions, sourced as they were in the dysfunctional aspects of her childhood and in repression, would very probably have been revealed as fantasy and remained unfulfilled.

Confidence and stability came, too, from the very closeness of her relationship with her mother. 'When your mother tells you all the time that you are loved and that you are the best at everything you do, you gain tremendous self-confidence.'

The only child may be a spoiled child, and getting one hundred per cent of her mother's attention and approbation there was a danger of this for Barbara: 'I was very spoiled, I was a very adored child,' she freely admits, but her character was not weakened by Freda's strategy. On the contrary, on account of the educational and disciplinary aspects, it empowered her. It made her unstoppable.

The discipline told first in little things at home. 'I always made my own bed, was always taught to hang up clothes and put my clothes away . . . and from when I was six or seven had my chores at the weekend. . . . I had to help with the washing up, scraping the potatoes and taking the eyes out, and I was allowed to dust the sitting room as long as I was careful with her ornaments and things. I went on errands, went to the corner shop and things like that. And I used to scour the steps. I had to do a line around the steps. My most favourite thing was helping my mother to bake, weighing the fruit, and I always got a bit of dough which was filthy when it went into the oven and it came out terrible, as hard as a rock.'

Marking the edges of front steps of a house with a donkey stone (light brown chalk) is an ancient ritual dating back to Celtic times to ward off evil spirits, though in commentaries on the working-class North it is generally taken to be a token of respectability, hard fought for in the terraced streets of industrial England.

Later, sometimes as often as twice a week, Freda took her to the Picturedome: '. . . she wanted to go, and had no one to leave me with. I think I learned a lot from the movies. I know that I thought Nelson Eddy was the greatest thing since sliced bread – wonderful musicals.' Eddy had a hugely successful partnership in the 1930s with Jeanette MacDonald. By the mid-1940s there were more than fifty cinemas in Leeds, all changing their programmes twice a week. Barbara was 'absolutely blown away in the cinema, watching people like Cary Grant and Ginger Rogers and Fred Astaire,' admitting in 2003 on *Desert Island Discs*, 'I think I fell in love with glamour then, because it was a time when movie stars really were glamorous and elegant and beautiful, and where are they today?'

The programme's presenter, Sue Lawley, was on absolutely the right track when she followed up Barbara's comment by asking, 'Fell in love with it in the sense that you wanted to write about it, or that you wanted to *have it to yourself* . . . ?'

Barbara stopped in her tracks and said only, 'I'm not sure . . .'

Movie glamour was a contemporary extension of the aristocratic world to which Freda was introducing Barbara in the Yorkshire stately homes. It was presumably bound up, too, in the acting element of her theatrical education. Very quickly, moviedom was integrated into her image of self and that of people who mattered to her, especially the man in her life – well-groomed Winston Taylor – who may not have had an aristocratic drop of blood in him, but could *be* the part as well as his namesake, Robert Taylor – another of Barbara's movie heroes – and was the idealised measure against which all her boyfriends would be assessed:

His looks dazzled.

It was his colouring that was so sensational. The gleaming black hair and the black brows were in marked contrast to a light, creamy complexion and cheeks that held a tinge of pink like the bloom on a peach; he had cool green eyes, the colour of light, clear tourmalines, fringed with thick black lashes. His eyes and his skin were the envy of his sisters – and most other women.

Matched to the striking colouring and handsome profile was a superb athletic body. He was exactly five feet nine and a half inches tall, well muscled, firm and taut and without one ounce of fat or flab on him.

Immaculate at all times, Vincent considered himself to be a bit of a dandy, loved clothes, wore them with flair and elegance. He cut quite a swathe wherever he went, especially on the dance floor, where his easy grace and good looks showed to such advantage.

Act of Will

Barbara's love of movies took Freda's strategy on, modernising it, giving it roots in a world beyond the fantasy bubble her mother had created and within which they dwelt, by connecting it to popular culture. This became her period. In the late Fifties and Sixties she would avoid pop culture like the plague, and the Forties and early Fifties would remain her period for the rest of her life, because it fitted her style. It was a modern extension of Frederick's style, which Freda nursed in her and which was Barbara's identity long before her contemporaries had found theirs.

Other elements of popular culture of the period conspired to influence her self-image. In her early teens, Barbara read Kathleen Winsor's *Forever Amber*, in which heroine Amber St Clare is prepared to go to any lengths (lying, stealing, even whoring) to get what she wants. 'I remember that I could not put it down,' Barbara wrote in a foreword to the Penguin edition in 2000. 'In fact, I read it so quickly that I immediately reread it to be certain I hadn't missed anything important.' What transfixed Barbara when she read it as a teenager, besides the detail of Restoration England, was that Amber *reached* for what she wanted, as Freda encouraged Barbara to reach. 'As a teenager, I was determined to get ahead. There was no way I was going to end up slaving in some textile factory, married off and perpetually pregnant like so many Yorkshire girls of my generation.'

By this time the Taylors had left 38 Tower Lane, the bucolic sanctuary of Barbara's earliest years, and moved a quarter of a mile east to No. 5 Greenock Terrace – 'One sitting room and kitchen, two bedrooms upstairs and an attic,' said Barbara as we approached the house. 'I remember it very well.' Greenock Terrace is in the thick of the brick-built terraces where Alfred and Esther Taylor dwelt, a stroll away from Christ Church School and the church, and still a long way from

153

the low end of Armley, the prison with Wortley Cemetery.

When we got there she was astonished that the house had only a few small steps leading up to the front door – how had donkey-stoning them so preoccupied her? The reality, a rundown terrace sadly unloved, disturbed comfortable memories and was dealt with instantly with the observation that at least No. 5 was well kept. It was the only one in the street that was. 'There was a fish-and-chip shop in the next street,' she said. 'My mother used to say, "Run around to the fish-and-chip shop." And next door lived another family called Taylor, and a girl called Barbara, no relation!'

I told her that when Graham Greene travelled the world, he was habitually met with the news at hotel receptions that another Graham Greene had just left, as if he were constantly one step behind his doppelgänger. The difference here was that Barbara was confronted by hers every day; she even attended the same school. 'They called her Barbara "Poppy" Taylor because she was born on Poppy Day. She had a disabled sister in a wheelchair, damaged at birth. She went to Thorsby Girls High School, I remember, got married and moved to some seaside place.

'Oh, but none of it looked like this, because everyone was so proud of their front steps and so on. I'm glad our house looks nice. It was a community then. I'll tell you something, we played in all these streets, and they all had window boxes and they were all pristine. I used to run down this way [via Ridge Road] to my church . . .'

A couple of years after the move, in 1945, Barbara left Christ Church School and enrolled at Northcote School on nearby Town Street, a private establishment independent of the State system. We find our way to the site, but look in vain. 'It *was* on Town Street and it was facing Keene's Dairy. What a memory I've got, Keene's Dairy! The school was an old house with a garden behind a wall, a high stone wall, a little like a manor house. There were trees. They must have knocked it down! Is that the Barley Corn? My father used to go there for a drink. These shops didn't look so garish then, they didn't have all the signs outside, they were very old-fashioned, you know. It was *all* more like a village then . . .'

The absence of the school building was soon forgotten in the search for another. Barbara attended Northcote for three years, between 1945 and 1948, following which she and her parents moved from Greenock Terrace to a house next door to the school, and Freda cooked meals for the teachers. So now we were looking for No. 148. 'This is the house! My mother used to take those lunches into the back door of the school. You see, a door opened from the school into our back-yard. And it didn't look so shabby! Ours was all one house, and I think they have made it into two – 148 and 146. It was all one house, surely, or have they built one on? That was the front room and you entered there where it says 146. That was the sitting room and the dining room, with the kitchen at the back. There were three bedrooms upstairs. It was all brick, not all concreted over like this . . .'

Not brick but grey pebbledash confronted us. It seemed that from somewhere an apology was in order, but the scene did not bend to her will. Barbara was twelve in 1945. The decision to go to Northcote was all Freda's of course, but had not been taken easily. A private school meant fees, a uniform. When Barbara's grandmother, Esther Taylor, rebuked Freda for suggesting such a thing, telling her it would give Barbara 'big ideas', Freda had retorted: 'Nothing's too good for our Barbara.'

I asked her what the school brought to mind. 'We wore green gymslips, yes, a green gymslip with a white cotton blouse, a green tie with a yellow stripe, and a green coat; and in summer

155

green dresses, cotton, green and white (I don't know, maybe it was striped) and panama hats, cream panama hats with ribbon – green with yellow – and we had green skirts and white cotton shirts, if we wanted with the tie. Oh, I remember things like the little plays, the parties, the church service and the present-exchange we used to have before Christmas when the senior classroom smelled like a perfumery from all the pot pourri and scented presents we used to give each other. Lots of people come up when I'm signing books in Leeds and say: "Hello love, you don't know who I am, do you, but we know who you are." And I'm looking at a woman who used to be at Northcote School with me.'

June Kettlelow sat next to Barbara at Northcote: 'Exelby was my name then. Barbara didn't live next door at that time, she lived at the Greenocks. We were the same age, her birthday was a week before mine. The school took children in from five, but the boys could only stay in the primary department, they left at eleven or something like that. We stayed until we left at fourteen or fifteen. It came in just at that point that you had to stay at school until you were fifteen. The morning was General Studies and then in the afternoon was typing and shorthand, and French. The headmistress was Mrs Harrison. Miss Smith taught the commercial subjects. Our form teacher was Mrs Cox. I didn't go until I was eleven, until I failed my eleven-plus. That's when I went. I think there were three classes from then on, probably fifteen or something in each class.'

Shirley Martin, another contemporary of Barbara at Northcote, told me: 'It was a lovely little school, a very small prep school and commercial college, from five upwards. A particular friend of Barbara, Alma Franks, lived across the road. At eleven, if you were a boy you might go to West Leeds High School or somewhere like that. From eleven or twelve everybody mixed together, all ages. You weren't divided up into grades. You'd do General Studies in the morning and then in the afternoon you'd do commercial, shorthand, typing certificates. In those days it went up to fifteen. Barbara left at fifteen and so did I. You could go on after fifteen, but if by then you already had 100, 120 words a minute shorthand and typing, why bother?'

Shirley, who works for solicitors Irwin Mitchell in Leeds, also remembers the Taylors living next door to the school, because her family lived next door to them. She remembers in particular how proud Barbara's mother was of her. 'I remember our mothers talking in the garden, over the fence, and it sticks in my mind her mother always talking about "*our Barbara*". She was a lovely person, very, very friendly, and it was a lovely school. I cried my eyes out when I left. No shortage of jobs secretarial-wise, but you needed good qualifications.'

Barbara, however, had no intention of becoming a secretary, nor indeed would Freda have countenanced it. Barbara is adamant that Freda had only university in mind for her, and although I first saw that as a measure of the dangerous level of fantasy in Freda's plans for her daughter, there being no precedent for a child going to university from Northcote, changes in education at the time might well have suggested to her that it was within Barbara's reach.

Post-war, the concepts of a vocational training and academia were no longer mutually exclusive, and higher education for families who would not earlier have been able to afford it, or indeed have it in their sights, was becoming a reality.

The 1944 Education Act had made secondary education compulsory and available free to all, and there evolved three categories of secondary school. Alongside grammar schools came technical schools (including those biased to commerce – like Northcote – or art) and so-called secondary modern schools (for pupils who had failed the eleven-plus and did not elect to go to one of the technical schools).

The notion of schools with a commercial bias had arisen out of a London-based network of 'central schools' established as early as 1911. When these were transformed into technical schools in 1945, a number of recognised secondary schools (many girls' schools among them) followed suit, offering thirteen- to fifteen-year-olds a two-year course that included shorthand, typewriting, bookkeeping and commercial practice. Northcote was one of these, and although the practice never became widespread (and Northcote itself would not survive), Freda's opinion that Barbara might go on to university was perfectly feasible in this context of change, the chief consequences of which lay in a socio-economic spread of university applications, and a massive increase of state scholarships and local authority awards towards that end.

Alan Bennett benefited from the latter, as he writes in *Telling Tales* (2000). Acceptance by a university meant that 'a boy or girl was automatically awarded a scholarship by the city . . . My education – elementary school [Christ Church], secondary school [West Leeds High School], university [Exeter College, Oxford] – cost my parents nothing, their only sacrifice (which they didn't see as a sacrifice) that by staying on at school beyond sixteen, I'm not bringing in a wage.'

Between 1900 and 1909, six new universities – Birmingham, Liverpool, Manchester, Sheffield, Bristol and Leeds – had been created. Three of them, including Leeds, had come out of science colleges, where training for jobs was uppermost. As early as 1900 Joseph Chamberlain inserted into the charter of Birmingham University (created in that year) a faculty of commerce, and now Leeds, Manchester, Durham, London, Bristol and Southampton had followed suit.

As it happened, none of this would matter to Barbara, who

had journalism, not university, on her mind, as indeed did another youngster living elsewhere in Leeds – Keith Waterhouse – with whom she would shortly be meeting. Waterhouse was born in 1929, four years before Barbara, and still writes his award-winning column for the *Daily Mail* today. His first considerable literary success came with *Billy Liar* (1959), a loosely autobiographical novel about a North Country undertaker's clerk who lives in a world of fantasy in preference to bleak reality. In collaboration with his friend Willis Hall, who also grew up in Leeds, Waterhouse made it into a play starring Albert Finney, and then a film starring Tom Courtenay and Julie Christie. The work reached far beyond the industrial North that spawned it, but no doubt found many sympathetic northern ears for the very reason that Billy himself, and others who suffered the traumas of a northern childhood, needed the fantasy element just to survive.

Like Freda Walker's, Waterhouse's family had fallen into the hands of the Board of Workhouse Guardians when he was a young boy. His father, a costermonger, died penniless after borrowing money to start a business and spending it instead on drink and the horses. The bailiffs stripped the family home of furniture, and his mother was reduced to hiding her prized possessions – a pair of fairground vases and a harp zither – from the means-test man in the coal-hole and airing cupboard. Later, the family was moved to a council estate on Halton Moor, not far from Temple Newsam. In *City Lights* (1994), Waterhouse tells us that he, like Billy, did work for an undertaker's – J. T. Buckton & Sons of Leeds – 'whose motto was "We never sleep", to which the sniggering rejoinder from everyone who heard it was, "Maybe not, but the customers do."'

He'd seen the job listed at the College of Commerce in Woodhouse Lane, Leeds, which he attended after failing the entrance exam to Cockburn High School, 'Leeds's answer to the grammar school on the other side of the tracks,' he calls it. The school ran courses like those that Barbara was studying at Northcote. The shorthand and typing would serve Waterhouse well in journalism, but he excelled in particular at commerce, because it was taught by making use of real-life, Leeds-based example: 'We were talking about the actual arrival of actual bales of wool on actual barges on the River Aire and the Leeds and Liverpool Canal, and their progress through the named mills of Leeds, such as the Perseverance or the Albion, to the named clothing factories of Leeds, like John Barran's, who invented the ready-made tailoring industry, or Montague Burton, the Tailor of Taste, to known retail outlets like Marks & Spencer, who had got their start in Kirkgate Market, or the Mutual Clothing company.'

Barbara will have learned all this too at Northcote, and readers of *A Woman of Substance* will appreciate how she put to good use her knowledge of the clothing industry in Leeds and the great Leeds landmarks of the industrial era (such as the multi-national retailer Marks & Spencer, which began here as a penny bazaar), and also how the entrepreneurial spirit of the times tied in with her burgeoning desire to get on in the world on her own account and not as a slave 'in some textile factory', a desire she made plain in the character of Emma Harte and in the character of Leeds, as she saw it:

> *Leeds was then, and still is, a lusty and vital city, and the streets on this busy Friday were, as usual, crowded with people rushing about their business. Tram-cars rumbled out from the Corn Exchange to all parts of the town and outlying districts. Fine carriages with prancing horses carried elegant ladies and gentlemen of distinction to their destinations. Prosperity, that sense of self-help and independence, nonconformity, hard-headed Yorkshire*

shrewdness and industriousness, were endemic, were communicated most vibrantly to Emma, so that she was instantly infected. And the rhythm and power of the city only served to consolidate and buttress these very same characteristics so intrinsic in her, for with her energy, tenacity, and zest, her obstinate will and driving ambition, she was, without knowing it, the very embodiment of Leeds. This was undoubtedly the place for her. She had always felt that to be true and now she was absolutely convinced.

She made her way decisively to Leeds Market in Kirk-gate, an enormous, sprawling covered hall composed of an incredible conglomeration of stalls selling all manner of merchandise imaginable — pots and pans, kitchen utensils, china, fabrics, clothes, foodstuffs to be bought and taken home or eaten there, including jellied eels, meat pies, mussels, cockles, cartloads of fruit, fancy cakes and toffee apples. She stopped at the Marks and Spencer Penny Bazaar, her attention riveted on the sign: Don't ask the price, it's a penny! Her eyes roved over the goods on display, so easy to view, open to inspection, so well organised in categories and so cheaply priced. She tucked the information at the back of her mind, her eyes keenly thoughtful. The idea of this Penny Bazaar is simple, yet it is exceedingly clever, she said to herself. Emma lingered for a moment longer, inspecting the goods, which included almost everything from wax candles and cleaning products to toys, stationery, and haberdashery, and then, still reflecting about the bazaar, she moved on. It was well turned two o'clock and she was conscious of a growing hunger gnawing at her. She bought a plate of winkles and mussels from the fishman's stall, lavished them with vinegar and pepper, ate them with her fingers, dried her

hands on her handkerchief, and set out for North Street,
where the tailoring shops were located . . .

A Woman of Substance

'Mummy and I used to go to Leeds market, the famous covered market burgeoning with fruits and flowers, cockles and mussels, on Saturday mornings,' Barbara told me. 'We used to go on the tram from Armley. I would buy daffodils and narcissus. It's near the Corn Exchange, the tram took us right there. A great treat for me was to go to the shellfish stall. Mussels, winkles, oysters and all sorts of shellfish like that with vinegar. Pickled vinegar. The first Mr Marks [of Marks & Spencer] was a pedlar and he peddled his goods on a cart and then he had a stall in Leeds market, which had the sign: "Don't ask the price it's a penny". So I had Emma go into that market.'

In *A Woman of Substance* we learn that Michael Marks was a Jewish immigrant, who came to Leeds from Poland. The idea of the M&S Penny Bazaars spreading from Leeds market all over the city and to other cities, making it a national chain, sets Emma's ambition alight.

There is a popular misconception, which fills readers' letters to Barbara, that Emma's flagship store in *A Woman of Substance* is modelled on Harrods in London. 'It is most definitely not,' stresses Barbara. 'I borrowed the building, but not the family story . . . But we did film two books there, *A Woman of Substance* and *Hold the Dream*.'

The description that links Emma's store to Harrods in people's minds comes relatively early in the novel and concerns the food halls: 'To Emma, the food halls would always be the nucleus of the store, for in essence they had been the beginning of it all . . .' Anyone who has been to Harrods Food Hall will likely conjure up an image of it as they read on.

However, anyone who knew Kirkgate Market in Leeds as

162

Barbara and Keith Waterhouse did, will prefer to bring the food halls in Emma's store to ground there. Waterhouse's father, Ernest, sold fruit, vegetables and unskinned rabbits from a horse and cart, so Keith knew this massive Victorian market well, and worked out early on that it was laid out in the form of rows or avenues, each with its own speciality. 'On one row,' he writes in *City Lights*, 'I could find the crumbly Cheshire [cheese] that my mother always bought ... On another would be hams swinging from brass rails and stuck with cloves like mapping pins; and below them white enamel trays piled high with chitterlings, polonies, Yorkshire ducks, stand pies, Aintree pies, whist pies, pork sausages, black pudding, white pudding, boiled ham, tongue, corned beef, sliced roast beef, salt beef, brawn, pigs' trotters – the same robust fare, in short, as could be found in any of the dozens of pork shops dotted around the city, which then as now could rival any Soho delicatessen or even Harrods Food Hall itself.'

Interesting, because it is Emma's fascination for the charcuterie department that brings her back to Leeds and the very roots of her ambition: 'When she came to the charcuterie department, a sudden mental image of her first shop in Leeds flitted before her, at once stark and realistic in every detail. It was so compelling it brought her to a standstill.'

Waterhouse claims that his desire to become a journalist, which he rehearsed in a paper of his own making as a very young boy, was moved by an image of himself as Edgar Wallace, the writer of such thrillers as *The Four Just Men* (1905), *The Crimson Circle* (1922) and *The Green Archer* (1923). Wallace was a journalist through and through, beginning his working life on Fleet Street (until the late 1980s the nucleus of British journalism) at the age of eleven, selling newspapers at Ludgate Circus. Waterhouse maintains that he was only three years old when he saw the picture of Wallace that

163

so influenced his future. It showed his hero 'wearing the peaked cap of a Reuters' war correspondent when covering the Boer War.'

Barbara was also determined to be a journalist and had in her mind's eye a strong journalistic image to which she aspired: 'At different times, depending on what movie I'd seen with Mummy at the Picturedome, I thought I was Rosalind Russell in *Front Page* or Jimmy Stewart in *Call NorthSide 777*, which was a newspaper story. I wanted to be a foreign correspondent or maybe a crime reporter.'

Freda, who so wanted her glittering prize to go to university, knew she had lost Barbara to journalism when she turned up wearing what looked like a brand-new raincoat that had been dragged around the back garden in an effort to confer on her something of the status of dogged reporter.

Being a journalist was not her principal ambition, however, more a means to an end. 'What I wanted to do was write books. I remember saying to my mother – whatever the words were – I want to write novels, I want to write books, I want to write stories when I grow up; and she smiled. And that was after I sold my first short story when I was ten to a children's magazine. I don't think I meant to sell it to them. Mummy sent the story in and they bought it. They paid me seven shillings and sixpence. I bought Daddy some handker-chiefs and Mummy a green vase, and put the rest in my moneybox.

'Daddy was always saying to me, "When I'm rich I'm going to buy you a pony." I don't know why he said it. I didn't want a pony actually, but he was always saying it. Anyway, he was never rich, so I never got the pony. But I wrote the story about a little girl who does get a pony. It was only about three or four pages, and three months later we got a letter and a postal order for seven and six. I said to my mother, I don't care about the money, I want to see my name there. So there was obviously that ambition and ego about writing already in me, wasn't there? I think my destiny was sealed that day. I said to my mother, "I'm going to write books one day." And she said, "Oh, that'll be nice." And then she said something to me that has always remained in my mind. She said, "You have to live life a little before you can write about it." She was right. Where better than in the reporters' room of a newspaper to do that? So the next best thing to writing novels, in my mind, was to become a journalist.'

Barbara loved reading, writing and telling stories – 'I was a big library girl. I had three library cards at one point' – and she had been writing them since she was seven: 'I always told stories – to my mother, to my doll – and wrote a lot of stories and poems and things, which my mother kept and I now have in New York, all yellowed with time.' Then, when she was twelve, her father bought her a second-hand typewriter. 'Now I wrote all sorts of stories and put them in a folder and stitched the folder so that they were held together firmly, then hand-painted the name of whatever book it was. I say "book", I mean just a dozen pages typed very badly.'

Soon, 'from about thirteen to fifteen I was taking books on journalism and on Fleet Street out of the library.' Barbara began writing little bits and pieces for the *Armley & Wortley News*, one of a group of newspapers within the Leeds Guardian Series, which also included the *North Leeds News* and the *Leeds Guardian* itself.

The *Armley & Wortley News* was published south of the river. The editors seem to have taken a leaf out of Fagin's book, encouraging gangs of city kids – budding reporters all of them – to deliver news on a daily basis, often for free. The papers were at a lower level than the three main Leeds papers, the *Yorkshire Post*, the *Yorkshire Evening Post* and the

Yorkshire Evening News, and, not unlike the freebie papers of today, ran on a shoestring. On the *Leeds Guardian* the editor was also the compositor, wearing an apron and lining the characters up on the hot metal printing machine, as well as being the main reporter, advertising executive and sub. This was the paper for which Waterhouse reported as a lad, after he heard that someone else at his college was doing it. Eventually he was offered a regular job for the princely sum of ten shillings a week. Unfortunately, his mother needed him to earn more, so he took that job in the funeral parlour instead.

These editors were willing to consider for publication any items that their reporters might bring in. Waterhouse's editor gave him some good advice – 'Remember that names make news.' When Barbara went to see the editor of the *Armley & Wortley News*, he was non-committal but encouraging to the fourteen-year-old. 'I said, "Can I do some local stories?" and he said, "Well you can, love, but I don't know if I'll use them."' It was enough. Barbara's uncle and aunt, Don and Jean Taylor, remembered taking her to a fair in Armley one time, and her running up to them to ask for change to make a telephone call. She had just witnessed a crash on one of the rides and wanted to file her story to the *Armley & Wortley News*.

Then one day at Northcote, her typing and shorthand teacher, Dorothy Smith, told her that she could get her an interview with the *Yorkshire Evening Post*. 'She knew the woman from the typist pool.' Barbara went home and told Freda that she had an interview on the *Post*, but not what job it was for. This was 1948, and Barbara was fifteen. Her mother still had it firmly fixed in her mind that she would get her daughter to university, but this was an interview with the respected *Yorkshire Evening Post* and Barbara played her cards tight to her chest. 'So it was sort of like, "Well if I don't like it – let me go and see if I like it." You know, she wanted to go to the interview *with me*! I said, "You can't! My mother cannot go with me to a job interview!"

'And then of course I got the job. I came back and she said, "I can't believe they've given a fifteen-and-a-half-year-old a job in the reporters' room." I said, "Oh, Mummy, it's in the typist pool." Well that did it, she was furious! To cut a long story short, I don't remember any rows about it. I remember long discussions and me pleading, and me finally convincing her to let me at least try the job. If I didn't like it, if it wasn't working out, I promised that I'd go back to Northcote and then go to Leeds University.'

Freda was not happy. Barbara had begun to place her own hand on the tiller of her fortune. With her love of glamour reflected in her passion for Hollywood movies she had given her mother's 'educational programme' a contemporary twist which may have seemed frivolous to Freda and possibly even reminded her of the pitfalls of her own mother's ambition. Her insistence on Barbara taking the contrary, academic route should be seen in that context of concern. Also, within the immediate family, Barbara's new outlook seemed to be associated not at all with Freda, but with Winston, and the way he looked and dressed and organised his priorities. Barbara was already beginning to operate outside the bubble Freda had made for them.

Barbara's awareness, whether conscious or not, of this subtle movement of relationships and ideas within the family, and her mother's feelings about it, can be seen in the central issue of the relationship between their fictional counterparts in *Act of Will*, where Audra's deep desire is not university but that her daughter Christina, who is a talented artist, realises her God-given gift. When Christina takes the commercial option, starting a fashion business instead, Audra accuses her

167

of compromising her 'gift for creating beauty', and unfairly reminds Christina of the sacrifices she has made for her, while Vincent tells her, 'You've just broken your mother's heart.'

The interview must have caused quite a stir at school because June Exelby remembers it to this day. 'I remember her going for this interview to the *Yorkshire Evening Post* and Mrs Cox saying, "You make sure you take the job in the typing pool, because once you are in, you are in."'

Mrs Cox was right, but it really didn't matter what anyone said about taking or not taking the job. Barbara was showing that at fifteen she could handle herself, she was able to determine the best strategy to get what she wanted, in particular steering clear of confrontation at home. 'I talked my mother into it; I talked them both into it. There weren't rows because I was a little scared of my father; I would never argue with my father because he was very strong-willed.'

Barbara had decided she was going to work on the paper, she was confident that she would soon rise out of the typing pool. More significantly, she had shown that once her mind was made up, nothing and no one could stop her.

CHAPTER TWO

Getting On

'And where's all this scribbling going to get yer?'
'On to a newspaper ... Maybe even the Yorkshire
Morning Gazette *... Stick that in yer pipe and smoke it,*
Emma Harte.'

A Woman of Substance

Barbara's job on the *Yorkshire Evening Post* took her daily
into the centre of Leeds, to Albion Street, the then site of
the newspaper group the Yorkshire Conservative Newspaper
Company. Later, the entirely fictitious Yorkshire Consolidated
Newspaper Group would figure in Emma Harte's empire, but
in the real world, the editor of the *Yorkshire Evening Post* was
one Barry Horniblow, his name almost ridiculously apt in the
context of a tabloid revolution to which he lent no small
endeavour. 'He had come up from Fleet Street,' Barbara
recalls. 'The *Yorkshire Evening Post* in those days was a broad-
sheet and he turned it into a tabloid like the London *Evening
Standard*.'

Top newspaperman Arthur Brittenden recalls Horniblow as
'the whiz kid who had come from London. Horniblow was
going to come and set the place alight and I suspect he got
bogged down by Yorkshire conservatism. He was succeeded

by Alan Woodward [an in-house alternative: Woodward was the news editor when Barbara arrived]. But the paper had an amazing circulation. It made all the money that kept the more prestigious morning paper, the *Yorkshire Post*, afloat.'

Keith Waterhouse, a reporter on the *Evening Post* when Barbara arrived, describes Horniblow as looking like 'an unusually bright bank manager', but notes that he was also one for giving youngsters a chance. The conservative daub might well owe something to his mode of dress, which demonstrated an attention to appearance applauded, as ever, by Barbara: 'They didn't like him because he was Savile Row from top to bottom,' she told me. 'Obviously if he was in the newsroom he was in shirtsleeves, but his clothes were impeccable, while the others were tough newspapermen from the North with their sleeves rolled up. Barry was white-haired, well-dressed, always impeccably dressed.'

As typist on the *YEP*, Barbara worked mainly for the advertising and circulation departments and was expected to take dictation. Her first day was a disaster, 'something I would never care to live through again. Not only could I not type the letters because I was so nervous but I couldn't even read my shorthand back. Fortunately, all were standard letters for the advertising department, so the other girls helped me. However, I was still typing away at seven o'clock and everybody had gone home and finally, after the tenth time on the last letter, I sighed with relief. Then, as I was leaving I saw the wastepaper basket full of very expensive notepaper engraved with the Yorkshire Conservative Newspaper Company – never mind printed, it was engraved! And I thought, I'm going to get fired for wasting all their stationery. So, I took a handful, went into the ladies' room, put a match to it and threw it down the toilet. Well, the blaze was so enormous I didn't know how to put it out. I kept flushing the toilet, and finally I put the blaze out. I then thought, well, I might get fired for wasting their stationery, but that's better than getting fired for being an arsonist. So I took the rest of the paper, smoothed it out and made what was a full wastepaper basket seem much less, before hiding the rest in the bottom drawer of my desk! I went to that drawer every night of the week, transferring paper from it into one of my mother's shopping bags, and brought it home and burned it.

'I hated that job, loathed it. I wanted to leave, I really did, in spite of my ambition. But I couldn't because I had this paper there in the desk. I thought they'd come and get me. What they'd do to me I couldn't imagine. I was very young and naive . . .'

In time, as the whirlwind in the typist pool began to settle, Barbara turned her mind to developing a strategy that would make her a journalist as soon as possible. Part of the week every girl in the typing pool had to spend time in the copy room, sitting in a little booth with a typewriter and head-phones, taking down copy phoned in by *YEP* reporters. 'Suddenly, the telephone would ring,' said Barbara, 'and the telephonist would say it was Keith Waterhouse, or whoever – Keith had brilliant red hair, very unruly – and I would type his dictation. You sat there typing as he dictated. You just put Keith Waterhouse at the top and when he'd finished he'd say goodbye and you'd take off your headphones and go into the newsroom and drop the article on the sub-editors' table, which thrilled me because it meant that I got to see the reporters sitting nearby.'

One day, Barbara realised that there was nothing to stop her feeding her own stories through the sub-editors in the same way. 'So I came to write my stories, used the same paper, put my name at the top – Barbara Taylor – and dropped it down on the sub-editors' table. And they ran three or four, and when it came to payday the question went up as to the identity of the writer of these stories and where to send payment. The Accounts Department said, "Barbara Taylor. Is she a stringer that we have in Doncaster or Harrogate or somewhere?" Suddenly, the penny dropped and they realised that I was the new girl in the typing pool. And so it all came out, and the editor was intrigued by this girl whom nobody knew, and he sent for me. This was when I first met Barry Horniblow.

'He said to me, "So, you want to be a journalist?" I replied, "Oh, yes sir, but I don't just *want* to be, I am *going* to be." This amused him. This was a man from Fleet Street and he liked that in me. I was about sixteen. Fools rush in where angels fear to tread . . . I was so nervous I worked my toe into the carpet and lost my shoe, and had to ask whether I could retrieve it from under his desk. He said, "What have you written?" and I told him I had written for local newspapers and all of that. He said, "Oh, some time bring me your cuttings book." Then I said, "Thank you very much, sir," and off I went.

'Then, at lunch time, I ran all the way to City Square, took a tram home to get my clippings book, dashed into the house and my mother said something like, "Oh, good, you've been sacked!" She didn't want me on this paper, she wanted me at Leeds University. Panting, I got the clippings book, took it back and went to his office. Mr Horniblow, who could see I was out of breath, said, "Barbara what is it?" and I told him what I'd done. He was intrigued, and that was the beginning of my journalism career.'

The editor made her assistant to his secretary, and within six months she was a reporter, graduating to women's-page assistant and then women's-page editor at eighteen. She was barely a woman herself, but she wrote the lot anyway: cooking, fashion, personalities. Two years later, 'Fleet Street beckoned!' as Barbara put it.

172

So meteoric a rise would not have been possible without a large dose of journalistic skill, which she crafted painstakingly with the help of Keith Waterhouse, as well as a personality to carry it off. Also, as we shall see, her rise coincided with an unprecedented media interest in women as a consumer force, in their attitudes and opinions, and by women's own interest in the kind of thing she was writing. It was a mixture of talent, style and publishing environment, which would set her fair into the third millennium and not only in journalism. Immediately, what was most striking to those around her was Barbara's style, the way she carried herself off.

Arthur Brittenden, nine years older, was working for the *Yorkshire Post*, same building, same group of newspapers, when in 1948 the fifteen-year-old Barbara appeared on the scene. Brittenden would rise to become editor of the *Daily Mail* (1966–71), then director of Times Newspapers and later director of Murdoch's News International. He has seen her only once since she left Leeds for London, but managed a crystal-clear recall of the winning spark in her. 'She was very winning, very appealing. I think people who are going to get on tend to show it at some time along the way, and she most certainly did. When she was there in 1948 and '49, I was on the *Yorkshire Post* – it was the prestigious morning paper of the group until it lost some of its character when it absorbed the *Leeds Mercury* [the paper, incidentally, on which Emma Harte gets her brother Frank a job in *A Woman of Substance*]. Barbara was a secretary then. We had these "moments", which I still remember nearly fifty years on, when we came upon each other in the canteen . . . very romantic. That was how I first remember her. We used to sit at a table having a cup of tea, but there was always something *different* about her. Of course, she was very attractive. Unbelievable. But she had this something different about her and it was very appealing. I

173

suppose, looking back, she must have been – I mean one reads about how ambitious she was and so on, but there was never anything aggressive or unpleasant or whatever about her. It was always terribly likeable. I think we were *all* in love with her probably.

'One other encounter stays with me – actually the last time I saw her. I was on the *News Chronicle* in Fleet Street. She had only recently come to London, and I remember coming down in the lift one day and going through the front hall, and there she was, sitting for some reason in the front hall with another girl. Why on earth would I remember with such clarity walking through the front hall of the *News Chronicle* fifty years ago and seeing Barbara there? But this was the thing about her, that you *did* remember. There was something always special about it – electric!'

I asked Leeds-based Bobby Caplin, a friend of Barbara since 1949, to describe her as she had been when they first met. 'She was a beautiful girl, absolutely *beautiful*. If you can imagine Barbara at sixteen years of age, a size eight or ten, slim, long hair –'

'Dangerous lady?' I teased.

'Very, very,' Bobby laughed. 'She was at the *Yorkshire Evening Post*, ended up on the women's page. In Leeds in those days there was a meeting for coffee on a Saturday morning at Marshall and Snelgrove, one of the top Leeds stores, a minia-ture Leeds Harrods in those days. There was also a lovely bar called Powolny's, a college bar in the centre, and we all met in there.

'Barbara had ambition, nothing wrong, but she had a good time, Barbara did. She was always very ambitious and she liked the nice things in life, as we all did, and she went out and got them. She never hurt anybody intentionally. I say that because at that age we are young, we want the nice things.

'But in those days it was a different life. The places weren't there to go to in Leeds, and we didn't have the money either. There was the odd coffee bar, but it was a different era. I mean, the trams were still running and I can remember when the cinemas were allowed to open for the first time on a Sunday evening, but had to close at nine thirty. So there wasn't much social life; well, there was a social life but it was in people's homes rather than going out.'

'She'd have been very popular with the boys, then?'

'Very. She was very, very attractive and she's always de-manded the best. Even at that age. She went out with a very close friend of mine, came from a wealthy family, very good-looking, born with two golden spoons in his mouth, lovely guy, that was the sort of guy she went out with. Ronnie [Sum-rie] is two years older than me. At seventeen these girls didn't want to go out with an eighteen-year-old, they wanted to go out with at least a twenty- or twenty-one-year-old. I mean we didn't have cars in those days, but Ronnie had a car, and he was a catch. He was known as one of the top catches in Leeds. And he took Barbara out, but then she moved, went to live in London.'

Barbara would need all of her feminine wiles to establish her position in the male-dominated *YEP* offices, especially after whiz-kid newcomer Horniblow backed her. 'After he'd seen the stories in my clippings book, he promised to get me shifted [out of the typing pool], and when he didn't, every time I saw him in a corridor I said, "It's Barbara, when will you move me?" So, one day I was told that I was going to be a part-time secretary. I was to help his secretary, Marion Greaves, and she hated my guts. An older woman, she resented me, she didn't want me there. I used to help out in the morn-ings. I had to be there in the morning and I had to be there when she took her lunch. The minute she'd gone to lunch Mr

Horniblow would come out and say, "Everything all right?" and I'd say, "Oh yes." '

But in fact Miss Greaves was only giving Barbara letters to type, when the editor had wanted her to get experience on the women's page, known as the Kay Boughton page. Why it was so named has been lost to time. As Barbara explained, 'There was no such person. It was a name, and the secretary, Miss Greaves, did it. It was really only a column. After probably about a month or two, he [Horniblow] said, "You're not getting any experience, are you, Barbara?" And I said, "No, I'm not." And he said, "Well things are going to change." I would say I'd been on the paper eight or nine months when he said one day, "Instead of coming here to work and help Miss Greaves, I'm going to have you in the reporters' room. I think you'll get some better training there." So, suddenly I was in the reporters' room, and one or two people were a bit suspicious . . .

'Edgar Craven was the chief sub-editor then, and they [the subs and reporters] thought Barry Horniblow was fancy pants. So, when he took this young girl under his wing, can you imagine? It never occurred to me at the time, but looking back I realise they had suspicions about him. They must have had. No? Wouldn't you think?'

'Were you having an affair with him?' I asked.

'No I wasn't. I know enough to know myself, I was not silly like that and he was a much older man. It was like a professor that you would adore. He was enchanted by my attitude. More than thinking it was sexual, they [in the newsroom] thought I was his spy. I know that because I've laughed about it with Keith. Barry loved young people, he was very focused on Keith so he put me with him – his desk butted up to mine, that's how they were arranged. So Keith was always helping me with my copy. He taught me to write for newspapers . . . Working on a newspaper at the age of sixteen and not a little village newspaper . . . a major, tough, provincial newspaper, a daily, being in that newsroom with a lot of newspapermen on the police beat, at court, inquest court, as a junior reporter. It was a great experience.

'Stanley Vaughan was the crime reporter, he used to drag me with him. I tagged on behind him – he spoke with an American accent and never took his hat off. I grew up in that newsroom. The nature of the work gives you exposure to life, sometimes life in the raw. Also, one has to have the human element in a news story. You can't just write about the land-scape or a room setting in a novel and not have it peopled with real people. I became conscious of the human element in stories when I was a newspaper reporter because that's what it's about, isn't it? A newspaper story is only interesting if it's about people, their tragedies, their dramas, their heartbreak – that's what I'm dealing with [in the novels]. I'm dealing in human emotion and, according to my French publisher, I am able to put it down on paper in such a way that I touch a nerve in the reader.'

The newsroom was L-shaped. To the right of the door was the long side of the 'L', a long table big enough to seat around six sub-editors: 'waistcoated or cardiganned figures, alter-nately red-faced or sallow-complexioned, pot-bellied or con-cave-chested according to whether they drank Guinness for strength or milk for their ulcers; but all of them chain-smoking . . .' as Waterhouse described them. Opposite the door sat the news editor, and beyond were the face-to-face desks of the reporters. The passage of copy was from reporter to deputy news or news editor, who having read and passed it would shout, 'Boy!', bringing the copy boy over to take it to the subs' table, who would prepare the copy for press.

'I was very much in love with newspapers,' says Barbara,

'and being a newspaperwoman, a newspaperman I should say, even down to wanting a dirty trench coat, which indeed I got. My mother accused me of having taken it off in the garden and rolling it around in the dirt to make it look used, but I didn't, it just got dirty.' The raincoat episode appears in another guise in her novel *The Women in His Life*, where Teddy shouts to her young charge Maxim, 'Don't trail your new raincoat on the ground!' as he is saying his sad goodbyes to the most important woman in his life.

Waterhouse recalled Barbara as 'an ambitious sixteen-year-old . . . apt to burst into tears from time to time when bawled out for not yet knowing her job to perfection, and I became her hand-holder-in-chief. Little did she know that it was a case of the blinded-by-tears leading the blinded-by-tears.'

'Keith was very sweet to me,' Barbara says. 'We'd go to the Kardomah Café and have beans on toast. I remember he told me not to cry in the newsroom, to cry in the ladies' room instead.'

Being slapped down by Ken Lemmon, deputy news and later news editor, would have been a new experience for Barbara, perhaps a salutary one, and the tears that flowed, which Waterhouse mopped up with such care and attention that it won him Barbara's hand at a Press Ball at Leeds Town Hall 'in a dinner jacket hired from Rawcliffe's where my mother had bought my school cap', might have damaged a less purposeful girl. There is a clear line between self-belief and delusion, and Barbara kept herself on the right side of it by sheer hard work, determination and discipline. She often stayed late in the office because she liked to get her desk cleaned up. And later, if she went for a drink with the others at one of the pubs favoured by the *YEP* reporters – the Pack Horse, say, or Whitelock's – she showed her self-discipline by not staying long. 'Frank

Shire, deputy news editor at the time, said: "You're smart, Barbara, you stay for one drink and then you leave." I'd have a drink, buy them a round and then go.'

Other characters among the *YEP* journalists included 'whimsical' Con Gordon, the feature writer, and the leader writer, Percy, 'who kept a cottage piano in his office which he would play for inspiration in his ceaseless fight against the Attlee government,' according to Waterhouse. Idly I mention to Barbara that Waterhouse is a socialist – odd then that he worked for the Yorkshire Conservative Newspaper Company and writes now for the *Daily Mail*, also a right-wing paper. 'Yes,' said Barbara, 'and do you know what he sang walking through the Yorkshire Conservative Newspaper Company? "The Red Flag"! He did all these crazy things. He walked once from Land's End to John O'Groats for a story. And on the other evening paper – the *Evening News* – was Willis Hall [Waterhouse's collaborator on *Billy Liar*, and most famous perhaps for his play *The Long and the Short and the Tall* (1958)]. And do you know who else was on the *News*? Peter O'Toole! I didn't know him well. Keith says I did and that I don't recognise him because he had acne in those days and didn't look so good, and that this big movie star, this gorgeous hunk with blond hair and bright blue eyes used to say, "D'you want to come t'pictures, Barbara?" in his Yorkshire accent, but that now he's very posh and he speaks like *this* ... and he's Lawrence of Arabia. Keith used to say all these terrible things to me when *Lawrence of Arabia* came out. He'd say, "Oh, you really blew it when he wanted to take you t'pictures." By then we were both in Fleet Street, [and we] went to a party given by Sam Spiegel for *Lawrence of Arabia* and met Omar Sharif ... but I don't remember Peter well, though I do remember there was a sort of pimply youth that used to hang around. It may have been Peter O'Toole.'

The Leeds newspaper scene does indeed have an extraordinary history of turning out famous people. Former *YEP* trainee Nick Clarke of BBC's *World at One* recalls a colleague being bawled out for poor timekeeping. 'The editor warned him: "You idle bugger – you turn up on time or one day you will have to make music for your living."' He couldn't have been more prescient: the poor timekeeper was Mark Knopfler, who would found multi-million-pound rock group Dire Straits.

At seventeen or eighteen, Barbara's social life tended to revolve around the friends she met regularly on a Saturday morning at Marshall and Snelgrove on Commercial Street: 'You could go through the circulation department, and there it was across the street. I would go across at 11.30 a.m. when I took my lunch hour. I'd meet Bobby Caplin and Ronnie Sumrie and various girlfriends. I didn't think of these boys as Jewish.' But they were among the second-generation Jewish clothiers who now ran Leeds industry. Many of the families had fled west out of Russia and Poland to escape persecution and stayed rather than continuing on to America, their original destination. It was a father–to-son inheritance.

Today, Barbara's readers will recognise this strain in her fictional account of the Kallinskis, a family of Russian Jews from Kiev. In *A Woman of Substance*, Emma rescues the patriarch of the family, Abraham, from a gang of hooligans in a Leeds street and he becomes a stalwart friend.

In the earlier twentieth century there were indeed Jewish ghettoes, and hooligan attacks on Jewish people, but Barbara developed a particular personal interest in the persecution of the Jews during the controversial and highly publicised trial of Adolph Eichmann in 1961, shortly before she met Robert Bradford, by origin a German Jew, when she was a journalist in London. In *A Woman of Substance*, after his rescue, Abraham takes Emma into his household, where she meets his

180

wife, Janessa, and their sons, David (who will later propose to her) and Victor. He tells her that most of the Jews in Leeds came 'to escape the terror and harassment of the pogroms directed against us,' an event which Barbara alludes to in her second novel, *Voice of the Heart.*

Meanwhile, in Emma's company we get a description of the Leeds ghetto, and naturally of the Kallinski tailoring shop, where she will come to work, and altogether it is one of those moments in Barbara's fiction for which she became famous, because it is researched with intricate precision. In this case we end up with a complete knowledge of the divisional labour-system invented by Jewish tailor Herman Friend, whereby a suit would be made up in parts by different tailoring outfits, according to the dictates of a factory. Friend worked with, and made famous, the John Barren factory, 'the first ready-made clothiers to start in Leeds after Singer invented the sewing machine'. He wrought a revolution in the industry, we learn, 'and helped to put Leeds on the map as the biggest centre of ready-made clothing in the world.'

Discussing all this with Abraham's son David 'on that hot August night in 1905', a friendship is born that will endure through the Emma Harte series: 'Together they would climb, in their own individualistic ways, struggling out of grim poverty, fighting all manner of prejudices, reaching for bigger and better things, and in their rising and their reaching they would carry the city with them.'

All of this was bred in the bone of the lads of Leeds, boys like Bobby Caplin: 'We were hard-working people in those days,' he remembers. 'I have an elder brother who was a little bit cleverer than me, who ended up as a project director at the World Bank in Washington, helped to design the first-ever computer in the UK. My late father started the family business and it was up to me to go into it. I was never good at school

... I was struggling, I was never an academic and I can honestly say that the happiest day of my life was the day that I left school. But I was never the boss's son. I went in to the factory at ground level and had to go to Technical College to get my City and Guilds. Ronnie [Sumrie] was also in the clothing business. Sumrie Clothes of Leeds was one of the finest clothing companies in the country, very famous. He never had to work. He was a good-looking guy, sat in the showrooms when people came.'

Alan Bennett mentions Marshall and Snelgrove in *Telling Tales*, but never felt at home there like Barbara did. He writes that he feared 'imminent exposure' when dragged in there by his mother: 'Marshall and Snelgrove's is a provincial outpost of a store in London's West End, so has a certain metropolitan grandeur, the carpets thicker, voices more hushed and fur coats much in evidence, and though the menu caters to Mam's core requirement of tea and toasted tea-cakes, tea-cakes come under an EPNS dish-cover, the proper manipulation of which is an additional hazard to the terrors of eating out.'

Bennett describes the 'floor show', which was a fashion feature of the restaurant, 'resident mannequins prowling the aisles between the tables modelling outfits on offer on the floor below in the couture department.' This was perfectly in tune with Barbara's love of glamour and also with an opportunity that had lined up for her. When Horniblow made her a reporter, he also hired Madeleine McLoughlin from the *Manchester Evening News* as women's page editor, and soon Barbara was to become her assistant. 'Suddenly Kay Boughton's column became Kay Boughton's Women's Pages, and the secretary was the secretary and no longer writing her page, or her little bit, and there was Madeleine installed as women's page editor instead.'

Barbara remembers that 'A lot of the fashion was how to

182

make six dresses out of two pieces of curtain or fabric or towel.' And Billie Figg showed me a photograph published in one of the Amalgamated Press women's magazines of her in a hat made out of a man's old tweed coat. 'It was still war-time "make-do-and-mend" days even in the early 1950s. Rationing went on for quite some considerable time after the war, food rationing until 1954. Isn't it funny,' said Billie, 'what was inculcated in us then never leaves you. I cannot waste the ends of soaps, I have to dig them into the next end. When I see youngsters in offices throwing away huge reams of paper that these machines spit out I'm appalled because I keep every sheet of paper that's got a plain white back to write on. You just can't get over it.'

No wonder Barbara had been so frantic at the paper wastage on her first day. Now she had justified her self-confidence, however, and found herself in something of a growth industry. What enabled Barbara's rise, what got her out of the North, with its relative lack of opportunity, was not only her singular desire to excel and a developing journalistic skill, but a combination of these and the opening up of the women's market, of which Horniblow's enhanced Kay Boughton women's page on the *Yorkshire Evening Post* was but one sign.

In the early 1950s, Billie was working on the PR side of an advertising agency – Napper, Stinton, Woolley in Great Chapel Street, off Oxford Street in London – with women as her main target. She recalls: 'There was this terrific feeling for the first time that women could get somewhere. People talk about the early 1950s as a depressed time – grey – but it wasn't. Things could only get better. Also, suddenly, women were being listened to. Women were the new market. Their opinion was being sought.'

Billie had professional experience of the old regime, when being a woman held considerably less promise. She cut her journalistic teeth on the Amalgamated Press in the mid-1940s, a group of magazines, 'all of them now defunct – *Woman's Pictorial, Mother and Home* and *Woman's Journal*. It was the company that the famous publishers Cecil King and Hugh Cudlipp later bought up and made into IPC. It was a marvellous place in Farringdon Street, with this rather ornate doorway and attendants in uniform.'

What is first relevant is why, in 1946, Billie left the company. 'The thing was that I got married and it was still the time that it wasn't done for the wife to work. As a matter of fact, until a very little way beforehand you weren't *allowed* to be married and work at the Amalgamated Press if you were a woman. There was a woman there then who was still pretending she was unmarried!

'Jack and I got married in 1946 and the practice must have been relaxed pretty soon after that. I was there in '44, '45, and right up to our marriage that woman was still masquerading under "Miss". I think it was a protection of men's jobs really. I remembered accepting at the time that once you've got married you've got a bit of support behind you and you mustn't occupy a job that could be given to a man. I remember accepting and understanding that,' said Billie.

Interesting, then, that Vincent's objection to Audra going out to work in *Act of Will* was not just a North Country thing. Indeed, Billie's husband Jack, who was brought up on a working-class estate on the edge of London in Dagenham, home of the Ford car industry, said, 'My mother never went to work except later on in the war when a lot of people did. I don't think any woman up and down our road went to work.'

'It was still part of the culture that the wife stayed home,' confirmed Billie. 'After a little while I got very restive and took a part-time job locally, over the road. Even so, one asked permission in those days,' she said, cocking an eye at her husband.

Realising too late what a career prospect she'd let slip at Amalgamated Press, Billie became determined to get back into journalism and found it almost impossible. 'I had to do it by steps . . .' and the job with Napper, Stinton, Woolley was an important one of these. 'I thought, I'll try and get in sideways. I couldn't get straight back into magazines, it was really tough. I'm talking 1950 . . . you couldn't easily get into magazines then.'

All of which attests to Barbara's ingenuity in landing a position as fashion editor on *Woman's Own*, though she had what was required, a column on a paper whose banner read 'Largest Circulation in the Country' – and a respected Fleet Street referee in Horniblow, if indeed they called upon him to provide one, for Barry Horniblow left Leeds some time before Barbara. There was also clear precedent for transfer from the Yorkshire newspaper group into Fleet Street. Arthur Brittenden had moved to the now defunct *News Chronicle*, and Waterhouse to the *Daily Mirror*, a move he describes with great wit in *Streets Ahead*, the compulsive sequel to *City Lights*. Fact was that a successful spell on the *YEP* was as good a setting-off point as any for a journalist, and had even been seen by Horniblow, of course, as anything but a step back in an already booming Fleet Street career.

Billie, on the other hand, was scuppered: 'My record was terribly broken, I hadn't got the cub reporter thing, had missed university because of a bout of TB and didn't even have much freelance. I'd very little and there didn't appear to be any vacancies. But I got in to the agency as secretary to Leslie Stinton, then I drove him mad. Bits of copy for baked beans, on anything the agency was selling, he'd find on his desk each morning.'

185

She soon realised, however, that she had in fact landed herself in the right place at the right time. 'The office had a woman doing what they used to call "editorial" then, and they decided that they'd like to expand, and they brought in a chap named Ted Jones.'

It was through Ted that Billie and Barbara first met. 'Oh, Ted was quite significant in our lives. He came from J. Walter Thompson, a journalist who'd got into advertising and PR. He just had a *sense* of a story and we did such exciting things.'

Finally worn down by these bits of copy coming from this girl who was supposed to be his secretary, Leslie Stinton decided to despatch Billie to Ted Jones's department. 'We had one hell of a lot of fun. As I said, women had become a market; people wanted their opinions. Suddenly women were being listened to. They'd actually say things like, "Let's have a meeting and get all the women in and hear what they say about it." I was made head of a women's section in the PR department.

'Now, Ted, when he'd been at J. Walter Thompson, had had something to do with fisheries and he'd been up to Hull and Barbara had covered a story for the *Yorkshire Evening Post*. It so happened that Ted had an eye for a pretty girl and he and Barbara had obviously struck it up, so she'd got his name. Barbara was sharp on getting her contacts ready for when she was going to come down to London, and so she'd arranged to call on Ted at his advertising agency when she came down. So, he brought her into the office at some time around 1951, 1952 – before she worked for *Woman's Own*. She would have been eighteen, nineteen.

'Barbara and I had already had dealings, without knowing each other, because she'd been using my stories in her column – the Kay Boughton column. I'm eleven years older than Barbara, so there was quite an age gap, but a friendship blossomed.'

I asked Billie whether she could remember the day Barbara first walked into Napper, Stinton, Woolley. 'I can picture it very easily. What I remember is someone very slim and pretty, and with laughing eyes, and my thinking, what a Scandinavian-looking person. My knowledge of Barbara now is putting her in a skirt to calf-length, but leaving me very conscious of very slim legs and . . . would it have been stiletto heels?

'Barbara Goalen was the great model at that time, and Barbara dressed to look like that. We wore hats and little white gloves. Couture ruled fashion, the ready-to-wear and the prêt à porter hadn't yet exploded. It was couture, Givenchy and the people that dressed Audrey Hepburn – those were our fashion icons then, and in London people like Victor Stiebel, and Hartnell, out of our reach but they were the icons who set the style you copied. You've made me realise that in those days Barbara dressed at a lower price-level than the couturier, but in those terms.'

The Kay Boughton page in early 1952 demonstrates Barbara's by then energetic but easy journalistic style: readable, well-informed, it captures her preference for elegance, sophistication and the feminine, well in advance of the more youthful, individual influences of the ready-to-wear revolution of the later 1950s.

These were the days when the public looked to royalty as well as Hollywood to set the tone. Long before she made spicier column inches in her regal Mustique period, with flamboyant villain John Bindon and other riotous crew playing court, Princess Margaret was about to become one such fashion muse, and Barbara was well on top of it. 'A new note in the Spring trends,' she writes, 'a hint of *the Princess line*, who is so much in the news just now. She was one of the first to adopt the flowing fan line used in coats and suits, and to favour velvet . . . Some say Princess Margaret will be the new

187

leader of fashion, following in the footsteps of her much-admired aunt, the Duchess of Kent. With an engagement in the air that is more than possible.' Perceptive stuff. When Anthony Armstrong Jones plighted his troth shortly after-wards, the couple became the Prince Charles and Princess Diana of their day.

By 1952, a year before she left for London, the fashion scene had taken Barbara far from the humdrum realities of home. On 6th February 1952, she was on the case at soon-to-be-crowned Princess Elizabeth's couturier Hardy Amies's London show, introducing us to 'the pyramid silhouette', and recommending 'wider hemlines worn over stiffening petticoats . . . for fun and games'.

The job also took her to the Paris Collections, and a surprise meeting with Barry Horniblow. 'I remember going to Paris for the first time when I was about seventeen. I went with the WPE [Madeleine McLoughlin] to cover the fashion shows. Paris totally overwhelmed me, I thought it was one of the most beautiful cities – it's still one of my favourite cities – and I came back and sat down at my typewriter and started to write the story of this ballet dancer [she called her fictional heroine Vivienne Ramage] who lived in a garret in Paris and was very poor. It was all very dramatic and I think actually it was probably something like *Camille*. I got to about page ten and thought, No, I have a feeling I've read this somewhere before.'

Camille had been filmed in 1937, starring Greta Garbo and Hollywood heart-throb Robert Taylor, who reminded Barbara of her father. The name of her heroine, Vivienne, was the feminine form of that of her older brother, who had so tragic-ally died as a baby.

'Madeleine McLoughlin wanted to take me to the Paris shows because some boyfriend she had was going to be there, and I could do some of the work. I was quite happy to go, but I realised what she was up to when she said, "You'll have to sit in for me at times", and I said "Why?". And she said her boyfriend was coming. I think I was quite smart even then because I remember saying to her, "Well, I don't know if my mother will let me go. I could talk her into it, but you've got to promise me that we'll go and see Barry Horniblow when we go to London. So she said, "Yes," but on the way there we didn't. So I blackmailed her – if we didn't see him I'd want my name on what we'd written, and of course I'd written most of it.

'On the way back, she said we'd go. I think Barry was on the *Sunday Sketch* or one of those Sunday tabloids, and I *kept* saying to her (I can be a bit of a nag), "Did you make the appointment?" I remember she finally said, "No, but don't worry, it'll be fine." And I said, "How are we going to get in?" She said, "We'll go through the Circulation Department." Circulation, two girls. I was seventeen so she must have been twenty-seven, maybe a bit older, and an Irish woman – from Manchester. She's chatting, "Hello, how are you? How are you chaps?" They just let us in as if we owned the place – in and up the stairs. She kept saying, "We've an appointment with Mr Horniblow, which is the way?" And finally we were outside his office door with some woman trying to put us off . . . and he came out. And I said, "Oh, Mr Horniblow!"

'There he was, white-haired and elegant. And he said, "Barbara!"'

Fade and cut. 'You *must* have been in love with him,' I said.
'No I wasn't, because . . .'
'A crush?'
'No, it wasn't a crush, it wasn't romantic, it was like . . .'
'Adulation?'
'Yes, a good word, adulation.'
At the start of 1952, in spite of a keener focus on the

women's market generally, prospects for manufacturers in the fashion industry were far from rosy. In January that year the Kay Boughton column reported that women's clothing manufacturers were facing gloomy trade prospects – eighty per cent had had to cut staff in the last three months. Barbara blamed this on 'world conditions . . . depression in the seven and eight o'clock BBC News bulletins . . . enough to put any woman off a day's shopping.' The saviour would be her hero, Winston Churchill. He, having been ousted by Labour in 1945, when all fell quiet on the Western Front, had, in 1951, been welcomed back as Prime Minister, and would continue to serve until 1955. '. . . On Mr Churchill's talks in America, the fashion industry's future lies,' Barbara told her readers. Meanwhile, if you couldn't find the money to dress yourself properly, you should 'Make the most of your voice! . . .

'An attractive voice is as important as appearance,' she wrote. 'While you will be admired for your clothes, you will be *remembered* for your voice . . .' It is an interesting thought, apt coming from Barbara, whose voice one does indeed remember. To broadcaster Richard Whiteley, who met Barbara for the first time in the autumn of 1979 when he interviewed her for Yorkshire Television's magazine programme, *Calendar*, her voice speaks of integrity: 'What I found most refreshing was that she hadn't adopted a transatlantic accent, despite having lived in the States for much of her adult life. She has retained a very crisp and rich English speaking voice.'

Barbara is skilful with her voice: she understands that it is a key instrument of style and she will modulate it accordingly and instinctively, so that she is probably not even aware she is doing it. In a crowded room at one of her book launches she wears its formal tones with all the confidence of a sophisticated woman of the world. Then, one-to-one, it can soften to reveal the vulnerable woman beneath. Sometimes – perhaps when she relaxes into her old self or is speaking to a rootsy Northern journalist like Michael Parkinson – the Yorkshire accent appears, as three years ago Simon Hatterstone of the *Guardian* ungallantly reported: 'The longer we talk, the flatter BTB's vowels become,' he wrote. 'As the hours pass, I lose sight of the huge-haired New York caricature and find myself talking to Barbara, the bluff, likeable Yorkshirewoman.'

Barbara had worked her voice strategy out years earlier. In the Kay Boughton column, in her youth, she developed a seven-point strategy for an appealing voice: avoid bad thoughts, read aloud for ten minutes every day, control the breathing, make it even, flowing, low, sonorous. Copy a voice you admire, but don't mimic the voice – 'Avoid artificial accents, they are unconvincing and insincere!'

It is a mark of the style of Emma Harte in *A Woman of Substance* that she relinquishes her Yorkshire accent as soon as possible, and the girls at Kallinski's Leeds tailoring shop tease her about it: '"Talking like cut glass," they called it.' Emma smiled and didn't take offence, so eventually it stopped, though they never quite became accustomed to her beauty or her air of breeding.'

Today, journalists are wont to assume that Barbara's mode of speech – 'the pleasing tones of the cosmopolitan,' as one put it – are 'the result of having lived her life in many different countries', but Billie Figg recalls that Barbara did not have a pronounced regional accent when they first met in London in 1950–1, though you could 'hear it softly behind'.

Regional accents were not part of the requisite accoutrements of success in the early 1950s, before *Angry Young Man* 'anarchism' took hold, domestic realism came in and working-class values were fêted in so-called kitchen-sink dramas, popular later in the decade.

When Barbara set forth from Leeds in 1953, it was

191

unthinkable that anyone should read the news on radio with other than an arching, upper-crust English accent. But soon Alan Bennett would make a living partly out of the multifarious working-class characters that a Yorkshire accent suggests to him. Bennett does admit, however, that in the Fifties when he went to Oxford University, he, too, made an attempt to hide it. 'Then it came back, and now I don't know where I am.' In *Telling Tales*, he recalls a particular problem he had in a public recitation of a sentence by Oliver Goldsmith:

> *Of praise a mere glutton he swallow'd what came*
> *And the puff of a dunce he mistook it for fame*

Bennett died when the last line came out first as 'the paff of a dunce' and the second time as 'the poof of a dance'.

Barbara has been known to indulge her Yorkshire accent when returning to her roots, and at other times has been described as – and alternately accused of – putting on a 'power-packed Alexis Colby accent' (this was in 1997) and having 'an almost stagey upper-class English accent'. The fact is that different accents can be called on to send definable signals to different audiences. Empathy is a powerful tool in communication; why not use such a tool if you have the confidence and credibility to do so?

Barbara might say that her experimentation is a sign of the actress in her, but her husband Bob, who was pulled up by the roots from his home culture at five years of age and can claim a more cosmopolitan life-history even than Barbara, makes no attempt to experiment in this way. Born in Germany, he was brought up in Paris, lived for many years in America and has an accent which encompasses every stage of the hybridisation process, relieving any possibility of confusion with a sardonic twinkle in his eye.

192

None of this is of idle curiosity, for accents have tradition-ally been connected with identity, and, as we have seen, iden-tity is the 'problem' at the root of Barbara's novels, slipped subliminally to her by Freda as the one to solve. It is significant, but not surprising, that Barbara's range of accents should attest this.

Uprooting from a home culture as strong as that of York-shire took guts for a twenty-year-old girl in 1953, but had a clear rationale. Roots not only feed a plant, they tie it down. Keith Waterhouse also left but took his Yorkshire roots with him, wearing them like a badge. He was still alluding this year to his childhood job in a Leeds undertakers in his column in the *Daily Mail*, and has retained, perhaps even built on, his Yorkshire accent, with its very particular, honest-working-class reference points.

On the other hand, Freda made a point of taking Barbara out of the local, post-industrial terraced world of Leeds as a child and into the rural landscape of North Yorkshire, where Barbara mined ideas, lessons of history and values with a wider relevance, ideas which set her up for the incredible inter-national journey on which she would embark. In particular, she mined the abstract idea of beauty – the special spirit of places like Middleham, Studley Royal and Haworth – which impart a sense of the sublime that the industrial revolution had squeezed out of the urbanised working-class in the North with almost Judgement Day finality.

Keith Waterhouse avoided any opportunity to visit the countryside as a boy, his preferred environment being one with 'not a blade of grass in sight'. When called up to National Service at an RAF base at Wombleton he likened the North Yorkshire moors to 'Noel Coward's Norfolk, very flat,' as if he had never set eyes on the real moors, often precipitous. 'The North Yorkshire moors,' he writes, '. . . even in the driest summers give out such an impression of being marshland that one expects Magwitch staggering across them through the mist.' The fictional setting, where the escaped convict Magwitch appears and terrifies Pip in *Great Expectations*, was in fact based by Dickens on Cooling in Kent. Nothing could be further in imagination from the wild, expansive Yorkshire moors.

Again, Alan Bennett casts a few frogs, a crayfish and the sight of a lizard on a rock as '*it* as far as nature in my childhood is concerned', concluding that 'with nature, as in other departments, adults *pretended* . . . [in order to] conceal the fact that nature is dull.' Both Bennett and Waterhouse made their way in the wider world still spiritually confined within the cobbled world of bricks and mortar in which they had been brought up, trading on it sometimes ironically, sometimes to point of caricature.

The Yorkshire countryside to which Barbara's mother turned her was something else. It awakened in her daughter an appreciation of landscape as character – as if it were over-written with the exploits of its history and the emotions and values redolent of those who had steered it. Being part of this landscape was Barbara's birthright, it reflected something in her and became part of her identity long before she saw the narrative potential and returned to it to articulate her woman of substance.

Imbued with the character of the sublime Yorkshire moors – indestructible, everlasting – her self-belief heightened by Freda's conviction that she was capable of anything, that there were no limits to what she could achieve, she moved out of childhood and shed her working-class skin with the courage and vanity of a Neville, the persistence of a Ripon, and the vitality of her maternal grandmother, Edith Walker, to press her game to the end.

The industrial revolution was over. If rising in the world was flying in the face of the working-class rubric, there was no longer a sense of it in Barbara. Prevarication went against her instincts and against everything her mother had told her. Freda had led her back in time to eighteenth-century William Hazlitt's idea that self-belief is the mother of opportunity.

CHAPTER THREE
The Jeannie Years

'There's nothing you can't have if you try hard enough, work hard enough and strive towards a goal. And never, never limit yourself.'

Audra in *Act of Will*

The Fifties were an unforgettable era. The decade saw the exploding of the H-bomb, the coming of TV to the masses, the first exploration of outer space, the beginnings of the affluent society and the start of a period of teenage-powered rebellion against existing attitudes to sex, class, authority and good taste. It was the time of Elvis Presley, Marilyn Monroe, James Dean, Burgess and Maclean, and the papers were full, too, of Khrushchev, Castro and Suez. Feminist writer Simone de Beauvoir used the term 'women's liberation' for the first time in 1953, and, two years earlier, Irish writer Leslie Paul wrote *Angry Young Man*, providing the soubriquet for a radical school of writers that included John Osborne, Kingsley Amis, Alan Sillitoe, Colin Wilson, John Braine and others.

In fact, as we shall see, there were large parts of this Fifties agenda that Barbara never experienced, partly because the 'good taste' and elegance elements that were up for replacement were endemic to her style, which it never occurred to her

196

to change. Nevertheless, the era was an unforgettable experience for her, if quite unlike most people's. It began in austerity and finished up with a Hollywood liaison.

In 1950, twenty Dunhill cigarettes cost three shillings and sevenpence (about 17.5p in modern coinage); a large loaf of bread, two shillings and a penny (10p); a pound of cheese, a shilling (5p); and a bottle of gin, thirty-three shillings and ninepence (about £1.70). On the face of it, that sounds like a cheap shop, but of course salaries were far less then too. A political columnist on the *Sunday Pictorial*, a national newspaper, was earning £1500 in 1950. Would he be earning forty times as much today? If so, his shopping bill today, at 1950s prices, would be £7 for the packet of Dunhills, £4 for a large loaf of bread, £2 a pound for cheese, and £68 for a bottle of gin. Quite clearly the whole product-cost balance is different today, and there are many more things to spend our money on, but clearly, too, things were not cheap more than half a century ago.

When Freda delivered Barbara to her small flat at 44 Belsize Park Gardens, NW3, between the Finchley Road and Haverstock Hill, she was going to find it hard to make ends meet on a salary that will have been considerably less than £1500 per annum. Typically, the burden was eased by Freda. Like Christina in *Act of Will*, the rent was paid by her mother: 'My parents supported me when I was in London,' Barbara has said.

This might be assumed to have had an effect on the frame of mind in which Barbara began her decade in London. Indeed, she makes plain in *Act of Will* that there is a price for Christina to pay for Audra's self-sacrifice on her behalf. Christina is driven by what she now perceives as 'the crucial debt she owed her mother ... She must repay it. If I do not it will weigh heavy on my conscience all the days of my life, she thought.

197

And that I could not bear . . .' Later, Christina says: 'I have a terrible need, a compelling need, to bring ease and comfort to her [Audra's] life. I want to give her the kind of luxuries she's never known . . .'

This was not, however, the case with Barbara, at least at this stage of her life. 'Of course I was very good to her, financially and in other ways, but I never felt there was a huge debt to repay. I wanted to succeed because I knew it would please her and that it would be wrong somehow if I didn't succeed . . . even at Northcote Private School, for instance, I always knew that I had to, that I couldn't waste time or dawdle or not pay attention to my school work, because I knew that that would be terrible. She expected the very best from me.'

Barbara has also said that she wasn't conscious until some time later of the sacrifices her mother had made for her. Nevertheless, she felt a need 'to prove that I was the best, and that all my parents' love and devotion had come to fruition.' This in no way restricted her or repressed a desire to do otherwise: 'I *wanted* to please her,' Barbara stressed.

It would have been completely out of character for Freda to burden Barbara with thoughts of the debt she owed her, or what it meant to slip quietly into the background when it was time for Barbara to be parted from her. 'My mother never said anything to me at the time, but later I found out that the day I left for London she took a doll I was fond of and put it in her bedroom. She kept it there until she died in 1981. I think she was trying to hang on to a part of me. When she died I found her handbag and my childhood library cards were in it – the handbag in current use! – and I found her diary, all her diaries, and read an entry that said: "Barbara has gone to London today – all the sunshine has gone out of my life." It was all it said on the page. It made me feel incredibly sad.'

Earlier, Barbara and her mother had traipsed around

London looking for a suitable place to rent. Had they come by car they would have had little difficulty parking, for, as Billie Figg recalls, many of the Second World War bomb sites were being used as car parks, 'and it was easy to park anywhere. Bombed-out London also became a rich bed for wild flowers.' Billie Figg, born in North London to Londoners who then moved to suburban Woodford Green, close to where she lives today, became Barbara's lifeline: 'She immediately treated Jack and me as friends. She would often ring, we would go out together, and she'd come down here, wouldn't she, first of all.' Jack remembered helping out at her little flat, putting up a curtain and rail to hang her clothes on. 'It was, as far as I remember, a living room, a bedroom, kitchen and bath,' he said. 'She was certainly living on her own. Looking back, I think we were just slightly stable anchorages in London, a place that was strange to her. She used to telephone us a lot, and so without complimenting ourselves there must have been some sort of comfort in that, whereas maybe quite a few of the other people she got to know, well . . . perhaps . . . they weren't married?'

Back in 1953, what *Woman's Own* meant to Alan Bennett's mum was columnist Beverley Nichols 'writing about his gardens and chronicling the doings of his several cats', Nichols' life seeming to her one of 'dizzying sophistication'. What it meant to broadcaster Sue Lawley was 'lots and lots of romantic fiction . . . all those soppy love stories. Did you never want to write one of those?' she asked Barbara on *Desert Island Discs* fifty years later. Listening to the radio programme miles away I could feel Barbara bristling before she answered, 'No, and I never read them actually, to tell you the honest truth.' Lawley stemmed the flow from her lancing cut rather unconvincingly with, 'They were very good.'

Her snipe was in fact off target because it would be quite

wrong to suggest that Barbara's lifestyle in London between 1953, when she became fashion editor of *Woman's Own*, and 1963, when she married film producer Robert Bradford and left for New York, bore even the remotest resemblance to that recommended by either Beverley Nichols or those romantic story writers. It came far closer to the one described in Barbara's novel, *Voice of the Heart*, which speaks to us of destructive relationships in a glitzy but ruthless world of show-biz, which it took Barbara but a short time to inhabit.

When she arrived at *Woman's Own*, Barbara discovered that she was in fact one of several fashion editors, and at once did what was required to adapt to their style, the young novice showing just how apprehensive and keen to swim with the stream she was: 'When I got there these women in the Fashion Department all wore hats, they came to work in them! So I went out and I bought a green hat, a pill-box we would call it today, but I always felt rather foolish because I'd worked on a newspaper before in a dirty old trench coat.'

A certain Patricia ('Triss') Lewis was the beauty editor. Brunette, sultry, sophisticated, Triss was extraordinarily attractive and far more experienced than Barbara. Straightaway Triss recognised the special spark in the twenty-year-old and took Barbara under her wing, introducing her to the fast-track world in which she operated from her Kensington flat.

'There were all these good-looking *men*!' remembers Barbara. 'We were a group. There was me and Triss, and because they were all show-business writers they dragged us to the nightclubs and to cocktail parties and that sort of thing.'

The men in the group, somewhat rakish and all of them much older than Barbara, included Roderick Mann, the then show-business columnist of the *Sunday Express*, ex-Fleet Street political columnist Frederic Mullally (*Tribune, Sunday Pictorial*), who now ran a Mayfair showbiz PR company –

both eleven years Barbara's senior – and their two compatriots, Logan ('Jack') Gourlay (*Express*) and Matt White (*News Chronicle*), also showbiz columnists (both now dead). 'The dashing men about Fleet Street,' as Barbara referred to them. An A&R man from Decca Records, Bunny Lewis, also tagged along.

Partly on account of the preoccupations of the members of this gang, what Barbara landed up in was not the traditional journalistic scene she supposed would envelop her in Fleet Street, but pure showbiz. 'My friends were mostly in the theatre or movies. I was in a much more writing/movie/show-biz world than I was in any other world when I lived in London.'

Before long, the celebrity culture swallowed up Triss, who was plucked from the Beauty department of *Woman's Own* to become a high-profile personality columnist for the *Daily Express*. 'They called her the Champagne Girl,' recalls Mullally. 'She was given a column of her own by Harold Keeble, who was the great assistant editor of the *Express*, the man who boosted all the great women journalists of his time. He invented page three of the *Daily Express* when it was at its peak. In terms of production, illustration, choice of pictures, choice of columnist, he was more than just Arthur Christianson [the famous editor]'s assistant, he was a dominating influence, the fresh wind that blew through the paper. And he *created* Patricia Lewis. Now, she and Barbara were very close. They were the two "glamour girls" of Fleet Street. Fantastic! Beautiful! They went around the West End together.'

Said Barbara: 'It was a particularly glamorous period, the Fifties and the early Sixties, lots of Hollywood stars around, and parties, openings, premières. When I look back, I feel very nostalgic for London as it was then ... so different from today.' Small surprise, then, that eventually Triss would marry classical actor/screen star Christopher Plummer, and Barbara would wed film producer Robert Bradford.

Mullally seems to have been something of a mover and shaker in the scene. He teamed up with (and married) Suzanne Warner, legendary film-maker Howard Hughes's representative in Europe. 'I left the *Sunday Pictorial* under Hugh Cudlipp, where for three or four years I'd had a column called "Candid Commentary", Suzanne left Howard Hughes and we formed this PR operation together in Mayfair. It was at that time that I first became aware of Barbara Taylor. The two girls, Barbara and Triss, were riding high, you see? There was an inevitable gravitation between my position and their interests. You have to remember that I was a more senior figure to these two girls in those days. I'd done my Fleet Street stint. I'd done my column. I'd written two books, political non-fiction. To them I was an attraction to be with, and I had my entrée to all the clubs of London. So there was no problem about calling them, they would call me. You know, "What are you doing today?" And I would invite them. I had a great situation in Hay Hill, Mayfair, two floors of a building. Everything happened around those two floors. On one floor I had my PR office, a staff of maybe twelve people, and above that I had my personal apartment. I would give parties there. And round the corner were all the places ... I would go out every evening, and it would be cocktails ... All the glamour was in that square mile. A little overlapped occasionally into Soho and into Pimlico, but basically it was Mayfair.

'There were not clubs as we talk about clubs today. These were places where you did not show your face before seven o'clock unless you were perfectly groomed and perfectly dressed. You did not show your face if you were a man unless you were superbly groomed and suited. There was a cocktail hour in Mayfair then, as it has never existed since and will probably never exist again. Full of style and grace. This hour was seven till dinner, seven till eight. Mayfair buzzed in about six or eight venues within that square mile, where the crème de la crème suddenly appeared, the actors, the cinema people that wanted to be seen. Most of these restaurants in those days, like the Caprice, had a bar, a tiny bar. You crowded in at about seven o'clock. You didn't even get where you were to sit. You were there in what was a kind of party atmosphere. And gradually the people drifted into the restaurant and the bar would go empty. But for that one moment between seven and eight you were shoulder to shoulder, cheek by jowl with the most glamorous models, the great Barbara Goalens of that time, who were not like the models of today. That was the scene; it was a very glamorous scene. Celebrities were celebrities. Today if you appear on television for five minutes you're a celebrity. In those days they were major, major stars. Actors, models . . . They got that much publicity, now it's flooded. There were only three, four, five magazines in those days.

'It's the forgotten era. As you came out of the war years into the Fifties, and rationing began to ease up, we were aware of Hollywood and were aware of Hollywood stars, trickle by trickle coming into London, into Les Ambassadeurs, into Ziggi's club . . . and we treated them as kind of nice pieces we were cultivating and would like to shake hands with, but we were not influenced by them at all. There was a British film industry that was appreciated around the world at this time – the Ealing comedies, and Sidney Box and one or two other greats. It was alive, and I was one of those involved with it. My company had clients like Sinatra, we had from top to bottom, we handled everybody, films by people like the Carrerases, you know, father and son, down to the lower end of it, which was the Hammer Horrors. We made the singer David Whitfield, managed him, got a deal with the A&R

man in Decca to get the right songs. But it was tough, it was only just beginning, papers were fewer and smaller, and for a while the major Fleet Street newspapers were totally anti-PR. Express Newspapers had a ban on PR altogether – you were not allowed to mention the name of a club at which someone had appeared last night. Puffs were out, it was very tough. So we had to be more ingenious.

'In the Mayfair bars and clubs it wasn't principally a Hollywood scene, it was an elegant *English* scene, though Hollywood actors came. What happened was that there was an infusion from America. Any Hollywood stars that came to London were attracted by that particular London scene, that very elegant London scene. So they would be there, they would be in and out of these various little places where we met, and I would meet Errol Flynn for the first time in my life. I would meet Douglas Fairbanks. They swanned in and out of our London scene, to which they were attracted by its elegance.

'The great clubs of the day were the Casanova, run by Rico Dajou, the Don Juan (Rico had the two clubs, they were in Grosvenor Street), Les Ambassadeurs – downstairs restaurant and upstairs nightclub (called the Milroy and run by big John Mills – not the actor), and Ziggi Sessler's at 46 Charles Street, now Mark's Club. The Caprice of course, a restaurant, but a top restaurant for the other people, not even really journalists, the Mayfair crowd. They had a wonderful head waiter. I wrote about him in one of my novels – *Danse Macabre*. If he hadn't chosen to be the greatest maître d'hôtel in London he would have been an archbishop of the Catholic Church. He was that wonderful a person. Annabel's came later.'

I recognised names out of Barbara's novels immediately. Early on in *Voice of the Heart*, Francesca Cunningham, Katharine Tempest and others meet for dinner at Les Ambassadeurs in Hamilton Place, between Piccadilly and Park Lane.

It's also a favourite lunch-time haunt of stylish film star Victor Mason, who is a model of what Katharine, Francesca, and, of course, Barbara consider a man should be: handsome, rough-hewn, suntanned, 'massive across the chest and back', clothes of the finest quality, chosen with panache, slacks with knife-edge creases, a man who cares about appearance. Francesca's family, the Cunninghams, whose aristocratic seat is Langley Castle in Yorkshire, also have a house in Mayfair's Chesterfield Street, which runs between Curzon Street and Charles Street, where Ziggi Sessler's 21 Club was. Maximilian West first walks onto the pages of *The Women in His Life* from his 'imposing house on the corner of Chesterfield Hill and Charles Street', and I knew that in the 1980s and '90s, when Bob Bradford was filming in England, he and Barbara themselves had an apartment in Charles Street. Annabel's, in Berkeley Square, which Barbara describes in the same novel as 'the chic-est of watering holes for the rich and famous, where the international jet-set rub shoulders with movie stars and magnates and members of the British Royal Family', actually opened during the summer before she decamped with Bob to New York in 1963. They are still members today. Berkeley Square was, and is, also home to bespoke yacht builder of distinction, Camper & Nicholson, where, in *The Women in His Life*, Maxim commissions the 213.9-foot *Beautiful Dreamer* for his wife Anastasia. And, as fate would have it, Barbara's English publisher (HarperCollins) was sited within the square mile at No. 8 Grafton Street where they first published her.

When she first came to the area at such a tender age in the early 1950s, Barbara could have had no idea that this was to become known worldwide as multi-million-selling author *Barbara Taylor Bradford's Mayfair*. Today it is, in every sense, her village:

205

Maxim came out of the imposing house on the corner of Chesterfield Hill and Charles Street and stood for a moment poised on the front step . . . Pushing his hands in his pockets, he forced himself to stride out, heading in the direction of Berkeley Square. He walked at a rapid pace along Charles Street, his step determined, his back straight, his head held erect. He was dark-haired with dark-brown eyes, tall, lean, trimly built . . .

He circled Berkeley Square, dodging the traffic as he made for the far side, wondering why Alan needed to see him, what this was all about . . . Oh what the hell, he thought, as he reached the corner of Bruton Street. Alan's been so special to me most of my life. I owe him . . . we go back so far, he knows so much – and he's my best friend. Crossing the street, his eyes focused on the Jack Barclay showroom on the opposite corner, and when he reached the plate-glass windows he paused to admire the sleek Rolls-Royces and Bentleys gleaming under the brilliant spotlights . . . He walked on past the Henley car showroom and Lloyds bank, and pushed through the doors of Berkeley Square House, the best commercial address in town and a powerhouse of a building. Here, floor upon floor, were housed the great international corporations and the multi-nationals, companies that had more financial clout than the governments of the world. Maxim thought of it as a mighty treasury of trade, for it did hundreds of billions of dollars' worth of business a year. And yet the buff-coloured edifice had no visible face, had long since blended into the landscape of this lovely, leafy square in the very heart of Mayfair, and most Londoners who walked past it daily were hardly aware of its existence. . . . Maxim crossed the richly carpeted, white-marble hall,

and nodded to the security guard, who touched his cap
in recognition.

The Women in His Life

I asked Barbara which her favourite haunts were in the early days. 'I went to the Mirabelle a lot in the 1950s, when it was in its heyday. The maître d' was Louis Emanuelli – Welsh, of Italian descent. Louis was later the man at Annabel's. He was a great pal and would do anything for me. Les Ambassadeurs nightclub, the Milroy, was always full of Hollywood stars, producers, etc. The famous bandleader there was Paul Adams. I particularly liked the Arlington Club on Arlington Street, a tiny club in Mayfair, but an "in" kind of place with a bartender called Joe who was a character. And yes, we went to Ziggi Sessler's. And there was an actors' club somewhere off the Haymarket called the Buckstone, a minor club, but I used to go there and so did a whole group of actors – Richard Burton, Christopher Plummer, Peter O'Toole and Jason Robards – Richard Harris as well, all dedicated drinkers. I used to love that club . . . I believe it's still there . . .

'At this time I became a friend of Jeannie Gilbert, an American woman who was press officer for the Savoy, Claridges and the Berkeley. She was from Kentucky and not to be confused with another Jean, who was press officer at one point, before my Jeannie, I think.

'Jeannie had a little house in Minerva Mews in Chelsea and gave lots of parties with guests, sometimes authors – famous writers, like James Baldwin, the black writer. He was around at that time. Unfortunately she had a habit of introducing them sometimes by nickname. One occasion I remember in particular, I was talking to this man whom Jeannie had introduced to me as Snips, or something like that, and he was chatting away about writing. I said I wanted to be a novelist

207

and he started talking about books. Finally he said, "You're a journalist, you're very young." – He was a rather portly man, middle-aged. – "You're rather young and pretty." I said, "You know a lot about books, what did you say your name was? Jeannie introduced you as Snips." And he said, "Oh, I'm Thornton Wilder." He wrote *Our Town*! Famous American writer! I was terribly embarrassed when I heard his real name.'

During the Jeannie years Thornton Wilder might well have been in London promoting his play, *The Matchmaker*, first out in America in 1954, to be adapted nine years later as the musical comedy *Hello, Dolly!* Mention of the literary novelist James Baldwin reminds us that Jeannie was dealing with a variety of 'product', their common denominator simply that they were staying in one or other of her hotels. Baldwin was born in Harlem, his fiction exposed taboos in courageous and disarming fashion and concerns both homosexuality and the plight of his disaffected people. His first novel, *Go Tell It on the Mountain*, was published in 1953, and caused a tremendous stir.

'There were also always famous movie stars floating around,' Barbara continued, 'people such as Victor Mature, Eddie O'Brien and other name stars of the day, also high-powered movie producers at her parties, such as Ilya Lopert (he made *The Red Shoes* among other films), Gregory Ratoff (also an actor), Arthur Krim of United Artists, and screenwriters Jack Davies and Michael Pertwee, who were partners . . . Other actors around at the time were Tyrone Power, and Sean Connery, as well as Lyndon Brooke (son of Clive Brooke) and Terry Longdon – both English. Lois Maxwell was another friend through Jeannie; she was the first and most famous Miss Moneypenny in the James Bond movies. Jeannie had a way of gathering all kinds of people, from the UK and the US. Eventually she married the Broadway producer David Merrick, after returning to New York in the early 1960s.'

Mullally remembers Jeannie as 'the best press officer the Savoy ever had, and she did what most PR people never succeeded in doing, she became friends with every columnist in Fleet Street. At any time you could drop in to her little suite of offices on the ground floor of the Savoy – two rooms – have a drink, ask questions and she would answer them. She cultivated and seduced our columnists. I was one of them. She was a great PR, wonderful; her model was a girl called Jean Nichols, a PR of the Savoy who came before. Jean married the author Derek Tangye, who went to live in Mousehole in Cornwall. They grew daffodils together and he wrote all those books, *A Gull on the Roof* and all that. Jean was another beauty; tragically she died of cancer.'

Jeannie Gilbert never had anything to do with Barbara's work, but became her friend and another means by which she came out into this glittering world. 'Because Jeannie did PR for those hotels, she'd always be having cocktail parties. If there were six actors in town and she wanted a pretty girl around, she'd call me up. I *could* ring her up and say, "Could I do an interview with . . . ?" But I was never a celebrity journalist. I did a couple, Dominguin, the bullfighter, for the *Evening News*, and a few movie stars, and the odd writer, but I was really doing the women's page. Celebrity journalists weren't quite "in" in those days, I don't think anybody ever got called a celebrity journalist.' Nevertheless, quite clearly, the Fifties opened the door for them.

What so astonished Billie Figg, and will have appealed to Jeannie, when Barbara stepped into a room full of her clients, was the way this twenty-year-old girl from the North 'felt very at home with all these famous people, as if it was only a matter of time when she'd be one of them.' Said Billie's husband, Jack: 'I noticed that she'd get into a room and suddenly you'd find that everybody was looking her way. It was the laughter in the eyes, they had enormous magnetism, they were *full* of laughter.'

'Her style is such,' continued Billie, 'that the feeling I always had was that Barbara, from her school days, had wanted to be famous and wanted to be successful and without thinking about it just had an inborn acceptance that she could, and that whatever she turned her hand to she would be good at. She had enormously high expectations of herself and a lot of assurance. It didn't cross her mind that anything she did would be second rate and she knew that she would work hard to ensure that it wasn't. I don't think she ever thought, I'm not very good at anything. I think it was built in. She says that her mother built that in to her, doesn't she. It was very obvious to me what a fuel it is to have that kind of confidence. It never struck her that she would in any way be boring to any of these people or that they would feel that, established though they were and up-and-coming though she still was. It didn't cross her mind that they wouldn't be entirely dazzled by her, and indeed they were, because Barbara had, without knowing it, a very good repartee, and she had the kind of huge sense of humour and comeback that makes very good conversation for men as well as for women. Certainly there was an immediate rapport in any conversation with men, and not just because she was sexually attractive.'

Soon Barbara's diary was studded with appointments at the smart places of the day, as some of those whom she met at Minerva Mews became part of her life. For example, she got to know Jack and Dorothy Davies particularly well, and from the mid-Fifties would live in a flat next door to them. Jack was the father of child star John Howard Davies, who played in *Oliver Twist* when he was nine, and was in *Tom Brown's Schooldays* at eleven, later directing *Fawlty Towers*, *Mr Bean*, *Reginald Perrin* and *The Vicar of Dibley*, among other comic masterpieces. Jack Davies had worked as a staff writer on the Will Hay films of the 1930s, before making his name on the semi-satirical *Doctor in the House* movies, hugely successful in Britain, the first of which was released as Barbara arrived in London, in 1953. It starred Dirk Bogarde as medical student Simon Sparrow, who runs foul of consultant Sir Lancelot Sprat, played by the robust figure of James Robertson Justice.

Mention of Justice jogged Mullally's memory of occasions when the actor joined him and Jack Gourlay at the Colony Club in Berkeley Square for an extended session, reminding us what a significant, often devastating, role alcohol played in the fast lane in the 1950s. 'In those days it was bottle parties after 11 p.m. You had to have your own bottle, due to licensing laws. Whichever club you went to, you bought a bottle of whisky the night before and they marked it up as to how much you'd drunk and put it in a cupboard with your name on it. In the Fifties that was it! You would go back to your club at eleven o'clock and they'd send for your bottle and put it on your table. You couldn't order another bottle that night because it was past the hour. That was the crazy licensing laws in those days. So, there I am with Jack in the Colony Club, we're with the owner – we only ever drank with the owners. And Jack and I were both after birds all the time. The owner would summon a girl from the show . . . Then, about 3 a.m., Jack would stagger off and I'd be left with Justice, and I'm living in South Audley Street, having split with my wife. I've got rooms. All I can do is offer him a couch. Biggest mistake of my life. I offer him a couch before he got the mail train to wherever the next day. Mistake, because when this giant of a man started to snore, the whole place shook . . .'

Of those she met at Jeannie's, Barbara also got to know Victor Mature well. I knew this because Billie Figg had mentioned an occasion when she joined them for dinner. Then,

211

on a separate occasion, I happened to ask Barbara what the derivation was of the phrase 'the whole enchilada', which readers of Voice of the Heart will know movie star Victor Mason uses repeatedly when he wants to say that something is 'the complete works', an enchilada being a tortilla stuffed with a variety of things. Barbara replied: 'That was a saying used constantly by Victor Mature. He always said, "the whole enchilada". I really didn't know what it meant until I came to live in America.' A coincidence not only of initials, VM, and first name of Victor, made me wonder how far I was expected to stretch similitude between movie star Victor Mature and movie star Victor Mason in the novel. Victor Mason falls in love with Katharine Tempest's best friend Francesca, where-upon Katharine hurts Francesca terribly by suggesting falsely that he is the father of her baby. Francesca, being Yorkshire born, a writer and something of an ingénue in the fast, showbiz scene in which she found herself, had always struck me as drawing on the vulnerable side of Barbara, fresh from the North and less experienced than her new showbiz friends. Katharine was altogether the more complete operator, gener-ous and loyal on the surface, but, as Barbara described her to me, 'a woman of great calculation, ambition, a degree of ruthlessness, and self-justification to a certain degree ... Always doing things for her friends, but somehow they all seem to serve her own ends.' Katharine dominates the book even when she's not on-scene, just as, quite clearly, the experi-enced Jeannie Gilbert was an important part of Barbara's life at this time. I knew that there had been a falling out between Barbara and Jeannie in later years. Did Francesca and Kath-arine represent the young Barbara and Jeannie respectively? Had there been something between them and Victor Mature? Had I stumbled on a real triangle of passion?

Barbara confirmed only that Victor Mature had met her via

Jeannie – 'He actually was a boyfriend of Jeannie's, that's how I knew him.' But she denied that she had an affair with him or that, like Victor Mason in the novel, Mature had been the cause of bad blood between her and Jeannie: 'I had quite a number of male friends who were not my boyfriends, if you know what I mean. Most people would think that if you knew a man that you were sleeping with him, but that wasn't the case.'

In an extended interview with Allison Pearson in 1999, Barbara was quoted as saying, 'I interviewed all sorts of men – movie stars – and they tended to chase me.' Pearson told us: 'The PR at the Savoy introduced her to Omar Sharif and Sean Connery. There is a photograph of Barbara and Sharif taken at this time and you suddenly realise that all those ridiculous clichés about beautiful people on which her novels rely were, for the gorgeous young girl about town, the height of realism. I congratulate Barbara on the heroic restraint clearly required to not sleep with Omar Sharif and I get a wry, knowing look. "I wasn't prim, Allison." "Weren't you?" "Not at all, but I was cautious. I wasn't a big sleeper-arounder. I was scared of getting pregnant. Mind you, I'm not saying I didn't sleep with anyone before Bob, but I worried what people would think, mostly my parents."'

This is so telling a comment, too apparently unlikely a claim (given the distance that now separated Barbara from her parents) not to be true. There was an affair with the photographer Terry O'Neill, who would later, after his marriage to Faye Dunaway, become involved in Hollywood and movie producing. 'But mostly I seemed to go out with actors. In fact, Mummy often teased me, said I always fell for the pretty face. My father would ask, "And what's this man's intentions?" like some Victorian, and I'd laugh and answer, "I don't know, Daddy, but my intentions are to have a career.' The parental involvement reminds us how young and inexperienced Barbara was, and, however independent she seemed, how tied to her roots she remained at this early stage. Freda's strategy for her daughter's rise in the world had left little to chance in matters appertaining to her mother Edith's fall.

Sex is often bound up with ambition, however. It is itself a Hollywood cliché – 'All the little girls were scalp hunters,' said Frederic Mullally – and Barbara's youth, beauty and laughing eyes might have been in danger of steering her into the path of trouble of a sexual nature, if trouble it would have been deemed. The first oral contraceptive – the Pill – would not become available until 1961, but that doesn't seem to have reined in the expectations of Mullally and the other male members of the gang, for whom, if they are to be believed, sex was always available, if often in rather seedy fashion: 'It was Mount Royal or White House (Regent's Park),' said Mullally. 'Those were the two places where you didn't have to bring your luggage with you.'

It may be significant that overt sex is a relatively unimportant aspect of Barbara's novels. She was once asked about this in interview, and replied: 'Well I do labour over the sex scenes, yes, a lot. I do a lot of rewriting. I try really to work from the point of view of the emotions of the people involved in the scene, rather than describing parts of the anatomy or using dirty words or being that explicit. It was shocking at the time of Harold Robbins when Harold started writing very, very explicit and rather dirty sexual scenes. It's really feelings that I'm writing about.'

Billie Figg recalled Barbara seeking security in the group culture. 'There was no shortage of boyfriends,' she told me, 'but they weren't a terribly important element. She was something of a loner herself, in a way, in that she was "*getting on*", she was fashioning herself, and she was a great crowd person.

214

She would go round in a crowd, she loved collecting people and creating groups.'

Lois Maxwell, who went to many of the parties at Minerva Mews, remembered this in particular about Barbara – her gang mentality at that time: 'Barbara was always beautiful and vivacious and full of mischief, and I am sure she has forgotten, but one day I was walking along the Kings Road with the man I thought I was falling in love with, and all of a sudden there was a dreadful whistle and who was on the other side of the road but Barbara and the two Jeannies and various other pals of ours, and they all looked at me, and she started to sing in high dulcet tones, "Love and marriage . . ." – and "Love and marriage . . ." followed poor Peter and me all the way down the Kings Road. I was blushing and he was a little bit put out about it. But I don't think this incident ruined anything because he did ask me to marry him and I did . . .'

There was safety in numbers, and Barbara preferred relationships that kept her ambition the commanding focus, rather than any sort of instant satisfaction, although that didn't mean that sex was off the agenda. Memorable relationships were those that were chummy and kept her career centre-stage, as had always been the case at home. Later, she would date film director John Berry, who was, like so many other of Barbara's male friends, significantly older than her. Born in 1917 of a Polish Jewish father and a Romanian mother, Berry had been a member of Orson Welles's legendary Mercury Theatre from 1937, and in charge of it until 1943 when he followed Welles to Hollywood, directing a number of films including *He Ran All the Way*, a thriller but also something of a statement on American life, starring John Garfield. 'It was a film noir,' Barbara recalls, 'and the last movie Garfield ever made. He died just after it was completed, at the age of thirty-nine.'

At the House Un-American Activities hearings into Communist Party influence within the film industry, Berry had been blacklisted and couldn't work in the States. In 1951 he had directed the documentary that supported those accused of communist ties, *The Hollywood Ten*, and had gone to live and work in Paris before arriving in London, where Barbara met him. 'He came to live in London in order to write a script with his great pal Ted Allan, the Canadian playwright/screenwriter, author of *Lies My Father Told Me*, among other plays and films. These two were hilarious together, and played lots of pranks on me, but we were great chums.' She went out with Berry, 'but ultimately we became just great pals. He and Ted loved having me around, probably because I fell for all their jokes and pranks, and cooked dinner for them . . . although they insisted they were the better cooks! . . . The two of them had sort of taken me under their wing.'

Again it is the group culture, safe, fun, and *useful*: 'What a lot I learned from those two men! They were both in their forties . . . and I was by then, what, about twenty-seven? It was like going to a theatrical movie school. I remember many of the discussions about *Oh, What A Lovely War*, which Joan Littlewood produced.' The famous stage musical about the First World War later became a film, but is known far and wide as having been composed by Joan Littlewood with her fellow artists in her Theatre Workshop at Stratford in East London. Ted Allan is generally only credited with coming up with the title, but apparently he was more completely involved with its genesis. 'Ted had written the original treatment,' Barbara told me, 'and there seemed to be quite a lot of dissension between him and Joan about the credits.'

In pursuit of her ambition to become a novelist, Barbara took every opportunity she could to talk to writers and to seek their advice. Jack Berry and Ted Allan encouraged her.

216

'I was writing a novel called *Florabelle* at the time ... they pushed me to finish it. Although neither of them liked the title, they did select one from a list – *The Things We Did Last Summer*.' She told everyone she met that she intended to become a novelist, and no doubt many of them gazed into her laughing eyes and, like Thornton Wilder, gave her all the encouragement she needed.

She was far from being alone in this ambition. In the gang, Roderick Mann published novels from the early 1960s; one, *The Account*, was actually published by Barbara's own UK publisher and involves a PR lady for a hotel who is not unlike Jeannie. Meanwhile, Mullally was in print even earlier with his fiction – 1959 – and has a dozen novels to his credit.

Barbara also became good friends with the writer Cornelius Ryan, whom she thinks she met at a film screening and who Mullally remembers as 'a movie photographer for one of the studios, a very sociable guy'. In fact, from 1941 to 1945, Ryan worked as a reporter for Reuters and the *Daily Telegraph* covering World War Two battles in Europe and the final months of the Pacific campaign. His first book, *The Longest Day*, was published in 1959 and sold four million copies in 27 editions, before being made into a film by Darryl Zanuck in 1962. His second book, *The Last Battle*, was published in 1966, and he finished a third, *A Bridge Too Far*, in 1974 while terminally ill with cancer.

With O'Neill and Ryan both being photographers, we should not be surprised to see the profession figuring in an important capacity in Barbara's novels. I asked her if Clee in *Remember* had been based on O'Neill, but no: 'Not really, although Clee was awfully good-looking too ... I made him out of whole cloth with a little bit of Robert Capa thrown in. And like Capa, he was a war photographer. In fact, so were the two male leads in *Where You Belong*. I enjoy writing about newspapermen and women, and photographers, because I know them well.'

The great thing about Ryan's books is his painstaking research and attention to detail. He built up a 7000-book library, and kept 'four or five hundred of the absolute best at my fingertips – a synthesis of the perfect World War Two library', as he told Barbara in an interview in 1968 for an article that appeared in a syndicated column she then wrote called 'Designing Woman'. She and 'Connie', as friends knew Ryan, became very close.

Barbara's own research is a strength already noted – be it the divisional labour-system of the Leeds clothing industry in *A Woman of Substance*, or a highly detailed salad and omelette-making scene in *Where You Belong*, or the procedure of an English coroner's court in *Hold the Dream*, or her meticulously drawn real-life environments: her Manhattan interiors, Yorkshire millworkers' cottages, French châteaux, or the dark waterways of Venice which inspired *A Secret Affair*.

She said to Billie Figg after publication of *Voice of the Heart* in 1984 that she suffered terrible embarrassment after she'd given the manuscript in to her American publisher, Doubleday. Her editor had critcised it for being too minutely descriptive. But her English editor, Patricia Parkin, who was with Barbara from the start and still edits her today, said: 'Don't you dare change that! That's what people want.' Time has proved it. 'I think that being a journalist has helped me greatly in many areas – not just the observation of people, but also in research,' says Barbara today, but she is happy, too, to acknowledge a debt long ago to Cornelius Ryan in this.

'We became good friends, before I knew Bob,' she told me. 'He taught me a lot about research. There was something of the teacher in him. He was always lecturing me – if you really want to write books you've really got to be serious about it

and you've got to write so many pages a day and so on . . .'
There may be something of Ryan in Nick Latimer in *Voice of
the Heart*, who gives Francesca five Ds of which every writer
should be aware:

> *Dedication, discipline, determination and drive. You've
> got to be obsessed with a book . . . And there's another
> D. D for desire. You've got to want to write more than
> you want to do anything else . . . [And] there's a sixth D,
> and this one is vital. D for distraction, the enemy of every
> writer. You've got to build an imaginary wall around
> yourself so that nothing, no-one intrudes. Understand
> me, kid?*

'We liked each other a lot and he was a bit of a mentor,'
Barbara remembers of Connie Ryan. 'He was very Irish. He
actually introduced me to the agent Paul Gitlin who represented me when I sold *A Woman of Substance*. We stayed
friends off and on, and he and Bob became good friends, and
then Connie got very sick with cancer, as you know . . .

'Dick Condon also became a good friend of mine. He wrote
The Manchurian Candidate, and I met him through Jeannie
Gilbert when Dick was Head of Public Relations for United
Artists. He wanted to be a novelist, and he and Cornelius Ryan
encouraged me, said that I really had to stick at it. They both
were singing the same song.'

Condon's *The Manchurian Candidate*, a political assassination thriller set in North Korea and America, prophetic and of
great social and political significance, came out in 1959 and
was a storm of a hit as book and film. Earlier Condon had
written a play (*Men of Distinction*, 1953) and a novel (*The
Oldest Confession*, 1958) and went on to write many more
of both. Back in the Fifties, at least until *The Manchurian*

Candidate met with such success, he was one of the wider group of Jeannie's people.

'That was the Wardour Street PR guy,' recalled Mullally, 'drinks for journalists and so on. That's how he would have met Barbara and myself and everybody else. Now, Richard went on, to everyone's surprise, to become a very good novelist. And I followed him and we met in Mexico. I went to live in Mexico in the late 1950s, rented a villa. Richard had just done two big things with two novels. And on my way up through Mexico City I had a little party . . . Wonderful character, big, expansive character, dominating the communications, and we're sitting in a little circle in my sitting room and he's telling one of his stories, or we were telling a story against him, and there comes a point when in the middle of his story he falls over from his chair and there's a very slim little barrier between him and eight floors to Mexico City. I promise you I leapt across the table and caught him, stopped him from going through. Now everybody starts roaring with laughter, but Condon, who did not find it funny, got up on his feet, went into the bathroom and stayed there for about a quarter of an hour. I knew what was happening. He had lost it.'

Every occasion in the Fifties seems to have been attended by drink and cigarettes. There were none of the health scares, they were the essential accoutrements of style, and alcohol in particular became reason for deep unhappiness and the kind of tragedy that Condon narrowly avoided. Barbara didn't start to smoke until she was twenty-nine, and with her *Yorkshire Evening Post* strategy well in mind, alcohol never lured her into serious problems. But it did for both Jeannie and Triss in the end. Their respective marriages, to David Merrick and Christopher Plummer, both ended not only in divorce but in tragedy on account of it.

In Jeannie's case there was a legal tussle over her child.

Merrick got custody after Barbara and other friends were sub-
poenaed to appear in a court of law on his behalf. As a result,
she and Jeannie didn't speak for years until, as Barbara recalls,
a mutual friend 'asked me if I would see her because she was
actually dying and I said OK. We had, the three of us – this
other woman and me and Jeannie – a very nice dinner and
then she said, "Can I come back and talk to you?" and the
minute we were alone she said, "Why did you testify against
me in a court of law?" I said, "I didn't testify against you, I
testified for your child."'

There is a section at the beginning of *Voice of the Heart*
inspired by this moment when two sometime very close friends
– Francesca and Katharine – who haven't spoken for years
come back together, Katharine returning to seek forgiveness
from all the friends she has hurt. 'Jeannie wasn't the model
for Katharine,' says Barbara, 'nor did she do anything to hurt
anybody like Katharine did, who actually ruined Francesca's
life by lying. But that section was the whole thing with Jeannie
Merrick, as she had become, and it was the idea of a friend
trying to become ... it was two friends *becoming* friends
again.'

In Triss's case, her very marriage to Christopher Plummer
was cast in tragic circumstances, as Mullally recalled: 'One
night they were driving back to London and opposite Bucking-
ham Palace there was a terrible crash. She went through the
window and her whole face was taken apart. She had to go to
the top plastic surgeon who did the job in those days for the
Airforce in one of the great hospitals for reconstructive surgery. They literally put her face back together again. I saw her
after months of surgery and she had wires coming out of her
head and scars all over her face. She nearly died. They were
not married at that point. He then said, would you marry
me. So, in between surgeries, they married. It was a bad way

to get married and their marriage was never a success. I knew them both at that time ... She followed him wherever Christopher was making movies: Madrid – they stayed in the next apartment to me there – New York. Then they divorced. He had bought a house in Park Street, Mayfair. She gave a New Year's Eve party at Park Street where she invited ... Barbara wasn't involved at this stage, must have been in America ... Triss invited two or three big Hollywood stars, there was haggis and whisky ... It was a disaster. I was invited with my then girlfriend. The butler was drunk. The Hollywood stars did not like the haggis, didn't know what it was all about. And dear Triss kept boozing and trying to forget what was happening. That was a typical post-Christopher event. But that house, a big townhouse in Park Street, was left to her as part of the divorce proceedings, which she flogged and has probably sustained her.' Triss retreated to Brighton and died in October 2003.

Barbara, Mullally and Mann are the three survivors of the gang today, and it is interesting to look at what it was in Barbara's make-up which may have ensured her survival. First, she didn't lose sight of home. Her mother would come and stay at regular intervals, and Barbara made a habit of going back to Leeds whenever she could. It was important to her to keep the two worlds in some way co-existing. Subconsciously she attached truth to Freda and Winston, to her upbringing, to what they had done for her. The glitzy world in which she was moving she knew to be all about appearance, and she was not ready to agree wholeheartedly with Oscar Wilde that 'Truth is entirely and absolutely a matter of style,' even if the next few decades would be a push-pull matter of indecision on that score.

Whenever Freda came to stay, Barbara made a point of taking her out with her. 'What was very nice, I always found in Barbara,' said Billie, 'was that every so often she'd have her mother come up and stay in London with her and she took her to all the places she frequented. You hear what I'm saying? All those places, whether her mother appeared to fit or not she was going to take her and she came.'

In *Act Of Will*, when Christina goes to London, her mother Audra lives for the times when she returns to Yorkshire or Audra goes to London to see her. When Barbara returned to Armley it was always a tremendous occasion, but one that also emphasised the difference between their two worlds, as Barbara's Ripon-based childhood friend Margery Clarke remembers: 'Barbara always brought her parents presents, like hampers from Fortnum and Mason ... Her father, being a Yorkshireman, found out how much they cost and shook his head and said, "Now then, our Barbara, they've seen you coming!"' A Fortnum and Mason hamper comes to Francesca's rescue in the making of a meal in *Voice of the Heart*. There is caviar, pâtè de foie gras Strasbourg, aged Stilton cheese with port, and three tins of turtle soup – goodies not exactly compatible with Freda and Winston's simple diet, one suspects. Billie remembers having Freda and Barbara for dinner one evening and serving a vegetarian meal, quite daringly radical even in suburban London in the 1950s: 'I suppose it was a bit ingenuous of us. Freda ploughed through this meal and was saying, "Yes, it is very interesting this," but at the end of the meal she really couldn't contain herself and she said that she was not one hundred per cent for it. She said, "Daddy likes meat!"'

These were little markers, a few of many that showed that however much Barbara might want to keep her past, present and future as one, it was going to be difficult. It is a difficulty to which she admits even today, although she couldn't survive if she didn't return at regular intervals. 'It's just that when I

go there I am a totally different person than I am in Manhattan, than I have become. I have to go back into the Barbara
people knew, and it is difficult sometimes.' On another
occasion she brought Freda home a squirrel coat, something
she had always told Barbara she wanted, a kind of symbol of
what her life had been sacrificed for. When Barbara gave it to
her, Freda said, 'Where will I ever wear *that*?' The story seems
to resonate with the differences between Armley and Mayfair,
and Barbara made a point of telling me, 'You must remember,
I became a completely different person when I went to London.
I lived there for ten years . . . I'm very far removed from Leeds
now.'

In some ways, however, Barbara did not change. She never
became trendy, for example. In fact, it is interesting to see just
how inured to outside influences was this traditional, stylish
Mayfair scene into which Barbara had fallen with as much
relish as if she had been coming home – which, in a sense, she
had: home to the fantasy world of Edith Walker, which she
was fast making real. Only just across Regent Street from
Mayfair, into Soho, things were happening that would change
the world, and yet Mayfair remained almost disdainfully
oblivious. 'All the glamour was in that Mayfair square mile,'
said Mullally. 'A little overlapped occasionally into Soho and
into Pimlico, but basically it was Mayfair. Soho was crumby.
Nothing was happening in Soho. It was just crumby.'

'What about the French Pub and all that?' I protested.

'Soho wasn't dressed up for cocktail Mayfair,' he replied.
'You'd go slumming in Soho.'

Soho was London's Bohemia. During the war, artists, the
military on leave, intellectuals, black-marketeers, prostitutes,
pimps and local working people made merry together in its
pubs every evening, the painter Nina Hamnett occupying a
central role and acquiring almost mythical status. 'After the

The Albion Street home to *The Yorkshire Evening Post*, where Barbara got a job.

'And where's all this scribbling going to get yer?' says Emma Harte to her brother in *A Woman Of Substance*. 'On to a newspaper,' Frank replies. 'Maybe even the *Yorkshire Morning Gazette*... Stick that in yer pipe and smoke it, Emma Harte.' The *YEP* had an amazing circulation. It made all the money that kept its sister, the more prestigious morning paper, the *Yorkshire Post*, afloat. This was 1948, Barbara was only fifteen, but she was on her way.

Within six months she was a reporter, graduating to Women's Page Assistant at 17 – shown here at that age, standing in for a professional model who hadn't shown up for a fashion shoot.

Barbara at 19 and already Woman's Page Editor. She was barely a woman herself, but she wrote the lot anyway: cooking, fashion, personalities.

Barbara's meteoric rise would not have been possible without a large dose of journalistic skill, which she crafted painstakingly with the help of Keith Waterhouse, also on YEP at the time, shown here with Willis Hall (left), who worked on the Leeds *Evening News* and with whom Waterhouse adapted his first novel, *Billy Liar*, for stage and screen, the film starring Tom Courtenay and Julie Christie.

Barbara in 1953, aged twenty (just before she went to London to work as a Fashion Editor on *Woman's Own*), wearing a grey taffeta dress – 'Look at the pleating! It came from a very expensive gown shop in Leeds.'

Peter O'Toole with Omar Sharif in Lawrence Of Arabia. O'Toole was another who worked on the Leeds *Evening News* before getting his break. Barbara met Sharif at a party in London given by Sam Spiegel, but any relationship with O'Toole was only ever alive in Keith Waterhouse's imagination.

Barbara met film producer Robert Bradford (pictured on location in Hong Kong) in the London home of screenwriter Jack Davies in 1961. They were married in '63, and suddenly she found herself in Beverly Hills, New York and Paris.

The irony was that, through marriage to Robert Bradford, Barbara no longer needed the ambitious drive that had set her apart. She needn't have bothered with a career at all.

Barbara's mother, Freda Taylor, a long way from home on Fifth Avenue, New York, during one of her visits to her daughter, and with friends of Barbara, Ruth and Leslie Leigh – 'What was very nice,' says Billie Figg, 'was that she took her mother to all the places she frequented ... whether she appeared to fit or not.'

Barbara, the writer. It was, in the end, an imaginative return to her roots in the landscape of Yorkshire that enabled her to write *A Woman of Substance*, the novel that defined her.

In her forties, Barbara was going back into her roots more, but it was an enlightened return: 'I remember her coming back to do research into the old mills for *A Woman of Substance*,' says Leeds-based Bobby Caplin. 'She had gone away a provincial girl and come back a very, very smart and intelligent person. It was great to see. But basically she was still Barbara.'

fall of France, the York Minster, already a popular watering-hole with the Bohemians,' wrote Judith Summers in her book entitled *Soho*, 'became the unofficial London headquarters of the exiled Free French.' It became known as The French Pub, its host Victorienne Berlemont, whose name, since the First War, meant something to squaddies all over the world.

Then there was Dylan Thomas's famous watering-hole at the Café Royal on Soho's western border, and drinking clubs beating the tight licensing laws – the Horseshoe in Wardour Street, the Byron in Greek Street and the Mandrake in Meard Street. 'Since the turn of the twentieth century London's so-called Bohemians had been associated with Soho and [on the other side of Oxford Street] Bloomsbury and Fitzrovia.'

In Barbara's day, during the 1950s, Soho was once more a magnet for the eccentric, the creative, the unconventional and the rebellious. 'To the young especially, Soho is irresistible, for it offers a sort of freedom,' wrote Daniel Farson in *Soho in the Fifties*. 'When I arrived there in 1951 [to take up a job with *Picture Post*], London was suffering from post-war depression and it was a revelation to discover people who behaved outrageously without a twinge of guilt and drank so recklessly that when they met the next morning they had to ask if they needed to apologise for the day before.' Soho was a place for characters and conversation, for ideas and revolution. It no more mattered whether you had wealth here than it did at La Coupole, or on the left bank of the Seine where poets, painters, intellectuals and revolutionaries gathered in similar fashion in Paris.'

Names like Augustus John and the by then alcoholic Nina Hamnett gave Soho media currency, along with journalists such as *The Spectator*'s Jeffrey Bernard, painters Francis Bacon, Robert Colquhoun and Lucien Freud, the writer Colin MacInnes (*Absolute Beginners*, 1957), playwright Frank

Norman (*Fings Ain't Wot They Used T'Be*, 1959), and jazz musician and writer George Melly. Gaston Berlemont now had the York Minster, and the famed drinking club the Colony Room in Dean Street had opened in 1948; its soon-to-become-legendary owner was the eccentric, warm but ruthlessly selective Muriel Belcher, for, as in the key meeting-places on the other side of the tracks in Mayfair, you had to be a member.

A defining mark of 'Fifties Soho', however, was the coffee-bar scene, where, as Summers writes, 'for the price of a cup of frothy coffee, Teddy-boys, Rockers and skiffle fans could sit for hours behind a steamed-up window listening to the latest Elvis or Chuck Berry hit on the jukebox, accompanied by the loud hiss of an espresso machine.' Soho coffee bars were the music Mecca for the young, and youth was, for the public at large, what the Fifties was all about.

As it happened, Billie Figg wrote the first article about Britain's first home-grown rocker, Tommy Steele. It appeared in *Picturegoer* in 1956 and was headlined, '*The* Coffee-Bar Sensation'. Steele, the piece tells us, was discovered by John Kennedy in the *Two I*'s bar in Soho. When Kennedy, who ran a picture agency and happened to be in there drinking coffee, heard him sing, he immediately arranged for a Decca A&R man to come and listen to him. Kennedy became Steele's manager, and Steele took the charts by storm. After that, Soho, and the *Two I*'s in particular, was the place to be discovered. Tommy Steele, Terry Dene, Adam Faith and Cliff Richard could be seen performing here in their earliest days. As Bruce Welch of The Shadows recalled: 'If it was good enough for Tommy Steele it was good enough for us . . . we almost lived there . . . If we were lucky we'd play downstairs four nights a week, from seven till eleven – mostly Buddy Holly and Everly Brothers numbers. It was a small place, very hot and very sweaty, with a tiny eighteen-inch-high stage at one end, a microphone and a few old speakers up on the wall . . . always packed.'

When Kennedy flipped through a sheaf of photos of Tommy Steele during Figg's interview, he stopped at one showing his charge playing a gig at London's swanky Stork Room, and said: 'But Tommy's not interested in Mayfair Society. He wants a girl just like his mum.'

The comment showed agent Kennedy's nose for a good photo caption, but it also pointed up precisely the breakaway nature of the scene Tommy Steele was setting. Glamour was no longer the dream. Hitherto, Mayfair Society had been the thing. It, royalty, Hollywood and couture fashion were all the news on the society and women's pages. But now a music scene was about to erupt onto the pages of newspapers and magazines which would change all that. Hollywood kept producing stars like Ava Gardner, Elizabeth Taylor, Rock Hudson, Grace Kelly and Stewart Granger in the old glamorous tradition, but the male iconography of the period belonged to the late Marlon Brando dressed in leather and sat astride a motorbike in *The Wild One* (1953), or, in mid-decade, to the smouldering, challenging youthful features of James Dean in *East of Eden* and *Rebel Without a Cause*.

Barbara left London on the cusp of a change that dealt a killer blow to Mayfair glamour as the dream of the young. Brian Epstein first heard The Beatles at the Cavern Club in Liverpool in October 1961, the year that Barbara met her future husband. In 1963, the year that she married and left England for America, they topped the charts for the first time with 'Please Please Me'. Beatlemania was upon us. The world would never be the same again. Yet, for Barbara, despite being a decade younger than Billie, change was never on the cards. She continued to wear conservative, classic, smart clothes, and her taste in music was still her father's – Eartha Kitt, Lena

Horne, Frank Sinatra and Tony Bennett. 'Just as she was conservative with clothes, there's a parallel with her taste in music,' observed Billie. 'At that period the other music that was square dancing, which had come over from America, bop and bee bop and all those sort of things . . . they were nothing to do with Barbara.'

Far away from all this, and yet so close to its centre geographically, was the glamorous Mayfair tradition to which Barbara belonged, and for which it now seemed Freda's educational programme had prepared her. On the face of it, as far as Barbara was concerned, what was established in Leeds was built on in London, not discarded, and contrary to her insistence that she was a completely different person when she went to London, there was no great change. 'But,' said Billie, 'the person *did* change. So, the appearance and the person were not the same.'

'On the outside,' Barbara says, 'people see a part of me and that's the part I allow to be seen. I think I'm a very shy person in many ways. I think the profession I've chosen should tell people that. It's private, it's an interaction between the author herself and characters. My typewriter is my psychiatrist . . . the Barbara Taylor Bradford complexities I save for my work.'

We have seen that identity is her constant theme, and this matter of truth and appearance is always worked out in the context of identity in the novels. From the moment she left the world that she identified with truth – the world of her childhood, her mother and father – up until the moment she rediscovered it in writing *A Woman of Substance* and *Act of Will*, Barbara has, so it seems to me, been on a pilgrim's progress of challenges, many of them self-imposed but without which her life would have been, for her, unbearably ordinary. In the process, her style has not changed, but over the years the person on the inside has been transformed, and the whole gamut of experience has allowed her to view what she started out with and where the trail has led her in a telling new light – the light in which her fictional characters glow. So, she can give us characters who are forgers, like Camilla Galland in *The Women in His Life*, who believes we can only ever have an unnatural identity, that we create a life out of more or less conscious choices, adaptations, imitations and plain theft of styles, names, social and sexual roles, that we write our own scripts and live by them. And she can give us others alongside them who are deeply centred on their natural, 'real' selves, like Anastasia in the same novel, or like Francesca in *Voice of the Heart*, and still others who, bereft of their own identity by dint of fate, create themselves whole new natures, like Maxim West, only to find there is something missing, there is a vacuum where his real self should be.

Barbara was, however, still a long way from this level of perception. At this stage of her life, in the 1950s and early 1960s, she was writing a script of her own to live by, while still drawing on the values of her youth for that script, which gave her a resilience that others around her – Jeannie, Triss – didn't have, and which inured her from the danger of becoming sick with celebrity narcissism, which did for so many in that showbiz world, even though one might think that Freda had inadvertently prepared her for it. Barbara used the self-assurance Freda gave her; she found security in Freda's conviction of her perfection, her superiority over others, her extraordinary qualities, but always knew it was something she needed to prove to be true for herself.

'One of the charming things about her,' remembered Jack Figg of Barbara in the 1950s, 'is that she never took all of that glamour very seriously. She did appear to, she was involved, but there was always a twinkle in her eye, as if to say, isn't this fun, all these people, who are not really part of us, but we are joining them.'

Now, the impression Jack has of this could conceivably have been one that Barbara intended him to have – another example of a persistent desire for psychological unity in the diverse scenes (Armley, suburban London, glamorous Mayfair) in which she moved. For, as Jack is the first to admit, he did not fall into the Jeannie-person category. On the few occasions he did find himself 'standing about with a wine glass in hand and chatting at one of Jeannie's bigger receptions', he made 'quite a lot of booboos. I remember once I was speaking to a chap about Norman Wisdom [the slapstick comedian] and I said that I thought Norman Wisdom was very, very clever, very funny, but his material was absolutely appalling and I felt that he needed better writers, one thing and another. And then Billie took me aside and said, "You are talking to Jack Davies, he IS Norman Wisdom's scriptwriter." '

An ability to rumble the celebrity culture was the best protection any girl in it could hope for. A sense of humour, an ability to laugh at herself, and to find genuine friends beyond her circle, would give the necessary objectivity – and in this the Figgs were so important. They regarded themselves as definitely off the celeb circuit: 'Sure I was a journalist,' Billie said, 'but I was a journalist who was coming home every night to a suburb. She'd obviously got these several circles of chums, which is healthy.

'I remember, Barbara often used to come down in a pale-blue Ford Zephyr that she had rather early on, a lovely pale-blue Zephyr, very stylish, and she'd come down to our other house, which was even more suburban than this. She stayed for a few days and I was away working, and Jack had just taken a new job – a travelling job . . .'

Jack seized the opportunity to develop what was, after all,

his story. 'Yes, Barbara was staying, and Billie I think had gone to Paris, to the Collections. So this chap, who was showing me the ropes, said, "Can we go somewhere quiet and go over some paperwork?" So I said, "Let's go home." He said, "Will your wife be there?" I said, "No, she's in Paris at the moment." So that was all right. So, we were sitting in this bay window we had, going over this paperwork, and Barbara roars up the drive in this pale-blue Zephyr with a mink stole that she used to trail around, and she had her own key and she came in, and I said, "Oh, Barbara, this is Maurice Brown," and Maurice Brown's eyes were popping out of his head. Barbara, who was a great prankster, her eyes twinkled and she said, "I'm going upstairs, darling, I'll be up there if you want me." Maurice was waiting for an explanation as to who this other woman was. Nothing was ever said. I just left it. But I'd arranged to meet Maurice Brown the next morning somewhere, at some station, and Barbara said that she would take me and drop me off, which she did. There he was waiting as we roared up in this pale-blue Zephyr, and she got out and put her arms round me and drew my head down and kissed me full on the mouth, and put her leg back as people do, and said, "Bye darling, see you later!" Well, I just sat in the car with Maurice and said, "Good morning, Maurice," and off we drove.

'That was typically Barbara really. It was the way she twigged immediately that there was a very funny situation to be made out of this.'

Although Jack is wont to describe himself and his wife as 'loose-end people ... someone for Barbara to lean back on', it is quite clear that they were as close then as they are today. They were close enough then for Barbara and Billie to go on holiday together to Paris, and indeed to indulge, on a few unlikely occasions, in the Figgs' sport of camping.

'We were campers,' said Billie, 'and we went away every

weekend at that time, with a tent and our car. If Barbara was at a loose end, as happened on some of these weekends, she came with us and we brought a little tent for her, a separate tent. I can tell you, she didn't stay under canvas. When we got there the first time, she looked at this thing we erected . . . Well, it was all right during the day, but then when we decided to prepare for retirement (it was quite early when we made the beds) there was this sudden scream and she shot out saying there were beetles in there. We said, "Oh yes, that's part of it, you often get beetles, you just brush them out." But no: "Oh, I'm not staying there!" So we had to go and find her a pub to stay in. On another occasion she came to a little piece of land my father had down on the River Crouch in Essex and we all camped on there. She was very tickled because my father had said, "It will be very nice to see you on our little estate!" When we got there, there was this tiny bit of land that we pitched our tent on. They were fun times. We roamed over half of Essex with Barbara.'

Typically, Barbara turned the Paris holiday with Billie into something of a journalistic coup. There never was a distinction between the worlds of work and play. As far as Barbara was and still is concerned, they are a seamless whole. 'We agreed that we'd like a holiday, a week in Paris,' Billie recalled. 'We just thought that would be nice to do. We took a hotel, which was a pretty run-down sort of Left Bank hotel. It had a bed in the wall, you had to press a button and the bed comes down. Somehow it worked out that Barbara was going to be on the side near the window. There was a fire escape outside which went right past the window, and I remember Barbara saying, "Oh, Billie . . . what if somebody comes down that fire escape? What if some man breaks in?" She paused and looked across at me and said: "I'm first!" And then she said, "I don't mean I want him first!" I shall never forget that. Did we laugh!

Anyway, it was a pretty run-down kind of place, but Barbara had fixed with a man she knew, called Escarti – he worked for a film company – to get an interview with Ingrid Bergman. This was an incredible coup because she'd only just come in out of the cold, as it were, after the seven years' banishment that Hollywood had treated her to. She'd run away with Rossellini seven years previously and America had shunned her. [She had left her husband Peter for director Roberto Rossellini and ignored the moralistic machinations of the Motion Picture Association of America to bring her to heel. Incredibly, Senator Edwin Johnson declared she should never again set foot on American soil.] No American film company would deal with her. But now, only that year [1956], they had starred her in *Anastasia*, and the British Press – us, if we got it – would be the first to interview her. It was a scoop! We did get it! Amazingly, it was agreed to! We fixed up that *Woman* magazine would take the article.

'Bergman was at the Théâtre de Paris playing in *Tea and Sympathy*, and we turned up about ten minutes before she did. She came in looking very ordinary with her hair in pin curls and a scarf over them, frightfully mumsy really. The story we wanted – the sort of thing that everyone wanted at that time – was Bergman's philosophy of life. Barbara and I each had some questions ready. We sat in a place full of sofas, so many we had one each, I remember, and we just fired questions at her as we went. She answered them all.'

The article begins with a comment about Bergman's characteristic lack of affectation, that she was a complete natural, not into the appearance culture – 'it is impossible to imagine that what she says is part of a pose,' and knowing that 'is important when you are talking to an actress,' wrote our two reporters. Then Bergman tells them about the early childhood loss of her mother – she died when Ingrid was two – and of her father, who died when she was twelve. Truth, appearance, loss, loneliness, the very themes that would, years later, dominate Barbara's novels, and already dominated her life now. The positive theme of the interview is about courage . . . the courage to do something with your life and not be 'put off by other people's advice or opinion . . . the courage it takes to stand on your own feet and do what you think is right . . . It is a duty,' says Bergman, 'each of us owes to the rest of the community as well as to ourselves . . . You have only to look about you to see a world full of people with chips on their shoulders. They wanted to do something with their lives but were put off by other people's advice or opinion. And so they feel cheated. They are impossible to live with because they have built up resentment within themselves. If I had not gone ahead and studied for the stage in Stockholm, my disappointment would have poisoned my whole life.'

Barbara and Billie hung on her every word. Thinking 'of the people who felt they could have been artists or writers but were afraid of the insecurity,' as they wrote, 'this aspect of Ingrid Bergman's philosophy of life became clearer.' For Barbara in particular, this was completely on song. She must have left the theatre walking on air, more determined than ever to move her own plans along.

By the time she interviewed Ingrid Bergman, Barbara had been out of *Woman's Own* for two years. In 1955, at twenty-two, she had been hired as columnist and celebrity profiler by Reg Willis when he became editor of the *London Evening News*. At that time it was the capital's largest circulation paper. 'I was working with a woman called Gwen Robyns, who you must have heard of because she wrote many books about Grace Kelly. She was married to a Dane or a Swede and she was this plumpish, jolly, nice woman. She was our boss and it was a room full of women. We did all sorts of features, and I used to be sent out to do stuff for the diary page, too.'

As usual with Barbara, the move forward was serendipity; she got the job in the course of what today might be termed networking, but was for Barbara simply an evening out. 'I met Reg Willis when he was features editor. It was probably a movie thing. Roddy Mann was probably there and Jack Gourlay and Matt White, and my little group of people that we all went around in. I noticed there were mostly men there with me, and I remember I had a blue hat, a knitted beret; it was a sort of bluish purple, but it had sequins that were like long tails; it was a glittery beret and very pretty. And that's how I met Reg. He came over and he said, "I love that hat, who are you?" And I said, "I'm Barbara Taylor." We chatted and he asked me who I worked with, and I said, "I'm with *Woman's Own*, but I really want to get back on a newspaper, I'm beginning to hate this." And that is how Barry Horniblow's name came up. I told him about *YEP* and my great editor Barry Horniblow, and we spoke about how Barry had gone out to South Africa, and eventually he said, "I'll see you around and if you ever – you know my number, it's the *London Evening News*, give me a call some time." Then, of course, I kept running into him and finally one day I did call him and said I really would like to come and have an interview for a job on his newspaper. So I went down to see him. I remember Reg interviewing me about my experiences in Leeds and what have you, and then he took me in to see the editor and when we left the editor's office, Reg said, "How long notice do you have to give, how many weeks?" So I said I'd have to find out, and he said, "Well, better give your notice anyway." Now nobody had offered me the job, the editor didn't say he was giving me a job. Reg just said, "Don't worry about it, Barbara." Then I found out I had to give two weeks' notice to *Woman's Own*, and when I told Reg, I don't remember how I said this, but I was nervously saying to him, "Are you sure I've got this job because the editor didn't tell me that I had the job." I wanted to hear it from the boss man, not the features editor. He said, "Barbara, I promise you, it's really all right." So, I gave my notice in with trepidation, and then I got a letter from the *News* and it was signed, "Reg Willis, Editor". He knew that the other chap was on the way out. So that's how I moved to the *London Evening News*.

'Then one night, after I'd been there maybe a couple of years – I often stayed late because I liked to get my desk totally cleaned up and do all the things that get neglected if you've been out doing stories, and Reg often used to look in – on this day he came in and said, "I see you're still here, Barbara, come on, I'll buy you a drink." And I do remember him standing in the doorway with a funny look on his face. He had some papers in his hand and I don't know what the words were – it's too long ago – but I must have said something like, "Is something wrong?" I knew . . . I always could read people. And he said, "I've just got something on the wire service, Barbara." He looked at me and said, "There's only one way to say it – Barry Horniblow just died." Well, of course, tears . . . I got sort of choked up. I didn't start sobbing or anything. I'd got tears in my eyes and I started to cry and he said, "Come on, I'll take you out for a drink." And I don't recall going for that drink, but maybe we did – El Vinos, somewhere like that. Horniblow . . . of course I worshipped him . . . Although then he was white-haired, he must have been a man in his fifties when I was fifteen, or maybe in his forties, I don't know.'

How far the death of Horniblow widened Barbara's perspective on what was going on in her life I cannot say, but he had started her off professionally, had been the first impetus, and his passing may well have encouraged her to look hard at where she had taken herself since. She had built up a great deal of journalistic experience, and contacts, too – perhaps now was the time to make a play for something more her own, or to buckle down and realise her ambition to write novels. Had she done so, it would have been no surprise to her friend, Billie: 'I always knew she was fiddling about with plots [for novels of her own] and trying them. She was very up on all the new books coming out. She had a knack, the books she chose – I remember Bud Schulberg's *What Made Sammy Run* in the early Fifties, things like that. They were the big best-sellers.'

However, instead of pursuing her ambition to write novels, as Keith Waterhouse, Roderick Mann, Frederic Mullally and Richard Condon were doing, and as Ingrid Bergman would certainly have advised her to do, Barbara looked for independence in what she already knew. She went the journalistic route with a new newspaper, which turned out not to be a good idea.

The paper was a weekly, geared to appeal to Americans living in London (there were some 80,000 at the time), and to American troops stationed in the UK. No doubt the thrill of the start-up appealed. As Barbara recalls: 'The staff was small and we all had to pitch in,' she as woman's page editor. An American, Bill Caldwell, was the first Editor. He had the distinction of appointing a youthful Bob Guccione, an artist and cartoonist before he launched *Penthouse* magazine.

Embellishing a story she had recently given to Anthony Haden-Guest (*New York Magazine* and the *Observer*), Barbara recalled the day Guccione first stepped into the office: 'I saw this man in reception when I went to lunch,' she told me. 'He was still there when I returned.' When she asked the receptionist who the visitor was and what he wanted, she replied 'Robert Sabatini Guccione. He's waiting to see the editor, but he doesn't have an appointment.'

Barbara, yielding to courtesy – 'It's so impolite to leave someone waiting for two hours' – conceded that she had 'better have a word with him'. The receptionist grinned, 'Oh yes, he just said he'd like to talk to the beautiful strawberry blonde,' and delivered the Sicilian to her door.

They became friends. After Guccione had made *Penthouse* a success, Barbara would attend dinner parties at his New York mansion, the walls hung with his collection of Van Gogh, Matisse, Renoir, Chagall, Degas, Modigliani, Picasso . . .

Meanwhile, on their first meeting he showed her his 'rather clever cartoons, with a feeling of Jules Feiffer about them', and she set up an appointment for him to meet Bill Caldwell. 'Bob told me he could also write, and that he had an idea for a political column called "Foggy Bottom",' she recalls. Caldwell hired him on the spot the next day. 'Everyone on the paper liked Bob Guccione,' Barbara remembers, 'and his column became very popular with the Americans.' His work routine was rather singular, however: 'He wouldn't come in until one or two, but he would stay there very late. He often worked all night.'

It was while Guccione was so doing that *Penthouse* began to take shape in his mind. Caldwell had by this time been replaced as editor by Derek Jameson. 'After he had been there a few months, Guccione tried to press sexy material on him,' Barbara said. 'Derek, a real dyed-in-the-wool newspaperman, declined. He told Guccione: "Look, we can't put tits and arse on our front page. We'll all end up in the nick!"' Given that Jameson would later become editor of the tabloid *News of the World*, and editor-in-chief of the *Daily Star*, his protestations must surely have been influenced by the prevailing, rather different market perception of *The London American*, though one can't help wondering if a move, however tentative, in Guccione's direction might have enhanced the newspaper's appeal to Americans, who were already being well prepared for it by Hugh Hefner. 'Later, Bob brought in a dummy of a magazine he wanted to start. It was beautiful. He was very professional in everything he did. I said, "Bob, it looks great, but isn't it a total copy of *Playboy*?" He said, "If there's one, there's always room for two." And he was right.' *Penthouse* went on to sell five million copies a month at its peak.

The London American fared less well. It ran for sixty-six issues between March 1960 and June 1961. '*The London American* lacked advertising revenue,' Barbara states today, 'and this in the end was the cause of the paper's failure. It couldn't justify its existence. The owners had other business commitments. So they finally lost interest. The paper closed down . . . I felt sorry for some of the people who were without work, and we were all sad to see it disappear. We'd all enjoyed being together, there had been a lot of camaraderie.'

Barbara returned to freelance work, moving in behind a desk in Billie Figg's 'funny little office on the fifth floor in Covent Garden, overlooking the *My Fair Lady* show in Drury Lane', as Billie herself described it: 'At that time I had a PR company with Shirley Harrison, the author. Barbara used to come and use a desk, not as part of Shirley's and my business but as a friend.' This was, without doubt, a low point for Barbara, the demise of *The London American* a terrific blow. 'By the end of the Fifties I think she was feeling disappointed that she'd not brought off anything very big,' recalls Billie. 'I remember, and Barbara agrees with me, she had an uneasy period. What she was up to was not meeting her aspirations.'

'I did feel out of sorts,' Barbara admits. 'I was rather irritated with myself, disappointed that I hadn't written a novel. That was my dream.'

Initially, she had begun to freelance for a Belgian magazine, specialising in celebrity-type interviews, while at the same time working on *Florabelle*. Soon, however, another reason not to pursue her declared ambition to be a writer of fiction would present itself and she would grab it. She was offered a job with a magazine called *Today*, a title from the IPC stable for whom Billie had once worked. 'I was a big admirer of the editor of *Today*, an energetic and talented American called Larry Solon, so I took the job immediately . . . I enjoyed it there, and perhaps that's why *Florabelle* never got finished. I was back in journalism full-time again.'

Then, not long afterwards, fate played Barbara a winning hand. Jeannie Gilbert, who by this time had made New York her home, was staying with her fiancé, the Broadway producer David Merrick, at the Beverly Hills Hotel in Los Angeles, where she ran into an old friend, a movie producer by name of Robert Bradford, who was waiting to meet a colleague for lunch at the pool where, famously, the Hollywood glitterati met. As he was about to go to London, Jeannie told him he must call on her best friend, Barbara Taylor. But she didn't have her telephone number, so she scribbled on a piece of paper the number of Barbara's then neighbour.

Barbara had recently moved to a swish address in Bryanston Square, Marylebone. 'Jack Davies and Dorothy, his wife, had a duplex apartment there,' she told me, 'what you would call a maisonette in London. You went in on the street level and then they also had the downstairs that opened on to a garden. I had the garden apartment next door to them, and next door to me was Sean Connery with a garden apartment . . . Dorothy was an interior designer, and we sort of did it up together. It was warm, cosy, with a living room, kitchen, bathroom and bedroom, just right for a single girl.'

Bob Bradford takes up the story: 'When I got to London I was inundated with work on a movie called *The Golden Touch*, a costume picture about the Louisiana Purchase [the transaction in 1803 which saw the sale of the French-speaking Mississippi state by Napoleon I to the US for $15 million].' There couldn't have been a more apt title for the project that would bring Bob and Barbara together, though it was a couple of weeks before he came across that bit of paper with Jack Davies' number on it: 'I phoned it and they invited me over. It was a Saturday night. They had promised to get Barbara in for a drink, but when I arrived there was no sign of her.'

Many miles away in Gloucestershire, Barbara was attending a friend's birthday party, unaware of what had been cooked up. Jack and Dorothy had failed to make contact. But they and Bob got on well and that night they asked him to join them for dinner in Soho. 'I did, and then when the bill came Jack discovered he'd forgotten his chequebook. So I lent him some money. Dorothy then insisted I come to lunch the next day, so they could repay the loan, and she promised to have Barbara for lunch as well.'

Barbara drove back from the country early that Sunday morning. No sooner had she parked outside her flat when she heard the phone ringing. 'It was Jack asking me to come to Sunday lunch. I explained I had a deadline to meet [but] he brushed it aside, said a friend of Jeannie's was in town, that he was a handsome man with lots of charm. I just laughed, explained I couldn't miss the deadline . . . Then Dorothy called and explained what had happened the night before. They would be embarrassed if I didn't show up . . . So I said, "Oh all right, but I can't stay long. *I have a deadline*!"'

When she made her way next door she was told to expect three people, one of whom would be this Bob Bradford. When the three arrived, there was an attractive red-head in a green suit, a smaller man, somewhat nondescript, and another man. It was quite obvious to her that the taller of the two men just had to be with the red-head, and that Bob – the one for her –

was this smaller nondescript fella. As it turned out, she was wrong and not to be disappointed. The woman was Pat Lasky, the smaller man screenwriter Jesse Lasky Jnr, who was working on the script of *The Golden Touch*, and 'Bob just came across the room with Dorothy and was introduced. He sat down next to me, started to talk about Jeannie, and how he had run into her in the Beverly Hills Hotel. We got on immediately: he was so warm and friendly, and he had the loveliest brown eyes, they were kind. I can remember thinking what a nice man he was, and also how attractive he was as well. I'd been led to understand that we were having lunch at the maisonette, but Jack announced that we were going to a restaurant. And so, half an hour later, we went to a nearby Indian restaurant. I'll never forget Bob leaning into me, whispering in my ear that he wasn't too fond of Indian food. Neither was I.'

What had been Bob's immediate reaction to Barbara? 'She was twenty-seven, twenty-eight. What did I see in her? I can't really say. She was very pretty, an attractive young woman. She was bright, and I put a lot of value on intellect, intelligence. She was *outcoming*, and I guess that really caught my attention. After lunch I asked her what she was doing. And she said, "Nothing."'

'That's true,' said Barbara. 'I'll never forget Dorothy's face. It was a picture, as silently she mouthed, "*What about the deadline?*" Of course, the deadline was forgotten. I went with Bob to the movies, and we've been going to the movies ever since.'

CHAPTER FOUR

Change of Identity

'*Most nights he lay awake, prowling the dark labyrinths
of his soul, seeking meanings for his life and all that had
happened to him.*'

Maxim West in The Women in His Life

In *Voice of the Heart*, people are not who they appear to be.
Immaculate, wealthy superstar Victor Mason started life as
'Victor Massonetti, construction worker, the simple Italian-
American kid from Cincinnati, Ohio', and tempestuous leading lady Katharine Tempest has also changed her name. She
was born Katie Mary O'Rourke in Chicago. They have both
become something else by their own efforts; they have risen in
the meritocracy, which the western world is in the process of
becoming, and – so it seems – have changed their names to
signal their new identities.

For Katharine in particular this becomes quite an issue in
the novel, her lover Kim reading her concealment of her true
identity as deception, a sign that she is incapable of true love.
Then we learn that indeed her decision to change her name
signals something dangerously repressive, a desire to blot out
her past. In Katharine we get to understand the effect that dire
childhood experience can have on a girl. This is surely what

Barbara sensed, however subconsciously, about Freda when she said of her, 'I believe that my mother always had a great sense of loss, in fact I know that she did,' although Barbara did not at the time know the details of that loss.

The particular childhood problems Katharine suffered were quite different from Freda's, although childhood loss is the common denominator. The reason the fictional Katharine changes her name is that as a child she was exiled from home by her father, sent to boarding school in England because of her influence over her brother Ryan, an influence which is benign, but at odds with their father's plan to turn Ryan into a politician. She isn't even allowed home for the school holidays. Then, she is abused by her father's business partner. 'In the ensuing days she began to realise how much that horrifying childhood experience had scarred her, what a devastating effect it had on her adult life.' In effect, Katharine has been cut off from love, cut off at her roots, and she changes her name in an effort to make a fresh start.

But it isn't possible to wipe the slate clean by changing your name, whatever cosmetically salutary effect it may have on self-image, and we see the results of that in the novel. There is more to identity than appearance. A name change will not suffice; this is very much the thematic material on which the author is sharpening her claws in *Voice of the Heart* and in *The Women in His Life*. Identity is again the theme, her message to her readers being, in her own words, 'to know who you are and what you are'.

If Barbara came to this theme subconsciously on account of her mother Freda's sense of loss, it was also brought home to her when she met Robert Bradford, her future husband. Like Barbara an only child, Robert Bradford was born in 1930 of a German Jewish family living in Berlin. He was three years of age when the Nazis seized power and at five he was taken out of Germany with a cousin and placed with a French family in Paris. 'His father was dead. He never saw his mother again,' Barbara told me. 'He prefers not to talk about it.' The final tragedy was that after the end of the war, Bob's mother, who did survive, went to America to find him, only to die shortly before they could be reunited. 'His mother took him out of Germany in 1939 and subsequently she herself got out but she couldn't get to him because the Germans had occupied France. Eventually when she got to America, because of the confusion she still could not find him. When he finally made it to America in 1946 she had died a few weeks before. I cried quite a lot when I wrote *The Women in His Life*. I suppose I thought of Bob as a child. I think he found the book very haunting, very moving.'

Exiled, cut off from his root culture, Bob changed his name, just as in *Voice of the Heart*, Nick's great-grandfather, also a German Jew, changed his when he emigrated to America. If we need confirmation that Barbara's preoccupation with identity owes something to her relationship with the man she loves (as well as to the mother she loved), this is surely it.

Fellow Jew Bobby Caplin makes the point that name change is common among this generation. 'Well, I've got a friend, his real name is Gerald Goldstein. At school he changed it to Jerome, then when he left school he changed it to Goddard. Who knows, having come through the holocaust maybe you don't want to be reminded . . . I mean it's not just a name, there's a whole lot of other baggage that comes with that. It's all very well saying, well I wouldn't do it, but we haven't been through those circumstances, where you do not want to be acknowledged as a Jew, which I can thoroughly understand.'

I was not surprised to learn from Barbara that Bob was always very tender with Freda. They cannot have had a great deal in common, but I could imagine an unspoken, subliminal acknowledgement of their shared sense of loss, similar because it had to do with identity, aloneness, being dispossessed. Just as Freda was drawn back to Ripon time and again in search of her lost world – the shadowy figure of her father and the environment in which her whole incredible childhood had been played out – so, in *The Women in His Life*, the character who comes closest to Bob in all the novels, Maximilian West, is drawn back to the places of his childhood loss, too:

> *The lure of childhood, he thought, how strong it is with me . . . is it because I lost so much when I was a child . . . had such irretrievable losses? Do I come back to Paris and Berlin in the hopes of finding something which escaped me long, long ago?*

Maxim's eternal quest is successfully completed in the novel. He comes to articulate the nature of his childhood loss and finds the something he did not know about himself, about his parents, about the love he thought he'd lost.

Another thing that Freda and Bob had in common, of course, was their love for Barbara, and, coming as they did from this similar ground of childhood loss, one is bound to speculate whether they drew something similar from her in recompense. Barbara, of course, was Freda's saviour. She will not stint in heaping praise on her mother for her encouragement and love, but it is also true that Barbara gave Freda a reason for living. Most significantly, she was, in her very character, an impressive expression of the very identity lost to her mother – her style, as I see it, an amalgam of the sparkle of Edith and the sheer persistence of a Ripon.

Is it also true to say that Barbara compensated Bob's childhood loss? Certainly she came to root his work in her own strong Yorkshire culture (nine films would be shot based on

Barbara's novels). It is not chance that many of Barbara's heroes and heroines call upon her Yorkshire-rooted values in order to make their decisions and to make their judgements. Bob, bereft of family and alienated from his home culture, will surely have benefited from the stability of so certain a cultural influence.

In the fact of their meeting in Bryanston Square lies an eerie coincidence, which pulls Freda and her mother Edith's story into the picture of Bob and Barbara's relationship from day one. For Bryanston Square is the very place where Edith's Studley Royal contemporary – her rival in Frederick's affections, the Most Honourable Constance Gladys Marchioness of Ripon – died less than half a century earlier. She was occupying No. 13 at the time. Barbara lived at No. 5.

Of course, loss was the last thing that Barbara read in Bob's brown eyes on the day they first met; nor did he need any help in the direction in which he saw himself going. He was already successful in the film industry, he was a sophisticated, cosmopolitan figure, and plainly, in that sense, he knew where he was coming from. He was the epitome of the suave Hollywood producer, with all the outward accoutrements that she had always deemed essential in a man. He had style, he dressed well and he gave the impression of being able, like a fine batsman, to deal with the swiftest of balls at his own pace, a master of timing with no small amount of wit.

Friends of Barbara could see immediately that they shared something in their personal style, something of the glamorous tradition, but then one also commented, 'I think my feeling, when I very first met him, was that he was someone I couldn't access, if you know what I mean.' This was, I think, the key comment. There was an emotional detachment. At one and the same time one can see this as an alienating factor (perfectly consistent as a response to the sadness of his own childhood

247

loss) and as a key element in his success as a modern man. There is no way you can be successful in the high-powered world of movie finance, where tens of millions of dollars are at stake, without a certain detachment, inscrutability even. 'And he was a major player,' said Bobby Caplin. 'He is certainly very shrewd, and I think in the part of the film business he was involved with, you have to be.'

Jack Figg had a taste both of Bob's inaccessibility and of what he presumed to be the tacitly structured nature of the film business when once Barbara was away and suggested Bob take him and Billie to dinner. 'Barbara wasn't with us,' said Jack. 'I was flattered. We went to a place – I think it was in Charles Street – more or less opposite the flat they had there. It wasn't a restaurant, it was just a big black door.'

'Mark's Club?'

'That's it. The door was opened by a chap with white gloves and we were shown in. We sat at a table and I was making small conversation, "Nice place this, etc." Billie and I suddenly felt very uncomfortable and nothing happened, and I was saying, "Do you want us to order, Bob?" "No, no, no," he said. And we sat there and the atmosphere was charged with some sort of tenseness and after a while I just said, "What's happening?" He said, "Well we're in the wrong place, we're down in the B Room. We've got to go into the other room." And there was a little flight of stairs leading up into it. He said, "That's the A Room." So we said, "What does it matter we're in the wrong room, aren't we all right here?" "No," he said, "And I can't be seen here. If I'm seen here everybody will start talking about Bob Bradford, did you see him there, etc."'

There is a prerequisite in business to get behind the face that is presented, to be accepted somehow into the inner sanctum, and personal style may be as important as the product you are selling in order to achieve that. Also, there is no way you can carry off the big deals, or handle the prima-donna personalities, unless you are utterly straightforward. You can be brutal, so long as when it comes to the crunch you are straightforward. Otherwise, no one will ever deal with you again. If, at the same time, you are able to disarm your interlocutor with a sense of humour, you are in, and Bob has this very nice, quiet, dry sense of humour, which Barbara loves and made a point of stressing to me: 'No day goes by when he doesn't make me laugh.' That is something for a wife to say of her husband after more than forty years of marriage. In 1961, others were soon shown this side of him, as one close friend recalls: 'I discovered, as we would see him more and more personally, that he's very easy to get on with and rather warm – and I thought, how odd!'

The period of Barbara's first meeting with Bob was one in which she couldn't have failed to be aware of the nature of his childhood loss, for their meeting occurred at precisely the moment that the world was examining the dispossession of German Jewry at the hands of Adolf Eichmann. 'I was stupendously aware of what had happened to the Jews of Europe under the Nazis,' Barbara said. 'I suppose because it was the persecution of innocent people. Though I was always into history . . . I couldn't bear the thought of it.' There could have been no more powerful a delineation of the loss that Bob had suffered (to be so tenderly described in Barbara's novel *The Women in His Life*) than the real-life re-enactment of the horrors of the Holocaust daily in the English newspapers as Barbara first met her future husband.

SS-Obersturmbannfuhrer Karl Adolf Eichmann, head of the Department for Jewish Affairs in the Gestapo from 1941 to 1945, chief of operations in the deportation of millions of Jews to extermination camps, was brought to a controversial and highly publicised trial in 1961. It lasted from 2nd April to

249

14th August. Eichmann was pronounced guilty, sentenced to death and, on 31st May 1962, hanged in Ramleh Prison.

Stepping over to the table, she grabbed the paper, stood staring at the headlines and the photographs, her eyes widening with shock, her face freezing into rigid lines of horror.

Names of places leapt off the page at her. Ohrdruf . . . Belsen . . . Buchenwald. The most fearful words stabbed at her eyes. Death camps . . . atrocities . . . inhumanity . . . extermination . . . Jews . . . millions murdered . . . genocide.

She lowered her eyes to the pictures. They stunned and horrified her, so graphic were they in the foul, inhuman story they told of the most unspeakable brutality and cruelty, a terrible testament to the pitiless torture and mass murder of innocent people.

Teddy (Theodora Stein) in *The Women in His Life*

Barbara was drawn deeply into the reportage. It had a marked effect on many, especially those involved with Germans. Bob Bradford told Barbara little about his own history, but the Holocaust recurs frequently in her novels. 'I do have strong feelings about it because I'm a child of Europe,' she explains. 'I grew up in England during the war, and my husband Bob was born in Berlin and taken out of Germany to Paris when he was five, because he was Jewish. Unlike many German Jews who didn't ever believe they would be hurt or touched, this family did. His mother got out, but all the aunts and uncles disappeared.'

Just how deeply Barbara mined the emotional strata of Bob's early life in her book was shown years later: 'I was somewhere in Ohio to give a little talk, most likely Dayton. Seats had been set out in a bookshop, you know the kind of thing, you could have a coffee, a soft drink. And this woman came and sat down and put her copy of *A Woman of Substance* on the table. She said: "I'd like you to sign it, but I haven't read it."

'"Don't worry," I said (as I do). "Millions have."

'She then pulled the book from her bag and said that she had been going to Switzerland by plane when *The Women in His Life* had just come out. She tried to read it on the plane, but figured from the opening that it was another business story like *A Woman of Substance*. She'd got to Zurich and her husband had become involved in a lot of business meetings so she had picked the book up again. "I picked it up in desperation!" she said (which was good to know!). "When I got into the part in Germany when Maxim is a small boy I became enthralled . . . the part when his mother and Teddy got on the train and said goodbye [to his family] . . . when she worried about the Germans on the train" – it was full of uniforms – "and Maxim was saying all sorts of Jewish things about what he had been eating. When I read this," she said, "I had cold chills, and when they got to Paris I felt tremendous relief . . . I started to cry and I cried and cried for hours and I didn't stop. You see, I was taken out of Germany like that and I had my mother's jewellery stitched into my clothes too. And you, your book, was a catalyst for me. I had not been able to cry since I was taken out of Germany in the 1930s. You're not Jewish, are you?"

'"No," I said.

'"How can you understand?"

'"Because I am a human being," I said.'

The women in the title are those behind her hero, Maximilian West – his mother, his grandmother, the woman that subsequently brings him up (Teddy), his first wife, his daughter, various women who have helped to form the man or are

251

important to the man. As he grows up, he wonders whether his dear parents are still alive. After the war, Teddy goes to Berlin literally to unearth Maxim's identity in the bombed-out rubble of the city, searching among the *Trummerfrauen*, the rubble women, who play their laborious part in the rebuilding of the city, counting out the bricks they retrieve at the storage depot every Saturday afternoon, to be paid accordingly. It is among the rubble that Teddy finds the Russian princess, Irina Troubetzkoy, a friend of Maxim's parents. She takes Teddy into her squalid cellar room below ground and informs her that Maxim's parents have gone to their death, his father in Buchenwald, his mother in Ravensbruck. When, finally, his worst nightmare of their murder is realised, 'Maxim suddenly understood that the sadness inside him would never go away. It would always be there. For the rest of his life.' His loss becomes the driving force in his life. He makes a pact with himself that he will become a dollar millionaire by the time he is thirty, and does so with time to spare.

Exiled from country and family at so impressionable an age, brought up by another family in Paris, his relations themselves victims of the death camps and his mother lost to him even after she had passed over to freedom, Bob, too, was deeply alone, bereft of loving parents, but also of the value-system that's part of the baggage of a national or religious culture. Like many other German Jews, he felt dispossessed as much of his German as of his Jewish heritage by the Nazi onslaught. Likewise, in *The Women in His Life*, Maxim's father, Sigmund, and mother, Ursula, are from 'great and ancient families', real Germans, as well as being Jews. Bob would need courage to attain the inner sense of unity on which a personality can normally count when rising out of the culture of his birth to make his way in the world.

Somehow he found it, and worked his way into the movie

business. He was a protégé of Jesse Lasky Sr, founder of Paramount, and was employed at one stage by the Hal Roach Studios in California, famous for stars such as Charley Chase, Will Rogers, Harold Lloyd and Laurel and Hardy, and for seeding the careers of the likes of Jean Harlow, Janet Gaynor, Fay Wray, and Boris Karloff, before their post-war TV production of classics like *The Lone Ranger*, Groucho Marx, Abbott and Costello, *The Life of Riley*, and *The George Raft Show*.

The late 1950s found Bob in Spain as Executive Producer for Samuel Bronston Productions. Among his film credits from this era are *John Paul Jones*, *King of Kings*, *El Cid*, *Fifty-five Days at Peking* and *The Fall of the Roman Empire*. From this time, producer Samuel Bronston, a sometime official Vatican photographer, developed Spain into a European capital of movie-making out of massive studios near Madrid. But when, in 1964, the big-screen epic *The Fall of the Roman Empire* failed to meet audience targets demanded by its enormous budget, Bronston went bust, was sued in court and forced out of production, still owing $4 million as late as 1975.

The Sixties saw Bob as Executive Vice President and CEO to Franco London Films S.A. in Paris, making among other films *Impossible Object* with Alan Bates and *To Die of Love* with Annie Giradot. By then Barbara and Bob were married, and there was a period in the late 1960s to early 1970s when they lived together in the French capital at the Plaza Athenée, an elegant hotel off the Champs Elysée that appears in *Voice of the Heart* and *The Women in His Life*, and where she and Bob stay regularly even today. 'I'm a creature of habit,' says Barbara, 'I always stay at the Plaza Athenée because for months we lived there when Bob was running a film company in the Sixties.' It is where Maxim, his mother and Teddy stay when first they flee Berlin to Paris. *The Women in His Life* is redolent of those days, Bob's childhood in exile, places which he came to share with Barbara from the early years of their marriage:

The lights changed to green and Maxim crossed the Place Saint-Michel and headed towards the Rue de la Huchette. Within seconds he was sauntering down that narrow old street, experiencing a sense of nostalgia as he glanced around. Here on his left was the hotel Mont Blanc, where he had stayed on a couple of occasions, and immediately opposite was the El Djazier, the North African night-club which they still frequented sometimes, going there to drink mint tea, ogle the exotic belly dancers, and eat couscous with harissa, the hot piquant sauce which blew his head off, but which he nevertheless enjoyed. And a few yards further along were the famous jazz joints, where some of the American jazz greats came to play and musicians of all nationalities to listen, as did he from time to time.

He paused when he saw the Rue du Chat Qui Peche. It was only a little alleyway, but he had never forgotten this street because the name had so delighted him when he was a child. 'It means the Street of the Cat Who Fishes,' Mutti had said, translating the French for him. Filled with glee, he had laughed out loud, tickled at the idea of a cat who fished. They had been on one of their outings, he and Mutti and Teddy. 'Investigating the quaint bits of Paris,' Mutti had called their wondrous excursions, and ever since those days this picturesque area had remained a favourite, and he often returned to walk around these narrow cobbled streets, to browse in the bookstores and galleries.

Barbara enjoys and nurtures her association with France in her books, most recently in 2002, of course, in *Three Weeks in Paris*. In 1994 she attributed her success as an author there to being one of the few foreign writers who paints an accurate picture of the country 'and actually gets the Eiffel Tower in the right place'. In *The Women in His Life*, Paris is a magnet to Maxim West, because he is the fictional persona of Bob Bradford as a child, and Monte Carlo is where later he moors his magnificent yacht, *Beautiful Dreamer*.

In *To Be the Best*, Paula and Emily go to Monte Carlo to meet their cousin Sarah, who has been living up the coast, near Cannes, for five years. Barbara used to holiday in Cannes before she even met Bob. In *Her Own Rules*, Meredith Stratton's quest for her mother takes in parts of France within Barbara's best-remembered experience. Havens Incorporated – the American-English-French group of upmarket inns, hotels and châteaux, which Meredith inherits – has its Paris office in rue de Rivoli, which is where eligible architect Luc de Moutboucher lures her to his château Clos-Talcy between Talcy and Menars in the Loire.

A Sudden Change of Heart is set in Connecticut, New York and Paris. In *Angel*, the heroine, Rosalind Madigan, returns us to the Loire Valley, to a band of country running 'from Orléans to Tours . . . through a verdant landscape known as the Valley of Kings', and gives us Montfleurie, 'the most magical of all the Loire châteaux'. Rosie looks into the dust of history for the spirit of the place, 'where once violent battles had raged when Fulk Nerra, war lord, predator and ruler of the area, had stalked this valley'.

In *Hold The Dream*, Emma Harte's will includes her Avenue Foch apartment in Paris and a villa at Cap Martin in the South. Both Paris and Cannes were of course part of Barbara's life before she met Bob, and in *Act Of Will*, her young alter ego, Christina, spends three days with her friend Jane in 'the lovely little town of Grasse . . . situated in the Alpes Maritimes high above Cannes,' where the painter, Fragonard, was born, famous for its gardens and Gothic cathedral. They go to the perfumery there when they're developing two fragrances, Blue Gardenia and Christina.

Was it here, too, that Barbara began seriously to paint? 'I remember we called in one night to have dinner with her,' Jack Figg told me, 'and she greeted us with absolute excitement – "I've learned I can paint!" She had just returned from a holiday in the South of France, where she'd met a Portuguese man called José, and she painted him at home. Suddenly she realised she could paint! She had these pictures expensively framed and they looked absolutely terrific.' Such is the power of France for this writer.

Bob Bradford's childhood experience may have been less happy there, at least to begin with, but together they have enjoyed many happy times in Paris since. In any case, in 1961 his life was set on an upward trajectory. If being cut off at the roots from his family had amounted to an existential challenge, he had by this time met it with success in his film projects. If you had asked him, he might have said that his idea of 'roots' wasn't Berlin or Paris, but the things that led up to the work he was doing now, things which, like tributaries of a river, flowed together to make his life what it was. It is a significant conceptual difference to the traditional roots metaphor, which carries with it the burdensome possibility that precepts of your birth culture can tie you down. Bob's past had ceased to exist for him, he had only a future – a feeling perfectly in tune with his new girlfriend's own ambitious nature and drive for autonomy – particularly welcome following the failure of *The London American*, which had temporarily sapped that drive.

If there was a feeling of disappointment that she hadn't

measured up to her own demanding aspirations in the late Fifties, by the early Sixties, with Bob, 'there was an infusion of new excitement,' as Billie recalls, 'and then she went off to live in New York, which had got far more "go". Although London had got a lot of "go" in the Sixties it wasn't Barbara's scene. But New York was, and they loved her.'

Bob was the tonic she needed. Life for such men is all about flow, change, movement *to* somewhere – the future. The modernist concept challenges the old idea of rootedness and static identity and, indeed, the whole notion of loss (which is why it appealed to him). In place of roots and family and community, there is freedom – man constantly on the hoof, constantly *in change*, man whose environment of airports and hotel rooms delivers the extraordinary *emotional detachment* that Barbara's friends had noted about Bob, while in his future-orientated projects he sought to deliver the sense of unity within that we all require – his autonomy, his values, his identity.

Maximilian West, the hero of *The Women in His Life* and the character whom Barbara has said, 'I truly love as if he exists,' and who, unusually for her, she brought back in another novel, *A Sudden Change of Heart*, is just such a man. His personal style is immutable, cast out of the materials of loss that his genetic and historical background has engineered, but he, like Emma Harte before him, makes it an uncompromising philosophy of life, which women in particular find irresistible.

At the start, Maxim *is* his projects. He has become identified with them. We are not sure what, if anything, he is beyond the performance of them.

His work comes first. It always has and it always will. It consumes him entirely. I know that now. He's not

257

normal, you know, not when it comes to work. Maxim is beyond a workaholic, Mother. They haven't invented a name for a person who works the way he does. Around the clock.

He has the stamina of a bull, and the most extraordinary concentration.

He is managing director of Westrent and Westinvest at twenty-five, with a goal to make his first million by thirty. Business is Maxim's discipline and only real pleasure. He has the accoutrements of success and the women to go with it, but considers them to be 'so much folderol'. Personal autonomy and cool self-sufficiency characterise his power over others, which is specifically contrasted with the boot-in-the-face power of the Nazis: 'strutting, arrogant, vulgar and bloated with self-importance'.

Maxim's brand of power is 'dangerously attractive' to women and at the start of the book excites his PA, Graeme Longdon. What turns Graeme on is something to do with the clinical, undemonstrative efficiency with which he attends to his projects – 'the intellect, the brains, the drive, the energy, the ambition and the success'. This may be the author telling us what she found exciting about Bob's lifestyle. What's for sure is that, at the start of their relationship, she didn't bring him to Yorkshire; rather, she let him take her further away from her roots, to New York. Barbara couldn't have been more ready for it at that time.

New York is an environment which fairly crackles with the belief that anything is possible. Bob Bradford took Barbara physically out of England, where making money and having ideas above your station were still frowned upon (particularly in the case of young women), into an environment where these things are a patriotic duty. A journalist once put it to Barbara that her personal claim never to have felt inadequate is rare in England. She replied: 'Don't you think that's why I live where I live today?'

At that stage in their lives, when Barbara had been left feeling dissatisfied and unfulfilled after the demise of *The London American*, and irritated with herself, disappointed that she hadn't yet written a novel, she would have found Bob's altogether pragmatic approach attractive. One can sense in her response to it the ambitious daughter Christina's need for a pragmatic set of values at a similar stage in her life in *Act of Will*. Christina's mother, Audra, has a perception of a *moral hierarchy* against which she deems her daughter, who has given up her art and become a commercial dress designer, a failure. But in *The Women in His Life*, in Maxim's modernist 'project-culture', there is no such moral hierarchy. Instead, project-goals achieved are used to elaborate a value system to replace the deep-truth culture in which Audra lives in *Act of Will*. There are no absolute values, no ultimate rights or wrongs, there is only project and the value of action, which is defined in terms of where action will lead. The Ten Commandments are justified on the altar not of Judaism or even on that of Yorkshire working-class culture, but on the altar of pragmatism – you abide by them for no other reason than that if you don't, no one will deal with you again.

In Maxim's world, morality has a *cash value*, which doesn't mean that you do whatever makes the most money in the short term, any more than his project-orientation suggests that he is mad for making money (which he is not – he is at ease with money but not overly impressed by it). Cash value in the moral context means, simply, pragmatism – what a decision will flow on to. The point in being morally pure is that no one will trust you if you are not.

Once the heart is taken out of morality and replaced with

the head, art and commerce are on a level pegging. There's a particularly pertinent moment in *Voice of the Heart* when Victor Mason gives Francesca (the character who appears most like Barbara) an antique copy of *Wuthering Heights*. Because Emily Brontë's work was always central to Barbara's aspirations as a novelist, the idea reminds us that in this high-powered milieu, art is a collector's item, a thing's cash value is not considered to be at odds with its artistic value, nor less intrinsic to it.

It is no longer 'better' to write a great novel or paint a beautiful picture than it is to put together an elegant deal. In the world of film (art at its most commercial), Maxim's flair as a financier is likened to that of an inspired artist. His father-in-law Alexander Derevenko observes, 'I can no more explain to you the creative impulse, what it is inside a painter that makes him capable of producing a breathtaking work of art ... than I can explain to you what it is inside Maxim that enables him to put together an incredibly successful company or a stunning deal.'

It may seem amazing that such a man as Maximilian West should consider marriage at all, let alone to sensitive Anastasia Derevenko, whose life is centred on the deep-truth-culture that his modernist approach is set on replacing. But then, as we later discover, Maxim is not so secure in his new project-identity as his outward display suggests, and it is the play between his philosophy (modernist) and Anastasia's more meaningful deep-truth philosophy, with which the book ultimately deals.

As for Bob, he seems to have been a similarly unlikely candidate for marriage in 1961. At thirty-three he had already been married before, and was well suited for all the reasons I have given to an autonomous, self-sufficient, single lifestyle. I remember Barbara telling me that her friend, the writer

Cornelius Ryan, had said to him: 'If you don't get on and marry Barbara, *I will*!' It had been a joke, but the point is made that Robert Bradford had not been looking for marriage when he was swept off his feet by Barbara Taylor. That, however, is precisely what they did. Bob and Barbara were married in London on Christmas Eve 1963.

CHAPTER FIVE

Coming Home

'The city of his birth and childhood. It had forever pulled him back, and he had always believed it held a secret for him. It had. The secret had been revealed to him today.'
Maxim West in *The Women in His Life*

Having a man behind her is what Edith Walker had lacked in her intended rise in the world, and 'a man behind her' is what Emma Harte in *A Woman of Substance* is given by her creator to get her project going.

The man Emma Harte selects to expedite her business is Joe Lowther. 'Quite by accident, when she had been shopping she had seen *it*. The shop. *Her shop* . . .' The shop, the first in her empire, is in Town Street, Armley, and it is to let. Joe Lowther is the name of the landlord she should contact. Emma goes to his home and offers him a deal. Against his better judgement – Lowther doubts that she has enough experience to make a go of it – he accepts her offer, partly because Emma has the temerity to put hard cash down on the table, and partly because 'he was drawn to her. Dangerously attracted to her.'

Lowther's considerable portfolio of property, which 'included eight shops in Town Street, a row of cottages in Armley, several terrace houses in nearby Wortley and . . . two large plots of land near St Paul's Street in Leeds itself,' was built up by his mother and her mother before her. When 'his ancient great-aunt' dies he inherits an additional £150,000, a large house in Old Farnley and 'four commercial properties in the centre of Leeds.'

Pursued by Lowther, Emma agrees to marry him, even though she doesn't love him. She is honest enough with herself to see that in marrying Lowther she is 'cheating him' of love, but that doesn't stop her. It is a bad match. Even the physical side of the marriage is unsatisfactory. The marriage is simply part of Emma's wider business strategy, although she tells herself that she needs him to 'protect her and Edwina' (her illegitimate daughter, you will recall, by Edwin Fairley). We have to conclude that she is using Lowther and that her action is completely unethical, but Emma then exercises her extraordinary business skills to turn the Lowther properties into an enterprise beyond his wildest dreams. Hers is an intricate and powerful strategy, which she executes with ruthless precision. We can only marvel at her performance.

There are some aggressive, apparently feminist traits in Emma Harte, particularly in her treatment of Joe Lowther, and one is tempted to hail her as something of a heroine of the feminist movement, which was gathering pace during Barbara's own rise, peaking as she wrote the novel. In 1953, the year that Barbara uprooted from home for London, Simone de Beauvoir first coined the phrase 'women's liberation' in her book, *The Second Sex*. In 1963, the year she uprooted from London for New York, Betty Friedan set the feminist fuse alight in *The Feminine Mystique*. Seven years later came *The Female Eunuch*, Germaine Greer's bitter landmark examination of women's oppression. Then, leading up to publication of *A Woman of Substance* in 1979, came associated bestselling novelists like Judith Rossner (*Looking for*

Mister Goodbar, 1975), and upfront commercial ones like Erica Jong (*Fear of Flying*, 1973) and Judith Krantz (*Scruples*, 1978), their all-woman themes leaving Jacqueline Susann standing, and characterising 1970s New York women as forceful, funny and free – figures epitomised by Diane Keaton in movies such as *Annie Hall* and *Looking for Mister Goodbar*.

None of this was quite *Woman of Substance* territory, however. When Barbara picked up her pen in 1976 to write her first novel, she did not pick up the feminist gauntlet as well. Emma, like many of her other heroines, is ambitious, disciplined and self-possessed. She can be ruthless, and is when crossed; she wants to win, and she is not averse to using her feminine wiles, but she always needs a man behind her. In the novel, she says: 'Being underestimated by men is one of the biggest crosses I have had to bear . . . [but] it was also an advantage and one I learned to make great use of . . . When men believe they are dealing with a foolish or stupid woman they lower their guard, become negligent and sometimes even downright reckless. Unwittingly they often hand you the advantage on a plate.' She uses men, but she does not get her kicks out of crushing them or castrating her male lovers. I was reminded of what a commentator once wrote about Margaret Thatcher: 'Her femininity added a frisson of sexuality to one's engagement with her and disturbed the public-school code of conduct and decorum formerly operating within the all-male preserve of the party's higher echelons.' Henry Rossiter, Emma Harte's financial manager in *A Woman of Substance*, is a paid-up member of just such a code of conduct and his loyalty to it is clearly disturbed by the allure of a woman whose 'mind was logical and direct. She did not think in that convoluted female way . . .'

A Woman of Substance showed Barbara's female readers how to go out and take up the opportunities that the feminist revolutionaries had opened up for them, but Barbara was not arguing the politics of feminism or any other movement. She already had her vehicle – the style to which she was born. She believes we all have this, if only we can find it, as Freda found it in her. The novels do not tell women what to believe, only to know themselves, thereby to put themselves in control of their own destinies. Movements are out. Feminism neither appeals nor appals. Barbara advocates not feminism but a brand of existentialism in which the feminine principle is preserved: 'I think that you have to *do it yourself*. I did it myself and Emma Harte did it herself . . . and it can be done without being abrasive.'

When Barbara married Bob in 1963 she was nowhere near ready to pen the character that would make her her fortune, however. It would be thirteen years before she did. So, what happened in the intervening years to bring Barbara to Emma Harte?

Marrying Bob meant weighing anchor altogether – on her family, on England, on the whole culture of her birth. A picture on the author's website captioned 'The Bradfords in Morocco on the set of *Impossible Object*' suggests that she slipped effortlessly into Bob's world, travelling with him on location. The film, made in the early years of their marriage by the company Bob ran, Franco-London Films, was based on a novel by Nicholas Mosley about a writer who finds it difficult to distinguish fact from fiction. When a journalist asked Barbara whether she had had difficulty in adjusting to life in the Manhattan glamour world, she was able to reply, quite truthfully, 'No, I'd been in it in London.'

But there were real differences in her life, which cannot have been effected without some measure of emotional insecurity. Contact with Yorkshire and friends like the Figgs was necessarily now limited. 'We lived in New York in Manhattan and in California in Beverly Hills. Bob had the apartment in Beverly Hills before we were married, so we went backwards and forwards, and then he gave it up because he ran a film company in France, so I commuted from New York to Paris and stayed three months in Paris then went back to New York for a month.'

Also, the irony was that, through marriage to Bob, she no longer needed the ambitious drive that defined her personal style and set her apart. She needn't have bothered with a career at all. She was able for the first time in her life to buy antiques of her own, and began to design and decorate their homes, capitalising on all those childhood trips with Freda to the country houses of Yorkshire. Bob saw to it that she wanted for nothing. Right from the beginning of their marriage he took a practical interest in the clothes that Barbara wore. In Paris, she was introduced to Ginette Spanier, the *directrice* of Pierre Balmain, who became a friend. 'I was very much into *haute couture*, but only a couple of pieces a year. Later, many of my clothes were by Pauline Trigère, the great American designer who was French born. Pauline and I were great friends until the day she died. She made the kind of clothes I love. Very sleek, very tailored, no frills and flounces . . . I also get clothes from Place Vendôme in London. Most of the things I choose there are by Italian designers. The owner, Seymour Druion, buys his collections in Rome and Milan, and picks out things for me which he knows I'll like and which Bob will like as well – dark colours for winter, no patterns; pastels, especially blue and pink, for summer.'

So, back in 1963 she had already made it. Becoming a multi-millionaire in her own right nearly two decades later apparently required little adjustment: 'I have always had quite a good standard of living and it hasn't made all that much difference,' she was able to say when asked what she was doing with her royalties from *A Woman of Substance*. 'I bought some English antiques and paid too much for them in New York, but the rest is simply invested carefully. I already had two fur coats and I didn't want any more. How many fur coats can you wear at one time?'

But, of course, the difference was that back in 1963 Bob was sourcing the finances of 'the whole enchilada'. He was strong, with definite ideas about the way he liked things to be. How far did he exercise control? How did the balance of power work then?

In 1994, the *Orlando Sentinel* quizzed Barbara as to how she and Bob had got on during their thirty years of marriage. She replied, 'We're both very bossy, so we lock horns a lot. So he calls me Napoleon, and I call him Bismarck. At Christmas in Palm Beach I saw an embroidered cushion that carried the words: "Napoleon lives here, I married him". And I bought it and crossed out "him" and put "her". And I gave it to Bob. Well, I've resigned my generalship now, I've come down to a lieutenant colonel.'

This story is retold in numerous interviews, and it happened so long ago that no one is quite sure whether Barbara bought the cushion and crossed out 'him' or Bob bought it and crossed out 'her'. But on one occasion she did speak plainly: 'I always think Bob's controlling and I know that I like to control.' But it is usually she who gives in: 'I say: "Oh, to hell with this, it's not worth arguing about."'

It was in these pre-*Substance* years, too, that Barbara wrote the trio of books to which I have already alluded and which seem to suggest an uncharacteristic compliance: *How to be the Perfect Wife: Etiquette to Please Him, Entertaining to Please Him* and *Fashions That Please Him*.

'Yes, I laugh about that these days,' Barbara told Sue Lawley on BBC Radio Four. 'They sold like crazy and, having written

A Woman of Substance about this warrior woman who goes out to conquer the world, people have teased me about it, especially the press who have managed to dig up these books and say, but Barbara this is terribly opposite, and I say, well I meant it when I wrote them ... now my attitudes have changed.'

She meant it when she wrote them, so what does this tell us about Barbara as young wife? Was it a period in which she luxuriated in pleasures of which most women dream, or one in which she was fighting to retain her self-respect and autonomy? Was it one in which she learned to play Lettice Keswick in *Everything to Gain*, 'a woman a lot like me ... a homemaker, a cook, a gardener, a painter, a woman interested in furniture and furnishings and all those things which made a home beautiful'? Or was it Emma Harte to the rescue in the mid-1970s when marriage threatened to cast Barbara forever in the role of second fiddle? One novel in particular plays over the whole range of possibilities for just such a woman in her situation.

The two really interesting wives of Maxim West in *The Women in His Life* are Anastasia Derevenko and actress Camilla Galland. After Anastasia's chance meeting with Maxim in Paris they fall deeply in love. Maxim gives her everything she could possibly want materially, but in time it is not enough. 'Maxim could not give all of himself to her.' We are not talking impotence here on a sexual level, rather on an emotional level. A dam holds back Maxim's emotions, and the block is to do with his childhood loss, to do with his being cut off from his roots by the war. For all his commercial genius, his successful projects, there is, deep down, a vacuum where his real self should be.

He cannot give Anastasia what she truly wants, for what she wants is *him*, and he is out of touch with himself. One night on their luxury yacht, after a spectacular party at which she has worn his gift of a diamond necklace, she felt 'something cracking and splintering inside her . . . "That's all I am to you these days, isn't it? The giver of your parties, the decorator of your homes, the wearer of your diamonds," she exclaimed coldly.'

This is a crucial point in the novel. Two sides of a coin are made to face one another. On the one side is Anastasia, whose nature encompasses a deep sense of truth and beauty and love; on the other is Maxim's project-obsessed psyche, which provides their riches but denies her access to his true self.

Maxim reels at her onslaught. Anastasia accuses him of infidelity, even though she knows he is a faithful husband, and she leaves him in the early morning. She attacks him because her womanly intuition tells her that what's missing is the crucial element of life: love. Maxim's script is not rooted in his true nature, he is still running away from who he is, which is why he is afraid to let anyone in, even the woman he could truly love.

In his second wife, Maxim finds someone who is able to meet him on his own terms. He and Camilla Galland live a kind of parallel existence, each engrossed in their own projects and deriving a shared exultation in their mutual success. 'If you marry me,' says Maxim to Camilla, 'I wouldn't want you to give up your career . . . I need plenty of space. In fact I must have it in order to do my work properly. I don't want you clinging to me, making me the core of your existence. I have to travel a great deal, and I hope you understand this. Of course you can come with me on the extended trips. I'd love it, love to have you with me. But not on the short, quick trips. They're too hectic, and I'm always locked up in meetings. I don't want distractions. Or to be deflected from what I have to do – because I am worrying about my wife. I've always had great direction, concentration. I can't change.'

Theirs is to be a project-marriage. There are to be 'ground rules', and we are conscious that the rules are set by Maxim not by Camilla, his very name a synonym for a rule of conduct. Nevertheless, Camilla is happy at the prospect: 'I have to work, Maxim, just as you have to ... they'd take me away in a straitjacket if I didn't.' She feels like 'the luckiest woman in the world', and then fate steps in to end it – Camilla breaks her neck by falling down a steeply pitched basement staircase.

Maxim replaces her with Adriana Macklin, who, like him, is consumed with business projects. He becomes unhappy, in fact he becomes impotent, though not with beautiful blonde Blair Martin, who wears pale-green silk pyjamas by Trigère, and lives in Sutton Place, Barbara's own apartment, over-looking 'the East River and a portion of the 59th Street Bridge'.

We shouldn't get too sidelined by matching up the bio-graphical elements, which have been scattered across the canvas so that no real-life colours attach to any one character in particular. What we are dealing with here is a theme which does have relevance to real players, indeed to us all, and has to do with life in the modern world in which, too often, truth is deemed merely to be the opposite of a lie.

Camilla and Adriana share Maxim's thoroughly modernist outlook. For all three, project – making things happen – is all. There is no great depth to their relationships, or if there is, as might have been possible in the case of Maxim and Camilla, both parties agree that it is not the priority. Emotional compli-cations have been eradicated by ground rules, or, in the case of Adriana, by the fact that she is a similar operator to Maxim. But, as Anastasia knows, they cannot be so easily dealt with. Anastasia haunts Maxim, and Adriana attacks him for always 'flinging that ex-wife of yours in my face'.

The rest of the novel leads up to Maxim's moment of truth. He is softened up for it by an accident, which delivers the crucial volte-face that 'there are more things in life than big deals.' He returns to his roots, finds his birthplace in the post-war rubble of Berlin, 'the city of his birth and childhood, [which] he had always believed held a secret for him'. Finally, he learns the secret, and 'the sadness inside him slipped away'.

Appropriately, when he discovers his true identity (a complete surprise, in which Barbara cleverly discovers the absolute value of love at the core of the concept of identity) he becomes whole again in Anastasia's arms just as the Berlin wall comes crashing down and his homeland is made whole again too.

Barbara and Bob had themselves crossed into the Eastern sector of the city in 1986. The trip had sparked the idea for the novel. 'I had always had this compulsion to go to the East zone,' she said at the time. 'So we went through Checkpoint Charlie and when we were there I had this flash in my mind's eye of a woman in a white satin evening gown in the style of the Thirties, blonde and very ethereal. Somehow I knew that her name was Ursula [the name of the woman we believe to be Maxim's mother], and I asked Bob if he had ever mentioned anyone of that name. But he just kept telling me it was my writer's imagination!'

The dam against emotion had been holed, the message was once again about identity – to remember where you came from, because 'it defines who you are.' By the time Barbara wrote *The Women in His Life* it had been revealed as the lesson of her own life, for it was Barbara's imaginative return to her very deepest roots in the landscape of Yorkshire in the mid-1970s that had enabled her to write the novel that would define her. Her return brought her back to the values of the landscape of Yorkshire to which her mother had introduced her as a child, 'a sense of honour, duty and purpose', the need

271

for 'integrity in the face of incredible pressure and opposition' and 'not only an honesty with those people who occupied her life, but with herself'. It is for this reason that her woman of substance is not quite the model of modernism, which at face value she seems.

So what led Barbara back?

For her part, soon after marriage to Bob, Barbara decided that she must have a project of her own and said, like Camilla Galland: 'Bob was busy being a movie producer – so if I didn't work, where would all my boundless energy go? I couldn't just sit at home and do nothing, I've never been one of those ladies who lunch and I loathe shopping.'

But she did not immediately buckle down to writing a novel, her true ambition, the one that would define her, which now for the first time she might have embarked upon unimpeded by want. Instead, she pursued a freelance journalistic career, writing about the homes of the famous in a syndicated interior-design column called 'Designing Woman' – first for *Newsday*, then, moving with editor Tom Dorsey, for the *New York Daily News*, and finally the *Los Angeles Times*. Her column went across America to 185 newspapers and she wrote it for twelve years. She also wrote a number of interior design books. I have seen them in her drawing room, now beautifully bound in leather, including the bestselling *Complete Encyclopaedia of Homemaking Ideas*.

Appearance, design, beauty – an arena in which the aesthetic and the commercial are indistinguishable – all absolutely in tune with the world in which she was now moving, and all the time she was learning, building up her knowledge base in an area that would, as it happened, prove useful for the novels, for she realised how compellingly she could write about the most exquisite artefacts of European origin – her favourite Biedermeier and Art Deco furniture and Impressionist paintings. America was looking to Europe in this arena, in which indigenously it could not of course compete. Barbara was unmistakably English, and brought up to the task. As she said: 'Mummy gave me this eye for antiques. She taught me to look.'

The design column and books became her project, which ran parallel to Bob's in the film world, and so began a pattern of life that in time would bring Bob and Barbara project-bound together in the marketing and filming of her novels, something Maxim and his wives never quite achieve in the novel. Today, this is the pattern of their lives. Bobby Caplin summed up the position well: 'I don't think either of them would have been as successful without the other.' Bob tends to get what he wants, he is a tough negotiator, but what he wants is now what Barbara wants.

They are a formidable team, now that their project is a shared one. Bob has engineered some unbelievably good deals for Barbara in America, involving many millions of dollars. Her personal wealth was quoted this year in the Rich List as £95 million (currently around $170 million), which puts her at 419th position in the world.

But this was only one side of the coin, the other was the fulfilment of those contracts. Her first English publisher, Mark Barty-King, became seriously concerned in the early days that she was working so hard, never seemed to go out because she did little *but* work. 'She was riding high on a worldwide reputation and there was always such big pressure to produce something,' he said.

The picture that came to mind was of the miller's daughter confined by the king to her room and spinning gold thread out of straw, the beautiful girl working away, working away, spinning a golden yarn according to magical directions from Rumpelstiltskin. But Bob would have none of it.

'*A Woman of Substance* was her first book. She sweated the

book for two years, twenty-four hours a day. I used to go to Hollywood to work and she worked twenty-four hours a day on that book. It's unbelievable and thank God she's not doing that any more. She doesn't have to . . .' But Barty-King remembers that the second book, *Voice of the Heart* (the finished edition, 928 pages in length), was another draining experience. Certainly it was not published until 1983, and yet Barbara had a contract and started writing it more than six months before *Woman of Substance* was published in 1979.

In an interview with Richard Whiteley, she once set out the regime that her life had become by this time: 'It's the salt mines. I do ten or twelve hours a day, seven days a week. I get in here [her study] at six o'clock in the morning. I'm wearing a pair of cotton trousers and a tee shirt, winter and summer, in winter I put a cardigan on, no make-up, no jewellery, just my glasses, very underdressed. And I edit what I finished yesterday, and at about seven thirty to eight o'clock I take Gemmy [the dog] out. So I've already done about two hours' work. I start at six a.m. in here, and then I work till noon, and then she [the dog] has her lunch and I have a salad and I take her round the block, bring her back, and go back to work till 6 p.m.'

I asked Bob how Barbara would relax. 'She doesn't like cocktail parties because you stand around and talk nonsense, totally idiotic stuff, and she'd rather sit at home and read a good book. She loves to read and to think and to work, and she likes to go out to dinner and she likes to be with close friends and relax. Particularly when she's working on a book, she doesn't want to sit around with people who talk nonsense. They have nothing to say but blow hot air and for her it's a total waste because her head is in the book, thinking, thinking out the plot.

'Although what we do keeps us apart we are not shut out

of each other's lives. When Barbara is shut up in a room writing for days, I am her window on the world, telling her what I've seen, who I've met. She tells me about her next chapter and uses me as a sounding board. I am very proud of her and what she has done. I'm a very secure individual, unworried by her fame and fortune. I had mine before her. She was my back-up and now I'm hers. My greatest relaxation is long weekends. I leave our New York apartment, take a plane and fly to Miami or Puerto Rico. I swim, I read and be away from people. You can't be a happy couple if you are always in each other's way. Barbara doesn't like the sun; I do. When she is "off the book" she travels with me.'

From the start of their marriage, Bob was himself project-bound in a difficult and highly commercial industry, and it was ever a two-career household with no idea of working nine-to-five. In the early 1970s, still some years before she started work on the novels, Barbara made a move to escape from the relentless urban vortex into the country, and one is minded how much she must have been missing her regular trips home to Yorkshire that she had always made from London. In 1971 she persuaded Bob to look for a country retreat in north-west Connecticut. She told *Architectural Digest*: 'Our search began when my husband, Bob, and I were guests of conductor-composer Skitch Henderson and his wife Ruth, at their house in New Milford. We were instantly entranced by the region, seeing elements of England and Europe in its scenic, sweeping beauty composed of rolling, tree-covered hills and shining lakes.'

Readers may recognise the connection in four of her novels. *Everything to Gain* (1994) began a series of books with narrative set in Connecticut, continuing with *Dangerous to Know* (1995), *Her Own Rules* (1996), and *A Sudden Change of Heart* (1999). She accords it great significance in the novels.

In *Hold the Dream* it is the place selected for the long-awaited coming together of the Hartes and the O'Neills. Emma Harte's granddaughter Paula first realises she loves Shane O'Neill, grandson of Blackie O'Neill (Emma's dear friend from the start of *A Woman of Substance*), when she is visiting Shane's converted barn in the country town of New Milford, near where Bob and Barbara's new home would be sited. In *Everything to Gain* it is where Mallory Keswick retreats to consider her future before she yields to the pull of the Yorkshire moors, where her late husband grew up and where she gets her idea about what to do.

'I fell in love with it the moment I saw it,' says Barbara of the house they finally bought, 'a wonderful old Connecticut colonial, classically elegant in its design, surrounded by ancient maples and smooth green lawns flowing down to a large pond. Dominating that pond, and adding to the decidedly pastoral feeling of the property, were two regal white swans floating on its surface against profuse pink water lilies.

'The first thing we did was hire Litchfield architect Paul Hinkel, whose work we had seen and admired. Paul is an authority on colonial architecture, and since he was nearby he could supervise the construction daily. He was quick to understand our requirements, agreed with us about the restoration and remodelling, and made other good suggestions. After much consultation and endless refining, Paul presented plans that turned a thirteen-room house into one with eighteen rooms, plus a wine cellar and storage space. He also redesigned the guest cottage.'

They could not have settled for less. However, this weekend retreat did not come as a release valve during the early years of their marriage. Though the search began in 1971, it would be twenty-one years before Barbara and Bob made the purchase. 'He bought it for me in May 1992, for my birthday,' said Barbara. It followed a spectacular deal Bob made with her American publishers, at the time the biggest author contract in history.

So, in the meantime, there was still no weekend escape. And even after they did buy it, 'We didn't go there much,' said Barbara.

'It wasn't Bob,' as Bobby Caplin pointed out, 'that home in Connecticut – magnificent, quite unbelievable, but that certainly wasn't Bob. That was one hundred per cent Barbara. Bob is a city man. This was maybe one of Barbara's dreams, with the lake and the swan. They were in the middle of nowhere!'

The Connecticut episode delineated once more Bob's pragmatism on the one hand, and Barbara's unfulfilled emotional needs on the other. 'Ten thousand square feet, enormous house,' Bob said when I asked him about it. 'It was wonderful! Paradise! But it was too big for us and we were never there, it was just draining us of money. The pool men, the tree doctor, the gardeners, every day there was something else. And it was two hours' travelling, over two hours to get there.'

'I think Barbara loved it,' I began. 'I remember . . .'

'She loved the country, she is a country girl, she is English, but I mean she is also a very practical lady and she works best at her home in New York.'

That may be so, but the novels give an impression that Connecticut meant something important to Barbara in a creative, imaginative sense. When she writes about the countryside around the Litchfield hills, the brilliant skies of the region are described in a manner not dissimilar to that in which she refers to Yorkshire. When I took this up with her she agreed: 'Connecticut has that very special kind of light, a clarity of light that I talk about in *A Woman of Substance*, *Hold the Dream* and *Voice of the Heart* – the part set in Yorkshire –

because it's like Northern light and it seems to emanate from some hidden source like the light in some of Turner's paintings. It's also rather an undulating countryside – I like moors and hills, though it's not, of course, the moors.'

That light of Yorkshire is a symbol of creativity – Barbara says as much when she finally returns to it in imagination to write *A Woman of Substance*. It is 'quite extraordinary. It's almost as if it comes from another source. A writer must turn inward in order to write – everything comes out of me and this is what I remember, that extraordinary clarity . . . that love of the light.'

Their work was indeed their life and vice versa, and Barbara would never have wanted it any other way. Bob knew that and in that sense Bob knew best. But there was this emotional side that did need an avenue if Barbara was ever to achieve her expression. Was part of her being denied? It was an unlikely bedding ground for a traditional family, of course, and there would be no children.

I detect a sadness that she and Bob didn't have children, but also no shadow of regret at the lifestyle they chose, which they are both so good at. 'I don't have any regrets at all,' said Barbara twenty years ago, and she says the same today. 'I don't think you miss a person that you haven't known. Bob and I got married in 1963 and I was then thirty years old. We didn't want to have children immediately. We said, "Well, maybe in a few years." Then somehow it was suddenly too late. It's the luck of the draw. I'd have loved children but I didn't have them. I've got a wonderful marriage, I've got Bob, who is a great supporter of mine in every way, as I hope I am of him, and we've got to be content.'

There is a touching moment in the old nursery at Pennistone Royal, towards the end of *Hold the Dream*, where Paula, Emma Harte's granddaughter, sings her twin children, Tessa

278

and Lorne Fairley, to sleep with 'The Sandman' song, for
which Barbara proudly claims authorship. 'I wrote that! That's
my creation!' she says excitedly. 'I also wrote a children's book
and edited a couple for a publishing house. They had bought
the most beautifully illustrated book I've ever seen for children.
It was in Czech – each page had an illustration and a poem
in Czech, and they asked me to write a little poem for each
of these wonderful illustrations, and I thought, well, why not?
I drove Bob crazy! I'd ring him up at the office and say,
"Just listen to this for a minute." And he thought something
wonderful was coming –'

> The sandman has the swiftest wings and shoes that are
>   made of gold;
> And he comes to you when the first star sings and the
>   night is not very old . . .

'– Bob would say, "Do you mind, I'm in a meeting!" It was a
lovely children's book, I enjoyed doing it; it was a challenge.
It wasn't that we set out not to have children, you know. And
I didn't say, "Oh I'm going to have a big career." I had a
miscarriage and I never got pregnant again.'

I talk to Bob about Barbara's relationship with her own
mother, the love she gave her, and he says, 'That's why Barbara
is so keenly interested in working with children today and
with literacy problems. She works with Literacy Partners in
New York, a charity that raises money and opens centres to
teach people to read. She is on the Madison Council of the
Library of Congress in Washington, and has worked with the
president's mother Barbara Bush and First Lady Laura Bush
on literacy and the National Book Festival. Over eighty million
Americans can barely read, or can't read at all . . .'

She is also on the committee of PAL, a children's charity

(the letters standing for Police Athletic League) which has, at different times, attracted Barbara Bush and Hillary Clinton to fundraising luncheons. 'PAL is a charity devoted to underprivileged children in the New York area,' Barbara told me. 'It was started in the 1920s to get poor kids off the street, but it is run more by business people today. Recently we opened one centre in the Bronx, a tough area of New York, and I sat in on a session and listened as fifteen-year-old girls talked about wanting to get their boyfriends out of the Latin Kings, a gang. "The only way you can get out," one girl said, "is to commit suicide or they will kill you because they don't want you to leave – they won't let you leave."'

It's a world Barbara drew on in *Everything to Gain*, where Mallory Keswick's husband and children are fatally shot in a tough area of New York. Their killer had been smoking crack cocaine.

Again, in her charitable work in the UK, children are the focus. She is on the board and is a trustee of PACT – Parents and Abducted Children Together. Again it is her fundraising capability that is to the fore. The charity was started by Lady Meyer, wife of the former British Ambassador to Washington. Catherine Meyer is a good friend of Barbara, and it was she who asked her to become involved. 'Catherine's children were abducted by her first husband,' Barbara told me, 'and I can't imagine how she lived through it. It must have been harrowing.'

In this context, suggestions, much proffered, especially by female journalists, that Barbara's dogs are substitutes for her unborn children seem almost pathetic, although it is perfectly true that she has, perforce, made her beloved bichons frises part of the family, and she explores the entire history of the breed in *Voice of the Heart*. If you ever get to meet pretty, fluffy Beaji and Chammi, do not refer to them as poodles or they'll have you. Barbara finds it mean of reporters to question her affection for her dogs, given that most pet owners shower affection on their animals. It is, after all, the point, isn't it? Beaji and Chammi have a very good lifestyle and they know all Barbara's secrets, for they sit under her desk when she is writing and thinking and working out characters and stories aloud.

When her first bichon frise, the late Gemmy Bradford, fell ill in 1987, Barbara flew home on Concorde straightaway. 'I went immediately to the vet, where the housekeeper had taken her. She was operated on, and miraculously lived. She died when she was twelve. She wrote a lot of books with me. Gemmy was short for Gemini, even though she was a Scorpio. My parents were both Gemini and so is Bob, and my agent. So I'm surrounded by Geminis.' Astrologers may be interested that Barbara herself is Taurus, and Taurus and Gemini are the two most likely signs for members of the Rich List, followed by Aries.

Beaji and Chammi undoubtedly enable Barbara to express a side of her that otherwise gets scant exercise, but it is of course the writing that provides the real fulfilment. Why then did it take so long after marriage to Bob, which had, after all, removed the need to earn money by other means, to start the novel that would make her name?

It was not for want of trying. She worked on four novels during the period up to 1975–6 – four false starts. 'I didn't like them, I'd get halfway and be bored with it, and I thought, if I'm bored then obviously the reader is going to be bored, and I'd put it away and start another one. I did this four times, four different novels before I got the idea for *A Woman of Substance*. I had been trying to write romantic suspense like Helen MacInnis.'

Helen MacInnis wrote espionage thrillers with romantic

sub-plots, which benefited from her extensive research skills into political events of the regions in which the novels are set. After her husband was assigned to intelligence work in the British army during the Second War, there was even suspicion that she had inside information. Once again, Barbara would not have been blind to the fact that research was a key element in the bestselling mix. By the time of MacInnis's death in 1985, more than twenty-three million copies of her novels had been sold in America alone, and they had been translated into twenty-two languages.

However, the reason why Barbara Taylor Bradford came to write the eighth most popular novel in the history of the world had nothing to do with an editorial analysis of what was currently selling well, and everything to do with leaving that side of her thinking alone.

With the distance of time, her own development away from the person she had once been, and the geographical distance from home that marriage to Bob entailed, and with a growing sense of frustration, as I have outlined, Barbara began thinking about her past and her family more.

'I only really began to understand my mother's life when I was in my forties,' she told me. Barbara turned forty in 1973, three years before she put pen to paper on her first published novel. She had begun to talk to her father about his relation-ship with Freda. 'I know that she rejected him constantly. No, she didn't talk about it, but I knew about it somehow when I was in my teens and the only person I discussed it with was my father. When she used to come and stay with me in London, when she first arrived she'd be saying, "Your father is terrible" and he is this and that, and she'd be sort of running him down – and after about a week, when she was supposed to stay for ten days, she said, "Oh, I really have to go home, your daddy can't manage without me and I miss him," and I looked at her in amazement. I never actually said to her, "Why did you stop sleeping with him?" but once, when I was in my twenties, I did say, "Why are you like that with Daddy? Sometimes you seem so cold with him."'

This was the occasion Freda told Barbara that she had wanted only her, that if she had had other children she would not have been able to give her everything, and that she was glad that Barbara had not been a boy because then she would have gone off more with her father. It must have begun to come clear to Barbara, now that she was thinking about these things with the benefit of some distance and objectivity, that there was more to all of this than met the eye.

So, in her forties, Barbara was going back into her roots more, and recognising that Freda was a person in her own right, not simply a mother with whom she was joined at the hip, and she was set on the road to realising that Freda's extraordinary mothering was a response to some deep loss of her own. Although she did not talk to Freda about this, maybe pieces of the mosaic – the 'bits and bats that I picked up over the years' – began to fall into place. Her subconscious will have set to work, and because her childhood, her parents and the landscape of her beloved Yorkshire were so far distant, these things would soon begin to occupy her imagination as well as her thoughts.

'There is something about Yorkshire which is deeply ingrained in me and I am moved by the beauty. It stirs me inside. I had this joy in it as a child, but I didn't know I had it. Being away from it, on this side of the Atlantic, and looking across the Atlantic in my mind's eye, I see it with great perception . . . There's also a lot of nostalgia in it, that yearning, the memories of my childhood are bound up in it, so to me it's very emotional . . . A writer draws on memories.'

The objective frame of mind that enabled these perceptions

was made available by more than the physical distance that now stood between Barbara and home. She was looking out from the window of a completely different culture and her emotional side was beginning to understand that she had left something important behind.

Then, at some time in the mid-1970s, Barbara read an interview with Graham Greene in *Time* magazine, which suggested a bridge between her writing and what that something might be. 'Greene said, "Character is plot."' One sentence, which marked a turning point in her approach to writing fiction. 'It made me understand what writing fiction is all about. I realised why I'd gone wrong in those four books that I'd started and never finished. It was because I'd come up with a plot and then tried to fit people in. Character is destiny. Develop character, *then* the story comes.' It fits, of course, the message of the novels: once you know where you are coming from, the 'narrative' way forward is clear.

As in life, so in fiction. 'We are what we are, character is what he or she is; that is what creates their lives. We live our characters, don't we? One lives one's basic character. If you're a weak person then you're going to have a quite different life from someone who is strong. Then, also, adversity can develop your character or shatter you. So many of the strong characters grow from adversity. I'm fascinated by the indomitability [of people]. Life is hard, it's always been hard, it doesn't get any easier and it's not important that life is hard, what is important is how we overcome that adversity, or the adversities we have in our lives. I think I'm writing often about courage, conflict, inside oneself and between people.' As in fiction, so in life, for Barbara herself grew from Freda's adversity, because Freda would not have mothered her in the way that she did, had Freda not suffered her own and Edith's loss.

So, marriage to Bob, which took her away from her roots,

ironically served also to refocus her on them. Yorkshire was the bit of deep-truth reality and it produced Barbara's first offspring together in Emma Harte, her unique character forged in the landscape of Barbara's youth.

Marriage to Bob also served to internationalise the arena in which the woman of substance would operate, which is why, unlike Catherine Cookson, Barbara is able to write about people and place beyond her homeland, and appeal to readers beyond the country that formed her. By the time she came to write her first novel, Barbara was able to call up the spirit of place that imbued character not only among the crags of Ramsden Ghyll, where she ran as a teenager 'on her beloved moors high above Fairley village', but also among Manhattan's skyscrapers, 'a living painting of enormous power and wealth and the heartbeat of American industry'. Cut off in her Manhattan eyrie, free from the cultural baggage of either Yorkshire or America, she was in touch with the spirit of both. So it was that when she came to write *A Woman of Substance* it featured a woman in tune with Manhattan, but with her roots in the gritstone hills of the Yorkshire moors.

Barbara's imaginative homecoming – her writing of novels which mine the two cultures in this way – was an emotional experience of rediscovery, and determined that the books would themselves have strong emotional and psychological dimensions. It meant that she would put her characters in touch with their true selves, as she had done herself, releasing them from the value-vacuum of their modernist world by returning them to their roots, often to Yorkshire, her own home culture, but in Maxim West's case first to Berlin, where he discovers that his real parents are not even the German Jews he had supposed to have been lost to him by the war. Then, finally, Barbara returns him to the love of the one woman, Teddy, who has been a true mother to him but is not even a blood-relation, thereby returning to us the notion of absolute love at the root of identity, the value which will henceforth organise rich and powerful Maxim's life along other than pragmatic lines, and to which he will refer for his values. This was the point of the homecoming.

One morning in her early forties, Barbara called up her subconscious and provoked her imaginative homecoming with a series of questions: 'Well, you haven't liked these four books you started and you put them away. So, what do you want to write and where do you want to set it? And what kind of book would it be? And what is it about, basically? And who are the characters?

'I didn't ask the questions out loud, obviously, [although] I do that of myself today. Walk around muttering to myself. But my answers were: set it in England because I'm English and I know the English. No! Set it in Yorkshire, because you really know the Yorkshire people. I want to write about a woman who makes it in a man's world . . . when women weren't doing that, at the turn of the century maybe. I realised as I answered these questions and wrote them down on a yellow pad that what I was describing . . . was a saga. I was really talking to myself about writing a traditional, old-fashioned saga.

'That's what I did write. And when I realised I was going to write about a woman who makes it in a man's world when women weren't doing that, and that she'd be a business-woman, I thought I wanted to write about a woman who becomes a woman of substance. And I looked at that and I knew at once that I had my title, and I also knew that this would be the novel I would finish. I thought it was a damn good title, especially since it can have two meanings, sub-stance: money, and the development of her character. You know, certain people didn't like this title! And I said: "Well, I'm not changing it. I love this title. And I didn't ask your

286

opinion."' Barbara laughs. 'I mean, Bob loved it, but people are funny, you know? They think they know better than you if they're in the business. I said: "I will never change this title."

'A week later I had written a twelve-page outline and I was on my way. For the next three months I sat at my desk in our Manhattan apartment creating Emma Harte in my imagination ... and I dug back into my memory for the countless details about Upper Armley and other parts of Yorkshire where the story is set. After I had written 190 pages I went back into the past in the novel. I was suddenly at the turn of the century, when Emma was a little girl. I swiftly realised that I had to make a trip to Yorkshire to discover more. I telephoned my father and asked him to look for old books on Armley and Leeds in the local library, which he did. It was rather fortuitous for me that my mother had been clipping out a series on old Leeds, which the *Yorkshire Evening Post* was running at the time. I talked at length to my parents, relatives and older friends and spent hours in Leeds Public Library studying old copies of Leeds newspapers, histories of Leeds and Yorkshire, interviews with local people who had worked in the mills, and visited old mills in Armley and Stanningley, drove into the Dales, and tramped around Ripon, Middleham and Studley Royal.'

Bobby Caplin remembers this period well. 'I remember Barbara came back to Leeds doing research into the old mills for *A Woman of Substance*. She had gone away a provincial girl and come back a very, very smart and intelligent person. It was great to see. But basically she was still Barbara. This friend of mine had a birthday party during the time she was here. This was Ronnie Sumrie, her old boyfriend, and Barbara contacted Ronnie as an old friend (he was well married at the time) to come and explain to her about manufacturers and to introduce her to the top mill-owners.'

There's a story attached to Barbara's meeting with her old boyfriend after so long. 'I ran into Ronnie on a Saturday, on a Pullman train from Kings Cross. I was coming up to Leeds to see my parents, and to do more research for *Woman of Substance*. It must have been about 1976. Ronnie and I found ourselves sitting in the same carriage; it was the restaurant car. I was further down but facing him, though I didn't immediately recognise him, and then this most enormous girl sits down with him. And it's a totally empty carriage. And I see this flick of horror enter these eyes and he looked across at me, those blue eyes full of horror. I thought, My God it's Ronnie Sumrie! And I stood up and sort of edged towards him and said, "Aren't you Ronnie Sumrie?" And he leapt to his feet and said, "It's Barbara, isn't it?" And as the train was pulling out I answered, "Yes," and so we sort of embraced in this rolling carriage, and I asked, "Do you want to come and join me?" "Oh yes, I do!" he exclaimed. And he was sort of polite enough and gentlemanly enough to say to this woman, "Oh would you excuse me. I've met an old school-friend." We have often laughed about that.'

It was a homecoming that determined her future as a novelist and would in years to come be reciprocated by the city of her birth. In 1990, two years after the scholarly Brotherton Library in Leeds had been granted their request to become the Keeper of the Barbara Taylor Bradford archive (her manuscripts sit next to ones by Charlotte Brontë), she was honoured by Leeds University with the degree of Doctor of Letters, presented by the Duchess of Kent. (Five years later, the city of Bradford would bestow on her an honorary DPhil.) The great coincidence was that Alan Bennett was similarly honoured at Leeds on the very same day – 12th May 1990. They hadn't set eyes on one another in more than half a century, and had no notion that they had attended the same school until broadcaster

One of Barbara's favourite shots of her husband, Robert Bradford, with the book that realised a dream. Between 1984 and 1999 Bob set up no fewer than nine TV mini series based on Barbara's books, producing them himself (*Hold The Dream, Voice of the Heart, Act Of Will, To Be The Best, Remember, Everything To Gain, Love In Another Town, Her Own Rules* and *A Secret Affair*). The impact of such television exposure – six hours in the case of *A Woman Of Substance*, starring Jenny Seagrove and Deborah Kerr – was astronomical in terms of book sales.

Left and above: Jenny Seagrove as Emma Harte, the woman of substance shown here with Liam Neeson, who played Blackie O'Neill.

Right: Deborah Kerr takes over as the older Emma, with Sir John Mills as Henry Rossiter, Emma's financial adviser.

With Lindsay Wagner as Paula O'Neill, Emma Harte's granddaughter in *To Be The Best* is Sir Anthony Hopkins, who played Jack Figg, named in the novel after Barbara's great friend since the '50s: 'When they came to write the screenplay they decided to turn him into a much bigger character, cast Sir Anthony to play him, and made him the hero of the film!' recalls Barbara with delight.

Barbara with the stars of *To Be The Best*, based on the third book in the *Woman of Substance* sequence. Fiona Fullerton and Lindsay Wagner flank their creator. Standing are Christopher Cazenove and Sir Anthony Hopkins

Victoria Tenant as Audra and
Kevin McNally as Vincent in
Act Of Will. Barbara loved
Victoria Tennant's portrayal
and the fact that she made a
point of sitting down and
discussing the character with
Barbara in New York.

Barbara's beloved Bichon
Frises, Beaji and Chammi
(above), lay claim to being
part of the family.

Alan Bennett and Barbara receiving the degree of Doctor of Letters from Leeds University in 1990, thereby going some way to satisfying Freda's wish for her daughter to attend university.

Barbara with Reg Carr, the then Keeper of the Collection of the scholarly Brotherton Library in Leeds. In 1988 the library became the Keeper of the Barbara Taylor Bradford archive. Her manuscripts sit next to ones by Charlotte Brontë.

Barbara Taylor Bradford by Lord Lichfield. Her name now synonymous with the character she created: when she stands up in public she is the woman of substance.

Richard Whiteley pointed it out backstage, 'and,' stressed Barbara, 'we are both the same sign – Taurus!'

Bennett at once slipped comfortably into character, his ear for dialogue, on which he had made his reputation, as deft as ever. 'Alan was so funny,' said Barbara. 'When we were getting robed, he suddenly looked down, looked at his feet, and he said disconsolately in this broad Yorkshire accent: "I wish I had known it was going to be this posh. I've got dirty suede shoes, Barbara." I followed his gaze and he did have on a rather mucky pair of stained suede shoes, and there we were being put in these velvet robes! Tremendous talent, and his father a butcher in Tong Road!'

The idea of *A Woman of Substance* had brought Barbara home. Suddenly, yesterday was now, the past was about to become a part of her. It was the psychological unity she needed to move forward. Acknowledging her anchor in Yorkshire had given Barbara back her centre.

CHAPTER SIX

An Author of Substance

'No matter what the publishers do, they'll never stop this book. It'll go through the roof.'

Inevitably, with a book about a character who had so many parallels with that of her creator, Barbara was an important element in its sale to publishers. 'When I came up with the idea for *A Woman of Substance* I wrote a twelve-page outline and showed it to Bob,' said Barbara. 'After he read it, he said it was a great idea, but that I was undertaking something quite enormous . . . the story of a woman's life from childhood to old age. I agreed, and he nodded and said, "You'll do it." He always had confidence in me.

'Although I had an American agent, Paul Gitlin, who had been introduced to me by Cornelius Ryan and had sold my design books, I decided to show the outline to my English agent, George Greenfield.' Greenfield was a leading literary agent in the rather traditional agency of John Farquharson, and had represented Barbara since she was a journalist in London. He liked the outline, but the first bite he got was ironically from an American – editor-in-chief Betty Prashker of Doubleday, when she was in London scouting for new authors. They'd had lunch, she'd told him she was looking for traditional sagas, and he had given her Barbara's phone number in New York. A week later Barbara received a phone call from Carolyn Blakemore, senior editor of the firm, who invited her to have drinks with her and Prashker.

'When we met, they were very cordial,' Barbara remembers. 'They told me they wanted to see the outline as soon as possible; I agreed to messenger it down to Doubleday the following morning. Within two hours of receiving it, Carolyn was on the phone telling me, "It's the best outline I've ever read, bar none! When are you starting work on it? When can we see pages?" I remember saying, "It's now early June, I'll give you at least one hundred pages the first working day after Labor Day Weekend."'

In early September Barbara kept to her promise, actually delivering 192 pages, 'which was the whole of Part One and the beginning of Part Two. Three days later she called me and said she loved it . . . They wanted it. I told her to get Paul Gitlin to make the deal. That was in about 1975–6. I delivered the book two and a half years later.'

In London, in the meantime, Greenfield had shown the original outline to Mark Barty-King, the then editorial director of Granada Publishing, part of Lord Bernstein's media empire, which included the Granada TV company. Then, by chance, Barty-King bumped into Barbara on a street corner in New York City. He already had the treatment and was intrigued, so, when he was in town on other business and caught sight of Greenfield walking along with a beautiful blonde companion, he was pleased to discover that she was none other than its author. 'I was so pleased to meet her,' recalls Mark, a handsome six-foot-four operator whose nickname in publishing circles at the time was Captain Marvel. 'She was an attractive woman of course, but what was palpable when she spoke about the manuscript of *A Woman of Substance* was her determination. Call it Yorkshire grit or whatever. I had no picture in my mind of her until that day, but I went back to England determined to buy that book, which we did for what was then the highest advance ever paid for the right to publish a book in the UK – £55,000.

'What swayed me was her cold certainty that it would be a success – it was determination rather than enthusiasm. There's a subtle difference. You trusted her judgement. She did not for one moment doubt that it would be a success. The only question was whether one wanted to be a part of it, and there was no question as far as I was concerned.'

When I described the scene to Bob, he replied, 'That's the way she's always been. You cannot take her off the track; once she's on the track she's like a locomotive . . . I mean, she goes. She puffs away and it's very hard to move her or to stop her.'

In due course (1982–3), Bernstein sold the publishing side of Granada to the old English firm of William Collins and it became known as Grafton Publishing, with Mark still heading up the editorial side, Patricia Parkin as chief editor (fiction) and Ian Chapman, from the Collins side, its managing director. Before long (1989), they with the rest of Collins would be bought by Rupert Murdoch for £320 million, and renamed HarperCollins. 'We met Ian Chapman,' Bob recalls. 'Ian became a great fan, he was a great champion of Barbara's because he was running the company. So, everything seemed to be rolling along, everybody was with us. She has never had any real problems with the UK publishers, I must say. They are the same today as they were then, though the executives have changed, except Patricia Parkin. Patricia, she's unbelievable.'

When Bob saw the enthusiasm both in New York and London, he became involved in the book's marketing, but there was some delay before publication. 'I finished *A Woman of Substance* and delivered it to Doubleday, New York on May fifth, 1978,' remembers Barbara. 'I know it was May fifth because it was just before my birthday. It was very long, it was 1520 pages, the weight of a small child. Then they took a year to publish it, which was perfectly normal in those days. We had to cut three hundred pages without it showing,' she laughs. 'Try it! You've got to do a page here and a page there and five pages there and you usually do description and I actually did lose three minor characters whom we felt we could get rid of. That took them a few months. Then the Doubleday editor Carolyn Blakemore said to me: "We still haven't lost enough pages. There is one chapter where Emma's not in it." So, she took out a chapter in its entirety.'

The dropped chapter was about the First War. 'And I said, "My favourite chapter!" It wasn't really my favourite chapter, but I wanted her to feel bad because I really didn't want to cut any more pages. Anyway, she said: "It lifts out and you don't know it's gone." I read it and it was true: Emma waves goodbye to Joe Lowther and Blackie on the railway station, they go off to war and the next chapter is six months later, she's there with her children, she's going to work, she's doing everything she normally does. Then of course one of the reviewers said: "Bradford goes into great detail about" – whatever – "and yet hardly touches on the First World War!" But there is a funny story attached to this. When Bob was making the film of *Hold the Dream*, which was the second book in the *Woman of Substance* trilogy, some magazine in London asked me if I would write them a short story about the Emma Harte family and I said: "I can't. I'm in the middle of a novel . . ." They came back and said: "Are you sure you haven't got anything? It doesn't have to be about the Hartes. Anything at all will do. What about a short story?" And I told them I didn't write short stories any more. Only when I was a child. And they said: "Oh! We'll have one of those." "*No*," was my answer.

'But then Bob was at the studio in London and he called me and said: "Try and find them something. You must have something somewhere because they're driving me crazy and it's very important that we get this big spread about the mini-series." So as I'm talking to him, I said: "Oh, Bob. There's the lost chapter." I sent it. They ran it. And then the British publisher said: "Why have we never seen this chapter! We want this chapter. We're putting it in the book." And they had reason to bring out a new edition, which said: "For the first time, the missing chapter", with a big medallion on the jacket. Then they did the same in America, they put that chapter in the book. It is a good chapter, actually, and it was one that I liked because it was the war.

'So, they took a year with *Woman of Substance*, it was delivered May '78 and within a few weeks I had also sold them the idea of *Voice of the Heart*, so I actually embarked on that book without knowing that *A Woman of Substance* was going to be successful. The first novel came out in May '79 in the US, and by that time I'd worked since the previous September on *Voice of the Heart*.

'In England it [*Woman of Substance*] was published in 1980. This delay had something to do with Mark holding it back because he wanted to get a deal with the book club.'

Mark remembers serious problems in getting the jacket right. The jacket of a book is of course crucial marketing territory, and the marketing on a first-time author is key. If the publisher goes light on that aspect you don't stand a chance. Bob didn't think twice about pitching in on Barbara's behalf. 'She really put everything she had into that book and when it was finished I'd already taken over a great part of her promotion, supervising the marketing of it, working with the

American publisher. I was always concerned, right from the beginning, that I didn't want her to be buried. When you are a new author coming in they don't want to spend the money unless the book takes off right away. If you're not careful, they can put you on the list and let the book ride by itself – good luck! So, I was very attentive to that, and she was on the road for the book. She did ten cities promoting the book, and I made sure that when I wasn't there, there was always somebody with her. I mean, it was the first time round, you know?'

I asked how Doubleday took to that. As a mass-market-paperback publisher myself at that time I knew that with no track record in Barbara's career it would have been deemed interference on Bob's part. Most husbands of new authors would have been shown the door.

'The publishers,' he began. 'They get a little nervous. They don't always like me.'

'I haven't heard that,' I said encouragingly.

'I don't know,' he said, flicking his eyes up at me, wondering just where I was coming from. 'I think that they are a little nervous of me.'

'How did you handle it?' I asked, cutting to the chase.

'I am always very polite and have no case against them. The point is that I do push. I was up in Doubleday and I met with the marketing people, advertising people, and if I saw this wasn't moving I came in there every day, made my phone calls till I got what I wanted. Barbara and I, we had a good relationship with Doubleday, a friendly and good relationship once she got going. In the meantime I had to ask to see advertising and marketing. You know, the guy working on *A Woman of Substance* and the next one, *Voice of the Heart* . . . I saw them failing in terms of marketing. I didn't see the energy there, so I was up to see the Vice President of the marketing and advertising. I'm afraid I used to drive the guy nuts. "What's happening? Show me what you're doing." I couldn't understand what they were doing. He was telling me, "Well, we're going to do publicity [as opposed to advertising]. That was nothing. So I finally had to go down the hall to see Nelson and get him on board and get him to give them more money.

'So I was with Nelson Doubleday. This was in 1979, when Nelson Doubleday owned the company, and I was sitting with Nelson in his office. Not too many people do that. I always remember he took off his jacket, I was off with my jacket and he said, "Let's talk." I said, "Listen, I treat a book like a motion picture, so let's start from that point of view. I know you don't have thirty million dollars to spend on a book, but let's start from that point of view: you are selling a motion picture, that's the way you tackle it. He was telling me that he'd allocated about fifty thousand dollars to promote the book and, you know, I said he had to do more. He didn't like it, but I think down the line he quite admired me for my persistence.

'We also – that's me and Barbara – took a big risk because I spent money. I believe my statement is correct when I say that I was the first person in the publishing business in the United States to take full-page ads in the *New York Times*. There was nobody before me who did that. I started it and I'm sure there were heads turning in the publishing industry thinking this publisher is spending a lot of money, because at the time we didn't tell anyone. I didn't want the publishers to be embarrassed, so we wouldn't tell and the publisher always got the credit, they got their name on the ad, but I paid for it. I did it on most of the books. Sometimes the publisher paid. Every book Barbara has ever written has had a full page in the *New York Times*.

'*Recognition* factor. It explodes on the scene, people see the name, they see the book, I don't know whether they buy the book or not, but the key point was the media in New York, the media in Washington. They saw. They saw that huge advertisement and you didn't have to call them and explain to them what you were talking about, they knew it, they had all seen it. I also bought spreads in the *Publishers Weekly* [the trade magazine of US publishing], so everybody in the business would know.

'My key market, target market, where work was needed, was always the United States, never England. I hired PR firms in England, but I never got into the advertising. I only did it in the United States because you've got three hundred million people to collar there.'

As Bob was only too aware, everything follows on from the American market. First publication happened in hardcover from Doubleday in 1979. There was so much noise that paperback publishers were alerted, and foreign publishers took *Publishers Weekly* and saw the spreads Bob had placed. He was doing the bedrock business, from which everything, even interest from the movie industry, would flow. And as one by one people picked up on this interest and made offers for rights in their territories and markets, the whole thing began to snowball. He knew what would happen and never doubted for one moment that the product could carry the interest that followed.

'I didn't know it was going to be a bestseller; in the US nobody knew. In fact, the hardback only sold 55,000 copies, but 55,000 copies in 1979 was a damn good shot for a hardback book when you are unknown. Like 250,000 today.'

Separately, I asked Barbara what she remembered about the book's first publication. 'In hardcover there was a rush for the book in some of the big cities like New York and Florida, and in Texas and California where I went to promote it. Surprisingly, for not a lot of money being spent by the publisher, it did get to the middle of the bestseller list, about number five or six, and did very well for a first novel with very little money spent either on advertising or promotion. They did do an ad, which was very clever. The person who did it actually ran all the Doubleday stores because in those days Doubleday had bookstores of their own, and he was so convinced it was going to be a huge bestseller that he did this ad saying that if you don't like this book we guarantee your money back. They had no returns on it. I remember going to one bookstore in Atlanta and the buyer said to me, "No matter what Doubleday do, they'll never stop this book, it'll go through the roof." It did well, and then it was bought for paperback by Avon.'

Avon was Bob and Barbara's prize for all the hard work. With the paperback publisher's interest came another Bob into Barbara's life. 'Bob Wyatt discovered the book; he was the one who bought it for Avon from Doubleday for quite a lot of money in those days,' she remembers. 'They paid about $400,000 for paperback rights. It's not a lot today, but in those days it was. He discovered the book; the book somehow came to his desk, they bought it and he did a very clever commercial for television where they had an actress dressed up as Emma, as a little maid, with a white apron and a frilly cap. I remember the commercial very well because first of all it was a very gloomy moorland setting and then it became a village street with cobblestones and the sound of horses, and a carriage and horse going down. Then suddenly it cut to this – well, actually to the inside of a grand room and a young girl carrying a tray piled up with a silver pot and all that. Then it flashed some other actors' faces and there was a voice-over which said: "A great family saga in the grand tradition. The story of the servant girl who became a world power . . ." or whatever it was. So, it was Bob Wyatt and whoever he worked with who made the book – he remained my friend for years. When the book came out it quickly went up to Number One on the paperback list and it stayed on the list for more than a year. Everybody talked about it staying there so long, it just never went off the list. It sold three and a half million copies in that first year in paperback. That edition came out in the summer of 1980 and it came out in England in hardcover at that time, too.'

The ball was rolling. In 1985, with publication of *Hold the Dream*, Barbara's third novel, her English publishers were boasting that '*A Woman of Substance* has been in the top twenty listings since its publication.'

Wyatt remembers the Avon TV advertisement as 'a first for fiction at least', but he was wary of taking too much of the credit for spotting the book's potential. 'Page Cuddy was also important to *Woman of Substance*. She was maybe the first reader on it,' he said. 'And that whole women's fiction thing was Nancy Coffey's at Avon. She invented it.'

Avon, through the 1970s and early '80s, had been enacting a publishing revolution, putting out romantic historical novels with explicit sex scenes known in the trade as 'bodice-rippers'. Authors such as Kathleen Woodiwiss, whose first novel, *The Flame and the Flower*, they published in 1972, and Rosemary Rogers (*Sweet Savage Love*, 1974) were sweeping the market with sales in the millions.

These authors had a huge impact on the marketing of women's fiction in paperback form in both America and England, where they were marketed by a company called Futura. Hitherto, the big-selling fiction titles on a paperback publisher's list had been bought in rights deals from hardcover publishers. Paperback publication had relied on publicity generated by the hardcover a year earlier. But now paperbackers were publishing their own original lead titles and had learned how to market and promote them in far more bullish fashion than any hardbacker.

Particular attention was paid to the cover artwork. Wyatt remembers how Coffey gave the Woodiwiss bandwagon its initial momentum with a 'certain look, a very specific, hand-drawn typeface which became the standard for the industry. No one had done things like this until then. After the soft Woodiwiss books came Rosemary Rogers, and another entire look and feel was established for that writer.'

A Woman of Substance was no more akin to *The Flame and the Flower* or *Sweet Savage Love* than was Maeve Binchy's *Light a Penny Candle*, first published in England in 1982, but both Maeve Binchy and Barbara benefited from the powerful marketing techniques that had been honed in the publishing of these and other similar books. Maeve Binchy was actually marketed by the paperback publisher of Woodiwiss and Rogers in England, and Avon would not have had the sophisticated conceptual reach via cover artwork and TV advertising for *A Woman of Substance* had they not built up to it with these precedents.

So, everything was falling into place. What had seemed good from Doubleday, now seemed fantastic from Avon, but it was still only the start. Between 1984 and 1999, Bob set up no fewer than nine TV miniseries based on Barbara's books, producing them himself (*Hold the Dream, Voice of the Heart, Act of Will, To Be the Best, Remember, Everything to Gain, Love in Another Town, Her Own Rules* and *A Secret Affair*). The impact of such television exposure – six hours in the case of *A Woman of Substance*, starring Jenny Seagrove and Deborah Kerr – was astronomical in terms of book sales. But Bob was not simply building sales of books any more. He was creating a brand.

Very wisely he kept out of the first film, retaining only approvals of locations, stars and script on Barbara's behalf: 'I didn't want to be the one to produce *Woman of Substance* as it was her first book. Since it contained so much of her Yorkshire childhood and youth, she was very emotionally attached to it, so I backed off making the miniseries, believing we would have terrible arguments as she was so emotionally involved. I flew over six or seven times while the series was being made in England and watched at arm's length. There wasn't a great deal of financial experience on the part of the producer, and there was another company, a British company, involved. So when I stepped in they had somebody to talk to about business. We were able to pull it together. I gathered some money and we pulled it together and got out.

'And, yes, the film promoted the book. There was a line-up of about 165 independent stations in the United States who locked together as a network for two nights. Very unusual. The independent station in New York was WPX; in Chicago it was The Tribune. 165 stations, it was enormous. They were all independent programming, but they locked into CBS, ABC, whatever, to show *A Woman of Substance*! Very unusual. We were going against the networks!'

Both he and Barbara became more closely involved in *Hold the Dream*, which would be transmitted the following year (1985). It again starred Jenny Seagrove (this time as Paula Fairley, Emma's granddaughter), Stephen Collins as Shane O'Neill and Deborah Kerr as the ageing Emma Harte. 'Barbara had just finished the book, *Act of Will*, when I flew her into London and talked her into writing the script for *HTD*,' Bob said. 'Four writers had not been able to give me the screenplay I wanted. I was in trouble. Barbara had her mind still full of *Act of Will* and practically had to start writing *HTD* again. It was unbelievable. I've known a lot of writers, seen some of the great ones in Hollywood. She'd never set foot in a studio in her life, at least not to work in. She'd never written a script. I put her in a studio – Shepperton – locked up in a room with a script typist. Her routine was 6.30 a.m. to 9 p.m., seven days a week for six weeks. We were already in production. I had over one hundred people waiting to get a script, a proper script. The typist who I had for her, who was typing the script, waiting for dictation, she said, "I think, Barbara, we've got to start doing something here. You can't just sit there and think! We have to start somewhere, so what's the location, where are we, where do we start?" Barbara says, "We are in Yorkshire. Outside scene, country." The secretary says, "Who's talking to whom?" That's the way the script was written. It turned out to be a brilliant miniseries. And Barbara was amazing.

'She did it twice – she did it with *Hold the Dream* and *Voice of the Heart*.' The latter, starring Lindsay Wagner, James Brolin, Victoria Tennant and Honor Blackman, was produced by Bob in 1990. Later came the film of *Act of Will* (Liz Hurley, Victoria Tennant, Peter Coyote), which Barbara remembers as more madness: 'Bob had a line producer at that point called Aida Young, and together they got a script of *Act of Will* developed, but for some reason he suddenly found himself with two miniseries going – he was doing *To Be the Best* (with Lindsay Wagner, Anthony Hopkins, Stephanie Beacham, Christopher Cazenove and Fiona Fullerton) at that time, and had to go to Hong Kong to shoot it. Naturally, Yorkshire Television didn't want to do *Act of Will* – God knows why, I've never been lucky with Yorkshire Television – so he let it go to Tyne Tees Television.

'He is extraordinary. He not only makes the movies of my books but he manages my career. I do have a literary agent for the publishing, Mort Janklow, but Bob's the one who deals with the publishers on a day-to-day basis. My only sadness is that he's never been able to get a TV deal for *The Women in His Life*, which is my own favourite novel among those I have written. He's got a script that's fantastic, and they say "No", and so he doesn't even bother to show it any more. "Oh, but it's not going to work, it's a little boy and what actor wants only part of a role," etc., etc.'

Altogether, advances and royalties from Barbara's novels total over £50 million, and she is reputed to be earning at the rate of £8 million a year still today. Following the original deal with Doubleday for *A Woman of Substance* and shortly afterwards for *Voice of the Heart*, and their subsidiary paperback deal with Avon, a contract was drawn up in April 1983 for $8 million with Barbara's three English-language publishers – Doubleday, Bantam and Collins – for three new books, which would be *Hold the Dream*, *Act of Will* and *To Be the Best*.

Three years later, with only *To Be the Best* not yet published under this contract, another three-book deal with leading US publisher Random House was signed for £6 million, which Bob recalls was a deal Mort Janklow (at that time the world's most successful agent) put together. This covered US rights for *The Women in His Life*, *Remember* and *Angel*.

Then, in 1989, Rupert Murdoch decided to take over Collins in London and publishers Harper & Row in New York, and turned the whole thing into HarperCollins. An executive called George Craig, who had been Collins' chief accountant in London, went over to New York to run the new HarperCollins company there as Chairman of the Board. Three years later, in 1992, Barbara, who was already with HarperCollins in London of course, made a move from Random House to the US side of HarperCollins, the sister company of her London publisher. The idea was that she would have one English-language publisher with a massive, cohesive, worldwide marketing effort.

The move caused quite a stir for a number of reasons. Ex-*Sunday Times* editor and husband of Tina Brown of *Vanity Fair*, Harold Evans, was head of Random House at that time, and he was not best pleased. Dialogue reported in the press showed that Evans was having to swallow the bitter pill that his old boss at the *Sunday Times*, Rupert Murdoch, was taking his author. The real story, as far as Barbara was concerned, was never told, however, and makes more fascinating reading. Naturally, Bob was in the middle of it.

'The new contract with George Craig. That was a historic event, that contract,' Bob told me.

I said I had a report to the effect that in 1992 a contract was drawn for £17 million between Barbara and HarperCollins.

'There was a contract. First of all it wasn't in pounds, it was in dollars, and I will tell you exactly the way this happened. That's a very interesting story. I had made a picture called *Voice of the Heart* and I showed the picture to George Craig. I said, "Look, I want you to see it. I want you to see all the things I'm doing." He'd seen *A Woman of Substance* and *Hold the Dream.* So George said, "Bob, this is a terrific series, really great, looks fabulous." I said, "Well, actually I have a favour to ask you," and he said, "Well, what's the favour?" I said, "The favour is I want to sell this series to Fox Studios [in Los Angeles]." And he said, "What do you want me to do? What exactly are you thinking?" I said, "I think this series is going to be terrific for Barbara's books and I want to sell the series to Fox Television because you are both part of News Corp, so it's a natural." He said, "What did you have in mind?"

'I said, "If I can be quite honest with you, I would like Rupert Murdoch to see this film . . ." And he said, "Look, the man is flying all over the world and he never looks at films."

'Anyway, here's what happened. I got a call two weeks later when we'd just finished editing the film (George had seen the rough cuts). It was just finished. I get a call from George. He says, "What are you doing today? You're not going out of town?" I said, "No, I'm here." He said, "You've got an appointment with Rupert Murdoch at 5 p.m. today. Hire a private studio and get the film over there. He wants to see the film at 5 p.m. sharp."

'I almost fell over. So, I was suddenly scrambling, getting a copy of the film, hiring the studios . . . At one minute to five, in walks Rupert Murdoch with his wife, Anna. I couldn't believe it. We sit in the theatre and I just explained to him a little bit about the history of how I made the film and he looked at me and said, "Where's Barbara?" I said, "Well, she's home writing, she's working." He said, "I'm not seeing the film unless Barbara comes here, I want her in this room." So I had to call Barbara at home. I spoke to her. I said to her, "I don't give a damn whether you are in blue jeans or a sweater, you must come, we are not running this film until you get here." So Barbara scrambled to get dressed in a hurry and came over. She arrived and we sat in that screening room for three hours. Three hours and thirty minutes – it's a four-hour miniseries with the commercials. I'd made it in thirty-five milli-metre. I wanted the quality of the best motion-picture you can get. So I showed this in thirty-five millimetre to Rupert Murdoch. It was just unbelievable in this room on the big screen, and he sat there enthralled.

'After we'd finished, he just said, "*Amazing* piece of film, fantastic, I love it." So then he said, "I am taking you to dinner." It's a good thing Barbara had changed. "I'm taking you to Le Cirque." It was unbelievable, we walked in. Cham-pagne. He stayed there till about eleven . . . eleven thirty with Anna.

'I didn't discuss the next move with him, I was careful. The next day George called me and said, "We are flying to the west coast, I'm flying with you. I've arranged a meeting, I think you've got two or three days to get it organised and I have the studio organised. A meeting with Fox, with the whole group at Fox. Without Rupert, but with all the heads of departments. George had called the head guy and talked to him and he said, "You want me to look at this film? What do you think, I've got nothing else to do? I don't look at my own films here in the studio. I give them three minutes, ten minutes, and you want me to look at the whole thing with you? You're crazy." George said, "Well the big boss asked me to call and he wants you to look at it because I think it's something we want to talk you about – putting it on the Fox network."

'George wouldn't let me go by myself,' Bob continued. 'We arrived at Fox and there are fifteen, twenty guys in the conference room, every head of department showed up. At the meeting there was a lot of conversation. Yes, we're going to like this, yes we're going to do this. But, to make a long story short, nothing happened. Later, much later, they turned down the film. Finally, they said, "Well, it's a woman's programme, our network is a new network [which it was at the time], and we are looking to attract a young male crowd. Anyway, they didn't turn it down that day and George and I flew back to New York together, and that is where history was made.

'We were sitting together having drinks. We were friends, we were talking publishing, pictures and publishing – what he was going to do with the company, all his plans. And then George said to me, "What's with our contract with Barbara? Since we're talking business, what's with the contract?" At the time our contract must have been very close to expiration. I said, "Well, George, frankly I haven't figured out all the things but I think Barbara is entitled to get more money." He asked

306

me my opinion on a lot of things. I said, "Primarily I want to make a deal that equals her talent and what she's going to contribute to the company. You know, she wears the uniform of HarperCollins, she's totally HarperCollins, and I think it needs to be given consideration." So he said, "Look . . ." He took out his pen and started to write on a paper napkin, and said, "All right, let's work." So, we started doing the figures and he made all kinds of calculations – and George was a good calculator, he's an accountant by trade – we started working figures on a napkin, and we made our contract right there. It was for $24 million, the biggest contract ever written in the history of publishing at that time. The contract was initialled by me. I think Barbara has the napkin. That was our deal. England and America, English-language rights, but not foreign language. Right there on the plane, on the napkin. Mort knew nothing about it, and certainly he was startled when he heard about it.

'From then on there were all kinds of rumours, resentment and jealousy; nothing directed at Barbara, because they all loved Barbara. They said, "This guy Bradford, he's a killer," and the publishers were apprehensive with me. Marketing guys said, "If this guy comes up we are going to be in trouble." Which wasn't the point. My job is only to protect Barbara and to promote her career.'

CHAPTER SEVEN

Hold the Dream

'You will never die, Emma. We are both going to live for ever and ever, here at the Top of the World.'
Edwin Fairley in *A Woman of Substance*

Many to whom I have spoken during the research for this book come back to the question of what has made Barbara's career so incredibly successful. Clearly, good management and marketing have helped to make her books big bestsellers. Unusually, if not uniquely among novelists, her name is synonymous with the character she created, which suggests a certain iconic status. When she stands up in public she *is* the woman of substance.

Barbara's first novel was published at the right moment for reasons apparently outside the author's intent. 'It was looked on as a saga,' Barbara remembers. 'It was the first time anybody had written a matriarchal dynastic saga.'

It was a new twist on a highly respected genre that had been enjoying a resurgence of interest. The family saga had been popular since the 1920s. John Galsworthy's *The Forsyte Saga* was first published in 1922, the second part of the family chronicles, *A Modern Comedy*, followed in 1929, and, in 1931, a further collection appeared called *On Forsyte*

Change. John Galsworthy established the Edwardian family saga as a genre of its own and had many imitators. In the 1960s an English television production, *The Forsyte Saga*, became such popular Sunday night viewing that vicars changed the time of Evensong to accommodate it. It also swept North America.

The Edwardian era was very much in vogue for another reason, too. *The Country Diary of an Edwardian Lady*, a naturalist's diary for the year 1906, found first publication in 1977. It remained at No. 1 on *The Sunday Times* bestseller list for an unprecedented sixty-four weeks, and, again, was similarly popular in America. Then, in 1978, *Dallas*, the family saga of the oil-rich Ewing family, started life as a five-part TV miniseries and ran for thirteen seasons. Small surprise, therefore, that the two biggest audience hooks in Avon's TV advertisement for *A Woman of Substance* were 'saga' and 'Edwardian'.

But the woman of substance herself is a woman of our times with universal significance. The opportunities of which Emma Harte avails herself are those later made available to many as working-class exploitation was eased by socialism and socialism gave way to meritocracy, with an emphasis on the individual, self-respect, self-belief, autonomy and self-sufficiency.

Emma Harte, woman of substance, was in the vanguard of this great change. She led the way in her fictional world, and in her we see clearly the character that will be required to survive and prosper in the modern world in which we all now live. That is what has made her so popular and inspiring. In an interview in the early 1980s, Barbara said: 'I hadn't realised how much people find Emma's story inspirational. I've had almost two thousand letters. Many of them said, "Emma Harte has been an inspiration to us – we lead this kind of life, very tragic, very difficult, and if Emma can get through it, we can." The strange thing is that they spoke of her as if she is a living person.'

Meanwhile, in the more modern novels, such as *Voice of the Heart* and *The Women in His Life*, there is a keener focus on the psychological trauma attached to having arrived, having uprooted and risen in the modern world. Weighing anchor altogether can leave modern man bereft of a sense that there is anything beyond the projects he works on so obsessively in his effort to get somewhere. Once again, Barbara could write about it because she had come this route: 'These characters couldn't have been created without my own development,' she has said. She can write about it because her project has become her.

She resolves her characters' trauma in a return to their roots, but this return is no retreat – Barbara doesn't see herself as Yorkshire, any more than she sees herself as American. Her life has taken her out of belonging to any one culture. Hers is the emotional circle we all need to make, to live and be loved, to break free and grow to a position of autonomy and independence, from which we can independently assess, use and not be used by the values on which we were bred and which formed us. Is there any better recipe for fulfilment in life?

There is, however, one other – I think central – reason for the novels' incredible success, which has to do with the work of the subconscious which generates unawares the ideas that claw at a writer's inner self and drive his or her best fiction.

When Barbara first read in the manuscript of this book that her own mother, Freda, and two of her siblings were illegitimate, and that Freda had been confined with them in Ripon Workhouse from the age of only six, she was deeply shocked. Bob had walked into the room at the crucial moment of her reading this and, seeing her turn white, he had asked what on earth was wrong. For a moment she had been speechless.

When I began researching Barbara's life and works with her full cooperation she told me that *Act of Will* in particular was very autobiographical, a lightly disguised story of her parents. Her mother Freda's fictional persona is Audra Kenton in the novel, a woman born into a middle-class family, who lives in a long, low, eighteenth-century manor house called High Cleugh. However, Audra's father is not who she thinks he is. Adrian Kenton, 'a shadowy figure in Audra's mind', had died in 1909, when she was only two years old. It is subsequently revealed that Adrian was not her father at all, that she is illegitimate and that her natural father is 'Uncle' Peter Lacey, who has been 'benefactor and protector' of Edith and her children – Audra and her two brothers, William and Frederick. When Lacey dies in 1920 after being gassed in a First World War action on the Somme, Edith follows soon after, dying of a broken heart. Orphans Audra and her brothers are then split up. Frederick and William are shipped to Australia. Audra goes to work at the Ripon Fever Hospital and later goes to Leeds to work as a nanny and meets Vincent Crowther, who is Barbara's father Winston Taylor's fictional persona. They marry. They have a son, Alfred, who dies in infancy from meningitis. They then have a daughter, Christina, on whom Audra focuses her not inconsiderable energies to the exclusion of everything else (including her relationship with her husband) to ensure that Christina lives the life that fate denied her.

Just how closely autobiographical this novel is should now be evident to readers of my book, and the question arises as to why Barbara should have been so stunned when my research made the match between fact and fiction more complete. Clearly, her mother Freda told her a lot about her background,

311

enough to write the novel, but withheld the fact that Edith was not married when Freda was born, even though, in the novel, Edith is not married to Audra's father. So why did Barbara find herself inserting in this overtly biographical novel this true-to-life theme, which had been deliberately withheld from her and which she did not know to be true? The answer that suggests itself is that the story demanded it – in the context of the story it had to be so. Audra's abnormal ambition for her daughter needed a motivation. The tragic death of Alfie was not enough.

Oddly, however, Audra's illegitimacy is more or less immaterial to the story. It is there but little happens as a result of it. What motivates her extraordinary education of Christina is her loss of status – the fact that fate (the death of her parents) has robbed her of her true position in life. Audra's coming from a different world (being 'a lady born and bred') is the key factor in her relationship with her husband. It is what Vincent must overcome to make their marriage work. She goes regularly to gaze on High Cleugh, her 'memory place' in the novel, just as Freda went to the real High Cleugh (in reality a window onto Studley Royal) – but there is no suggestion that it is Audra's illegitimacy that has robbed her of her status.

In that sense, Audra's illegitimacy isn't even a very necessary ingredient of the story. So why *is* it there? The answer has to be that the issue of Freda's uncertain paternity lay buried deep in Barbara's subconscious and she couldn't help it rising to the surface when she began writing about Freda's father. Was it planted there perhaps by Freda when Barbara was small? Things stick in our minds as children, particularly when they represent a threat, as mention of the words 'fancy woman' in relation to Barbara's father did when she was a young child. Perhaps someone once said something about Freda's illegitimacy and she attached it in her mind to the sense of loss inherent in her mother's personality, which she admits always to have noted, blanking it out until she returned to her roots as writer, and then up it came.

This subconscious memory or feeling was evidently of significance to the writer right from the start. Seven years before *Act of Will* she had used the theme in *A Woman of Substance*, making it the principal motivation of her narrative. Emma Harte has an illegitimate child, Edwina, by aristocrat Edwin Fairley, who refuses to recognise her. It is this that motivates Emma's incredible story, and, I believe, Freda's too.

This is the grit in Barbara's oyster: her characters and plots are secreted in layers around the fundamental psychological problem of identity which Freda's childhood raises, so to produce the pearls of her success. The magic is that the conscious mind is not involved. The issue of Freda's identity at the root of the novels is a piece of the past – Freda and Edith's past, and now Barbara's past too – which rose up and appeared in present time, unbidden.

A Woman of Substance launched the whole theme of identity and loss that permeates so many other of her works, and which, to the benefit of their sales, mirrors a universal sense of loss and alienation that modernity engenders in those who cut themselves off from their roots, the values of home. There is this desire in almost all her main characters to find out or remind themselves where they are coming from, who and what they are, even in Maxim West, who comes closest to Bob, who as a German Jew was dispossessed of his identity as a child. It is *the* theme of Barbara's works and it is, too, *the* theme of her life, because her return to her Yorkshire roots in the novels is, as I have shown, an affirmation of her own identity. The process of its uprising from Barbara's subconscious (from the past unbidden) is akin to the process of the past rising up into the present to which she warms in places she visited in her

313

childhood, like Middleham Castle and of course Studley Royal, which she characterises as déjà vu. This, again, is caught up in Freda's problem of identity, her need to belong, expressed in the intense way she – orphan child of the York-shire she so loved – held these places when she took Barbara to them as a child, determined that her daughter would inherit their mettle, which she felt was rightfully hers.

Ten days after Barbara first read my manuscript, and many transatlantic phone calls later, she accepted what I had dis-covered and concluded: 'I don't know what I write. You know, I don't. You have found out what I don't know. You have analysed why I do what I do, but I never knew I was doing it.'

Mother and daughter had been unusually close, their re-lationship exclusive. 'No one knows what went on in that house with my mother. . .not even Daddy,' she said to me. 'I don't think my father knew any of this because my father and I got very close after I married and moved to New York, and we had some very intimate talks, but he actually never said anything about her past, only that she had a heart the size of a paving stone.'

If it is really true that Freda kept so much secret from her own husband, she must have been a troubled as well as a determined woman, unless of course Edith had told her that her father was the Marquess of Ripon and she had learned not to share the information because no one believed it. It is also possible that Freda did tell Winston, but he chose not to share it with his daughter because he didn't really believe it either, or didn't think Barbara would, or it had become an issue that upset the balance of his family (as it does in *Act of Will*).

All that we know from Barbara is that she did not know.

In the days before genetic fingerprinting, a birth certificate proved a person's identity. Finding no father's name appended, that only half an identity is known and recognised, can be

314

traumatic, and uncertainty can drive an adopted child to seek out its natural parents. In Freda's case there appears to have been no attempt to find out who her father was, which again suggests to me that she already knew.

Barbara found it difficult to accept the evidence of the work-house archival records because 'my mother loved Ripon so much, and she always spoke happily and lovingly of her child-hood.' But the evidence is cast iron, written by hand at the time in the official record books of Ripon Workhouse, and Freda's workhouse experiences do not discount the possibility of happiness at other times. If, as I think likely, Edith was 'looked after' as the Marquess's mistress up to 1910, when she was first listed as a pauper in the workhouse with her children, there would have been material happiness during Freda's first and most impressionable years – years when, as she told Barbara, she made frequent trips to Studley Royal. There would also have been love all around, for if Edith had three children by Frederick she would have loved him, and almost certainly he would have loved her. In the first flush of Edith's relationship with Frederick we have a credible image at which all Freda's subsequent education of Barbara was aimed. That image, regenerated by Freda's natural desire to recapture who she really was, is surely what drew her back with Barbara to Ripon, and time and again to the Studley Royal Estate after Studley Hall had been destroyed, as well as to other stylish, historic and aristocratic domains which helped to identify what kind of man she believed her own father to have been. Of course she told Barbara they were happy times. They were everything she had desired and lost, and now they were everything she desired for Barbara who she perceived rightly had inherited from Edith the fighting qualities necessary to reach far beyond the situation of her birth. The telling point about Freda's 'educational programme' for her daughter –

315

force-feeding her 'books and art and furniture and style' – is its uniqueness. No other mothers of Barbara's friends did anything similar for their daughters, and the desire to put Barbara where Freda felt she herself ought to have been explains why she did it.

Freda's longing to be who she really was drew her back to Ripon from Armley and from the loving arms of the Taylor family at the least opportunity. Indeed, as in *Act of Will*, it set her in opposition to her mother-in-law, Esther Taylor, who was concerned that Freda was giving Barbara ideas above her station. But it also engineered this extraordinary 'educational programme', preparing Barbara to resolve the problems of identity and loss Freda suffered, and be what Edith might have been. The nature of Freda's love was to give *and* take; for Barbara became the unwitting instrument of resolution in her mother's loss.

Pulling herself out of the dismay that my revelations had initially caused, Barbara began quite naturally to be concerned about what people might think. 'I don't want the press to say that I have grandiose ideas about my background when I have always said that I come from a working-class background in Leeds,' she said, and urged me to verify the link with the Marquess of Ripon. It is not possible, at this distance of time, to verify the identity of Freda's father, given that Freda's mother never married him, his name does not appear on Freda's birth certificate, and there is no DNA evidence available. Equally, it is not possible to discount the likelihood of Freda's father being the Marquess of Ripon either. In fact, it doesn't matter whether Frederick Oliver Robinson actually was her father. It doesn't matter if the evidence is circumstantial. What matters is that we can perceive the fiction arising out of the problems of identity Freda experienced as an illegitimate child, which she believed included loss of status, and how these problems determined the environment in which Barbara was brought up and so fed through to the novels. The happy ending is that Barbara's novels did resolve these problems.

It is nevertheless interesting to see how closely we can push the link with the Marquess, and logically we should start with the birth certificates themselves. To recap, Edith chose not to name the father of her first three children on their birth certificates. Refusal to do so was unusual even in cases where the true name of the father was unknown, because it openly declared the child's illegitimacy, and in those days illegitimacy was a terrible stigma, which a mother would do anything to avoid. I quote an official Ripon illegitimacy return, where in every single case a father is named, whether or not the nomination is authentic. I show that this is consistent with the case of Catherine Cookson, born illegitimate around the same time as Freda, whose birth certificate carries the name of a man believed to have been her father, even though Catherine's father had known her mother, Kate, for a very short time (by Kate's account they made love only twice) and had long left her in the lurch by the time Catherine was born. Finally, it is in research terms unusual to find no father mentioned on a birth certificate, for a name not to have been conjured up from somewhere, especially on three consecutive certificates, as was the case for Freda and her siblings Fred and Mary. So much so that it begs explanation.

The unusual nature of Edith's omission of the father's name suggests a definite decision on her part not to invent one, as if the truth about Freda's paternity was barred to her, that she had agreed to withhold the truth but that she was not prepared to compromise the truth with a lie. Second, it is implicit that Edith didn't name the father because she wished to protect his reputation, and therefore that the man had a reputation worth protecting, that he was a prominent man, a man of position

317

whose public reputation would suffer were his 'indiscretions' made known. The three children are bound to one father by the fact of his not being named on all three certificates (as well as by the telling choice of their names, to which we will come). It suggests that each birth was not a one-time fling, that Edith was enjoying an ongoing relationship with one man, whom she didn't name because she loved him enough not to want to blacken his reputation.

It is logically possible that the father was an ordinary married working man, as Barbara pointed out to me, but this is, I think, unlikely. First, would an ordinary working man have been able to persuade Edith he was worth waiting for through three births and the passage of seven years, from 1904 to 1910? And would such a man have refused to detach himself from his spouse? (The Marquess, a highly visible Catholic, had much to lose.) Again, how would an ordinary working married man have afforded to keep Edith?

Photographs of Edith in fine clothing, and the image handed down of her as 'the beautiful Edith Walker', plus the fact, passed down to Barbara, that Edith didn't work because she didn't need to, are consistent with the father of her children being Edith's benefactor and protector (again like Audra's father in the novel). Being pregnant so often would have made it impossible for Edith to hold down a job of the sort that she had been doing prior to 1904 – no one would have hired her as a domestic – yet she does not appear as a fully fledged workhouse inmate until after the birth of Mary in 1910 when the affair is abruptly curtailed. She could not have avoided the workhouse earlier unless she was supported – either married (which we know she was not) or being kept by the children's father. Again, Freda's very obsession with the lifestyle of the aristocracy, a main plank of Barbara's education, steers us away from the idea that her father was from the working classes.

This is all reasonable argument. All in all, it is likely that her suitor was a man of substance, someone able to support a mistress. So, we come to the question as to which man of substance in the locality is a likely contender.

Edith's first two illegitimate children, Freda and Fred, were given names derivative of their father's name as a kind of lifeline to their true identity. It was a typical thing for a single mother to do. In A *Woman of Substance*, Barbara chose to name Emma Harte's illegitimate child Edwina, after her aristocratic lover Edwin Fairley, not because she knew or suspected that Freda and Fred were illegitimate and had been named after a father called Frederick, but because the practice was recommended to her as typical.

There was only one contender among local families in village-sized Edwardian Ripon. And not only was Frederick a family name of the Robinsons of Studley Royal for hundreds of years, but their family seat was a magnet for Edith and Freda in Freda's childhood, and later Barbara in hers, and is a key focus of the novels. Finally, of course, Edith's naming her third child Mary ties down the father's identity more successfully than if she had chosen to name her Frederika, because Mary had been the name of Frederick Oliver Robinson's dear sister, who had died in infancy.

Of course, it is possible that the very visibility of the Robinsons led to Edith naming her first three children after his family as a sort of joke, but it seems unlikely that such a joke could be made to run for seven years. By naming her children Freda, Fred and Mary, Edith was, I am certain as certain I can be, pointing the finger of paternity at Frederick Oliver, and she made sure he had nowhere to hide by desisting from inventing a fictitious name for their birth certificates.

There was, as I have shown, access and opportunity for such a liaison. Despite the huge social gulf, there were certain points of contact between rich and poor in the Ripon community, among the most likely being places of entertainment, theatres, music halls and inns. Edith was living in the thick of these in Water Skellgate. The nearby music hall and theatre, the Victoria Hall, was the ideal venue for an ambitious girl – or indeed anyone – to mix with 'the other half'. Closer still was the Palace Theatre, a casino and a permanent funfair. Then, opposite the music hall, was the Masonic Temple, where the Marquess's very own Lodge held meetings throughout the year, and round the corner was the popular inn, the Unicorn, owned by the Marquess and a venue he would have attended regularly as the Masons used to gather there after meetings. The Masons were involved in providing support for the poor. Frederick Oliver was patron of the Ripon Home for Girls (ironically his own wife seems to have drawn him into this). All of which brought him in person into the poorer parts of Ripon that I have described.

Edith gave birth to Freda not at home but in an apartment in Water Skellgate, where the Studley Royal Estate owned numerous properties. Although the ownership of Freda's birthplace, 9A Water Skellgate, is not known, it was given over to the Irish League, an organisation of support for the many Irish workers in Ripon and very likely one sponsored by the Estate. Then there is Frederick's own need. His wife, Gladys Lonsdale, had all but deserted him, taking a long lease on a house in Sussex and becoming part of a racy arts set in London. Someone like 'the beautiful Edith Walker', with a love of glamour and a natural vitality, would have made an attractive antidote to such a loss.

Then there are the dates, which bring the business to ground in actual events in Edith's and Frederick's lives, the ups and downs of Edith's fortune coinciding with crucial dates in the lives of the Robinsons of Studley Royal. It is clear that when

Edith's long stint in Ripon Workhouse began in 1910 she was no longer operating under the protection of her lover. Something had happened to curtail their relationship. Is it pure coincidence that the first Marquess died the previous summer? Was this occasion for son and heir Frederick Oliver to tussle with his conscience and put his new responsibilities as Marquess before his love for the woman who had borne him three children? Or had the philanthropy of the first Marquess sustained Edith, charity which, with his death, simply dried up? What seems certain is that the Robinsons closed the door on any contact with Edith Walker for the time being, and handed the matter over to the State.

While still incarcerated in the workhouse, Edith married John Thomas Simpson in what looks very like an arrangement to settle the matter for good, at least to all appearances. Edith travelled from the workhouse to Ripon Cathedral to marry him, and travelled back again. Why did she not move in with Simpson straightaway if this was a love match? Why, when her next child was born in 1912, did the registrar enter different addresses for father and mother – Edith in Bedern Bank (another property owned by the Estate), Simpson in Bondgate? Simpson was a labourer, six years younger than Edith. Was he perhaps in the employ of the Estate, a solution that would distance the problem of Edith by keeping her in a more profitable fashion and at arm's length? There have been Simpsons in Ripon for hundreds of years. The name takes us back to the John Simpson who, as I said, was head gardener during John Aislabie's landscaping of the Studley Royal Estate, and in Edith's time the family gardening business was still kept up by a John Simpson who lived with his wife Elizabeth at Cemetery Lodge in Kirby Road, until 1919 when George and Lavinia Simpson took over.

Then there was Edith's sojourn in Morpeth, miles away

to the northeast in Northumberland, which would have distanced Frederick's little 'problem' still further – geographically speaking – and which once again provides a connection with the Studley Royal Estate, the Morpeth 'solution' later becoming a major part of the Studley Royal charitable portfolio, a whole project where workers could relocate, live in their own house and work a piece of Studley Royal land in far-off Northumberland. The connection is further boosted by the discovery that here, at Winton House, was also the headquarters of the Freemasons, in which the Robinsons of Studley Royal were leading lights and through which Frederick may well have funnelled the 'solution' in the first place.

Barbara was staggered when she read about Edith's sojourn in Morpeth. 'I saw "Morpeth" and I remembered immediately Mummy saying that when she was little – I don't know how old she was – she used to go to Morpeth.' Freda's half-sister Frances was born there in 1914. This period could well have been another particularly enjoyed by Freda, for if Edith's marriage had begun as stratagem, it appears that she and Simpson were now (perhaps as a result of the Morpeth experiment) close. Certainly when they returned to Ripon they continued to live together, taking up lodgings at Yorkshire Hussar Yard, where Edith's sixth and last child, also called Edith, was born in 1917.

It is likely that Freda did not live with them in the Yard, and one should remember when considering how happy Freda told Barbara she had been in Ripon, that despite the roller-coaster ride Edith led her, she was part not just of Edith but of the extended family of Walkers who lived in the city, particularly those living at 8 and 9 Bedern Bank. Besides, up to 1910, along with the patriarch John Walker there was Edith's brother Thomas, his wife Frances, son Jamie, and three daughters, Lillie, Elizabeth and Minnie in residence. Elsewhere in Ripon, until his death in the First War, lived another brother Joseph and his wife, Ruth Matilda, who had been witness at Edith and Simpson's wedding. Then there was Gabriel Barker, either Freda's Aunt Elizabeth's or Aunt Minnie's child, who also lived in Bedern Bank for some time, and very probably there were others that have eluded my researches. Ripon, with its winding streets and strong family continuity, would have seemed a magical playground even allowing for the workhouse years.

Being the eldest, Freda would never suffer the ultimate degradation of deportation, when, in 1924, it all fell apart. In that year, three of the children were put in the hands of the NSPCC, who delivered them to Dr Barnardo's Homes, thence to Canada and Australia. Why in 1924? Surely because Frederick Oliver Robinson, Second Marquess of Ripon and the last of the line, died the previous year and Edith's protection finally came to an end. No one else could afford to look after them.

Subsequently, John Simpson would decline and meet an early death in Ripon Workhouse, but the spirit of Edith found its medium in Freda, who went forth to ensure that the emotional circle of her mother's life was as yet far from complete, and that her dream would at last be realised.

A lifetime later, Freda herself died, but only after the publication of *A Woman of Substance*, as if to announce that her job was done. She saw the book become a success; she saw the two ends of the circle meet. The midwife's feelings are never thought of in the joy of a mother's first birth, but Freda's part in this is nurse *and* mother. The scope of her life took her from the Edwardian era into which she and the woman of substance were born, right through to the modern world in which people like Maxim West operate. Her feelings following her daughter's success are unimaginable, and one must believe

323

that she had a few things to say quietly to her mother, Edith Walker, whose ambitions lie at the fount of all this.

'My parents died in 1981 just as I was at the end of *Voice of the Heart*,' Barbara told me. 'They were both alive to see *A Woman of Substance*. I remember after my father died in the October I said to Mummy that I wanted to bring her to New York, but she insisted on staying where she was. I suppose an old person doesn't want to be moved. She said, "Just do me a favour, go back and finish that book, I want to read it." But she died before it came out, just five weeks after he did. She was seventy-seven, and I think that she had decided she didn't want to live any more.

'The housekeeper rang me one day and said, "Your mother's really not well," and the doctors came and were going to put her in hospital. I said, "Can she come to the telephone?" And Brenda said, "Let me go and see if Rose and I can help her to come to the telephone." There was another lady there, a neighbour called Rose. I heard the phone go down – of course this is six o'clock in the morning in New York, six thirty because it was about eleven thirty in Yorkshire. They were ages and I'm then saying, "Hello?" and there isn't anybody there – and eventually Brenda came back, about five minutes later. She said, "I don't really know how to tell you this, Barbara, but your mam just died."

'I couldn't believe that she'd just died. I didn't want to believe she was dead at all. Then I thought she must have died earlier and that they had talked it out between themselves because they hadn't known how to tell me. I felt so helpless. Later I found out what they told me was true. Mummy had been sitting in the chair and she had smiled at Rose. Apparently Rose said, "It's me, Mrs Taylor," and my mother didn't answer her, but finally she looked at her and she smiled, and it was a very radiant smile and she had extremely blue eyes.

Rose said, "If you could have seen your mother's eyes, they were shining." She told me this when I got to Leeds. Apparently Rose sat and held her hand and talked to her for a few minutes and Mummy looked at her and said, "God help me, please have God help me." Rose said, "He will if you're asking him." She smiled again and Rose said the eyes seemed bluer than ever, and she said, "She smiled at me a third time and her face was so radiant, I don't know what it was that she saw." And Brenda was there – this lady who'd looked after Mummy – and Brenda said, "Well, what she saw was Winston" – my father – ". . . she was happy when she died."'

Emily stared at Paula. Tremulously she said, 'I've always been afraid of death. But I'll never be afraid of it again. I'll never forget Grandy's face, the way it looked as she was dying. It was filled with such radiance, such luminosity, and her eyes were brimming with happiness. Whatever it was our grandmother saw, it was something beautiful, Paula.'

Hold the Dream

ACKNOWLEDGEMENTS

Principally I would like to thank Barbara Taylor Bradford and Robert Bradford for their friendship and generous contribution to this book; also Amanda Ridout for suggesting the idea in the first place and Patricia Parkin for performing the requisite balancing act to see it through.

Among those who kindly agreed to be interviewed I would like to thank especially Billie Figg, whose journalistic skills were rigorously applied to our discussions, and who, unprompted, offered me some of her own early interviews with Barbara, as well as many personal memories. My thanks are also due to her husband, Jack, whose gentle humour added a very welcome dimension to the book. I would also like to thank Doreen Armitage, Judy Blanchland, Mark Barty-King, Arthur Brittenden, Bobby Caplin, Margery Clarke, June Kettlelow, Roderick Mann, Shirley Martin, Frederic Mullally, Beryl Thompson, Vera Vaggs, Richard Whiteley and Bob Wyatt.

Once it became clear that the book would require as much detailed archival research as interview I fell into the hands of a number of professionals in institutions and libraries in the North of England, to whom I am very grateful. In particular I would like to acknowledge Doris Johnson, whose scholarship in archival science and incisive appraisal of material was not only useful to the book, but taught me a lot along the way.

I would also like to thank the following: *Architectural Digest*; Lonnie Ostrow of Bradford Enterprises; Chapel Allerton Hospital; Mr B. C. Smith of the de Grey and Ripon Masonic Lodge; Mr P. Allen of the Morpeth Masonic Lodges; Granada Television; Guardian Newspapers; Leeds Library; Mike Yaunge and Ros Norris of the Ripon Local Studies Research Centre; Morpeth Library; Northallerton Library; Northallerton County Archives; Northumberland Record Office; Observer Newspapers; Adrian Munsey and Samanta Elliott of Odyssey Video; Probate Sub-registry: York; Public Record Office: Kew; Registrar: Leeds, York, Harrogate; Christine Holgate of Ripon Library; *Ripon Gazette*; Ripon Workhouse; Scarborough Library; West Yorkshire Archives (Sheepscar and Wakefield); Yorkshire Archaeological Society; Louise Male of the *Yorkshire Evening Post*; David Hartshorn at *Yorkshire Post* Archives; and Doreen Stanbury and Pat Smith for their typing.

There are, in addition, some crucial literary sources, works that I have acknowledged when quoting from them in the text: in particular, by Keith Waterhouse – *City Lights* and *Streets Ahead* (Hodder Headline, 1994 and 1995); Alan Bennett – *Telling Tales* (BBC Worldwide, 2000); Paul Murray Kendall – *Warwick the Kingmaker* (Allen & Unwin, 1957); Jack London – *The People of the Abyss* (Journeyman Press, 1977); Anthony Chadwick and Beryl Thompson – *Life in the Workhouse* series (Ripon Museum Trust, 2003); Jim Gott – *Bits & Blots of T'Owd Spot* (Crakehill Press, 1987); Judith Summers – *Soho* (Bloomsbury, 1989); and Daniel Farson – *Soho in the Fifties* (Pimlico, 1993). Other works which have informed my text include *Lancaster and York: The Wars of the Roses* by Alison Weir (Pimlico, 1998), *Oliver Twist* by Charles Dickens

(Penguin Classics, 1966 etc.), *Wuthering Heights* by Emily Brontë (Penguin Classics, 1965 etc.), *Backing Into the Limelight: The Biography of Alan Bennett* by Alexander Games (Headline, 2002), *Empty Cradles* by Margaret Humphreys (Doubleday, 1994) and *Neither Waif Nor Stray: The Search for a Stolen Identity* by Perry Snow (Universal, 2000).

I would like to acknowledge all the picture sources. Maps are by Ordnance Survey and photographs are by or from Ripon Library, the Robert and Barbara Taylor Bradford Photographic Archives, Leeds Library, Yorkshire Tourist Board, Skyscan Balloon Photography, English Heritage, Cris Alexander, Lord Lichfield, the *Yorkshire Post*, Columbia Pictures and Odyssey Video.

While every effort has been made to trace copyright sources, the author would be grateful to hear from any unacknowledged ones.

NOVELS BY
BARBARA TAYLOR BRADFORD

1980 A Woman of Substance
1983 Voice of the Heart
1985 Hold the Dream
1986 Act of Will
1988 To Be the Best
1990 The Women in His Life
1991 Remember
1993 Angel
1994 Everything to Gain
1995 Dangerous to Know
1995 Love in Another Town
1996 Her Own Rules
1996 A Secret Affair
1997 Power of a Woman
1998 A Sudden Change of Heart
2000 Where You Belong
2001 The Triumph of Katie Byrne
2002 Three Weeks in Paris
2003 Emma's Secret
2004 Unexpected Blessings

INDEX

PS 3552 .R2147 Z68 2005
Woman of substance / Piers
Dudgeon

STREET STORIES
The World of Police Detectives

STREET STORIES

The World of
Police Detectives

ROBERT JACKALL

HARVARD UNIVERSITY PRESS
Cambridge, Massachusetts, and London, England

2005

Library
University of Texas
of San Antonio

Copyright © 2005 by Robert Jackall

ALL RIGHTS RESERVED

Printed in the United States of America

Library of Congress Cataloging-in-Publication Data
Jackall, Robert.
Street stories : the world of police detectives / Robert Jackall.
p. cm.
ISBN 0-674-01709-9 (cloth : alk. paper)
1. Criminal investigation—New York (State)—New York—Case studies.
2. Detectives—New York (State)—New York. I. Title.
HV8073.J23-2005
363.25′097471—dc22
2004060793

**Library
University of Texas
at San Antonio**

For my students at Williams College
With great appreciation and affection

CONTENTS

STREET STORIES
The World of Police Detectives

I

IN THE FIELD

The call came into the Midtown North squad at 2340 hours on October 17, 1991, just as detectives were packing it in for the night.[1] The evening before, some kids had done a stickup of a twenty-four-hour delicatessen at 55th Street and Lexington Avenue. The deli's owner was an Arab man, accompanied that night by another sandwichmaker, a man who had his own shop in Brooklyn. The robbery went bad. The kids shot and seriously wounded the deli owner and killed his companion. Then they fled in a van. The deli owner said that three Hispanic boys had done the robbery, but his descriptions were vague. A couple of leads had not panned out. A local derelict named "Moose" claimed to have seen the whole show; but after extensive interviews, squad members discounted his story as the fabrication of someone looking for a handout from cops.* A truck driver dropping off early editions of the morning papers at a

---

* Moose is a pseudonym. Throughout this book, pseudonymous names are placed in quotation marks at their first usage.

newsstand saw a van screeching away heading west on 55th Street, but he was unable to get the plate number.

All the detectives were out of the squad room, so I answered the phone. The caller, sighing and weeping, plunged immediately into his story, the words pouring out one on top of the other. It was "Harry," from Brooklyn, who had read in the newspaper about the death of the sandwichmaker. Harry, who had bought his lunch from this man for years, had overheard one of his employees talking about the murder, a boy named "Ramon," who was like a son to Harry, and sometimes babysat his children. When questioned, Ramon had told Harry that he had heard "Apples" talking about the stickup. Apples had been shot and wounded in the left arm by one of the two men in the deli. He checked into Wyckoff Hospital in Brooklyn, claiming that black robbers had mugged him. Apples also said that two other boys had gone into the deli with him. Another had stayed in the van, a vehicle that Apples had "borrowed" the previous Friday night from his workplace, an auto-leasing shop in Manhattan. Harry was calling Midtown North to ask if the police wanted to find and question Ramon.

Detectives George Delgrosso, Alex Renow, Robert Chung, and Pete Panuccio drifted back through the squad room on their way home, only to find themselves tumbled into a long night of interviewing Harry and searching for Ramon. The detectives caught up with Ramon the next day, and he quickly named the whole crew. Apples, Roberto, and "Sonny" had gone into the deli; "Dexter" had stayed in the van. Detectives found Dexter first, accompanied by a beautiful girl. When the detectives asked Dexter if she was his girlfriend, he said: "Who, her? Hell no, I just fuck her." Dexter led detectives to Apples and Roberto. Both admitted their participation in the robbery and put the gun in the hands of Sonny.

A day later, after getting a warrant to search 16-year-old Sonny's apartment, detectives found a financial statement from Beth Israel Hospital in Manhattan and tracked him down there. He was visiting his mother, who had just given birth to another son. Sonny at first denied that he was in fact Sonny; then he denied any knowledge of the crime. Then he said he had thrown the gun off the Williamsburg Bridge. In the end, Sonny confessed to the shooting, arguing that the two Arabs had started all the trouble by pulling a gun instead of allowing themselves to be robbed peacefully. Sonny had his 7-year-old nephew deliver his 9-millimeter weapon to police in a brown paper bag.

A random crossing of lives leads to the overhearing of a conversation that prompts an anguished call to the police. Detectives track down a witness who heard his friends bragging about a robbery. The witness names the culprits. The youths admit their participation in the crime and give up the shooter. The shooter confesses. Suddenly savagery committed in the dead of night becomes illuminated.

The Warrant Squad of Central Robbery, headed by Detective Sergeant Dennis Boodle, routinely crashed apartments all over the city at 0500 hours every morning, searching for long-time absconders from multiple bench warrants. Hitting locations at such early hours raised the likelihood of catching wanted criminals, who are invariably night people, but did little to endear squad members to the families or neighbors of their quarries. The work carried dangers, even when the warrants were for misdemeanors; many sections of the city resembled armed camps. Some absconders managed to beat the system for years, by learning how police think and then planning dodges and evasions accordingly.

As 1991 was coming to a close, with many wily absconders still in the wind, warrant squad detectives decided to think like criminals. They came up with Operation Jackpot. They rented a large office on the second floor of the Port Authority bus terminal at Eighth Avenue and 42nd Street and blocked out the floor-to-ceiling windows with blue paper. At the door, they erected a huge sign reading Casino Club. They sent out letters to the fifty most wanted absconders in the city, informing them that they had won $50 and a free round-trip bus excursion to Atlantic City to go gaming. The winners were invited to come to the Casino Club on December 20 to collect their money. All winners who appeared were also included automatically in a grand drawing for a Sony television.

On the appointed day, Detective Vinnie Valerio donned a tuxedo, slicked back his jet-black hair, and waved a huge cigar while effusively greeting people who came to or passed by the Casino Club. Detective Tony Gonzalez wore a Santa Claus outfit complete with white beard. He rang a huge bell so vigorously throughout the afternoon and early evening that it eventually caused blisters on his hand. Inside the room, Detective Stacy Weiss acted as the receptionist, checking winners in as they arrived. Behind her, Detectives Imani Booker and Debbie Lawless, posing as secretaries, pounded away on ancient police department manual typewriters. Christmas carols, featuring "I'll Be Home for Christmas," played in the background.

After Detective Weiss checked the letter that a winner gave her, she motioned to Detective Sergeant Boodle, who congratulated the winner and then quietly consulted the manila folder that contained mug shots, arrest records, and warrants. Sure that he had an absconder in hand, the sergeant escorted the winner to the "cashier's office" in a back room to receive his reward. (When one winner asked Detective Weiss about the Sony television, she responded: "Oh, well, that's for the grand jury. Uh, I mean, the grand drawing.")

The waiting area in the main room contained several chairs. Companions who accompanied the winners lounged there while waiting for their friends to collect their money. Lieutenant John Walsh and several other detectives, an Associated Press reporter, and I, all posing as winners waiting for the bus to leave for Atlantic City, chatted amiably about inconsequential trifles, engaging one another and winners' companions in discussions about the weather, the recent renovations of the Port Authority building, the lack of reading materials in the waiting room, the vagaries of the free bus schedule, and the kinds of gambling one could do in Atlantic City. Several photographers from local newspapers snapped pictures of the winners, claiming they were for promotional purposes. The air of normalcy in the waiting room gulled other winners entering the Casino Club.

At one point, two men came into the club to claim winnings. A young man with a decorated fade haircut slapped his companion on the back and said: "Well, Mr. Jones, get your $50 there, and I'll see you outside. I got to go out cuz this here cigarette smoke is botherin me." Boodle quickly realized that the man about to leave was the wanted culprit. Two detectives quietly escorted him to the cashier's office, while Boodle told his bewildered friend to go home.

Inside the cashier's office, detectives, supervised by Detective Sergeant Ed Keitel and later accompanied by the Associated Press reporter and myself, waited to arrest, search, and handcuff the winners, who were then hustled out a back door to a paddy wagon. Most of the winners were incredulous at being duped, and chagrined to have been made into suckers. One said: "You mean there ain't no trip? This whole thing a fake? This here a hoax?" Another said: "Do I still get my $50?" Some were mildly indignant: "Well, you coulda come to my house and I'da come wit you."

One woman, a search of whose person revealed seven bottles of crack, was downright outraged by the police deception. When the Associated Press journalist asked her if she had been surprised by the operation, she lashed out: "What do you think this is? Get the fuck outta here, you silly bitch. What country is this? Is this South Africa? Is this Nazi Germany? What country is this? You all are the worst mothafuckas on the face of the earth. I hate you fuckin mothafuckas—all you police, scientists, and whatever the fuck you is—you is the lowest scum and slime on the face o the earth. What the fuck is you waiting for? Who is your president these days? Who is the fag? What is your president's name, or prime minister, or prime rib roast mothafucka, whatever the fuck he is. How does you get out of this country? How does you go to another world? Why don't you arrest each other? If you think you ain't gonna get slammed by everything out there that all of us is, you got another think comin."

But such outbursts just enlivened the festive atmosphere at the Casino Club. Detectives delighted in the reversal of roles. The ethos of the street is to get over on others, to make saps out of marks and exult in their humiliation, to outwit and thus defy authorities. The Casino Club sting allowed detectives to play on the larceny in the hearts of criminals, which led to their undoing. Reporter, photographers, and fieldworker became active participants in the deception, as the price of admission to watch detectives get over on those who make a living by getting over.

Detective John Bourges of the 34th squad faced a puzzle. Beginning in January 1992, detectives in the precinct had caught an unusually large number of violent crimes, including several murders of extreme brutality. The spree started with a wild nightclub shooting by a major drug dealer. Then a kid was shot twice in the head and run over with a car for good measure. Several seemingly random drive-by assaults and other wanton shootings followed in short order. One trail of blood led back to the apartment of the mother of a key suspect. A pitched gun battle in the middle of a busy street left 9-millimeter shells so thick that detectives had to kick them out of the way to avoid tripping. A restaurant where drug dealers congregated was fire-bombed, and several wild car chases compounded the mayhem in the streets.

Bourges had caught a few of these cases. In his investigations, the names of some key local players repeatedly surfaced. Curious about how far that pattern extended, he collected files from scattered cases caught by different squad detectives, all involving street violence, guns, or drugs. He discovered that the same players appeared again and again in a whole skein of incidents, sometimes as suspected assailants, sometimes as victims. He made up a large wall chart, with pictures, case outlines, and known links among the players. It was a brilliant first attempt to make sense of seeming chaos, and an electrifying moment for a fieldworker trying to understand criminal investigators' ways of knowing.

Working with Detectives Garry Dugan of the Manhattan North Homicide Squad and Mark Tebbens of the 40th squad in the South Bronx, Bourges learned that an unsolved highway murder of a college kid in the 30th precinct was linked to the crowd of suspects. So was a murder in Brooklyn and a notorious quadruple homicide in the Bronx. The seemingly wanton violence in the 34th turned out to be part of a three-borough-wide street war between rival factions of the same gang of drug dealers, along with a sideshow war between one of these factions and an insurgent crew trying to muscle in on business.

In an unprecedented development, the district attorneys of New York, Kings, and Bronx counties agreed to consolidate their cases under a previously untested 1979 state conspiracy statute and to try the case in Manhattan. I sat on the bench with Justice Leslie Crocker Snyder during much of the joint trial of nine members of the Wild Cowboys gang—a unique opportunity to see a case from initial violence on the streets all the way through the tangle of the legal system.

These episodes were three among many defining moments during my long fieldwork with New York City police detectives. I began with a set of questions about the nature of public investigations into events that have crucial social consequences: How do officials charged with important investigations determine "truth"? What are the structure, social psychology, ways of knowing, and occupational ethics of official investigative work? Such work is critical to the functioning of any highly organized society, and an enormous amount of it gets done in the United States. The 9/11 Commission examined the plots that hatched the atrocities of September 11, 2001, and the institutional failures of different U.S. agencies that contributed to the terrorists' success. Congressional committees probe scandals in the financial and pharmaceutical industries, in the White

House, or in Congress itself. Federal Aviation Agency investigators conduct exhaustive studies in the wake of every airplane crash, to determine who or what was to blame. New York City water-main inspectors unearth whole streets whenever a pipe bursts to ascertain the cause of the break. And police detectives investigate crime, particularly violent crime.

In all of these instances, investigators try to establish particular "truths" on the basis of which "responsibility" for actions can be assigned. Different occupational groups develop their own ways of determining what constitutes truth in their worlds and, consequently, what institutional structures or what particular persons are going to be blamed or rewarded for specific actions. So this book, as part of a larger intellectual project, examines the institutional and organizational contexts in which crucial decisions about what constitutes the truth of matters are made;[2] and, in the paradigmatic case of detectives who seek to discover who did what to whom, it asks how such truth gets transformed into public proof in the judicial system.

I began fieldwork with the NYPD in September 1991 at a temporary Midtown North stationhouse on 42nd Street near Tenth Avenue. Midtown North, the sprawling, variegated 18th precinct, is a cross-section of both high and low New York City, and my time there introduced me in a bracing way to the world of urban policing. Police make the city safe for everyday gullibility, foolishness, high-spiritedness, and anonymity—for respectable deals, romantic liaisons, and the lawful and unlawful pursuit of tawdriness. Early on, I spent days accompanying a uniformed police officer on his beat at the western edge of the city's old "Tenderloin" district, seeing through his eyes the shifting intricacies and intrigues of street players and coming slowly to understand how dedicated uniformed cops thwart crimes every day through sheer cunning and detailed street knowledge.[3] But when robberies, homicides, and other acts of violence shatter this protective shield, uniformed officers call in detectives to interview witnesses, interrogate suspects, and piece together the story behind the yellow tape guarding the crime scene.

In broad daylight and into the wee hours of nights, I followed the four-days-on-two-days-off schedule of one of the Midtown North detective squad's three teams.[4] I accompanied detectives as they knocked on doors in tidy apartment houses and flophouses alike to glean knowledge about crimes. I watched them investigate "legitimate" fronts that provided markets for the spoils of robberies, or laundered cash, or screened insurance scams. And I rode with them as they delivered prisoners to Central Booking's holding cells and sat with them watching the Manhattan Criminal Court's all-night sessions, the famous lobster shifts that, during those years, arraigned hundreds of criminals each night, seven nights a week, in rapid-fire succession.[5]

A couple of months later I started fieldwork with the Central Robbery squad of the New York City Transit Police (NYCTP), then a separate authority from the NYPD. Located in the New York State Parole Office building on 40th Street between Eighth and Ninth avenues, Central Robbery caught all robberies on the trains and buses in the five boroughs. The squad was divided into teams that focused on warrants, token booth holdups, gun robberies, and violence "in the hole" by youth gangs. Squad members regularly went to the several transit districts in each borough to pick up suspects detained by local uniformed cops. Robberies by youth gangs took detectives into high schools throughout the city to interview victims and to apprehend suspects, who, with their attorneys, jammed the squad room and its adjoining offices.

The transit police taught me all the routines and details of police procedure, indeed all the basics of detective work. I learned how detectives control crime scenes and canvass the surrounding area. I observed detectives interviewing witnesses, recording statements, establishing the identities of suspects, and interrogating them. I watched scores of lineups, peopled with suspects and "fillers" whom I had seen recruited from the nearby Port Authority terminal. I accompanied detectives as they got witnesses to court, shaped and managed "cases," and assisted prosecutors in constructing trial strategies. Rich stories about criminal investigations, impossibly complicated to my ears at the time, were the regular stuff of detectives' conversations. I worked steadily with all the Central Robbery teams until early January 1992 and continued to associate with the transit police through the authority's merger with the NYPD in April 1995.

In early January 1992 I began working with the three teams in the NYPD's 34th precinct detective squad. At that time, the precinct extended from 155th Street to the Spuyten Duyvil at the upper tip of Manhattan, and from the Hudson to the Harlem rivers. It was the NYPD's largest geographical precinct, its busiest, and its bloodiest. I continued this fieldwork daily through the summer of 1993 and then twice a week during the following academic year, with continuing contact for years afterward. Squad bosses and detectives gave me full access to all phases of the squad's work, even as they attempted to make their own sense of the strange fieldworker in their midst.

Unlike journalists, police buffs, fiction writers, criminal psychologists, forensic anthropologists, or movie and television script writers, I came to work with detectives day after day, night after night, for several years, in order to understand the structure and meaning of their work. I learned the intricacies of detectives' cases as well as they did. Eventually, the police created a new role in their world to explain to themselves what I was doing there: I became known as "the Professor." The tag stuck and was quickly adopted by police throughout the city. The police called me the Professor even in my presence, and the Professor became a character in many of detectives' new stories.

I was the outside observer admitted to the theatre-in-the-round drama of their occupational world. The Professor also became, along with everyone else in this workaday setting, the butt of the constant gags, rough banter, and gotchas that punctuate squad work. Once I was handed an "evidence box" that contained a facsimile of a still-twitching bloody hand. I was also expected to participate in fooling other detectives and uniformed cops. In one duping, I extolled the Malaysian beauties who graced a former vice-squad detective's photo album of memorable prostitutes. The Professor's seemingly innocent appreciation of female charms helped the detective gull visiting officers into unwitting expressions of sexual attraction to these men posing as women, much to the entire squad's raucous merriment. One of my own gags was sending the squad a color photograph of myself locked behind the bars of an Alcatraz cell. This portrait of the Professor hung in the 34th squad room for years.

My long fieldwork with the police took me in several different and unexpected directions. At the beginning, to capture the rhythm of various squads' work, I followed ongoing cases from the time detectives took them on complaints through their final or semifinal dispositions. With the 34th squad, this included an analysis of all one hundred murder cases that the squad caught in 1992. During slack periods, I also reviewed the squads' files of old cases that detectives singled out as being particularly interesting. Usually, but not always, detectives targeted homicide cases or violent subway robberies, some of which included arson.

All the while, I interviewed detectives and had countless informal conversations with them about their work.

I always began these talks with the specific cases at hand, either new or old. I wanted to learn detectives' habits of mind and how and why they attached significance to specific details of their investigations. These interviews and conversations invariably led detectives to tell me stories about other cases they had caught and handled. Detectives' occupational consciousness is much more narrative than analytical. They organize the multiple realities they confront in their work through stories—stories that nearly burst with complicated details. Whenever files on cases were available, I asked detectives to retrieve the paper. Once I had read the dossiers, mastered the recorded details, and got the central plot reasonably straight, I reinterviewed detectives about their investigations, trying to get behind the official written versions of events. The extraordinarily detailed character of detectives' work, along with their distinctive narrative organization of it, made my work with them singular in my own field experiences.

At detectives' urging, I accompanied them downtown to 100 Centre Street for their consultations on cases with assistant district attorneys, and eventually for their testimony at hearings and trials. Through detectives' good offices, I met prosecutors in all six trial bureaus at the District Attorney of New York (DANY). I eventually extended my fieldwork to include prosecutorial work.

Beginning in the summer of 1992 and throughout much of the following spring, I spent one day each week in the Early Case Assessment Bureau (ECAB) with junior assistant district attorneys (ADAs) from Trial Bureau 50. Always supervised by bosses, often by the legendary chief of the bureau, Assistant District Attorney Warren Murray, the ADAs assessed cases and prepared specific accusatory instruments, depending on their interviews with uniformed police and detectives and, frequently, with victims and accused culprits of crimes. These instruments also assumed narrative form, as ADAs anticipated presentations to judges or possibly juries.

Occasionally, the grinding routine of ECAB interviews generated electrifying sparks. One day, while I watched, a rookie assistant interviewed a young man from the lower reaches of the 30th precinct about his second armed robbery arrest. Faced with long prison time and hoping for more lenient treatment, the suspect described in detail several street hits in his neighborhood that he had eye-witnessed, all committed, he said, by the same tall, lanky, hooded assassin, a man whom he knew well. The story of the rookie's unexpected bonanza of information made her a star among her fellow ADAs on duty that day. District attorneys, like detectives, revel in stories that place them at the center of action.

Events in the 34th precinct led me in another unanticipated direction: to detectives' investigation of violence emerging out of the drug trade. By the early 1990s Washington Heights had become the city's hub for wholesale distribution of Colombian cocaine. It was also a retail bazaar for coke, crack, and heroin buyers from throughout the city and upstate New York, as well as New Jersey, Delaware, Maryland, and Virginia to the south and Connecticut, Massachusetts, and Rhode Island to the north. Drug dealers in Washington Heights regularly sent vast sums of money to the Dominican Republic, often in ingeniously illegal ways. They also laundered drug riches through the purchase of various restaurants, bodegas, beauty parlors, travel agencies, *casas de cambio,* car repair shops, and car dealerships. These high-cash-flow "legitimate" fronts washed dirty money, while often providing comfortable hangout spaces for gang members.[6]

Washington Heights exploded in fiery riots in July 1992 after Police Officer Michael O'Keefe, a member of the precinct's aggressive anticrime unit, Local Motion, shot and killed Kiko Garcia in a desperate struggle over a gun. Massive demonstrations besieged the stationhouse. Cars were overturned and buckets of cement were thrown from rooftops at firemen and cops, as angry crowds demanded O'Keefe's head. American-flag-burning demagoguery both from local Dominican politicians and from citywide technicians in moral outrage focused media attention on the protests. Mayor David N. Dinkins rushed to comfort Garcia's family and then had the city pay for his funeral in the Dominican Republic.

When the detectives discovered videotapes demonstrating incontrovertibly that Garcia was a drug dealer, they argued convincingly that the riot had been largely orchestrated by other drug dealers. Narcotics traffickers were, and remain, savvy in manipulating the long-term political and cultural alliance between elites and criminals—an alliance brokered by voracious media reflexively drawn to dramatic images of racial and ethnic confrontation—in order to bring aggressive cops to an enforcement standstill. A thorough investigation by DANY exonerated PO O'Keefe that September, and the small army of police on duty in the 34th precinct forestalled further threatened civil unrest.[7] But seeing the riots first-hand in the company of the police, I realized that any study of New York City police detectives, particularly in Washington Heights, had to examine drug-related violence. And it had to take a square look at how criminal investigators see New York City's quasi-tribal politics.[8]

Throughout my fieldwork, I systematically collected detectives' stories, and stories within stories, trying to understand the internal logic of each tale and how one tale led to another. The first book that resulted from this work was *Wild Cowboys: Urban Marauders & the Forces of Order*—a kaleidoscopic set of interconnected stories aimed at capturing the narrative-rich consciousness of detectives and prosecutors as they investigated related outbreaks of violence in Manhattan, the Bronx, and Brooklyn in the early 1990s.[9] To reflect investigators' always-incomplete-at-any-moment understandings as they try to unravel a bewildering skein of events, I wrote the book as a "broken narrative." Like detective work itself, it starts at the bloody end of these seemingly discrete, disparate stories and moves back in time and space to their common beginnings, through all the fits, starts, confusions, dead ends, and sudden put-it-all-together insights that characterize criminal investigation.

Detectives' distinctive ways of knowing, their moral rules-in-use, and the meaning they assign to their work are all embedded in and emerge from the endless stories they tell. *Street Stories* takes readers once again into detectives' world, with tales about the way New York City police detectives investigate violent crimes in an unruly metropolis. The more complicated the case, the more intricate the story. The longer detectives work, the more stories they tell, and retell, and the more these stories trigger their own and colleagues' memories of more stories. Detectives' stories include their assessments of the stories told by others—by witnesses, culprits, and fellow detectives. They lay bare detectives' occupational consciousness—their self-images, assumptions, investigative techniques and craft, moral judgments, and attitudes toward all the players in their world. Detectives' stories reveal their aspirations, sensibilities, hopes, resentments, and fears. Their stories also illuminate dark corners of modern society.

No single book can possibly capture the full range of big-city criminal investigations, especially in New York City at the peak of violent crime in the city's history. But the stories presented here, together with three analytical essays, describe the typical encounters and intricacies of detectives' work, then and now. The book's stories detail both the routine and the bizarre crimes that detectives encounter, and introduce the range of criminals that detectives regularly meet in the course of their investigations. They dwell on the sharp moral conflicts detectives must negotiate and the sometimes strange byways of legal procedures, and they show how chance case assignments can affect career opportunities in the police hierarchy. These street stories convey detectives' sense of responsibility for their cases, while exploring some of the social and psychological consequences of lives spent investigating mayhem within a bureaucratized framework.

To do their job, investigators must piece together fragments of information to discover who committed relatively opaque criminal acts. In rare cases, detectives find fingerprints or other traces of identity at crime scenes that lead them to suspects. But usually they work with knowledge picked up off the street—some of it self-serving, some downright false, some right on the money. They analyze relationships between known criminals and their associates to map out networks of culprits or follow a chain of linked incidents. They search the vast web of municipal, state, federal, and private bureaucracies for information, carefully cultivating reliable and trustworthy sources. Detectives look for people, sometimes for months, sometimes for years, peering behind masked identities and legitimate façades, following wormholes through shadowy underworlds to find those who witnessed, knew about, or committed crimes.

Above all, detectives talk to people. They unpack the common-sense worlds of criminals, the unspoken assumptions, and the intricate associations of the street. They assess the plausibility of the stories they hear from witnesses and informants and the credibility of those stories in courts of law. With the wiles and snares of hunters, they interrogate suspects, seeking to bring their prey to the moment when the need to confess crimes outweighs the dangers of self-betrayal. When they succeed in gaining self-betrayal—through cajoling, feigned friendships, surrogate paternal concern, intimidation, or simple human empathy—they then betray their subjects to a relentlessly impersonal and thoroughly bureaucratized criminal justice system.

As public officials, detectives labor in the interstices of procedure-driven, semi-military police organizations, district attorney offices, and courts, a world of splintered jurisdictions and scattered information, intense competition for prestige, vying hierarchies, and arcane, hair-splitting distinctions. These bureaucratic behemoths fracture authority and knowledge and make absurdity the constant bedfellow of rationality. Within this milieu, detectives labor to transform their hard-won street understanding into convincing public proof that fixes responsibility for crime.

The knowledge that detectives gain emerges, as often as not, out of deception and a willingness to bypass or bend procedures. These methods pit them against the necessarily upright public self-images and rhetorics of the organizations in which they must make cases. Both official and self-appointed watchdogs persistently scrutinize the way detectives work. The roguishness of police detectives often breaks the gear-grinding institutional gridlock of the criminal justice system, even as it makes detectives and their behavior constantly subject to attack.

As representatives of established social authority, detectives become lightning rods in American society's tempests over authority and its proper uses. Police guard a social order that they had little hand in forming. Even when they give voice to the plight of victims who can no longer speak for themselves, or rush into ugliness and danger from which others flee, their efforts become subject to an endless, concerted public acrimony over what kind of social order shall prevail. Their beleaguerment helps shape a remarkable occupational solidarity that binds them one to another.

2

LOOKING FOR SHORTY

Detective Sergeant Ed Keitel wanted Shorty. For weeks, a one-man crime wave had terrorized passengers on the 1 and 9 trains in northern Manhattan. The robber stuck a gun in people's faces while taking their jewelry or money. The scores of victims who complained to the New York City Transit Police all described their assailant as a light-skinned Hispanic man of slender build, about five feet six inches tall, with a distinctively sallow, pock-marked face and close-cropped hair. Most victims said he robbed alone, but some put an accomplice with him. Some of these described his companion as a strung-out white woman in a cap who looked like a boy, others as a nondescript but clearly worse-for-wear black man.

At 2000 hours on November 12, 1991, Jeff Aiello, the bear-like senior detective of the Central Robbery Squad, returned to the squad's 40th Street headquarters between Eighth and Ninth avenues. The windowless, stiflingly hot first-floor complex was leased from New York State Parole, which occupied the rest of the yellow-brick building. Aiello shouted with exultation—he had "Tyre" in tow, a raggedy man collared for a mugging by city police and picked up by Aiello at Central Booking downtown.

Tyre was looking to trade. He said he had robbed three times with a small, skinny guy named Shorty on the 1 and 9 lines, along with a skinny white girl named "Anna," who knew how to get cash for credit cards. Shorty, Tyre said, had a funny-looking face with marks all over it. He always got on the train at 191st Street and Saint Nicholas Avenue and either rode uptown to 225th Street or downtown to 157th Street, robbing on his way. Sometimes he robbed again on his way home to 191st Street, where, just east of Saint Nicholas Avenue, he frequented two drug spots, one to cop crack, the other to smoke it. Tyre said that Shorty also haunted another drug spot at 175th Street between Audubon and Saint Nicholas avenues. After wolfing down a bag of burgers purchased by Aiello, Tyre announced that he was ready to take the cops to Shorty's uptown hideout.

Detective Sergeant Keitel was a fresh face in the Detective Division. After a few years on uniformed patrol, he had spent many years in plainclothes chasing sophisticated out-of-town pickpockets who come into the city to work its sprawling underground world. Keitel ordered six detectives to saddle up. A laconic Lucky Strike chain-smoker, he retained vivid memories of jungle combat in Vietnam, as well as a moment in the transit police's cramped underground office beneath Times Square when an arrested culprit, just frisked by another officer, suddenly pointed a loaded gun in his face. Keitel insisted that all Central Robbery detectives wear bullet-proof vests. In a van with tinted windows, Tyre, cuffed in the backseat, and Keitel, along with Detectives Aiello, Zeke Lopez, Carl Nuñez, Jimmy Nuciforo, Ed Vreeland, Detective Sergeant John Dove, and myself, took a

ride up to the wild and wooly 34th precinct, looking for Shorty.
Two other detectives followed in a car.

Tyre guided the cops to Fairview Avenue, a steep, crescent-
shaped, canyon-like hill, sloping downward from Saint Nicholas
Avenue to Broadway. Halfway down Fairview, Tyre pointed out a
vacant lot tucked between seven-story buildings on the south
side of the street. The lot ran uphill into inky opaqueness. Next
to the high stone wall at the back of the lot, Tyre said, stood
three shacks. Shorty's crib was in the middle. The outline of an
electrical transformer peeked above the buildings. When the detectives slid open the van's doors, the steady hum of high volt-
age current filled the 2200-hours dark night. Led by Keitel,
the detectives stormed the sharply graded lot. They discovered
couches and beds in the three well-appointed shacks. Hotplates,
televisions, and videotape recorders tapped electricity from the
transformer, which supplied a subway station located on the far
side of the stone wall.

Meanwhile, Tyre sat cuffed in the van with me. When I asked
him why he was giving up Shorty, he snorted derisively. Then
he blurted out: "Fuckin Shorty's the fuckin scum o the fuckin
earth." It turned out that Shorty had crossed Tyre. Once when
they went to a friend's house to smoke crack, Tyre had unknow-
ingly dropped his bag of drugs, only to have Shorty pocket it.
Then Shorty began robbing other crack smokers—social misbe-
havior that caused the owners of the 191st Street crack house to
prohibit him from entering their establishment. Tyre explained
the rules at issue: "Shit roll downhill, man, and Shorty, he shit
on his own peoples. You can rob all the peoples you wants on the
trains, but you don't rob the peoples you smokes crack wit." But
then Shorty really stepped over the line. In their last job together, while Tyre and Shorty were robbing people on a subway

platform, an old man, scared out of his wits, had a heart attack and collapsed, half of his body on the platform, and half sticking out over the track. Tyre said: "Ain't no way to treat old peoples. I gots a grandaddy too. Wooden want him treated like at."

The detectives returned to the van, panting from the exertion of the trip down the sheer hill, swearing loudly as they scraped excrement from the soles of their shoes. In tow was a small, slender woman, eyes darting, dirty face twitching, who said her name was "Alice." But no Shorty. Keitel ordered Alice into the car with two detectives. Then Keitel got into the van and asked Tyre about the girl. Tyre said that he knew her as Anna. He went on to say that Shorty, a man of fixed habits, could only be in a few other places. So Keitel ordered the detectives in the car to wait on Fairview Avenue with Alice, who had agreed to identify Shorty when he returned. Everybody else piled back into the van and headed down to 191st Street and Saint Nicholas Avenue, Shorty's home base. Detectives Lopez and Nuñez posted themselves inside the train station. All the others, including myself, sat in the van on the east side of Saint Nicholas, waiting for Shorty.

The sidewalks pulsed with near-midnight energy on that unseasonably warm late-fall night: surging crowds, boys in baggy pants and baseball hats hitting on leather-jacketed, bangled girls wearing blazing magenta lipstick and bright red shoes, all moving to the deafening throb of merengue music blasting from bodegas and gigantic speakers in the open trunks of BMWs. Money and drugs changed hands in plain view up and down the block. Tyre knew all the players. Spotting his girlfriend talking to just about everybody, he said that he hoped she was asking his whereabouts. Tyre then mentioned to detectives that Shorty might be carrying a fake gun. Detective Aiello responded that, if

Shorty went for the gun, the police would treat it as a real weapon, regardless. Better a dead Shorty, Aiello said, than a wife living off his life insurance policy, dancing her nights away with some NYPD bozo named "Ricardo" and saying "Jeff was such a nice guy."

Aiello spotted a huge hulking man strolling down the teeming street who reminded him of a ghost from his past. He mused about the incident. He was policing a train when a similarly gigantic man, dead drunk, boarded, leaned over a woman, and began fondling her breasts. Aiello immediately grabbed the man's right arm, ordering him to unhand the woman. The man casually swung his right arm back, as if he were swatting a fly, and went back to work on the now-hysterical passenger. Heels over head, Aiello rolled backward like a bowling ball, banging into the train's shelf-seats as he went. Leaping to his feet, he raced toward the man, nightstick held high. He hit the assailant on the shoulder with such force that he broke his stick in two. The baton's end piece cracked the reinforced glass of the train window. Unfazed, the man turned and charged Aiello, wrestling him to the ground in a mighty bear hug, squeezing his life-breath away, until the train reached the next station and five uniformed officers, alerted by the conductor, piled in and rescued Aiello.

No Shorty. After an hour's surveillance, Keitel ordered the van to pick up Lopez and Nuñez from the train station and then directed the whole team to return to Fairview Avenue. There, in order to conceal Tyre's new role as informant, the detectives told Alice they had just nabbed Tyre on the street. A few detectives remained on Fairview to wait for Shorty. The van, with both Alice and Tyre in the backseat, slowly cruised the drug spots on 191st Street, as well as all the other tangled streets in the area, dipping down to the 175th Street hangout. No Shorty anywhere. After another hour, Keitel radioed the unit on Fairview to call it a night. As the van turned left on Broadway at 191st Street to head downtown, both Tyre and Alice simultaneously yelled: "There's Shorty!" They pointed to a man standing at a phone booth on the west side of Broadway, smoking a cigarette.

The driver, Detective Vreeland, made a precipitous U-turn on Broadway and then halted the van. Aiello, Dove, and Lopez clambered out, raced across Broadway, seized the slightly built man on the phone, slapped his smoke to the ground, and cuffed him. But when the detectives brought the man across Broadway to the van, both Alice and Tyre said it wasn't Shorty; he was too dark. The detectives released the man and got back into the van. Aiello asked Alice how she could mistake someone else for the man she was sleeping with. But Alice shrugged and said that the man had looked like Shorty in the dim light. A few minutes later one of the detectives cracked: "Well, we sure scared the shit outta that fuckin guy. Maybe we shoulda said: 'Pardon us, sir, but here's your aromatic cigarette back.'"

At headquarters, I watched Aiello and Lopez interrogate Alice. She confessed to doing two robberies with Shorty. The first, she said, happened totally by chance. Shorty had asked her if she wanted to blast. She readily agreed, and they ducked into a nearby building to light up their crack pipes. As an old lady entered her first-floor apartment in the building, Shorty yanked her bag right off her shoulder, throwing her violently to the floor. The bag contained a large number of credit cards that Alice admitted selling later. But, Alice said, she protested to Shorty at the time: "Are you outta you fuckin mind? This lady's gotta be 110 years old." She added that Shorty regularly victimized another old lady every time she got her welfare check.

The second robbery happened downstairs. Shorty and she had

planned to stage an argument on a 1 or 9 train. Shorty was supposed to jostle a victim and grab his wallet. But then before they got on the train, Shorty stuck a gun in some guy's face right at the token booth and got exactly one dollar. Alice said she jumped the turnstile in disgust and walked away. But Shorty followed her onto the platform, admonishing her to be careful because she could get arrested for fare-beating. Alice gave no specifics for any of these crimes. Only the incident on the train platform fell within NYCTP jurisdiction. But detectives possessed no paperwork on such a robbery and therefore had no case. No complainants, no formalized complaints, no crime, even though Alice confessed to crimes. That night, Keitel put Alice in two lineups for other cases where complainants pointed to a robber matching Shorty's description and working with a female accomplice. But they got no hits, and so they cut Alice loose in time for an early breakfast.

The next night, Detective Jimmy Nuciforo and Detective Sergeant John Dove went to Shorty's shack twice, first at 1800 hours, then later at 2130 hours. No Shorty. But the detectives did find a man named "Jamaica" at Shorty's shack. Jamaica said that Shorty knew the police were looking for him, so he was making himself scarce. The detectives then hit the crack houses on 191st and 175th streets, causing customers to scatter, racing down fire escapes or over rooftops. The owners of both establishments expressed great dismay at the police visits, but they directed their anger principally against Shorty for provoking the unwanted attention. One owner mentioned a jewelry shop on Audubon Avenue that, he said, fenced gold from all the local robbers. Maybe Shorty went there.

The next afternoon, Keitel and two of his men visited the jeweler and described Shorty to him. The jeweler denied knowing anyone who matched the description. So Keitel reminded him how the world works: either the jeweler phoned the transit police immediately the next time he saw Shorty or he could expect the police to shut down his fence and put him in jail for receiving stolen goods.

Robbery has always constituted the main crime in the New York City subway system.[1] Indeed, robbery, major crimes emerging out of robberies, or mayhem often accompanied by robbery have catalyzed key decisions about policing underground. The public outcry at the early-morning robbery-murder on May 14, 1936, of 54-year-old Edgar L. Eckert, an executive of the Rogers Peet clothing company, prompted Mayor Fiorello H. LaGuardia to create the official position of Special Patrolman (Railroad) to police the subway.[2] Eckert had been working late at the Rogers Peet 35th Street and Broadway store to clear a basement flooded by a sudden storm. He was manually strangled in a mezzanine men's room of the Eighth Avenue and 42nd Street Independent Subway station for his gold pocket watch and company medallion.

The city's Board of Transportation employed the initial twenty-one patrolmen essentially as private guards, supervised by a police captain from the NYPD. More transit officers were added gradually over the years, with officers taken from the city's regular police list and assigned to the subways. In 1947 the transit patrolmen received peace officer status. In 1953 they were designated police officers with full police powers and obligations within the City of New York—the same year that New York State established the New York City Transit Authority. In 1955 the New York City Transit Police Department was established as a separate police authority, though its officers were not granted full police powers for all of New York State until 1964. By that time, the NYCTP boasted about 900 officers.

A series of subway assaults in 1964, almost all with strong racial overtones, paved the way for major increases in the number of transit police officers. Some assaults emerged directly out of robberies. On May 31, 1964, twenty black youths robbed and terrorized IND elevated train passengers in Bensonhurst, Brooklyn. The same day, five black youths stabbed a white 17-year-old on the IND train for refusing to hand over $10 and play his radio for his assailants. On July 17 roving bands of black teenagers beat and robbed two white men in different incidents on upper-Manhattan IND and IRT trains. Other assaults had no ostensible motive except perhaps forcing others into submission. For example, on May 31 four young men wielding a meat cleaver intimidated a motorman and passengers on two BMT trains near Prospect Park in Brooklyn. And on July 26 two young black men were arrested in the IND 4th Street station for terrorizing fellow passengers.

Civil rights leaders, including Dr. Martin Luther King Jr., urged public understanding of the problems facing Negroes. Roy Wilkins, executive secretary of the National Association for the Advancement of Colored People, argued that such violence by blacks was caused by "the deprivations of slum life" and by the "bitterness and frustration which all Negroes feel at the continued denial of equal opportunity everywhere and at the unpunished beatings and killings of Negroes . . . in the Deep South." Other black leaders excoriated the mainstream press for reporting "delinquency" by blacks when there was plenty of unreported violence committed by gangs of white youths. Almost all these leaders insisted that adding more police was no answer at all to the violence committed by some black youths. Nonetheless, public clamor about safety on the subways, spearheaded by a June 1 editorial in the *New York Times* arguing that "the basic danger in this city is the existing peril to the life and property of peaceful citizens," led Mayor Robert Wagner to increase the size of the transit police force to 1,077 officers by early July 1964.

Harlem streets exploded in late-summer riots beginning the night of July 16, sparked by NYPD Lieutenant Thomas Gilligan's shooting of 15-year-old James Powell, who, the officer said, confronted him with a knife after Gilligan tried to break up a neighborhood dispute.[3] Riots engulfed Harlem and Bedford-Stuyvesant for six nights. In the fall, the boiling cauldron of racial unrest spilled downstairs. On October 23 gangs of black high school students rioted on subway platforms in both Harlem and Brooklyn, terrorizing train passengers. On November 1 roving bands of black youths jostled and robbed passengers on Manhattan's IRT line. Vandalism increased throughout the system, and the subways—"everybody's second neighborhood"—were increasingly seen as a hostile and menacing milieu. In early February 1965 the Transit Authority reported that violent subway crimes had increased by more than 50 percent in the previous two years.

Then, on March 12, 1965, a racially-tinged robbery-homicide catalyzed growing public fears about safety in the subways and prompted a major increase in the transit police force. Just before midnight, three black youths boarded the Manhattan-bound A train on the IND line at Broadway-East New York in Bedford Stuyvesant, Brooklyn, and started harassing all the passengers in the car. In particular, they hit on three young black girls, who ignored their advances. The youths began intimidating other passengers, demanding money and cigarettes. They prodded 17-year-old Andrew Alfred Mormile, who had been dozing. Mormile woke up and tried to walk to the next car. The youths blocked his path. A shoving match ensued. One of the youths drew a knife and lunged at Mormile, stabbing him first in the face and then in the back of his neck. Mormile fell to the floor, bleeding profusely, his death observed by other passengers too paralyzed by fear to intervene. The youths fled the train at the Nostrand Avenue station. They were later apprehended and convicted with the testimony of the three girls.

Motivated by sensational media coverage of this event, amidst a rising tide of subway violence, Mayor Wagner promised to post an armed police officer on every train and every platform—some 465+ stations—between 2000 and 0400 hours. To keep this promise, he increased the ranks of the transit police by 700, bringing the force to over 2,720 men and women by the end of 1965.

A new kind of robbery plagued the subways in the early 1970s. Partly in response to the nearly ceaseless wild-west banditry of city buses, the Transit Authority instituted an exact fare policy for surface transportation in September 1969. Almost immediately, robbers went downstairs and began to rob subway token booths in record numbers. Between April 1970 and April 1971 there were 771 booth robberies. Crazed drug addicts, looking for quick cash, carried out the great majority of these haphazard affairs. But organized professional crews of four men—the drop (gunman), the bagman who collected the loot, the lookout who watched for police, and the wheelman who drove the getaway car—committed many booth robberies. Crew members split the take proportionally, based on street perceptions of the relative danger, legal liability, and nerve required for each task.

A booth robbery could net as much as $5,000 cash and an

equal amount in subway tokens that could be sold to bodegas at a 30 percent discount. Bodega owners resold the tokens at full price "for the convenience of customers." The transit police instituted stakeout teams that lay in wait in darkened stations for booth robbers to strike. Shootouts occurred regularly. If a transit police officer apprehended a robber, he received one vacation day for meritorious service; if he shot a robber, he received two days vacation. The Transit Authority's policy reflected the contemporary public attitude toward police shootings and the regrettable consequences of banditry.

After the early 1970s flurry, the number of booth robberies dwindled, eventually declining to about 100–150 a year into the early 1990s. Token-booth clerks with inside knowledge of the system or romantic relationships with robbers, or both, set up some of these robberies. One clerk joined with a motorman and other accomplices to rob more than a million dollars from token booths between 1970 and 1976. More typically, token-booth robbery crews made marauding, Jesse James–style raids on easy-target stations.

One group called the Black Hoods robbed more than sixty token booths in their home territory of the South Bronx between October and December 1974, netting, however, only about $1,000 a job because they took only the cash on the counter and ignored the valuable boxes of tokens in the booths. Similarly, in late 1991 a drug dealer named Jake from 118th Street and Lenox Avenue needed cash in a hurry because he owed his supplier a lot of money. Armed with Uzis, Jake and his crew carjacked women drivers, used the stolen vehicles in token-booth robberies, and then sold the getaway cars to one of the many chop shops on White Plains Road in the Bronx. Jake and his crew also took only cash, ignoring the tokens.

When transit police detectives finally ran Jake to ground, they openly admired his cool toughness, his refusal to speak with them beyond polite niceties, and especially his unwillingness to rat out his companions—qualities that carry a premium in the detectives' own world. Although detectives were happy to have Jake in custody, they nevertheless despised the accomplice who gave him up. But Jake's public persona collapsed when he was nailed in a lineup. Facing an eight-to-ten-year stretch because of his previous convictions as a robber, Jake hanged himself on Rikers Island.

In the early 1990s the famous and much more professional "Macdougal" gang, whose members trained for robbing booths by lugging weighted bags up and down stairs, robbed eight Brooklyn stations in less than two months. According to Detective Billy Courtney (famous for poster-size photos of "the ideal lineup": a refrigerator, a sheep, a fat lady, and a lone subway robber), "Kevin Macdougal," a family man and a gentleman crook, ran his crew like a detective squad boss. Displaying unfailing politeness to the detectives who arrested him, Macdougal, though an armed robber, did not use violence wantonly. Indeed, he pistol-whipped and expelled a crew member who had shot a token-booth clerk in a robbery gone bad. Macdougal exhibited shrewdness and careful planning as well. He rigged a rope ladder from the roof of his apartment building, giving himself a ready back-door escape if needed. He always targeted subway stations where one could see the street and the tracks at the same time. Macdougal's underlings, however ("Dillinger," "Machine Gun Kelly," and "Baby Face Nelson"), gained reputations as wild men. Baby Face was also a rat. He betrayed Macdougal and the others to the police for a $5,000 reward, before going out and robbing a token booth all on his own.

The J-line gang plundered stations up and down the J line in Brooklyn. "Rasheem," the gang's lookout, had helped rob the token booth at the Norwood Street station on February 5, 1989. Arrested and convicted of that robbery, Rasheem spent two years in Sing Sing before being sent to a medium-security prison in Pennsylvania. The first day there, he walked off and went back to the city, where he hooked up with some associates. On October 2, 1991, he returned to the Norwood Street station. When the token-booth clerk—the same clerk Rasheem had robbed in 1989—re-entered the booth after a break, Rasheem put a gun to his head, pushed into the booth, and seized cash and two boxes of tokens.

With the proceeds from that job, Rasheem bought some cocaine in "weight." He took it to the Carolinas, where he sold the coke at the normal out-of-New-York four-to-one markup. With the profit from that venture, he bought guns and returned to New York, where he sold the guns at a five-to-one markup and used that profit to buy more weight. When he returned to the Carolinas, police arrested him for drug-dealing and sent him back to New York on an escaped prisoner warrant, based on a photo-array identification by the Norwood Street token-booth clerk.

Jeff Aiello and Billy Courtney greeted Rasheem like a long-lost brother, in a display of fellow-street-warrior camaraderie that included bear hugs and high fives. Aiello laughed that during one year he had spent more time with Rasheem than he had with his wife. The huge, easy-going, straight-talking Rasheem bantered with the cops and with me. "Whatchoo wanna know, Professor? You wants to know why I does what I do? Professor, does you have any idea what my share was when we was ridin high on the J line? Just my share, Professor? I was making $8,000 a week. And

I could fuck at will. Why should I take a straight job? That wooden make no sense at all."

Professional robbers like himself lived large, the way life was supposed to be lived, Rasheem said. In addition to investing his money in his drug business, he had gold-and-diamond rings for every finger, gold chains—herringbones, Gucci links, and Bismarks—real gold ("You wear fake, everybody know it and talk bout you"), gold sleeves for all of his teeth ("Just like th rappers, man"), a wardrobe filled with Polo clothes, leather jackets, and shearling coats, several BMWs and Mercedes Benzes, and more fur coats for his wife than he could count. And he drank only "Moat" (Moët-Chandon) champagne for breakfast, lunch, and dinner, and in great quantity before any robbery. Rasheem held to strict pricing standards, insisting that he never accepted less than $850 for a box of 1,000 subway tokens, worth, in 1991, $1,150 face value. Only someone who was desperate, he argued, accepted less than that because there were any number of "legitimate" bodegas ready to pay top dollar for stolen tokens.

After his discourse on the merits of robbery as a career choice, Rasheem asked Courtney and Aiello for a favor. "Fellas, I been straight wit you and I wants some help. I wants to go back to Sing Sing." Both of the detectives expressed surprise that Rasheem wanted to return to the fearsome maximum-security prison. "Man, you can buy anything you wants on the yard at Sing Sing. It like being on the street in New York, same prices. Sides, there a little girl Correction officer at Sing Sing who like to take the young boys up to the infirmary at night for $100 a ride. I lookin forward to gettin some of that again."

Although guns predominate as booth robbers' tools of choice, some use other weapons to persuade clerks to open their doors, particularly after the Transit Authority bullet-proofed a hundred of its older stations. Railroad clerks Oscar Williams in 1977 and Regina Reicherter and Venezia Pendergast together in 1979 were burned alive when robbers squirted liquid petroleum into their booths and then casually tossed in lighted matches. Railroad clerk Harry Kaufman suffered the same fate in November 1995. All of these clerks suffered grievously before dying from severe burns. Several others over the years have endured serious burns and narrowly escaped death.

Members of the famous Mankowitz gang, headed by Brian Mankowitz from Middle Village, Queens, preferred to use pickaxes for their ten 1987 token-booth robberies, smashing and shattering the Plexiglas booths while promising to do similar work on clerks' heads unless available cash and tokens were handed over. Police say that one clerk dropped dead of a heart attack when he saw the gang coming down the stairs toward his booth carrying their farm implements. No one could or would identify the members of Mankowitz's crew. The gang was finally stopped by Police Officer Vinnie Valerio, a powerfully built man shaped like an egg atop tiny, extremely nimble feet that regularly danced him to the front line of officers ready to crash into apartments after armed suspects ("Professor, when we go through that door, stay behind me cause I want you to write that fuckin book"). Valerio shot and paralyzed Brian Mankowitz in the middle of a robbery at the Elderts Lane station on the J line on December 7, 1987, after surprising the pickaxe king from his hidden stakeout.

With nearly two-thirds of the transit police force deployed every night, daylight hours became more attractive to subway criminals. In 1973 daytime subway robberies and larcenies exploded, mostly purse and jewelry snatchings. So did the number of assaults, rapes, and brawls. Some of this statistical bulge in daytime crime came about because officers fudged their arrest times, under administrative pressure to show decreases in nighttime crime. The falsification of records led to a major scandal. But the subways had, in fact, become more menacing during business hours. Police expressed particular concern about youths who were committing an increasing number of crimes in the afternoon and early evening hours. By 1974 two-thirds of all felonies were committed between 1200 and 2000 hours. This led to a sharp redeployment of transit police officers, now 3,600 strong, to daylight hours.

The department also carried out significant "anticrime" operations using undercover plainclothes officers to catch criminals in the act. On the time-honored police premise that criminals cause crime, police posed as drunks or out-of-town rubes, among many other guises, to lure criminals into robbery or grand larceny. Critics savaged the decoy program from the start, arguing that police were "making crime" instead of preventing crime. The NYPD countered that the decoy program did indeed prevent crime by locking up inveterate predators.

The late 1960s and early 1970s brought New York City and the nation the Black Liberation Army (BLA), an offshoot of the Black Panther Party for Self-Defense. The BLA aimed to free all "Afrikan" people from racist oppression. To do so, several of its members attacked police officers.[4] On June 5, 1973, transit PO Sidney L. Thompson tried to arrest a fare-beater at the 174th Street and Southern Boulevard elevated IRT station in the Bronx. The fare evader's companion shot Thompson fourteen times. Thompson managed to shoot the man he had originally stopped in the throat and left ankle. Detectives arrested that man, Victor Cumberbatch, at the Bronx Lebanon Hospital and later apprehended Thompson's killer, Robert Hayes aka Seth Ben Yssac

Ben Ysrael, and another man after a shootout in a BLA apartment.[5]

Despite what the *New York Times* on January 21, 1976, called "rising terrorism" in the transit system—fears prompted by marauding youth gangs sometimes wielding shotguns, sometimes clubs and knives, who held subway cars at bay and occasionally hijacked buses—New York City's 1975 fiscal crisis caused dramatic cuts in the transit police force, along with every other city agency. Through attrition, hiring freezes, and outright layoffs, the city reduced the force to 2,600 officers, a level maintained until 1979.

The period was notable for several one-man crime waves. An 18-year-old who had been arrested fifty times beginning at the age of nine specialized in riding between cars and, as trains left stations, snatching purses or chains from women standing on platforms, occasionally yanking people to the ground and dislocating shoulders. This muscular would-be-boxer worked the Lexington Avenue line at Grand Central Station and the Seventh Avenue Line at Times Square, taxiing between the two terminals for convenience. He was said to be personally responsible for 15 percent of subway crime in the late months of 1976, credit that should probably be shared with several other young men to whom he had taught his peculiar craft.

On September 29, 1977, police finally caught a 16-year-old girl and 17-year-old boy for robbing forty women on the trains within six weeks. Police apprehended a man for burglarizing Anne Picyk's apartment on July 8, 1978, after snatching her purse in the subway to get her address and keys and then throwing her off the platform into the path of an oncoming train, which she miraculously dodged. The youth confessed to three similar incidents in the previous month in which all the victims, unlike the

lucky Picyk, had been seriously harmed. The famous Willie Bosket killed two men in the subway in separate incidents in March
1978. After a five-year prison sentence for the two murders, Bosket continued an active criminal career. Arrested again in the
early 1980s, he claimed to have stabbed 25 people while doing
over 200 robberies and, overall, more than 2,000 crimes. While in
prison, he set his cell on fire several times and attacked his keepers nine times.[6] The transit police used such cases to argue stren-
uously that relatively few culprits were responsible for the vast
majority of subway crimes.

The chaos downstairs began to peak in early 1979 with rashes
of extraordinary violence. Youths regularly rampaged through
subways late at night after leaving discotheques. The occupation
of token-booth clerk was considered extremely dangerous, not
only because of the fiery deaths of Reicherter and Pendergast
but because, in addition to normal shootings, several clerks were
head-bashed or stabbed (or both) by robbers. Subway passenger
Reilly Ford was burned to death, and in a separate incident
down-and-outer Michael Starkman was set afire on a Brooklyn-
bound train on March 1, 1979. Several passengers were nearly
thrown off platforms, and on February 25 Good Samaritan Yong
S. Sou was hurled to his death in front of an onrushing train by
a mentally ill patient who used his newfound freedom under
New York's deinstitutionalization program to menace subway
passengers in Greenwich Village. In the moments before his
death, Yong Sou had tried to dissuade the young man from pes-
tering another passenger.

The incident presaged a flood of down-and-outers, derelicts,
and mentally ill who poured into the shelter of the subway system in the early 1980s. Dignified as "the homeless" by their advocates, they panhandled aggressively, under an umbrella of court

edicts declaring their activities an expression of constitutionally protected free speech. With the city still strapped for cash, Mayor Edward Koch did not increase the number of transit police officers. Instead, he ordered a massive reorganization of the department, redeployment of existing personnel, and overtime work for all officers. For a good part of 1979 transit police officers worked twelve hours a day to combat crime in the subways and allay public fears about the safety of the system. For a time, city police officers were assigned to the subways to aid the transit police.

In 1980 crime in the subways soared more than 70 percent over 1979. Gold-chain snatching abounded, and the police discovered an entire network of jewelry stores that routinely fenced, and sometimes melted down, the stolen items, often selling the precious metal back to the thieves as jewelry or gleaming tooth-sleeves. One robber roamed Queens subways slashing riders with a meat cleaver; another seriously wounded Alexander Hudson at Brooklyn's Botanical Garden station with a bow and arrow. Throughout the system, marauders with guns, knives, and link-chain whips terrorized riders. And everywhere in the city, subway cars displayed graffiti—usually crude (Teddy-bear love Teddyette 4-ever), sometimes grotesque, inventive, playful images or self-portraits, including cartoon figures in riotous color, some in three dimensions, some with legends ("We will all just fade away"; "The children of tomorrow can't love this world if we the people of today destroy its beauty before they even see it"; "How can we destroy and kill ourselves while our killers stand alive and waiting . . . STOP THE BOMB"). The displays usually had tags (street names like Taki 183, Cornbread, Cool Earl, Lee, Samo, T-Kid, Mad and Seen [the Partners in Crime], Dust, Json, Kase, Dondi, Vulcan, and Futura), "writers" searching to "get up" on steel-canvas trains to display their prowess and defiance to the world. Throughout the 1970s city officials, invoking middle-class fears of the seeming visual chaos underground, had unsuccessfully waged war on the audacious subway graffiti artists, who risked life and limb to enter subway lay-up yards in the dead of night to vie with their peers all across town. Emerging artistic elites in downtown Soho adopted the youngsters as avant-garde heroes who seized public spaces for self-expression. But even after spending millions of dollars on police overtime and on train clean-up, the MTA's rolling stock continued to display fantastic and sometimes compelling images from worlds apart. The repainting of trains merely provided fresh canvases for these budding artists and writers. The MTA solved the problem only in 1989 when the city began to buy easily washable stainless steel trains. Graffiti receded dramatically in the subway, but it remains a major problem on New York streets.[7]

Violent crime in the subways, as well as vandalism costing more than $30,000 a month in broken train windows, continued to surge throughout 1981. Mayor Ed Koch increased the number of transit police officers by 850 in 1982, bringing the force to 3,343, close to its pre–fiscal crisis strength. Over the next three years, physical conditions in the subways continued to deteriorate, generating great public outcry, media lament, and alarm among city officials. But reported crime downstairs, including robberies, leveled out. This period was marked, however, by two sensational incidents, each of which became symbolic flashpoints in the city's ongoing racial tensions for years to come.

On September 15, 1983, Michael Stewart, a black 25-year-old graffiti writer, was arrested in the Union Square station in a melee with several transit police officers, all white. He died thirteen hours later in Bellevue Hospital. His death was variously attri-

buted to acute intoxication (his blood alcohol level was .22, compared with a New York standard of .10 for legal intoxication),
blunt force trauma, cardiac arrest, and asphyxia. Three transit
police officers were tried for criminally negligent homicide, to
wit, beating Stewart to death. Three others were tried for perjury. All were acquitted, amidst strident accusations about rampant racist police brutality against blacks. Stewart became a
patron saint of the emerging hip-hop movement, his death im-
mortalized in a painting by Jean-Michel Basquiat that portrayed
cartoon-like white police officers beating a black Christ figure.[8]
To this day, denizens of the hip-hop world as well as many artistic elites refer to Stewart's death as a murder.

Then on December 22, 1984, Bernhard Goetz, a white man,
shot four black youths who, he claimed, were trying to rob him
in the subway. A jury acquitted Goetz of several charges of attempted murder, assault with a deadly weapon, and one charge
of reckless endangerment. He was found guilty only of possession of an illegal handgun.[9] Years later a Bronx jury found him
liable in a civil trial for permanently paralyzing one of the four
youths.

Violence associated with the burgeoning drug trade, particularly in crack, first hit New York City streets in 1985 and then
spread throughout the nation the following year. Major subway
crimes rose precipitously, causing the city to increase the transit
police to 3,800 officers. A year later, in November 1987, the department cited the strength of its force as the reason for an 8.7
percent decrease in major crimes. But conditions continued to
worsen downstairs. Droves of homeless drifters, many mentally
ill or crack-addicted, made subway trains, platforms, and even
tunnels their homes. Aggressive panhandlers stalked the trains,
demanding money from cowed passengers. The Appellate Term

of the New York State Supreme Court, much to the delight of advocates for the homeless, ruled on March 12, 1987, that a statute outlawing loitering in transit centers was unconstitutionally vague, thus prohibiting police, for the time being, from clearing public spaces. The New York State Court of Appeals later reversed the Appellate Term's ruling, but the floodgates had been opened.

Late at night, the Eighth Avenue Port Authority Terminal, Grand Central Station, Pennsylvania Station, and stations throughout the subway system became night-of-the-living-dead bivouacs. Tents and makeshift sleeping bags littered the floor, and squalor was everywhere, as undressed or half-dressed men wandered aimlessly through public spaces now claimed as their own. Their dull eyes brightened only when detectives, with the promise of a quick five or ten bucks that might make the recipients the next robbery victims, scoured for lineup fill-ins, eliciting alternate responses ("I looks black but, honest to God, Officer, I passes for Spanish"; or "You outta you fuckin mind you think I gonna stand in a fuckin lineup so's you can pin a fuckin robbery on me"). Trains all over the city became rolling sleeping cars, as the drugged or drunken made six-foot-long benches into beds and filled the air with the bouquet of the streets. Brazen fare-beaters leapt over turnstiles ("How come I should pay to ride this shitty system?"), while thieves sucked tokens out of stuffed-up slots or, in plain view of terrified clerks in their booths, "popped" token boxes to steal gray canvas bags bulging with the morning rush-hour's take.

The disorder downstairs exacerbated the anxiety of middle-class riders and caused grossly exaggerated estimates of the prevalence of subway crime. The transit police found themselves battling on two fronts: against criminals themselves and against plummeting public confidence in the safety of the trains that carried over three million passengers every day. Only in May 1989 did MTA chairman Richard Kiley announce that the transit police would begin enforcing rules against aggressive begging on trains or platforms and sleeping on train seats. But then advocates for the homeless took the matter to federal court; and on January 26, 1990, federal judge Leonard B. Sand voided the MTA's no-begging policy, ruling that poor people have a constitutional right to beg under the First Amendment.

Only after four months of chaos, during which increasingly intimidating and threatening panhandlers read excerpts of Sand's decision from printed cards to subway passengers before accosting them for money, did federal Court of Appeals judge Francis X. Altimari, speaking for a 2–1 majority, overrule Sand's decision, saying that subway begging had become "nothing less than assault." The legal wrangling paralyzed effective enforcement of ancient anti-loitering and anti-begging laws until November 1990, when the United States Supreme Court let stand the decision by the Court of Appeals.

In the meantime, the transit police efforts to control violent predators ran into serious trouble. Members of the elite decoy squad were once again posing as inebriated, and then arresting culprits when they snatched chains or jack-rolled seeming drunks for their wallets. Accusations past and current about "dubious" arrests by the squad, especially arrests of black and Hispanic citizens—including accusations of frame-ups, "aggressive enticement," and planting evidence in order to meet quotas for collars—led bosses to disband the unit in early December 1987. The issue was: When does legitimate undercover work cross the line into entrapment? That same month, arrests plummeted by nearly 38 percent as all officers hunkered down under the media maelstrom provoked by the allegations.

Later, Robert M. Morgenthau, the district attorney of New

York, dismissed allegations of false arrests brought against the decoy squad for want of sufficient credible evidence. Morgenthau's office then became the target of accusations of burying evidence of abuse.[10] Federal prosecutors pursued the case and eventually won convictions against two transit officers for violating the civil rights of eight black, Hispanic, or Asian men by arresting them falsely for sex-abuse claims brought by white women complainants. This was perhaps the first use of what became a pattern of federal prosecution of police officers in response to well-orchestrated choruses of civil rights advocates. The decoy unit began operations again under strict new guidelines in March 1991.

In 1988 subway crime rose by 10 percent, with robberies soaring by 21 percent—sharp annual increases that extended, with monthly fluctuations, throughout the next several years. Freelance gunmen like Shorty did most of these robberies. Reports of gun robberies averaged between 1,500 and 2,000 a year, with an estimated 4:1 ratio of actual incidents to reports. Token booths once again became principal targets. Horrific, widely publicized crimes made riders more apprehensive than ever. On June 4, 1988, railroad clerk Mona Pierre was roasted alive by a man who poured flammable liquid into her Halsey Street-Wyckoff Avenue Station token booth in Bushwick, Brooklyn, and then lit her up when she refused his demands for money. Copy-cat robbers assaulted three more token booths with flammable liquid within the next week, almost killing railroad clerk James Madden in one attack.

Nearly a year later, on March 15, 1989, at 0050 hours at the old rickety wooden train station at Intervale and Westchester avenues in the Bronx's 41st precinct, just as the uptown train pulled into the station and passengers streamed onto the platform, three young men and one woman approached the token booth of the railroad clerk, a 48-year-old woman, and pushed a note through the small token window that said "Give us the case [of tokens] or we gonna burn you down." The clerk tried to push the note back when she heard one of the robbers yell: "Let's burn the bitch." At that point, the four splashed a liquid on the booth and set it on fire. The clerk dashed out and escaped, but the whole wooden train station burned to the ground in a spectacular conflagration. Exhaustive work by Detectives John Cornicello and Jeremiah Lyons resulted in the arrests of four suspects and eventual guilty pleas.[11]

Robberies by groups of youths—a perennial subway problem since the 1960s—became rampant in the late 1980s as teenagers born to postwar baby boomers and 1960s-era immigrants peaked in numbers. The transit police called robberies committed by five or more youths acting in concert "multiple perpetrator robberies" or, more colloquially, wolfpack robberies. Subway wolfpack robberies reached their zenith in 1990 and 1991, with a thousand incidents in each of these back-to-back years with the typical estimated ratio of unreported to reported incidents of 4:1.[12] By early 1990 the threat to other youths alone had become so severe that School Chancellor Joseph Fernandez asked for special police-patrolled subway cars to escort youngsters to and from schools, particularly in Brooklyn. The transit police, by then the sixth largest police authority in the nation at 4,000 strong and under the new leadership of William Bratton from the Boston Police Department, initiated the Central Robbery Squad (CROB). It was the brainchild of roly-poly, mustashioed Detective Lieutenant Jack Maple (always clad in bright suspenders and bow tie, complete with spectator shoes and black bowler) and his whip, Detective Sergeant Tommy Burke, a Brooklyn boy,

whose remarkable articulateness suggests the quality of a Bishop
Loughlin Memorial High School education in the early 1960s. In
addition to gun and booth robberies, Central Robbery focused
on wolfpacks, adopting a policy of never closing a case until all
robbers in an incident were arrested. This organizational stance
was a source of great pride to the transit police and distinguished the force from its hated "Big Brother," the New York
City Police Department.[13] The privately stated policy of the transit police was to make the subways so inhospitable to crime that
criminals would choose to commit their depredations upstairs.

———

At 2100 hours on November 15, 1991, Detective Jeff
Aiello called Detective Sergeant Keitel at CROB headquarters
and asked him to come to the Bronx to aid in an apprehension.
Aiello had caught a wolfpack robbery that had occurred earlier
in the evening on the platform of the East 180th Street station. A
gang had attacked and robbed three 17-year-olds, taking about
$100 in cash in addition to jewelry. One victim, who was black,
had a gold chain snatched. The robbers beat him into uncon-
sciousness and then kicked him in the ribs while he was down
and out. They seized a gold watch with diamonds from another
victim, also black. The assailants then knocked him off the platform and onto the tracks. The third victim, who was Hispanic,
managed to smack one of his assailants before the gang ran off.

The victims reported the incident at the Transit Authority's
District 12 at East 180th Street and Morris Park Avenue in the
48th police precinct. The transit police immediately drove the
badly beaten victim to the hospital for treatment. Then a plainclothes anticrime cop accompanied the other two kids on an
escort around the immediate neighborhood in a van with dark-

tinted windows. In a local chicken shack, the pair spotted the assailant whom the Hispanic victim had smacked. The cop arrested the culprit, 19-year-old "Tiko," and brought him back to District 12, along with the two complainants.

Aiello arrived and began to interview Tiko, telling him that, if he chose to take the full weight of the robbery and assault charges, Aiello intended to bury him. Tiko crumbled fast and gave up his whole group, telling Aiello that the gang hailed from the Melrose section of the Bronx, usually going north to rob where they were less likely to be recognized. Aiello pressed for more detailed information. Tiko then mentioned that one of the ringleaders often parked his black Mustang in front of the neighborhood video store on Tinton Avenue near East 163rd Street.

Keitel, accompanied by Detectives Zeke Lopez, Roger Fanti, and myself, arrived at District 12. Keitel and Aiello discussed the available options. The police could have Tiko show them exactly where the gang hung out and identify the participants in the robbery/assault. This approach required lineups back at the district. Or the police could take the complainants out on an escort, identify the culprits right in the street, and go straight to arrest. Because it was already midnight, the detectives decided on the latter course. Just as everyone headed for the street, the anticrime cops brought the badly beaten victim back to his friends, much the worse for wear and practically immobile with bandages around his fractured ribs. However, the victim leapt at the chance to pile into the van to accompany his two friends and the rest of us in hunting for his assailants.

The police van drove up and down the streets of the Melrose neighborhood for almost three hours with no luck. Suddenly, the three complainants yelled that the three kids entering a candy store were among their assailants. The detectives parked the van and waited. But the suspects lingered in the store, playing a video game. So Aiello and Keitel accompanied the complainants into the store. But these were different kids wandering the streets at 0300 hours.

A few minutes later, the complainants spotted another youth in a yellow sweatshirt who, they said, resembled one of the gang. But a closer inspection eliminated him as well. Detective Lopez drove repeatedly past the video store on Tinton Avenue, but there was no black Mustang in sight. Around the corner on East 163rd Street, a large group of youngsters clustered in front of a bodega. Keitel told Lopez to drive slowly. A complainant insisted that one of several young men in front of the store, this one wearing a green sweatshirt and a green hat with a pompom on top and sucking on a lollipop, was part of the gang. Lopez stopped the van. Detectives Aiello and Fanti jumped out, grabbed the suspect, and brought him close to the van's tinted glass. Aiello asked: "Is this one of the guys?" The complainant said yes; but his two companions waffled. Aiello asked the complainant who had made the identification: "How sure are you?" He said: "Seven out ten shots." Aiello responded that that wasn't sure enough and let the pompom kid go. Another complainant announced repeatedly that he was keen on finding a short guy with curly hair and also a heavyset guy with dreadlocks and a wispy peach-fuzz goatee.

The van kept circling the neighborhood, with the cops looking into the chicken shacks, the candy stores, and arcades. Children, large and small, were everywhere at 0330 hours, in bare, ruined shells of once-grand buildings or on empty moon-surface lots. But no good suspects. Getting restive, Keitel ordered the van back to the district. On the way, Lopez drove past the video store one last time. Now a black Mustang was parked in front. Lopez circled the area quickly and parked a block away from the Mustang on Tinton Avenue and East 161st Street, but with a clear view of the video store and car. Keitel ran the Mustang's plate; it came back to a girl with a nearby address. So the police sat in the van and waited. Their vehicle immediately became an object of great curiosity. Although the street was pitch dark, the scene was lively. Teenagers alternately made out with each other on a nearby tenement stoop or kidded around on the sidewalk, making sure to come close to the van to try to peer inside; elders hung out of sixth-floor windows staring steadily at the van; passersby made obvious detours to scout the van. But no one went near the Mustang.

Suddenly, the complainant who had been hospitalized spotted another teenager wearing a green University of Miami jersey walking down Tinton Avenue toward the van, on the opposite side of the street, his back to the Mustang and video store. The complainant yelled: "Boom. That's it. I forgot completely. One of em was wearin a Canes jersey and I think that's him." Aiello and Keitel pressed him to be sure. The suspect had an angelic face and seemed much younger than the complainants' earlier descriptions of their assailants. But the complainant insisted that the boy in the Canes jersey was one of the gang. By this time, the would-be Hurricane had turned right at the corner of East 161st Street and was walking west, away from the van. To follow him across East 161st Street meant losing sight of the Mustang on Tinton. Keitel made the decision to go after him, and the van lumbered slowly down the street, tailing the quarry.

The suspect headed toward the door of a ramshackle building on the northeast corner of Trinity Avenue and East 161st Street. Deafening music blared from the ground floor of the building, one of the many unlicensed social clubs in the area. The detectives quickly leapt out of the van and grabbed the suspect before he could enter the building. They brought him to the van and asked the complainants if he was one of the gang. Two complainants immediately said yes; the third, the kid who had been knocked onto the tracks, was hesitant. But Keitel ordered angelface cuffed and put into the van. Keitel then told Detective Fanti, standing at the door of the social club, to demand that everyone come out slowly into the street.

Fanti held the club door open and ordered everybody out. One by one, teenagers straggled out of the club. With shouts, the complainants identified a young man wearing a baseball hat. When Aiello and Lopez brought him over to the van, he looked startled when he saw the complainants' faces through the open windows. The complainants taunted him as the detectives cuffed him. Then another youngster in a green jersey walked right past the van, pulling up his sleeve and ostentatiously showing a gold watch with diamonds on his wrist. The complainant who had lost his watch screamed: "That's my watch." Aiello grabbed the kid and cuffed him. More youngsters drifted out of the club; the police grabbed all of them and showed each to the complainants. The complainants identified one more assailant, whom the police also cuffed.

Suddenly, the scene exploded. An irate crowd poured out of nearby buildings into the street and swirled around the van, badmouthing the cops, while both berating and yelling encouragement to the youngsters being arrested. Within a few seconds, hundreds of neighborhood people filled the entire street. Aiello put a 10–18 over the radio (police officer needs assistance). The crowd grew and became more vociferous: homegirls and homeboys in baggy pants and sweat shirts shouted obscenities in street-rap rhythm, chests puffed out, hands chopping air; old men with rheumy eyes yelled and spit at the "goddam DTs" (detectives); early-thirties mothers of the teenagers under arrest screamed at the police not to take their sons away. All the detectives by this time had drawn their weapons. The street had become a tinder box.

Fanti put a 10-13 (police officer in distress) over the radio, upgrading Aiello's original call. Just then a heavyset man with dreadlocks sauntered out of the club with a huge pit bull on a leash. He walked through the milling crowd heading east on East 161st Street, virtually unnoticed, except by the complainants. The boy who had been kicked began to scream: "That's the guy. That's the guy who kicked me." Keitel yelled to Aiello to arrest him. Aiello hurried down the street and, with his gun trained on the pit bull, cuffed the man at the corner of Tinton Avenue. The crowd pressed forward, drawing tighter and tighter circles around the arresting officers. Suddenly wailing sirens, at first distant, then rushing closer, broke the tension. The cavalry arrived, complete with paddy wagon and several police cars, a full six minutes after Aiello's first call for help, ninety seconds after Fanti's upgrade.

With Keitel shouting orders, uniformed officers from the 40th precinct took the four teenagers already arrested into custody for delivery to Transit's District 12. On the ride back to the station, Keitel yelled at everyone, but especially at Aiello, for not wearing a vest. "I don't care if you don't arrest any of these mutts, but I want you to come home alive at the end of the day. That situation needed just one spark. One spark. I don't want to have to say this again."

Back at District 12, an old black man came into the house, his eyes blackened, his face and head battered and cut. He told

Aiello and Lopez that he had heard on the street that the DTs had locked up a heavy-set kid with dreads. He said the same guy had robbed him and beaten him earlier that night in front of the social club at East 161st Street and Trinity Avenue. The detectives told the old man that, if he could identify his assailant in a lineup, they would be glad to charge him with another robbery. The old man cordially thanked the police but said that he planned to shoot the boy himself.

———————

Youngsters who participate in wolfpack robberies think of themselves as part of a "posse," the name borrowed from the fearsome Jamaican gangsters who, Uzis at the ready, roamed Edgecombe Avenue in upper Manhattan in the mid-1980s. The Jamaicans in turn borrowed the term from the western movie shoot-em-ups that gangster wannabes watch by the hour. But in a symbolic inversion, Jamaican posses and those who adopted the name prided themselves on being outlaws, instead of newly deputized assistants to lawmen. Youthful posses still roamed the subways in 2004, though in fewer numbers than in earlier years. Typically, they beat up their peers. But they also threaten other "vics" or "herbs" with box cutters, knives, screwdrivers, clubs, or claw hammers, all for leather, Eight-Ball, or Starter jackets, baseball hats, gold chains, earrings (often ripped right from the victim's ear lobes), flashy watches, or sneakers that are standard-issue attire for many teenagers in the city. Posses also go after the occasional big score—a "print" (a visible wad of money) in the pocket of some hapless rube who, when approached by the posse's scout to see if he is "Five-O" (an undercover cop), makes the fatal mistake of betraying the startled eye-darting fear that invites aggression.

Many robberies by youth posses become male initiation rites
("I beat the shit out th kid because, if I dint, Rafael ud call me a
pussy on the street"; or "I not wanna hit the kid but just wanna
fit in, I just wanna not get called a faggot or that I a girl").[14]
Youths who get enticed into such rites resist giving up their
comrades when apprehended, because the whole point of such
crimes is making friends. The organizers of these rituals, however, routinely betray others to save themselves. Other wolfpack
robberies bind youths to one another by attacking rival gangs
or by punishing fake claims to gang membership. Thus on May
10, 1991, "Basher," "Killer John," and "Lunatic," all claiming to
be members of the Decepticons' Lost Boys legion from Crown
Heights, ran into a teenager on the subway who claimed falsely
to belong to the gang while on the shuttle train at Prospect Park
in Brooklyn.[15] Lunatic told the story in his own hand:

"Me and 'Carlos' was coming back from the park got on the
train at Caton Ave met some Decepts on strain. This boy said he
a Decepts Killer John call us. He said he was a Decepts we said
are you a decepts he said yes we ask what leand [legion]. He don't
no so we start hit him Killer John slice he than we though
hem off the train. I was not the gay that cut hem. I was punch-
ing hem and kick once I was Carlos and Killer John 'Fangreen'
Basher and 'Rob' I wasn't the one that went in his picket. I didn't
get or see any money one of the guy had it the money [quotation
marks added]."

Incessant struggles over honor and street reputation produce
other wolfpack robberies. Street youths who can claim little distinction in the established arenas of American society make respect and disrespect the main framework of their lives. In New
York City in the late 1980s and early 1990s, the trains became an
important milieu where black and Hispanic youths, in particu-

lar, tested each other's mettle. Of the 1,002 reported youth-posse subway robberies in 1990—handwritten into Central Robbery's huge bound ledger by officers who recorded victims' reports of the time, place, and circumstances of the assaults on them, along with their descriptions of the assailants—exactly two descriptions by victims pointed to assailants who were not black or Hispanic, and these two descriptions were ambiguous.[16] Many robberies were precipitated by insults or the smallest slights, real or imagined: fearful glances, baleful looks, careless words. Other robberies were inflicted precisely to humiliate perceived adversaries and to bolster reputations or self-images.

Growing up in a city that mainlines on celebrity and fame, youths who feel obscure often seize whatever is at hand to dance, if only briefly, in the bright lights. The enormously popular dance clubs of Manhattan from the 1970s through the 1990s, both legal and illegal, drew not only glittering social elites but youths from worlds apart, dressed to the nines, claiming glory through vibrant good looks, dancing or musical skills, or chic dress. Notorious uptown drug dealers regularly frequented the downtown club scene. And subway robberies often financed club excursions. On September 1, 1990, for instance, while two youths kept lookout for the police, six boys surrounded the Watkins family from Provo, Utah, who were in town for the U.S. Open tennis tournament and were waiting for an E train at the Seventh Avenue and 53rd Street station. The youths lacked Roseland Dance Center's $15 cover fee. So one boy slashed the father's pocket and stole his wallet, with $200 and credit cards. When the mother tried to intervene, she was punched in the face and kicked while on the ground. When 22-year-old Brian Watkins reacted to protect his mother, Yul Gary Morales stabbed him once with a chrome-plated butterfly knife. Brian continued to pursue

Morales but collapsed near the turnstiles. As he lay bleeding to death in his brother's arms, with his father murmuring a last Mormon blessing over him, Brian's assailants used the family's money to go dancing a block away. After the gang was apprehended and discovered to be members of a group named FTS (Flushing Top Society), dedicated to subway graffiti and disco dancing, Morales's street mates back in Queens painted a mural in tribute to their now-famous homeboy, known on the street as Rocstar.

Youth posse robberies often resemble wanton, senseless melees. For instance, in a handwritten statement one robber described an attack on the 3 train at Saratoga Avenue in Brooklyn on October 23, 1991: "We was all going train station myself 'Randy' 'Sean' 'Opie' 'Forever' we saw a kid on the train Then Randy start punch the kid punch the kid in the back then Forever punch the kid in the Face. Opie kick the did [kid] off the train sutter. Then we all walk throw the train and saw anther kid Randy punch kid in face start kick he. I punch the kid in the face Than people I don't know start hitting the boy. A girl pick up the boy thing [quotation marks added]." Violence, not words, expresses marauders' most important self-images, even as they shrewdly attempt to use words to insist that they only engage in assault, not in far-more-harshly-punished robbery.

Sometimes youth posses attack adults for real or imagined insults, affairs that are often racially tinged but always fraught with the tension and thrill of dominating others. In a handwritten statement, a culprit described an incident on the Q train on May 13, 1991: "We first was going to the beach after school. So we when down to the beach. And we when to chach the Q train to go to the beach And . . . the train was stoping Avery 2 memets and we was gething sick of that so we when to the next wagon

and my the kid saind the man in that chear look at him so some
of the gays wen to the man and they stared we him sain bad
words to him and meacking him flinch. So the gay gat up and
stud stio [still] and looking at every bady so he tall black kid
snoftem [socked or knocked? him] first and so "Jeimy" noftem
agan. And to more gays snoftem. And I pod my hands in his face
and when against the door I her every Bady sain Kill the white
man Kill the white man sad Kill the White man one time and I
stop and every Body keep on sain et I when to seat whi my Frean
'Eduardo' the train stop and the man when out the train to
more minutes the Cup came in . . . [quotation marks added]."

For all their amateur air, subway wolfpacks provide their
members with essential experience, at least for those who decide
to pursue criminal careers. Wolfpacks provide a social arena for
participants to test and display their "balls" in confronting another person, seizing property by force, and sometimes beating
the victim wantonly and savagely to show that they are "down"
with each other. Occasionally, wolfpack robbers engage in op-
portunistic sexual assaults on female robbery victims. In such a
context, participants often dare one another to act wildly. For instance, on November 14, 1991, four boys—"Zombie," "Two-Z," "J-
Boy," and "Harley"—approached a man on the downtown 5 train
between 96th and 86th streets. The other boys dared Zombie to
hurt the intended victim. Zombie pulled out his gun, pointed it
at the victim's head, and pulled the trigger three times.

Two-Z described the scene: "At 12 noon in the afternoon I met
Zombie at the corner store on 169 Washington Avenue and just
chilled out until 6:00 That's when we all met up . . . Then we
were 6 deep we started walking and we planned to go this girl
named 'Sissy's' house so we got on the train We rode the two
train from Prospect Ave to 149 Grand Concourse and caught the

5 train and everybody dared Zombie to pick a herb or take the dare so the Zombie said yeah OK then he pulled out the gun and clicked it three but the bullet didn't come out so he put the gun away and hit the man on time stopped then hit him again that's when the man started to bleed he pulled out a napkin to wipe his blood and then he got up Harley pulled him back down and kicked him in the mouth [quotation marks added]."

A reputation for wildness and for wanton cruelty shields one from others' encroachments. Few people wish to risk their lives to confront someone whose whole manner devalues life. But among those who wish to move on in the occupation of robber—encouraged, police argue, by youthful experiences of seeing little connection between action and consequences—wildness is not held in high esteem. At levels beyond the wolfpack, robbery is a relatively rationalized operation; like most occupations in modern society, it places a premium on control and self-control. Consequently, some career robbers become habituated to the subway because they favor a closed moving environment where victims have nowhere to go.[17] Such habituation extends to choice of victims. Subway robbers, working individually or in pairs, routinely target victims from particular ethnic backgrounds. Black robbers prefer new immigrants, easily spotted by their cultural uncertainty. Mexican and Chinese immigrants are special favorites because they possess slighter physical builds, and many, robbers know, are illegal aliens who are unlikely to complain to the police about being victimized. In addition, black robbers think that Chinese immigrants cannot identify them in lineups because "All we niggas look alike to da Chinks." Other robbers track what cops call the "wounded buffaloes" of the street—the feeble or the old. Many rob only elderly women, or women shepherding children. Whatever their choice of vic-

tims, robbers typically employ thoroughly rehearsed, indeed rit-
ualized, methods of operation, as well as formulaic verbal commands that signal and frame a robbery: "Up gainst the wall,
muthafucka!" "Do you know what time it is?" "This is a stickup;
hand it over." "Everybody be cool; this here a robbery."

Dedicated robbers take pride in their work. Detective Sergeant
Tommy Burke once conducted a lineup that included a suspect
photo-identified by a witness along with five fillers. After her
first look, the witness told Burke that she thought the person
who robbed her was number 3, but she wasn't sure. She asked
Burke if all those in the lineup could say what the actual robber
had said when announcing the robbery and, with some embarrassment, she whispered the formula to Burke. Burke gave the
order. Filler number 1 said in a mild voice: "Yo, muthafuckas,
this is a stickup." Filler number 2 said in a meek voice: "Yo,
muthafuckas, this is a stickup." The third person in the lineup,
indeed the actual suspect, strode up to the microphone, stuck
out his arms, and said: "YO, MUTHAFUCKAS, THIS IS A
STICKUP!" He then turned around and, with a condescending
look, indicated to the other guys that his was the right way to
announce a robbery.

If robbers are successful in their trade, and especially if they
are able to develop the "hardness" essential for occupational
longevity, they stick to tried-and-true routines for many years.
Among those who do robbery for a living, hardness becomes the
most highly prized occupational virtue.[18] Hardness means, above
all, a mental toughness and the ability to project such a thor-
oughgoing ruthlessness that victims become compliant and
nonresistant, enabling robbers to control completely the peculiar social interaction called robbery. Such control of the situation prevents robberies from "going bad." Robberies that do go

bad are usually thought to be caused by a victim's resistance to the robber, which creates a situation that is "out of control," typically producing an escalation of violence.

One learns such managerial skills only through experience. Because of constant practice as evinced in robbery rates, Brooklyn robbers are generally considered "harder" than those from other New York boroughs. A prosecutor tells the sad tale of a Brooklyn robber who had teamed up with a Manhattan robber to do subway work. One robbery went bad when the Manhattan robber lost control of the situation and then shot the victim. "The worst mistake I ever made in my life," the Brooklyn robber told the prosecutor, "was teamin with that New York nigga." Indeed, the general aphorism among Brooklyn robbers in comparing themselves with those from their sister borough is: "New York niggas, they be soft; Brooklyn niggas, we be haaaard!"

The police finally nabbed Shorty. On November 18, 1991, he visited his fence to sell some gold. The jeweler did his civic duty and called the detectives at his local Transit District 3. Shorty turned out to be a predicate felon already convicted of two previous felonies, including shooting at two police officers in the 10th precinct. After twelve lineups at Central Robbery (with eight positive identifications) and four lineups at Transit District 3 (with three positive IDs), Shorty was housed on Rikers Island awaiting further proceedings on his case. The transit police obtained an order to produce Shorty from Rikers for twenty-one more lineups at Central Robbery on December 4, 1991. The event was intended to be signal, with brass from up and down the transit police hierarchy in attendance, as well as media representatives. Even in the world of subway robbers (who tell detectives that they do between 30 and 35 robberies for every one in which they are caught), Shorty was a prodigiously prolific robber.

On the morning that a prisoner faces a legal proceeding, the Correction Department begins processing orders to produce prisoners for court at 0200 hours and puts its charges on buses heading for the city four hours later. Prisoners destined for Manhattan are taken to Central Booking, where they are held on the twelfth floor of the new Tombs building adjacent to 100 Centre Street, the location of Manhattan's criminal courts, until the detectives in one or another command fetch them and accompany them to their appointments. After the legal proceedings, prisoners are returned to Central Booking and remanded to the custody of Correction, which transports them back to Rikers.

Detective Jeremiah Lyons from the NYCTP's Major Case Squad was to pick up Shorty from Central Booking and deliver him to Detective Aiello for transport uptown to Central Robbery along with "Morgan," Shorty's accomplice in three robberies. Lyons called Detective Sergeant Keitel at 0900 hours to check in and reconfirm plans. Lyons was at Central Booking waiting on Correction officials, always slow and deliberate in their work. Most of the complainants had gathered at Central Robbery by mid-morning. Shorty's Legal Aid lawyer showed up a little later. The transit police brass drifted in late in the morning and chit-chatted with the Central Robbery detectives. Detective Aiello went downtown at 1100 hours to meet Detective Lyons in order to fetch Shorty.

At 1300 hours, Aiello returned with Morgan in tow, but no Shorty. According to Correction, Shorty had not yet arrived at Central Booking. Detective Lyons, a vital, personable, and excitable man, was, according to Aiello, beside himself. Lyons had confronted the Correction officer at Central Booking, but to no effect because nobody knew where Shorty was.

Things began to get tense at Central Robbery. But everyone made allowance for the usual mishaps that occur in transporting prisoners. Prisoners use any number of scams to avoid being produced, including declaring that they are sick and need to go to the infirmary. Detective Sergeant Keitel called the Rikers infirmary. No Shorty. Next, because buses dispatched from Rikers Island make several stops distributing prisoners to different jurisdictions, each of which requires transfer of custody, Keitel called Correction at Rikers and had them review each transfer for all the buses dispatched that day. But Shorty appeared on no Correction roster for any bus at any stop heading off the island.

The detectives waiting at Central Robbery began to chat about the regular lawsuits that prisoners institute against all law enforcement officials and the timidity of judges in cases that inevitably draw the attention of the New York media. Everybody acknowledged that the prisoners run Rikers Island, especially when regular city budget crises force a slowdown. Moreover, who could blame Shorty? Detectives readily noted that, if they themselves had already been identified in eleven lineups, they would do everything possible to dodge twenty-one more complainants. And everybody noted the grim irony of Shorty's elusiveness both on the streets and in the system.

No Shorty. The awkwardness of the situation increased by the minute. In the meantime, detectives from Transit District 3, who had collared Shorty and Mason on the tip from the jeweler, arrived at Central Robbery and asked: "Where the hell is Shorty?" Keitel made several more calls to Rikers. The Correction officer at Rikers insisted that his log showed that Shorty had left the building. That was all he knew. He assumed that Shorty had boarded the bus and was safely delivered to the Correction facility at Central Booking. Maybe, the Correction officer suggested, Shorty just was not answering to roll call at Central Booking. Did the detective receiving Shorty at Central Booking know him? Keitel ordered Aiello to go back downtown with a mug shot of Shorty for Lyons. With the photograph, Lyons went up and down the entire holding pen looking for Shorty among the dozens of prisoners waiting for processing. No Shorty.

Where was Shorty? The afternoon wore on. Shorty's lawyer announced that he wasn't staying after 5 P.M. Detectives wondered if the lawyer had known that Shorty planned to stiff them. All the transit police bosses sat in solemn assembly in Lieutenant John Walsh's office, the awkward silence broken only occasionally with caustic remarks. Lineups seemed less and less likely with each passing minute unless, one boss quipped, the police arrested the complainants. Bosses dispatched Detectives Don Mounts and Vinnie Valerio to test the mood of the complainants. The detectives reported that, although all the complainants were burning mad, sixteen of them remained so angry at being robbed on the subway that they were willing to stay until Shorty showed up. The detectives sent out for food and drinks for all the remaining complainants and made special arrangements for some, including transporting a singer to walk his dog and then to attend a long-sought audition.

Correction finally found Shorty at 1645 hours. He had never left Rikers Island. Instead, he had spent the day sitting quietly in the yard after refusing to get on the early morning bus. Whoever had drawn up the order to produce him had neglected to check the box allowing Correction to use necessary force.[19]

3

WHEN THE BALL FELL

Detective George Delgrosso had almost nothing when he reported to the Midtown North squad at 0800 hours on January 1, 1987. No case taken by Nightwatch. No police paperwork. No established crime scene. No ambulance reports. No hospital reports. Times Square and all the midtown watering holes had hosted several hundred thousand New Year's Eve merrymakers the night before, and the usual drunkenness, disorderly conduct, pick-pocketing, muggings, assaults with deadly weapons, and fisticuffs had kept two thousand police officers and all the hospital emergency rooms on the island busy until dawn. Detective Delgrosso had only 71-year-old Jean Casse of Toulouse, France, retired insurance broker, father of three children, grandfather of five, in St. Luke's-Roosevelt Hospital, along with his distraught wife, Renée, a retired physician who spoke no English. Jean Casse lay near death with a broken skull and brain contusions, fractures of the thyroid cartilage and the cervical spine, and multiple contusions on his head, face, and neck.

Midtown North, the sprawling 18th precinct of the New York

City Police Department, extends at its southern end from the Hudson River and 43rd Street to Eighth Avenue, then from 45th Street and Eighth Avenue to Lexington Avenue, all bordered on the north by 59th Street. The precinct jumps with action every day. Well-organized rings of high-class prostitutes roam the Sixth Avenue hotel corridor looking for midwestern businessmen flashing Rolexes ("I can spot a John a block away"), whom they French-kiss with atropine-smeared lipstick to immobilize them and turn them into easy pickings. Con men offer a checking service (with stamped receipts) at Saint Patrick's Cathedral for tourists' valuable cameras and video equipment ("No photography is allowed in the house of God"), while fake priests persuade naïve young girls to relieve the devilish pressure in their saintly loins so they can continue God's work without distraction.

Jewelers on 47th Street between Fifth and Sixth avenues provide a market for the diamond rings, gold necklaces, and earrings snatched from terrified subway passengers by young robbers, who prefer payment in recycled gold bangles, ear studs, or tooth sleeves. They also make a market for diamonds or jewelry taken by more seasoned robbers who track traveling diamond dealers and seize their leather bags stuffed with six-figure merchandise. African immigrants stake out the square in front of the Plaza Hotel at 59th Street and Fifth Avenue, their tables laden with fake seventeen-jewel watches that tempt long lines of honest citizens looking for too-good-to-pass-up deals. Taxis swerve across Fifth Avenue in reverse to pick up dressy fares, sometimes hitting jaywalking pedestrians. Other taxi drivers refuse to leave midtown's bright lights to take fares to the dark corners of the city, even menacing the occasional intrepid passenger who insists on his inalienable right to be taken to Highbridge or Coney Island. Insouciant messengers and fitness-minded bankers alike barrel through red lights on bicycles, scattering pedestrians like tenpins.

In an Eighth Avenue salad bar, a homeless drunk helps himself to fresh spinach with his bare, bloody hands after killing his drinking buddy in a blind stupor. Gypsy fortunetellers dazzle tourists from the hinterlands with flashes of bosom and promises of future furtive delights, all the while making wallets vanish with expert sleight-of-hand. At the southeast corner of Eighth Avenue and 42nd Street, the Lost Tribes of Israel, dressed in full biblical regalia and quoting prophesies from the Holy Book, promise salvation for all peoples of color, while hurling racial slurs at passersby condemned by their pale skin to eternal damnation and hell-fire.[1]

Just across the street, young men and women, boys and girls from all over the country disembark Greyhound buses and pour out of the Port Authority terminal into the bright lights of the Deuce, seeking thrills in smack, crack, blow, speed, smoke, or crank, and in gay bars or porn theaters along Eighth Avenue. Street honeys, mostly girls from uptown, ply their trade on Eighth Avenue as well, while stud hustlers ("I always be the fuck*er*, never the fuck*ee*") prowl the streets and bars for rough-trade seekers, whom cops often find trussed and robbed, sometimes killed, in the flophouses that crowd the western end of the precinct. Flophouse owners, fearful of the strength of the predators under their roofs, regularly scorn willowy or short female officers dispatched to quiet disturbances, demanding policemen who can solve problems quickly by knocking heads ("I wants fuckin *men* po-lice!"). And bands of young robbers from Brooklyn ("Manhattan make it, Brooklyn take it") prowl midtown corridors, looking for just-paid city workers, theatergoers, shoppers, thrill seekers, or gawking tourists whom they can corner, punch senseless in "strong-arm robberies," and relieve of their money.

Delgrosso knew a lot about Brooklyn robbers. In the 1970s, after a stint on foot patrol in Brooklyn's 94th precinct, he worked in the Neighborhood Stabilization Unit on Muggers' Row between Ocean Parkway and Prospect Park in the 70th precinct, catching hundreds of street and push-in robberies, muggings, and store heists, once getting stabbed in the back by the brother of a boy he was arresting in the midst of a melee. Later, he did anticrime work in Brooklyn's 75th precinct in East New York, always one of the city's busiest and most violent precincts, and then in that precinct's task force on youth gangs.

After a tour ghosting buy-and-bust narcotics operations on Manhattan's Lower East Side, Delgrosso was promoted to the detective squad of Midtown North's D team in December 1985. He worked with Robert Chung, who made his bones by infiltrating Chinese tongs; Alex Renow, a career Midtown North veteran, first in uniform and then in the squad, a sage observer of the human comedy that police see unfold every day; Pete Castillo, who brought several years of federal task force work to the squad; and Pete Panuccio, another veteran of the "People's Republic of Brooklyn" and something of a specialist in its colorful and telling street language ("Man's heart done seized up and the parimutuels had to jump-start it").[2]

During Delgrosso's first year in the squad, he worked on several homicides caught by other detectives. When Jean Casse died from his injuries at 1030 hours on New Year's Day, 1987, Delgrosso had caught his first homicide.

———————————

At 1100 hours, with the help of Sergeant Frederick Sachs of the 17th precinct, a police interpreter, Delgrosso got the original story from Renée Casse at the Plaza 50 Hotel on East 50th Street. Mrs. Casse told Sachs and Delgrosso that she and her husband had never visited New York before. They had arrived on December 27, 1986, with seven other French men and women, for a sightseeing and shopping trip. On New Year's Eve, the entire group went to the New York City Ballet's evening production of *The Nutcracker* at Lincoln Center, which ended at 10 P.M. They ate a late dinner at Scarlatti's restaurant on East 52nd Street near Madison Avenue and then returned to their rooms at the Plaza 50 to change shoes and drop off shopping bags before heading to Times Square to ring in the New Year. But the Casses had trouble keeping up with the younger members of the tour group. They suddenly found themselves alone on a very crowded street. As bells began to ring, people started shouting and singing and the crowd pressed around them. Renée told Jean that it was midnight and they should return to the hotel, but Jean insisted on continuing to Times Square. Renée had her arm on Jean's, when suddenly she found herself losing her balance. She fell down and hurt herself. She got up quickly but, to her horror, saw Jean prostrate on the sidewalk, bleeding from the nose and the mouth. A man was kneeling on Jean, holding him around his neck and smashing his head on the sidewalk near a large flowerpot in front of a restaurant. Renée threw herself on the assailant, pulling him by his hair and ears. The man took Jean's wallet, which contained about $500 in one-hundred-dollar bills, and then fled east on 52nd Street.

At 1300 hours, Delgrosso and Sachs took Mrs. Casse and an American friend of the Casse family to 52nd Street. Mrs. Casse pointed out the scene of the assault, just under the awning of Ben Benson's steak house, a well-known New York City restaurant, at 123 West 52nd Street. But Ben Benson's, indeed the whole street, was dark on New Year's Day, the quietest Thursday of

1987. Delgrosso ordered uniformed officers to cordon off the area in front of the steak house.

After returning Mrs. Casse to her hotel, Delgrosso went to St. Luke's-Roosevelt Hospital, where he tracked down one member of the ambulance crew, Patrick Powers, an emergency medical technician from Saint Clare's Hospital, a West-side sister institution. Powers said that he and his partner, Michael Vaughn, whose day job was teaching biology in public school, were stationed on Seventh Avenue at 53rd Street on New Year's Eve. At midnight, from their truck, they watched the ball fall. Suddenly, they received an emergency call for a man shot in the head on 52nd Street between Seventh and Sixth avenues, right around the corner from them. Despite the wall-to-wall crowd, it took only a minute for the EMTs to drive down Seventh Avenue and then east on 52nd Street. All intersections were cordoned off, and blue police barricades stopped all other traffic. Police officers cut a slot through the surging crowd for the ambulance.

On 52nd Street, jammed with milling pedestrians and more than a dozen police officers, two uniformed cops directed them to the north side of the street, halfway down the block to Ben Benson's steak house. The technicians found a heavy-set gentleman with a gray beard lying on his back, his feet pointed toward the street. The man breathed rapidly and, with each breath, blood spurted profusely from his nose. He also bled from his right ear. His pulse was eighty-four, his skin moisture normal, his skin temperature cool, his color pale. Powers and his partner shone a flashlight into the man's pupils but got no flicker in return. Police officers swung a huge potted plant out of their way so that the technicians could turn the man on his left side and clear his airways. That was when Powers and Vaughn saw matted blood on the back of the man's head, though they were unsure of its source. The technicians stayed at the scene for about three minutes and then raced the victim to St. Luke's-Roosevelt Hospital, the nearest medical facility, where they left him with nurses in the trauma room of the emergency ward.

Delgrosso went from the hospital back to the Midtown North station house and combed through the log of radio calls that police had received the night before. The log gave him the sector that handled the call. However, before going off duty, the sector cops had done no paperwork that might provide more details of the assault. Delgrosso then contacted the Department of Transportation and found some traffic cops and meter maids who were on duty in the Times Square area on New Year's Eve. But none of them had seen the assault on Casse. At 1450 hours he returned to 52nd Street with Detective William Schachtel and other members of the Crime Scene Unit. The unit took photographs of the entry to Ben Benson's, particularly the still-visible bloodstain on the sidewalk and the tilted potted plant. Back at the house, Detective Alex Renow suggested that Delgrosso check with Central Booking to see if anyone had been arrested in the Times Square area with C-notes (one-hundred-dollar bills) on his person. Delgrosso took a ride downtown to Central Booking at 100 Centre Street and reviewed all the arrests for robbery in the Times Square area on New Year's Eve, about fifteen in all, but officers had confiscated no C-notes from those arrested on New Year's Eve or morning. Calls to Central Booking in Brooklyn and Queens also produced nothing.

When Ben Benson's opened the next day, Delgrosso found three witnesses to the assault. James Head, known as Mike, had been working tables near the front windows of the steak house. Suddenly, around midnight, Head saw twenty or thirty black kids creating havoc in the street, pushing and knocking around several pedestrians. Head told the maître d' to lock the restaurant's door. Then he saw, as if in slow motion, an elderly man falling down, straight down, right on his head with a thump, just between the two large planters underneath the awning that stretched from the restaurant's front door to the curb. The man was wearing a camel-hair overcoat. His feet were pointed toward the street. Head and the maître d' raced outside and screamed at the three or four kids surrounding the old man, one of whom straddled the man from his knees to his thighs as he went through his pockets. While another waiter brought out tablecloths to cover the victim and keep him warm, Head went back inside the restaurant and called 911, claiming, in order to get a quicker response, that a man had been shot.

The maître d' echoed Head's account, as did Richard Farrar, another waiter who had been upstairs changing his clothes. Farrar looked out of the upstairs windows and saw about two dozen black kids running in circles on the street. To his right, he saw several kids surround a girl in a white fur coat and grab something from her. To his left, he saw three kids push someone to the ground; he could see the victim's legs from the window, but his view was partially obscured by the sidewalk awning. Farrar went downstairs and joined the maître d' and Head.

After the EMS team removed the victim from the street, restaurant employees doused the pool of blood that had gathered on the sidewalk with buckets of water and detergent. Police officers on the street did not tell them to stop. Delgrosso asked the steak-house manager for a list of all his employees as well as credit card slips for all the restaurant's late-evening New Year's Eve customers in order to locate other potential witnesses. But the manager could produce only two credit card slips for the entire evening, claiming that all the other customers had paid in cash.

Back at the station house, Delgrosso found PO Robert Giannetta, one of the officers who had responded to Head's call about shots being fired. Giannetta said that he had just come on duty at 2330 hours and was sitting in a radio car with his partner, PO Dan Danaher, on Broadway at 54th Street facing south when the 911 call came across the air at 0005 hours, just after the ball fell. The officers drove to 52nd Street and found a man lying on the ground with a sheet over his body. A female police officer and three plainclothes police officers, one of whom was Sergeant Sheerin, were huddled around the prostrate victim, along with the victim's wife and a restaurant employee. The three plainclothes officers left the street heading east. When the EMS team arrived, Giannetta helped turn the large man over and then place him in the ambulance. To do so, he had to move a barrel that served as a planter. Giannetta said that he did not safeguard the scene.

Delgrosso immediately tracked down the plainclothes assignments for New Year's Eve. Sergeant Sheerin had headed a detail from 2330 to 0800 hours, policing the area from 44th to 66th streets, from the Hudson River to Lexington Avenue. Sheerin supervised two officers, one of whom was Michael Bachety. Bachety told Delgrosso that Sheerin's team had parked their unmarked car at 55th Street and Seventh Avenue because it was too congested to drive. The officers walked down to 50th Street on Seventh Avenue, where they watched the ball fall. As the officers walked back uptown to their car, a large group of black youths thronged past them, heading north up Seventh Avenue, then turning east on 52nd Street. The plainclothes team followed them, staying about six feet behind the youths on the south side of 52nd Street. Suddenly the youths broke into a run.

Across the street, the officers saw a man lying on the sidewalk, with a woman kneeling next to him. The man lay perpendicular to the street, his feet ten feet from the curb, his head a little to the west of a restaurant's awning. After radioing for a sector car and a "bus" (an ambulance), Bachety and Sheerin ran east on 52nd Street to see if they could catch up to the youths they had been following. But they found no trace of the youngsters. When Bachety and Sheerin returned to the crime scene, the ambulance was just pulling away. The next morning, after learning that Casse had died, Bachety went to the Medical Examiner's Office at Bellevue Hospital, where he identified Casse's body, number 87–18.

Several New York dailies ran articles on Casse's murder on January 2, 1987. Though replete with inaccuracies, the articles prompted a couple to come forward that day to relate to Delgrosso what they had seen on 52nd Street on New Year's Eve. "Bob Brown," who used to work in New York City in one of the big financial houses, had flown back from Los Angeles to spend New Year's with his friend, "Janet Yost." Brown gave a detailed statement. He said that he and Yost had had dinner on New Year's Eve and then went to a party at NBC studios on Sixth Avenue near 50th Street. Afterward, the couple strolled up Sixth Avenue and then decided to walk over to Seventh Avenue to see the ball fall. They turned west on 52nd because the police had not cordoned off that street. As they walked along the north side, they saw a group of kids running toward them, yelling and screaming, striking anyone in their path. Brown saw the kids hit two men. Brown and Yost were surrounded two or three times by youths shouting at them and menacing them, but they were not hit.

Suddenly, to his right, Brown saw a young woman in a white coat being chased under the overhang of a building on the north side of the street and knocked to the pavement. Immediately, Brown turned Yost around and headed east with Yost on his right. To his left, only twenty-five feet away and just a few seconds after he had seen the assault on the young woman in the white coat, Brown saw a tall, husky young man strike an elderly man with his right fist with great force. The man's feet went straight out from under him and his head hit the pavement with a ghastly crack. Brown said that the assailant then jumped two steps to his left, exhibiting enormous excitement. Brown could not see the assailant's face because his back was turned. Brown kept moving, but he looked back and saw someone on top of the old man, while the "hitter" was sprinting toward Sixth Avenue. Brown added that all the kids involved in the assaults were black. They kept yelling: "Just the whites!" Yost heard them saying: "Get the whites!" She caught only a glimpse of the assailant. She described him as tall and skinny, with a close-cropped Afro, and wearing a light gray jacket.

By this time, Delgrosso had a complete scenario of what happened to Jean Casse and his wife. The incident had all the earmarks of one of the Brooklyn-style strong-arm robberies that plagued midtown Manhattan during the 1980s. But he had as yet no accurate description of the assailants and no further leads.

The first break came in the early afternoon of January 3. PO Michael Paccione of the Midtown North robbery squad interviewed a young man from Brooklyn arrested with three accomplices for a strong-arm robbery on January 2 in front of 56 West 47th Street. Robbery squad officers prowling the streets in an unmarked car had actually eye-witnessed the as-

sault. Paccione asked the young robber if he knew about a rob-
bery/murder in Times Square on New Year's Eve. The boy said
that on Thursday, January 2, while at the Albee Square Mall on
Fulton Street in Brooklyn, he had asked an acquaintance if he
was going uptown to "get paid"—Brooklyn robbers' phrase for
their "work."[3] The acquaintance said: "No, I think I caught a
body when the ball fell." Paccione pressed for a name, but the
boy said he knew this man only as Smokey. He added that James
Walker, one of the people with whom he himself had been collared, might know more because Walker had been with him
when Smokey made the remark.

Paccione immediately called Delgrosso. Delgrosso tracked
down James Walker, who was also still in custody. Walker had
been around the block, chalking up arrests for robbery, attempted robbery, chain snatching, and assault. Walker claimed
that he had had been at home on New Year's Eve until after the
ball fell. Then, at around 1:30 A.M., he went into Manhattan and
ended up at the Latin Quarter on Broadway and 48th Street,
where he saw Smokey.[4] Walker knew Smokey from the Albee
Square Mall, a meeting place for Brooklyn robbers who go to
Manhattan on Thursdays and Fridays to get paid. There, he said,
they exchange gossip, street lore, and stories with one another
and boast about their exploits. They also choose up three- or
four-man robber bands on the criterion of perceived criminal
prowess. They then carve up Manhattan, assigning specific lo-
cales to specific bands, in order to reduce competition and avoid
unwanted police attention.

Walker knew Smokey by that name and by another street handle, Catfish. Walker said that Smokey stood over six feet tall and
weighed about 230 pounds. He acknowledged that Smokey had
said he "caught a body when the ball fell." Delgrosso pressed

him for the meaning of that phrase. And Walker said that Smokey thought he had killed someone at midnight on New Year's Eve.

Delgrosso took Walker to several police units that housed collections of photographs of assailants, including the 75th precinct's robbery unit and anticrime unit, Brooklyn's CATCH unit, the New York City Transit Police robbery squad, and ending up at the Manhattan CATCH unit, then housed in the 20th precinct on West 82nd Street.[5] Walker looked through scores of photographs but found none of the young man he knew as Smokey. He did find pictures of several of Smokey's associates, however, including Jerry Sanders aka Drak and David Warren aka Young. Delgrosso took down both names, writing a note for his file that Drak's typical modus operandi (as recorded in police notes that accompanied his picture) was to rob on the streets and then duck into a subway to escape apprehension. Walker told Delgrosso that David Warren, in particular, hung out with Smokey at the Albee Square Mall and that he had seen Warren at the Latin Quarter on New Year's Eve with Smokey.

Delgrosso pulled Warren's sheet and discovered that he had been arrested for strong-arm robbery a few months before in the Midtown South precinct with another Brooklyn teenager named "Chico." After PO John Bane had retrieved Chico from his house, Delgrosso grilled him about Warren. Chico admitted being arrested with Warren in September. Chico said that Warren lived in the Pink Houses in East New York, Delgrosso's old stomping grounds. Chico acknowledged knowing Smokey as well, though he did not know his real name. Smokey also lived in the Pink Houses on Crescent Street. Chico could not give an exact address, but Smokey could always be found in the Albee Square Mall on Thursdays, Chico said. It was payday for city workers, and Smokey would be putting together a team to go uptown to rob.

Delgrosso needed Smokey's real name and exact address. So he ordered James Walker to telephone David Warren and find out how to reach Smokey. Walker did as he was told, and the unsuspecting Warren provided Walker with Smokey's phone number. Delgrosso did a reverse search, which came up with an address on Crescent Street.

Sergeant Wally Zeins led a team of cops and detectives to Brooklyn to find Smokey, accompanied by James Walker in a van with dark-tinted windows. The unit camped outside the address on Crescent Street and waited. At 2000 hours a hulking young man came out of the house, accompanied by a girl. Warren immediately identified the person as Smokey. Smokey and the girl started walking toward the Linden Boulevard area. The police car followed the couple closely for a few blocks, and then the detectives jumped out and surrounded the pair. A detective told Smokey to get against the fence to be tossed. Sergeant Zeins asked him to come to the local station house, which he agreed to do. Smokey's girlfriend later claimed that the cops had put guns to Smokey's head and back. To protect Walker's identity, POs Maloney and Crowley responded to Zeins's call and took Smokey to the 75th precinct station house in a separate vehicle. The police did not arrest or handcuff Smokey. Smokey called his father from the station house and told him to come down. Smokey turned out to be 19-year-old Eric Smokes.

Delgrosso interviewed Smokey at a desk in the 75th precinct squad room beginning around 2230 hours. Delgrosso told Smokey that he was investigating numerous assaults and robberies in the Times Square area on New Year's Eve. Delgrosso asked to take Smokey's picture, explaining that he intended to show the photo to witnesses to an assault. If he were identified, Delgrosso explained, Smokey might be in trouble. Smokey agreed to be photographed. After using the Polaroid camera, Delgrosso settled into the interview. Where, Delgrosso asked, was Smokey on New Year's Eve? Smokey said that he went to see the movie *Heartbreak Kid* with some friends. Delgrosso expressed disbelief that Smokey had been at the movies on the biggest night of the year. Smokey hemmed and hawed a bit. Then he said that he had mixed the nights up and that, actually, he had gone to Manhattan at about 9:30 or 10:00 P.M. with some friends for the New Year's bash. He said that he was with David Warren and others, including a boy named "Ned Davis." He claimed that he took the F train, walked up Sixth Avenue, and ran into some other kids from Brooklyn, eventually ending up at 47th Street and Broadway, near the Latin Quarter.

Delgrosso asked Smokey where he was when the ball fell. Smokey replied that he thought he was at 47th Street and Broadway. From there, he and his friends walked downtown toward Madison Square Garden at 34th Street and Seventh Avenue. Delgrosso asked Smokey if he had gone any further uptown than 47th Street, but Smokey insisted he had not. Delgrosso asked if Smokey had seen any robberies while he was walking around. Smokey denied seeing any, although he said there was a lot of craziness in midtown that night. Smokey did say he had seen a fight and a shooting at 39th Street and Seventh Avenue around 12:45 A.M. Smokey claimed that he went back to Brooklyn with the same friends at about 2:30 A.M.

Smokes's father arrived during the interview. Although Smokes was old enough to be interviewed without having a parent present, Delgrosso allowed Smokey's father to remain.

The cat-and-mouse interview continued. Delgrosso mentioned that some kids had robbed an old man, who had died several hours later, but Smokey denied knowledge of any such incident. He asked if Smokey knew anybody who had made a good score on New Year's Eve. Smokey said he knew no one who got lucky that night. Delgrosso showed Smokey some photographs that he had brought with him from the CATCH unit. Smokey picked David Warren's picture out of the bunch and gave Delgrosso Warren's address. Delgrosso then asked Smokey if he hung out at the Albee Square Mall on Fulton Street. Smokey said that he went to the mall from time to time. Delgrosso told Smokey that the police heard that Smokey had said that he could not go to Manhattan because he caught a body when the ball fell. Smokey indignantly denied making any such statement and demanded to know who had lied to the police about him. Delgrosso continued: Had Smokey punched anyone that night in Times Square? Smokey denied it. Delgrosso asked if anybody in the group accompanying Smokey in Times Square had done a robbery on New Year's Eve. Smokey said no.

At midnight, leaving Smokey in the squad room with his father, Delgrosso went to David Warren's house on Linden Boulevard in another city housing project. Warren's mother was upset to have the police visit her home at midnight. But after Delgrosso showed David's picture to her and told her that the police were investigating assaults in Times Square on New Year's Eve, she fetched David from a neighbor's apartment down the hall. David acted sullen and withdrawn. But he acknowledged that he had been in Times Square on New Year's Eve with Eric Smokes, Ned Davis, and some other friends. Warren's mother asked if Delgrosso wanted to talk with Ned Davis, in whose apartment David had just been. When Ned arrived, Delgrosso spoke to him privately. But Davis denied being in Times Square on New Year's

Eve. When Delgrosso pressed him with the statements of both Smokey and Warren, Davis reiterated his protests vehemently and said that he had been home watching television with his parents.

Delgrosso asked Warren to come with him to the station house, and both Warren and his mother acceded to that request. Delgrosso tossed David for a weapon before they got on the elevator but did not handcuff him. When they reached the 75th precinct station house, Delgrosso put Warren in an interrogation room in the robbery unit. Smokey was still in the squad room down the hall. The two young men could not see each other.

Delgrosso went over Warren's statement with him in detail. David said that he had gone to Manhattan with some friends on New Year's Eve, got off the train at 42nd Street at Sixth Avenue, and walked up Sixth Avenue and then over to Broadway at 48th Street, near the Latin Quarter. When the ball fell, he and his friends were in that area, Warren said. Then they walked downtown to Madison Square Garden at 34th Street. Warren said that he had seen a lot of pushing and shoving but no robberies. Warren insisted that he had never gone above 48th Street. Delgrosso asked Warren if he had seen Smokey hit anyone that night, but Warren denied that he had.

At this point, Delgrosso lied to Warren. He told Warren that Smokey had admitted hitting somebody on New Year's Eve. He asked Warren if he had seen Smokey hit an old man. Warren said no, but then he said that he had indeed seen Smokey hit somebody, but it was a young man, not an old man, and the assault occurred, Warren said, at 41st Street and Sixth Avenue sometime after the ball fell. Warren insisted that neither Smokey nor he had taken any property from the man Smokey hit, an emphasis that told Delgrosso that Warren was well aware of the legal distinction between simple assault and robbery. Instead, David said, a young Hispanic man named "José" whom they had met at the Latin Quarter had taken property from this victim.

Delgrosso went back to the squad room and confronted Smokey with Warren's statement about the assault at 41st Street and Sixth Avenue. Smokey admitted the assault. But then he too insisted that he had not taken any property from the man he struck. Instead, he said, someone named José from Ninth Avenue and 50th Street took the man's wallet. Delgrosso pointed out to Smokey that, under the law, hitting someone so that someone else can pick his pocket is considered robbery. But with a notion of responsibility characteristic of the street—one is responsible only for actions that one directly and personally performs—Smokey insisted that he had not robbed anyone.

Delgrosso returned to the robbery interrogation room and continued his interview with Warren. He told David that the police knew all about the robber shape-ups at the Albee Square Mall on Fulton Street, suggesting that Smokey acted as a ringleader in organizing robber crews to go to Manhattan. Warren, who had only the one previous arrest with Chico for robbery, admitted that Smokey and he had gone into Manhattan about a dozen times since the summer and that he had seen Smokey hit people. But, working with the same notion of responsibility asserted by Smokey, he insisted that he had never seen Smokey take property, and he denied any personal responsibility for robbery of any sort. He also stuck to his story about New Year's Eve, arguing that he and Smokey had never been above 48th Street. Delgrosso warned David that he and Smokey stood to be arrested if any witnesses to New Year's Eve robberies in Manhattan picked their photographs out of photo arrays. With that, at around 0130 hours, Delgrosso drove Warren back to the projects, dropping him off a block from his house because David did not want neighbors to see him getting out of a readily identifiable police vehicle.

In the meantime, Smokey told Sergeant Zeins, who was sitting with him and his father in the squad room, that he just remembered that a homeboy named "EZ" was counting C-notes the day after New Year's. According to Smokey, EZ had said that he "did a poppy." Zeins reported the information to Delgrosso, who laughed when he heard that Smokey had miraculously remembered such a vital piece of information. But he went to Smokey and asked him for help in finding EZ. Smokey said that EZ hung out at the Albee Square Mall. Smokey then left the station house with his father at 0200 hours on January 4, a free man. But Delgrosso now had his picture, along with the Polaroid shot he had taken of Warren at the station house, to show to the witnesses of Casse's murder.

Before leaving Brooklyn, Delgrosso rousted Ned Davis out of bed in the early hours of January 4 and went over his story. Davis now sleepily acknowledged that he had in fact been in Times Square on New Year's Eve. He said that he, Smokey, David Warren, and another kid had arrived at 42nd Street in Manhattan between 11 and 11:30 P.M. They walked around the Times Square area with several friends until midnight. When the ball fell, they were, he said, at 44th Street and Broadway. Then they walked downtown to Madison Square Garden. The atmosphere was tense. Cops with radios watched them steadily and followed them. They walked uptown to the Latin Quarter at 48th Street and Broadway and checked to see if they had the requisite $25 per person for admission. But they were short. So he and David Warren went back to Brooklyn together. He admitted lying to

Delgrosso earlier, pointing out that he usually lied to the police until he had a clear idea of what they wanted. Delgrosso sensed that Davis was lying now, but he could not shake his story.

Delgrosso also talked again with James Walker. He insisted on his own alibi for the early hours of the New Year but now acknowledged that he himself had gone to Manhattan many times with Smokey and Warren to rob. Smokey always played the strong-arm puncher. Walker and Warren went through the felled victims' pockets. If anyone pursued them, Smokey punched them out too. Walker admitted that he had robbed with Smokey and Warren at least a dozen times in this manner, probably many more, but they were never caught. After robbing, the crew often went to the Latin Quarter, where they met lots of girls from the projects in Brooklyn. Walker also said that he knew EZ, and that EZ was still robbing people in Manhattan. Delgrosso became convinced that he had two of the right players in Smokes and Warren. But what police think they know and what they can prove in court are often two different things.

The witnesses to the assault on Jean Casse could not give Delgrosso the proof he needed. Renée Casse could identify neither Smokes nor Warren. Things had happened too quickly, and Mrs. Casse, in a state of panic and shock, had never seen the hitter in the first place. Mike Head, the waiter from Ben Benson's steak house, could not pick Smokes or Warren out of a photo array. It had been chaotic in the street, and Head had never seen the hitter's face. He had only gotten a glance at the back of the man rummaging Casse's pockets. Janet Yost's description of the hitter as tall and skinny did not fit Smokey—big, strapping, and muscular. And Bob Brown had seen the face of neither the hitter nor the man on top of Casse. Delgrosso could not make a case with these witnesses.

Over the following few days, Delgrosso continued working the Albee Square Mall connection. He reviewed scores of arrest reports for New Year's Eve robberies committed by wolfpacks from various Brooklyn projects. Most of these assaults had occurred in midtown and Upper East Side precincts, but a few took place further uptown. One New Year's Eve wolfpack robbery caught Delgrosso's eye, a rip-off with a hammer in the 25th precinct in East Harlem about an hour after the Casse assault. Police charged "Charles Dawson" from the Fort Greene projects as one of the assailants in that robbery. Delgrosso added Dawson's picture and those of Dawson's associates to his pile of photographs to show witnesses.

Delgrosso got his second break late on January 7. He received a call from a detective in Brooklyn's 88th precinct who was dating a woman from the Fort Greene projects. The detective said the woman had discovered a press clipping on the Casse murder in her 16-year-old son's room. When she asked him about it, her son told her that he and his friends had seen the murder. He was with some homeboys from the Fort Greene projects, he said, when he saw the body on the ground; they all fled when a man came out of a restaurant toward them. The boy's mother, fearful of asking for more details, suggested that he call the police with this information. Instead, her son went to his grandmother's house because his nana would hide him from the police. The woman told her detective boyfriend that her son's name was "Michael Edwards."

Detective Jimmy Kennedy of the Midtown North squad picked up Michael Edwards at his grandmother's house at midnight on January 8 and brought him to Midtown North for questioning, telling him only that the police wished to speak to him about something that had happened on New Year's Eve. At the station house, Kennedy revealed the real reason for the discussion. Shocked that the police were questioning him, Michael at first denied all knowledge of the murder. They discussed his friends. Michael said that he hung out with his cousins, "Bill Johnson" and "Bobby Johnson." He admitted that he regularly went to the Albee Square Mall. Kennedy checked Michael's record and found that he had been arrested in 1986 with Bobby Johnson for criminal mischief—bending bars in a subway station—but the case was later dropped. Bill Johnson, a big, husky young man, had only a misdemeanor arrest for fare-beating on his record.

Kennedy asked where Michael and his friends had been on New Year's Eve. Michael said that he, Bill and Bobby Johnson, "Shawn Green," and two other kids from the Fort Greene projects had gone into Manhattan around 10 P.M. They wandered around Times Square until the ball fell and then went back to Brooklyn. Kennedy said that Michael knew more than he was disclosing. Kennedy suggested the possibility of Michael going to jail for withholding evidence. But that threat did not faze Michael. Kennedy said that kids go to Albee Square Mall to team up for robberies in Manhattan. Michael said that he went to the mall to shop. Kennedy pointed out that Bill Johnson certainly fit several witnesses' descriptions of Casse's assailant. The detective said that he was going to get photographs of Michael and all his friends and show them to other witnesses to see if they put Michael and his crowd on the set.

Kennedy's last remark seemed to trigger apprehensiveness in Michael. At 0345 hours, Michael gave Kennedy the following story: He and his friends were at 50th Street and Seventh Avenue when the ball fell. They did, in fact, wander over to 52nd Street, heading east, following crowds of kids from Brooklyn. They formed a line on the north side of the street, Bill Johnson leading, Michael five feet behind him, and Bobby Johnson five feet behind Michael. They had been on 52nd Street for no more than thirty seconds when, suddenly, just to their left, Michael said he saw a lady getting mugged. The boys kept walking. Then, again to his left, from a distance of fifteen to twenty feet, Michael said that he saw an old man fall like he had been hit. The man fell right in front of Ben Benson's steak house. Michael also saw an old lady in a fur coat at the scene.

Kennedy pressed him on what else he saw. Michael said that a guy named Smokey, whom he knew from the Albee Square Mall, was walking east toward Sixth Avenue after the man fell. And he saw another guy, whom he recognized from the mall, bend at the waist, lean over the man on the sidewalk, and go through his pockets. Michael said that he, Bill, and Bobby kept walking east on 52nd Street, went around the corner, saw other people whom he did not know, and then returned to 52nd Street. The ambulance had just arrived and the street was crowded with police. Then he and his friends went back to Brooklyn. When Kennedy showed Michael photographs, he picked out Smokey as the hitter and David Warren as the person rummaging through the pockets of the man on the sidewalk.

The detectives kept Michael in the house while they raced back to the Fort Greene projects to pick up the Johnson brothers and Shawn Green at 0500 hours. The three boys made extremely reluctant witnesses. Shawn Green, himself tall and husky, admitted to being in Times Square on New Year's Eve, and he even admitted being on 52nd Street right after the ball fell, but he claimed to have seen nothing. The detectives pressed him, but Green did not budge from his story. At 0600 hours, Kennedy and Delgrosso interviewed 19-year-old Bill Johnson, whose arrest for jumping a subway turnstile turned out to be complicated by his persistent failure to show up in court on that minor offense, resulting in the issuance of two bench warrants for contempt of court. Bill admitted being in Times Square with his brother, Bobby, and with Michael Edwards. But he insisted that that he had done nothing wrong and that he wanted nothing whatsoever to do with a police investigation.

The detectives confronted him with Michael Edwards's statements about the assault on Jean Casse on 52nd Street, but Bill remained adamantly silent. Then, the detectives pointed out to Bill how closely he resembled the descriptions by witnesses of the man who slugged Casse on 52nd Street. They suggested that they might charge him with the homicide unless he aided them in identifying those responsible. Bill agreed to talk, but only because he felt that it was entirely possible the cops might charge him. He pointed out that the police officers who had followed the Brooklyn boys on 52nd Street started smacking the black kids, not any white people. He had no expectations of fairness from the police. He said that he did not want to stay at the station house any longer. But Kennedy and Delgrosso insisted that he give a statement.

So Bill admitted seeing the old man getting hit on 52nd Street from a distance of eight to ten feet away. His brother, Bobby, was with him. The hitter was a big guy, he said, who swung with his right, perhaps his left, hand and then crossed the street. Bill said that he saw other kids going through the man's pockets; one of them jumped up with his fists closed and ran east on 52nd Street. Bill recognized both the hitter and the kid who went through the victim's pockets. He knew them from the Albee

Square Mall in downtown Brooklyn on Fulton Street. The hitter
was named Smokey. Bill did not know the other one's name, but
he knew his face. The detectives showed him photographs, and
Bill picked out David Warren.

Sixteen-year-old Bobby Johnson initially told the investigators
that he had no idea who hit the old man. But when the detectives pointed out that his brother, Bill, had put Bobby at the
scene and that Bill's size made him a likely suspect on the basis
of other witnesses' descriptions of the hitter, Bobby told them
what he saw. Once the crowd of Brooklyn kids reached 52nd
Street, Bobby saw a young woman to his left get pushed against
a wall and robbed by four or five kids. He saw an old man leave a
restaurant that had an awning. The old man was, he said, standing under the awning facing the street with his back toward the
restaurant. A Brooklyn guy named Smokey, or Catfish, whom
Bobby knew from playing basketball in the schoolyard at 46th
Street and Clermont Avenue, hit the man with a roundhouse
swing of his right hand. The man fell backward with his face
up. Smokey walked away immediately, off the sidewalk into the
roadway, heading east on 52nd Street. Four or five other guys,
one of them kneeling over the man, went through his pockets
and then ran when a man came out of the restaurant. All the
while, Bobby was only about twenty feet from the action. When
the detectives showed Bobby the photographs, he picked out David Warren as the person kneeling over Jean Casse.

Detectives and paddy wagons headed over to Brooklyn not
only to get the prime suspects but also to pick up other boys
from the projects who might have been involved. At the formal
lineups the afternoon of January 8, Michael Edwards, Bill and
Bobby Johnson, and Shawn Green (who agreed to look at the
lineups despite his adamant denials of having seen anything on

52nd Street on New Year's Eve) all picked Eric Smokes and David Warren. The Johnson brothers and Shawn Green also picked out Charles Dawson, who, they said, was one of several youths surrounding Casse after Smokey hit him. Smokes, Warren, and Dawson were charged with felony murder. The charges against Dawson were later dropped because the office of the District Attorney of New York (DANY) decided that the Johnson brothers' testimony placing Dawson at the crime scene was insufficient evidence to tie him to the actual assault and robbery of Casse. DANY also discounted Shawn Green's lineup identifications of Warren, Smokes, and Dawson because Green's earlier denials of seeing anything on 52nd Street undercut the credibility of his testimony. Dawson was later convicted of felony murder arising out of yet another robbery in Brooklyn.

After arresting and charging Smokes, Warren, and Dawson, the police found another young Brooklyn man in jail on robbery charges who admitted witnessing the assault and robbery of Casse. "Corky Jones" had gone to Franklin Lane high school with Smokey and had known him for several years. He also knew David Warren by the name Young God. Jones had gone to Times Square on New Year's Eve, had ended up on 52nd Street, and had seen Smokey hit the old man in the face. He then saw Warren going through the man's pockets. He claimed that Smokey actually took the man's wallet and then both Smokey and Warren fled east on 52nd Street, before cutting through a passageway between blocks to head uptown. Jones, who had a considerable record of his own for chain snatching, robbery, and possession of burglary tools, was terrified of Smokey's wrath against himself, his wife, and young daughter if he testified to what he had seen. On Rikers Island, Smokes had told Jones that if Jones testified against him, Jones was going to get "persecuted."

In eight days, Delgrosso's investigation had gone from nothing to arrests and indictments. At trial before Justice Richard A. Scott of the New York State Supreme Court, Pamela Jordan, the defense attorney for David Warren, and John Avazino, who represented Eric Smokes, argued that the remarkable dispatch with which the case came together suggested the insufficiency of the evidence against their clients. The attorneys argued that Delgrosso had fallen into the honey pot when James Walker was arrested for robbery. Walker, in their view a walking definition of reasonable doubt, had led Delgrosso, falsely, to Eric Smokes and David Warren. From that point on, the defense argued, Delgrosso developed tunnel vision on the case.

Moreover, the defense attorneys argued that Delgrosso and the other detectives working the case had intimidated and deceived Michael Edwards and Bill and Bobby Johnson into betraying Smokey and Warren to the police under threat of getting pinned with the robbery-murder themselves. Here, Avazino stated, one sees "the true conduct of the police." He railed against the detectives' deception of one witness after another, pitting them against each other and against the two defendants. It was no accident, both attorneys argued, that Edwards and the Johnson brothers eventually had to be arrested as material witnesses and coerced to testify against their clients at trial. The detectives' and prosecutors' reliance on the likes of James Walker and Corky Jones, known predators, revealed not only the weakness of the state's arguments but also the corruption at the heart of a criminal justice system that gives breaks to felons to make cases against boys who were guilty of nothing more than high-spiritedness on a traditional night of celebration. Further, one of the people's two civilian witnesses, Janet Yost, saw a tall, skinny kid assaulting the old man, a description that certainly did not fit the burly Eric Smokes.

Assistant District Attorney Michael Goldstein, who worked the case with ADA Susan Axelrod, argued that one takes witnesses where one finds them. Michael Edwards, Bill and Bobby Johnson, and Corky Jones all knew Smokey and Warren, all were on 52nd Street on New Year's Eve right after the ball fell, and all, however reluctantly, named Smokes as the hitter and Warren as the robber the night that Jean Casse was punched senseless and killed. James Walker, Smokey's and Warren's companion in a dozen robberies, directly heard Smokey admit killing someone when the ball fell. Walker was allowed to plead guilty to a felony with probation in exchange for his testimony, a regrettable exigency that, for better or worse, is the way the criminal justice system works.

The discrepancies among witnesses were typical of any criminal case and were not, in any event, substantial, Goldstein argued. The defense attorneys wanted things both ways. When the evidence from different witnesses against their clients coincided, they claimed collusion between witnesses to frame their clients. When different witnesses gave slightly different accounts of a situation, the attorneys claimed vast exculpatory gaps. Finally, Smokes's threats against the witnesses were reason enough, Goldstein argued, for Edwards, the Johnson brothers, and Jones to fear retaliation against themselves or their families so greatly that they had to be coerced to testify.

In the end, a jury convicted Smokes and Warren of the murder and robbery of Jean Casse. Warren was sentenced to fifteen years to life imprisonment. Smokes, sentenced to twenty-five years to life imprisonment, was enraged at the verdict and shouted out a long string of statements protesting his innocence. "They can kill me right here man, but I ain't do it . . . I didn't even do this. I don't want to hear that bullshit any more. I didn't do this shit . . . These mothafuckas lie, none of them seen shit. They all seen what that lady seen, she didn't see me do it. I didn't kill anybody . . . I told you they would convict me on this shit. They just mad because [I] knocked somebody, they don't have nothin, nothin I did, I didn't do this . . . I didn't do this. You know I was going to be convicted of it, you unnerstan, they want somebody, they didn't even see me. I tell them this. These dudes stand up and lie, this justice system is so unbalanced . . . This is all bullshit. Be good for what? I am finished. I am finished, man."[6]

As it happens, Eric Smokes received a bachelor's degree in behavioral science from Mercy College at Sing Sing Correctional Facility on June 2, 2004.[7]

Instead of an exercise in tunnel vision, Delgrosso's investigative work on the Casse homicide was so thorough that it aided the investigation of the robbery and murder of John Gelin a year later on June 15, 1988, at 1118 Sixth Avenue. A young man approached Gelin underneath some scaffolding on the east side of Sixth Avenue near 50th Street. The assailant attacked from behind, hitting Gelin in the head with his fists. Gelin died of his injuries five days later. His assailant ripped a considerable wad of money and a gold bracelet from Gelin's pocket and fled south on Sixth Avenue. An engineer's helper who worked on Sixth Avenue saw the assault and chased the culprit down the east side of Sixth Avenue, then down the subway stairs at the southeast corner of 47th Street and Sixth Avenue, then westbound underneath Sixth Avenue to the set of stairs leading up to the street on the southwest corner of 47th and Sixth. At the top of the stairs, the assailant paused and kicked his pursuer. But the culprit left behind an Atlanta Braves baseball cap, with its distinctive "A."

The week that Gelin was assaulted, Midtown North had a slew

of robbery complaints, many of them strong-arm jobs committed by kids from Brooklyn who shaped up at the Albee Square
Mall. Some of the robbers claimed membership in the Decepticons. Going with what he had, Detective Harry Bridgwood developed a huge list of the Decepticons, every one of whom had
been arrested at one time or another, a great many for robbery.
The focus on the Decepticons proved to be a blind alley. But as
Bridgwood pounded the streets looking and listening for any
news of young robbers who had recently made a big score in the
city, he heard the street name "Drak." In squad-room talk, Delgrosso told Bridgwood to go through the Casse file because he
remembered that the name Drak had surfaced in his investigation. Bridgwood not only found Drak's name but Delgrosso's
note about Drak's predilection for ducking underground when
in flight. Drak turned out to be a prototypically troubled young
man from Brooklyn, and a member of the A Team from East
New York, with arrests for robbery, assault, possession of a
deadly weapon, thievery, and violent opposition to arrest. The
engineer's helper, who had witnessed the assault on Gelin and
had pursued and nearly caught Drak, identified him immediately. To Delgrosso's great satisfaction, the helper's testimony
at trial sent another Brooklyn strong-arm robber down for the
count.

4

THE GIRL IN THE PARK

The squawk came into the 34th precinct squad room at 1202 hours on Friday, March 27, 1992. Two female joggers had discovered a woman's body in Fort Tryon Park, the densely wooded urban retreat at the north end of the precinct where the Cloisters, with its stunning collection of medieval art, stands. The apparent cause of death of the unidentified woman, the radio run stated, was a gunshot wound. The squad began its usual banter, always coarse when cases have dump-job dead-end earmarks. One detective cracked that the dead woman had probably also been jogging and had run into a branch with a protruding bullet.

Detective Tony Imperato caught the case. Imperato, a barrel-chested, darkly handsome man, grew up in the streets of the Bronx. From as early as he can remember, he had wanted to be a cop. When he first applied for the job in the mid-1970s, at the height of New York City's fiscal crisis, the list for the police department was closed. He became a correction officer for a time—dreaded work pervaded by the fear that one's charges will shat-

ter their cages and wreak their fury on their keepers. Imperato
finally got the call to join the NYPD in 1981. After the usual stint
in patrol, followed by work in the Narcotics Division in his old
boyhood stomping grounds, he became a detective in the 34th
squad in 1989.

In 1992 the 34th precinct extended from 155th Street to the upper tip of Manhattan, Hudson River to Harlem River, making it
the largest geographical precinct in the city. In October 1994
the NYPD split the old 34th precinct in two, creating the 33rd
precinct from 155th Street to 178th Street, river to river, while
the 34th precinct retained jurisdiction from 178th Street to the
Spuyten Duyvil. But in 1992 the old 34th precinct, along with the
28th precinct in Harlem, the 75th precinct in Brooklyn's East
New York, and the 40th precinct in the Mott Haven section of
the Bronx, was a "shithouse," one of the most crime-plagued
precincts in the city. It was the prototype of what cops call "killing grounds," areas of the metropolis marked by disproportionately high numbers of homicides, armed robberies, and other violent crimes. Most of the crime in the old 34th precinct emerged
out of Washington Heights, between 155th and 207th streets,
the residential and cultural gathering grounds for hundreds of
thousands of immigrants from the Dominican Republic.

Dominican immigrants—like immigrants of other ethnic
groups—work hard, strive to achieve a respectable place in American society, and abide by the law for the most part. But some
Dominicans choose to live outside the law, either as illegal residents or as active participants in a range of criminal activities,
often both. The few Dominicans engaged in crime consumed almost all the efforts of the old 34th precinct squad detectives.
Further, most victims of crime in the precinct were also Dominicans. The drug trade accounted for the vast percentage of vio-

lence in Washington Heights in the late 1980s and early 1990s. Beginning in the early 1980s, the area became the principal wholesale and retail distribution point for Colombia-produced cocaine smuggled into John F. Kennedy International Airport. Dominican immigrants acted as the principal retailers of Colombian coke as far north as Toronto and as far south as South Carolina.[1]

Imperato partnered with Detective Bobby Small, a chain-smoking, tough-as-nails former army sergeant who joined the cops right after returning from Vietnam. Small's colorful language reflected his Hell's Kitchen boyhood, where he grew up with the famous Irish thugs known as the Westies and learned the street smarts that served him well later in policing the 34th precinct.[2] On one sweltering day, after catching an old lady dead on arrival and a young man's carefully premeditated shotgun suicide, Small received a call from Patrol asking the squad to respond to 181st Street and Saint Nicholas Avenue. A chaotic scene greeted Small, Detective Eddie Cruz, and myself when we arrived. A large truck had jumped the curb and smashed into the window of an electronic equipment store. Videocassette recorders and televisions hung out the window. A yellow crime scene tape and fifteen uniformed officers separated 400 youths from the enticing gadgets.

A uniformed sergeant smoked nervously, pacing back and forth. Small approached the sergeant and asked why she had called the squad. Just then, a large, belligerent man pushed his way through the crowd of youths, demanding the release of his truck. The sergeant turned away and continued to puff on her cigarette. Small asked the man if the truck was his. The man said that he did indeed own the truck and that it was a refrigerated vehicle carrying perishable food that would quickly spoil if the truck remained inoperable. Small asked who had been driving the truck. The man responded that someone he hired had been driving, but when Small asked who that person was, the man responded: "¡Yo no se!"

Small laughed and asked the man if he expected the police to believe that he had hired someone he didn't know to drive his truck. But the man responded: "¡Yo no se!" Just then, a uniformed officer came up and whispered to Small that the truck had cut the legs off an elderly woman, who had been rushed to Harlem Hospital. Small again pressed for the name of the driver. Again, the owner proclaimed: "¡Yo no se!" Small immediately impounded the truck and ordered it parked in the sunniest part of the police parking lot at the station house until the owner produced the driver. Late that afternoon, the truck's owner came into the station holding a man named "José" by the scruff of the neck, declaring him responsible for the accident. The incident confirmed Small's cardinal belief about policing: in a world of radically opposed ideas about how things should work, only the energetic assertion of legitimate authority brings about a reasonable public order.

So just after noon on March 27, 1992, after the usual groans about missing lunch, several squad members rode up to Fort Tryon Park in separate cars. The park's narrow, tangled roads made for heavy going in the clunky Chevrolet squad car driven by Imperato. Finally, after several wrong turns in the thickness of the park's woods, Imperato and Small, accompanied by myself, spotted a uniformed officer who pointed up a steep footpath that led, he said, to the crime scene. Imperato parked the car, and we three men hoofed up the incline. At the top of the hill, an open field spread out. Across the field, to the right, was a set of stone stairs that led from the street down to the park. In the middle of the field stood a modernistic sculpture shaped like an oversized tuning fork; to the left, under trees, was a pagoda. Still further left, hidden from view, another set of stairs headed down to a path that led toward Broadway. The other squad detectives had already arrived at the field and were waiting for Imperato. Uniformed officers from the Emergency Service Unit were also on the scene.

A low wire mesh fence bordered the field at the top of the hill. A young light-skinned Hispanic woman, probably Puerto Rican, lay slumped with her back against the fence, facing down the hill. She had been shot in the face. Stippling marks—pinpoint skin abrasions caused by gunpowder from a gun fired at a distance of less than eighteen inches—scarred the left side of her face. A spent 9-millimeter shell nestled undisturbed in a nearby bed of old plane-tree leaves. The detectives tried to reconstruct the scene. Clearly, detectives said, the woman's killer had held the gun near her face with his right hand and fired. The shell ejected from the gun over his right arm, landing behind him to the right.

The woman was dressed in black fishnet stockings, a purple minidress, a short, tatty brown-cloth jacket, and new pumps with black chiffon bows. She had a ring on each hand and a watch on her right wrist. A gold post took the place of her right front tooth. Her garish makeup did not disguise or improve her unusual facial features. Detectives placed her at the margins of the oldest profession.

When the woman had fallen against the fence, her right leg had gone beneath her left, catching her purse between her legs. The detectives could not retrieve any identification it might contain without disturbing her body, and so they had to wait until the Crime Scene Unit and the technicians from the Medical Examiner's Office arrived and did their work.

In the meantime, Detective Imperato took Polaroid pictures

of the victim's face to show around the street. Detective Small interviewed the joggers, two Puerto Rican girls, "Eliza" and "Jean," both in their early twenties. They said that they had left Eliza's house a mile or so away at 11 A.M. and reached the park at about 11:30 A.M. They entered from Margaret Corbin Drive, went past the tuning fork monument, and came down the circular path. Before entering the park, they had checked the open field carefully, as they always did, to make sure that no one was around. The only people they saw were a young couple with a tourist book looking at the monument. Once in the park, they noticed something brown propped up against the fence, perhaps a cardboard box. But as they came around the bend, they realized it was a street person, perhaps sleeping. Then they stopped, came closer, and saw blood dripping from the person's face. They screamed and ran out of the park down to Broadway to the nearest alarm box, where they put in the call to the police. When that call got no response, they crossed the street and called 911 from a bodega phone. The Emergency Service Unit arrived shortly afterward. They told Small that they knew nothing else.

Curious onlookers, mostly young Dominican men, had already clustered in the pagoda at the edge of the park. Detective Small and I went over to talk with them. One man wondered aloud if the murder had anything to do with the nightly devil worship rites in the pagoda, always held by candlelight. Another young man laughed uproariously and asked if the speaker was so ignorant that he knew nothing at all about *santería*.[3] Detective Joe Montuori stood at the edge of the crime scene, smoking. Montuori is a former Brooklyn street kid, a Bishop Loughlin Memorial High School graduate who trained with the American Institute of Banking to become a loan officer. A drinking buddy persuaded him to take the test for the police department, on which he scored high. But he faced another hurdle: the NYPD's five-foot, eight-inch height requirement mandatory in the 1960s. So Montuori had himself stretched by a chiropractor and transported on his back in a hearse to the police physical exam. While most of his boyhood friends ended up either dead from heroin overdoses or in prison, Montuori became the legendary veteran of hundreds of murder investigations in the 34th precinct. He noted the complete absence of the media at this crime scene. "No one gives a fuck about this place," he said.

The Crime Scene Unit arrived, headed by Detective Chris Fortune. She cleared everyone out of the way and took extensive photographs of the scene with her Nikon camera. She then drew a detailed map of the surrounding area. The detectives chatted with the uniformed officers about local "pross" operations, but none of the cops had seen the dead woman before. Detective Fortune gently chided her male colleagues for some derogatory remarks about prostitutes. "It's just a job, guys." The men shifted uneasily, not used to being reprimanded by a woman, let alone by a striking red-haired colleague. The technicians from the Medical Examiner's Office arrived at 1300 hours and took the woman's body temperature rectally. The thermometer registered 97 degrees. Bodies cool at a rate of 1–2 degrees an hour, depending on the surrounding air temperature; the day was mild. The technicians officially pronounced the woman dead, fixing the time of death at 1130 hours, just about when the two girls said they had entered the park to jog.

Detective Imperato finally retrieved the dead woman's purse. In it, he found several pieces of identification, mostly cheap facsimiles with various names. He also found a work release card from Correction that named her as Yvette Tirado, with an address in the 43rd precinct in the Bronx. Several Bronx phone numbers were scrawled on an index card, with no names attached to the numbers. Just then, a high-ranking boss arrived and discovered footprints near the woman's body. He ordered photographs of the prints to be taken because he knew of a murder that had been solved the previous year through footprints. But one of the uniformed cops pointed out that the Emergency Service Unit had been stomping around the body in that area and that the footprints were probably theirs, an observation that prompted several head-turned guffaws.

————————

On the ride back to the house, Detectives Imperato and Small wondered aloud how a Puerto Rican prostitute from the 43rd precinct had ended up over in the largely Dominican 34th. Homegirls, like homeboys, don't usually wander far from their blocks, and Puerto Ricans and Dominicans stay out of each other's yards. Back at the house, Imperato made the requisite notifications to the police hierarchy and then called Correction. The Correction officer told Imperato that Yvette Tirado had a record as a petty thief, occasional armed robber, drug addict, and prostitute. A quick check of her sheet revealed a long list of arrests for grand larceny, forgery, robbery, and drug possession dating back to 1985. She had just been given work release from upstate after serving several months in prison for holding small amounts of cocaine and possessing burglary tools.

Imperato took down her address in the Bronx and the Bronx phone number of her nearest relative, her sister, whose name was listed as "Mercedes." But when Imperato called Mercedes's number, the woman who answered denied that she was Mercedes or that anyone there went by that name. Imperato called all of the numbers on the index card that he had found in Tirado's purse.

He got no response at any of them until the last call, when he finally reached a woman who said that her name was "Maria Gonzalez." Maria denied knowing anyone named Mercedes. However, she said, her own sister's name was Yvette Tirado. Imperato told her that Yvette had been hurt and asked her to come to the station house. Gonzalez became upset and said that she herself had broken ribs, but that she would come to Manhattan as soon as she could.

At 1645 hours Maria Gonzalez arrived with another woman, who gave her name as "Alicia Serrano." Maria, who walked only with great difficulty because of her heavily bandaged ribs, told Imperato that Alicia was also related to Yvette. Imperato, accompanied by Detective Gennaro Giorgio and myself, took both women into the larger of the squad's two private interview rooms. The women were apprehensive. As gently as possible, the detectives informed them that Yvette had been hurt badly. The women demanded to know her condition. The detectives glanced at each other and then told the women that Yvette was dead.

Maria Gonzalez began weeping and wailing. She embraced Alicia, screaming at the top of her voice: "My sister is dead! My sister is dead!" She fell to the floor, thrashing her body around the room, uttering mournful cries, while Alicia and the detectives tried to comfort her. After a minute, she composed herself, sat up, and in a calm voice asked the detectives how it happened. The detectives told her that Yvette had been discovered in Fort Tryon Park and that she had been shot to death. Gonzalez paused, and then asked what Yvette was wearing. The detectives described her attire. Gonzalez pressed for more physical particulars. When the detectives mentioned the gold post in Yvette's mouth, the two women looked at each other and asked to see a picture of the dead woman. Reluctantly, the detectives showed them the Polaroid photos that Imperato had taken in the park.

Suddenly, the women jumped up and hugged each other, with Gonzalez exclaiming: "My sister is alive! My sister is alive!" Gonzalez explained that the photograph did not picture her sister, whose name was also Yvette, but instead her sister's good friend, Yvette Tirado. Her own sister, Gonzalez said, regularly used Yvette Tirado's name as one of many aliases and, like the dead woman, was known by different names to different people. The detectives, their eyes used to the shadowlace of street identities, scarcely blinked. The women quickly said that they could take the detectives to the street that the dead woman frequented and could, in fact, introduce the police to the dead woman's real sister, Mercedes.

At 1745 hours Detectives Imperato and Small, accompanied by myself, took Gonzalez and Serrano to Manor and Watson avenues in the 43rd precinct, one of the hubs of the drug trade in the South Bronx. Gonzalez said that she could find Mercedes on the street there. During the ride over the Cross Bronx Expressway, Gonzalez lamented about Yvette, her own sister, who, she said, spent all her time on the street going with men, doing anything to get drugs. She hoped that Yvette Tirado's death might bring her own sister to her senses. As the squad car approached Manor and Watson, both women announced that they did not want to be seen in a police car. Imperato dropped the women off two blocks away from the intersection to look for Yvette Tirado's sister. Detective Small admonished them not to reveal anything about Tirado's death. Gonzalez and Serrano ran down the street in search of Mercedes.

The detectives pulled the squad car up to the corner of Manor and Watson and awaited their return. They surveyed the street scene. Eight uniformed police officers strolled the block, which was thronged with hundreds of people. The area's famed open-air drug market, only one of many throughout the city during the mayoralty of David N. Dinkins, was in full swing, with transactions taking place only a few feet from the officers. The drug dealers all knew that uniformed officers were forbidden to make drug arrests, a task restricted to special narcotics units as a presumed shield against corruption, or, as many police officers thought at the time, a policy enacted by the Dinkins administration to keep arrests down in largely black and Hispanic areas of the city, the core of the mayor's constituency. Still, the dealers practiced a symbolic furtiveness to prevent open confrontations with the cops.

Three uniformed officers, POs Robert Johnson, Carlos Perez, and Christopher Weston, made the squad car, strolled over, leaned in the window, and asked the detectives what brought them to such an idyllic spot in the Bronx. Detective Small showed them the Polaroid photo of Yvette Tirado. Suddenly, previously scattered sparks of knowledge combusted, as all three officers spontaneously shouted bits of information: "I know her, she's out here all the time"; "That must be the woman snatched off the street this morning"; "There was a robbery at a crack house last night and this must be connected." The cops said that one of the guys involved in the snatch was on the street only a few minutes before. In fact, one of the cops had just "disconned" him (issued him a summons for disorderly conduct). His street name was "MuscleMan."

The officers also said that word on the street was that a black man had witnessed the snatch and had then been severely beaten by the kidnappers. One cop said that the kidnappers had told the black man: "You'll never see that girl again." But another cop said: "No, no, he told his own mother: you'll never see that girl again." The cops then dispersed, one to find the witness, the other two to find MuscleMan. A quarter of an hour later, the cops returned. The witness was not around, but the two officers had MuscleMan in tow. The detectives put the burly teenager into the back seat of the squad car.

Just then Gonzalez and Serrano returned, followed by a strikingly beautiful full-figured blond who announced herself as Yvette Tirado's sister, Mercedes. Several young men followed Mercedes, but kept their distance from the squad car. Imperato told her she needed to come to the 34th precinct to help her sister. Mercedes stood in the street, hands on her hips, looking skeptically at the police. Detective Imperato repeated: "Your sister needs your help." Mercedes said that she would find her own way to Manhattan. She left with a flourish, trailed by the coterie of young men. Detective Imperato drove back to Manhattan with MuscleMan sitting nervously in the back seat, kept company by Detective Small. During the ride, MuscleMan asked the detectives: "What's this about?" Imperato responded: "If you've done nothin wrong, you got nothin to worry about. Have you done anything wrong, MuscleMan?" MuscleMan didn't respond.

The detectives arrived back at the station house at 1850 hours, the same time as Mercedes and her entourage, consisting of her other sister, "Gladys," her brother "Enrique," and her tall, gangly boyfriend with an infant in his arms. Detective Imperato assigned Detective Pete Moro to interview MuscleMan. Moro, once in a platoon that was pinned down for four days by enemy fire— only one of many harrowing Vietnam wartime experiences he rarely mentions—joined the police department in 1980 and went straight from the police academy to the street wars of the South Bronx. He later served there in the Narcotics Division. He came to the 34th squad in 1988, where he quickly developed a repu-tation for self-effacing patience in interrogations and careful, methodical investigative work. MuscleMan gave Moro a series of alibis for the entire day of the abduction, and all of them seemed to check out.

Moro surveyed other information about MuscleMan. He had been arrested eleven times for holding narcotics with intent to sell, and once for robbery. He seemed to be one of the principals in a major drug operation near Manor and Watson avenues, one that was well-known to the Street Narcotic Enforcement Unit (SNEU) at the 43rd precinct. Detectives noted with marked dis-favor that MuscleMan, whatever his toughness on the street, mewed and whined in custody. The detectives suspected that MuscleMan had guilty knowledge about the abduction; but with nothing specific on him, they cut him loose.

Detective Imperato, with Detective Giorgio and myself, went into the sergeant's office off the main floor of the squad room, where Mercedes sat with her retinue. When Imperato broke the news about Yvette's death, pandemonium broke out in the tiny room. Mercedes swept the sergeant's desk clean of all belongings and threw herself on the floor in a fit of hysterical anguish, thrashing, kicking, and rending her garments. Her brother, sis-ter, and boyfriend, along with Detective Giorgio, tried to restrain her, but their efforts only provoked further displays of grief and attempts to destroy fixtures in the sergeant's office. Detective Imperato stood at the edge of the room, arms folded, looking impassively and unblinkingly at the tumultuous scene.

Gradually things quieted down, and to confirm the identifica-tion Imperato showed Mercedes and Enrique the Polaroid photo of Yvette in death. Mercedes, who became calm just as quickly as she had exploded, mumbled: "I'll bet it was those two mothafuckas she was with." Giorgio immediately asked her whom she meant, but Mercedes clammed up. The detectives asked if Yvette went with men for money, but Mercedes vehemently denied it, saying that Yvette would rather jerk (rob) them. Besides, she said, Yvette had a wife, "Martha." In response to queries about whether Yvette owned a brown beige coat, Mercedes said that she and Yvette had traded coats the night before, with Mercedes loaning Yvette her own long black leather coat. Mercedes then said that she had heard on the street that Yvette had been robbed of the leather coat. Mercedes suddenly directed her brother to find out from the police how to recover Yvette's body, and she abruptly left the squad room with dramatic flair. A few smitten detectives talked a lot about Mercedes in the next several days.

PO Carlos Perez of the 43rd called Imperato at 2030 hours. He said that the cops had been unable to find the black man who had witnessed the abduction. However, a routine SNEU sweep had netted one of the street dealers from the drug operation at Manor and Watson who, Perez said, had information about the homicide. The police had arrested the dealer on the street that day for felony possession of twenty-six jumbo vials of crack stuck to a skein of tape. He had tried to ditch the skein under a car when he saw the plainclothes SNEU cops coming down the street, but he was too late. Later, the cops pulled people out of the holding pen, one at a time, and asked: "Something happened on the street today. Do you know anything about it?" When they came to the street dealer, he asked immediately: "Is the girl dead?" Imperato, Small, and I raced back to the Bronx, reaching the 43rd precinct station house at 2100 hours.

The jammed-up street dealer, going by the name of "Tennessee," was a 39-year-old man whose prolonged crack use made him look 65. Tennessee readily admitted that he pitched crack on the street, taking most of his $40-an-hour wages in trade (one $10 bottle for every fifty-bottle "bundle" sold, four bundles per hour on average). Detectives Imperato and Small, observed by myself, first interviewed Tennessee in the SNEU office on the first floor of the station house, where cops crowded into the room. Tennessee told his story in a low, mumbling voice, but forthrightly and with great observational detail, focusing only on actions and words, taking the police step-by-step into his world.

At the start of his story, he said that he had seen "Cuba" leave the street driving the car that abducted the woman, who had been bound and gagged. He saw Cuba return to the street about two hours later and clean some debris out of his car, putting what looked like cloth, twigs, and branches into a trashcan on the corner of Manor and Watson in front of the street's hamburger joint. The detectives immediately interrupted the interview to talk with 43rd precinct cops about getting that trashcan back without attracting any attention, although everyone acknowledged that after nearly twelve hours there was slim hope of finding anything worthwhile. The SNEU lieutenant called a friend in the Sanitation Department and explained that if the police went to pick up the trashcan everybody would be in the wind. The lieutenant's Sanitation Department friend agreed to make a night run through the area, pick up the can, and deliver it to the station house. Detective Imperato asked the cops to take Tennessee back upstairs to the squad room, where he could go over Tennessee's story in greater detail and take a written statement.

In the hall, the SNEU cops briefed Detectives Imperato and Small on the local drug-trade scene. The SNEU cops had been after the guys in the drug operation near Manor and Watson for a long time. The key local players, they said, were "MuscleMan," Cuba, Midnight, "Fatso," and Johnny, all of whom were managers for the drug spot. The cops were delighted that "these guys now have a body on them." The owners of the business were "Tony" and "Frankie," apparently brothers, both of whom hailed from Audubon Avenue around 170th Street in the 34th precinct. They produced jumbo red-topped bottles of crack that sold for $10 in the Bronx. All the regular pitchers, like Tennessee, were crack-heads.

The dead woman was well known to the department as a crack-head and sometime prostitute. At 0940 hours on Friday morning, she had made a 61 (a formal complaint) to PO Jones on the street about being robbed of her coat by Midnight. Later, a black man told other police officers about a girl being snatched and shoved into a car. The black man had witnessed the kidnapping and then been beaten by the same bunch. Only when those cops saw the detectives from the 34th precinct show up on their street with pictures of a murdered woman had everything suddenly come together. Imperato mentioned that he had had to cut MuscleMan loose. He asked the SNEU cops to arrest him again soon.

After pulling Tennessee's sheet, Detective Imperato went upstairs to begin taking his statement, accompanied by Small and myself. Tennessee already had three felony arrests for sale of narcotics, four for possession. He faced big time for his arrest that day with felony weight. Imperato had him repeat his statement several times. Each statement produced more details and slightly different nuances of meaning. Finally, Imperato committed the statement he elicited from Tennessee to writing and had Tennessee sign it—standard police procedure when interviewing witnesses to major crimes.

Imperato's style of interviewing typifies that of a great many detectives. Imperato always pushed hard, giving no quarter, respite, or rewards. He held everything said in suspension, waiting to see if it merited credence. He distinguished between mopes and mutts, both street types that he grew up with and knew well. When he interviewed mopes, he offered them tough choices, threatening to place the entire weight of crimes on them, bringing them to the point where they were ready to give up things easily, the distinguishing mark of a mope. Even when they lied, mopes wanted to be accommodating. ("At's the best troof I can come up wit, Officer.") But as soon as mopes yielded and showed signs of weakness, it was time to push back ferociously, offering no concession, satisfaction, or approval, increasing the pressure on them to the breaking point. Mutts were harder, cockier, tougher, more self-assured, fully aware that the police could never be their friends, that the game between them and detectives was always deadly serious. Their very demeanor announced an emotional remoteness behind an impenetrable shell. They lied to police on principle, even about simple matters.

Tennessee turned out to be a mope. He said that he was at the crack house on Manor Avenue in the third floor apartment on or about 11 P.M. on March 26, 1992. Yvie (Yvette) and a girl named "Missy," as well as several male drug dealers, were also there, along with a scar-faced woman named "Mariana." Yvie and Missy kept eyeing Mariana. Tennessee, recognizing a robbery in the making, motioned to Mariana, took her into the bathroom, and explained the situation. Mariana asked Tennessee to get her out of the apartment safely. But suddenly Yvie burst into the bathroom and stood between Tennessee and Mariana. Tennessee told Yvie to let Mariana alone. Then a man named "Jack" came into the bathroom and told Tennessee to mind his own business. Yvie pulled a knife on Mariana, and at that point, Tennessee said, he left the bathroom and went to the living room. Tennessee saw Mariana leave the apartment. Yvie and Missy followed her but returned in about five minutes. Yvie offered Tennessee five dollars so that he could buy some crack. Tennessee refused, telling her he had his own crack. Shortly after that, Yvie and Missy left the apartment again and this time did not return.

Later on, Tennessee said, he heard a heavy knock on the door and thought it was the police. As he raced toward the fire escape in the kitchen, the front door to the apartment opened. Tennessee ducked into the bathroom and locked the door. He heard Cuba's voice saying: "Who done it?" And then he heard Cuba smack someone, demanding information from her. Cuba tried the bathroom door. When he discovered it was locked, he demanded that it be opened or he would break it down. Tennessee opened the door and faced Cuba. Mariana was at Cuba's side. Cuba asked Mariana if Tennessee had robbed her. Mariana said no. In fact, she said, Tennessee had tried to help her. Two girls whom she didn't know had done the robbery, Mariana insisted. Cuba then made the rounds of all the patrons of the crack house. Eventually, one of the crack-heads gave up Yvie and Missy as the robbers.

Tennessee said that the next morning at about 8:30 A.M. he saw Yvie and Missy walking down Watson Avenue toward Manor Avenue. Suddenly, Cuba and Midnight accosted the girls and took both of them into a building on Manor Avenue. A short time passed and then Tennessee saw Cuba leaving the building carrying Yvie's black leather coat; Midnight was carrying a brown jacket. The two girls followed them out of the building. Cuba and Midnight left the block. Tennessee saw two uniformed police officers standing at Manor and Watson avenues. He saw Yvie approach them. She talked with one of the police officers. Tennessee assumed that she gave him information about the leather coat being taken. He guessed that she also gave information about the drug house on Manor Avenue because he saw her point to the building. The police officers wrote something down and then left the area.

Later that morning Tennessee saw Cuba, Midnight, and Johnny forcibly take a struggling Yvie down the front steps of the drug house on Manor Avenue and put her inside Cuba's sky-blue Caddy. Midnight rode in the back and Cuba drove. Johnny remained on the street. After a while, Tennessee saw Cuba's car parked on Manor Avenue. Cuba got out of the car and opened the rear passenger side door. He threw a multicolored piece of cloth and another object that looked like a piece of wire or a twig into the trashcan on the corner. Cuba then returned to his car and drove away.

Tennessee worried aloud that he had to have some document that he could show people back on the street, some piece of paper to prove that he had been arrested but then was sentenced to time served, or some such thing. Otherwise, he said, the people on the street would figure he had talked to the police. Imperato and Small mused briefly about Tennessee's odd faith in the credibility of official documents in a world far removed from procedural niceties.

While Imperato was taking yet another iteration of Tennessee's statement, Detective Small and I went downstairs to the

SNEU office. Sanitation had just returned with the trashcan. Officers and bosses crowded into the tiny office to watch Small fish through the can. He found a few small bags of garbage, paper plates, and Big Mac containers. But he also found a long wool dark blue scarf, with red and yellow stripes at one end. And there were some pieces of brown packaging tape. One piece was small and circular, appropriate for binding a person's hands. Had this been used to bind Yvette's hands? The other piece of tape was larger, broken, and matted with human hair. Had this been placed around Yvette's head as a gag?

The police officers who were jammed into the room marveled at their luck. Rarely do the stories of street-murder witnesses get corroborated with forensic evidence. Small brought the objects upstairs to show to Tennessee, and Tennessee identified them as those that he had seen Cuba put into the can. The detectives then put Tennessee into the squad-room holding cage and went out on the street with the SNEU to do surveillance on the Manor Avenue house in order to grab Midnight, Cuba, and Johnny.

Surveillance is grueling, tedious work, paced entirely by others' comings and goings. Over several long hours, amidst the usual rough banter of men together in the middle of the night, the SNEU cops briefed the detectives and me on drug operations on Manor Avenue. First, the cops said, a woman controlled the heroin trade on the street. Indeed, she headed a major heroin distribution network in the Bronx. Although the cops had arrested her many times, they could never make anything stick because, like many dealers at her level, she never traveled armed nor did she ever directly handle the dope.

Second, the street crack dealers receive their bundles of fifty bottles from their managers on a long skein held together with a piece of broad scotch tape. When they make a sale, they cut as many bottles as they need from the skein with a razor. They have to work so fast to keep up with the trade that they inevitably cut their hands with the razor, and they put them in the exterior pockets of their hooded sweatshirts to stanch the blood flow.

Third, the heroin dealers receive their bundles bound with rubber bands, ten packages of dope to a bundle. As they open a bundle to sell packets, they keep the rubber bands on their thumbs as a way of accounting. When the cops roust either group, the first thing they look at is dealers' hands. Crack dealers' hands are always sliced up; heroin dealers always have rubber bands on their thumbs.

Fourth, despite constant complaints about the quality of the crack in jumbo red tops, it still seemed a good buy at $10 a big bottle, compared to the regular-sized green tops around the corner for $5. The cops pointed out that this particular drug operation, relatively small potatoes compared with others in the Bronx, netted about $25,000 every weeknight, week in and week out, and on weekends the total take was about $125,000.

The long night dragged on. At 0230 hours, a boss radioed Imperato and Small and ordered the detectives back to the 34th precinct station house, much to their frustration. Precinct commanders are judged by how well they keep overtime costs down, even more than by their troops' adroitness in solving crimes. The SNEU cops continued the surveillance.

"Midnight" came back to the street later that night. The SNEU cops scooped him at 0400 hours and transported him to the 34th squad holding cell. Imperato was notified at home, and he arrived at the squad at 0830 hours to interview Midnight. Showing him the death picture of Yvette Tirado, Imperato told Midnight that a witness tied him to the girl's murder. Midnight had one chance, and one chance only, to get out in front of things. Otherwise, Imperato swore to Midnight that he would bury him.

Midnight, a 22-year-old Dominican, oversaw street sales at the Manor Avenue drug house. His version of events confirmed the broad outlines of Tennessee's story but with some crucial differences. Midnight said that a customer named "Mariah" complained that she had been robbed by two women in front of the crack house he managed. She lost two rings, a hair clip, and $90. Midnight did some detective work of his own, and, through street informants, identified the two robbers, a Spanish woman and a black woman. Along with Cuba, Midnight confronted the two women and demanded that they pay back the $90. The drug house could not afford to get a reputation as unsafe for its customers.

Midnight told the Spanish girl that she had to sell the leather coat she was wearing to recover the $90. The girl refused, arguing that the coat belonged to her sister. But Midnight insisted. The Spanish girl left in a huff and returned shortly with the $90, which Midnight tucked away to give back to Mariah. Later, Midnight looked out the window of the crack house and saw the same Spanish woman speaking with police officers at the corner of Manor and Watson. He also saw one of his own employees, Miguel, walk past the woman and police officers. Then Miguel came up to the apartment and excitedly told Midnight that the girl was talking to the police not only about being robbed of her coat but about the drug spot and the stash apartment.

Midnight was troubled. He had to act. He waited until the cops left the street and then he went downstairs and, accompanied by Miguel, confronted the Spanish woman. He told her in no uncertain terms that what she had done was wrong. He had not robbed her. Instead, he had only made her pay for her own robbery of one of his customers. Besides, what did she think she was doing, identifying him to the police or talking to police about the location of the crack house and stash apartment? It's simply not right to tell police such things, Midnight explained. The girl denied giving the police an accurate description of Midnight or telling them about the drug house. She then stormed off into an apartment on the ground floor of the crack house building.

Miguel, Midnight said, had listened to the whole exchange. He told Midnight: "Somethin got to be done; we got to kill her." Midnight said that he told Miguel: "I ain't killing nobody. You got to kill the girl." Midnight said that Miguel took charge. First, Miguel went to get Johnny and told him to go into the apartment and get the girl out. Then Miguel told Cuba to get the car. Then Miguel went upstairs to an apartment and got the 9-millimeter gun. And then Miguel came down and went into the ground floor apartment where Johnny and the girl were.

Midnight said that he stayed in the lobby. He saw Johnny and Miguel drag the girl out of the apartment, her hands bound with beige tape. Miguel held the woman while Johnny put more tape around her mouth. Miguel and Johnny forced the girl into the back seat of the four-door Caddy that Cuba was driving. Miguel sat in back with the girl. Midnight rode in the front passenger seat, while Cuba drove. Johnny stayed on the street.

They took the Cross Bronx Expressway to Manhattan, all the way to the Henry Hudson Parkway, where they headed downtown, getting off at 158th Street. But no sooner had they left the highway than Miguel told Cuba to go back uptown to Fort

Tryon Park in Washington Heights, a remote place with lots of murders and overworked cops—in short, a good place for one more dumped body to get lost in the shuffle. Once at the park, Cuba drove around the museum and took the road that led toward the highway. Miguel told Cuba to stop the car just before the arch in the middle of the street. Miguel forced the girl out of the car and across the street, walking with her toward the three-foot wall that bordered the park.

Midnight got out of the car and went to the wall and looked down into the park. Midnight said that he watched Miguel and the girl walk down the path and then up some stairs until they were briefly out of Midnight's view. He walked up Broadway toward the museum to keep them in sight. Meanwhile, Cuba had turned the car around and driven to the parking lot, opposite the museum.

Midnight saw Miguel pull the gun out of his waistband. The girl slumped to the ground near a fence and raised her hands in front of her face. Miguel fired a shot into the air. Then he stepped toward the girl, stood directly over her, and pointed the gun at her head. He shot her twice. Miguel then walked around the edge of the grassy circle, went back down the stairs, and headed up the path to the road. Midnight said that he and Miguel piled into the car, and Cuba drove back to Manor and Watson avenues, where the three split up. Miguel later told Midnight that he had stashed the gun.

Detectives didn't accept Midnight's story at face value. They noted his studied effort to present himself as a passive participant in the scene, one who simply followed Miguel's lead, a typical strategy of those interrogated about serious crime. And if Miguel, who was not mentioned at all in Tennessee's account, had in fact shot the woman, detectives knew that Midnight didn't see himself as responsible in any way. But, under the guise of getting suspects to tell their side of the story, detectives can usually elicit exculpating tales, like Midnight's, that in the eyes of the law are actually confessions.

———————

Apprehension is an essential part of investigation. After talking with Midnight, the detectives' primary task became to find the other players in order to interrogate them. Detectives Imperato, Small, and Montuori took Midnight in the surveillance van and went over to Manor and Watson avenues, looking for Johnny, Miguel, and Cuba. They parked on Manor Avenue in front of a fire hydrant. Meter maids gave them three parking tickets during the long morning's wait. When each player appeared on the street, first Miguel, then Johnny, and then Cuba, Midnight identified him through the van's tinted glass, and the detectives radioed Detective John Bourges, who was waiting at the 43rd station house. Bourges dispatched plainclothes cops in an unmarked police car, who swooped down on the street, arrested each culprit, and took him to the 34th precinct. By mid-afternoon everybody was there, sitting in separate rooms.

Detective Imperato prefaced the interviews by reminding each player of the intersection between the law of the street and that of the criminal justice system. Both worlds run on hard calculations of self-interest. Friendship and loyalty mean nothing. Personal survival means everything. The only issue is: does one betray others or does one get betrayed?

Johnny, a 21-year-old Puerto Rican man, wrote out his statement in English. Johnny used most of the strategies afforded by the common law's insistence on intention as an essential part of crime. First, he portrayed himself as an unwitting participant in an event that he misunderstood. "Friday morning I came down from my girls house. I went to the game room to play some machines. But the machines was shut of so I went into the sotre [store] and bought a pack of Newports and I was talkin to the owner and I over here the guys talkin about same thing they were up set because this girl had did some thing a[nd] I went out side to mind my own busness I so I was waitin for my girl in the corner to come down with her baby to go around. So they told me that they was going to scare that girl I seen her around the area. Ive heard that she liked me too. So I said your crazy she is a drug user so they said we gonna scare her like evyone else lets bug out I said allright just to scare her ok. they said ok."

By stressing the disparity between his own intentions and what he thought were the intentions and eventual actions of the rest of the crew, Johnny exculpated himself and set the stage to inculpate his friends: "so they were takin [talkin] to her in the first flor so I came in the apartment we all start to take [talk] to her they were saying why did you do that that wasnt right. they said your were only to scare her for she wasnt have to do it do it again just like we do to all the drug users. its just a fake they said so I said ok lets scare her. & they said put the tape on her hands. so I put tape on her hands and mouth laughin at the time they take her out to put her in the car. When they put her in the car I went up to the car and closed it I [illegible] down in the car window and the all left."

Johnny suggests to Imperato that his nonchalance about the whole affair points to his true state of mind: "so I went back to the corner to wait for my girl to come down. later on me and some friend went to the movies I didn't bother to ask what happen because he was with his girl when we got back I saw Midnight and I asked him yo what happend to this girl he said yo we scared the shit out her. She got out the car and ran off she wont do that again. so I said cool I see you whenever. I left to my friends house with friends and girls. We [illegible] night at his house I came back to the block. I didn't see anyon around so I went into the game room to play street fighter."

Finally, Johnny asserts his shock at Yvette's fate, an expression of moral revulsion that proclaims his own innocence, even as he betrays his friends: "& a friend of mine came in and said yo I herad that they took Midnight in this morning I said why she told me for some murder. I said that what stop playing with me ok she yo for reel they take him in I keeped playing my kids game I was saying to my self oh my god the didn't no it cant be true they said it was going to be a joke just to scare her. My friend came and told me yo John they just took Miguel I said why she told me I don't know. So I when to find out what happen when the cop car pulled up and hand cuffed me. So when I was takin [talkin] to the Detective he told me why I was her for and it kind of shooked me because for me it was just a joke to scare her the girl that I set the tape her name Evette and the other peopl in the car was Miguel, Midnight, and Cuba."

Miguel, a 16-year-old Guatemalan boy, told a different story than either Midnight or Johnny. Miguel said that he worked as a lookout for drug dealers. That Friday morning, he had seen a Spanish girl talking with three uniformed police officers at the corner of Manor and Watson avenues. He went looking for Midnight, the manager of the drug spot on the street, and found him in the second-floor hallway of the building. Miguel asked Midnight why he was hiding upstairs instead of being out on the street. Midnight told him that the Spanish girl had robbed a customer of $90 and two rings and, as payback, Cuba and he had taken her leather coat. Midnight said he wanted no beef with the police who, he had heard, were on the street.

Miguel went back to the street. Another man approached him and told him that he had overheard the girl telling the police about the drug apartments on Manor Avenue. Miguel hung out in the local bodega until he saw Midnight come outside. He asked Midnight if he knew about the girl giving up the drug apartments to the police. Midnight said yes. Midnight told him that he had just spoken to the girl in the first-floor apartment of the drug building.

Cuba arrived in a sky-blue Cadillac. In the building's lobby, Cuba, Midnight, and Miguel talked about what they should do with the girl. Just then Johnny came into the lobby. Midnight told Johnny to go talk with the girl and cool her out so she wouldn't leave. Johnny was told that the girl was going to get hit for what she told the police. Cuba said that she should be shot or killed, but Miguel could not remember which. Johnny went into the apartment, followed by Miguel, Midnight, and Cuba. They asked the Spanish girl what she told the police. She insisted that all she told the police was that she wanted her jacket back. But Miguel confronted her about what she was heard telling the police. The girl remained silent. Miguel asked her why she told the police. The girl did not respond.

Cuba gave Johnny the tape to tie the girl's hands. Miguel, Midnight, and Cuba held the girl while Johnny tied her hands. Midnight told Johnny also to put tape around her mouth. Cuba left the apartment first, unlocked the car, and scouted the street. Then Miguel and Johnny dragged the girl out of the building and put her into the back seat of the car; she resisted every step of the way. Midnight followed, bringing her jacket, shoes, and purse. Miguel sat in back with the girl. Midnight sat in the front passenger seat. Cuba drove.

They went first to 158th Street and Henry Hudson Parkway. They had intended to drive to the lighthouse under the George Washington Bridge, but the roadway was blocked. Besides, they noticed two Parks Department employees. So they drove north on the Henry Hudson Parkway to Fort Tryon Park. They circled the museum and then headed back toward the highway. Cuba stopped just before the tunnel. Midnight told Miguel to take the girl out of the car, saying that he would catch up with them. When Midnight caught up, the three walked down the path into the park, staying to the right. The path led them to some stairs. They walked up the stairs and stopped at the top by the fence. Midnight told Miguel to grab the girl and hold her because he was going to shoot her. But Miguel said: "Are you outta you fuckin mind? You gonna shoot *me!*" Then the girl threw herself to the ground and covered her face with her arms.

Midnight wanted to shoot the girl in the forehead, so he pointed the gun at her. But he didn't fire. Miguel said that he told Midnight: "Gimme the gun!" Midnight handed him the gun and Miguel shot the girl in the head. Miguel saw the girl's head nod; blood came out her nose and mouth. But she was still moving. So Miguel pointed the gun at her again and fired, hitting her in the face. Midnight took the gun from Miguel. Midnight cleared it by ejecting two bullets, which he told Miguel to pick up. Then they walked back to the road and met up with Cuba. They all drove back to Manor and Watson avenues. Miguel last saw the gun in the glove box of the Cadillac. He threw the two bullet-shells down a sewer at Manor and Watson. Midnight drove away in his red Mitsubishi Conquest. Cuba drove away in his blue Caddy. Miguel went to his girlfriend's house and told her what had happened.

Miguel recounted his story with boyish animation and pride. Both to the detectives, and later on videotape to the district at-

torneys, he spoke with particular forcefulness about how he
seized the gun from Midnight's hesitant hands and acted decisively to shoot the woman. Such initiative and nerve are virtues
widely admired on the street. Detectives had no reason to doubt
that Miguel was indeed the shooter of the bunch. But who was
the man who had told Miguel that he had overheard Yvette telling the police about the drug apartments on Manor Avenue?
And why had Tennessee not mentioned Miguel in his version of
events? Could it have been Tennessee himself who had betrayed
Yvette to Miguel and thus to her fate?

Cuba, a 23-year-old Cuban man, gave a still different story.
Early Friday morning, around 7:30 A.M., Cuba said he was driving a friend's 1983 Cadillac at Manor and Watson avenues when
Midnight came over and said he needed to borrow the car because he wanted to go out with a girl. Just then, Cuba said,
Miguel walked over and started talking to Midnight. They told
Cuba to wait right there, and they went into the drug-spot
building on Manor Avenue. About twenty minutes later Miguel
and Midnight came out of the building holding a girl by her
arms; Johnny was next to them. They all seemed to be pushing
the girl. Midnight told Cuba to start the car. The girl got into
the back with Miguel. Midnight got into the front seat with
Cuba. Midnight told Cuba to drive to Manhattan. During the
trip, Miguel talked to the girl. Then, Cuba said, in the rear-view
mirror he saw the girl give Miguel a blowjob.

At one point, Cuba said, he drove under a highway. Midnight
told him to stop, but then both Miguel and Midnight told him
it was too light there. They told Cuba to get back on the highway. Then, Cuba said, when they got up to Fort Tryon Park, they
told him to make a turn into the park and keep driving. Finally,
Cuba was told to stop. Miguel and Midnight took the girl out of

the car and told Cuba to park the car. Then, Cuba said, Miguel and Midnight took the girl down some stairs. About fifteen minutes later, they returned without the girl. They were walking fast. Cuba asked what happened to the girl. Miguel and Midnight said they were going to fuck the girl, but she ran away. Both of them got into the car with serious looks on their faces. Cuba was told to drive back to the Bronx, where he dropped Midnight and Miguel off at Manor and Watson avenues.

Cuba said that he then went home. Later that night, he heard from people on the street that Midnight and Miguel killed a girl, using his car. But when Cuba saw Miguel and Midnight the same night, they said it was not true. Then, on Saturday morning, Cuba heard on the street that Midnight had been busted for murder. Cuba ran into Miguel and told him that news. Miguel told Cuba that he had shot the girl because she talks too much. Cuba said that he told Miguel: "You should not have killed that girl."

The detectives viewed Cuba's determinedly exculpatory account as skeptically as they had Midnight's, particularly in light of Tennessee's and Miguel's accounts of Cuba's role in Yvette's abduction. Two weeks later, Imperato tracked down the young black man who had witnessed the abduction and had then been beaten up. That witness told Imperato that he had seen Cuba come out of the building, look both ways, and then give a high-sign to others, who then dragged a struggling woman out of the building to Cuba's car. Still, detectives chose to believe wholeheartedly Cuba's description of Miguel's back-seat sexual assault of the woman on her way to her death. And they had Miguel's confession to the actual shooting.

In the meantime, a woman named "Carmela" marched into the 34th precinct station house and boldly demanded the return of her Cadillac, which, she claimed, had been stolen. She produced two keys to the car used in Yvette Tirado's kidnapping as proof of her ownership. Imperato asked her if she knew or had ever heard of Midnight, Miguel, Johnny, or Cuba, or several other of the players on Manor Avenue where she lived. But Carmela denied any knowledge of the people that Imperato named. She just wanted her car back. Imperato informed her that her vehicle was now evidence in a kidnapping/murder case and she could forget about seeing it for at least a year. Carmela stormed out of the squad room.

Imperato and Small argued strongly to DANY that this case should go to trial. The heinousness of the crime and the evidence against the defendants demanded it. At pretrial hearings the sitting New York State Supreme Court justice, when she saw the videotaped statements of Johnny, Midnight, and Cuba, and particularly that of the flushed, bright-eyed Miguel, declared the murder an "execution." Moreover, another eyewitness had come forward, though reluctantly. He had been having sex with another man in the woods immediately adjacent to the murder scene and had actually seen the shooting. At the hearings, when the state produced Tennessee, arrested for yet another drug violation, the resolve of the defendants' lawyers crumbled.

The defense attorneys had planned to take advantage of New York State's prohibition on using evidence obtained from one codefendant against another codefendant without independent corroboration. But Tennessee put Cuba, Midnight, and Johnny on the set of the kidnapping, and Miguel proudly boasted of the murder itself. Tennessee's failure to mention Miguel's participation in the abduction, possibly because he himself had betrayed Yvette to Miguel, became moot. The lawyers nearly climbed over one another to reach the bench to arrange pleas. All the defendants pleaded guilty to felony murder. Midnight, Johnny, and Miguel received fifteen years in prison. Cuba received twelve years.

At the start of the police investigation, Yvette Tirado's death was a prototypical "uptown murder." It represented one of hundreds of crimes that cops call "public service homicides," where victims' ways of life and habits of mind help fashion their own deaths, where today's witness is tomorrow's assailant and yet another tomorrow's victim. And Detective Imperato noted the circle of street justice in this case: a robbery provokes a retaliatory robbery, which led to contact with the police, which prompts fatal sanction. Still, the terrifying circumstances of the woman's death, underscored by Cuba's account of Miguel's sexual assault on her when she was completely helpless, transformed Yvette into an honorary innocent victim. Any derogatory comments about her ceased the moment that Detective Imperato came out of the interview room and told his brother officers what Cuba had said. Yvette Tirado became known, once and for all, as "the girl in the park."

5

SQUAD WORK

During a typical four-day tour of duty between 1991 and 1993, a Central Robbery squad detective of the New York City Transit Police might catch one armed robbery by a single assailant specializing in sticking up subway riders for cash. In addition, he might catch two or three multiple perpetrator robberies by five or more youths working together to appropriate new sneakers, earrings, Eight-Ball jackets, or other in-vogue accoutrements. In each of these years, specialized units within Central Robbery handled roughly 160 cases of token booth robberies, while the Warrant Squad spent the wee hours of the morning searching all five boroughs for the hundreds of wayward souls who had ignored Desk Appearance Tickets (also known as "Disappearance Tickets"). The transit police's Major Case squad, based at the Metropolitan Transit Authority offices on Jay Street in Brooklyn, handled murders, rapes, and sexual abuse cases in the subways and buses throughout the city.

During a typical four-day tour of duty in that same period, a Midtown North squad detective of the New York Police Department might catch two complaints of assaults by rough-trade preying on men with uncontrollable sexual desires; several complaints by tourists relieved of their money and dignity by flim-flam artists who construct too-good-to-be-true deals that never fail to reel in suckers; a complaint from a local bar of extortion by men the size of taxi cabs, armed to the teeth, and speaking in raspy voices; a DOA (dead on the detective's arrival) in one of several Eighth Avenue flophouses; and a slashing outside one of the trendy clubs that dot the precinct. He might also catch three or four homicides a year, ranging from the stomping-to-death of a sushi chef for casting a longing glance at an attractive Italian girl, to a 45th Street dump-job of a heavily insured upstate cross-dresser done in by his wife and her lover.

During a typical four-day tour in the upper reaches of Manhattan, an NYPD 34th precinct squad detective might catch two aggravated gun assaults; an armed extortion of a bodega owner balking at his creditor-controlled installation of illegal slot-machines in his store; a complaint about a knife-wielding derelict; a shake-down of a bookie running the illegal Dominican lottery; a jumper off the George Washington Bridge; a report of a caveman living in the Ice-Age-old mica caves of Inwood Hill Park; and a drug-related homicide.

Throughout the city, rapes and other kinds of sexual abuse, sexual abuse of children, trafficking in women, drug possession, drug sales, and even robberies and burglaries, although initially reported to the Midtown North and 34th precinct squads, were usually handed over to specialized NYPD units located either at the precincts or, more often, in other organizational wings of the sprawling department that then numbered more than 30,000 officers—by far the largest police force in the world.

Detectives in all of these units constantly scrambled to keep up with their caseloads. For instance, the eighteen squad detectives catching cases in the 34th precinct, who were organized into three teams of six, each received about 300 complaints per year in the 1991–1993 period, far in excess of those handled by detectives in sleepier precincts. Of such complaints, each squad detective "took" between 200 and 250 cases a year, that is, he or she decided to investigate those complaints. In addition to the eighteen "catching" squad detectives, four senior detectives on the precinct's homicide team assisted junior colleagues in investigations of murder cases. The squad also housed three senior detectives "on a steal" from the Manhattan North Homicide Squad (MNHS), a specialized unit whose members assist detectives in precincts north of 59th Street in investigating recent and old homicides. Other members of MNHS regularly visited the squad as well.[1] The pace of complaints and cases taken in transit's Central Robbery unit and in the Midtown North precinct was only slightly slower than that in uptown Manhattan.

Against such a frenetic pace, in the open-but-confined, metal-desk-cluttered forums of squad rooms, detectives then as now talk constantly about the Job. The Job is, first, to investigate and establish responsibility for crimes and thus help contain the forces of disorder. Second, the Job is to negotiate the sprawling interconnected bureaucracies of the criminal justice system and the concomitant thicket of rules, regulations, and laws that impede the investigation of crimes. These two aspects of detectives' work clash constantly. Investigating crime means successfully working the streets, but that task requires habits of mind and practices that are completely different from those instituted, valued, and rewarded in police bureaucracies and in the legal system.

What are the organizational frameworks of detective work? Squad detectives work cases. Cases are bureaucratic entities—official representations of real-life incidents that prompt complaints. The bureaucratization begins with the complaint itself. Uniformed officers on the street, or complaint-desk officers at the station house, "take" complaints from complainants. The basic rule in New York City is "No complainant, no crime." The NYPD does not accept third-party complaints except in the case of a death. When that happens, the first uniformed officer on the scene becomes the complainant, standing in for the deceased person. The officer at the complaint desk gives each complaint that she receives (from uniformed officers or directly from complainants themselves who come into the precinct) a number on a UF (uniformed force) 61 form in the sequential order in which she receives them. She writes up a terse account of the complaint and sends it forward to the uniformed desk sergeant for disposal. The desk sergeant reviews all complaints, decides which ones require an investigation by the precinct detective unit (PDU), marks a time on them, and forwards them upstairs to the squad room.[2]

Squad detectives work a four-two schedule—four days on duty, two days off duty. The first two days of a tour are from 1600 hours until midnight, followed by a turnaround beginning at 0800 hours until 1600 hours for the next two days, followed by a swing of two days off. The schedule thus moves through the days of a calendar week, to compensate for the weekly rhythm of crime, typically heavier on weekends and sparser in the beginnings of weeks. Detectives catch cases according to schedules constructed by each team in a squad. Typically, a detective is responsible for whatever cases the desk sergeant sends to the PDU during his team-assigned catching hours, say, from 1600 hours to 1800 hours. Homicides are the exception to the normal catching order because each detective in a team catches murders in a sequential batting order. When a detective catches a homicide, he goes "off the chart," that is, he catches no other cases for four days so that he can investigate the murder while, presumably, leads are hot.

Once a detective takes a case on a complaint, detective bosses, first sergeants and then lieutenants, review the case several times at intervals of seven, thirty, and sixty days. Each case must be "closed," though only those that are "cleared" by arrest or by exceptional clearance (discussed below) count as hits in the detective's batting average. All other case closures count as outs, with the exception of those later determined to have been inappropriately assigned to the squad; these so-called walks don't count either way. A detective's batting average is just one ingredient—though an important ingredient—in promotion reviews. Once caught, a case belongs to a detective until it is resolved one way or another. In investigations of homicides, which have no statute of limitations, this produces at least formal investigative continuity, sometimes over long periods of time, and ensures the chain of custody of whatever evidence is obtained in an investigation.

The catching system generates a proprietary sense about cases. This is reflected in detectives' use of the possessive adjective in referring to victims or incidents: "*my* girl in the park" or "*Bobby's* shooting on 160th" (where Bobby is the detective who caught the case, not the victim or the suspect). Rarely are detectives removed from cases once they catch them. Occasionally bosses, to assuage the anxiety of being dependent on subordinates' work when outcomes are uncertain, break the catching order and assign certain cases to experienced detectives, but for the most part this happens only when the crimes involved are sure to draw media attention. When detectives die, retire, or get transferred to other jurisdictions, other squad detectives inherit their cases and become responsible for them. Practically speaking, though, this applies only to unsolved major crimes because bosses insist on the closure of all other outstanding complaints.

The division of every squad into three teams, each with its own time chart planned a year in advance, creates obvious barriers to squad-shared knowledge. Only the few squads whose teams regularly meet with one another to discuss ongoing cases see the larger patterns in criminal activity within their precinct. But even within teams themselves, the catching system, with its allocation to individuals of responsibility to resolve cases, unintentionally fragments knowledge about cases. Consider the instance of two detectives who had served as uniformed partners together, who had each saved the other's life on different occasions in the middle of wild melees, who had worked on the same squad team for years together, their desks next to each other. During the frantically busy year of 1992, one detective interviewed a man as a witness to a homicide and then sent the witness on his way. Meanwhile, his teammate was looking all over Washington Heights for the exact same man as the known gunman in a celebrated triple murder. The detectives discovered the mixup only by accident.[3] Such incidents spur momentary impulses for reform, specifically calls for thorough communication of the details of all cases in a squad.

The catching system's fragmentation of knowledge can and often does produce a bailiwick mentality—a narrow focus on resolving one's own cases, without attention paid to colleagues' work or to the larger problems facing a precinct, let alone the city as a whole. This bailiwick mentality, typical of large organizations in general, characterizes all police authorities in New York City. Before the New York City Transit Police merged with the NYPD in April 1995, its avowed aim was to drive all crime out of the subways, off the buses, and into the streets. Even today, the Metro-North police work mainly to keep the Westchester County and Connecticut-bound commuter trains and Grand Central Station crime-free. The Amtrak Police labor to make Pennsylvania Station safe. NYPD precinct commanders try to force crime out of their own jurisdictions and into other precincts or boroughs. And the privately stated aim of top NYPD bosses is to run all criminal activity out of New York City and into New Jersey and Connecticut.

Crime itself will always be with us, police say, because some human beings enjoy transgression and, in some cases, evil. In detective squad rooms or, for that matter, in the entire police department, one finds few believers in the inherent goodness of humankind, or in social explanations for criminal violence, or in the perfectibility of human society. The important thing, depending on one's rank and level of organizational accountability, is to make one's own watch, one's beat, one's tour, one's precinct, one's borough, or one's whole city inhospitable to crime and criminals.

But the exigencies of investigating crimes limit and penalize too much self-absorption. Detective work depends on teamwork coordinated by the detective who catches a particular case. No one detective alone can conduct even the mandated, let alone innovative, steps in, say, a homicide investigation. The time-consuming tasks include canvassing apartment buildings, streets, family members, and known haunts of victims and suspects alike; combing motor vehicle registrations, telephone and credit card records; and arranging for ballistics comparisons or, more rarely, fingerprint analysis. Each detective has to rely on his team members, who trade their time for the expectation of similar help when they catch cases that require such assistance. Of course, detectives who do not cooperate with their colleagues on their cases can expect no cooperation on their own. For better or worse, squad work is teamwork.[4]

When teammates shirk the reciprocal arrangements of teamwork, they become the butts of fiercely barbed humor. A detective on one team was thought to be so notorious a slacker that, when she tripped and fell one day in the squad room and was transported to the hospital for a claimed injury to her back, her teammates set up a crime scene, complete with yellow tape, and then wrote up a complaint charging a large paper clip found on the squad-room floor with the crime of assaulting a police officer.

The catching system and the proprietary sense about cases that it generates shape subtle norms of team etiquette, particularly for how partners should behave. A good partner does not give unsolicited advice to the lead detective in a case. He pitches in willingly to help when asked, going out readily with his partner even when he is busy with other matters. If requested to do so, he promptly writes up reports on the investigation and adds them to the case file. He keeps quiet during interviews when his partner has assumed the lead. A detective who adheres to these norms can expect the same unassuming cooperation from his partner when he is working his own case. Those who do not conform to this etiquette find themselves working alone.

At the same time, a few senior detectives, usually those in the borough-wide homicide units, sometimes subvert the catching system by stealing juicy cases from the junior precinct squad detectives they are supposed to be aiding. This happens with the silent blessing of assistant district attorneys, who want seasoned witnesses in the courtroom for high-profile prosecutions. Such outright theft always causes profound resentment, as does more routine appropriation of credit—in casual squad-car conversation, bar talk, or rare public appearances—for solving others' cases.

Although bureaucratic incentives and occupational group norms generally ensure detectives' attention to their cases, who catches a case actually does matter. Ability, drive, and dedication to duty are always inequitably distributed in large organizations, and the police force is no exception. Some detectives mean well but are incompetent, unable to winnow the wheat from the chaff in an investigation. Some detectives are just lazy, a phenomenon that plagues all big organizations, perhaps especially civil service bureaucracies dogged by the near impossibility of firing people for cause. Some detectives actively dodge work by surreptitiously altering the desk sergeant's marked times on UF 61s, thus dumping difficult complaints onto their colleagues. Other detectives go through the motions of an investigation, even on homicide cases, producing impeccably orderly case-file folders that are, however, empty of all real information.

Detectives acknowledge that the catching system creates a roulette that enables some culprits to get away with serious crimes, including murder. But shirkers are the exceptions. Most detectives work at their cases, major and minor, faithfully and assiduously, some even spending money out of their own pockets to pry information from informants. The highest occupational virtue in detectives' world is dogged persistence.

Detectives talk constantly about their bosses because bosses shape their work in decisive ways. The New York

City Police Department may be one of the last workplaces in the United States where the use of formal titles—sergeant, lieutenant, captain, or the colloquial title "boss"—is expected and observed.

Every detective squad has at least one, sometimes two detective sergeants. In addition, most have a detective lieutenant; only a few squads are supervised by sergeants. Because there are only twelve detective captains in the whole city, each responsible for large geographical areas and several precincts, most squad detectives know their captains only from a distance. Even more distant are the top police brass, the inspectors and chiefs who turn out of One Police Plaza downtown.

All promotion to the civil service ranks of sergeant, lieutenant, and captain—the commanders of police officers on the street—proceeds "up the blue," that is, through the uniformed ranks on the basis of standardized written examinations. Most detective commanders rise through the uniformed ranks and get assigned to the detective bureau without any experience as detectives. A working detective who takes and passes the sergeant's examination and gets called off the waiting list when a position opens has to go back "into the bag" (uniform) for at least a year, with no guarantee that he will ever be invited back into the detective bureau as a supervisor. If he is in fact invited back into the bureau as a detective sergeant and he then takes and passes the lieutenant's examination, he has to return to uniform once again, with no guarantee of ever making it back into the bureau. In short, the dominant ethos of the police department emerges out of the semimilitary world of the uniformed forces.

The people who become detectives' bosses enforce police department rules and procedures covering virtually every conceivable situation that officers of every rank face, and they place a premium on officers' mastery of these sanctioned skills, behavior, and knowledge. Detectives regularly find themselves supervised by bosses who are skeptical, suspicious, or downright afraid of detectives' ability to do street work, a skill that requires bending rules and subverting procedures in order to obtain the information necessary to make cases. Detectives return the suspicion, especially in the case of commanders who excel in test-taking but are thought not to have "made their bones" on the streets. In addition, many policewomen in the New York City Police Department retreat from the streets to take desk jobs as soon as possible after leaving the police academy. They also far exceed their brother officers in test-taking skills, besides being the beneficiaries of relentless affirmative action pressures for more women in higher ranks.

From the viewpoint of street warhorses, "house mouses"—both male and female—increasingly govern the police department. Typically, the less street experience such superiors have, the more they insist that their subordinates adhere closely to established procedures. These rules proliferate in response to every crisis, as bosses scramble to close loopholes in existing regulations. In such a world, cops or detectives who aggressively pursue criminals or who, for that matter, demonstrate initiative beyond standard operating procedures become liabilities to ambitious bosses. Indeed, some detectives claim that bosses worry far more about detectives' possible violations of procedures than they do about crime. Some young cops just coming on the Job, and indeed some young detectives just promoted to the squad, take "Don't get involved" as their motto, for fear of being hammered by bosses who insist on rules for the sake of rules.

Bosses' allocation of overtime frequently makes or breaks cases. Even when detectives are in hot pursuit of a murder suspect, they need their precinct boss's approval to work longer than an eight-hour tour of duty on any given day, at time-and-a-half pay. Bosses in the high-crime, low-profile areas of the Bronx, Brooklyn, or Queens generally extend overtime as needed to solve serious violent crimes. But in Manhattan completely different rules apply, and these reveal the NYPD's organizational premiums.

Top police bosses in Manhattan—inspectors and chiefs—know that sooner or later a high-profile case will jam them up. A serious assault in Central Park, statistically the safest police precinct in the city, or a bold robbery or murder in the Midtown North precinct, a shopping mecca visited by millions of tourists a year, will immediately command worldwide media attention and place top police bosses under intense scrutiny. Top bosses in turn force that pressure down the line to precinct squad commanders, who demand that detectives work around the clock if necessary to produce sound-bite-sized stories out of the inevitably messy tangles of crimes and criminal investigations. Such cases simply have to be solved, and solved promptly, at almost any cost, to preserve the credibility of the police department and, more important, the careers of top bosses. Those bosses reward precinct squad commanders in the darker corners of Manhattan who, whatever the carnage in their own jurisdictions, can keep their detectives' overtime down, against the inevitable rainy days that threaten important publics' perceptions of order in the bright lights and of the competency of top bosses. Detectives in the killing grounds of Washington Heights or Harlem find their work subordinated to the exigencies of top bosses' careers—an organizational practice that sanctions callous attitudes toward crime victims in the low-profile corners of the city.

But even as they strive to keep down overtime costs and get

their charges to adhere to regulations, bosses need detectives who know how to spend time on some cases and not on others because case management is taken as a key index of their own administrative prowess. No matter how rule-oriented precinct bosses might be by temperament, they have to compromise rules in order to get their own work done, even if only by turning a blind eye toward detectives' legerdemain. The busier the house, the more bosses value those detectives who know how to get rid of cases.

For instance, in spring 1992 neighbors in a Washington Heights building heard two brothers tangling in a terrible fight. One brother left but returned shortly afterward. The brothers' fight resumed, moving into the building's corridor, where one of them got shot in the chest with a .22 caliber gun. Police later found the weapon on the pavement outside the building. The brother who was not wounded reported the incident to the police, saying that while he and his brother were arguing a robber rushed into the building and shot his brother. Detectives visited the wounded brother in the hospital. He claimed that he was completely drunk and the next thing he knew he had been shot. The detectives pointed out that numerous witnesses had heard the wounded man wrangling with his brother and asked if it was, in fact, his brother who had shot him and then thrown the gun outside. At that point, the victim asked them not to investigate the case any further. Despite the violence of the incident, always dangerous to the maintenance of public order, the detectives closed the case immediately, laughing about "two no-good mutts in a family quarrel," pointing out that some of their colleagues, who confuse police work with social work, might have interviewed everyone in the family and written a book on the moribund case.

Bosses come to depend on such cut-to-the-chase practicality. They are especially happy when presented with an opportunity to get rid of several cases simultaneously through the administrative mechanism known as the exceptional clearance (EC). Under police department rules, a detective can take an exceptional clearance if and only if police have enough evidence to arrest a suspect and know where the suspect is but, for reasons beyond the detective's control, such as the suspect's death or his taking refuge in a country that does not extradite wanted criminals, the arrest cannot be effected. In practice, police apply the mechanism only to the most serious violent crimes, such as homicides, rapes, and robberies. An exceptional clearance simultaneously closes a case, relieves bosses of the necessity of further review, eliminates paperwork, and gives the equivalent of an arrest (a "hit") to the officer handling the case and to his boss. Moreover, should a culprit be extradited or new evidence be adduced involving others in a crime that has been cleared to one person, police can reopen the case.

But the procedure has a dark side. One boss relied heavily on a detective to EC opaque homicides through the magic of statements by Spanish-speaking informants whom no other detective had ever met. And in the late 1980s and early 1990s, uptown Manhattan hosted a number of badmen involved in skeins of wild drug-related violence. "Guerrero" settled scores with his drug-business rivals and disloyal associates alike by approaching them openly on the street, backed up by gunmen in a livery cab behind him, and then drawing on his prey in Old West fashion, leaving them to die in the streets as an object lesson for anyone else who might challenge him. When someone beat Guerrero to the draw on a Caribbean island paradise, and his death was confirmed, detectives attributed several homicides to him, to their bosses' great jubilation. No one looked too closely at the details of those cases. As long as witnesses mentioned the name Guerrero as being at or near murder scenes, detectives took exceptional clearances. All useful bureaucratic mechanisms create invitations to abuse.

Sooner or later, even the most rule-oriented boss gets jammed up by the vagaries of the street and turns to a detective to "make a case come out right." This usually occurs when police officers are likely to be accused of wrongdoing for decisions they made in the heat of action on the streets. This does not mean that bosses encourage detectives to falsify the facts of incidents. But it does mean that bosses sometimes strongly suggest that detectives who catch such difficult cases give prominence in their written reports to those facts, observations, and interpretations that bolster rather than undercut fellow officers' explanations of events in order to guard against the possibility of inevitable later inquiries going south.

But some police bosses live by the book and demand that everyone else does as well. The foibles of such upright souls become fodder for detectives' stories. Detectives tell the story about the boss who became obsessed with missing equipment and demanded an immediate accounting for anything misplaced. One day the boss passed a detective's desk and noticed that she had not put the squad's Nikon camera back into the property room. The boss took the camera off the detective's desk and hid it in his own. When the detective returned, she panicked at the camera's disappearance. To cover the loss, she went out immediately and bought a new camera out of her own pocket and put that camera into the property room as if it were the squad's equipment. Two days later, the boss casually asked the detective where the squad's camera was. The detective went to the property room, retrieved the camera she had bought, and showed it to her boss. At that, the boss took the original camera out of his desk and asked what it was. The dumbfounded detective was summarily flopped back into uniform.

An Internal Affairs Bureau field associate observed a top-grade detective taking a drink while on duty, and he informed the squad's lieutenant of the crime. The lieutenant grilled the squad's sergeant for his opinion on the detective's character. The sergeant said that the detective in question was a man of great integrity and valor whom he had known for twenty years. While one couldn't condone his taking a drink on duty, one had to measure it against an impeccable two-decade record. The lieutenant flopped both the sergeant and the detective back into uniform.

Some bosses obsessively assert their authority, earning themselves the scathing ridicule that detectives reserve for the self-important. Detectives laugh about the squad boss who upbraided another detective visiting the squad for using the phone without the boss's permission, and about the inspector who ordered a full-scale investigation, complete with an all-but-useless fingerprint dusting by the Crime Scene Unit, of his departmental car stolen right in front of the station house and recovered a few blocks away, in all likelihood joy-ridden by local youths who found the keys the inspector had left in the ignition. Transit detectives howl at the boss who ordered his sergeant to call in a handwriting analyst to root out the culprits who disrupted a Christmas party. Some detectives had covered the plaque of the "detective of the month" trophy with a paper sticker containing the names of detectives whose work they thought demonstrably better than the detectives singled out by bosses for praise. After a confrontation between the boss and ringleader (who was identified by his peculiar scrawl), the boss banished the detective to Coney Island, as far from the detective's home as one could get and still remain in New York City.

Bosses who seize center stage at crime scenes, stomping on potential evidence and barking officious orders to detectives about how to conduct a proper investigation, always become the objects of squad-room laughter. Bosses who don't or won't temper the normal brusqueness of police department supervision with good humor and understanding open themselves up to especially barbed humor. Detectives in one squad kidnapped their boss's beloved plant, on which he lavished more attention than he paid his human charges. They mailed the plant back to the boss leaf by leaf, once accompanied by a photograph of the plant with a gun to its stalk.

Detectives think that bosses who don't seek out the hardearned knowledge of street veterans and consequently blunder into a public relations disaster deserve any embarrassment that comes their way. No detective who worked in northern Manhattan has forgotten the foolhardiness of Police Commissioner Lee Brown's 1992 unannounced good-will tour of West 160th Street to herald his Community Policing Initiative. A videotape of the visit featured Commissioner Brown unwittingly shaking hands with local drug dealers and hitmen posing as law-abiding residents, and thereby becoming fodder for street lore—"Get over on the commissioner, man, ain't nothin to get over on the precinct po-lice."

Finally, bosses whose rigid adherence to regulations makes their squads forgo simple human pleasures come in for special criticism. One fabled day in February 1992, several members of the 34th squad and I were leaving court in downtown Manhattan when we encountered a crestfallen man on the down elevator who was carrying a huge package redolent of garlic and seafood. Detective Joel Potter suggested that from the delicious smell of the package the man should be going up instead of down. The man turned out to be a confidential informant for the district attorney squad. He told the detectives that his wife had made thirty pounds of shrimp scampi and rice for the squad, but the boss would not allow his men to accept it for fear of violating Knapp Commission regulations. Potter suggested that the men and women of the hard-working 34th squad were worthy beneficiaries of the DA squad's loss and would in the bargain be grateful for his wife's cooking. So that evening the 34th squad and the Professor enjoyed the fattest, most luscious shrimp in anyone's memory, as the Saga of the Shrimp Scampi and the Bonehead Boss spread like wildfire across the city's station houses.

More serious issues arise when struggles over who controls an investigation become obstacles to solving a case. Early Saturday morning, December 28, 1991, Sergeant Keith Levine of the Communications Division was heading home following an evening spent with two fellow musicians, both civilians. As the three friends drove east on 57th Street from Ninth Avenue at 0217 hours, Levine looked out the back-seat window of the large black car and saw a man dipping a card into an automated teller machine in the lobby of Manufacturer's Hanover Bank. Another man stood beside him, gun in hand. The gunman then pistol-whipped the victim, crossed the street in front of the car occupied by Levine and his friends, and quickly headed west toward Ninth Avenue on the north side of 57th Street.

Levine apologized to his friends but said that he had to act.

He told them to stay in the car, which by this point had made a U-turn on 57th Street. With his off-duty pistol drawn, Levine jumped out of the vehicle and walked quickly behind the robber. A quarter of the block toward Ninth Avenue, in front of the Pasta Roma Italian restaurant at 315 West 57th Street, Levine announced himself as a police officer. The robber turned quickly and fired at least two shots. Sergeant Levine was hit twice, in the chest and stomach. The gunman immediately headed south, crossing 57th Street and then cutting through a driveway between buildings toward 56th Street. Sergeant Levine died at Bellevue Hospital an hour later. A check of his 38-caliber weapon revealed that he had fired one round.

The police response to the crime was massive. Levine had worked for a year in Midtown North before being posted downtown, and he was liked by all the cops in his old command. The robbery victim had fled, never to show up. But, remarkably, a homeless man named "Charlie" who lived in the Manufacturer's Hanover branch where the robbery occurred came forward to tell the police that a man he knew as Jay had done the robbery. His unsolicited testimony was bolstered by a match of his rolled palm prints to latent prints found on a garbage can in the bank's foyer, as well as by surveillance videos that confirmed him as a habitual door-opener-for-money at the bank. Jay, Charlie said, hung out uptown on 116th Street. Detectives' canvasses of the midtown area produced information that a couple of girls had acted as lookouts during the robbery, and one of them had worn a white coat.

Charlie later identified Jay from inside a police surveillance van on 116th Street manned by Detectives Harry Bridgwood and Danny Rizzo. In that surveillance, Charlie had taken his time, pointed out and named several players on the street, and called the alarm only when he saw the man he knew as Jay heading toward the subway. The police apprehended him in the subway station. Jay turned out to be Christopher Lewis, a miscreant with more than thirty arrests, most for petty crimes such as fare-beating and sucking tokens out of subway tollgates. Lewis said that he had been at Rockefeller Center in the early morning of December 28, sucking tokens out of the toll gates in the Sixth Avenue-47th-50th streets subway station. He then sold the tokens, as was his practice, to a Metropolitan Transit Authority clerk for 50 or 75 cents. The clerk resold them, he assumed, for the going price of $1.15.

Lewis remembered a man in a brown uniform who watched him sucking the tokens and subsequently selling them to the clerk. Police later found a brown-uniformed security guard at Rockefeller Center who corroborated Lewis's story, placing him at Rockefeller Center about an hour before Levine's murder. Lewis went on to say that he had walked north and then west a few blocks, stopping at a McDonald's on the southeast corner of 56th Street and Eighth Avenue for a bite, and then heading up Eighth Avenue, on his way to 72nd Street and Broadway, where he met friends. He put himself at the scene of the robbery at the time it occurred, but he insisted that he had nothing to do with it or with any subsequent shooting.

When Lewis was put in lineups, several witnesses identified him, though with varying degrees of certainty. The witnesses included Levine's two musician companions and also two lawyers who had seen a man running past them in the driveway between 57th and 56th streets as they were exiting a taxi. One of these lawyers had served as an assistant district attorney in Brooklyn. Lewis was charged by the police and then indicted by the grand jury with the murder of a police officer. The *New York Daily News* ran a front-page story on the arrest and indictment, including a picture of Lewis.

But there were other culprits and a weapon still out there. Detectives continued the hunt on 116th Street. At a local coffee shop that served as a hangout for street denizens, the detectives heard about a young woman named "Yvonne," who regularly wore a white coat and fit midtown street witnesses' description of one of the girls who had acted as lookouts for the robbery at the ATM. When the police found Yvonne, she turned out to be a crack-head who couldn't remember where she had been early that Saturday morning. She didn't think that she had done the crime, but she admitted that she knew Jay from the street. She said that Jay was also known as Black.

Then Yvonne confessed to the crime. Who else could it have been, she asked? All the detectives seemed sure Jay had done it, and she often hung out with Jay on the street. And if the detectives thought she had done the robbery with Jay, well, then she must have done it. Yvonne's statements were rambling, circular, incoherent, and confused. Eventually, she implicated another woman named "Sharon." The police picked up Sharon on Yvonne's identification. Both women were indicted by the grand jury for the murder of Sergeant Levine. Most detectives in the Midtown North squad and the next-door robbery team were delighted at the prompt resolution of the case. The Sergeants' Benevolent Association gave a plaque to the Midtown North squad for solving the murder of one of their own.

But several Midtown North detectives, especially Robert Chung and Harry Bridgwood, who worked the case from the beginning, as well as George Delgrosso, Artie Swenson, Pete Panuccio, and Detective Sergeant Al Regenhard, were deeply skeptical about Christopher Lewis's guilt. Lewis wasn't known for the kind of violence that robbery entails. Even though his jailers downtown found rap lyrics in his cell about the killing of Sergeant Levine, Lewis seemed an unlikely cop killer. Bridgwood, a 23-year veteran who had taken hundreds of statements and confessions over the years, felt deeply uneasy with Lewis's interviews, based on the sixth sense that only long street experience produces. Lewis expressed surprise where Bridgwood expected a calculated dodge. He seemed calm where Bridgwood expected edgy anxiety. His story about being at Rockefeller Center sucking tokens an hour before the shooting checked out. And his description of his daily routine—express train from Harlem down to 72nd Street or Times Square, over to Rock Center to suck tokens, back to the West Side to catch the train uptown—made perfect sense for a West Harlem street guy. His presence at 57th Street and Eighth Avenue when the robbery occurred might have been just an unfortunate coincidence.

Further, neither Bridgwood nor Chung thought that the identifications of Lewis in the lineups were especially strong. Certainly none were unequivocal. Moreover, Yvonne and Sharon lived in their own little world. Sharon told police that she and Yvonne had been lovers but then Yvonne had rejected her. Why would Yvonne name her, Sharon asked, unless she had had a change of heart and wanted to set up a love nest in prison where they could get three squares a day and live together? Police wondered if these two women were actually robbers or two hapless souls seeking refuge from their harsh lives on the streets.

In the initial canvass in midtown right after Levine's slaying, Detectives Danny Rizzo and Artie Swenson, following tried-and-true robbery-investigation practices, had tracked the shooter's escape path to see if he had discarded anything in flight. Rizzo and Swenson found a blue jacket hanging over a wire outside the basement window of an apartment four feet below street level on 56th Street. The jacket had clearly been tossed over the six-foot fence that guarded the sidewalk cutout and basement windows.

The jacket's zipper was damaged and there was a hole in one of its collars. It contained a package of Kool cigarettes in one pocket and a matchbook with a phone number in the other, along with two subway tokens. The phone number came back to an apartment in the Bronx off the Major Deegan Expressway. When Detectives Delgrosso, Swenson, Panuccio, and Sergeant Regenhard went to the address, they found a woman in her sixties who worked as a cleaning lady at a television studio near where Sergeant Levine had been shot. The woman became upset at first because she assumed that the police were there about her job. The detectives assured her that she was not in any trouble. They then asked her about the jacket, but she denied having ever seen it. When told that it had contained a matchbook with her phone number, she said that it must belong to one of her two sons. She gave the police an address on 129th Street in Manhattan, one block east of the Park Avenue elevated train tracks. She said both of her sons lived there. The apartment had been hers, she said, but her sons turned it into a crack haven.

When the detectives got to the apartment, they found broken windows, unlocked doors, and no lights. One of the woman's sons, "Alexander," was there, but his obvious disorientation from substance abuse convinced the detectives that he knew nothing about the robbery-murder. When quizzed about the jacket, Alexander said that he thought it belonged to his brother, Butch McBryde, but he didn't know where Butch was. The detectives cajoled Alexander, demanding that he contact them when Butch got in touch.

Back at the station house, Delgrosso ran both brothers through the computer and discovered that Butch had a misdemeanor warrant, enough to hold him if the police tracked him down. Delgrosso visited the 129th Street apartment a half dozen times in the next couple of weeks, sometimes with Pete Panuccio, sometimes with Artie Swenson. Twice they found Alexander there, along with several crack addicts, whom the police chased out of the apartment. The police vowed to return night after night until they found Butch. Detectives Chung and Bridgwood also visited the apartment several times, but without finding Butch.

In the meantime, a struggle for control of the investigation had broken out, pitting Assistant District Attorney Elizabeth Lederer against a boss in the police department, but also several Midtown North detectives against a boss. Lederer, who became famous for her prosecution of five teenagers accused of assaulting the Central Park Jogger in two sensational trials in 1990, had worked closely with Chung, Bridgwood, Delgrosso, Rizzo, and other detectives from the first night of the Levine case.[5] As the detectives' doubts about Lewis's guilt deepened and as the intense social pressures and expectations detectives feel when investigating the murder of a police officer increased, they pushed the department, with Lederer's strong encouragement, for license to pursue full time the other leads they were developing.

But the case against Lewis, though circumstantial, was strong, with identifications from solid citizens as well as street people and confessions from two presumed participants. Although the police department never spares resources to catch cop killers, it seemed at the time as though Levine's killer was already locked up. And the number of complaints requiring investigation pouring into the always busy Midtown North squad had not abated.

Furthermore, police officers, not the district attorney's office, run the police department and its investigations, and in all bureaucracies some bosses don't appreciate talented, strong-willed subordinates, especially those who speak their minds.

An impasse developed. It focused on the kinds of questions that plague all organizations during crises: Who gets blamed, both within the police department and in the public eye, if Lewis turns out not to be the killer? Who gets departmental and public credit for finding the true culprit? How do the police find the right guy, but also get other ongoing work done? Finally, and most important, who is boss? Some detectives debated in the locker room whether they might have to end up going uptown on their own time to continue working the case. But this was dangerous because, without formal authorization, officers might bear individual liability should something go awry. In the end, after a memorable clash with a boss over these issues, Lederer went downtown to One Police Plaza and got a chief's authorization for Detectives Chung and Bridgwood, though not for other detectives, to go off the charts completely to do whatever was necessary to resolve the ambiguities in the Levine case.

A week later, Detective Panuccio was delivering a prisoner to Central Booking at 100 Centre Street. As he passed the bullpen, a prisoner called his name. It turned out to be Alexander, locked up for fare-beating in the subway. Alexander told Panuccio that Butch had showed up at the apartment. Further, he said, a guy named Black often came to the 129th Street apartment, sometimes with a girl, to meet with Butch. Black had a gun. Alexander told Panuccio that he had overheard the trio talking about the ATM robbery and the cop shooting. They said that the cops had locked up the wrong guy. When Panuccio showed Alexander a picture of Christopher Lewis and noted that Lewis was called

Black, Alexander said that he knew Lewis from the street, but the Black he meant was a different guy. Panuccio called Delgrosso immediately, who told Robert Chung of his partner's finding. With this information, Chung got Lederer to apply for a warrant to hit the apartment and arrest Butch. The detectives enlisted Alexander, newly released from jail, to alert them to Butch's presence in the house.

But when Alexander went into his building on his assigned mission, he disappeared, evidently through a rear exit, and the plan unraveled. The police waited. Days went by. No Alexander. No Butch. Just before the warrant's expiration ten days later, the detectives hit the apartment. With Emergency Service Unit support surrounding the block, they found and arrested Butch McBryde. After he was thrown to the floor, secured, and cuffed— all proper procedures when executing an arrest warrant and a search for a weapon—he looked up at Harry Bridgwood and said: "I'll tell you what happened." He quickly blurted out key details of the robbery at 57th Street and Eighth Avenue on December 28, 1991. And the tale that he told seemed like an alternate reality to the case already made against Christopher Lewis.

Butch admitted being at 57th Street and Eighth Avenue early Saturday morning, December 28, 1991, along with Irving Crumb, known as Everett, and with two girls, one named Michelle, who wore a white coat, and the other "Sally." With them was a man whom Butch knew as Black, to whom Butch had loaned his blue zippered jacket because of the cold weather. While Michelle, Sally, and Everett kept watch outside, Butch went into the ATM with Black, confronted a customer who spoke a foreign language, and robbed him. But, Butch insisted, Black had the gun, a silver 38-caliber with a six-inch barrel and black handle. It was Black who had been chased by the white guy along 57th

Street. Butch knew nothing about the shooting of the cop. When shown a photograph of Christopher Lewis, Butch said that he didn't know him. He also claimed that he didn't know Black's real name.

On January 22, 1992, detectives found Michelle, who confirmed Butch's story and named Black as one Michael Alston. Alston had gotten out of prison just a month before Levine's murder after serving time for his second conviction for homicide. All told, Alston had served barely ten years for two homicides. He was nowhere to be found. Michelle also told the police that she and Alston had sold the gun used in the ATM robbery to a nearby drug house for forty bottles of crack. One of the owners of the drug establishment told Bridgwood and Chung that, after buying the weapon, he hid it wrapped in two cheese-doodle bags in a garbage can near his building. He took the gun out at night to play with it. Just two evenings before the detectives approached him, he discovered the gun was missing from the can when he went to retrieve it. He assumed that Sanitation had picked up the gun when it collected the garbage.

The detectives immediately contacted Sanitation and ended up spreading out a couple of tons of the vicinity's recent domestic garbage on the 125th Street pier to look for the gun. No luck. It turned out that the drug dealer had been stashing the gun in a wire trash basket on the street—a refuse depot subject to different schedules than Sanitation's runs for domestic garbage. And Sanitation had already hauled that load of trash to the Staten Island landfill. The detectives had missed the gun by no more than two days. But a gunpowder residue check of Michelle's kerchief, in which she claimed she kept the gun before she and Alston sold it, came back positive.

On April 16, 1992, after an exhaustive search, the detectives finally found Michael Alston, who admitted shooting Sergeant Levine. Although Alston was much bigger than Christopher Lewis, the pair looked alike. Alston said that the policeman's bullet had indeed hit the zipper of the borrowed jacket he was wearing and then made a hole in the collar, causing him to abandon it. When Bridgwood asked him what was in the jacket, Alston replied: "Cigarettes, matches, anna cupla tokens," the kind of telling detail that convinces detectives they have the right guy. Christopher Lewis, Yvonne, and Sharon were released with apologies on May 6, 1992. Alston pleaded guilty to murder, Butch and Michelle to attempted murder, and Everett to attempted robbery.[6]

Detective Artie Swenson was transferred to another command. On the basis of superlative homicide clearance records, Detectives Delgrosso and Panuccio made contracts to transfer to the Manhattan South Homicide Squad, but their move to that prestigious unit was blocked, forcing the two detectives, after a year of waiting, to relocate to other precinct commands. At least one boss outran any mistakes made in the Levine case to climb high in the ranks of the NYPD.

———————————

The detective bureau itself, beginning in the late 1980s, has created internal obstacles to the development of the habits of mind that make for effective criminal investigation. Most important, recruitment to the bureau has become thoroughly routinized, lowering the overall quality of talent of those who enter the bureau. Detective work used to be organized as an apprentice system. Detectives are simply uniformed patrolmen on special assignment. They used to be plucked from the uniformed ranks on the basis of demonstrated initiative, organizational know-how, and investigative potential, as assessed by other detectives or by department bosses. Police officers had to demonstrate investigative abilities to superiors, who then became their advocates for promotion to the bureau.

For instance, PO Kenny Ryan of the Midtown North precinct pounded his beat from 45th Street to 47th Street, from Eighth Avenue to the Hudson River, for years. Ryan used the discretionary power of policing gingerly and humanely, but also practically, given the formidable process involved in collaring someone. A collar takes a cop off his beat for an entire day, leaving the street to those who would be wayward. Instead, Ryan became a fixture on his beat, drawing bright lines for acceptable and unacceptable behavior, giving all the players fair warning of the consequences of crossing those lines. He accumulated extraordinarily detailed ethnographic knowledge of the ways of the street and thumbnail biographies of all its players—the dazed shadowboxers, the rough trade, the drug dealers, and especially the working girls, the street's bellwethers. He got to know all the girls on his beat, learning all of their names, real and fake, the names of their men, even the names of their children, learning as well their particular habits, predilections, and fears. Ryan always treated the girls squarely, and in return they told him about the predators in their midst. He passed that information on to detectives, helping them break scores of cases. They, in return, became his champions, urging his promotion to the squad because in their estimate he was worth ten cops. Ryan eventually made the detective bureau.[7]

But now the path to the coveted detective's gold shield is for the most part governed by union contract. Many young officers are automatically made detectives after serving eighteen months in buy-and-bust undercover narcotics operations—work that demands nerves of steel, remarkable physical courage, and street smarts, but no demonstration of investigative promise. The total number of officers with detective rank in the NYPD ballooned in the last years of the twentieth century, reaching 6,900 detectives in the department by January 2000. Only about 3,000 of these were in precinct squads, affiliated robbery units, or transit districts specializing in subway crime whose jurisdictions overlapped precincts. The rest were in highly specialized units of one sort or another, such as narcotics, borough-wide homicide units, crime scene analysis, ballistics, or criminal identification.

Experienced detectives have no mechanisms to transmit their knowledge of how to conduct street investigations to the green juniors pouring into the bureau, except for occasional, thoroughly standardized courses in general criminal investigation or more specialized courses in homicide investigation. But most young detectives find such courses irrelevant and boring, and many view the senior men teaching them as "hair bags," out-of-touch old-timers intent on talking about their personal triumphs. To be sure, some senior detectives do adopt favored juniors and school them carefully in investigative techniques. These ties can be personally powerful and organizationally effective, but they produce resentment among detectives not selected for such exclusive mentoring. The endless storytelling in precinct squad rooms serves to disseminate some occupational lore in a more widespread and equitable fashion, especially about the subterfuges of criminal investigation. But as older detectives retire, they take with them vast, almost entirely untapped reservoirs of accumulated knowledge and experience.

The increased bureaucratization of the detective bureau has undercut any lingering notion that detectives might have of a necessary connection between ability, hard work, and promotion. Hooks—that is, family ties, patronage relationships, sexual liaisons, or other personal connections—have always been extremely important in the NYPD, reaching back to the days when the department was Tammany Hall's graft collector, discipline enforcer, and sometime fall guy.[8] No one gets choice assignments without what cops call a "contract," a reciprocal tradeoff of personal favors at high levels of the organization. And no one gets to make contracts unless he is part of interlocking elite social circles, with years of experience on the Job and a track record of reciprocating favors. No one becomes the beneficiary of contracts unless key people see him as a future player in those same social circles. The more elite the police unit at issue, the more complicated the contract because one must give or call in favors from more people. Thus, units such as the prestigious Aviation Unit, from which even the most experienced pilots are regularly excluded, or the Auto Crime division, a sure-fire ticket to lucrative insurance investigation after retirement, become essentially the property of cliques of social intimates who establish criteria for entry that only a chosen few can meet.

Contracts become particularly important in the detective bureau when promotion is at issue. Proven ability in solving cases has always been a sine qua non of moving up the steep hierarchy of the three grades within the detective bureau, though not always a sufficient reason. Of the 5,939 detectives in the New York City Police Department in September 2004, 661 were second-grade and 197 were first-grade. All the rest were third-grade detectives, whose annual base salaries ranged between $57,943 and $61,670. Second-grade salaries ranged between $66,414 and $69,300; and first-grade detectives earned between $75,524 and $79,547 per year.[9]

Some of the best and most seasoned detectives in the department, many in local precinct squads, languish their whole careers in third-grade status, overworked, undervalued, and poorly paid, while promotions to "grade" go to relatively inexperienced detectives brought by administrative fiat into the bureau and lucky enough to land in choice details where grade is a perquisite or in specialized squads with powerful bosses who have the clout to get their own charges promoted. Even spectacular success in solving high-profile crimes, once the pathway to first-grade status, no longer guarantees detectives anything except the envy of colleagues and bosses who themselves long for the bright lights that New York's ravenous media shine briefly on detectives who crack big cases.

―――――――――――――

The perceived exigencies and subsequent practice of detectives' work regularly clash with official rules, both administrative and legal. Take, for instance, the issue of informants. All detectives rely on informants for information. They therefore spend a lot of time developing and maintaining the intricate social relationships with informants, actual or potential, that enable them to elicit secret knowledge. Some informants see cops as father figures to whom they can divulge the wickedness all around them. Others barter information in exchange for investigators' turning a blind eye to their own criminal activities. Still others want an immediate payoff for information, in the form of money or bags of drugs left behind after a raid they help arrange. Still other informants want the police to protect them by locking up their enemies. Although informants are used most extensively in the netherworld of narcotics investigations, virtually all criminal investigations come to a standstill without informants of some sort.

The police department, acting on the insistence of the district attorney and the courts, requires detectives to register those informants whom they use regularly. Officially, this allows a detective to go before a magistrate by himself, without his witness, in order to obtain a warrant. It also avoids the real possibility of false accusations and manipulation of the criminal justice system by criminals. But detectives argue that, like most procedures, informant registration first and foremost protects bosses if things go wrong. Those with the deepest knowledge of criminal activity are the least likely to submit to official registration. The rule leaves detectives with the tricky problem of finding a way to utilize the most detailed and arguably the most reliable information they get from the street.

And sometimes police must protect informants even at the cost of big arrests. One detective tells the story about receiving a call from an "anonymous" informant, when he worked in narcotics, telling him that three dealers were leaving a particular building at that very moment. The detective and his partner rushed to the building, arrested the dealers, and seized 800 crack bottles, a bag of uncut cocaine, and several hundred dollars. But in court the judge threw out the arrests and suppressed the evidence of the seized drugs because, he argued, the "anonymous" call was insufficient cause for the police to stop and search the men. As it happens, the "anonymous" informant was the brother of one of the three dealers, a man whom the detective knew well. He gave up his brother against the inevitable day when the "weight" apartment would become the target of merciless robbers. The detective told the court the literal truth but could not testify to the whole truth, that is, his personal knowledge of the dealer's brother, for fear of injuring innocent people.

Detectives' relationships with informants are dangerous and complicated in other ways. If an informant on one case casually provides information about yet another case, and the detective pursues those leads without following carefully prescribed procedures, he runs the risk of being considered overzealous by bosses. The informant who gives a detective good information one day might betray him to the Internal Affairs Bureau or the district attorney's office the very next day if it is made worth the informant's while. Indeed, the police department often uses known informants for integrity tests. An informant might call a detective with an important piece of news about a case, but then mention that cops stole drugs from dealers while making an arrest. If the detective fails to report the latter allegation, whether he thinks it is true or not, he runs the risk of a visit from Internal Affairs. However tempting the regularly offered promises of sexual delights, one sleeps with prostitutes, always good informants, only at the risk of immediate betrayal, and one trusts junkies, reliably in the thick of things, only if one also believes in tooth fairies.

The Internal Affairs Bureau lays traps for police officers to enforce adherence to standard administrative procedures—the basis, in the IAB's view, of the legitimacy of policing. Thus, a lieutenant who was one precinct's integrity control officer received a complaint from a civilian who told him that cops had roughed up his son. The lieutenant spent the better part of an hour with the man, sympathizing with him and assuring him that his complaint would be addressed swiftly and justly. The following day IAB notified the lieutenant that he had violated procedures by talking in such a way with the complainant. Instead, IAB said, the lieutenant should have given the complainant a Civilian Complaint Review Board form and sent him on his way. In this view of police work, human kindness has no place in the station house.

Detectives in Internal Affairs face their own organizational exigencies. Once they open an investigation of a police officer or detective, they have to close it, just as squad detectives have to close their cases. This imperative sometimes produces perverse results. For instance, Detective Joseph Montuori recalls an incident during his investigation of a death by autoerotic asphyxiation. The parents had been so shocked at finding their dead son in feminine garb that they had redressed him and placed him in bed. Because the young man had died of strangulation, Montuori called the Crime Scene Unit. When CSU arrived, its detectives took photographs and dusted the boy's room. Eventually the parents acknowledged their alteration of the scene, and Montuori ruled the death accidental.

A few days later, however, the boy's mother called Montuori to report that her son's watch was missing; it had been on his dresser, she said. Montuori immediately checked the CSU photographs and, sure enough, there was a picture of the boy's watch on his dresser. Only police officers and the parents had been in the room. Montuori called IAB to report that a watch had been stolen by one of the police at the scene. IAB took the case and interviewed all the officers who had responded to the scene but had no success in uncovering the culprit. Montuori then received a call from the investigating IAB detective informing him that IAB was charging Montuori himself with the theft and asking him if he wanted to make a deal. Montuori pointed out that it was he who had initially reported the incident. Only when Montuori threatened to see the matter all the way through departmental trial, if necessary, did IAB drop the charges.

As the despised "Rat Squad," IAB has always had trouble attracting talented investigators, and the bureau developed a reputation for conducting poor investigations. Finally, in the mid-1990s, the NYPD instituted a policy of forcing top investigators into a two-year IAB stint in return for a free choice in their next assignment. Requiring premier investigators to serve a tour with IAB helped diminish police officers' antagonism toward the bureau, if not toward the idea itself of policing the police.

In the legal arena proper, attorneys and courts place a great premium on written documentation of the investigation of criminal events, an important procedural safeguard both to obtain evidence considered reliable by courts and to guard against false accusation, arrest, and prosecution. Precisely because of this emphasis, detectives know that everything they record in writing will be subjected to extensive public interpretation and reinterpretation. But every major investigation takes detectives down blind alleys. The street yields its secrets grudgingly. Some men and women crawl out of the woodwork to confess to crimes, but more often they accuse their neighbors or relatives. Young women employ precise legal formulas to accuse their parents of abuse in order to gain emancipation from restrictive traditional authority. Spouses file complaints of child abuse or sexual abuse against their mates when a divorce is imminent in order to have a legal record for leverage in custody hearings. Criminals offer information to jam up business rivals or nail enemies as payback for still other crimes. Informants provide seemingly reliable information that turns out to be triple or quadruple hearsay. Hot lines opened in major cases invariably elicit fantastic stories describing sins and crimes, whether real or imagined, that have never been reported to the police. Even an open-and-shut case goes through many twists and turns before its telling in court, and "final" adjudication rarely closes a major case permanently.

As detectives see it, their specific goal in, say, a murder investigation is to identify a suspect on the basis of available evidence and produce a coherent, compelling narrative, including a demonstration of the suspect's means, opportunity, and plausible motive linking him or her to the crime, a story that permits no alternate readings. Thus, detectives commit to writing as little as possible about their ongoing investigations until they know the final shape of their narratives, in order to minimize later alternative exegeses of their work. This practice, and others like it, make rule-oriented police bosses, courts, and defense attorneys deeply suspicious of detectives' work methods.

Prosecutors also have their doubts. They tussle endlessly with detectives in the course of the joint occupational struggle to transform street knowledge into legal proof. For instance, detectives argue that prosecutors' constant search for a smoking gun stems from their fear of circumstantial cases that rely on detectives' intuition about criminals and their ways of thinking. Moreover, many prosecutors, whose work demands a public face of truthfulness and propriety, worry about their dependence on an occupational group that relies on deception as a tool-in-trade.

A few prosecutors take haughty and scornful stances toward all police officers—sentiments that can destroy even important cases. For instance, Detective John Bourges caught the June 12, 1991, case of an off-duty police officer who was rousted at 2130 hours by the superintendent of his building with the news that a culprit was burglarizing the officer's car on the street. The policeman went out and ended up confronting a drug dealer borrowing a tire to replace a flat on his own identical-make automobile. The cop showed his shield and identified himself, whereupon the thief pulled out a gun and fired at the officer. The cop returned fire, hitting the thief once. The thief fled on foot, pursued by uniformed cops who had responded to a 911 call placed by the building superintendent. The officers found the thief's bloody shirt, but not the culprit, and called the Crime Scene Unit to recover it.

Later, one of the officers who had given chase was waiting in the area for a ride back to the station house when suddenly he spotted the thief crawling out of bushes where he had been hiding. Once again the chase was on, and the officer finally tackled the thief as he was breaking into an apartment window after going up a fire escape. The next morning, a resident of that building found a weapon under his own vehicle and handed it over to the police. The culprit had just been released from Attica after serving a term for attempted murder.

Bourges did a warranted search of the thief's apartment, which yielded some drugs, a photograph of the thief's girlfriend posing with money and a gun, and a .38-caliber bullet, which a ballistics test later determined had been ejected from the gun found under the car at the scene of the culprit's apprehension. The police-officer complainant identified the culprit in a lineup. The culprit gave a statement to Bourges arguing that the police officer who had originally confronted him in the street had not identified himself. Moreover, the thief argued, it had been his "partner," whose name he claimed not to know, who carried the gun during the attempted tire theft and who fired at the cop. The cop had meant to shoot at his partner, not at him, and it was his own bad luck that the cop was such a terrible shot.

Whom should a prosecutor believe? As it happened, the prosecutor who caught the case considered the police a "morally stunted occupational class," as he once told me—a deficiency he attributed to police officers' constant proximity to street violence. Such an attitude makes trust and subsequent decisive action difficult. As Bourges and other officers saw it, the prosecutor simply did not believe the police officers' accounts and therefore did not press the case. Indeed, although Bourges and other officers (including two auxiliary cops, two police officers

who had responded to the scene, two housing detectives, the police complainant himself, and crime scene detectives, as well as
witnesses from the complainant's building) went to court constantly, whenever requested, the prosecutor's lack of aggressive-
ness enabled the defense to stall the case again and again by asking for and gaining continuances. Months went by. Finally, one
day the prosecutor asked for a continuance, the people's first
such request. The defense claimed to be outraged by the delay
and demanded that the judge dismiss the case under the right-
to-a-speedy-trial rule. To police officers' bewilderment and chagrin, the judge did indeed throw out the case, and the culprit
walked.[10]

The prosecutor saw the matter completely differently. From
his perspective, the case against the accused was ambiguous and
indeed compromised by the question of whether the complainant off-duty police officer might have been inebriated at the
time of the original altercation. In his view, he made every effort
to bring the case to a closure acceptable to the police officers,
but the case was victimized by the unaccountable procedural vagaries of the court system.

Most prosecutors, despite their suspicions that cops bend the
rules and despite the discomfort this generates for officers of the
court, accept the moral trade-offs inherent in working hand-in-
glove with detectives, and they reject sanctimoniousness among
their colleagues or former colleagues.[11] For the most part, they
resolve their tensions by adopting a knowing-and-not-knowing
stance toward the ambiguities of detective work ("I don't wanna
know what happens at the station house").

The moral ambiguity of detectives' work clashes
with the necessity of courts to project a public image of upright-

ness, essential for the legitimacy of the judiciary. Courts claim to be impartial institutional arenas in American society whose purpose is to establish truth and fix responsibility for crime.[12] Both claims are increasingly problematic within a social order marked by epistemological wars that pit all against all and one rife with remarkable excuses and justifications to escape or mitigate responsibility.[13] Criminal investigators work with a minimalist concept of truth, that is, they seek to ascertain who committed specific actions defined as crimes. If they uncover a motive during the course of that search, it helps convince themselves and others that they have, in fact, ascertained who committed those crimes. But the basic quest for knowledge about who committed specific actions is always primary and encounters many obstacles.

Criminal violence is a relatively hidden human activity. Gaining even bits and pieces of knowledge about it requires painstaking work. Few people actually see criminal violence first-hand, and only rarely are criminal investigators among them. Most violent acts have few witnesses: the culprits themselves, their now-violated or dead victims, civilian witnesses usually so terrorized or confused that their perceptions are jumbled or limited, neighborhood residents, some of whom are civilians too frightened to come forward, some of whom are the indirect beneficiaries of criminal activities, and some of whom are themselves criminals who reveal information only when they can trade it for their own advantage.

Moreover, when witnesses are themselves criminals, one sees in stark relief the central dilemma of criminal investigation. One of the street witnesses to the infamous 1991 quadruple murder in the Bronx's 40th precinct gave Detective Mark Tebbens a vivid eyewitness account and key identifications of two of the several shooters. But he was useless in court because he made his living riding the trains in an ankle-length leather coat beneath which he carried a sawed-off shotgun to encourage people to part with their money.[14]

Further, criminal violence proceeds from a social world that has its own peculiar ethos, that is, its own rationality, institutional logic, and moral rules-in-use. Many career criminals, drug-traffickers foremost among them, see violence as a standard-issue occupational tool. With one's rivals, the kill-or-be-killed rules of warfare obtain. With the employees of associates, such as couriers carrying narcotics or money, the rules of predatory opportunity apply because robberies and murders of others' underlings can be blamed on the regrettable chaos of the streets. With witnesses to crimes, the rules of expediency sometimes dictate a dead-men-tell-no-tales prudence. With informers, the rule of "snitches get stitches" simultaneously punishes informers and cautions would-be informers. With one's peers, and their women, the rule of respect reigns supreme, and the smallest slights, wayward glances, infringements on personal space, or untoward words, whether real or imagined, often provoke murderous responses.

Such a world places a premium on daring and cruelty. Only those whose reputations inspire dread get ahead and simultaneously fend off trespass. Violent criminals almost always boast about their crimes to one another in order to boost their reputations ("A murder ain't a murder until you talk bout it on the street"), a habit that gives still other criminals rich material to trade for their own advantage when betrayal time comes around. The world of the streets makes for minds uncluttered with the burdens of middle-class existence. Street players have vast tracks of time to watch other people, discern their strengths and weaknesses, and measure opportunities for predation. Even to talk with criminals, let alone to understand their worlds, detectives must temporarily set aside their own moral frameworks and grasp criminals' habits of mind and moral rules-in-use from the inside out. Such chameleon-like ability at moral alternation makes those judges and prosecutors who become moralizers (one of the many hazards of their own professions) deeply uneasy.

Investigators gather information from witnesses and transform some of it into sworn testimony. In the process, investigators continually assess the credibility of men and women who tell them stories and the plausibility of the stories they tell. How trustworthy is this witness? How will a jury appraise his or her character? What incentive does this witness have for telling this story instead of another one? Is the story consistent with other information gathered independently? Does the story ring with the logic of the streets, however improbable its twists and turns? Criminal investigators expect people to lie to them. This expectation puts a premium on tough-minded skepticism, often bordering on cynicism, an occupational virtue having less to do with temperament than practicality.

At trial, prosecutors regularly call on detectives to "put in" cases. When forensic evidence is available and useful, crime scene detectives describe the collection and analysis of fingerprints, blood samples, semen, hair, carpet fibers, footprints, or other traces that criminals left behind. Far more typically, they troll various bureaucratic nets to track the suspect's movements and tie him to the time and place of the crime at issue. The nets include vast police records of summonses, arrests, or jail time, but also multiple public records documenting lawsuits and insurance policies, as well as the use of passports, telephones (both land lines and cell phones), subway fare cards, EZ passes, automated teller machines, credit cards, computer networks, and banks. Detectives describe the typical procedures and results of witness identification through photo arrays and lineups. They frame the testimony of civilian witnesses who will be introduced later in the trial, or they introduce statements by accused criminals. These include admissions about foreknowledge of a crime, outright confessions, or even exculpatory statements that, with requisite contradictory circumstantial evidence, can undercut a suspect's credibility. But the key issue, always contested, is how did the detective obtain information from civilian witnesses or statements from accused criminals? More often than not, to defense attorneys' delight, prosecutors' chagrin, and judges' skepticism, the answer is through deception or subterfuge of one kind or another.

Detectives regularly use ruses of all sorts to outfox both civilian witnesses and criminals. Civilians refuse to cooperate with investigators for many reasons. Kinship ties command primal loyalty in many communities, whatever the depredations of one's relatives. In some minority groups, racial or ethnic solidarity far outweighs any adherence to universalistic criteria. Sometimes civilians are beneficiaries of the river of cash that big-time criminals, particularly drug traffickers, send flowing through a community; so are car dealerships, automobile-repair shops, night clubs, restaurants, jewelry stores, travel agencies, and bodegas. Sometimes civilians are simply scared stiff. They know the merciless retaliation of street violence and they know that the legal system, which for the most part requires policing to be reactive not preventive, affords little protection against it. And many civilians want no part of the vilification of their own characters that testimony in criminal matters always invites.

Joe Montuori talks about his worst experience with a witness, one that haunts him to this day: "We had a dump-job in Fort Tryon Park and we got nowhere with the case. Several years later, we hear from the FBI in Baltimore that they had an informant who not only witnessed the murder but who could tell us the name of another witness to it. Apparently the dead guy had ripped off a Baltimore gang for $60,000. So the boys come up from Baltimore lookin for this guy, they find him, kidnap him, and put him into a van. He had been hangin out with two people, the guy who became the FBI informant and a girl. To save their own lives, they're both ordered to shoot into the thief's body, after the gang has already killed him. They torture the girl with an electric drill to get her to do this."

"We go to trial with these two witnesses. I had my doubts about the guy's credibility. He's an ex-Baltimore cop turned mutt. And at trial he breaks down and cries on the stand and the jury thinks it's an act. But the girl has no criminal record whatsoever. And, independently, she gives us the exact same story, almost verbatim, of events leading up to the thief's death. Eventually, it all gets down to her. She didn't want to testify. I spent hours and hours tryin to convince her that this was the right thing to do. But she says: 'No, I'm gonna get killed.' But I finally convinced her. We became friends and she had faith in what I was tellin her that everything was gonna be okay. I get her to court and she had to get up and identify these mutts one at a time. She gets through two identifications and breaks down on the stand. The judge orders a five minute break to let her regain her composure. She runs out into the corridor outside the courtroom where I'm havin a cigarette. She asks me for a smoke, she's shakin and cryin. I put my arm around her and tell her: 'Everything's gonna be okay. They can't hurt you.'"

"Just then the defense attorney, a big-money guy, comes out

into the corridor, sees this little huddle, with my arm around the girl. Back in court, he begins to rant and rave that I was tellin her who to pick out, that I was feedin her information about the crime, that it was all a setup. So even though she identifies the other three guys, now her credibility is in jeopardy. The jury acquits all five defendants, all big-time drug dealers. And I'm left with a witness who's scared to death and won't believe ever again anything I tell her."

The reluctance or fear on the part of civilian witnesses to testify against violent criminals represents a danger to democratic institutions just as ominous as systematic police brutality. Unchecked criminal violence leads just as surely as state violence to a society of bystanders. The care, protection, and encouragement of witnesses constitute a major part of investigators' work. Detectives fashion arguments to persuade witnesses of the necessity of civic duty, even though they themselves have often ceased to believe in it. They simultaneously cajole and subtly threaten witnesses on the edge of criminal activity. They assure civilians that their testimony will bring about convictions and therefore protection against retaliation, knowing full well the crap-shoot unpredictability of the criminal justice system.

The apprehension of criminals is itself a crucial part of criminal investigation because the unraveling of who committed specific actions defined as crimes often depends on criminals' own statements to police, even if those descriptions are not entirely accurate. Detectives coax, wheedle, insult, frighten, bully, and tease information from suspects. If a detective can elicit an admission from a suspect that he was "down for the robbery, but not for the shooting," the detective can then nail the suspect for felony murder.

Or to obtain a confession, detectives often feign more knowledge than they have, suggesting the availability of witnesses even when none exist. They hint at possible betrayal from accomplices, even when this is unlikely, making suspects choose between betraying or being betrayed. They conjure up images of accomplices fondling suspects' girlfriends while suspects themselves languish in prison. Or they sow doubt in suspects' minds about possible residues of hard forensic evidence, even when there is none to be had. To elicit a denial of some crucial fact in the face of firm evidence to the contrary, thus establishing *mens rea* or "consciousness of guilt," detectives engage criminals in long meandering seemingly off-the-trail conversations that they have planted with traps.

Detectives help those criminals who are willing to talk to them construct self-serving accounts for their crimes ("You gotta give em an out"). These enable criminals to sustain and project valued self-images even while confessing depredations to men in suits and ties or, later, making requisite plea statements in the alien legal world. Detective Robert Chung's 1988 work with "Roberto Rodriguez" provides a classic case of the construction of a justification for murder.

Roberto had just gotten out of jail and was staying in the small, squalid apartment of two men, "Aristos" and "Joseph," both heavy drug users who had been lovers for fifteen years. One day Joseph had to go to court, and when he returned Aristos was not at home. Roberto told Joseph that Aristos had gone out to score drugs. When Aristos did not return, Joseph became alarmed and went to the Midtown North precinct to file a missing person's report. Aristos's family, profoundly worried by their son's disappearance, came to New York and stayed in the apartment for three days. In the meantime, Roberto supplied both coke and heroin to Joseph, and the two began to sleep together.

A persistent smell began to pervade the apartment. Joseph cleaned the apartment several times to get rid of the odor, but to no avail. Finally, the smell became overpowering, and Aristos's body was discovered under the only bed in the apartment, the same bed on which Joseph and Roberto had been bouncing around. Aristos, whose body was wrapped carefully in black plastic garbage bags, had been stabbed repeatedly with a large knife.

In his interview with Chung, Roberto gave bits and pieces of events during his time at the apartment but did not speak about the stabbing itself. Finally, Roberto admitted having sex with Aristos. From his time in the Midtown North precinct, home to hustlers from all over the country, Chung knew well that male Hispanic hustlers' street code for homosexual sex is always to penetrate, never be penetrated. One can thus preserve a self-image as macho and heterosexual. Chung gave Roberto his out: "Look, I can see how it happened. I mean I wouldn't want to take it up the ass either with all this AIDS stuff going around." Roberto leapt out of his chair yelling: "That's it! That's it! He tried to grab me! He wanted to fuck *me!* I ain't no faggot!" In court, Roberto pleaded guilty to stabbing Aristos, claiming that he had no idea what came over him and that he had acted out of fear that Aristos meant him harm. The public-record versions of a great many crimes are highly sanitized in a similar fashion.

When investigators flip criminals to turn on their accomplices, they engage in the profound moral quandary commonly called a "deal with the devil." How many concessions can one make, say, to an admitted murderer in return for information against other murderers? Moreover, how will a jury assess the credibility of a criminal who has been given material inducements for his version of events? Police investigators, perhaps because they see the results of horrific violence first-hand, vehemently oppose deals with the devil. But prosecutors, who must grapple in open court with the irrationalities and vagaries of a legal system where witnesses recant regularly and sometimes disappear altogether, where the credibility of cops' testimony is weakened by recurring police scandals, and where judges sometimes unmake well-made cases with procedural rulings, see such deals as the necessary price that virtue pays to vice in order to achieve any justice at all.

But some prosecutors take perfectly well-made cases, sometimes even for vicious crimes, and knock down charges, accepting pleas to lesser offenses even when culprits more than deserve to be prosecuted to the full extent of the law. Whether motivated by fear of losing in court, or stage fright, or the desire to maintain collegial relationships with defense attorneys against the day when they enter private practice, or orders from bosses under political pressures, or the sheer burden of impossibly heavy work loads, or victims' lack of moral standing, virtually all prosecutors are haunted by the ghosts of nonprosecutions-past and worry that today's plea bargain may turn out to be tomorrow's crazed killer.

Detectives establish emotional relationships with some criminals, particularly with youths who have had no prolonged interaction of any sort with adults before being taken into custody. Even with hardened criminals, detectives occasionally joke around about the absurdities of the underworld or the legal system, or the idiosyncrasies of particular judges, all sources of endless wonderment. But in the end, detectives betray suspects—all of them—to a thoroughly impersonal criminal justice system, violating all the normal expectations of social relationships. In private, detectives occasionally rue their role in such betrayal, or question the larger meaning of a system built on betrayal.

("We're really in trouble. What kinda society is it where a kid gives up everything that matters to him for the price of a bucket o' chicken?") But in public at least they are unapologetic about the moral ambiguity of their work.

They argue that all criminals are thoroughly deceptive anyway ("They all lie all the time"). And they assert that the fantastically bureaucratized system in which they labor would collapse of its own weight if they did not bend procedures or circumvent them entirely. Deception grounds whatever truth detectives can hope to attain, and they delight in the discomfort that such an anomalous role causes for those whose need for the appearance of moral probity outweighs any sense of justice.

The legal system's dependence on the morally ambiguous role of criminal investigators confers no privileges on detectives themselves, however. When it comes to formal proceedings, the watchword in detectives' world is: "It's always the detective who's on trial."

6

STREET WORK

To investigate crime, detectives must master the logic of the street, in particular the ways, wiles, habits of mind, moral rules-in-use, indeed the entire life-worlds of criminals. They try to achieve this understanding by telling and retelling stories that present both the wildly improbable ins-and-outs as well as the humdrum routines of life on the street.

Detectives' stories draw from their experiences in uniform or in plainclothes anticrime duty as well from their investigative work. A major theme of their tales is the unpredictability of New York City streets. Detective Angel Morales recalls the day he was driving a patrol car alone through Central Park when a young woman hailed him, claiming in a foreign accent that she was lost. Her destination was the Museum of Natural History at 79th Street and Central Park West, but she had gotten turned around and was on the east side of the park. Morales told the pretty Amsterdamer that he was heading west and invited her to hop in the rear seat for the ride across the park. Stopped at a traffic signal in the park, Morales turned around to chat with the woman. When he turned back to resume driving, a man stood in front of the car wearing only a woman's long pink nylon slip, a gray jacket, and a woman's hat.

The man had his arms folded and stared fixedly at Morales. Then the man walked toward the rear of the car on the driver's side, giving Morales a wide-eyed look as he passed, ending up with his back to the police car. Ignoring the traffic signal, Morales steadily watched the man in his rearview mirror when, suddenly, the man spun around facing the car, simultaneously pulled a gun, and, holding the weapon in both hands, went into a firing crouch. Morales immediately yelled for the young woman to dive to the floor, as he swerved the car to his right, braked, and leapt out of his seat. Morales drew his own weapon, shouted at the man to drop the gun. But the man seemed to be in his own world and held his battle-ready stance.

Not a day passes that Detective Morales doesn't thank the Lord for his good eyesight. He thought he saw something red in the barrel of the man's gun, and so he held his fire, again commanding the man to drop the weapon. When the man did so, as a prelude to meekly surrendering, the gun turned out to be a black plastic toy. The young woman crawled out of the police car, thanked Morales profusely for the ride, but said that she had decided to walk.

Things are never as they seem. Right beneath the surface of seemingly placid social life are ticking time bombs, and one must be ever alert to their detonation. Detective Morales recalls another time that he had landed a choice detail in Central Park on a lovely summer day. "I'm just arrivin at my post at 0830 hours near 59th Street and Fifth Avenue, and had turned into the walkway that takes me into the Zoo area, when a young man comes runnin toward me. The weather's warm, but this guy has on a long-sleeve shirt and a tie and he's carryin his coat, prolly walking through the park on his way to work on the East Side. He's badly scratched, and his right sleeve is torn and hangin only by a thread with all the rest bunched up down by his wrist. He looks panicky and I'm immediately suspicious because of his appearance. He tells me that he's been attacked by someone in the park. So I go with him to find the attacker when suddenly he says: 'There she is.' I look up and see a middle-aged lady, reasonably well dressed in dark conservative clothes, who's walkin toward us with her hands folded inside a lightweight coat."

"Now I'm even more suspicious of the complainant once I see this here supposed attacker. So I go forward to talk to the woman while tryin to keep an eye on the man so that he won't be behind me unobserved. Suddenly the woman leaps forward toward me with both hands raised, barin huge long nails and tryin to rake my face! I'm lucky to grab both of her wrists at the same time and we begin a wild kinda dance. She's extremely strong and for several minutes I can't overcome her. Meantime, she's tryin to kick me in the balls. Finally, I get my foot behind her and manage to trip her. I fall on top of her still holdin both hands. By this time, she's tryin to bite my face and is kickin even more ferociously. A crowd gathers around us and, finally, a construction worker asks me if he can help. Between the two of us, I manage to get the woman cuffed. I look around, the complainant's gone, some of the people in the crowd are yellin police brutality, and askin how come the cop had to throw this poor woman to the ground. *See, you never know what you got.*"

Even when officers know street players personally, the streets can explode in a heartbeat. Detective Joe Montuori talks about the time he was working in plainclothes in Brooklyn right after the 1968 assassination of Martin Luther King, Jr.: "I was tight with a bunch of kids in the neighborhood and I considered many of them my friends. But right after King got shot, my partner and I are in the middle of a group of these kids tryin to talk them out goin on a rampage and one kid says: 'We can't trust these guys. Let's get these guys.' So my partner and I turn around and start to walk right outta the circle and all the kids, including the kids I know well, begin to pelt us with stones. My partner gets hit in the head and I end up draggin him to safety behind the uniformed lines."

Part of detectives' experience of unpredictability on the streets stems from the nearly superhuman strength of many of its denizens. Detective Gennaro Giorgio recalls a typical encounter: "There was this ham and egg place at 71st Street and Broadway that had a direct line to the 20th precinct because the owner was so good to cops. Anything happens, he calls, and cars are there immediately. One day, this great big guy comes in, eats, and refuses to pay. The owner calls the cops and several cars arrive right away. Then the call becomes a 10–13 [police officer in distress], and by the time I get to the joint, there's eight cops bouncin off this guy, and he's pickin up cops and throwing 'em halfway across the room. We pile in and finally by sheer weight of numbers we subdue him on the floor and he says: 'Okay, that's enough.' He had been in Dannemora for twelve years, four in solitary. Had gotten beat with aluminum bats covered in rubber up there. So this was a short workout for him."

Apprehension situations, in particular, often provoke wild behavior that detectives remember for the rest of their lives. Detective Montuori recalls going with two other detectives to a suspect's apartment to arrest him. "I go in first and peek around the corner and see a shadow. Suddenly, a machete comes whistlin so close that it whispers my cheek as I pull away. Then the guy we want comes chargin out the apartment with the machete raised high and begins chasin me. One of my partners is huddled against a corner and shoots the guy as he runs past him and down the stairs after me. I keep goin down, turning around on the landings to shoot the guy, while my two partners are shooting at him from above. But the guy kept comin after me. Finally, after three flights of stairs, he stops, with six bullets in him."

The expectation of violence is an integral part of street players' life-worlds. When violence does in fact occur, even searing pain seems to be experienced as routine, thus mitigating trauma that others find catastrophic. Detectives never cease to wonder at the physical and psychic resilience that life on the streets seems to confer on its habitués. In the summer of 1991 Detective Austin Muldoon went to Columbia Presbyterian Hospital to see an elderly overweight woman reported by police officers to have been stabbed a dozen times. When Muldoon found the woman sitting alone on a gurney in the crash room, she had a twelve-inch kitchen knife jammed all the way through her throat, presumably left in place by doctors until they could prepare her for surgery. Muldoon asked her what happened. The woman said: "Whatchoo mean?" Muldoon said: "Well, you might not be aware of it, but there's a knife sticking clean through your throat." The woman answered as if Muldoon had reminded her of something, pointed at the knife, and said: "What? This? This ain't no thang."

There was also "Ramon," who had been badly wounded in his left side, the bullet ripping up muscle tissue and barely missing his kidney and spleen. On August 17, 1992, Detectives Bobby Small and Tony Imperato, accompanied by myself, went to Columbia Presbyterian Hospital to talk with the victim. Ramon

told the officers a story they had heard hundreds of times in the
34th precinct—that he was walking down Broadway minding
his own business when some guy he had never seen before shot
him. Ramon then told the doctors that he was checking out.
The doctors pleaded with him to remain in the hospital because,
they told him, he faced the great danger of peritonitis if his
wound was not properly treated. But with blood already seep-
ing through his bandages, Ramon stood up, wrapped a sheet
around his naked body, signed a release form, and walked slowly
but upright out of the crash room. Doctors, nurses, police officers, and I watched with mouths agape. Later, Small learned
from Ramon's mother's boyfriend that Ramon had been asleep
on a couch in a drug apartment on Academy Street when robbers ripped it off and casually shot the young man on their
way out.

For months in that same year a photograph hung in the 34th
squad room to remind everyone of the events of May 13. At 1600
hours the squad responded to a shooting at 174th Street and Audubon Avenue, in front of a known drug-sale location. Blood
splattered the sidewalk, spent shells littered the gutter, and
crowds in a gala mood surrounded and gaped at the roped-off
crime scene. The first uniformed officer on the scene told detectives that he had found the victim on the sidewalk, shot several
times, twice in the head. He was "likely" and had been rushed to
Columbia Presbyterian Hospital. Detectives Marta Rosario and
Danny Medina stayed to secure the scene, while Detectives Pete
Moro and Billy Siemer, accompanied by myself, hurried over to
the hospital's crash room.

Though it was already jammed with uniformed officers, other
detectives, and doctors, the room was hushed, indeed somber.
Tubes protruded from the comatose victim's nose, mouth, and

body. Blood soaked through the heavy bandages on his skull. Detective Moro asked the attending doctors about the victim's chances. The doctors said that few victims survive wounds as serious as this young man's. Moro then asked the doctors' permission to take a photograph of the victim to show around the street. When they agreed, Moro nudged closer to the victim's bed and asked the nurse changing the dressing on one of the victim's wounds to move briefly so that he could take a picture.

No sooner had Moro said "picture" than the victim sat bolt upright, pulled the tubes off his face, mugged for Moro's camera, and then reached for the nurse's derrière. Although his body resembled Swiss cheese, the shots to his head had apparently splintered and circled instead of penetrating his skull. The young man gave his name as "David Limón," and he chatted amiably with Moro and Siemer for a moment, while claiming that he had no idea who might have shot him. The name, of course, turned out to be fake, and two days later when Moro returned to the hospital to see David Limón, he had already checked out and was nowhere to be found.

Despite the unpredictability of the streets, most stories told by detectives in northern Manhattan and the city's other killing grounds focus on the regularities of criminal occupations and the violence that emerges out of criminal trades. Detectives tell scores of stories about robbers who repeatedly follow the same routine in plying their trade, even using the same exact words each time they rob. A veteran robbery detective talks about a case that came to colleagues in his old squad: "There was this new pattern there that the guys in robbery hadn't seen. A tall guy in a ankle-length leather coat kicks in the door of a saloon, pulls out a sawed-off, and yells: 'All right, muthafuckas, up gainst the wall! This here's a robbery!' When they couldn't catch up with the guy, they went to one of the old retired guys, Sullivan, and told him about the MO and, right away, Sullivan says: 'That's "Joe Brown." That's what he does.' So they pull Joe Brown's sheet and it turns out he's in prison, so it can't be him, they think. But they call Correction and find out that Joe Brown is on work release every weekend. They check the robbery dates and, sure enough, every robbery is on a Saturday night."

"So they go visit Joe Brown in prison after he returns from a weekend leave and put the robberies on him. And Joe Brown says: 'How'd you figure it was me?' And the detectives told him that they had talked to Sullivan. And Joe Brown says: 'Oh yeah, ole Sullivan. He locked me up ten years ago. How's he doin?' Then the detectives asked Joe Brown how come he was still doin the same thing that he was doin when Sullivan locked him up, makin it easy to figure out it was him. And Joe Brown looks at them and says: 'Hey, this is what I do.'"

Detectives' success in solving crimes depends on criminals' habituation to routine. But the constant repetitiousness of much criminal occupational behavior leads some detectives to admire criminals who exhibit ingenuity, enterprise, or a spark of creativity in their efforts. Typically, however, such sentiments are reserved for scam artists, who, instead of physical violence, use their wits to gull victims or find and manipulate glitches in bureaucratic systems to enrich themselves.

Detectives tell myriad tales of extreme violence as a tool-in-trade: drug dealers hog-tied, gagged, and head-shot; a face pushed into red-hot electric-stove burners; creditors, rivals, and partners assassinated. Even marginal players are tortured for information. In spring 1992, for example, uniformed officers brought a dazed young drug dealer to the squad room with a steel bit protruding from his skull. Robbers had used an electric drill to persuade him to reveal the location of his boss's stash.

Hitmen and the workings, vicissitudes, and logic of their trade figure prominently in detectives' stories. Detective Mark Tebbens talks about one of the many spin-off cases that his Wild Cowboys investigation revealed: "'Blanco' and 'Moreno' were partners in the drug trade, but they began to get on each other's nerves and suspicious of each other. One day they both end up shoppin in the same clothing store. And they come round a rack of coats and see one another and both fall to the floor and pull their guns, pointin at each other. They don't shoot, but each takes out a contract on the other. Now the contract that Moreno takes out on Blanco gets sublet, twice, each middle-man taking a cut, and ends up over to Cypress Avenue in the Bronx. So three of the boys there hop on their motorcycles and head off to Bushwick in Brooklyn to kill Blanco at this car garage that he manages."

"Well, they burst into the garage and there are four guys there and the killers don't know which one is Blanco. They hadn't checked. So they kill everybody. Thing is, none of these four guys there were Blanco and now Blanco's raised up and knows that there's a hit out on him. And Moreno's nose is out of joint and he complains that he spent his money and didn't get what he wanted. So the word goes down the line again and the three killers say they'll fix it because they don't want to be known as fuckups, so they get a picture of Blanco, follow him from Washington Heights down to Bushwick, and whack him there. And everybody's happy except Blanco."

In the 34th precinct, many tales point to the complicated betrayals and bloody struggles for power typical of drug organizations. Such tales bristle with thickets of names that confuse and bewilder most outsiders. But detectives navigate such terrain easily, because fixing identities in a world where everyone has multiple, documented aliases constitutes the heart of criminal investigation. Here is Detective Garry Dugan's recounting of such internecine strife, based on information from a hitman named "StreetSweeper," after StreetSweeper himself was arrested for two murders in Queens and for plotting to kill an ace detective in Brooklyn who had unraveled a drug operation: "'Bingo' owned Blue Tops [a crack operation]. He gets into trouble with the police and has to go to the DR [Dominican Republic]. He leaves his business with his brother 'Bongo.' Bingo comes back from the DR, but Bongo refuses to give him his business back. By then, Bongo's got a strong security force, anchored by a guy named 'Big Joe.' But then Bongo gets into trouble with the police and *he* flees to the DR, leaving all his spots with Big Joe and his managers. But Big Joe's unhappy because he's sending Bongo close to $35,000 a week in the DR, while he's only getting $1,000 for himself."

"In the meantime, Bingo sees his chance and approaches Big Joe and offers him a contract to kill Bongo for $40,000. Big Joe refuses at first. Bongo comes back from DR and meets with his security force. Big Joe tells Bongo that he wants a bigger cut, so Bongo gives him a [drug] spot and they agree. Big Joe distrusts Bongo because he gave up the spot too easily and thinks that Bongo will send someone to kill him. In the meantime, Bingo renews the contract on Bongo and now Big Joe accepts. He tries to kill Bongo in New York, but can't find him. Bongo has fled again to the DR. Big Joe sends 'Louie Louie' to the DR to kill Bongo. But Louie Louie can't get the job done, so Big Joe sends a guy named 'Ace.' Together with Louie Louie, Ace tracks Bongo down at his girlfriend's house and kills him and, with Louie Louie, returns to the States via Puerto Rico. In the meantime, Big Joe makes his move and takes over all the spots in the operation, indicating to his men that if Bingo makes any trouble he'll kill him too."

"One day in early 1992, Big Joe goes over to Washington Heights to kill 'Swordfish,' a paid killer, cuz he's got an ongoing beef with him. He's accompanied by Louie Louie, 'CatnMouse,' and two other guys. They stake out the restaurant on 183rd Street where Swordfish hangs out, and get themselves some food from another restaurant in the meantime. They're joined by a blond girl who often spends time with Louie Louie. Louie Louie decides that he wants to take care of some other business, namely collecting ten grand from a guy for yet another murder that he did in the DR. So he goes into a nearby building to find this guy. The blond girlfriend worries aloud that Louie Louie needs backup. Big Joe motions to CatnMouse to follow Louie Louie and back him up. But by this time, unbeknownst to Big Joe or CatnMouse, Louie Louie has pulled up his hood. CatnMouse thinks that Louie Louie is Swordfish, the guy that they are there to kill in the first place, so he shoots Louie Louie. Everybody races into the building at that point."

"Big Joe is furious with CatnMouse for shooting Louie Louie, his good friend, so he shoots CatnMouse twice, once in the chest and once in the leg. The building empties with a lot of other shooters shooting and then the cops arrive. In the middle of all of this, Bingo takes over the drug spots. Later, Ace gets killed by 'Brillo' and a guy named 'Pato' because Ace had gotten fresh with Brillo's wife. They dump Ace's body in Queens. Pato used to be Big Joe's driver whenever they did a hit. But then Big Joe and Brillo kidnap Pato and kill him because they find out he was working for Bingo, spying on Big Joe."

Dugan salts the story with skepticism because his source, StreetSweeper, has carefully removed himself from the action. Stories from other informants have StreetSweeper himself killing CatnMouse at Bingo's orders, and later killing Pato. Moreover, if Bongo's assassins were in fact Louie Louie and Ace, then the three Dominican policemen who were arrested and convicted for taking the job on hire were framed. But the story rings with authority because it recounts chaotic wildness coupled with the nonchalant, routine use of lethal violence, both typical for Dominican drug organizations between 1985 and 1993.

Indeed, the boldness of criminals in the 34th precinct became legendary. On January 11, 1992, a man named "Chamorro" came to the station house at 0400 hours demanding to see Detective Eddie Cruz. Cruz had locked up Chamorro for smashing "Chi-Chi" in the head with an aluminum baseball bat in December 1990. But Chamorro, who admitted the assault, pleaded self-defense, claiming that ChiChi set his German Shepherd on him and he feared for his safety. Cruz had several witnesses who placed the dog attack several hours before Chamorro's assault on ChiChi. Chamorro eventually beat the charge with testimony from fourteen street mates who backed up his story of the timing of the dog's attack. Detectives at the station house became apprehensive as Chamorro restlessly paced the floor demanding to see the cop who had arrested him. When they finally searched Chamorro, they found a Tech 9-millimeter pistol with 40 rounds of ammunition.

And, in July 1992, at the high point of the riots that shook the precinct that summer, a group of homeboys poured gasoline down the hill of 183rd Street leading directly to the station house and set it on fire, destroying several cars parked on the hill and bringing flames right to the doorsteps of the house.[1] Shortly afterward, much to everyone's mixed outrage and merriment, even the quartermaster of the NYPD refused to have needed clerical supplies delivered to the 34th precinct, on the grounds that the wildness of Washington Heights criminals made routine police business unsafe.

In a world where violence is the first resort to settle problems, witnesses holding season tickets to the mayhem on the streets sometimes have trouble keeping things straight themselves. Detective Gilbert Ortiz, assisted by Detective Joe Montuori, caught the homicide of one Pete from the Bronx in early January 1992. Pete and "Buster" made their living sticking up drug dealers. One of their victims was "Romero," the boss of a drug ring remarkable even among drug rings for its internecine strife, double-crossing, skimming of funds, and trading information to other drug dealers, and even to the police. Romero complained about the robbery to a hitman named "StickMan," telling Stick-Man that he wanted Buster dead. StickMan told Romero that he would kill Buster for $3,000, and Romero agreed. But then StickMan got into an altercation with Pete in a drug apartment in front of several witnesses. The fight spilled out to the street, and StickMan shot Pete. He ordered Pete thrown into a car. Two girls who had witnessed the shooting on the street decided to go along with StickMan for the ride. StickMan carted Pete to Washington Heights, criminals' favorite dumping ground for corpses. On the way to Manhattan, Pete, despite his serious wounds, began talking, so StickMan shot him again to finish the job. Stick-Man then demanded $3,000 from Romero because he had gotten rid of half of the robbery team plaguing his operation, but Romero gave StickMan only $1,000 because he had especially wanted Buster, not Pete, dead.

In the course of the investigation into Pete's death, Montuori

interviewed the two girls. They began going over the circumstances of a shooting in a car that they had witnessed, but nothing fit what Montuori and Ortiz had already learned from other
witnesses about Pete's death. The girls were talking about still
another shooting that was news to the police. It turns out that
StickMan had felt slighted by Romero's refusal to pay him the
$3,000, but he concealed his anger. He persuaded Romero to
take a ride with him along with the same two girls. During the
ride, StickMan turned around from the driver's seat in the car
and shot Romero for not paying him what he considered his
due. Montuori, Ortiz, and the two girls had a good laugh together after they sorted out the mix-up.

Detectives often laugh with certain kinds of criminals, usually
those self-conscious enough to tell stories that comment simultaneously on the overlapping absurdities of their respective
worlds. Here is one gunman's tale about his trip back to the Dominican Republic, at a time when the exchange rate was one
American dollar to fourteen Dominican pesos: "I hadda get
outta New York so I took my girl and flew to the island. Because
a the flight, I hadn't brought a weapon so when we get to Santo
Domingo, I feel kinda naked. So I ask around where can I get a
gun, and the second guy I ask offers me a nice Beretta with some
ammo for Five-Hundred American. So I buy it and I feel better,
cept now I start to worry what if I get picked up with a gun
down here and no permit, because I don't got no lawyer on the
island. So I see this guy in military uniform with a lot of medals
on him and I axe him do I need a permit to carry a gun down
here. And he says a permit'll cost you Twenty-Five American. So I
say okay, and he motions to an old lady on the street with a grocery bag. She comes over and he tears a strip of paper offa her
bag and writes me out a permit with my name on it and signs it
Colonel something."

"So now I got a gun and a permit and it's hot and we're thirsty so we go to a bar. It's late afternoon and the bar's packed with people. I order two beers and the bartender tells me that's Eighty-Four. And I'm shocked at the price, eighty-four dollars for two beers! And I know he's trying to take advantage of me cuz he sees I'm a Dominican-York. So I pull out the Beretta and stick it in his face. And all of a sudden everybody in the place pulls out a gun and dives to the floor pointin at me and at each other. And everybody stays like that, not movin, like freeze-frame, then the bartender runs into the middle of the floor, with his hands above his head, yellin: '¡Pesos! ¡Pesos!'"

Detectives tell and retell such stories with gusto because of the resonance with their own sense of futility in controlling guns, of being burdened with procedures in a world of exigency that mocks legal niceties, of the corruption of law enforcement in the homelands of many immigrants to New York and the conse-quent delegitimation of police everywhere, and of the funda-mentally different sets of norms, expectations, and taken-for-granted behavior in the underworld.

In their musings about how criminals view the world, detec-tives also tell tales of vindictive domestic violence, such as the case of the young woman who threw lye mixed with liquid choc-olate on her lover's face. She was considerably surprised when detectives asked why she had used the chocolate. "Chocolate make the lye stick better!" she said. She then proceeded to give the police a detailed recipe on how to mix the right proportions of chocolate and lye in order to inflict maximum damage to a man's face.

There are countless tales of men assaulting women, though one man's moralizing tale about a friend suggests that chivalry of a sort is not entirely dead. ("I mean, it okay that he shot her cuz these bitches need to be taught a lesson. But you don't shoot a woman *in the face.*") Such stories underline detectives' regular entry into social worlds that run on unconventional moral premises and rules-in-use. A few detectives, working-class intellectuals such as Joe Montuori, become fascinated with the kaleidoscopic view of multiple moralities that their work affords and are delighted when they come across criminals with whom they can dispassionately talk trade.

Finally, detectives tell stories of the codes of honor and respect on the streets, and the fatalistic bravado that they often trigger: "A guy gets chased by the cops for sellin dope and he tosses his gun in an alley before the cops catch him. He tells his younger brother where the gun is and the kid goes and gets it. The same day, the piece-of-shit heroin dealer who runs the operation starts slappin around a woman in the middle of the street. The woman's the mother of one of the kid's friends. So the kid tells the dealer that ain't no way to treat a lady, and he pulls out the gun. The dealer starts walkin toward him and says: 'You wanna shoot me? Go ahead and shoot me. You wanna shoot me? Go ahead and shoot me.' He says it three times. So the kid shoots him and he dies in the street."

Detectives' narrative forays into the workings of the criminal mind are unvarnished by disclaimers and sentimentality. Here is a story about amateurs trying to break into the big time: "Didja ever hear the story of the Apple Dumplin Gang? 'Manuel' and 'Ricky,' two Spanish kids, decide they wanna be big-time taxi robbers. So they hail a livery cab and get in the back seat. Manuel pulls a gun and Ricky announces the stickup. The driver snaps on the automatic door lock, starts drivin like mad, and they end up in front of the precinct station house. They have to shoot their way outta the car. They figure they made a mistake gettin in the car in the first place."

"So, next time, they both come up to the driver's side of a livery. They wanna show the driver they're serious, so Manuel shoots him in the leg, and Ricky announces the robbery. The driver steps on the gas and squeals down the street, leavin the kids standing there. Manuel shoots after the car and shatters the rear window. They figure they fucked up by not surroundin the cab."

"So, next time, Manuel comes up on the passenger side of a livery, and Ricky on the driver's side. It's hot outside and the car's windows are open. Manuel sticks the gun in the passenger window, Ricky announces the stickup, and Manuel tries to shoot the driver. But he misses and ends up shootin Ricky, right in the chest. The driver takes off, leavin Ricky on the ground and Manuel standin there. Manuel goes over to Ricky. Ricky says to Manuel: 'Yo! You shot me, Manuel! You shot me!' And then he dies. And that's the story of the Apple Dumplin Gang."

Others might lament such violence as senseless or tragic, but to detectives it makes perfect sense that the streets treat bumbling ineptitude mercilessly. In their view, stupidity turns all tragedy into parody.

———

Squad-room horseplay mirrors street work. Horseplay breaks the routine of listening to essentially similar complaints and especially suspects' endless recitation of almost-identical excuses and justifications, improbable explanations, and outright lies. Especially deadening are complaints by victims who then refuse to press charges. Some detectives amuse themselves by playing jokes on suspects when the cases against them are clearly moribund because of complainants' unwillingness to testify.

For example, detectives told one suspect that they were putting him into a lineup and that he could choose any number he wished. The suspect picked number five and then was put in a room completely alone, with the number five perched on his lap. After a while the detectives then came into the room and told the young man that the complainant had picked number five from the lineup. Bewildered, the young man responded: "Is this the way you always do it?"

Detectives told another suspect that they were going to subject his fanciful story to the station house's new lie detector machine. They sat the young man in front of a photocopier, which they had rigged in advance, and then revisited his account of the incident. Every time the suspect completed a sentence, the detectives punched the start button of the photocopier, and out popped a sheet with the word "Lie" on it.

Detectives told yet another suspect who was thought to have been involved in a shooting that the station house now had sophisticated methods to determine at a glance whether someone had fired a shot. One detective then put on a pair of sunglasses and, after carefully adjusting them to make sure that no harmful rays assaulted his own eyes, examined the suspect's hands under a common fluorescent lamp. He declared that the suspect's hands did indeed reveal telltale traces of paraffin.

Horseplay sometimes celebrates the occupational lore and experiences that bind detectives to one another. In spring 1992 a ring began running an unusual scam in the 34th precinct. A man hailed a gypsy cab and directed the driver to the precinct station. Upon arrival, the passenger announced that he was a police officer and needed the keys to the car for official police business. He ordered the cabbie out of the vehicle and then drove it to a nearby gas station, where a woman sold the car to customers for $200. She even signed the sales slips with the name "Angela Y. Davis," a classy touch that amused old-timers who remembered the nationwide womanhunt for the real Ms. Davis when she was a celebrated fugitive.[2] Detectives brought a man and a woman whom they suspected of being part of the ring into the station house and placed them in the large interview room. The man had bought one of the vehicles from the woman, but it was unclear whether he knew the car was stolen. And so the detectives waited and watched the pair through the one-way window-mirror.

The man and woman quarreled briefly. The man remained alert, wary, and anxious, pacing the floor of the room all the while. But the woman slumped down on one of the benches and fell into a deep sleep. After removing the man from the interview room, the detectives gathered and held a mock court session as the woman slumbered. Several held their left hands over their hearts, right hands raised high, as they addressed an imaginary judge: "Your Honor, case closed. Observe the sleep of the guilty." The phrase describes behavior readily observable in the pokey of every station house in the city. Outlaw veterans of the criminal justice system seem to know almost instinctively when the two-way con game that typifies most police–criminal interaction is up. They know when the police know they are guilty as charged, and this shared knowledge allows them to drop whatever pretense is necessary for maintaining the game. With the tension and anxiety of uncertain fates relieved, they promptly drift off to slumberland.

Squad members direct most of their horseplay toward their own colleagues, and most of the time it takes the form of "gotchas." In perpetrating a "gotcha," squad members poke fun at their comrades' idiosyncrasies, making their targets rueful victims of their own foibles. For instance, one detective regularly kept six sexual liaisons going at the same time. He calculated that more than six women on the string made scheduling too complicated, but less than six might give some of his girlfriends the wrong idea about his availability. His colleagues persuaded a refined, cultivated, college-educated comrade who can talk like the street at a moment's notice to call the philanderer and impersonate an enraged Puerto Rican husband, promising dire revenge against the detective for seducing his wife. The swordsman spent the rest of the day on the phone calling all his women, trying to discern whose husband knew of their romance.

Gotchas beget retaliatory gotchas. Detective Gennaro Giorgio, a much celebrated detective who loves the spotlight and thrives in it, had written a letter to *60 Minutes,* protesting what he felt was a prototypically hostile liberal attitude toward the dilemmas of police officers in one of its weekly shows. Not long afterward, Giorgio received a call from a man who said that he was a producer of the show and that he wanted to interview Giorgio at home. He told Giorgio that the program would call the deputy commissioner of public information to obtain permission, a routine task, he claimed, for *60 Minutes,* and then he would call Giorgio back promptly. After receiving the call, Giorgio exulted in the opportunity to confront the mighty program. But when the man did not call back, Giorgio realized that he had been duped.

Suspicion fell on police administrative assistant Marina Amiaga, the squad's beloved godmother, because in retrospect Giorgio realized that she had seemed overly solicitous while he was waiting for the producer's call. Giorgio plotted his revenge. Amiaga had just received an inquiry about her state income tax return, and so Giorgio, in cahoots with Lieutenant Joseph

Reznick, sent a letter to Amiaga's home, instructing her to call a "Mr. McGeever" at a local Internal Revenue Office about a serious problem on her federal tax return. Amiaga went into a panic, worrying that she faced a time-consuming audit. But then Detective Tony Imperato, who, as it turned out, had been Amiaga's accomplice in the *60 Minutes* plot against Giorgio, "gave up" the trick to Amiaga, earning the label of "cheese eater" (rat) for tipping off his co-conspirator.

This betrayal prompted Giorgio to plot against Imperato. The following Saturday, Imperato interviewed a culprit in the detective sergeant's office because all other space had been taken up with the interviews of suspected accomplices in a murder. When Imperato reported again for duty on Monday, the detective sergeant told him that a weapon was missing from his desk and had been found on the culprit's person down at Central Booking. Imperato protested vehemently that he had searched the culprit twice before dispatching him downtown. The sergeant told Imperato that his shield was on the line, but then weakly (in the opinion of the squad) gave the game away by cracking a smile.

When the occasion presents itself, detectives regularly play jokes on the desk sergeants, in retaliation for sergeants' sending them so many troublesome complainants. One day a desk sergeant forwarded a phone complaint to the Midtown North squad from a woman who demanded the arrest of another woman who, the complainant said, was beaming subtle sexual messages urging her to Sapphic delights. Detective Pete Panuccio told the distressed woman that for some unexplained reason she was on the wrong frequency and all she needed was a head tuner to adjust the signals she was receiving. Luckily, Panuccio said, the station house had just such a specialist on staff. With that, he switched her call back to the desk sergeant, telling the complainant to ask for the chief head tuner.

Detectives' intramural sport suggests their own internalized appreciation of criminals' transgressiveness, as well as their respect for those criminals who adhere to a code of silence. The diversions also mimic the cat-and-mouse games of criminal investigative work, a mix of skill, cunning, deception, lies, and bluffs. Most importantly, the rough banter in the squad room during downtime welds detectives to one another against those moments on the street when lives hang in the balance.

————————————

The emotional meaning of detectives' work depends on their construction of the moral status of victims. Suicides—numerous in the 34th precinct because men and women drive halfway across the country to jump off the George Washington Bridge (jumpers west of the Mississippi usually drive to the Golden Gate Bridge)—are often regarded with a level of scorn that those who work constantly with death by violence reserve for people who throw away life. ("What degree of difficulty was the dive?") Detectives resent the painful work of informing suicides' next-of-kin, not least because relatives of suicides almost always deny reality and insist that police open homicide investigations.

Other kinds of violence afford finer distinctions. Detectives shed few tears over the serious injuries or deaths of known robbers, drug dealers, or hitmen because detectives see physical catastrophes or sudden demise as inextricably linked to such violence-prone trades. Indeed, they take grim satisfaction in "street justice," most of which is administered by other criminals. Rapists get raped in prison. Drug dealers kill robbers.

Robbers kill drug dealers. Robbers kill other robbers and drug dealers kill other drug dealers. From the police standpoint, the perfect "public service homicide" is a drug-sale apartment with a dead drug dealer and a dead robber, each with the murder weapon that killed the other in his hand. Such occurrences were fairly common in the 34th precinct between 1987 and 1993. Many police in especially bloody neighborhoods come to see their mission as the geographical containment of violence so that criminals mostly kill each other and the larger public remains unaware of the carnage.

The sense of fair reckoning that detectives attach to such cases—and their judgment that the world is a better place because of it—contrasts sharply with the lack of proportionality between acts and consequences typical of the world of legal niceties in which detectives must work. Over time, however, any satisfaction detectives feel when the streets mete out just deserts is outweighed by the drudgery of cleaning up other people's messes. Moreover, detectives cannot encourage street justice. Indeed, they are sworn to actively discourage it, though this often brings results that are unsatisfactory to everyone.

For instance, Detective Kevin Walla caught a shooting of a young woman, the consort of "Cholo," a well-known drug dealer in his precinct. The woman's body was shot through with ten bullets, leaving her conscious but barely alive. When detectives talked to Cholo, he told them that he had heard from eyewitnesses, who had no intention of coming forward, that "Mingo," his archrival, had shot his woman to get at him. Cholo added that he intended to take care of Mingo in his own way. But Detective Walla, assisted by Detective Joe Montuori, cautioned Cholo that street vengeance could only lead to more trouble and that Cholo had to trust the legal system instead of the streets to do justice. Cholo told the detectives: "Look, I'm in the game, and I know you don't think much of me because of that, but I love this woman, she's my life."

So the detectives went to the hospital with a photo array that included Mingo's picture. They found the woman somewhat improved. Even though she was unable to speak because of tubes in her trachea, she picked Mingo's photo out of the array with eye blinks at the detectives' direction and identified him as her assailant. Walla and Montuori took her identification to the district attorney and were just about to go into the grand jury to obtain an indictment against Mingo when the squad called to notify them that the woman had died.

The assistant district attorney asked Walla and Montuori if they had obtained a "dying declaration" from the victim. The detectives told the ADA that they had not because at the time of their visit to the hospital the woman was not dying. Indeed, they did not mention the possibility of death to her for fear of precipitating a crisis that might lead to her death. But under state law, without a formal dying declaration, the woman's identification of Mingo was hearsay, leaving the ADA with no choice but to drop all plans to go to the grand jury. When Walla and Montuori informed Cholo that Mingo could not be prosecuted because of this legal technicality, Cholo asked in disgust, "What kinda system you guys got here?"

Detectives regularly use criminals' casual and often gratuitous violence, or the threat of it, to aid their own investigations. For instance, two detectives driving around the 34th precinct spotted a wanted suspect who had fired shots at a crowd, seriously injuring several bystanders, and then, while being chased by street cops, had thrown his still-loaded automatic weapon into a baby carriage holding an infant. The detectives jumped out of their cars and chased the culprit on foot. He ducked into a building with several drug apartments. By the time the detectives got into the building, the gunman was nowhere to be found. They did, however, run into a well-known drug dealer who operated one of the drug-spots in the building. The detectives informed the dealer that because the building was harboring a marauder, he and all the other local dealers could expect to be shut down that day. After walking back to their squad car and then driving back to the building, the detectives found the gunman lying on the sidewalk, beaten to a pulp, one arm with a compound fracture and the other out of joint.

Often, just reminding players of the shoot-first-ask-questions-later rules of the street encourages a change of heart. One way that detectives help improve a hard guy's memory is to suggest that they will "put a jacket on him"—that is, tell his street associates that he snitched—unless he actually does provide them with needed information.

Detectives' sense of the moral status of victims greatly affects their investigations. On January 31, 1992, a young black man was found dead at the rear of a notorious drug-sale building controlled by Dominicans on 162nd Street. There were no bullet wounds on his body, and it seemed that he had died from a fall. Detectives debated briefly whether he had simply slipped while trying to "step over" from the fire-escape ladder to a window sill in order to burglarize one of the many drug apartments in the building or whether he had been thrown from the window of one of those apartments after a failed burglary or robbery attempt, a common-enough punishment for luckless predators on drug dealers. Scores of young Dominican men crowded both sides of the decrepit building, mocking the dead young man as well as the cops, screaming obscenities that some uniformed officers returned in kind. But in the absence of any ballistics evidence and in the face of the ¡yo-no-se! stonewalling typical of Washington Heights, the case was effectively closed "before he hit the fuckin ground."

Another detective uptown caught a no-witnesses, few-clues case of a DOA. A man had gone over the ledge of a short cliff. A bag of burglary tools was found next to his body, and his sheet revealed that he had indeed made his living this way. Because there had been a series of reported car burglaries in the same area, it seemed most likely that the man had been caught in the act of jimmying someone's vehicle and had been thrown off the precipice. The detective who caught the case asked the victim's mother whether her son had been depressed of late and, when she said that her son often felt down and out, the detective wrote up the death as an apparent suicide due to mental anguish.

Detectives take particular satisfaction when victims of crime mete out street justice. Bronx detectives found a young man floating beneath the Whitestone Bridge almost naked. A fingerprint match revealed several arrests for robbery. A check for local robberies yielded a series of stickups by a two-man team who preyed on couples parked in a nearby lovers' lane. Calls to local hospitals produced another young man shot in his backside. He confessed that he and the deceased had been robbing young lovers when, suddenly, one of their victims pulled a gun, shot him, and sent both robbers scurrying across the Whitestone Bridge. But a high wind caused his partner to lose his balance and plunge into the waters far below. The force of the fall stripped off his clothes. One detective quipped, "There *is* a God, and He was workin in the Bronx that night."

When 34th precinct detectives responded to a just-robbed

grocery store, they found the robber lying on the floor, shot in his rear end by the grocery owner as he was making his getaway. They laughed uproariously, even though medical technicians told them that the young man's bladder, spleen, and liver were all badly damaged. In another case, a robber pushed his way into an elderly couple's apartment, seized the old woman, put a gun to her head, and demanded money and jewelry. Her granddaughter fled to the back of the apartment and told her grandfather about the intruder, whereupon the old man grabbed his shotgun and raced into the living room. Just as his wife broke away from the robber's grasp, the old man blew the robber's head clean off. When detectives arrived on the scene, they asked the old man if it had been hard for him to shoot the robber. "Hell, no," he said, endearing himself to the squad forever, "it was just like shootin a big buck."

In December 1992 the 34th squad received a call about a shooting at a bank on Dyckman Street in the upper end of the precinct. The victim was reported as "likely" (though he survived), and the shooter was in hand at the scene. When the squad and I arrived, the uniformed police officers were cracking jokes and laughing loudly, while standing over the bullet-torn, bleeding body of the victim. The shooter, who was not in custody, was chatting amiably with the officers. Handing over his gun and his carry-permit, he explained that he owned a local business and had tailed his employee, who was carrying $30,000 cash receipts for deposit at the bank. As the employee started to enter the bank, the "victim" tried to rob him at gunpoint. The businessman said that he came up behind the robber, told him to stop, and then, when the robber turned around with a gun pointed at the owner, shot the robber in the shoulder. The "victim" turned out to be a career robber wanted for several similar heists. The detectives took the shooter back to the station house to record his statement and then congratulated him for his good citizenship.

Detectives almost always become quite emotionally involved in cases with innocent victims—those who are not engaged in illegal occupations or other activities that court violence and who therefore are thought to suffer through no fault of their own. Such cases raise the problem of theodicy, the justice of God, in sharp, unforgettable ways. The February 2, 1986, discovery in Fort Tryon Park of the body of 9-year-old Bertha Acquaah, battered, stripped, and then baked in an oven by her stepmother because she resented the tiny girl's attachment to her father, stirred head-shaking sadness among detectives for years. On March 3, 1991, a drug-gang hitman accidentally shot and killed 13-year-old Leideza Rivera as she happened to step in front of his intended target at a chicken shack in the Bronx. The incident provoked detectives' profound indignation, wrath, and determination to nail the shooter, a hope that was never fulfilled.

Urban marauders wantonly murdered a college student, David Cargill, for sport on a city highway on May 19, 1991, after a minor traffic incident. His death fueled exhaustive investigative efforts and eventually a conviction.[3] A boyfriend's battering murder of 2-year-old colic-afflicted Kenya McPherson on September 7, 1992, provoked detectives' icy rage against him and the girl's mother, who was on the street copping marijuana during the assault. After discovering her child near death, she slipped down to the local bodega to get a beer to sip with her smoke before calling the ambulance. The February 8, 1991, wanton stabbing to death of a hard-working, much-beloved building doorman, Willie Lantigua, provoked such unrestrained grief among the building's residents, both adults and children, that detectives were unexpectedly moved by his death.

The still unsolved murder of an unidentified 4-year-old girl, sexually violated, tied up, stuffed into a picnic cooler, and abandoned at a construction site near the Spuyten Duyvil in sizzling 1991 July heat triggered deep paternal grief among squad members and a formal adoption and burial of the child, christened Baby Hope by the men. It also set in motion a meticulous cataloguing of all missing children throughout the nation and the tracking of child abusers, pedophiles, and pornographers. The Smithsonian Institution reconstructed the murdered child's face from studies of her skull. After the sketch was publicized far and wide, detectives investigated hundreds of hotline phone calls with dead-end leads, conducted countless interviews with community residents, psychics, and distraught onlookers alike, but all, to detectives' frustration and despair, to no avail.

On January 30, 1992, a homeless man's discovery of the weathered skeleton of a 5-month-old fetus in the "salt mines" near the Hudson River triggered rueful silence among the officers and detectives called to the scene, as they wondered if the baby had taken a full breath on its own outside the womb, New York State's post–*Roe v. Wade* legal definition of human life, and therefore of civil rights. Even after the medical examiner's office declared the fetus to be a discarded fossil, probably from the Museum of the American Indian formerly located at Broadway and 155th Street, a mournful air hung over the squad as detectives pondered the long-ago fate of Baby No Hope.

The case spurred Detective Montuori to rummage through the 1986 files to retrieve the case of Baby Jane, a 30-week-old fetus whose mother was stabbed twice in her abdomen by her vodka-bloated boyfriend on August 22, even as the mother's two older children beat the boyfriend with a baseball bat. The knife stabbed Baby Jane in her right thigh and then severed the mother's umbilical cord. Baby Jane died instantly, much to the

mother's hysterical grief. Montuori wanted to charge the boyfriend with homicide. But, although the medical examiner conceded that the fetus could have survived outside the womb, the
district attorney, citing New York's legal definition of when human life begins, allowed no charge on the fetus's death. But to
Montuori and all his colleagues, this was not just a fetus but a
baby that they had named. The district attorney did allow a
charge of assault against the mother, but even that fell apart
when the mother testified to the grand jury that she had fallen
on the knife. Asked about the case after charges were dismissed
against her boyfriend, she said that she could always have another child but she might not get another boyfriend.[4]

A pint-sized 10-year-old Ecuadorian complainant galvanized
the whole 34th precinct squad room in mid-1992 with his crisp
description of how he fended off a would-be abductor, sending
detectives pell-mell to squad cars to scour the precinct with pho-
tocopies of the boy's hand drawing of the predator's face. And
the 1988 murder of PO Michael Buczek, detectives' brother officer, while interrupting a drug-house robbery prompted fierce
rage that spurred an international manhunt and a lingering
grief that lasted for more than a decade.

Detectives make even career criminals into honorary innocent
victims if they have died a particularly cruel or vicious death. A
drug-dealer tortured to reveal the location of drugs or money
before being executed, or a prostitute raped before being killed,
invariably prompts detectives to express moral revulsion. Innocent victims, whether real or honorary, allow detectives to assert
publicly their most valued self-images: defenders of the innocent
and avengers of the social order, self-images that are difficult to
assert or maintain in the dark and forbidding world of drug-related homicides where "everyone is guilty."

Detectives' stories also provide detectives with a self-dramatizing venue to display their skills, experiences, and self-images to one another in a striking way. Detective Gennaro Giorgio tells the story of his warning to "Zorro," one of three assailants of Columbia University law professor Wolfgang Friedmann, stabbed to death in a 1972 Amsterdam Avenue mugging because he refused to part with an heirloom watch that he had brought from Nazi Europe. Giorgio got Zorro to roll over on his accomplices for reduced time. But in prison Zorro regretted his betrayal and, after pumping iron and bulking up, put out threats on Giorgio's life. Giorgio tracked Zorro's release, followed him to his Manhattan residence, and told him that, if Giorgio ever so much as saw his sorry face again, he was a dead man.

Detective Joe Montuori talks about the day that he and a partner were returning to the house via the West Side Highway when Montuori spotted a man walking and jumping nimbly on the catwalks atop the eastern tower of the George Washington Bridge. Just about that time the radio squawked to the Emergency Service Unit about a possible suicide on the structure. Because they were on the spot, Montuori and his partner stopped their car and hopped on the iron-mesh elevator to ascend to the top of the bridge's tower. The howling winds that rattled the elevator cage terrified Montuori's partner and prevented both detectives from enjoying the sweeping panoramic view of the Palisades, New York's canyons, and the roadway beneath them. When they reached the top, they shouted over the wind to the young man to come down. But he refused, saying he might get in trouble. Montuori kept insisting, and finally the youth acquiesced. Once the cops had him safely cuffed in the cage, he told Montuori that he had felt lonely of late and simply wanted a quiet place to eat his lunch.

Detectives place a premium on knowing how and when to act decisively in the midst of chaos. Detective John Bourges tells about a fabled day when he, Pete Moro, and Tim Muldoon handled a wild shooting spree that resulted in a double homicide and four seriously wounded victims, a drug-related homicide, and another gun assault, all within a span of a few hours.

Finally, detectives see themselves, and present themselves to each other, as men and women unafraid of confronting and immersing themselves in the ugly underside of modern society, a world from which most people turn away, although in private detectives sometimes rue the emotional costs of such immersion. They pride themselves on their mastery of the sometimes extraordinary and often bewildering details of street life and on their ability to slip easily in and out of the peculiar moral frameworks of street players. Law enforcement is dirty, difficult, sometimes dangerous work, always poorly paid. The search for glory fuels more investigations than material rewards.

Everyone worth his salt wants to go after dangerous criminals because such cases present arenas in which investigators can demonstrate their prowess to their peers and to the world. With important exceptions, detectives exult in the danger of their work, in the heart-pumping excitement that only physical risk, the chase, and mortal combat afford. Detective Bobby Small describes the thrill of confronting and arresting armed robbers, men at the top of the criminal prestige hierarchy, honored even by police for their nerve. Detective Mark Tebbens recounts his intensive search for Platano, the most feared and elusive hitman and getaway wheelman in New York City. For years Detective

John Bourges prominently displayed on his desk a famous quotation from Hemingway: "There is no hunting like the hunting of man and those who have hunted armed men long enough and liked it, never really care for anything else thereafter"—a sentiment that aptly characterizes the best investigators.

7

WAITING FOR CHOCOLÁTE

The case looked like a ground ball. The uniformed cops had been sharp, especially in finding the parking summons in the apartment. Nightwatch had done a good job. A typical late-1980s Washington Heights story. Two wannabe-big-time dealers from Connecticut come to 160th Street in the dead of night to buy coke for resale back home. Wrong place, wrong time, wrong guy. Bang, bang. So, at 0800 hours on October 4, 1988, as he began his tour, Detective Austin Francis Muldoon III, known to everyone as Tim, had two teenage kids likely to die in Harlem Hospital, but miraculously he also had several possible witnesses.

Detective Muldoon reviewed the Nightwatch reports. PO Wilfredo Ocasio and his partner had received a radio run at 0445 hours directing them to 552 West 160th Street, where shots had been fired. There, in the rear of the first-floor hallway, outside apartment 3, the cops found two youngsters lying face down on the floor, their hands tied behind their backs with lengths of curtain material, both bleeding profusely from their heads. Two live rounds, both .22 caliber, and two spent cartridges, one .25 caliber and the other .22 caliber, were scattered in the hallway. After calling for buses to carry the two boys to the hospital and notifying the 34th precinct patrol supervisor of the assault, Ocasio entered apartment 3, the door to which was open.

The apartment contained four rooms and a bath, with the hallway running the length of the filthy, sparsely furnished apartment. One bedroom held a double bed, a television, and a bicycle. Another bedroom contained only a low table with several pictures of saints, adorned with dollar bills, surrounded by scores of lighted votive candles—a typical *santería* shrine favored by Dominican drug dealers. On a glass table in the living room stood a scale, with a box of aluminum foil next to it. On the kitchen sink Ocasio found a traffic summons issued to a New York-registered blue Audi for being double-parked on West 175th Street. Ocasio put the car's description and license plate number over the radio, asking any police officers who spotted the vehicle to detain its occupants for questioning.

When the Nightwatch team arrived at Harlem Hospital, the surgical resident on duty told the detectives that the young men, both still unidentified, were in extremely critical condition, with intracranial bullet wounds. The doctor held out little hope for their recovery. Unable to speak with the victims, the Nightwatch detectives went to 552 West 160th Street, arriving at 0645 hours. There they found a young man named "Nathaniel" hanging around the building where the shootings happened.

Nathaniel told the detectives that he and several other friends had taken two cars down from Connecticut to visit people at 170th Street and see New York. Then, Nathaniel said, two boys in the group, Warren and Paco, had gone off in one car with another boy, "Jimmy," saying that they would return in a few minutes. But an hour later Jimmy returned alone and said that Warren and Paco had disappeared. So Nathaniel, Jimmy, and a woman companion went to look for them. When they spotted the police cars on West 160th Street, Nathaniel looked into the hallway of the building where all the commotion was and saw a sneaker belonging to his friend Warren. He stayed on West 160th Street while the others went to the hospital to find out what had happened to their friends.

The Nightwatch team radioed other officers at the hospital asking that the rest of the kids from Connecticut be rounded up and taken to the station house. With minor variations, the whole crew corroborated Nathaniel's story. They also gave the police enough information to identify the victims: 16-year-old Warren Hodge and 19-year-old John Irizary aka Paco, both of New London, Connecticut.

In the meantime, the radio alert on the blue Audi paid off. Patrol officers spotted the car at 164th Street and Amsterdam Avenue at 0700 hours. The driver, a 23-year-old man named "César," had a woman with him—a waitress named "Oriana," he said, who had just finished her shift. He was giving her a ride home. Everything about César announced him as a street player. In any event, he had to explain to detectives how the traffic summons got into apartment 3 at 552 West 160th Street. So the cops hauled him and Oriana back to the station house.

After reading the reports, Muldoon visited the crime scene to get the lay of the land. West 160th Street had long been a drug supermarket in the most drug-saturated precinct in New York City. Male crack-heads stumbled down the block, desperately trying to hold themselves together as pitchers for one of the dozens of dealers headquartered on the street. Crack-whores trolled for trade, offering heavenly delights in exchange for a five-dollar blast. Muldoon strolled the street and talked with a precinct old-timer, "Mighty Joe Young," a huge man completely addicted to crack who lived in a cardboard box on the block. Mighty Joe Young acknowledged seeing a commotion in the early morning, but he was in such a fog that Muldoon discounted him as a possible witness. Muldoon went into the building where the shootings occurred, walked through the apartment, and then returned to the station house to have a quick go at César.

César gave a more elaborate version of his story about picking up Oriana, a story that she corroborated in a separate statement. But César also acknowledged that he had been in apartment 3 earlier that night. He said that, "before the shooting," he had noticed a heavy-set black man with a white female in the hallway outside the apartment. César "got the feeling" that the man was a "runner"—someone who steered customers to drug dealers—and that he was waiting for a dealer named Chocoláte.[1] César said that, "at the time of the shooting," he did not see these two people in the area. Muldoon ordered Oriana released and César detained as a possible participant in the shooting. He immediately returned to West 160th Street to look for the man named Chocoláte and his associates.

Chocoláte was nowhere to be found. But in the meantime, a stocky 26-year-old black man named "Pirate," who fit César's description of the runner, had walked into the 30th precinct station house and told detectives that something had happened on 160th Street and he had been there. Detectives called Muldoon, who asked that Pirate be transported to the 34th squad. There, Pirate acknowledged to Muldoon that he worked on the street and knew Chocoláte. Pirate had recently served time for strong-arm robbery and possession of stolen property. He lived in a building on upper Riverside Drive that police knew to be a honeycomb of robbers. Pirate's own family, including his father and several of his brothers, were well-known predators who preyed on anyone intrepid enough to stroll Riverside Drive at night.

Pirate gave Muldoon the following statement, written out by Muldoon and signed by Pirate. The specificity of the account attests to Pirate's actual observation of the events but also exhibits a skilled detective's extraction of information from a witness: "On the morning of October 4, 1988, at approximately 4 A.M., I saw two males approaching me on West 160th Street from the direction of Amsterdam. I recognized these two males because I had dealt with them on the previous Saturday night when I took them to a location where they had purchased a half-ounce of cocaine. When they saw me, they asked if I could get them the same thing as I had on Saturday. I took them to a location, but it was closed. We then went to 552 West 160th Street where we saw two male Hispanics standing in front. I knew both of these males from the street. One I knew by the street name of Chocoláte (male Dominican, 20s, 5 feet, 9 inches tall, slim, clean-shaven, dark skin, scar on left cheek, short afro, often wears a white hat with a black band). The other male I have also seen many times but did not know him by name (male Dominican, light skin, 20s, 5 feet, 7 inches tall, skinny, goatee, often wears a stone-washed jacket). We told them what we wanted and they took us into apartment 3 at 552 West 160th Street."

"Once inside the apartment, they began to frisk the two buyers. The shorter one [buyer] said in English that his friend was 'strapped.' When this got no response, he repeated it in Spanish. This alarmed the two sellers. The shorter buyer then took a small automatic pistol from the front of the pants of the taller buyer and handed it to seller #2. At this time, all five of us proceeded into the living room. The buyers were saying that they had the gun strictly for their own protection and that they had no intention of ripping anyone off. About this time, a female entered the apartment with some groceries (#3 female, Spanish, light-skinned, 20s, 5 feet tall, slim, white jacket, aqua blue pants). I believe this woman does the cooking and cleaning at this apartment in return for drugs. After this female entered, #2 gave the buyers' gun to Chocoláte and then left the apartment."

"After several minutes, the buyers became nervous, but Chocoláte told them to relax and that his partner was going to return with the drugs. When #2 returned, he was in the company of three other Hispanics, all of whom are known to me from the neighborhood. [Here Pirate described three men, one of whom was armed.] #4 was holding what appeared to be a 9-millimeter automatic. When I saw this, I began to say that the two guys were all right, and that I had dealt with them before. Chocoláte began to ask questions in Spanish that [one of the men] translated into English, and then translated back into Spanish. The questioning dealt with the reason why the two buyers had a gun."

"At this time, Chocoláte left the room and [then] returned with gold and brown cloth that #2 used to tie the hands of the shorter buyer and #5 used to tie the hands of the taller buyer. I said these guys were not rip-off guys. At this time, the girl [#3] offered to go to the apartment where the buyers had gone on Saturday, but then I remembered that apartment was closed. The buyers were then asked if they had any money on them and when they said that they did, #2 removed it from the short one's pocket and placed it on the table. There appeared to be several hundred dollars. Chocoláte and #6 then went into the back room where they spoke in Spanish. I went there and told them that the two guys were okay. But Chocoláte waved the small au-

tomatic in my face and told me to shut up. #5 then left the
apartment and I was told to leave as well. #5 was standing on the
stoop looking around. I told him that the two guys were okay,
but he told me to mind my own business. I then went home.
Later that morning, I saw a police car on the block. I asked
someone what happened and he told me that the two guys had
been found in the hallway and that they had been shot. I was
also told that the female crack-head had stolen the jewelry off
the short buyer after he had been left in the hallway."

Muldoon took Pirate to the 26th precinct station house to
view photographs of people in the narcotics trade. Pirate picked
out a photograph of the man identified in his earlier statement
as the #2 seller, Chocoláte's aide, arrested on September 8, 1988.
The man had given the police the name "Daniel" and an address
in Weehawken, New Jersey, which of course turned out to be bo-
gus. Pirate could make no further identifications, and he was released. Muldoon put out a wanted card on "Daniel." But he still
had only verbal descriptions of Chocoláte and the girl, who had
presumably witnessed the slaughter.

Muldoon went back to the station house and confronted
César. Muldoon told César that the traffic summons put him on
the set. That, coupled with César's outstanding record as a citizen, gave the police every reason to believe that he was involved
in the shootings. César then gave Muldoon a different story
than the one he had told earlier. He told Muldoon that he had
been sitting in his car with his brother's girlfriend in front of 552
West 160th Street at about 4:30 A.M. the morning of October 4.
There he saw seven other people whom he knew only by sight,
not by name, hanging around the building. Suddenly, Chocoláte
came out of the building and told the whole group that two
guys were trying to hold him up, that he had disarmed them,

and "that he knew what he had to do." César said that he told Chocoláte to think about it and not do anything stupid. But Chocoláte said that he was tired of getting held up and that he was taking the robbers up to the roof. Then Chocoláte went back into the building and into apartment #3, the door to which was directly visible from the street because the building had no front door.

A short while later, Chocoláte came out of the apartment holding a gun to a short man's head. César said that he yelled "Po-lice!" to bluff Chocoláte, but the bluff failed. Chocoláte tried to push the short man up the stairs toward the roof, but the man did not budge. So Chocoláte shot him right in the hall-way. César said that he grabbed his brother's girlfriend and fled. As they were running, he heard a second shot. He then went over to Broadway and drove a waitress home, had a few beers with her at a local bar, and was then grabbed by the police as he was driving yet another waitress home. César provided only a vague description of Chocoláte and reiterated that he knew the other street players on West 160th Street only by sight and in passing.

Muldoon had heard stories like César's a hundred times in his near-year with the 34th squad, stories in which casual, routine, taken-for-granted violence and the havoc it wreaks are far less important than the fundamental things of life such as drinking, selling dope, and chasing women. Muldoon knew that the interview was over and that as soon as César left the station house he would be in the wind. Still, César's story corroborated Pirate's account of the shootings. Muldoon had an impressionistic image of the morning's events on West 160th Street.

The next day, October 5, 1988, another witness added some touches to that picture. A man named "Pepe," accompanied by a friend, walked into the station house and told Detectives Louie

Bauza and Muldoon that he heard that César had implicated him in the shooting and he wanted to set the record straight. He acknowledged being at 552 West 160th Street in the early morning hours of October 4. He had gone there, he said, to visit a girl who lived in an upstairs apartment. As he entered the building, he saw several men near apartment 3 on the first floor. One of them had his head down and his hands behind his back. Chocoláte was pointing a black gun at another man.

Pepe said that he told Chocoláte to "leave those people alone." He went upstairs to fetch his girlfriend, but she told him to wait downstairs. He retreated down the steps and saw only the man with his hands behind his back. As Pepe headed toward the building's front door, he heard a shot, followed quickly by another shot. He fled toward Broadway and took a cab home. The police thought Pepe's story about visiting a girl extremely unlikely. Indeed, detectives had information that Pepe was running his own drug-spot in the same building. Still, they did not suspect Pepe in the assault on the two Connecticut youngsters because everything pointed to Chocoláte.

A week later, on October 12, 1994, Warren Hodge died at Harlem Hospital without ever regaining consciousness. The staff, reflecting the hospital's long-standing animosity to the NYPD, did not notify the police. Instead, the 34th squad received a courtesy call from the Medical Examiner's Office the following day with the news. On October 13 Muldoon reclassified the case as a homicide and went to Harlem Hospital to see if he could speak with the surviving victim, John Irizary, still listed in serious condition. The shot to his head had passed through his skull from right to left, leaving a metal fragment behind in his brain. Paralysis on his left side made his left arm and lower left leg and foot useless, and he suffered from unpredictable seizures. But Irizary had vivid memories of the early morning of October 4 and readily relayed them to Muldoon.

Irizary said that he had come down to New York from Connecticut with several other people in two cars, stopping to party on the way, snorting heroin and smoking some crack. He carried a .25 caliber automatic pistol. His good friend, Warren Hodge, a 6-foot 4-inch 16-year-old kid, was unarmed. When they reached Manhattan, Irizary, Warren, and Jimmy went to West 160th Street to buy three ounces of cocaine with $900 that Irizary was carrying. Irizary and Warren got out of the car to make the buy. Warren asked for the gun and Irizary gave it to him; Warren stuck the pistol into his waistband.

The street was packed with people, including several runners looking to guide customers to drug apartments. Irizary and Warren saw a runner whom they recognized from previous transactions. Without speaking a word, they followed him into an apartment on the first floor of a building. A man whom Irizary assumed to be the watchdog of the spot answered the door. He asked Irizary in Spanish: "Do you have a gun?" Irizary, who is bilingual, told the guard in Spanish that he did not. The watchdog then frisked Irizary. He then turned to Warren and asked him in Spanish if he had a gun. Warren did not speak Spanish at all, but he seemed to understand the question. He responded in English: "I got a gat," a common street-slang term for gun. The guard then frisked Warren. When he discovered the automatic in Warren's waistband, he excitedly exclaimed in Spanish: "He has a gun! He has a gun!" All the while he was yelling toward the back of the apartment.

The watchdog seized the gun from Warren's waist. Irizary immediately yelled at him in Spanish: "He told you that he has a gun! He just finished telling you!" But the watchdog said that, no, the big guy had told him no such thing. Irizary became aware of another man in the adjacent room covering them with a weapon. The guard went to the back of the apartment and came out with a dark-skinned man and a young Spanish woman, who called the dark-skinned man "Choколáte." Chocoláte had the pistol found on Warren in his hand. The guard told Chocoláte that the big guy had had the gun. Irizary tried to explain the linguistic mixup to Chocoláte, but Chocoláte was uninterested in hearing the story. He sat both boys down at opposite ends of the room and asked Irizary what was going on.

Irizary tried again to explain the situation, arguing that they had just come to buy drugs. Chocoláte asked to see their money. Irizary pulled out $900 and put it on the table. Chocoláte looked at the money, then at Irizary. Irizary began to panic. He told Chocoláte that he and his friend wanted no problems, to keep the money and the gun, or sell them the coke, but just let them walk out of the apartment in one piece. Chocoláte said no, and told the bodyguard to retrieve something from the other room. In a few minutes, the guard came back with cloth and, at Chocoláte's orders, tied the boys up, first Irizary, then Warren. Warren kept asking what was going on, but Chocoláte silenced him by pointing the .25-caliber automatic at him. Irizary told Warren: "They gonna kill us, man, we gonna die." He saw Warren beginning to weep silently. In the middle of this, the runner fled the apartment.

Chocoláte stood both boys up and walked them outside the apartment, with one hand on Irizary's bound hands, and the other wielding the gun. The guard followed, forcing Warren in front of him. Irizary's knees began to tremble and he felt weak all over. Chocoláte tried to force Irizary into the garbage-filled alleyway off the hallway. Irizary knew that if he went into the alley, Chocoláte would empty the automatic into him. He refused to go there. Chocoláte, holding the automatic at Irizary's right temple, said that he would kill him immediately if he did not go into the alleyway. And Irizary said: "Well, you gonna have to kill me right here." The next thing he heard was a shot and the woman screaming in Spanish: "Chocoláte, no! Stop!" Irizary fell to the floor, the woman still screaming at Chocoláte to stop, and he heard another shot. His friend Warren fell on top of him.

Muldoon showed Irizary two photo arrays put together from police files on the players from West 160th Street. Irizary picked out photographs of two people, one of a local hanger-on from the street and the other of "Daniel," Chocoláte's watchdog. He did not pick out a photograph of César. And Irizary did not see Chocoláte's face in either photo array. Nor could he provide anything more than a basic description of the woman he thought to be Chocoláte's girlfriend. Muldoon still had no idea who Chocoláte was.

The break came a week later. On October 18, 1988, PO Michael Buczek and his partner tried to stop three men fleeing from a drug robbery on West 161st Street. In an ensuing struggle, Buczek was shot and killed.[2] In the massive police investigation that followed, detectives across the city pulled in all their informants, demanding to know who had shot their brother officer. Two detectives interviewed an informant named "Ivan" who reported that a robbery gang preying on drug dealers had killed Buczek. The detectives took Ivan to Manhattan's CATCH Unit in the 20th precinct and had him review hundreds of photographs of street players in the low 160s, looking for faces of either robbers or their victims familiar to him.

Serendipitously, Ivan picked out a photograph of one Orlando Rodriguez aka Garis Abreu along with several other monikers. Ivan said that this guy had done the shootings of the two kids at 552 West 160th Street a couple of weeks earlier. The detectives investigating the Buczek murder immediately notified Muldoon. Muldoon made up a new photo array with Orlando Rodriguez's picture in it and presented it to Irizary, who was still in Harlem Hospital, on October 21, 1988. Irizary sat bolt upright in bed, pointed at the #6 photo in the array, and said: "That's Chocoláte. That's the guy who shot Warren and me."

The identification of culprits underpins all criminal investigation. Muldoon pulled Orlando Rodriguez's sheet. Chocoláte had a fairly typical record for a West 160th Street drug dealer: an October 1986 narcotics sale arrest; a May 1987 gun arrest; and a June 1987 violent assault arrest. Chocoláte had served only minimal time for these crimes, even though he had been on probation since his first offense. The murder of Hodge and the assault on Irizary fit what the police knew about Chocoláte well.

But where was Chocoláte? Identifying culprits is one thing. Catching them is another. Early street rumors had it that Chocoláte had fled the United States for the Dominican Republic. Muldoon put in a wanted card on Chocoláte and hunkered down for a long wait. Lists of wanted suspects comprise an integral part of the bureaucratic nets that aid investigators. An officer looking for a suspect sends a bulletin with all particulars to the wanted desk. If the wanted suspect gets caught in another net in New York City, and if the arresting officer in that case faithfully calls the wanted desk, as required, then the suspect's card will "drop," and the officer originally putting out the wanted notice will be notified. The system is designed for local use only, so arrests outside New York City will not produce a hit.

The premises of the system are that criminals break laws regularly and that sooner or later their own actions and habits of mind will lead them into legal entanglements, most often in the local area where they practice their trades. But the success of the system depends on officers' care in filing reports and making notifications and on the ability of the police to act quickly enough to make apprehensions before courts free suspects on bail.[3]

In 1989 the 34th squad heard a rumor that Chocoláte was locked up in Puerto Rico for assault. But when the squad checked, Chocoláte was not in custody on the island. Even as he shouldered his 250-cases-per-year workload, including a couple of dozen other homicides in northern Manhattan, Muldoon checked regularly to see if Chocoláte's wanted card had dropped. And he kept searching for the woman in Chocoláte's apartment who had implored the dealer not to shoot Hodge and Irizary.

After graduating from college, Tim Muldoon had followed his grandfather's and father's footsteps in becoming a cop. The New York City Police Department was not hiring in February 1979, so Muldoon joined the Nyack, New York, police force and became close friends with two brother officers, one black and one white, PO Waverly "Chipper" Brown and Sergeant Edward O'Grady. Both were later slain in the infamous Brinks robbery of October 20, 1981, conducted by black radicals and white Weather Underground terrorists seeking to finance the Republic of New Afrika.[4] Muldoon joined the NYPD in November 1979. While a cop, he completed two years of law school.

Muldoon served in uniform in the Bronx's 46th precinct until 1985, amusing himself by confiscating and bringing home to his little brother brass knuckles, serrated spring knives, and nunchakus from the denizens of the precinct, playthings that prepared his brother for his own subsequent police career. Guns, of course, went downtown. Muldoon partnered for much of his time in the 46th with PO Pete Moro. Muldoon and Moro shared many close calls, but none closer than the day they found themselves surrounded by a hostile mob after both officers had responded to a domestic dispute. The man of the house, already on the street by the time Muldoon and Moro got to the scene, refused to come with the officers and made threatening gestures toward them. Muldoon tapped the man with his stick to get his attention. The man went crazy, burst past the officers, and began running down the street.

Moro jumped on the man's back and Muldoon grabbed him around the waist, while the man determinedly plodded down the block. A full fifty yards later, the man attacked Moro with his teeth and lacerated his chest right through his uniform. Suddenly, a crowd circled the melee. Someone in the crowd tried to grab Muldoon's radio, but he held onto it and got off a 10-13 call. Central responded slowly because Muldoon sounded so calm. Meanwhile, the crowd began to chant: "Get their guns." Both officers found themselves simultaneously held and pummeled while tightly gripping their weapons as several hands grabbed at their holsters. The partners managed to fight off and scatter their assailants, with the assistance of responding officers. Later, Muldoon became friends with the domestic disputant, and whenever he drove by the stoop, the man, reconciled with his wife and holding a new baby, heartily waved to him.

Muldoon was assigned to the narcotics division in 1985, working in the 34th precinct, which was just then emerging as the drug gateway for the entire Eastern Seaboard. He earned his gold shield in 1987. In January 1988 he joined the 34th detective squad, serving on the same team as his old friend Pete Moro. Muldoon's irrepressible good humor, beaming from his stout figure and broadcast in his booming voice, quickly made him a welcome figure in the squad, even as his wide reading habits, his love of poetry (which led him to Seamus Heaney long before the Irish poet became a Nobel Laureate), and his devotion to Lou Reed's and John Cage's music set him apart.

New York City police detectives always attract attention in public gatherings, but Muldoon's engaging persona attracted more than most. Men and women gravitate to an authority figure who radiates bonhomie, and sometimes marginal characters see such a personality as holding the answers to life's riddles. Once, when Muldoon caught a hit-and-run case, he left his business card with the man who reported the incident. For years after that encounter, the man regularly came to the station house with a stack of papers neatly tied up in a bundle, everything from old telephone and electric bills to newspaper clippings on a wide variety of subjects. He always left the package with police administrative assistant Marina Amiaga, saying: "These are for Detective Muldoon."

Muldoon's boyish open face and ready smile also invited sometimes remarkable confessions. At 1700 hours on Monday, August 5, 1991, two men and a woman came to the 34th station house to report that their close friend, "Sara Long," was missing. One of them had had dinner with Sara the previous Thursday evening, parting with her at 10:30 P.M. But Sara didn't go to work on Friday, nor did she call in. They called Sara's roommate, Julian Cowell, on Friday. Cowell told them that Sara had returned on Thursday night at 11:00 P.M. He said that he had heard her take a shower on Friday morning and leave the apartment at her customary hour of 6:30 A.M. The three friends looked for her over the weekend, even visiting her apartment and speaking with Cowell in person. Then Sara didn't report for work on Monday. Sara's mother and brother-in-law, alarmed by behavior totally uncharacteristic for her, flew into New York from California.

Missing persons occupy a low priority in squad work. Detectives know from long experience that even the most unlikely people declare timeout from life's complicated games and go missing, sometimes for inexplicable reasons. Indeed, in New York City, a fantastic aggregation of functionally interconnected but discrete little worlds, hundreds of people take such timeouts every day by walking out of their normal worlds and into other ones. Most are found only when they want to be found. One day, for instance, the squad took a call from a distraught mother in Hackensack, New Jersey. Her grown daughter, "Lisa," had left home early that morning to take a bus to Manhattan for her special nurse's training course at Columbia Presbyterian Medical Center. But at 11 A.M. Lisa's supervisor phoned the mother asking where Lisa was. The mother panicked because, she insisted, Lisa was emotionally stable, drug-free, completely responsible, orderly, and always on time. Going missing was simply unlike her. Lisa's supervisor at CPMC gave detectives the same account of Lisa's character and also stated that there was nothing particularly stressful about Lisa's training. Because of these characterizations, detectives from both the NYPD and the Hackensack Police Department spent the better part of the day searching high and low for Lisa. Late in the day, Lisa's CPMC supervisor called the squad to announce that Lisa had been found. She had taken ill, his story went, and had spent the day at a friend's apartment on 94th Street. She had not thought to call anyone. Detectives dismissed this fable out of hand as they mused about the thousands of dollars of public monies spent searching for Lisa.

Muldoon listened to Sara Long's friends with a skepticism born of such incidents. Nonetheless, because of the reported alarm of Sara's relatives, Muldoon called Julian Cowell while Sara's friends waited in the squad room. Cowell repeated the same sequence of events that Sara's friends had related to Muldoon. Then Muldoon asked Cowell if he had heard any conversation in the apartment between the time Sara returned home at night and left for work the next morning. Cowell hesitated, and then said no. Muldoon asked him if that was an honest answer. Cowell said that he did not know how to answer Muldoon's question. Muldoon said that he could answer yes or no. Cowell hesitated again, before answering yes.

The oddness of the conversation made Muldoon uneasy. Was Cowell's social awkwardness just a personal quirk, or was it a signal of inner turmoil? Muldoon called the Missing Persons Unit of the NYPD to see if they had received any notification about Long. Then he called all the hospitals in upper Manhattan to see if she had been injured, and he called Correction to see if she had been arrested. He called Missing Persons again to make doubly sure that it had no reports, because Missing Persons is famous for missing persons. No word of Sara anywhere.

Muldoon asked Detective Joe Montuori to accompany him to Sara's apartment on Hillside Avenue in the Inwood section of upper Manhattan. After the detectives were buzzed into the building by Julian Cowell, they noticed a faint odor in the lobby, but the smell dissipated as they traveled to the third floor on the elevator. Cowell, an engaging 23-year-old computer programmer, admitted them to the large apartment he shared with Sara Long. The detectives asked Cowell about the odor in the building, which was slightly stronger in his and Sara's apartment, and Cowell said that he had not cleaned Sara's cat-litter box as he should have in the prevailing ugly August weather. After asking and receiving Cowell's permission, the detectives searched Long's room thoroughly. In addition to the faint odor, they noticed a strong smell of pine disinfectant. But Cowell seemed completely cooperative and reasonable. The apartment was generally neat and orderly. The detectives left the apartment and went to the basement and discerned no odor. Moreover, there was no longer any smell in the lobby.

Muldoon and Montuori headed toward the squad car. Suddenly they stopped and looked at each other. Muldoon told Montuori: "Joe, you taught me to trust my instincts. I don't know about cat shit. But I do know about dead bodies. Something's wrong. Let's go back." They headed back to the building and up to Cowell's apartment. Again, Cowell was completely cooperative. The detectives asked Cowell for permission to look at his room and he agreed. Cowell apologized for the messiness of his relatively tidy space. In the corner of the room stood a large black duffel bag which, when Montuori tried to lift it, was quite heavy. Montuori asked Cowell what the duffel contained. Cowell responded that it held wet towels that needed washing. Cowell excused himself and went to the kitchen for a glass of water and then returned to his bedroom.

Montuori asked if he could look inside the duffel and Cowell agreed. Montuori dragged the duffel to the middle of the room, unzipped it, and found a green plastic garbage bag inside the duffel. He reached his hand inside the bag and felt something wet and squishy. When he withdrew his hand, it was covered with a reddish brown substance. Cowell said that he didn't know what that was, that it shouldn't be there. Muldoon noticed that

Cowell had begun sweating profusely. Montuori said nothing, but went to the kitchen, washed his hands, retrieved a butter knife, and came back to the bedroom. Using the knife, Montuori continued to open the garbage bag and a terrible odor flooded the room. He turned and asked Cowell: "What's in the bag?" And Cowell blurted out: "It's her body. The rest of her is in the garbage in the basement."

While Montuori supervised the Crime Scene Unit's work and Detective John Bourges rescued Sara Long's limbs from the garbage that had already been carted outdoors, carefully labeling and reassembling the pieces of her once beautiful body, Muldoon, joined by Detective Joel Potter, took Julian Cowell's statement back at the station house. Cowell spoke with great equanimity and directness. Muldoon and Potter listened patiently to him for hours, without revulsion or judgment, as Cowell methodically explained his version of events. Muldoon wrote it down.

———————————

Cowell told the detectives that he and Sara had lived together as roommates for two years.[5] That Friday morning, Cowell said, he took a shower at 6 A.M. after a long, sweaty night of computer programming in the 100 degree heat. Sara knocked on the bathroom door and asked Cowell to hurry it along. But Cowell tarried a while to wash his hair. Sara barged in. As Cowell covered himself with a towel, Sara demanded that he get out of the bathroom, and then stormed out, slamming the door.

Cowell told the detectives that that he finished drying off and then left the bathroom, but encountered Sara just outside his bedroom. She lunged at him with a black-handled kitchen knife and slashed his face. Cowell stepped back and the towel dropped from around him. She was poised to lunge again. Crying out "Sara, don't!" Cowell ran into his bedroom and shut the door. He pleaded with her: "Sara, what's the matter? Why are you doing this?" No answer. "If you don't answer me, I'll call the police," Cowell said. But in trying to make an emergency call, he mistakenly dialed 199, the hotline in his native Jamaica. Then he heard the other phone extension being taken off its hook.

Cowell said that he crept out of the bedroom and down the hall, all the while saying: "Sara, calm down. What's the matter?" As he turned the corner, he saw Sara at her desk, phone turned upside down, still holding the knife in her right hand, and muttering. She stalked him and then lunged again. Cowell dodged Sara's thrust, grabbed her knife arm as she missed, and then throttled her neck with his left arm. The pair fought for the knife, still clutched by Sara. The knife slowly turned toward Sara and, still in her own hand, stabbed her at least twice. She went limp and Cowell held her up by her midsection. Suddenly, she came to life and struck at him again. Cowell pushed her knife hand toward her throat and she stabbed herself there. As she fell to the floor, Cowell wrenched the knife out of her neck. Sara cried out: "There, now you have done it!" and lay still.

Cowell pondered calling the police but feared that no one would believe he had acted in self-defense. As he sat dazed on the floor, the phone rang. On the answering machine, he heard "Marlene," Sara's close friend and boss. Sara was late for work, and Sara was never late for anything. Was everything okay?

Cowell wanted Detectives Muldoon and Potter to understand the difficult situation he faced. His apartment was now a bloody mess. What if Sara's blood dripped downstairs? How was he going to get rid of her body? And how was he going to explain her disappearance?

He had to clean up the blood first, so he changed into jeans. He put old newspapers down on the floor to soak up the liquid and wiped the apartment's entry with a shirt so that the blood wouldn't seep under the front door into the public hallway. He used a sponge mop to absorb the rest of the blood. Then he stuffed Sara's body into a green duffel bag and dragged her to the bathroom, where he put her into the bath tub. He continued to clean, first the floor, then the walls, then the furniture. The phone rang. It was Marlene again, and she sounded more concerned than she had been before. He returned to his cleaning. He put some rugs and clothing into the clothes washing machine, but the machine began to bang so loudly that he feared it might attract attention. He took all the rugs out of the machine and threw away those that seemed hopelessly stained.

The phone kept ringing. All calls for Sara. "Kirk," Marlene's husband, called several times and said that he was coming over to the apartment. Cowell began to panic. He had cleaned well, but what about Sara's body? He had to get rid of her body. He remembered images of Buddhists burning themselves in Vietnam. He'd burn her body, and all traces of her would disappear. So he placed newspapers all around and over Sara's body, still clothed in her nightgown, and lit the paper. Acrid smoke filled the bathroom, then the whole apartment, and began pouring out the open windows. The doorbell buzzed loudly. Cowell quickly changed his clothes and looked out his front door's peephole. One of his neighbors stood in the hallway looking toward his apartment. Cowell opened the door and told her that he had left a pot on the stove. Only a minute later, the building superintendent banged at his door, and Cowell repeated his story about a cooking mishap.

At this point, Cowell told Muldoon and Potter, he realized that burning Sara's body with newspapers wasn't going to work.

So he stuffed her back into his green duffel bag and dragged her body into his bedroom. He started to clean the bathroom. He heard a loud knock at his door, followed by Kirk's voice asking: "Sara, are you in there?" Then Cowell heard keys in the lock, so he opened the door. Kirk was accompanied by a friend. Kirk asked about Sara. Cowell replied that he had heard Sara shower that morning and then leave for work. Kirk and his friend had beers and then left. Cowell went back to cleaning.

The phone rang again. Marlene told Cowell that Kirk might return. Cowell stopped cleaning and deferred getting rid of Sara's body. He didn't want Kirk walking in on him while he was doing either. The phone rang again. "Mark" was calling from California. He was flying to New York the very next day and said that Sara had invited him to stay at her apartment. The phone rang again, and again, and again. Sara had a lot of friends who wondered where she was. One of her friends was "Bob," who told Cowell that he'd be over the next day, after he met Mark coming in from California.

Cowell dozed off. He awakened at 7:00 A.M. on Saturday morning and went to a nearby bodega, where he bought aerosol spray and a bottle of disinfectant. At home, he sprayed the apartment and mopped the floors with the disinfectant. He decided to burn Sara's body again, so he dragged the duffel bag back into the bathroom and dumped her into the tub. He searched his apartment for flammable non-smoky liquids. He settled on a spray can of cleaning oil. But that smelled horribly when he lit it. So he thought of rubbing alcohol. He went back to the bodega and bought four bottles of rubbing alcohol and a can of scented alcohol. He tested the concoction on his kitchen counter, but he could only raise a weak flame. Then he remembered that the Buddhist monks in Vietnam had used gasoline to burn themselves. If he used gasoline on Sara, Cowell thought, he'd have nothing left but an easily disposable skeleton.

Cowell went back to the bodega, bought a gallon bottle of Great Bear water, returned to his apartment and emptied the water, and then, with the container, went back out again to a nearby gas station to buy gas. But the station attendant required him to purchase a regular gasoline can in order to buy gas. Back at the apartment, Cowell experimented with the gasoline to see how effective it would be. He poured gasoline on Sara's head and lit it. The flame flashed down her body where gas had dripped and then back up toward her head, and, to his shock and distress, almost burned Cowell. Billowing, thick, sooty smoke filled the bathroom. Cowell opened a window and released the smoke, a bit at a time. He repeated the procedure three more times, but the smoke nearly overwhelmed him. He put Sara's now charred body back into the green duffel bag and dragged it back to his bedroom, along with the remainder of the gasoline. He tried cleaning the bathroom, but it was a complete mess.

Exhausted, Cowell started watching a police movie when he heard a knock on the door. It was Bob, with his wife and two boys in tow, accompanied by Mark. The wife and kids settled in to watch television with Cowell, while Bob and Mark went to look at an apartment. When they returned, Bob was aghast at the state of the bathroom. Cowell explained that a female friend had accidentally lit the day's newspaper with her cigarette while using the toilet. After Bob left with his family, Cowell and Mark spent the rest of the evening chit-chatting and taking more phone calls from Sara's friends.

On Sunday morning, Bob and Mark went back out apartment hunting once again. Cowell hurried over to a supermarket and bought some more air freshener, but on his way home he ran into an old boyfriend of Sara's who was hunting the neighborhood for her. Cowell dashed upstairs and sprayed the apartment with the pine-scented spray. Then he invited Sara's friend into the apartment and chatted with him for a while.

Cowell told Muldoon and Potter that he realized that he had to do something. He began to think of chopping up Sara's body and disposing of it in pieces. When Bob and Mark returned from house hunting, Cowell joined them in a thorough search of the neighborhood looking for Sara. Then, after Bob and Mark went off separately on other business, Cowell went to a hardware store and bought a power saw. He dragged the green duffel bag back into the bathroom and once again put Sara's body in the bathtub. The saw made a loud whirring noise. He tried to cut off Sara's head with the saw, but he couldn't cut through the neck bone. So he then took a hammer and metal ruler and tried to hammer the ruler through the bone. No luck. He took a large kitchen knife to the task, with no effect. Finally, the saw worked and he severed Sara's head from her body and put it in a trash bag. Cowell paused in his story and asked Muldoon and Potter if they had ever cut a head off a body. When the detectives said that they hadn't, Cowell said that it turned out to be hard work, much harder than people might think.

Cowell continued his tale. He told the detectives that he first sawed off Sara's left arm followed by her right arm, and placed them in trash bags. Then he sawed off her legs below the knees. But her body remained in too large a piece, so he sawed off what remained of her legs at the thighs. He stuffed these parts into trash bags and stashed the bags in his bedroom. He put Sara's torso in a blue suitcase, which he placed at the head of his bed.

Mark returned and made small talk with Cowell. Cowell occasionally jumped up and checked on matters in his bedroom and sprayed more air freshener. The odor kept getting worse. The suitcase was leaking blood. Cowell mopped up the blood with some cloths and placed the suitcase on its side, used more air freshener, and plopped down in the living room with Mark for the rest of the evening. He began to feel sick and developed a fever.

On Monday morning, Cowell and Mark drove downtown to a breakfast diner in the West 80s. Then Mark dropped Cowell off at the Deuce, where he bought two black duffel bags and returned home on the train. He bought more ammonia and disinfectant at the bodega. Cowell was alarmed to find Mark already back at the apartment, but he retreated to his bedroom and scrubbed it thoroughly. His fever got worse. Then Mark left on his apartment hunt. Cowell doubled up the trash bags that contained Sara's body parts and stuffed them with cardboard and papers to give them a natural appearance. He did a trial run with regular trash. Although he saw two men in the building, he decided that he had to act. He took two bags with body parts downstairs and placed them randomly in the basement, followed by another trip with one bag of body parts and one bag of blood-stained papers and cardboard.

Cowell was feeling sicker by the minute. Mark returned in midafternoon and got him some aspirin to break his fever. Then Marlene and "Michael" arrived. All three were walking around the apartment. Someone said: "It stinks," and started looking for the odor's source. Someone decided it was the cat litter that hadn't been changed in the brutally hot weather. Marlene and Michael went out to buy fresh litter. After they returned, all three went to the police station to report Sara missing. Cowell jumped up, sprayed the apartment with air freshener, and started cleaning the bathroom again.

Suddenly, Detective Muldoon called. Cowell found the conversation awkward, strained, and nerve-wracking. Then Marlene, Michael, and Mark returned from the station house on their way out to have dinner and to pick up Sara's mother and brother-in-law at the airport. Cowell had some uninterrupted time to finish the job. The suitcase had leaked badly. He took Sara's torso out of the suitcase and stuffed it into one of the new black duffel bags freshly lined with a green plastic bag. He rinsed out the suitcase in the tub with hot water. The smell of blood and decaying flesh permeated the apartment. He cleaned the tub, the floors leading to the bathroom, and the floor in his bedroom.

Suddenly Cowell heard the downstairs buzzer ring. He didn't answer. It buzzed again. He stopped cleaning and ran through the apartment spraying again. He answered the door through the intercom system. It was Detectives Muldoon and Montuori.

The detectives entered Cowell's apartment. They asked him about the smell. Cowell blamed the cat. When the detectives left, Cowell quickly resumed spraying the apartment. But the buzzer rang again. Muldoon and Montuori asked Cowell if they could look around the apartment. Cowell agreed. Montuori asked if the closed room belonged to Cowell. Cowell acknowledged that it did. Montuori asked if he and Muldoon could look inside. Cowell agreed, even as he realized that "it was coming down now. It was almost over." Montuori saw the duffel bag. He asked: "What's this?" Cowell said it was clothing. Montuori opened the duffel bag, revealing the green plastic trash bag. He asked again: "What's this?" And then, Cowell said, he told the detectives that it was Sara's body and the rest of her was in the basement.

The multiple knife wounds on Sara Long's body and the defensive wounds on her hands and forearms gave the lie to Cowell's version of events. But over the nearly seven hours that the confession took, Muldoon did not confront Cowell with the discrepancy because his confession of his deeds was legally quite sufficient. Muldoon allowed the young man a story that he could live with, listening patiently and sympathetically to the theatrical yet clinical, self-dramatizing, self-pitying account of the dilemmas that Long's body presented for him, offered entirely without fear and without remorse. When the detectives asked if he had done anything like this before, he responded indignantly: "No! I hate violence!" Joel Potter tried briefly to explore possible motivations for the psychopathic savagery of Cowell's actions, but Cowell steadfastly stuck to the story that he had told Muldoon.

Muldoon and Montuori worried endlessly about how a judge might rule on the issue of when custody of Cowell began and when, therefore, they were required to read him his *Miranda* rights before asking any further questions. Did custody begin after Cowell blurted out his deed at the apartment? Or did custody begin earlier when Montuori asked Cowell what was in the black duffel bag? Or did custody begin still earlier when the detectives asked if they could enter Cowell's closed room on their second trip back to the apartment? Could a defense lawyer successfully argue that Montuori knew that Long's body was in the bag, that Cowell knew that Montuori knew, that Cowell was in custody at that point, and Montuori was obliged to read Cowell his rights?

Neither suspect's nor policeman's subjective state of mind determines when custody begins and when *Miranda* warnings must be read. The legal test is "what a reasonable man, innocent of any crime, would have thought if he had been in the

defendant's position."⁶ But this definition leaves ample room for
contradictory opinions and rulings. On May 16, 1980, Montuori
helped investigate a case in which the timing of reading *Miranda*
rights became crucial. A couple reported a shooting in Fort
Tryon Park to Detective Harry Hildebrandt at the 34th precinct.
They had seen a white female with blond hair wearing a white
dress and carrying a large black bag fleeing the scene of the
shooting. Hildebrandt found Vincent Eckes dead on a park
bench, shot once in the head and once in the chest.

Detectives Richie Serpa and Montuori found two .38 caliber
cartridges near the bench, one spent, the other live. Their canvass of the park turned up four other witnesses who provided
them with essentially the same story that the original two witnesses had told Detective Hildebrandt. About two hours later,
Mary Ann Balint appeared at the station house in everyday street
clothes. She carried no bag or purse. Balint claimed to Detective
Serpa that she had been sitting with her fiancé on the park
bench when a black man leapt out of the shrubbery, a blaring radio in one hand and a gun in the other, said not a word, and
then shot her fiancé twice before fleeing. Balint said that she was
terribly upset by the incident. She ran home, changed out of her
nurse's uniform, and took her dog for a walk before coming to
the station house to report the shooting.

Serpa was unhappy with her story. But he had seen people in
shock do bizarre things. Besides, Balint said that she could identify the man who had shot her fiancé. So Serpa had her look
through the mug books of known black male criminals in the
area. In the meantime, Montuori and Hildebrandt were dispatched to Balint's apartment to verify her address, to speak
with her mother, with whom she lived, and to check out her
story. Balint's mother corroborated her daughter's version of

when she had come home, changed, and left for the station house. Montuori asked about the clothes that Mary Ann had been wearing. Balint's mother described them and readily produced them; Mary Ann's white nurse's uniform and shoes had bloodstains on them, and this also fit Mary Ann's story.

As the detectives were about to leave, Montuori spotted a black handbag on the dining room table. He asked the mother if that was the bag that Mary Ann had been carrying earlier that day. The mother said that she and her daughter actually shared the use of that bag and that it contained personal articles belonging to each of them. Both Montuori and Hildebrandt found this odd. They had never known two women, let alone a mother and a daughter, to share a bag. Montuori asked if he and Hildebrandt could examine the contents of the mother-daughter handbag. The mother agreed and, after removing some "women's things," emptied the purse onto a white sweater on the table. The contents consisted of entirely ordinary items: lipstick, comb, hairpins, and coin pouch. But the purse also contained a live .38 caliber round.

Montuori immediately called Detective Serpa and told him about the live round. Serpa asked Balint about the .38 caliber round in her purse. She responded casually that her purse had been open. One of the robber's rounds must have popped into it, she said, an explanation that Serpa found barely plausible. An automatic weapon with a jammed breech could eject a live round, and detectives had discovered one live round at the crime scene. But what was the likelihood of yet another live round ending up in Balint's purse? She kept looking through the photo books and, over the course of about an hour, offered other possible explanations for how the bullet found its way to her handbag. Finally, she told Serpa that she doubted that the police had actually found a bullet. Serpa told her that her mother had seen the detectives recover the round and suggested that she call her mother to confirm this.

Mary Ann did call her mother, speaking in a language Serpa didn't recognize. After the call, with Serpa asking her "What now?" Mary Ann said: "I'll tell you what happened. I did it, but first I want to see my mother." Serpa tried to read her *Miranda* rights at that point, but she insisted on seeing her mother first. After conversing with her mother privately, and then hearing her rights read, Balint made a full confession both to Serpa and to an assistant district attorney in a later videotaped statement. She also led the detectives to the waste basket into which she had dropped the murder weapon while fleeing.

Balint pleaded guilty to first-degree manslaughter but promptly appealed her conviction. She argued that she had not been given her *Miranda* warnings in a timely fashion and that her initial admission and all subsequent statements had therefore to be suppressed, along with the gun that was discovered in a search premised on her confession. And, with her mother now agreeing, she claimed that the purse belonged to her alone and that her mother could not give the police permission to search it without a warrant. The live bullet, as well as the gun, must also be suppressed. The Appellate Division of the New York State Supreme Court agreed with her and overturned her conviction. In a concurring opinion, one justice went so far as to assert that "police testimony invites doubt as to its trustworthiness and bears significant indicia of having been carefully tailored to meet a perceived constitutional requirement." The court ordered a new trial. But now prosecutors didn't have the bullet, the gun, or Balint's statements, and they had one judge proclaiming that the police account was mendacious. All they had was Vincent

Eckes's dead body. Mary Ann Balint was released from prison. For years, Montuori regularly ran into her at a diner on West 181st Street.[7]

Cowell's case ended up before Justice Harold J. Rothwax, who ruled Cowell's spontaneous statement at the apartment and his confession at the station house admissible.[8] In January 1993 Cowell's attorney was delighted to be offered a fifteen-year plea by the assistant district attorney, much to Detectives Muldoon's, Montuori's, and Potter's surprise and chagrin. Cowell's counsel also filed notice of appeal late that same month. In 1997, never having acted on his state appeal, Cowell filed a *habeas corpus* petition with federal court in Manhattan arguing that he had had ineffective counsel, had entered an involuntary plea not aware of its consequences, and had been denied his right to a state appeal of his conviction, a petition eventually dismissed by the federal court because Cowell had not exhausted his state remedies.

This precipitated a flurry of motions by Cowell in state court over the next several years, alleging inadequate representation at the time of his plea.[9] All of these motions were denied. Then in 2004 Cowell reiterated his earlier claims and argued as well that the spontaneous statement he made in his apartment as Detective Montuori searched through his duffel bag was the product of a custodial interrogation without the benefit of *Miranda* warnings. And because the detectives had not read Cowell his rights at that point, Cowell argued, not only should that statement be thrown out but also, as fruit of a poisoned tree, Cowell's subsequent full confession to Detectives Muldoon and Potter and all forensic evidence, including Sara's body and body parts, Cowell's cleaning equipment, and the condition of the bathroom and other areas of the apartment, should be suppressed by the court. In short, Cowell argued for a reversal of his con-

viction, the vacating of his plea, the suppression of all statements and physical evidence, and the dismissal of the indictment against him. One of the cases Cowell cited in his favor was
the court's ruling in favor of Mary Ann Balint.

In fall 2004 the Appellate Division of the New York State Supreme Court finally dismissed all of Cowell's claims, ruling that
the detectives had acted entirely lawfully in their inquiry into
Sara Long's disappearance. Moreover, the court asserted, that
"even if we were to find the initial inculpatory statement to be
inadmissible, we would find that defendant's subsequent oral,
written, and videotaped statements, provided after *Miranda*
warnings, were sufficiently attenuated from the initial statement
to be admissible."[10]

On November 24, 1990, Chocoláte fell into the net.
He was arrested on a weapons charge in the Bronx. When
Muldoon received notification that Choocoláte was in jail, he immediately telephoned an assistant district attorney on a Sunday
night, telling him that he had four witnesses to Chocoláte's
murder of Warren Hodge, three of whom were for the moment
in jail. But the assistant district attorney insisted on waiting until Monday, too long a delay to keep up with the Bronx's alacrity
in releasing violent felons back into the community. Chocoláte
was released on bail in the early afternoon just as Muldoon discovered that his fourth witness, John Irizary, now said that he
was unable to identify Chocoláte as his assailant. Without the
victim's identification, Muldoon's case was weak unless he could
get an out-of-custody statement from Chocoláte that tied him
to the shootings. Muldoon called in all his markers with his
informants until he discovered Chocoláte's hiding place and

eventually his car on the street. Muldoon sat on Chocoláte's car for hours waiting for him to return to it. But then he received notification from the 34th precinct that further overtime was denied, and he had to call off the hunt.

Then, Chocoláte shot two men in a bar in the Bronx, one of whom nearly died. Police officers apprehended him after a wild chase down the Grand Concourse in the 46th precinct, Muldoon's old beat. In custody, Chocoláte refused to speak with the police, and Muldoon's three briefly jailed witnesses were already in the wind. With no witnesses, he had no case. But at least Chocoláte was off the street for a while.

Muldoon had another drug-related homicide at 552 West 160th Street that made him uneasy about his pursuit of Chocoláte. On October 11, 1989, a young woman named Rita Bellamy from Brooklyn went there with her boyfriend to score drugs. Rita went upstairs to make the buy, while her boyfriend waited downstairs. Rita ended up strangled and thrown out of a third-floor window to her death. The boyfriend fled the scene. At first, it seemed that the dealers in the drug apartment had demanded sex from Rita in exchange for drugs. When she resisted, the story went, they killed her. But later, Muldoon interviewed a man named "Antonio" who said he feared for his life and was willing to talk. Antonio claimed that Ivan, who dealt out of 552 West 160th Street, had shot him. Muldoon remembered that it was a man named Ivan who had identified Chocoláte as Orlando Rodriguez to detectives investigating the murder of PO Michael Buczek. Antonio's connection of Ivan to Chocoláte's building closed that circle.

Antonio said that "Arames," with another man named "Gordo" standing nearby, told Antonio that he and Gordo had killed the girl. But Antonio later recanted his testimony. Muldoon knew, from still other cases, that Gordo was a psychopath, who sold beat (fake) dope to boot. He made Gordo for Rita Bellamy's killer but, with Antonio's defection, had no way to prove it. Years later, it turned out that the early story of sexual assault was wrong. Gordo had simply sold Rita beat dope, and when she came back to squawk, the crew had its customer relations department choke her and throw her out the window.

Muldoon waited and hoped that someone who had a falling out with Gordo would give him up in return for a favor from the police. Sure enough, Detective Joel Potter got a statement from an arrestee that Gordo had done Rita's murder; indeed, the informant said that Gordo intended to rob the girl of her money all the while. But then the informant got out of jail on bail and disappeared. Following Potter's lead, Muldoon kept working the case, looking at all the West 160th Street players. Eventually, the name "Samson" came up, who turned out to be Antonio's brother. Samson also gave up Gordo for Rita's murder. Gordo, in the meantime, was on trial for yet another murder. He was acquitted for lack of convincing evidence, and Muldoon arrested him for Rita's murder just as he walked out of the courtroom, thinking he was a free man.

Gordo went crazy at being sent back to jail to await trial for Rita Bellamy's death. But Samson and another witness that Muldoon produced fled to the Dominican Republic, and before long Gordo was back on West 160th Street selling beat dope. Muldoon consoled himself with the thought that sooner or later street justice would accomplish what the courts failed to do, and in a far more fearsome manner. But then Gordo went north to Newburgh, New York, with a crew to rob a bodega owner on his way to the bank. The crew members were caught near the scene, and Gordo, who was found hiding in a tree, was convicted and sentenced to ten years in prison for armed robbery.

In the opaque world of drug-related homicides, the central problem is always the witnesses. Over the years, Muldoon kept trying to put the case against Chocoláte together. But he either could not find witnesses, or, as in Irizary's case, the witnesses could no longer make necessary identifications. Even when he did locate witnesses, Muldoon faced the fundamental paradox of all criminal investigation: those with the deepest knowledge of criminal activity are usually criminals themselves, or they live, by choice, in criminal environments, making their perceptions and judgments suspect to people outside those social circles.

Furthermore, in New York City's highly compartmentalized and competitive law-enforcement world, one cannot always count on cooperation from other police units in finding witnesses and working cases. While Chocoláte was safely in jail and Muldoon was looking for witnesses uptown, DANY's Homicide Investigation Unit was quietly investigating the case on its own. At the time, HIU had much greater resources at its disposal than the average squad detective. Specifically, it could pull people out of jail or prison at will and interview them for information without the always-present restrictions of overtime and cost that frustrate many investigations at the squad level. Moreover, it could offer inducements to uniformed and undercover officers to bring interesting cases to HIU instead of to precinct detectives.

Unbeknownst to Muldoon, the Puerto Rican woman who was with Chocoláte at the time of the shootings had wandered into the 34th precinct on January 27, 1992. She wanted to talk about the events of early October 1988 on West 160th Street. Two undercover officers took her downtown to HIU instead of upstairs to Muldoon. "Chica" was a crack addict who had momentarily awakened from her drug-induced haze. She told HIU investigators a story similar to the ones that Muldoon had gathered from witnesses earlier, except that, she said, Pepe, the man who had voluntarily come into the police station to say that he was visiting a girl in the building when he happened on the murder-assault in progress, was actually talking to Chocoláte immediately before Chocoláte began to get excited and ordered the two customers tied up. At that time, Chica said, Pepe fled.

Moreover, she said that after Chocoláte shot the shorter Hispanic man, Chocoláte's worker shot the tall young black man. Chica insisted that she had pleaded with Chocoláte not to harm these customers, but he did not listen. Later, on June 30, 1992, HIU investigators pulled the runner, Pirate, out of one of his many stints at Rikers Island and interviewed him. Pirate told the investigators essentially the same story that he had told Muldoon earlier. The HIU investigators did not notify Muldoon about their interest in Chocoláte. When Muldoon learned about their surreptitious work, he was furious, all the more so when he discovered from street sources that Chica had returned to the twilight world of homelessness and crack addiction. Some street rumors had her dead, her head buried in a city park, though no one knew which park. Other street people warned Muldoon that, when and if he ever found her, he would find her brain fried beyond repair. But at least Muldoon now had her identified.

Finally, in 1996, Muldoon got transferred to the Manhattan North Homicide Unit, later led by his old 34th squad boss, Joseph Reznick, now an NYPD inspector. Muldoon finally had the time to make one last run at Chocoláte. Muldoon beat the bushes on West 160th Street and came up with César, who still hung out on the block and who, years before, had claimed to see Chocoláte shoot Irizary and to have heard a second shot as he fled down the street. Muldoon also found Pirate, the runner, who had witnessed the shootings. Between 1989 and 1996, Pirate had been arrested for robbery, including a violent assault on a bystander; criminal possession of a loaded firearm; fare-beating in the subway; violent assault with intent to cause serious injury with a firearm; and possession of narcotics with intent to sell. But with these two witnesses and the cooperation of a new prosecution unit investigating old homicides at DANY, Muldoon was able to get an indictment on Orlando Rodriguez aka Chocoláte for the murder of Warren Hodge and the attempted murder of John Irizary.

Moreover, Muldoon located the owner of the spot that Chocoláte managed, who admitted that Chocoláte had told him about the shootings right after the event. Muldoon began to put the case together with ADAs Steven Saracco and Stacey Mitchell.[11] But there are many slips between arraignment and trial. Pirate, the runner, disappeared, and the spot owner refused to testify. Once again the case against Chocoláte was in jeopardy.

Muldoon doggedly went back to West 160th Street to search for witnesses. He got some help from the system's bureaucratic net. On April 27, 1997, Chica had tried to sell crack cocaine to an undercover officer in the Bronx and landed in jail on felony charges. Chica had stopped using drugs five months earlier and, despite brushes with the law that her new occupation provoked, was trying to put her life back together. Chica told Muldoon the following story.

Two or three days before the shootings, she had been sitting on the second-floor landing of 552 West 160th Street. Pepe came into the drug building with a customer, a Spanish man about 5 feet, 10 inches tall. Chica said that Pepe sold drugs out of apartment 8 in the same building. Chica saw something bulging in the customer's jacket and she warned Pepe that the man might have a gun. Just as she said this, the man did indeed draw a weapon. Pepe and he struggled ferociously for possession of it. The man fled the building and jumped into an automobile with Connecticut plates.

A couple of nights later, Chica said that she woke up around 3:30 A.M. and went to West 160th Street to score some dope. She ran into Cho1coláte, for whom she regularly ran customers and translated English into Spanish during drug sales. Chocoláte asked her nicely to come into his sale apartment to sample some drugs. She went without hesitation because she had often done this for Choco1áte. But when they arrived at Choco1áte's sale apartment, Choco1áte put a gun to Chica's head. He demanded to know what had happened to his kilo of cocaine. Chica had no idea of what he meant and pushed the gun away. But Choco1áte said: "I'm not playin." Chica begged for a quick death. She heard the gun click. Then one of Choco1áte's workers, pushing the gun away from Chica's head, told Choco1áte that she had nothing to do with the theft. Someone else had snuck in and burglarized (plundered with stealth) Choco1áte's apartment, getting away with a kilo of cocaine. But no robbery (seizure of property from a person by force) had occurred. Choco1áte was in a rage because he had to answer for the loss of the cocaine to his own boss. He had no idea who had stolen his drugs, but he suspected his sometime workers or local street players, people who knew of his operation and who could observe his comings and goings.

Chica was now terrified and tried to leave the apartment. Choco1áte picked up an aluminum baseball bat and smashed her in the hip, hurting her badly. As she spun around from the force of the blow, she saw in an adjacent bedroom a woman friend of hers, also a crack addict, bound to a chair, bleeding profusely and screaming in agony. Choco1áte had carved the woman's back up with a broken windowpane, demanding to know what she knew about his missing kilo. Chica hobbled out of the apartment and down to the street when, suddenly, she saw a friend of hers, Pirate, a runner who often brought customers to Chocoláte. The runner had two kids with him, one tall and black, the other short and Spanish. They were heading into 552 West 160th Street. Just then, Chocoláte and his helper came out on the stoop and Chocoláte asked Chica to come back into the apartment to translate for him. Chica was hesitant but she did limp back.

Chocoláte and his worker led the way into the apartment, followed by the runner and the two customers. Chica brought up the rear. As she usually did, Chica told the kids to stand to the side to be frisked. Chocoláte's helper frisked the tall black kid first, then started on the short Spanish one. The latter, Chica said, told her in Spanish that he had a gun, but too late, because the helper had already seized it from him. The helper became agitated and ordered everyone into the living room, where Chocoláte was. Just then Pepe came into the room and talked to Chocoláte. By this time, Chocoláte had a gun pointed at the two kids and he was yelling in Spanish that they came from Connecticut to rob him, just like the guy a few nights before who tried to rob Pepe, and that he was going to "kill the two muthafuckas."

Chica told Muldoon that she tried to defend the two boys, pleading with Chocoláte not to hurt them, but Chocoláte had no ears that night. He told Chica to shut up. At Chocoláte's orders, the helper tied the boys' hands behind their backs. Chocoláte then ordered everyone out of the apartment. The runner left in a hurry, with Chica right behind him. But Chica stayed in the building, on the landing of the stairs. Chocoláte came out of the apartment with a gun to the short Hispanic man's head. Chica screamed that these boys had nothing to do with the attempted robbery of a couple of nights before and that they couldn't possibly have done the burglary of Chocoláte's apartment. But Chocoláte told her to shut up. Then Chocoláte pulled the trigger. Chica said that Chocoláte's face was angry. She saw the Hispanic man fall to the floor. Chica screamed: "Oh, my God!" She heard Chocoláte order that the other man be shot too. She turned away and heard another shot. When she looked back, the young black man was also on the floor.

Just then, "Angie," another woman friend of hers, came into the building and asked what happened. Almost immediately afterward, uniformed police arrived at the building, discovered the bodies, and asked Angie what was wrong with Chica, who was screaming. Angie told the cops that she and Chica had had a fight with their boyfriends. The police told both women to get out of the building. Both Angie and Chica fled immediately and went to an abandoned building nearby, where Chica wept while Angie comforted her.

Because Irizary could no longer make a positive identification of his assailant and because other witnesses had disappeared, refused to testify, or were so compromised by their own criminal involvement, Chica became the linchpin of the case against Chocoláte. Of course, there were loose ends. Muldoon still had no idea who Chocoláte's helper was. At the very least, that man was guilty of felony murder, if not the actual murder of Warren Hodge. And why had Chocoláte insisted on going through with the shootings despite vehement protests by a regular translator, a regular runner, and an upstairs business associate that the two victims were in no way connected with the attempted robbery in the building a couple of days earlier? Everyone knew that Chocoláte's kilo had been snatched in a burglary, not a robbery.

Was the assault on the two boys just Chocoláte's statement to his boss that he knew how to protect his spot? And what about the discrepancies between Irizary's and Chica's versions of events, specifically whether it was Irizary or Warren Hodge who had the gun when confronted by the drug dealers? Muldoon figured that Irizary, who had admitted bringing a .25-caliber weapon with him from Connecticut, put the gun on his friend at the showdown to avoid more trouble with the police.

But with Irizary's testimony about the circumstances of the night he was shot and Chica's testimony nailing Chocoláte for the shooting, Chocoláte was convicted of murder and attempted murder and sentenced to 37 years in prison. His accomplice was never apprehended.

The long time-span of the case enabled Muldoon to see transitions in street players' lives that are usually not visible to police, or are often overlooked by them. César got married, had two children, and became a taxi-cab mechanic. When Muldoon last spoke with him, he had stayed out of trouble for years. Chica, despite her brushes with the law, seemed to be on the road to complete recovery from crack addiction. Muldoon was most impressed with the self-transformation of the crack addict known as Mighty Joe Young, the giant who used to live in a cardboard box on West 160th Street. Muldoon thought there was an off chance that Mighty Joe Young might remember something about the night of October 4, 1988, but he held little hope of finding him. Street rumors about the man's fate abounded. He had been murdered; he had died of AIDS; he had moved to the Carolinas; he had become rich; he was serving a life sentence somewhere.

Eventually Muldoon did find Mighty Joe Young, although it was the unrecognizable former Mighty Joe Young who recognized Muldoon. In 1990 Mighty Joe Young had turned his whole life around. He stopped using crack, got married, became the father to two stepsons, found steady work, and maintained an apartment in a Bronx working-class neighborhood. Although he readily admitted his past, he looked back on his days as the wild man of West 160th Street as if he were viewing a movie about a strange man in another world.

8

TRACING THE PAST

Detective Edward Dermody of the 5th Homicide Zone arrived at 611 West 204th Street, apartment 5 on the second floor, at 1830 hours on March 5, 1974.[1] Mrs. Guadalupe Diaz, a 55-year-old widowed Dominican immigrant, lay dead on her bedroom floor, nearly perpendicular to her bed, disrobed from the waist down. Her hands were tied in front of her with white plastic clothesline. She had been sexually assaulted. Five small, superficial puncture wounds surrounded her right upper eyelid in a circle. Three superficial wounds marked her left upper eyelid. She had evidently been threatened with a sharp instrument, or perhaps struck by a pointed ring on the hand of her assailant, but these minor wounds had not caused her death. A brown electrical extension cord was wrapped twice around her neck, and the bones there were crushed from the ligature.

Mrs. Diaz held her lower denture in her right hand. A pocketbook was found emptied on the bed. A heavy odor of cigarette smoke filled the tiny room, and the ashtray on the kitchen table was glutted with Salem cigarette butts. A radio sat on a small table in the bedroom. In front of the radio, a half-filled bottle of brandy stood next to a used brandy glass, along with a pointed kitchen carving knife engraved with the legend "The Miracle Worker." Household records were strewn on the floor. In the adjacent living room, the window leading to the fire escape was open, as was the iron gate that protected it. The gate had been forced. But the gate's padlock was unlocked, undamaged, still in good working order, and sitting on the window sill. The kitchen was undisturbed and everything seemed in its place, except for a small drinking glass left out on the counter next to the refrigerator.

Dermody, a nearly two-decade veteran of the police department with twelve years as a detective, notified the city-wide Crime Scene Unit then turning out of the Police Academy on East 20th Street in Manhattan. Then he interviewed Rafael Diaz, who had discovered his mother's body. Rafael lived at 613 West 204th Street in a second-floor apartment directly across the courtyard from his mother's apartment. His apartment, in fact, mirrored hers. From his kitchen window he could see directly into his mother's kitchen, the first large room that one reaches after walking down a fifteen-foot corridor from the front door and past the small bathroom. Rafael told Dermody that his mother was a woman of extremely regular and tidy habits. She was a churchgoer who did not smoke or drink. After being widowed, she devoted herself to her children and grandchildren.

Like many Dominican immigrant women, she worked as a seamstress, sewing dolls at the Alexander Doll Company factory in West Harlem on West 131st Street from early in the morning until shortly after 4:30 P.M. Then she took the train home, usually reaching her apartment around 5:30. To signal that she had arrived safely, she always went to her kitchen window and waved to her son across the courtyard. He played with his daughter in his own kitchen every day after returning from work. Rafael regularly gave his mother time to wash up, then, using his own key, entered her apartment and had coffee with her in her kitchen around 6:15 P.M. That evening, Rafael told Dermody, he and his mother followed their daily routine exactly. But he knew right away that something was wrong when he went to her apartment because her front door was chained from the inside, something that had never occurred in all his years of evening visits to her.

The CSU team arrived at 2150 hours, headed up by Detectives Edward Meagher and Stephen Colangelo, veterans of crime-lab and crime-scene work. Meagher and Colangelo took thirty photographs of the crime scene, especially Mrs. Diaz's body but also every nook and cranny of her apartment. Using black powder on light surfaces and white powder on dark surfaces, they dusted the windows, the sofa, the walls, closets, bureau and tables, bottles, glasses, the refrigerator, the radio, and various other objects in the apartment for latent fingerprints—the invisible residues of perspiration from the sweat pores on the friction ridges of fingers and thumbs. The powder revealed ten partial prints, which CSU photographed and then lifted with transparent tape, affixing each print to a contrasting-color card.

Two prints came from the center lower frame of the inside of the living room's left window, which had provided entry to the apartment. One came from a wall adjacent to the same window. Two came from the carving knife found in Mrs. Diaz's bedroom, one on each side of the blade. One came from the side of the small drinking glass found on the countertop of the kitchen cabinet next to the refrigerator. And four came from the sides of the brandy glass found on the small bedroom table in front of the radio. Meagher and Colangelo labeled each print and marked it with the Crime Scene Unit run number, the date, and the location where it was discovered in the apartment. They then signed all the lifted prints. They also gathered up the papers strewn on the floor of Mrs. Diaz's bedroom and, together with the ten lifts, sent everything back to the Crime Scene Unit, where they had to write a report to accompany the materials before transmitting them to the Latent Print Unit of the police department.

Beginning that evening and in the days following the murder, Dermody and other detectives repeatedly canvassed all the residents at 611 West 204th Street and in all the buildings adjacent to it to see if anyone had noticed suspicious persons in the area on the evening of March 5. But no one said they had. The detectives went to Mrs. Diaz's wake and talked with everyone who attended, seeking to learn more about the woman's habits and routines. They met Mrs. Diaz's three other grown children besides Rafael—another son and two daughters—and tried to comfort them in their shock and grief. They learned that Mrs. Diaz had raised all four children alone because her husband had died only a year after they immigrated to New York in 1961. The detectives also visited the Alexander Doll Company and interviewed Mrs. Diaz's coworkers and checked her locker for any leads. They noted that she had punched out at 1647 hours on March 5, corroborating Rafael's description of her adherence to routine.

Dermody checked police records for similar sex crimes committed in Washington Heights in recent years, on the assumption that sexual predators almost always repeat their crimes as if following a script. Two crimes in particular caught his eye. About three months before Mrs. Diaz's murder, and in the same vicinity, an intruder had pushed his way into an apartment, tied up a 58-year old Hispanic child-care provider, stripped her, and knocked her out. But when detectives re-interviewed the victim of that crime, she claimed not to have suffered sexual assault. Further, her assailant had menaced her with a gun, not a knife. Nonetheless, the detectives proceeded with a side-by-side comparison of this victim's associations and habits with those of Mrs. Diaz, checking for similarities in their places of shopping, banking, entertainment, and churchgoing, or for overlapping social circles in their choices of doctors and friends. But the two women lived in different social worlds.

The second assault had occurred in Mrs. Diaz's building a year before. A handyman who worked and lived there, and was presumably known to her, had violently raped a woman in her apartment on March 1, 1973. But the police had arrested the handyman on his victim's identification the same day as the assault. Because strange things regularly happen in the criminal justice system, Dermody checked with the Department of Correction to verify the handyman's whereabouts when Mrs. Diaz was assaulted. But the handyman had been incarcerated since his arrest. No other sexual crimes in Washington Heights matched the assault on Mrs. Diaz.

Rafael Diaz called Dermody on March 20, 1974, to report a conversation with a close friend of his mother's. Mrs. Diaz had told her friend that she had been harassed several times in the summer of 1973 near the subway station at 207th Street and Tenth Avenue by a man who came out of a nearby bar and made advances toward her. But when the detectives interviewed the woman, she could provide no further details of the incidents, nor any description of the man who had reportedly bothered Mrs. Diaz.

The fingerprint analysis came back from the Latent Print

Unit. There were no usable fingerprints on any of the papers on Mrs. Diaz's floor. The ten lifts from other items in the apartment had been manually compared with the fingerprints of known sex criminals in uptown Manhattan and the Bronx and with those of all known push-in robbers and burglars in the area, but there were no matches.

Dermody and other detectives regularly returned to West 204th Street to complete their canvass of residents at 611 and adjacent buildings. On March 23, 1974, Dermody finally caught up with "Nikki Sterling," who lived on the first floor of 611 West 204th Street, directly below Mrs. Diaz's apartment. Sterling said that she had heard about the homicide but had no information that could help the detectives' investigation. But something about her manner gave Dermody pause. Sterling seemed nervous and evasive. Pressing her, Dermody queried her about friends who used her apartment. Sterling said that "Robert Tucker" and "Larry Tucker" were among the visitors to her apartment, and she provided Dermody with their addresses. On March 29, Dermody talked with Larry Tucker. He claimed that Nikki Sterling was his girlfriend and that he had been with Nikki at her apartment at 6:30 P.M. on March 5. He had heard about Mrs. Diaz's murder, but, he said, he knew nothing about it. He readily gave his fingerprints for comparison. Two weeks later, the Latent Print Unit reported that there was no match.

Over the course of the entire next year, in response to Dermody's submission of over a hundred additional sets of fingerprints, including those of Robert Tucker, the Latent Print Unit found no matches. Mrs. Diaz's killer had disappeared into the night, leaving only traces of his identity behind him and a family shattered and bewildered by grief.

The fingerprints at Mrs. Diaz's apartment were unusual. Despite remarkable advances in forensic techniques, which have generated an enormous following in popular culture, local police investigating violent crimes rarely find such clear forensic traces of identity. Precinct squad detectives address the vast majority of cases they catch by searching the far-reaching bureaucratic nets that freeze-frame suspects' movements, and sometimes actions, and allow police to pinpoint their whereabouts and associations at particular times. But most especially, detectives talk to people, witnesses, informants, and criminals themselves, continually assessing their credibility and the plausibility of the stories they tell. When it comes down to identifying culprits of particular crimes, they rely for the most part on witness identifications.

The most important kind of witnesses are those who incriminate culprits on the basis of their own direct observations of crimes being committed or on the basis of statements that culprits make to them. In both cases, police first try to match witnesses' recollections with the officers' own records of people who have already committed crimes. It is an axiom of police work that criminals commit most crimes and that a small, hard core of criminals commit the vast majority of crimes, grossly disproportionate to their numbers. The police force's extensive photographic records of criminals are the key tool in this identification process.

The photograph records of the New York City Police Department began in 1858 with a 450-ambrotype rogues' gallery that cops studied to learn the faces of their adversaries. In the late nineteenth century, the legendary first chief of the Detective Bureau, Thomas F. Byrnes, a formidable force in the city until he ran afoul of the Presbyterian Reverend Charles H. Parkhurst and the latter's Society for the Prevention of Crime, compiled a fantastic collection of photographs of every well-known criminal east of the Mississippi and had his men memorize their visages.[2] Most of these arch enemies of the public order were con men, forgery artists, and counterfeiters. If they were spotted by Byrnes's detectives below an imaginary line near the Wall Street financial district, they were summarily arrested and jailed on sight.[3]

During the twentieth century, the NYPD's photograph collection grew exponentially as mug shots were taken of everyone arrested, whether for misdemeanor or felony crimes. The sheer vastness of the collection soon posed enormous problems of classification. Each of New York City's five boroughs has its own centralized CATCH unit, housing photographs of everyone arrested in a particular borough, going back decades, catalogued by general descriptions and by types of crime. To the trained eye, the "oracle numbers" reproduced on the bottom of each criminal's photograph provide at a glance the person's race, sex, age, height, weight, and the crime for which he or she was arrested. This ancient system still exists, with hard copies of photographs of all persons arrested in each borough going back nearly forty years. To get some sense of the scale of such an enterprise, consider that, in Manhattan alone, there are more than 100,000 arrests each year.

Beginning in the early 1990s, the New York City Transit Police Department began using photo-imaging technology, starting with a database of 13,000 felons who had been arrested in the citywide transit system. After the NYCTP merged with the NYPD in 1995, under the aegis of former chief of transit police William Bratton, the NYPD adopted the new technology and began slowly transforming its vast photographic files into computerized images. Now detectives sit with complainants or witnesses in front of a computer and type in their descriptions of culprits. Ready-made photo arrays of possible suspects appear on the monitor. As of 2002, photographs of all culprits arrested since 1994 had been scanned electronically for storage in the database, with more added each year. The old hard-copy photograph files still serve as the primary identification files for those who committed crimes prior to 1994.

Visual identifications of criminals are still extremely important in everyday police work, although eyewitness identification itself has come under intellectual, governmental, and judicial scrutiny.[4] Moreover, given the long delays between arrests and trials due to the overcrowding of the criminal justice system, some accused culprits change their external appearances dramatically by "bulking up" on three squares a day, along with heavy weight-lifting regimes. Newly fat faces and muscular torsos make some witness identifications falter in court.

But what does one do when there are no witnesses to a crime, or when the only witness was the victim, who is dead? How does one begin to track and then identify the assailant?

Fingerprints were first used in a systematic way to identify people for administrative, not law-enforcement, purposes. Consider the practical problems faced by the likes of colonial authorities in sprawling, multiracial, multilingual 1850s British India. These authorities turned to fingerprints as they tried to winnow fraudulent claims by imposters to government pensions.[5] Or consider how industrialization, political upheavals, the collapse of estates, and urbanization shattered the routines of village life, exponentially increased migration and social mobility, and unraveled the community ties that made face-to-face recognition of others commonplace. Immigration officials in the United States and Argentina adopted fingerprinting to control certain types of undesirable aliens. In New York City, police and magistrates used fingerprints to identify, register, and regulate habitual offenders such as prostitutes, vagrants, mashers, degenerates, and other miscreants at the periphery of the social order.[6]

The ease of collecting fingerprints created a staggering demand for a storage and retrieval index that allowed comparison of prints. Modern fingerprint classification systems went through a stop-and-start-again development. Francis Galton, a sometime eugenicist, divided all fingerprints into "arches," "loops," and "whorls" and provided the foundation for most later indexing schemes. A British-India colonial administrator, Edward Henry, added a fourth pattern, called "composites." Henry also numbered the fingers and focused attention on differing ridge characteristics as the distinguishing marks of fingerprints. Later, Scotland Yard instituted a national standard of sixteen "matching points" of such ridge characteristics for identification through fingerprints. Around the same time, Juan Vucetich in Argentina developed a sophisticated version of Galton's system.

Competition between originators of classification systems, and their emulators, was often fierce. Different countries, indeed whole geographical regions, made fateful commitments to one or another classification system. In the United States, the patchwork of jurisdictions led to myriad choices, sometimes producing incompatible systems in adjacent bailiwicks. Moreover, unlike Britain, the United States did not adopt a fixed, cross-jurisdictional national standard of matching points to establish identity.[7]

The use of fingerprints for criminal investigation, specifically to link suspects to crime scenes, had a somewhat more sporadic development. In the mid-1850s Detective John Maloy of the Albany, New York, constabulary convicted a burglar who had left a bloody thumbprint on a piece of paper in a building he had illegally entered; Maloy was able to match the crime-scene print to an inked print of the culprit's thumb. In 1897 the Indian police charged an ex-servant with the burglary and murder of his former master, a tea-garden manager who had been found stabbed to death in his home. A bloody fingerprint found at the scene matched one of the ex-servant's inked prints taken when he was apprehended as a suspect in another crime. The magistrates hearing the case refused to convict the ex-servant of murder but did convict him of burglary on the basis of the crime-scene print.

Late-nineteenth-century European experiments demonstrated the possibility of lifting invisible (latent) prints from certain surfaces and later comparing them to inked prints. In both Europe and the United States, the early 1900s brought several successful prosecutions utilizing matches between latent and inked prints.[8] One of the most famous occurred in New York City in 1908, when Sergeant Joseph Faurot of the New York City Police Department, already celebrated for gaining a confession two years earlier from a master burglar after identifying him through his fingerprints, tracked down a suspect in the murder of a young woman using latent prints left behind at the crime scene. When Faurot confronted the suspect with the fingerprint evidence, he confessed to assaulting the girl while in a drunken rage.[9]

Each success further legitimated the use of fingerprints as crucial evidence in linking suspects to crime scenes. Indeed, early on, fingerprint examiners—the interpretive experts who elucidated fingerprint "matches" to judges and juries—claimed that, barring gross disfigurement, fingerprints are permanent and, moreover, that no two fingerprints are identical. These experts argued that every finger of every human being bears a papillary ridge arrangement that begins to form between the third and fourth fetal month. Barring accidental or surgical removal of those ridges or of the finger itself, the ridge arrangements on each finger stay with a person for life, providing a unique identifying mark.

Examiners classify fingerprints into three general groups of patterns—arches, loops, and whorls, each with several subcategories. In comparing fingerprints, examiners look at three basic ridge characteristics. The first is called a bifurcation, a ridge that runs along as a single line and then separates into two lines. The second is called an ending ridge, which runs along but then abruptly stops. The third is either a dot without direction or a short ridge. In addition, examiners study the relationships of these friction-ridge characteristics to one another to determine whether the ridge characteristics occupy the same relative position in the impressions being compared. Examiners seek to ascertain as many points of similarity as possible. Identifications are made by successfully comparing the number and similarity of these characteristics. The United States Federal Bureau of Investigation has made positive identifications with as few as seven points of similarity, although typically American prosecutors seek many more in order to convince juries.

Over the course of the twentieth century, courts in the common law countries came to accept examiners' claims that fingerprints are permanent and that no two fingerprints matched as a matter of practical, though not absolute, infallibility. Here, one makes a presumption of truth based on fingerprint experts' long experience without successful rebuttal to their claim that each and every fingerprint is unique. This position was forcefully argued by a Scottish judge in 1933:

> The value of finger-print evidence depends on the reliance which can be placed on the result of expert investigation and experience—in an immense number of cases, examined over a very extended period of years—to the effect that identity is never found to exist between the skin ridges on two different persons' fingers. This is what leads the experts to claim infallibility for the finger-mark method. I deprecate the use of the word "infallibility" in this connexion at all. What the experts obviously mean is, not absolute, but practical infallibility—that is to say, a presumption of truth, the reliability of which may be accepted, not because it is irrebuttable in its own nature, but because long and extensive experience is shown to provide no instance in which it has ever been successfully rebutted. All proof depends at bottom on presumption; even the evidence of two credible and uncontradicted witnesses who speak to the same occurrence is *probatio probata* [conclusive evidence] not because it is impossible that they should both be mistaken, but because of the high presumption that what two credible witnesses say happened in their presence actually did happen.[10]

But fingerprint experts themselves actually asserted a somewhat stronger claim. Examiners from both local police and federal authorities regularly argued in court and in print that their training, methods, and experience enabled them to discern the uniqueness of every print with which they were confronted, making their work more of a scientific enterprise than an inter-

pretive technical craft.[11] Only the most skeptical critics of fingerprint evidence dispute identifications ascertained by matching
ten inked or "rolled" prints with a previously obtained set of ten
inked prints. The crucial issue is, instead, the scientific validity
of matching latent prints lifted from crime scenes, which are
usually partial and often smudged, against sets of inked prints
taken by police from suspects in custody. The process that examiners typically use to do such investigation consists of four steps
called the ACE-V procedure: observation and *analysis* of distinct
patterns of a latent print; *comparison* of those latent print patterns with those of an inked print; *evaluation* of the compared
patterns to determine whether or not the latent and inked prints
were made by the same person; and *verification* of one's work by
another examiner who repeats the same process to verify, or not,
the initial examiner's finding.

On May 9, 1991, Detective Edwin Cruz received a
call from Detective Thomas Montero of the Latent Print Unit.
Beginning in January 1990, the New York City Police Department had begun using a statewide automated fingerprint identification system (SAFIS), originally developed by the Federal Bureau of Investigation in the late 1970s and adopted gradually by
several states in the 1980s and early 1990s. SAFIS photographs latent prints presented to it and then electronically searches its
own database for prints with matching friction ridge characteristics. Doing work with lightning speed that previously took untold man-hours, SAFIS compares thousands of already stored
prints per second, presenting detectives with possible matches.
Detectives then pull the records of suspects matched by the automated system and do manual verification checks. Montero

told Cruz that the Latent Print Unit had been submitting usable fingerprint evidence from old homicide cases to SAFIS to see if any matches were discovered. SAFIS had identified three possible suspects in the homicide of Guadalupe Diaz.

The manual examination narrowed the possibilities to one suspect, 36-year-old Herman Nathaniel Myers, who was 19 years old at the time of Mrs. Diaz's murder. New York City police had first arrested and fingerprinted Myers on May 14, 1975, in the Bronx for grand larceny auto, almost fourteen months after the Diaz murder. He had later been arrested in both Manhattan and the Bronx for burglary, possession of burglar tools, and possession of stolen property. A nationwide records search revealed that Myers was currently residing in North Charleston, South Carolina, where he had also had encounters with the law. Montero told Cruz that there were six initial matches: one match of a print of Myers's No. 3 finger (right middle finger) to the latent print lifted from one side of the brandy glass found on the small bedroom table in front of the radio in Guadalupe Diaz's apartment; two matches of prints from Myers's No. 1 (right thumb) finger and one match from Meyers's No. 6 (left thumb) finger to latent prints taken from the other side of the same glass; and two matches of a print of Myers's No. 6 finger (left thumb) to the latent prints found on the inside of the lower frame of the left living room window.[12] Montero told Cruz that his squad was still working on making other possible matches. Cruz immediately took the case to the squad commander, Lieutenant Joseph Reznick. Without hesitation, Reznick assigned the unusual case to Detective Gennaro Giorgio, the legendary homicide detective and the Big Daddy of the 34th precinct squad.

The son of Italian immigrants, Giorgio grew up on Bleecker Street in Little Italy in an apartment right above Zito's Bak-

ery. After exploring an acting career on the basis of his rugged
good looks, personal magnetism, and remarkable facial expres-
siveness, he joined the New York City Police Department in 1959
at the age of twenty-six and spent six and a half years in uniform
on foot patrol in the 20th precinct on Manhattan's West Side. In
those days, the department selected police officers to join the
Detective Bureau on the grounds of demonstrated potential to
become good investigators. In 1966, an incident that electrified
the precinct gave Giorgio an opportunity to show his stuff.

A huge intruder had surprised a young woman by climbing
across a board placed between buildings and entering her tenth-
floor apartment through her bathroom. He raped her while
threatening to butcher her baby with a large knife that he
brought with him. After he finished, the intruder started to leave
the apartment, but then pulled down his pants and demanded
another go-around. The victim went crazy and fought the intruder fiercely. He stabbed the woman twenty-six times, but she
miraculously survived.

The precinct detectives worked night and day on the case and,
with a police artist, managed to get a composite drawing of the
assailant from the woman. They printed up poster-size copies of
the drawing for distribution throughout the neighborhood. Detective Frank Leo brought a bunch of the posters for distribution on Giorgio's beat. Giorgio took one look and told Leo that
he knew the assailant, and that if the posters hit the street the
culprit would be in the wind. He pleaded with Detective Leo to
give him two days to find the rapist and bring him in. After con-
ferring with his squad's commander, Leo reluctantly agreed.

No luck on the first day. But late in the second day, with time
running out, Giorgio spotted the suspect sitting on a stoop.
Giorgio got out of his radio car, sauntered up to the man, and

struck up a conversation. After shooting the breeze for several minutes, Giorgio mentioned casually to the man that someone had dropped his name in connection with several robberies in the precinct. Would he be willing to come down to the station house and straighten things out? The man readily agreed, and Giorgio and his partner drove the man to the detective squad. The detectives promptly took him to the hospital, where the rape victim was wheeled out in a chair, bandaged head to foot. She immediately identified the man as her assailant. The squad locked him up. The very next day, the chief of detectives summoned Giorgio to his palatial office at police headquarters, then located at 240 Centre Street, congratulated him on his street savvy, cunning, wiliness, and ability to con criminals, and invited him to don the coveted gold shield.

After a five-year stint with the 26th precinct detective squad, Giorgio worked as a detective specializing in homicide investigations, first in the Fifth Homicide Zone from 1972 to 1979, then in the Manhattan North and Manhattan South Task Forces, and finally in the Homicide Team of the 34th precinct, all the while gaining a reputation as one of the premier interrogators in the New York City Police Department and a great mentor, often surrogate father, to young detectives.

On May 10, 1991, Giorgio called the North Charleston police department to inquire about Herman Myers. Detective Eurzin Douzart told Giorgio that Herman Myers was well known to the police there. Over the years, they had arrested him for domestic violence, possession of drug paraphernalia, and other offenses. Myers had just served four months of a one-year sentence and was currently performing community service. Giorgio asked Douzart to make sure that none of the North Charleston police raised Myers up by approaching him or, worse, detaining him.

Giorgio then contacted Assistant District Attorney Warren Murray, chief of the sixty-attorney Trial Bureau 50 of DANY. Murray, a former United States marine famed for running his bureau like a combat platoon, all the while inspiring profound loyalty in his well-trained young assistants, tries few cases because of his supervisory duties. But Murray wanted this unusual case not only because of its challenges but because Murray's most valued self-image as a prosecutor is identical to that of the best detectives. Murray's motto, preached regularly to his charges, is: "No truth, no justice." Murray thought that cracking Mrs. Diaz's murder could send a powerful message to the Dominican community in crime-ridden but generally uncooperative Washington Heights. The two men met a few days later to review the old case file and plot strategy.

Giorgio told Murray about two cases he had worked where fingerprints had proved crucial in eliciting statements. When Giorgio was with the Manhattan South Homicide Task Force, he and Detective Juan Medina, his suave, always impeccably dressed long-time partner, investigated the murder of 38-year-old Patrick Kehn, a tax and investment attorney with Shearman & Sterling on Wall Street. On September 26, 1981, Kehn was walking with his girlfriend on the promenade by the East River at Franklin D. Roosevelt Drive and East 10th Street in the 9th precinct. Three kids, one armed with a bat, another with a knife, surrounded the couple and pushed them onto a bench. Kehn and his companion readily handed over a watch, a gold chain, and his wallet with $100 in cash. The teenagers took the watch, chain, and the cash but discarded the wallet at the scene. One assailant then told Kehn: "I don't like your face." The boy with the bat slammed Kehn, and then the culprit with the knife stabbed him in the chest and abdomen. Kehn died on the spot in front of his companion, who became hysterical and unable to provide any useful descriptions of the assailants.

The case went down the usual blind alleys. A fireman had been jogging on the East River promenade ten minutes before the assault on Kehn and had been approached by a group of kids looking to rob him. The fireman dispersed the wolfpack and reported the incident to detectives at the 9th precinct, noting that he could identify at least one member of this crew. Precinct detectives showed the fireman their profiles of wolfpack robbers in the area and, with his photo-array identification, arrested a young man with a long list of robberies on his resumé. He was indicted for the Kehn murder. When Giorgio interviewed this suspect, he admitted the aborted assault on the fireman and admitted yet another robbery and what turned out to be a minor stabbing on Avenue D the same day as the assault on Kehn. But he denied any involvement whatsoever in the Kehn robbery-murder.

Despite pressure from the police department to close the Kehn murder with this suspect, Giorgio asked the forensic experts exactly what their search of Kehn's wallet had revealed. And it turned out that they had found one fingerprint on one of Kehn's credit cards that did not match Kehn's prints, nor the prints of the suspect in custody, nor those of any of that suspect's usual cronies. The indictment against the original suspect was later dropped, thanks largely to Giorgio's efforts. At that point, an informant came into the 9th precinct and told detectives that he had heard on the street that Leapo had killed Kehn. Giorgio and Medina checked every record available but could find no one nicknamed Leapo anywhere. But then another detective remembered a street kid named Leopoldo, which was close. The Latent Print Unit ran the lift from Kehn's credit card against the prints of several men named Leopoldo, and the match came back to Leopoldo Siao-pao, a 19-year-old karate expert with a long record of robberies.

When Giorgio and Medina caught up with Leopoldo in Brooklyn late on October 22, 1981, he acknowledged that people called him Leapo, but he denied any involvement in the assault on Kehn. When Giorgio said that he had evidence that put Leopoldo at the crime scene, Leopoldo smirked, thinking, he later said, that Giorgio meant that Kehn's girlfriend, a hopeless witness, had identified him. But Giorgio showed Leopoldo his memo book with his notation about the fingerprint match. Giorgio told him: "That is evidence." Leopoldo believed him and eventually admitted that he had stabbed Kehn. Still later, after the detectives pointed out the inequity of Leopoldo taking the weight of the charges alone while his comrades partied, he gave up the bat wielder who had started the assault, as well as the other accomplice. The batsman turned out to be the same informant who had given the police the name Leapo in the first place. Leopoldo's recognition of the persuasive power of fingerprint evidence prompted his confession and, in the end, his betrayal of his associates.

The second case was similar. Kathleen Williams, a 30-year-old vice president for a large bank, had just arrived in New York City on a business trip on September 22, 1982. Thirty minutes after she checked into the Waldorf-Astoria hotel, she was found dead, stabbed in the throat, in a stairwell on the 19th floor. She had apparently used the wrong elevator bank to reach her assigned room. When she couldn't find her room on the floor, she had either decided to descend the stairs back to the lobby, instead of using the elevator, or she had been forced into the stairwell by an assailant.

The police found one witness who had seen the woman in the hallway with what he described as a short, Mexican-looking man. Detectives had this witness view more than five hundred photos of employees, as well as watch all employees coming in and out of the hotel for more than twenty-four hours. But the witness became totally burned out and could make no identifications at all. Giorgio had come into the case late and had not been to the original crime scene. The physical evidence had yielded nothing, but Giorgio decided to look it over again. Ms. Williams had six cards in her purse with her name on them. In the upper right-hand corner of one of them, Giorgio thought he noticed a mark, and it turned out to be a latent thumbprint. The Latent Print Unit compared the lift from the card with everyone in the files who even remotely fit the witness's description of a short Mexican-looking man, more than two thousand possibilities in all, but their search yielded no matches.

As Giorgio and Juan Medina were leaving the hotel after another day of investigation, they saw two uniformed police officers near the lobby candy shop. The store had just been robbed. The clerk gave a description of a tall and husky Hispanic robber, about 6 feet, 3 inches tall, between 250 and 260 pounds. Giorgio called the Manhattan Robbery squad and asked the detectives there for lists of robberies in the midtown area, with a special focus on hotels. It turned out that there had been thirty-six robberies in the midtown area in smoke shops, candy stores, and clothing stores, many of which were connected with hotels. The victims were always women, either a woman working alone or two women working at night.

Giorgio and Medina had the Latent Print Unit run the thumbprint found on Williams's card against prints from the robbery squad's voluminous list of robbers who worked Manhattan. No match. But when the unit ran the latent thumbprint against the prints of the borough's known burglars, it was able to match the latent thumbprint to an inked thumbprint of one Juan Robles, 21 years old, 6 feet, 4 inches tall, weighing 245 pounds. Working with a photo-array containing Robles's mug shot, Giorgio and Medina identified Robles as the culprit in twenty-nine of the thirty-six midtown-Manhattan robberies. Evidently, Robles had graduated from burglary to robbery.

When Giorgio and Medina searched for Robles, they discovered that their quarry had other problems. He was out on bail, charged with burglary and attempted murder for entering the apartment of a former girlfriend and then firing a shot at her new suitor. New York State court rulings of the time dictated that detectives could not interview Robles about the hotel robberies or Williams's murder without an attorney present because Robles already had an attorney for the pending, albeit unrelated, case. So the detectives bided their time. Indeed, Giorgio posed as a probation officer in order to monitor Robles's guilty plea to a knocked-down version of the burglary and attempted murder charges in return for ninety days at Rikers Island.

The day that Robles left Rikers Island, Giorgio, Medina, John Johnston, and several other detectives waited across the bridge for him. Giorgio stayed hidden in a car and told the other detectives to go straight up to Robles and cuff him as soon as he walked out of jail. When they did this, Robles, as expected, went berserk. Giorgio then magically appeared and asked the other detectives why they had cuffed Robles. He ordered them to release Robles. Robles fell for Giorgio's con and quickly focused all his attention on Giorgio, asking him why he had been picked up. Giorgio told Robles that he was a suspect in several robberies and took him over to the 13th precinct to interview him.

During the interview, Giorgio never mentioned the homicide of Kathleen Williams. Instead, he went over all the robberies in the area, asking Robles if he had ever been in midtown hotels. Giorgio mentioned the Plaza and Robles allowed that he had been there once for a drink. When asked about the Sheraton, Robles said that he had stayed there once. When Giorgio asked about the Waldorf-Astoria, Robles said that he had heard of it but he had never been there. Giorgio went on to other hotels, the Hyatt, the Regency, coming back again to the Waldorf-Astoria. Again, Robles said that he had heard of it but had never been there.

Giorgio sensed that the interview was approaching that fleeting moment when a suspect's need to share his guilt with another person outweighs the dangers of self-betrayal. All Giorgio had to do was give Robles an out. So Giorgio said to Robles: "Juan, I've never met you before today. I don't particularly like you. But I also don't dislike you. I'm just doing my job. But you've lied to me here. I've got a fingerprint that puts you at the Waldorf-Astoria. Something may have happened at the Waldorf-Astoria. Sometimes a guy goes somewhere with no intention of hurting anyone and something happens." And Robles blurted out: "I didn't mean to hurt that lady" as a prelude to a full confession.

But the problem with the case at hand, Giorgio and Murray agreed, was that Herman Nathaniel Myers was no longer a young man, like Leopoldo Siao-pao or Juan Robles. Myers had been around the block too many times to be conned so easily. Eighteen years of silence made it unlikely that Myers would blurt out his guilt in order to cleanse his soul for murdering Mrs. Diaz. Giorgio and Murray mused about the case. In and of itself, the fingerprint evidence inside Mrs. Diaz's apartment was insufficient to convict Myers. He could offer any number of reasons for the presence of his prints there. Myers might, perhaps, argue that he and Mrs. Diaz were friends or even secret lovers who met at her home. Or he might argue that his job took him into her apartment. Or he might admit that he had entered her apartment illegally but claim he had left before she came home from work and knew nothing about any harm done to her.

Rapists rarely acknowledge their depredations. What Giorgio and Murray needed from Myers was statements that removed plausible excuses or reasonable justifications for his presence in Mrs. Diaz's apartment—statements that so contradicted the evidence of the fingerprints that they demonstrated "consciousness of guilt" to a jury. But the logistics of getting any statements at all were tricky. What strategy should Giorgio use to get Myers to talk with him in the first place? Under what cover should Giorgio conduct the conversation? At what point should Myers be read his rights? When should a New York court issue a warrant for Myers's arrest, a move that confers an absolute right to an attorney whether or not a suspect waives his right to have an attorney present during an interrogation with a police officer? In short, the groundwork of truth here, in the minimalist sense of establishing who raped and killed Mrs. Diaz, would have to be deception.

Giorgio needed to learn more about Mrs. Diaz in order to lay a snare for Myers. Working with Detective Louie Bauza, a barrel-chested young man whose buoyant, sunny disposition brightens even dark days, Giorgio tracked down Mrs. Diaz's four children, two sons and two daughters. On May 21, 1991, Giorgio and Bauza talked at length with the Diaz family.

Giorgio showed Mrs. Diaz's children a picture of Herman Myers; all stated that they neither knew nor recognized him. He asked them whether their mother smoked, drank, or had any boyfriends. Mrs. Diaz's children laughed sadly and told Giorgio and Bauza about their mother. She was an old-fashioned lady, they said. She had never smoked in her life. Neither did she drink, although she kept a bottle of liquor on hand for her sons' visits. She worked hard at her sewing job, all day, every day. She went to church three or four times a week, where she saw other older women who were her friends. She lived for her children and grandchildren. Her sons and daughters used to kid her about dating. They told her that she was still an attractive woman and they suggested that she see some men. But, her children said, Mrs. Diaz told them that she was too old for that sort of thing. She was content with her life as it was, whatever its difficulties.

Mrs. Diaz's children said that their mother's apartment building had had several robberies and such violence terrified her. Still, Mrs. Diaz did not trust banks. Despite her children's urgings, she kept money at home, as well as the jewelry that they all regularly gave her as presents. As a compromise, she always kept the gate guarding the fire-escape window in her apartment closed and locked, with the window itself firmly shut. They said that it was clear to them that the gate had been forced open. They were puzzled that some of Giorgio's questions suggested that he was pursuing the possibility that their mother had invited her assailant into her house. But Giorgio reassured them that he was just examining the situation from every angle in order to catch their mother's assailant.

Back at the station house, Giorgio made further preparations for his coming meeting with Herman Myers. At his request, the Diaz children had given him a photograph of Mrs. Diaz taken shortly before her death. Giorgio paired this with four other shots of Latinas, all about Mrs. Diaz's age. He also randomly selected two photographs of a criminal from the 34th precinct's mug books. He packaged these with the fingerprint report from the Latent Print Unit and with Dermody's investigative reports in the squad's brown manila folder marked Homicide/Rape. In the next several days, he conferred frequently with ADA Warren Murray, who drew up a felony complaint on Myers but did not docket it. Giorgio gave the complaint to Detective Bauza. Bauza was to take the complaint to a New York State Supreme Court justice for authorization and legal transformation into an arrest warrant once Giorgio contacted him from South Carolina. Giorgio also called the detective squad in North Charleston and asked them to invite Myers into the station for questioning on May 30, 1991. Myers was, Giorgio insisted to the South Carolina detectives, to be allowed to come in voluntarily.

Giorgio finally met Myers at 0900 hours on May 30, 1991. Myers had driven to the station house in North Charleston in his sister's car at the request of investigators Alphonso Scott and James Smalls, who had visited Myers earlier that morning. Giorgio, accompanied by Detective Angel Morales from the 34th squad, greeted Myers at the door. Myers stood 5 feet, 10 inches tall and weighed about 160 pounds. He had shrewd, quick eyes and a hard, calculating look. When Giorgio informed Myers that he and Morales were from New York City, Myers refused to shake their hands. Instead, he asked the detectives: "What's this all about?"

Giorgio began to spin the web. He told Myers that a man had named him as an accomplice in several burglaries in New York and that the detectives wanted to talk to Myers about those crimes. Giorgio showed Myers the photographs that he had randomly selected from the precinct's mug books, identifying the man in the photographs as "Leroy Jones." Myers said that he knew no one named Leroy Jones, that he did not know the person depicted in the two photographs, and that he had not even been to New York in several years. He paused for a few moments, and then he agreed to speak with the detectives.

Giorgio told Myers that because of the burglary accusations he had to advise him of his rights. After the reading, Myers waived his rights and agreed to answer Giorgio's questions without an attorney present. Giorgio once again mentioned the several burglaries in New York City in which Myers had supposedly been implicated by Leroy Jones. In connection with those crimes, Giorgio showed Myers the five photographs of Latinas that he had prepared, including a picture of Mrs. Diaz in life. Giorgio said that these women had been present in their apartments when the burglaries took place and that the burglars had tied them up. Myers said: "These are Latino women." Giorgio acknowledged that they were Latinas and asked Myers if he knew or recognized any of the women depicted in the five photographs. Myers declared emphatically that he did not. Giorgio asked him to view the photographs again. Again, Myers denied knowing or ever seeing any of the women. Giorgio asked Myers to initial the back of each photograph to acknowledge his statement.

Giorgio then asked Myers about his several arrests for burglary in New York between 1980 and 1983. Myers deflected the query, saying that he "scrapped" for a living, picking up metal objects on the streets and selling them to junkyards. He stressed that he never went inside apartments either by invitation or on his own to retrieve metal. Giorgio asked Myers in what parts of New York City he plied his trade. Myers said pointedly that he

never went above 190th Street in Manhattan to do scrapping because there were no metals on the street in that part of uptown.
Myers did admit that he was quite familiar with the area around
190th Street and Audubon Avenue since he used to hang out at
George Washington High School.

Giorgio locked Myers into his story. He again asked Myers if
he had any friends or relatives above 190th Street. Myers said
that he remembered a boyhood friend whose family lived on
190th Street between Saint Nicholas and Audubon avenues. But
Myers said again that he had never visited or entered any houses
above 190th Street. Indeed, Myers said, the only time he had ever
been above 190th Street was when he attended P.S. 52 at Academy Street and Vermilyea Avenue as a youngster. At that point,
Giorgio wrote a statement incorporating all the elements of
Myers's story. Myers, Giorgio, and Morales signed it. The investigators then took a break and got Myers some coffee. The three
men chitchatted informally for a while. Giorgio noted that
Myers smoked Salem cigarettes.

Giorgio resumed the interview. He showed Myers a copy of
Detective Thomas Montero's fingerprint report identifying the
fingerprints on the brandy glass and on the lower window frame
in Guadalupe Diaz's apartment at 611 West 204th Street as belonging to Herman Nathaniel Myers. Myers insisted that he
knew nothing whatsoever about a burglary in that location. He
went on to insist that he had never been at 611 West 204th Street.
He said that he did not know, nor had he ever known, Guadalupe Diaz. When asked again, Myers repeated that he had never
been at 611 West 204th Street. Again, all three men signed this
statement.

Giorgio left Myers with Detective Morales in the interview
room and telephoned Warren Murray. The two men reviewed the

status of the interrogation and Giorgio informed Murray about South Carolina's simple extradition procedure, which required only a suspect's waiver, not an appearance before a magistrate. But when Giorgio returned to the interview room, Myers suddenly said: "I thought you said this was about a burglary. How come the brown folder says Homicide/Rape?" Morales shrugged apologetically at Giorgio. Apparently, right after Giorgio had left the interview room, Myers had reached across the table and yanked the folder toward himself in order to read the writing on its cover.

Giorgio acknowledged that he was investigating a homicide/rape case that had probably started as a burglary. He suggested to Myers that perhaps he was only in Mrs. Diaz's apartment for the burglary and that an accomplice had committed the homicide and rape. Myers said: "Write that down." Giorgio noted that, if this indeed had occurred, he had to have the name of the accomplice. Myers said nothing. Giorgio wrote out the scenario as he had described it, complete with Myers's statement, but Myers refused to sign it.

The conversation meandered on as Giorgio continued to construct plausible accounts for Myers, hoping to find one that Myers might embrace. He suggested that sometimes burglaries get out of hand and unintentionally escalate into murder. But Myers did not respond. Giorgio then said that if Myers told the whole truth about the crime, the district attorney could extend him some consideration. Myers responded that judges impose sentences, not district attorneys. But then Myers said: "If there's one thing a person doesn't want to go to jail for it's R.A.P.E. [spelling the word out]. The inmates treat you different. Maybe I could go into protective custody." Giorgio pressed the advantage. He suggested that someone might be charged with burglary, homicide, and rape, but plead guilty to homicide to cover all the charges so that when he went to prison his "jacket" would only read "homicide." And Myers said: "You can do that?" Myers said that he would think over what Giorgio had said about possible consideration from a district attorney.

But then Myers shook his head and said that he could not explain how his fingerprints got into Mrs. Diaz's apartment. Indeed, he said, the police had obviously made a mistake because they could not be his fingerprints. A bit later, he said nonchalantly that he thought that his sister, "Carol Blake," lived in an apartment located north of 190th Street at one time. Abruptly, Myers asked if he could leave the station house and come back the next day to finish up the interview. But Giorgio said that they had to finish. Giorgio left the interview room again, called Louie Bauza in New York, and told him to take the already-drawn-up felony complaint before a judge. Bauza faxed the subsequently authorized arrest warrant back to Giorgio within a half hour.

Giorgio reentered the interview room and said that he wanted Myers to return to New York City with him the following day. Myers asked who would pay for such a trip? Giorgio produced a one-way ticket to LaGuardia Airport and placed Myers under arrest for the murder of Guadalupe Diaz. ADA Warren Murray presented *People v. Myers* before a grand jury on June 3, 1991. The same day, the grand jury returned a true bill indicting Myers of both murder and felony murder in the death of Mrs. Diaz.

———

Giorgio immediately began preparations for trial. He called Edward Dermody and informed him about Myers's arrest. Dermody, who had retired from the NYPD in 1976 haunted by Guadalupe Diaz's savage death and frustrated by his dead-end investigation into it, was delighted to hear the news and expressed his eagerness to help in trial preparations in any way possible. Then, out of the blue, Dermody asked Giorgio: "What about Nikki Sterling? Was she involved? I've always had a funny feeling about that girl." Giorgio told Dermody that he intended to look into Sterling.

A few days later, Herman Myers called Giorgio from Rikers Island. Giorgio asked Myers if he had rethought their discussion about getting consideration from the district attorney in return for telling the truth. But Myers said that he had no idea what Giorgio was talking about. He simply wanted to make sure that one of his family members had stopped by to pick up a coat and books that he had left in the North Charleston police station house when he was arrested. Giorgio asked him whether he knew a woman by the name of Nikki Sterling. Myers paused and admitted that he did. Giorgio pointed out that Sterling used to live in the apartment directly below Guadalupe Diaz's apartment at 611 West 204th Street. Myers said nothing. But later in the conversation he said that Nikki Sterling was, at one time, his girlfriend.

On June 14, 1991, Giorgio, together with Detective Tim Muldoon, tracked down Nikki Sterling, who was now working in the research division of a major corporation. Sterling warily agreed to talk with the two detectives in the company of one of her supervisors. Giorgio showed Sterling a photograph and asked if she knew the man depicted by it. Sterling said that she recognized him but could not remember his name. Giorgio informed her that the photograph was a likeness of Herman Myers, who was under arrest for the murder of a woman who lived directly above Sterling at 611 West 204th Street. Sterling said that Herman's sister, Carol Blake, lived at 613 West 204th Street. Carol had a key to Sterling's apartment. When Sterling stayed overnight in the Bronx with her boyfriend, Larry Tucker, Carol looked after her cats. Sterling recalled returning home after spending one night in the Bronx and finding her apartment dirty and her cats starving. She said that it was possible that Herman came to her apartment with Carol.

When Giorgio asked if Herman had ever been Sterling's boyfriend, Sterling vehemently denied any relationship of the sort. Giorgio wrote out Sterling's statement and she signed it. But then, as Sterling's supervisor went to photocopy Sterling's signed statement at Giorgio's request, Sterling grabbed it from her hands and tore her signature off the bottom of the legal-sized page. At that point, Sterling became indignant and belligerent and demanded to be left alone. Giorgio and Muldoon looked at one another in amazement, and Giorgio determined to check out her story.

Giorgio tracked down Carol Blake, Herman Myers's sister, in South Carolina. Blake's memory of events cast doubt on Sterling's story. Blake asserted that she had been good friends with Nikki Sterling and visited Nikki's house quite often to take care of her cats. She said that her brother Herman took at least five trips with her to Sterling's apartment to feed the cats. She remembered that Nikki's apartment was always stuffy and needed air. So she and her brother opened the window onto the fire escape. On those occasions, Herman sat on the windowsill with one leg in the apartment and the other on the fire escape. Blake also remembered visiting Nikki's apartment with her brother when Nikki was home. On those occasions, Nikki and Herman sat on the fire escape together and played music. Carol remembered that Herman had told her one day that he and Nikki were lovers.

She also stated that she, Nikki, and Herman were smoking a lot of pot in those days. Giorgio judged from her appearance and demeanor that Blake still used drugs, making her a poor witness at trial. Still, her acknowledgment that she and Myers had together visited the building where Mrs. Diaz lived, not once but several times, further undercut Myers's story. Giorgio also reluctantly came to the conclusion that he had no grounds for proceeding against Nikki Sterling because the law does not prohibit citizens from concealing their knowledge of heinous crimes. She became one of the many loose ends of police work.

In the meantime, Herman Myers became further entangled in the web of his own denials. Detective Angel Morales had finger-printed Myers on May 31, 1991, when he and Giorgio brought Myers back to New York City in custody. Morales also printed Myers's palms, routine procedure in major cases, and sent all the prints to the Latent Print Unit for further comparison of Myers's prints against those found in Diaz's apartment. On June 17, 1991, Detective Montero issued a report matching Myers's left palm to the latent palm print lifted from one side of the blade of the knife found in Mrs. Diaz's bedroom, and Myers's No. 10 finger (left little finger) to the fingerprint lifted from the other side of the knife's blade. Now eight prints put Myers in Mrs. Diaz's apartment. His denials of ever knowing her or of ever be-ing above 190th Street negated possible legitimate explanations of the traces of his presence in her apartment. To Giorgio and Murray, at least, Myers's denial, set against compelling evidence to the contrary, strongly indicated a consciousness of guilt.

In the end, *People v. Myers* came down to the fingerprints, to Myers's denials to Giorgio, and to ADA Warren Murray's skill in arguing a case based on circumstantial evidence. At trial, Rafael Diaz described the years-long regularity of his and his mother's nightly routine and how the evening of her murder was no different from any other of hundreds of nights, thus narrowing the timeframe of the attack on her to less than 45 minutes. Mr. Diaz also testified to his mother's personal habits, particularly her abstention from tobacco and alcohol. Then retired Detective Edward Meagher, formerly with the NYPD's Crime Scene Unit, flew in from his new home in a far western state to testify about his and his partner's work at Mrs. Diaz's apartment on March 5, 1974. He described in detail the discovery, collection, and preservation of the several latent prints in the apartment, clearly establishing the chain of custody of the fingerprint evidence over the years.

Detective Ronald Alongis from the Latent Print Unit testified about the uniqueness of fingerprints, even those of identical twins and those from different fingers of the same person. He stressed the "scientific" character of fingerprint analysis. He described the automated SAFIS system and its selection of Herman Myers as a possible culprit in the Diaz homicide. Murray then had Alongis focus on the Latent Print Unit's manual comparison of Herman Myers's inked prints against the latent prints found at the Diaz crime scene. To illustrate his unit's work, Alongis used a large blow-up of the inked print of Herman Myers's No. 3 finger (right middle finger) and a blow-up of the latent No. 3 finger print lifted from one side of the brandy glass in Mrs. Diaz's bedroom. He pointed out twelve points of comparison between the inked and latent prints that, he argued, matched exactly. He stressed that there were many more "matching points" but that he had stopped counting at twenty and plotted only a dozen for courtroom illustration. He concluded that, based on his expert analysis, eight latent prints found at the Diaz homicide scene matched corresponding inked prints of Herman Myers.[13]

Detective Gennaro Giorgio, dressed to the nines and with his customary aplomb, testified about his cat-and-mouse interviews and conversations with Herman Myers. Giorgio's rules for interrogation are simple and straightforward.

- Know the case from beginning to end, down to the smallest detail. Specific knowledge is the key to successful interrogation.
- Listen patiently to suspects. Never confront them in an accusatory way.
- At first, write nothing, taking in everything a suspect says without challenge. Then go back over the suspect's statement, writing it out carefully.
- Read it back to the suspect and have him sign it. Lock suspects into their statements, whether true or false.
- Then key in on inconsistencies in the statements or on aspects of the statements one knows independently to be false.
- Make careful notes of casual conversations with suspects. Sometimes suspects blurt out damning statements spontaneously at off-guard moments.
- Observe the suspect's demeanor carefully during the interview, especially when he is telling known lies. Make a mental note of any behavioral patterns that regularly accompany the known lies, such as facial tics, hand rubbing, head touching, turning away, licking lips, or displays of anger.
- Point out the lies without, at first, letting the suspect know how one knows he is lying. Ask the suspect why he is lying.
- Then point out some piece of actual evidence that contra-

dicts his story. Insistently but quietly demand an explanation for the discrepancy. If none is forthcoming, move on to the next discrepancy.

- If one has no tangible evidence on hand, use dodges, ruses, or tricks to elicit statements from suspects.
- At a certain point, offer the suspect an out—a plausible explanation, justification, or excuse for his depredation, suppressing all personal moral revulsion and clearly indicating that one understands and indeed empathizes with such a motive or account.

In short, let suspects convict themselves with their own words. Denials of guilt are as useful legally as admissions or confessions if one has independent evidence to undermine the denials and thus the suspect's credibility before a jury. By the time Giorgio had finished speaking with Herman Myers, Myers had become entangled in a web of actual dark deeds, denials, half-truths, and bold lies.

Nevertheless, the case against Myers was based largely on circumstantial rather than direct evidence. The law makes no distinction between direct evidence (such as eyewitness testimony, physical evidence, or admissions or confessions by suspects) and circumstantial evidence (which requires a disinterested observer to make a logical inference from facts and circumstances surrounding an event). But as all New York prosecutors and defense attorneys know, it is no easy task to get a jury drawn from the city's vastly contrasting social strata to agree on what constitutes a proper and legitimate inference even from compelling circumstantial details. Prosecutors and judges use homey analogies to describe the nature of circumstantial evidence and how one draws conclusions from it.

In prosecuting Herman Myers, for instance, ADA Warren Murray argued: "Suppose it were important to prove whether or not it had been raining. Assume a witness testified it was cloudy, but that it wasn't raining. The witness continued to testify that he walked into a store for several minutes [on] some business, and when he emerged onto the sidewalk, he saw that the pavement was wet, that small puddles had formed, and that several passersby on the sidewalk had what appeared to be water splotches on their clothing. [The witness] adds that it was not raining when he came out of the store. Based on that testimony, you can reasonably conclude that it had rained while the witness was in the store."

"In making [such] a determination based on circumstantial evidence, you must do two things. First, you must apply the usual tests of credibility to determine whether or not the witness told the truth about what [he] saw and heard. If you accept all or part of [his] testimony as truthful, then you must perform a second function. You must use your powers of reasoning and logic to determine whether those facts, which you accept as truthful, support the inference at issue [that it was raining while the witness was in the store]."

In *People v. Myers,* Murray went on to argue, if one believed Rafael Diaz's portrait of his mother and of his brief glimpse of her as she waved to him from her kitchen on her final evening, retired Detective Edward Meagher's testimony about the gathering and preservation of the latent fingerprint evidence in Mrs. Diaz's apartment, Detective Ronald Alongis's testimony about the science of fingerprinting in general and the matches between the latent prints found in Mrs. Diaz's apartment, Detective Gennaro Giorgio's account of Herman Myers's several statements, particularly his denials of ever seeing Mrs. Diaz or of ever being north of 190th Street in Manhattan, but also Myers's later casual statements faithfully recorded by Giorgio, then, logically, one could, indeed one had to, infer that Herman Myers was in Mrs. Diaz's apartment on March 5, 1974, for no good purpose. He actually murdered her or, at the very least, participated in her murder.

The trial jury agreed and, on October 15, 1992, convicted Herman Myers of both murder and felony murder. Myers appealed, citing error in connection with the SAFIS system and with the expert testimony on the fingerprint issue presented at trial. The Appellate Division rejected Myers's arguments and upheld his life sentence. The justices went out of their way to remark that Myers's arrest for the brutal murder of Mrs. Diaz after seventeen years "is a testimonial to the unrelenting perseverance of the New York City Police Department."[14]

Not long after Herman Myers's conviction, fingerprint examiners' claims to the scientific validity of their craft came under serious assault. First, actual and alleged evidence-planting by police undercut public confidence in the objective character of "evidence."[15] Second, tests sponsored by the International Association for Identification in 1995 demonstrated disturbing variability in fingerprint examiners' judgments on whether sample prints match.[16] Third, British examiners dropped their sixteen-point standard in January 2001, thus conceding that matching fingerprints depends not on a fixed set of measures universally applicable in all cases but on the interpretive skill of examiners in comparing prints. Fourth, nineteenth-century models of science have foundered on the rocks of the social constructionists' insistence that all knowledge is arbitrary, including knowledge derived through scientific procedures.[17] Fifth, the 1993 *Daubert v. Merrell Dow Pharmaceuticals, Inc.* decision by the United States Supreme Court revised the rule governing the admissibility of expert scientific testimony, thereby subjecting the admissibility of fingerprint evidence to multiple court challenges.[18]

In *Daubert*, the Supreme Court rejected a practice that most courts had followed for over half a century—namely, admitting expert scientific opinion only if it was based on a scientific technique that was "generally accepted" as reliable within the relevant field of scientific expertise.[19] The Supreme Court held that, in determining the admissibility of testimony by scientific experts, a trial court may consider among other unspecified factors: (1) whether the "theory or technique" is one that "can be (and has been) tested"; (2) whether "the theory or technique has been subjected to peer review and publication"; (3) "the known or potential rate of error [of the technique] and the existence and maintenance of standards controlling the technique's operation"; and (4) the extent of the acceptance of the technique within the relevant "scientific community."[20] In 1999, in *Kumho Tire Co., Ltd. v. Carmichael,* the Court held that the *Daubert* rule applies not only to scientific testimony but to all expert testimony, including that relating to technical knowledge.[21]

Between 1999 and 2001 fingerprint evidence survived fourteen separate court challenges under these new standards.[22] In every case, the courts denied demands by the defense (or the plaintiff in one case) to exclude fingerprint evidence and testimony. In two cases *(United States v. Mitchell* and *State of Georgia v. McGee)* the courts took judicial notice of the uniqueness and permanence of human friction ridges.[23] In *McGee* the court concluded "that despite numerous legal challenges in state and federal courts, the courts have held that fingerprint identification has reached a scientific stage of verifiable certainty" and that "fingerprint identification is reliable evidence."

Then, on January 7, 2002, the Federal District Court for the Eastern District of Pennsylvania found that the "evaluation" stage of the ACE-V procedure for fingerprint examination is inherently subjective and did not meet the *Daubert* standard for scientific or technical knowledge. Judge Louis H. Pollak took judicial notice of the uniqueness and permanence of fingerprints but ordered that fingerprint experts could provide only "descriptive" and not "judgmental" accounts of their analysis of prints in court. In short, they could not declare in court that, in their opinion, a latent print found at a crime scene *matched* an inked print taken in custody.

The defense bar rejoiced amid predictions that Judge Pollak's decision sounded the death knell for fingerprinting, the premier forensic investigative technique of the twentieth century. Indeed, the decision prompted several other challenges to fingerprint evidence. Observers also predicted that the decision would affect other forensic techniques such as ballistics, hair and fiber comparisons, and handwriting analysis. Some famous defense attorneys, such as Peter Neufeld and Barry Scheck, opined that only after the National Institute of Justice funded exhaustive academic studies of the "verification and validation" of fingerprint identification should it be admissible in court.[24]

But the United States Justice Department asked the Court to reconsider the decision. On March 13, 2002, after hearing live witnesses in his own courtroom, Judge Pollak reversed his prior decision and allowed the government to present FBI fingerprint examiners as witnesses at trial and to make evaluative judgments based on their analysis of the prints at issue.[25] The court stated that fingerprinting is a "technical discipline" instead of a science but that *Daubert* does not require expert testimony to be scientific testimony. Exhaustive studies should indeed be conducted to establish more firmly the "verification and validation" of the technical discipline of fingerprinting. However, to postpone judicial use of a forensic device that helps separate goats from sheep until knowledge is absolutely conclusive—an impossibility, by definition, in scientific matters—"would be to make the best the enemy of the good."[26]

9

A DEATH IN THE FIELD

The junction of Chauncey Street and Central Avenue exploded in pandemonium in the warm early evening of Friday, September 21, 1984. Police radio cars, red and blue lights flashing, flooded the area. Scores of uniformed police from the 83rd precinct roamed the streets, trying to find anyone who might talk to them about what had happened. But most neighborhood residents stayed indoors, peering at the wild scene from behind drawn shades. The eleven New York City Transit Police officers in the search party, their work done, clustered to one side of the scene, joined by several transit police detectives. Several of the men wept openly. New York City Police Department Commissioner Benjamin Ward, accompanied by New York City Mayor Ed Koch and an entourage of white-shirted top police brass, waded through the ragged six-foot-high horseweeds in the vacant lot below the Conrail lines that abut the Most Holy Trinity Cemetery, demanding explanations from NYCTP Commissioner James Meehan.[1] Members of the print and broadcast news media had already caught wind of the event on their police scanners and had begun to descend on Brooklyn's Bushwick section, adjacent to East New York.

Vincent Carrera of the NYCTP's Major Case Bureau's squad 12, known as the Hollywood Squares because of their flamboyant, unceremonious raucousness, stood ashen-faced amidst the tall weeds, his Falstaff-like figure shrunken, his booming voice muted, his magnetic personal intensity stilled. Detective Thomas Burke, one of Carrera's partners, pointed out the pair of handcuffs at the feet of the body, as well as the distinctive three-inch-long safety pin that fastened the leather case containing PO Irma Lozada's shield to her purse. Burke said: "Vinnie, look, she's got her cuffs out." Carrera flipped the leather case on the purse so that Lozada's shield 4721 showed clearly. Both detectives also noted that Lozada's 38-caliber standard-issue five-shot revolver was missing.

Carrera spotted flashes coming from the roof of the stark, gray-cement cemetery administration building at 675 Central Avenue at the edge of the lot. Gesticulating wildly, he ordered several of the uniformed cops to climb the fire escape and grab the photographer. The cops clambered up the rickety old iron ladder and apprehended the newsman. Carrera ordered the cops to throw down the newsman's camera. When the camera crashed to the ground, Carrera smashed what was left after its two-story fall. Carrera then ordered the cops to throw the man down too. When the officers hesitated, Carrera yelled even louder: "Throw him down! I want him on the ground. I want other photographers to take his picture when he's lying dead. And I want his wife to see him like that in the paper tomorrow." But the cops brought the trembling newsman down the fire escape. The newsman apologized. Carrera acknowledged that he had destroyed the camera and film, but the newsman, happy to be in one piece, made a quick exit without complaint. The news media later pegged Carrera as the wildest man on the scene, but one of Carrera's many mottos was: "Sometimes you gotta get wild."

Carrera grabbed his boss, Detective Sergeant Louie Cosentino, the sergeant's sergeant, a man, in Tommy Burke's words, "from the warrior class . . . who thought he could control anything by the force of his own personality." Years later, on October 2, 1994, Cosentino was left for dead on a Brooklyn street, savagely beaten by nine unruly youths when he tried to quiet them, an attack that he miraculously survived. Because all catching orders go out the window in a cop killing, Carrera asked Cosentino who was catching this case. Cosentino said: "You're the big fuckin hero, you do it."

Part of Carrera did not want the case. The trail was already as cold as the young woman's body. Even minutes after a homicide, witnesses go home, talk with friends or family, and decide not to get involved. Were this case left unsolved, Carrera knew that his eight-year career as a detective, indeed his twenty-year career as a police officer, would be finished. But in his heart of hearts, Carrera did want the case because of his rage at what the murderer had done, and because all detectives worth their salt want big cases, whatever their risks. He turned back to his fallen 5 foot 6 inch, 120-pound sister officer, curled into a fetal position on her left side, her blue-and-white-striped white polo jersey soaking up the pool of blood that streamed from the gunshot wound on the left side of her face and the lethal bullet crater in the back of her head, and he said: "Sweetheart, I never met you, but, by Jesus Christ, I'm gonna find out who did this to you if it's the last thing I ever do."

Vincent Carrera had always wanted to be a police detective, ever since he witnessed the authority that the "bulls" exercised over his bookie grandfather, even in the Joey Gallo–controlled Red Hook section of south Brooklyn where he grew up. While working at a supermarket on Knickerbocker Avenue in the 83rd precinct, he regularly needed police protection to make his twice-a-day receipt deliveries to the bank. One day, PO Joe Picciano, who was driving him to the bank, asked Carrera: "Whatcha gonna do with your life, kid? Gonna be a stock boy all your life? Here's an application to become a cop. Fill it out and send it in." Carrera took Picciano's advice and sent in the application. Just a few years later, on February 15, 1971, Carrera rushed up to the 41st precinct in the Bronx, the notorious Fort Apache, to console the widow of Detective Joe Picciano, killed in the line of duty while fingerprinting a prisoner who grabbed Picciano's weapon and shot him to death.

Carrera eventually got the call from the transit police in 1965, and for two years he led his district in arrests and the issuance of summonses. In 1967 he began a campaign for promotion to the detective division. But one of his supervisors urged that he not be promoted to detective because "Officer Carrera takes every incident that happens on his post as a personal affront." Carrera considered the remark to be the highest compliment a cop could get because, he said, he tried to treat every victim of a crime as if the victim were his own mother. All of his requests to become a detective went unheeded. Instead, in 1969 Carrera was assigned to the Booth Robbery squad, a stakeout team that interrupted in-progress robberies of token booth clerks, a crime that scourged the transit system throughout the 1970s. Carrera and his partners hid in tiny bathrooms or on the floors of clerks' booths or in porters' closets until, on planted radios, they heard the magic words: "Yo, this here a stickup." Then the officers raced out of hiding, guns drawn, ordering the culprits to freeze.

Carrera made several arrests during attempted token-booth robberies, receiving an extra vacation day for each arrest. He also shot two robbers in pitched gun battles underground, receiving commendations and two vacation days for each robber successfully shot. Later, he was assigned to the decoy squad, where, disguised as a Hasidim, thought by robbers to be easy prey, he tempted more than 1,450 muggers in just a few years to rip him off, all the while observed by brother officers ready to move in and make arrests.

Carrera finally made the detective squad in early 1978. He caught his first homicide in 1979, and the lessons that he learned from it served him for his entire career. The victim was a man going to work on a Friday afternoon in early September, carrying three bottles of wine and some snacks to celebrate his twenty-fifth year on his job with his office mates. The man was shot twice in the chest on the northbound platform of the Shepherd Avenue train station in the 75th precinct. The victim had no wallet when he was found, and there were no eyewitnesses to the murder itself. But a hardware store owner, who had worked on his street for twenty years, did come forward to say that he had seen two young men running past his store near Shepherd Avenue on Friday afternoon.

On Monday Carrera led a whole contingent of detectives from the NYPD's Brooklyn North Homicide squad and the transit detective division to canvass stores on Shepherd Avenue as well as passengers exiting the northbound train. On the platform, Carrera decided on the spot, much to the surprise of other detectives, to stop every passenger, not just a random sample of them, to make inquiries about what they might have seen on Friday afternoon. Two boys lingered behind the crowd after one train pulled away from the station. In exchange for the promise of a reward, they said that they had overheard two other boys at a numbers joint bragging that they had made some money by ripping off a guy in a train station.

With these witnesses' help, Carrera, together with Detective Artie Christiani, the NYPD police detective with whom he worked the case, grabbed "Smith," one of the two suspected assailants. Christiani immediately recognized Smith because he had interviewed him only the Monday before the Friday slaying while Smith was in custody for yet another robbery. But a New York State Criminal Court judge had set Smith free on Wednesday two hours before the legal deadline for obtaining an indictment against him, and just half an hour before the grand jury's indictment was issued. The public uproar over the judge's decision and the murder that followed in its wake eventually led to a lengthening of the time allowed for district attorneys to obtain indictments after arrests.

Smith immediately admitted being "down for the robbery," but he put the gun in "Jones's" hands. When the detectives nabbed Jones that night, they took him back to the station house and confronted him with Smith's statement. Jones laughed at the detectives, insisting that they had nothing and that Smith would never give him up. Carrera went to the room where Smith was being held and made a tape-recording of Smith betraying his friend. When they played the tape to Jones, he became furious and turned on Smith, insisting that it was, in fact, Smith who had shot the man at the Shepherd Avenue station after both of them had done the robbery. Of course, Jones saw himself as being in the clear for the shooting. Carrera became a life-long believer in the investigative value of tape recorders.

Carrera asked immediately where Irma Lozada's partner was, and police officers pointed over to a radio car. In the back seat sat PO Nat Giambalvo, weeping openly, bleeding from his mouth. Lozada had been a lovely, engaging, outgoing young woman, and an enthusiastic, dedicated officer, well-liked by all her colleagues during her three years with the force. Her death enraged her fellow officers. Giambalvo told Carrera that two cops who were especially close to Lozada had tuned him up—that is, thrashed him—at the crime scene because he, a 42-year-old, nineteen-year transit police department veteran, had been Lozada's partner for her first six weeks in the plainclothes anticrime unit, and veterans were supposed to take care of rookies. Carrera said to Giambalvo: "Nat, I got one question for you. Did you kill her? If you tell me that you had been in some kinda shooting and that it's possible that your friendly fire coulda killed her, then we stop all questioning and you talk to your attorney. Nat, did anything like that happen?" Giambalvo, a well-regarded street cop, said: "I did not kill her." Carrera responded: "Fine, then I don't wanna hear nothin about no attorney. I want to know what the fuck happened."

Giambalvo told Carrera the following story. He and Irma Lozada, whom everyone knew as Fran, had begun their plain-clothes duty at noon that Friday, patrolling the trains in East New York, looking out for chain- and purse-snatchers. After a few hours of work, Fran asked Giambalvo to accompany her to a shoe store in Bushwick to buy a new pair of sneakers. Fran's purchase took only a few minutes, but Giambalvo groused about her carrying the bag with her old shoes while they had work to do. Fran noted that they were due for a meal back at Transit District 33 at 2493 East Fulton Street at 1600 hours. She said she could leave the bag in her locker before they went back on duty.

At about 1545 hours the two officers boarded a southbound LL train heading toward East New York where the district office was located. They took seats in the middle of the train. At the Wilson Avenue station, Giambalvo got up and stuck his head out of the door, peering up and down the platform, a routine he had performed thousands of times over the years, usually without incident. Just as the train doors were closing, he saw a young man wearing a short-sleeved gray tee-shirt, blue jeans, and sneakers get out of the train hunched over, his back to Giambalvo. The man ran alongside the train and, as the doors closed, reached in with his right hand and came out with a yellow metal chain.

Giambalvo shouted to his partner: "Fran, chain snatch!!" Then he yelled "Police!!" to the conductor, who was operating the train from the same car. Giambalvo forced the train doors back open and ran after the thief. The young man bolted down one set of stairs in the elevated Wilson Avenue station. Giambalvo rushed down the other. But before Giambalvo reached the bottom of the stairs and ran into the mezzanine where the entrance turnstiles were, he heard a loud boom from the heavy wooden doors at the bottom of the stairwell, and then another loud boom from the wooden doors at the station's entrance. He jumped over the turnstiles, raced past the railroad clerk's token booth on his left, and hurried out the door, straight ahead to Moffat Street, which crosses Wilson Avenue at a right angle. But the thief was nowhere to be seen.

Across the street, on the south side of Wilson Avenue, Giambalvo, well known in the area as a cop after years of street work, saw two women sitting on the stoop of the first house on the south side of Wilson Avenue up from Moffat. He gestured toward them with open palms. One of the women pointed to her left, to Giambalvo's right, and then made a jumping motion with her hand. From years of experience, Giambalvo knew immediately how the thief had fled. He raced half a block to his right to the playground adjacent to the Audrey Johnson Day Care Center, heading north on Moffat Street toward Knickerbocker Avenue. Two groups of children were in the playground, shepherded by a young woman. Talking over the five-foot-high wall separating the center's property from the sidewalk, Giambalvo identified himself to the young woman as a police officer and asked her if she had seen a man jump over the wall. She said she had and that the youth had run through the playground, heading toward the Conrail tracks up the hill beyond the back fence.

Giambalvo then told Carrera that he glanced back and saw Lozada just reaching the corner of Wilson Avenue and Moffat Street. She had gotten stuck in the train. Only because the conductor saw her hand reaching out did he reopen the doors, allowing her to follow her partner. Giambalvo quickly explained the situation to her, telling her that, scores of times, he had been in chases exactly like this at the Wilson Avenue station. He told her that they had to split up. She should go down Moffat Street toward Central Avenue, then left on Central toward Chauncey Street where there was a vacant lot. The chain snatchers usually jumped over the wall of the day care center, then through a hole in the back fence, up to the Conrail train tracks. Sometimes they hid in one of six huge manholes alongside the Conrail tracks if they were being pursued. But in any case they usually came out down the hill into the vacant lot at Central and Chauncey. Giambalvo gave Fran the only radio between them, telling her to radio Central for assistance. The last he saw of her, she was running south on Moffat Street, radio in one hand, shoe bag in the other.

Giambalvo then went north, back to the day care center, jumped over the fence, raced across the playground, through the hole in the fence, and up the hill to the Conrail train tracks. He ran along the tracks until he reached the back of the Wilson Avenue subway station, looking all the while with his flashlight into the work-station manholes that lined the tracks, perfect hideouts for those on the run. Behind the station, Giambalvo told Carrera, a path juts off the tracks to the right heading along the edge of a jungle-like ditch, roamed by wild dogs and dotted with abandoned cars. Ten minutes later, Giambalvo climbed over a large mound of dirt and reached the vacant lot behind the cemetery administration building at Central Avenue and Chauncey Street. Nothing but tall horseweeds. No Fran.

Giambalvo was worried. He went around the building and then began walking west on Central Avenue. But he saw children riding their bikes, men and women sitting on stoops, listening to radios, and drinking beer. Nothing seemed out of the ordinary. He reached the corner of Central Avenue and Moffat Street. People were everywhere; nobody paid him any attention; nothing seemed amiss. When something happens on the street, people invariably disperse, avoiding any interaction with someone who is obviously a cop. Giambalvo breathed a sigh of relief, assured of Fran's safety.

Still, he walked quickly north on Moffat Street back to the Wilson Avenue train station. There, he asked the female railroad clerk if she had seen his partner, a white Spanish female with blond hair. The clerk said that she had seen a woman running out of the station right after he had but not seen her since then. At that moment Giambalvo spotted PO Thomas Birmingham, who had just come on patrol in the station's mezzanine. At 1610 hours, Giambalvo explained the situation to Birmingham and borrowed his radio. First, Giambalvo called Central and told the operator on duty that he could not find his partner. Then he tried to reach Lozada himself. "Shield 4721, Shield 4721, come in 4721." "Fran, where are you?" But he received no response. He called headquarters back to make sure that Central had asked cops in radio cars in the area if they had seen Lozada or if Lozada was in one of their cars. After a minute, Central responded that the inquiries had been made, but the responses were negative. Central began broadcasting over the transit police's two-frequency radio system: "Shield 4721, come in to Operations. Operations to Shield 4721, come in Shield 4721," transmissions that became burned into the memories of cops all across the city.

Giambalvo tried to reconstruct the situation. Perhaps Fran had missed the chain snatcher and thought that they were to meet back at District 33 for their meal. She needed, after all, to drop off her old shoes. Giambalvo gave Birmingham back his radio and hopped the southbound train to East New York, arriving at 1635 hours. At the station, he saw PO Michael Wasser, who originally had been scheduled to work with him and Lozada that day. He asked Wasser if he had seen Fran. But Wasser had not. He also ran into PO Gerry Howard, who asked "Where's Fran?" Giambalvo, now visibly upset, explained the situation. Then Giambalvo walked over to District 33, about five minutes away from the East New York train station. He reported the crisis to the desk officer and the executive officer on duty. Then he telephoned Central asking if Central had heard anything from

PO Lozada. Central now reported that a female officer was en route to District 33. Giambalvo sagged with relief. But when he inquired again a few minutes later, it turned out that the cop had been taken to the 83rd precinct. In any event, it was not Fran.

The desk lieutenant then told Giambalvo to take a marked car and do a search. Together with POs Wasser and Howard, who by that time had returned to the district, Giambalvo headed toward the Wilson Avenue station at about 1730 hours. There, three uniformed transit police officers joined them. While all the other officers scattered to different points, including different stations up and down the LL subway line, to question anyone they could find about the whereabouts of PO Lozada, Giambalvo retraced his steps the opposite way, heading south on Moffat Street, east on Central Avenue to Chauncey Street, through the high horseweeds in the vacant lot, over the mound of dirt, back along the path at the ditch's edge in the jungle, up onto the Conrail train tracks, and back along the tracks. No Fran anywhere. By now, it was past 1830 hours. More uniformed transit police officers arrived to join the search. Giambalvo was beginning to panic.

PO Theodore Shelto found her at 1900 hours. Shelto had gone through the vacant lot once, then up onto the Conrail train tracks back to the Wilson Avenue train station, then back toward Central Avenue through the jungle and its overhead foliage, and along the path next to the deep gully. He came again into the deep-weeded portion of the lot that he had already tramped through several times. That's when he saw her brandnew white sneakers, her blue-and-white jersey and jeans, and her bullet-shattered face. Several other officers had been searching in that same area, but they too had missed her body because of the horseweeds.

Giambalvo told Carrera that he had no idea how Fran got

killed. All he knew was what he had told Carrera, and that his partner was dead, and that his own career as a street police officer was finished, no matter how things turned out, because other cops would always see him as the guy who lost his partner.

While Carrera was talking to Giambalvo, the scene had gotten even wilder at Chauncey and Central. Carrera had called for Father John King, S.J., chaplain to the transit police. Father King, who was reached by radio at a social event on a boat on the Hudson River, arrived in Brooklyn by police department helicopter. He knelt on the ground beside Irma Lozada and administered the last rites to her lifeless body, crossing her eyes, ears, lips, nostrils, hands, and feet with the sacred oil, murmuring the ancient words—*per istam sanctam unctionem et suam piissimam misericordiam, indulgeat tibi Dominus quidquid deliquisti*—conveying her pardon for all the sins she had committed. Then he beseeched the Lord for her safe passage into the company of the blessed.[2]

The news media had arrived in full force, with lighting trucks flooding the now dusky street. Reporters were hammering the police bosses with questions, even though no one had any answers. Police officers were circulating all kinds of views, many of which originated from NYPD Commissioner Benjamin Ward's open arguments with NYCTP Commissioner James Meehan right in the field. Why did Giambalvo split up with his partner in a dangerous take-down situation? Why didn't both officers go to Central and Chauncey and just wait for the thief? No petty collar is worth risking an officer's life. The first rule of policing is to come home at the end of the day. Even assuming that they had to split up, why in God's name did Giambalvo send a rookie female cop to handle the confrontation with the thief?

Further, how could Giambalvo not have heard the shots that killed his partner if, in fact, he had been in the area searching for a chain snatcher? Maybe Lozada had gone shopping, and Giambalvo had gone gallivanting, or drinking, or both. Was Giambalvo making up a cock-and-bull story to cover up his own dereliction of duty? And what kind of two-bit equipment were transit police officers using? There had been plenty of reports of radio failures in the past few years. Maybe the girl cop would still be alive if the Metropolitan Transit Authority spent some money to equip officers properly.

The arguments captured the long-standing acrimony between the city police and the transit police: the NYPD, lumbering organizational goliath, proud to the point of arrogance, condescending toward the NYCTP, tunnel moles who spend their lives underground; the NYCTP, the "force on the move," innovative, aggressive, contemptuous of the NYPD's stodgy, hidebound methodologies, deeply resentful of Big Brother. Meehan argued back to Ward that the transit police, unlike the NYPD, patrolled alone in the subways. The transit police had no prohibition on single-person policing or apprehension. Therefore, splitting up did not carry the same taboo as it did with the city police. But Ward wanted to hear nothing about differing departmental practices. At the moment, what mattered was that the transit police had handed him, the mayor, and the city the worst possible nightmare: a policewoman killed in the line of duty.

Events of a few days later suggest the remarkable blame-time, finger-pointing tensions of the early evening of September 21. After a microphone-grabbing press conference that saw Ward and Meehan excoriate each other, much to the media's delight, Ward took drastic action. Intent on destroying Meehan's version of events and validating his own fixed view that Giambalvo was never in the crime-scene area, Ward ordered sixty "field associates"—officers in precincts who secretly provide information on their comrades to the Internal Affairs Bureau—to report to Chauncey Street and Central Avenue. The order created a sensation in the police department, and scores of police officers from all over the city came to Chauncey and Central to see the unmasking of the "fuckin rodents" who spy on their fellow cops.

At the scene, after the Emergency Service Unit had chopped down all the tall horseweeds nearly to the ground, Ward had thirty field associates posted at fifteen-foot intervals from the spot where Lozada was shot, up the hill to the Conrail tracks, and all the way back to the Wilson Avenue station. Then the ballistics unit fired a 38-caliber pistol just like Lozada's into a safe box. When Ward asked every officer who had heard the shot to raise his hand, officers all the way back to the Wilson Avenue station put their hands in the air. Ward then ordered the experiment repeated with another thirty field associates, with the same result. Choosing to ignore the sound-baffling effects of the six-foot-high horseweeds the day that Lozada was killed, Ward and many of his top brass became more convinced than ever that Giambalvo had not been at the scene.

On the night of the murder, several different officers reported these swirling disagreements to Carrera as the detective in charge. He had to test Giambalvo's story before he could proceed with the investigation. Carrera went back to the train station and interviewed the railroad clerk, Joan Cheatham. She told him that at 3:50 P.M. she heard a loud boom that caused her to look up. She saw a young man between eighteen and twenty years old, 5 feet, 11 inches tall, slim build, racing through her station. He jumped over the turnstiles and then rushed through the second set of doors and out of the station, then turned right on Moffat Street. Then, she said, she saw a man in a red-checkered shirt running through the station. The man paused a moment in front of her token booth, and she pointed out to the street. The man raced out of the station and turned right.

A moment later, she said, a young woman wearing blue jeans, a white polo shirt with blue stripes, and brand-new white sneakers came running out of the station. The woman ran up to her booth and turned the knob. Then she waved at the clerk and rushed out of the front door. The man in the checkered shirt returned to the front of the station and talked with the woman. Then both, she said, ran to the right, a detail that contradicted part of Giambalvo's story; but this is the kind of discrepancy that detectives expect to hear from different witnesses. One person's life-or-death detail is an insignificant moment in another person's day.

Then Cheatham told Carrera that the man in the checkered shirt returned about fifteen minutes later, out of breath. She asked if he had caught the guy. He said no. She said the man went into the train station and came out with a uniformed police officer. Both came up to her booth and asked to use her telephone. The man with a checkered shirt identified himself as a police officer and asked if she had seen a young woman running after him. Cheatham told him that she had. He asked if she had seen the woman since that time, but Cheatham said no.

Carrera went out to Wilson Avenue, which by then was teeming with people from the neighborhood as well as radio cars from the 83rd precinct and Transit District 33. He promptly found the two women sitting on the stoop of the first house on the south side of Wilson Avenue.[3] He took them inside the Wilson Avenue subway station, away from the prying eyes of the street, and talked to them. They insisted they had done nothing wrong, and Carrera assured them that they were not in trouble of any sort. They claimed that they had seen nothing whatsoever. Carrera pushed back, saying that a police officer claimed that he had seen them. Had they seen the cop? And when? They demanded not to be involved publicly. But Carrera told them that he had a cop killing on his hands and there was no way to promise them that. The two women had, of course, heard that a cop had been killed but were surprised to learn that it had anything to do with that afternoon's events. They told Carrera the following story.

They were waiting on the stoop for their daughters to come home on the school bus. They heard the door to the train station open with a boom, and then a slim-built young man wearing a gray or blue tank top, with a close-cropped haircut parted on the left side, raced out of the station and made a right turn on Moffat Street, heading toward the day care center. They insisted that they could not identify the young man. Then, they said, they saw a man dressed in a red-checkered shirt run out of the station. He motioned to them and they pointed to their left, his right, in the direction of the person fleeing. Suddenly, a young woman carrying a bag on her shoulder and a shopping bag joined him. They assumed that she was the victim of a robbery. Then the man pointed toward Central Avenue and the young woman raced in that direction while the man headed toward the day care center. The two women were dumbfounded to learn that the person they assumed was the robbery victim was in fact the dead cop.

From Carrera's perspective, Giambalvo was in the clear, whatever error of judgment the officer might have made in splitting up from a rookie partner. The railroad clerk and the two women from Wilson Avenue had corroborated his story in independent ways. When Carrera returned briefly to the crime scene, he informed Chief Meehan that there was no reason to doubt Giambalvo, adding kerosene to the already raging fire between Meehan and Ward. Meehan told Carrera that, on Carrera's assurance, he would back Giambalvo against Ward's tirades and accusations, but he warned Carrera that he better not be backing a ghost.

By then, with dark rapidly falling, radio cars from the 83rd, the 81st, the 75th, and the 79th precincts cordoned off a three-block-square crime scene, the largest in New York City history to that date. Cops sat all night in their cars to prevent anyone from entering the area and to tell local residents within the crime scene to stay in their houses. They were waiting for daylight when they could search the entire area for Lozada's gun, her shoe bag, and any evidence that might point to what had happened.

Carrera returned to the 83rd precinct station house at 179 Wilson Avenue, where he was greeted by Lieutenant Nicholas DeLouise, commander of the detective squad. DeLouise introduced Carrera to Detective Gaspar Cardi, who caught the case for the city police. Carrera remembered seeing Cardi at the crime scene, which, as it happens, Cardi had reached late because an earlier investigation had taken him to Queens. The two detectives shook hands and quietly promised each other not to allow their respective bosses' or organizations' animosity, or the news media circus, all of which had by then moved from the crime scene to the station house, to derail their investigation. Carrera briefed Cardi on his corroboration of Giambalvo's story and suggested that they focus their investigation on known chain snatchers in the area. Cardi agreed, but noted that Bushwick was notorious for larceny and robbery and that even such a focus meant scores of possible suspects.

In the meantime, DeLouise mentioned to Carrera that a female police officer wished to speak with him. But so did the news media, chiefs, captains, lieutenants, and other bosses from both police organizations, as well as transit and city detectives from commands all over the city waiting to be told what to do to investigate the death of one of their own. Adding to the clamor, an attractive young woman clad in a short skirt and halter top repeatedly called out to Carrera as he stormed around the station house barking orders. Hours passed as Carrera and Cardi made assignments and fended off demands. At one point, Commissioner Meehan grabbed Carrera and asked him what he had. When Carrera told him nothing, Meehan told him to get something in a hurry because his own career was on the line. DeLouise again asked Carrera to speak with the female police officer. And, finally, precisely because he had nothing and because the investigation was stymied at least until daylight when Crime Scene could begin its search of the area again, Carrera agreed to see her. The officer turned out to be the pretty girl in the scanty civilian clothes.

PO Deborah Barker had grown up in Bushwick, right on Moffat Street. She told Carrera and Cardi that an old friend from the neighborhood had approached her earlier that evening. Although her friend hated the police and had no direct knowledge of the cop slaying, he wanted Barker to know that it was likely that one of only three people did the crime. Carrera and Cardi immediately went with Barker to speak with her friend, who had been waiting for hours in the basement room used for detaining youths. The man was wary and hostile with the detectives. He asked about reward money should his information lead to the cop killer. Carrera said that Cop Shot, a private organization that works in conjunction with the Policeman's Benevolent Association, offers a $10,000 reward for information leading to arrests in shootings of officers. The man then told the detectives that three local guys did practically all the chain snatches and most of the robberies in the neighborhood and especially at the Wilson Street train station. Their street names were Kilo, Skeeter, and Jeter. Cardi went back to the squad room, sorted through the squad's nickname file, and came up with the three names, with photographs and addresses to boot. All three had records for larceny or robbery. He showed the photographs to the witness, who immediately verified that the photographs were indeed pictures of Kilo, Skeeter, and Jeter.

At midnight, Detectives Carrera and Cardi went back to the squad room, now packed with bosses from both of their jobs. They described what Barker's witness had told them and proposed picking up all three men immediately. But the bosses thought of scores of serious reasons for not doing so. The bosses worried particularly about judicial rulings then in effect in New York State that undercut long-assumed police prerogatives to enter homes even when invited to do so. And they worried about then-extant rulings that required suspects' attorneys in unrelated cases to be present during any police questioning. The debates among the bosses went on for more than two hours without being resolved. Finally, at 0200 hours, all the bosses left. And Carrera, Cardi, Cosentino, and Burke began the hunt.

The detectives first went to the address that they had for Skeeter, but it turned out to be Skeeter's girlfriend's mother's address. The mother sent the police to her daughter's house a few blocks away. When the police arrived there, both the girlfriend and Skeeter were home. The girlfriend readily told the detectives that she had been watching television around 4 P.M. the previous afternoon when Jeter knocked on her door asking for Skeeter, who was out. She said that Jeter, who was sweating and looked like he had just been running, showed her a chain and told her: "Skeeter shoulda been with me cuz I just ripped a chain off from the train station."

Skeeter then joined the conversation. He told the detectives that his girlfriend had mentioned her previous afternoon's conversation with Jeter to him. Later, he said, they watched the late-night news together. When they heard that there was a big reward for a cop killer, Skeeter and his girlfriend talked about dropping a dime on Jeter because the money looked so good.

Just at that moment, Kilo dropped by to see Skeeter and was dismayed to find his friend's house filled with police officers. The detectives took Kilo and Skeeter back to the 83rd precinct station house in separate cars and placed them in separate interview rooms.

The detectives spoke first to Kilo. Kilo had just gotten out of prison and was built like a tank. But when Carrera told him that he might go down for murder one—for killing a police officer—Kilo burst into tears, a hint of how fearsome New York State prisons are even to those who know them well. Kilo protested vehemently that he had done nothing wrong, had not been with anyone who did something wrong, and wanted nothing to do with the investigation into the killing. He refused to speak further with the officers.

But the interrogation of Skeeter went differently. Skeeter denied doing anything himself but said that, if the detectives gave him the advertised reward, he would tell them what he knew. After the requisite exchange of assurances, Skeeter said that Jeter had come to his house at 3 P.M. the previous afternoon and told Skeeter that it was his birthday and he wanted to make some money to celebrate. Skeeter told Jeter for the one-hundredth time that he did not like to rob on Fridays because an arrest meant a weekend in jail. So Jeter went off on his own to do a chain snatch. Later, when Skeeter heard on television that a late afternoon chain snatch had turned into a cop killing, and because his girlfriend had told him about Jeter's 4 P.M. visit to their apartment, Skeeter knew that Jeter had killed the cop.

Skeeter later vehemently denied making these statements to the police. Indeed, he claimed that a detective offered him $10,000 to put the murder on Jeter, an offer that he steadfastly refused because Jeter was his best friend.

But for the moment Skeeter and his girlfriend had provided detectives with evidence of a chain snatch by Jeter. Backed up by uniformed officers, the detectives hit Jeter's house at 0245 hours. His mother told the police that her son was at his girlfriend's apartment only a few blocks away. When the detectives arrived there at 0300 hours, the girlfriend opened the door and, from the six-month-old mug shot that he had in his hand, Carrera recognized Darryl Jeter sitting quietly in the living room. Carrera said that he wanted to talk with Jeter about anything that he might have seen or heard about the previous afternoon's events. Jeter agreed to come back to the station house.

When they got there, Carrera and Cardi ran into the boss of the homicide division of the Kings County District Attorney's office, accompanied by a young, black-mustachioed man wearing horn-rimmed glasses. Carrera thought that he was Groucho Marx reincarnated. But the boss introduced him to the detec-

tives as Assistant District Attorney Eric Seidel, that night's riding prosecutor. The detectives quickly briefed Seidel about their
interviews with Kilo and Skeeter and told him that they were
about to interview the prime suspect in Lozada's murder. Seidel
told the detectives that he was sticking around the station house
and to let him know when they wanted him to enter the case. Beginning at 0330 hours, after reading Jeter his *Miranda* warnings
and gaining his assent to being interviewed without a lawyer
present, the detectives asked Jeter to reconstruct his entire day.
Jeter told them the following story.

He said that he had slept at his mother's apartment on Thursday night. He got up at 6 A.M., showered, dressed, and went out
to meet his friend Kilo, who needed to go to the unemployment
office in Manhattan's Chelsea area. Jeter said that he and Kilo
stayed at the unemployment office a good part of the morning.
Then they went to Washington Square Park, where they hung
out for a few hours. They returned to Brooklyn around 1:30 P.M.,
and Kilo went to his house to drop off the papers from the unemployment office. In the meantime, Jeter ran into Skeeter, who
had been hired to rummage through some garbage to find registration papers stolen from the automobile of the local liquor-
store owner. Jeter joked around with Skeeter for a while when,
suddenly, he spotted his girlfriend talking with a guy whom he
disliked. He became angry at his girlfriend but decided not to
confront her at that time.

Friday was Jeter's birthday and he wanted to get some money
to celebrate. He suggested to Skeeter that they rob someone. But
Skeeter said no precisely because it was Friday and he did not
want to risk getting jammed up for the weekend. So, Jeter said,
he too decided to skip chain-snatching for the day. Instead, he
went to his sister's house, where he usually lived, and stayed in
his room all day. Around 7 P.M. he went uptown to the city with

his girlfriend, saw three movies in a Times Square movie house, played some games in a street arcade on the Deuce, and then accompanied his girlfriend back to her home. Shortly after they arrived at the girl's house, the police came and brought him to the station. Jeter concluded his statement by admitting that he had been at the Wilson Avenue train station on Friday. He said that he had done an earlier chain snatch of a piece that he thought had a diamond in it. He showed it to a knowledgeable friend who hung out at the station and knew what chains were worth. But the friend told him to "get that mothafuckin thing outta my face," which meant that the diamond was fake. So, Jeter said, he tossed the chain up onto the roof of the train station.

Carrera and Cardi settled in for a long night. They asked Jeter to tell the story again from the beginning. Jeter reiterated the same sequence of events. Then he told the story again, and again, and yet again, for more than three and a half hours, using almost exactly the same words to tell his tale. In the meantime, the detectives had uniformed officers search the roof of the Wilson Avenue train station, but there was no chain to be found there. When they confronted Jeter with this discrepancy, Jeter said that he may not have thrown the chain on the roof after all.

By 0800 hours the interview had not moved an inch beyond its start. Carrera, Cardi, and Jeter drooped with weariness. Carrera sat directly opposite Jeter, his head cradled in his hands. As soon as Jeter finished a telling of the story, Carrera told him to tell it again from the beginning. Cardi, leaning his chair back against a locker, listened to Jeter's droning repetitions of his story with his eyes closed. The detectives had studiously avoided any reference to Lozada's murder, fearing that if they directed the interview to that end, Jeter would immediately shut down and demand an attorney.

After yet another retelling of his story, and yet another command from Carrera to retell it once more from the beginning, Jeter suddenly asked: "About the lady?" Carrera and Cardi became immediately alert. Carrera said: "Yes, Darryl, tell us about the lady. That's what we want to hear." Jeter said: "Well, you din't tell me you wanna hear about the lady." He paused and then said: "I see the lady get killed, ya know, but I ain't kill her."

Then Jeter told the detectives the following story. He said that he had, in fact, snatched a chain from a man riding the train at the Wilson Avenue station. As he fled the station, he saw that he was being chased by a plainclothes cop in a red- or orange-checkered shirt. He ran out of the station and then jumped over the wall of the day care center, ran through the playground, and up the hill to the freight lines. He went along the freights trying to fix the snapped chain as he walked. Finally, he reached the top of the hill, overlooking the vacant lot at Central Avenue and Chauncey Street. He sat on the edge of the hill, where he kept working at fixing the chain. Then, he said, he saw a lady come into the vacant lot. A neighborhood friend of his, "Gerald," was also in the lot. The lady said that she was looking for a dog that had gotten lost in the tall weeds. Gerald began helping her look for it.

Suddenly, the lady pulled out a gun. She and Gerald got into a fight. Jeter said he heard the lady say: "Don't shoot." He heard Gerald say: "I'm tired of you all fuckin with me all the time." Then, Jeter said, he saw Gerald shoot the lady. Jeter told the detectives: "When he shot her, I ran. I din't want to stay there. I ran down the hill, through the bushes and up over a fire-escape ladder on the house next to the vacant lot, onto the roof." On the way up the ladder, he said he heard the lady say: "You don't have to do this." Then he heard a second shot. Jeter said that he saw a woman in the top floor apartment as he climbed the ladder. He went to the roof, looked over its edge into Central Avenue to make sure that nobody was looking, then descended though the building and went home.

Carrera watched Jeter closely throughout this statement. Jeter took deep breaths and kept swallowing deeply, licking his lips, visibly nervous. Detectives interpret such demeanor as evidence of internal turmoil. Carrera recognized Jeter's story as a confession in the making, but Jeter was still not ready to give it up. He pressed Jeter. "Darryl, Darryl, listen to me. Darryl, you're telling me somethin about what somebody else did. Darryl, you gotta tell the truth before God. If the lady got killed, and you killed her, you gotta talk, Darryl."

At the same time, Carrera sensed that, in this case, because of the length of the interrogation, he and Cardi needed more than the typical statement, traditionally written out by detectives and signed by a suspect, or written out by the suspect himself. He thought back to his first homicide at the Shepherd Avenue train station and his use of a tape recorder to ensnare one of the culprits. Privately, he suggested to Cardi that they tape-record Jeter's statement. Cardi was hesitant. A tape-recorded statement could indeed be invaluable, but it could also kill the case. Did the detectives want to marry themselves to whatever Jeter might say on tape? Written statements always provide detectives with more flexibility. But in the end Cardi agreed that a tape-recorded statement made by the suspect would be more powerful evidence in court in such an important case than detectives' handwritten summaries of the suspect's statements.

Shortly before 0900 hours, with church bells ringing out the time shortly after the tape began, the detectives tape-recorded a statement by Darryl Jeter. In that statement, in almost exactly the same words that he had used earlier, Jeter reiterated his story about being a witness to Gerald's shooting of the lady in the vacant field. But he added an epilogue to the story. Jeter said that Gerald threw the gun down in the field after shooting the lady. Jeter picked it up and took it with him. After Jeter finished his statement, and the tape-recorder was turned off, Carrera asked Jeter if he knew where the gun was. Jeter looked at Carrera, started to tremble, his lips moving without words. Carrera repeated the question, telling Darryl to let it go, to let it out. Jeter said that if he were allowed to make a phone call, he could have the gun at the station house in ten minutes. But the detectives said that they wanted to go and get the gun themselves. Jeter said that he would lead them to it.

Cosentino, Carrera, Cardi, and Detective John Medina piled into a squad car with Jeter, whom they handcuffed. Jeter directed the police to a debris-filled vacant lot at Central and Putnam avenues. He told the detectives that he had flipped the gun into the lot as he ran past it on his way home. While Detective Medina sat in the car with Jeter, Cosentino, Carrera, and Cardi searched every square inch of the one-hundred-foot-square lot for more than a half hour, lifting up the old furniture strewn everywhere, kicking through the bags of garbage, piles of human and animal excrement, dead cats and dogs, looking under every rock and behind every bush. No gun.

Carrera, Cardi, and Cosentino confronted Jeter. No one, they said, could have found a gun in that lot and brought it to the station house in ten minutes. Even three detectives had had no luck. There was no gun in that lot, they told Jeter. Cosentino and Carrera both climbed into the back seat, with Jeter in the middle. Angrily, Carrera demanded to know where Jeter had hidden the gun. Jeter told Cardi to get back in the car, and he directed the police to his brother-in-law's house on Woodbine Street. The gun was there, he said, under a heavy night table in his bedroom.

After gaining entry to the building, Cosentino, Carrera, and Cardi, with Jeter's brother-in-law's permission and cooperation, retrieved PO Irma Lozada's weapon from under the night table in Jeter's bedroom. Two of the gun's five rounds had been fired. The detectives drove to Central Avenue and Chauncey Street where the Crime Scene Unit had erected a temporary headquarters. They delivered the gun to the unit's lieutenant, who demanded to know just what the detectives thought they doing. Crime Scene had the absolute prerogative to retrieve evidence, he pointed out angrily. Moreover, Crime Scene should have photographed the weapon where it was found before moving it. With the lieutenant fuming about the lost opportunity for his unit to recover the weapon that had killed a cop, Carrera left the gun to be vouchered.

Back at the 83rd precinct station house, Carrera confronted Jeter again. Carrera pointed out that, earlier that morning, he had told the detectives that he had run away and gone up the fire-escape ladder after the first shot. And while he was on the ladder, he heard the second shot before fleeing through the building. But then in another statement, he told the detectives that he picked up the gun in the field. What had actually happened? How did he get the gun? Jeter began swallowing hard again and sweating profusely. Carrera urged him to let it out. And, with the tape recorder running once again, Jeter told the following story.

"Okay, me and Gerald we met up around 3:00 o'clock, somewhere around there, and we was goin on a train to get some jewelry. My job was to snatch the jewelry. His job was to make sure nobody come behind. So when the train pulled up at Wilson Avenue train station, the doors opened, that's when I got up to fake like I was goin out, but as the doors were closing that's when I snatched the chain. I ran down the stairs and I was headin out the train station and I was almost out and I looked back and I see this man in red. So I ran around the Audrey Johnson Day Care Center and jumped over their wall and went through the gate in the wall on the side of the building to go up on the freight. As I got up on the freight, I looked down on the street to see whether or not anybody was behind me. There wasn't nobody behind me. So I ran just so I could get past the train station, get in back of the train station. As I passed the back that's where I started lookin at the chain and everything and I fixed it. After I fixed it I put it around my neck and while I had it on my neck I started walkin towards Central Avenue."

"As I got to Central Avenue I noticed that I heard some sound on it. I heard some sound, two people talkin. So I looked and there was a white lady and there was Gerald. So they was talkin about a dog. You know, the dog . . . was lost . . . And you know she told Gerald that the dog had passed her, so when he turned around she pulled out her gun and she told Gerald, 'Don't move.' He turned around and that's how an argument started, you know, tellin him what's going on, this, that and the other. He was just speakin in general to defend himself. So they got into a scuffle and everything and the gun was away from her and I picked it up and I told her not to move. I had the gun and everything off the floor. So I went to hand Gerald the gun but he wouldn't take it unless he had something on his hands. So Gerald put his nylon over his hand and he took the gun. But I didn't know he was going to shoot her. I was just tryin to make time for myself."

"Then you know one shot rang out and it hit her. I seen the

blood on her face. And Gerald told me if we gonna go down, we gonna go down together, and he gave me the gun. I fired in her direction. I wasn't meanin to hit her. I don't still know whether I hit her or not and I just ran after I fired. I ran and I hit the fire escape. And I went up through the fire escape. As I was goin up a lady looked out the window and she looked me directly in my face. So I just kept on goin straight on up the fire escape and got on the roof. I looked down Central Avenue to see if anybody was down there. There wasn't. So I came back and went down the stairs of the building, inside of the building, and I went out through the front. I came out through the front and just ran straight all the way down to my house on Woodbine."

The detectives pressed for more details. Jeter said that he had sold the snatched chain to the SweetTooth Man for $70. He was going to split the money with Gerald, but he did not see Gerald the rest of the day, so he spent it that night on his girl-friend. The detectives began tape-recording this second state-ment at 1125 hours, ending at 1140 hours. On the two taped statements, Carrera called Jeter "Gerald" a total of eleven times. Each time, Jeter corrected Carrera, reminding him that he was Darryl. The detectives asked Jeter if he had given them permis-sion to enter his bedroom at his sister's house to search for Lozada's weapon in order to render any possible issues of do-main moot. Jeter acknowledged that he had given them permis-sion.

The detectives conferred privately about Jeter's statement and went over the sequence and the relative seriousness of PO Lozada's wounds. One bullet, surely the first to strike her, had entered her left cheek from a distance of about 18 inches, caus-ing stippling marks on her cheek. The bullet exited behind her left ear. The impact of this shot caused Lozada to collapse on her left side with her face to the ground. The medical technicians at the crime scene guessed that this wound by itself would have caused Lozada to bleed to death in great pain over a period of several hours, had she remained undiscovered. But the second bullet was fired into the back of Lozada's head with the gun pressed against her skull. The bullet exited through her mouth. The shot left gunpowder residues deep inside her, and, in the technicians' view, killed her instantly. For practical and legal purposes, the second shot was the death-dealing blow.

At 1145 hours, the detectives resumed the tape-recording. Specifically, Cardi asked Jeter some pointed questions about the sequence of the shots:

> *Cardi:* I am asking you, Darryl, that the statement you had made prior of you firing the second shot, are you sure of that?
> *Jeter:* I'm absolutely sure.
> *Cardi:* You fired what shot?
> *Jeter:* The second one.
> *Cardi:* There isn't any doubt in your mind?
> *Jeter:* There isn't any.
> *Cardi:* Gerald fired the first shot and then handed you the gun and said what?
> *Jeter:* We gonna go down, we gonna go down together.
> *Cardi:* And then you took the gun . . .
> *Jeter:* I fired in the direction of the tall grass.

Later, Cardi asked Jeter if he were willing to speak with a district attorney and make the same statement. But at that point Jeter stopped talking and asked for his lawyer.

———

Who was Gerald? Cops at the 83rd knew a local boy named Gerald whom neighborhood robbers regularly blamed for their depredations. Gerald was mentally retarded, the cops said, and would admit to the crucifixion of St. Peter as long as he was provided with the necessary details during questioning. Did he actually have anything to do with the murder, or had Jeter acted alone and was now trying to pin it on Gerald?

Other detectives found Gerald and brought him back to the station. When Carrera and Cardi met Gerald and saw how impaired he was, they doubted that this young man was involved in the murder at all. Indeed, they wondered if the police had brought in the right person. So they brought Gerald to the room where they were holding Jeter and asked Gerald if he knew Jeter. Suddenly, Jeter jumped out of his seat and said: "I can't do this to him." Jeter then blurted out a third statement to the detectives, one that they could not tape, nor have Jeter sign when they later wrote it down, because he had requested an attorney at the end of his second taped statement.

In this third and last statement, Jeter told the detectives that he had ripped a chain off someone's neck in the Wilson Avenue LL train station. Gerald was supposed to be a lookout. He said that he ran out of the station and saw that he was being followed by a white male in an orange shirt. He made a right turn, jumped over the day care center wall, and went up to the Conrail tracks. He ran down the tracks toward Central Avenue. He looked behind him but no one was chasing him. He started walking while fixing the lock on the chain and then put the chain around his neck. When he reached the vacant lot, he saw a lady standing in the lot. She told him that she had lost her dog and asked him to help look for it. He turned around and heard the woman say: "Freeze, don't move." She had a gun in her hand.

Jeter advanced on her, grappled with her, and they both fell to the ground tussling. Jeter said that he grabbed the gun away from her and shot her in the face. He saw blood on her face as he started to get up. She rolled over on her side and he shot her once in the back of the head. He put the gun in his pocket and ran to the rear of the building adjacent to the vacant lot. He saw Gerald entering the lot as he fled. He climbed up the fire-escape ladder. A woman came to a rear window and looked him straight in the face. On the roof, he waited a few minutes but did not see anybody in the street. He went down the building's inside stairs to Central Avenue and then fled home.

Following the lead in Jeter's third statement, Carrera and Cardi went back to 669 Central Avenue, the brownstone apartment building that bordered the field where PO Lozada had been murdered. They went to the third floor apartment on the field side of the building and interviewed the woman who lived there, "Hally Moore." The detectives asked Ms. Moore if she had seen someone running up the fire-escape ladder the previous afternoon. When Ms. Moore reluctantly admitted that she had, the detectives pressed her for a time of day. Ms. Moore said that she watched a soap opera faithfully every day, one that began at 4 P.M. Just prior to the start of her favorite program, she heard a noise out in the field. Then the opening music of her show came on. She heard a second noise. She went to the kitchen window that faces onto the fire-escape ladder. A young man came up the ladder, looked her in the face, and told her: "Get the fuck inside." She closed the window and then heard tramping on the roof of her building. She settled down to watch her soap opera, determined not to get involved in anything. The detectives thought that Moore's statement was valuable. Her memory of the opening music of her favorite soap opera fixed the time of the incident, a time that corresponded closely with Giambalvo's account.

But when the detectives brought Ms. Moore to the station house to view a lineup with Jeter in it, she did not pick anyone out as the person she had seen on her fire-escape ladder. The detectives were profoundly disappointed at her inability, or unwillingness, to identify Jeter, and they felt the ground of their case shift underneath them. However, while Moore was being driven home by other detectives, her five-year-old daughter asked her out loud why she had lied to the police. Ms. Moore began to cry, was brought back to the station house, and this time picked out Jeter as the person she had seen on the fire escape on the afternoon of September 21.

Ms. Moore's later open-court testimony differed in some details from the story that she told Carrera and Cardi on September 22. She said that, after returning home from work, she was in the kitchen making a sandwich in anticipation of her soap opera. She heard a noise in the field adjacent to her building, went to her back window, and yelled at whoever was in the field to get out of there. Someone yelled back at her to "mind your mothafuckin business." She went to the front window and saw nobody. Then she went back to her kitchen, looked out the window, and saw two people in the back yard. A white female was on the ground and a male stood hovering over her as the opening music of the soap opera started. She thought that the man and woman might have been having sex because this occurred regularly in the field. She saw the left profile of the man standing over the female from a distance of about thirty yards. She said that she had deliberately not identified anyone in the first lineup because of fear of retaliation.

While Carrera and Cardi were on Central Avenue, the Crime Scene unit informed them that PO Lozada's radio and her bag with her old shoes had been discovered in the back yard of 196

Moffat Street, an easy throw from the roof of 669 Central Avenue. Further, two other detectives had paid a visit to the Sweet-Tooth Man's store and confiscated a chain from him that he said he had bought from Darryl Jeter on Friday. However, the Sweet-Tooth Man claimed that he bought the chain from Darryl between 8 A.M. and noon. Later, the SweetTooth Man admitted lying to the investigators. He eventually testified that Jeter and his girlfriend had come to his store in the late afternoon of September 21. The SweetTooth Man said that he bought the chain from Jeter for $30. He steadfastly denied that he was a fence.

All the police brass were dumbfounded and delighted that Carrera and Cardi had, in less than 24 hours after the crime, not only arrested a culprit for PO Lozada's murder but had gotten several statements from him. ADA Eric Seidel and his bosses were initially unhappy at the procedural irregularity of tape-recording two of Jeter's statements. Seidel worried that defense attorneys could seize on this marked departure from detectives' normal practices to claim that the police had singled Jeter out in some kind of prejudicial way. But since the detectives had used a tape recorder, Seidel wished that they had been able to tape Jeter's third statement before he asked for an attorney.[4] Later, Seidel called as a witness an expert in voice spectrographic analysis who offered her opinion in court that the voice on the tapes was indeed that of Darryl Jeter. Jeter's lawyers had originally hired the same expert to prove that the voice on the tapes was not Jeter's but fired her when her report was not to their liking.[5] Whatever the complexities and ambiguities, Seidel had Jeter's two taped statements themselves, including Jeter's admission of the chain-snatching, the detectives' account of

Jeter's third statement, the SweetTooth Man's admission of buying the chain from Jeter, Moore's eyewitness testimony putting Jeter in the raggedy field next to her house, and the murder weapon found in Jeter's bedroom.

But the case became still more complicated when the medical examiner's official report on PO Lozada came back. The autopsy confirmed the sequence and seriousness of the two gunshot wounds that she had suffered. But it also revealed trace amounts of opiates and cocaine in the officer's bile and urine that pointed either to the ingestion of codeine in cough medicine or, more likely, to low-level narcotics use. Her brother officers knew Fran as a desirable, happy-go-lucky party girl in addition to being a dedicated, enthusiastic rookie cop. If she had been using illegal drugs in violation of strict rules and been discovered through the department's random drug-testing program, she stood to forfeit her police job. Such a scenario had nothing whatsoever to do with her murder. But Seidel, Carrera, Cardi, and everyone else connected with the case knew that it could be made to seem so. Seidel imagined vigorous defense claims that Lozada had been killed in a drug deal gone bad, while her partner, Nat Giambalvo, was off drinking, or worse. The criminal justice system is one of several institutional crossroads in American society where fantastic claims are regularly made and often honored.

Moreover, the report of Lozada's drug use escalated the open fighting between the bosses of the NYPD and the transit police. Police Commissioner Benjamin Ward blasted PO Nat Giambalvo not only for "poor judgment" but for "violat[ing] clear-cut rules of the department." Ward argued that Giambalvo had failed to search adequately the area where he had separated from PO Lozada and that he seriously erred by waiting an hour and a half to notify his own superiors and the city police department that his partner was missing. Commissioner Ward also excoriated the transit police, even as scores of pundits in New York papers criticized the mayor's office and anyone else who could plausibly be blamed for the radios carried underground by transit officers. But a thorough analysis revealed that at no time had PO Lozada radioed Central or anyone else for assistance, as PO Giambalvo had instructed her to do when the officers split up. Nor had Lozada responded to the scores of calls Central had made to her, all recorded: "Shield 4721, come in to Operations. Operations to Shield 4721, come in Shield 4721."

At one point, Ward's hostility to the transit police went so far that he ordered the chief of the Internal Affairs Bureau to summon Cardi and Carrera to his office and demand all of the detectives' investigative reports. Carrera, who was keeping the file because Lozada had been a fellow transit police officer, adamantly refused, arguing that the file contained the names of confidential informants and that IAB would blow the case to pursue its own agenda. The chief was beside himself with anger and astonishment that a detective would refuse his direct order, and he threatened to flop Carrera back into uniform unless he complied. When Carrera pointed out that he and Cardi were trying to make a cop-killing case, in the eyes of police officers the most important kind of case, the chief responded that his principal concern was to please the police commissioner.

Public debate about the homicide quickly assumed the warlike tones that mark most discussion about anything to do with policing and criminal justice in New York City. For instance, Mayor Ed Koch wrote an op-ed piece for the *New York Times* on September 29, 1984, describing his own first-hand experience at the gruesome crime scene, saying that it was an awful place to die. He recounted the suspect's previous repeated criminal record and noted that he was on parole when the crime occurred. He called for the reinstatement of the death penalty in New York State. Later, Koch said in reference to the deeply flawed parole system: "Our system of justice doesn't work." A few days later (October 8, 1984) a *Times* reader lambasted the mayor for his callousness, arguing that if Bushwick was an awful place to die, it must be an even worse place to live. The answer is not greater punishment, the writer argued, but a commitment to remedy the general injustice and inequity of American society that causes violence, a finger-pointing account for criminal depredations that has enduring contemporary resonance.

Still other pundits indulged in standard I-told-you-so hectoring about the dire consequences of police not following proper procedures. Female police officers, as well as civilian feminist advocates, argued that PO Lozada's gender had nothing whatsoever to do with her death, a position vigorously maintained even though Lozada faced a parolee desperate not to be rearrested, who towered over her by seven inches and outweighed her by fifty pounds. In this view, Lozada's death stood for the dangers faced by all police, not just those faced by policewomen.[6]

Jeter's 1985 trial was racially charged, a replay of long-held, bitter resentments felt by black Americans against the police and a rehearsal of arguments for trials to follow. Richard W. Foard III represented Jeter, funded by the International Committee against Racism. Foard argued that Darryl Jeter was never in the vacant lot, that he was the victim of a police frameup to conceal police corruption, police wrong-doing, and a bungled investigation. In this view, there had never been a chain snatching nor a hot pursuit of a thief. Indeed, as Seidel had anticipated, the death of Irma Lozada was, Mr. Foard suggested, an execution due to a drug deal gone bad. The person who was interrogated on the two tapes was not Darryl Jeter but Gerald, as shown by Carrera's repeated use of the name "Gerald" during one of the taped interrogations.[7] The police felt, according to Foard, that in such a big case they could not get a conviction on Gerald because of his mental impairment, so they pinned the crime on Darryl Jeter. Mr. Foard also argued that this cover-up by the police was "racially motivated," typical of a legal system that is "racist to the core."

Jeter's story at trial carefully matched the scenario that Foard proposed to the court and, outside the courtroom, to the news media who gathered in great numbers to follow the trial.[8] As he had said in earlier statements, Jeter described getting up, accompanying Kilo to the unemployment office, and running into Skeeter, who was rummaging through garbage to find the liquor store owner's stolen documents. But in this telling, the story changed from all previous statements. In this version, Skeeter eventually "left out" and Kilo returned. Jeter went with Kilo to the latter's girlfriend's apartment, where Jeter and the girlfriend smoked some reefer. After a while, Jeter left the apartment and walked around the neighborhood. Suddenly, at Central Avenue and Decatur Street, Gerald ran around the corner completely out of breath. Gerald told Jeter that he had just done a stickup and the police were right behind him. They talked for about thirty or forty seconds. Jeter said that he told Gerald: "Yo, you want me to take the gun and I'll put it up in my house till tomorrow." Gerald agreed and gave Jeter the gun in a skinny paper bag.

Jeter said that he went back to his sister's house, went to some lengths to avoid encountering her children, and hid the gun under a heavy night table. Later, his brother-in-law, at his mother's request, gave him twenty dollars because it was his birthday. He went out and walked around the neighborhood with his brother-in-law, bought another joint, and smoked it. He ended up at the SweetTooth Man's store. While he was playing a video game in the store's game room, his girlfriend came into the store. They played the game together for a while and then Jeter walked her home, reaching there at 7 P.M. They had agreed to go uptown that evening, but the girlfriend needed some time to get ready. So Jeter told her to be ready by 8 P.M.

In the meantime, Jeter walked around the neighborhood and ran into Skeeter and an acquaintance whose name he could not remember. They all smoked some reefer, and Skeeter told Jeter that the police had found a police officer's body in the train station and that detectives were going up and down Wilson Avenue talking to everyone about the murder. Jeter said that he had better get off Wilson Avenue because he was on parole and smoking marijuana. Then he "tricked" Skeeter and the acquaintance into walking him back to his girlfriend's house at 7:45 P.M. But his girlfriend still was not ready to go out, so Jeter fed his daughter. Finally, at 8 P.M. Jeter and his girlfriend went uptown, saw three movies in a Times Square theater, and then came back to Brooklyn around 3 A.M. As they were walking back to his girlfriend's house, they saw police radio cars on the block and detectives hustling Skeeter into a squad car.

Soon after they reached the girlfriend's house, police banged at the back door to her apartment. One detective entered the apartment and stood on Jeter's feet so he could not move. Then, he said, uniformed cops handcuffed and beat him, all the while yelling at him about killing the lady cop. Jeter said that at the station house he was kept in a pen in the squad room for almost an hour and then grilled by several detectives for hours about what he had done that day. He told them, he argued, exactly what he had done, except that he did not initially tell them about hiding Gerald's gun. Eventually, he did tell them about hiding the gun for Gerald. The detectives, he said, took him for a ride to the vacant lot at Central Avenue and Chauncey Street and then brought him back to the station house. The detectives took Jeter into another room and brought him face to face with Gerald, but the two did not speak. The police then confiscated Jeter's sneakers and his tank top and put him in a lineup.

Jeter steadfastly denied snatching a chain on Friday September 21, 1984, or even discussing snatching a chain with Skeeter. He denied selling a chain to the SweetTooth Man. He denied shooting the woman in the field. He denied taking the police to his brother-in-law's and sister's house, although he said that he did provide the exact location of the gun there. And he denied giving any taped statements, or indeed any inculpating statements whatsoever, to the police. Jeter's attorney brought several witnesses into court, including Darryl's sister, brother-in-law, and Skeeter, who, in varying ways, provided alibis that supported Jeter's in-court story. Moreover, the defense produced a 12-year old boy, "Robert," who claimed to be an eyewitness to the events and who described a mysterious man named "Sherrod," presumably a drug dealer, who, Robert said, struck the lady cop with a stick and then fled. Robert also stated that he had seen Sherrod with the same cop on other occasions, implying clandestine transactions. Seidel treated the youngster gently even as he discredited his story.

Darryl Jeter was convicted of murdering PO Lozada by a jury of six men and six women, black and white. He was also convicted of felony murder and of criminal possession of a weapon. At sentencing on June 20, 1985, Justice Thaddeus Owens denied Mr. Foard's motion to set aside the jury's verdict as incompatible with the weight of the evidence. ADA Seidel noted that Jeter's probation report listed eight arrests for crimes committed as a juvenile. The Juvenile Court records in 1981, according to Seidel, described Jeter as "beyond parental control and supervision" and as "a danger to the community." His adult record listed five crimes for robbery, grand larceny, and assault. The Probation Report described Jeter as "an exceedingly dangerous individual." Seidel pointed out that the report also noted that many of Jeter's crimes were committed against women.

Justice Owens sentenced Jeter to serve concurrent terms for the two murder convictions and a consecutive term for the weapons possession charge, a total of thirty-two and a half years. Jeter's conviction was upheld on appeal, though the higher court ordered that Jeter serve the sentence for the weapons possession charge concurrently with those for the murder convictions. At the sentencing, the courtroom broke into chaos, with representatives from the International Committee against Racism and the Progressive Labor Party shouting that the whole trial had been a racist frameup and displaying placards denouncing the entire criminal justice system. In the midst of the melee, Vinnie Carrera said to a television news reporter: "Irma Lozada, rest in peace."[9]

Five thousand police officers from the New York metropolitan area and from as far away as Buffalo had turned out for the inspector's funeral given to Irma Lozada on September 26, 1984, at the Holy Name Church at 96th Street and Amsterdam Avenue in Manhattan. Police Commissioner Benjamin Ward was conspicuous by his absence. As the Emerald-Irish War Pipe Band's bagpipes wailed "Amazing Grace" and muffled drums thudded slowly on the dreary, overcast day, police officers in starched blue uniforms and white gloves, standing five deep, saluted as the hearse carrying Fran, their fellow officer, sister, daughter, and friend rode up Amsterdam Avenue. Many wept openly. Father John King, S.J., the priest who had anointed the young woman as she lay dead in the field, told her grieving brothers and sisters that "No words, no wisdom will ever wipe away the pain of grief. But we can see how she came from God's hands and how she now returns to God's hands."

A great many officers remain haunted to this day by the terrible dilemma that Fran Lozada confronted when she suddenly came face-to-face with her assailant among the tall weeds of the vacant lot at Central Avenue and Chauncey Street on September 21, 1984. Jeter's own statements pieced together describe what happened. Fran's ruse of looking for a dog gave her the chance to get her gun and handcuffs out of her purse and to display her shield. But when she ordered Jeter to freeze, he advanced on her, wrestled her to the ground where he disarmed her, and then shot her in the face. Gravely but not fatally wounded, Fran begged for her life. While exclaiming his hostility toward police for regularly interfering with his own life and work, Jeter shot her in the back of the head and then fled. Given the size of her opponent, Lozada's only chance of surviving the encounter was to shoot Jeter straightaway when he disregarded her orders and moved on her. But shooting an unarmed man, then and now, means the end of a police career. The split-second hesitation made PO Irma Lozada the first, but sadly not the last, police-woman killed in the line of duty in New York City history.[10]

10

THE LONG ARM OF THE JOB

Detectives' shuttle between the streets and the courts, between the investigation of bloody mayhem and its rationalized processing, makes them outsiders of a sort in the vast apparatus of the criminal justice system, alert to its inconsistencies and irrationalities. But detectives are also police, agents of the state, symbols of authority, ultimate insiders with privileged access to hidden social arenas and forbidden knowledge. They become objects of fear, anger, and resentment. This double-sided role shapes the meanings of detectives' work, their images of the world, and their own self-images.

Detectives' work-world teems with highly rationalized agencies. The gigantic New York City Police Department, which since 1995 has included the housing and transit police, is splintered into 76 separate precinct commands, more than 60 specialist divisions, and an untold number of secret units. Another dozen police authorities also flourish in the city, ranging from the Metro-North Police to the Amtrak Police to the Port Authority Police, each with specialized jurisdictions. Every federal law enforcement agency—Federal Bureau of Investigation, the Drug Enforcement Administration, Alcohol, Tobacco, and Firearms, the United States Marshals Service, the United States Postal Service, and the several law-enforcement wings of the former Immigration and Naturalization Services, now taken over by U.S. Immigration and Customs Enforcement of the Department of Homeland Security (DHS)—has major operations in the city. The DHS, created in the aftermath of September 11, 2001, also has other extensive divisions operating in the city.[1] The city has five different district attorney jurisdictions, one for each borough of the metropolis, with each elected district attorney connected to a separate branch of the New York State Supreme Court. The city also has two federal prosecutor jurisdictions connected to federal courts, one in the Southern District in Manhattan and the other in the Eastern District in Brooklyn. Detectives come to see their investigative work as regularly subordinated to the exigencies of competition among and within this tangle of agencies and organizations.

Given the city's huge, constantly shifting population, its premier presence in the nation's financial, business, cultural, and intellectual arenas, its role as a gateway for legal and illegal immigrants, its vast criminal underworld mirroring the complexity of its upperworld, and its narcotics-trafficking and market-making in every kind of illegal goods, New York demands organized vigilance by police and prosecutors. But police and prosecutors also need crime. The more vicious the crime, the more innocent the victims, the better the opportunities to demonstrate prowess to bosses, peers, and larger publics. In such a world, detailed knowledge means power, and jurisdiction brings opportunities for garnering prestige, at least for certain kinds of cases.

The competition between the NYPD and federal police agencies for control of important cases is legendary. Detectives regularly savage federal agencies for their bloated budgets, their agents' lack of grounded understanding of how the city works, and especially agents' unwillingness to share information about criminal groups. But also within the NYPD itself, some detective squads regularly hide important information from other units. Sometimes even detectives in the same squad hold back informants who might help their colleagues' cases if sharing informants might jeopardize their own big cases. Borough-wide homicide squads steal good cases from precinct detectives, and bosses regularly appropriate credit for their subordinates' hard-slogging work on the streets.

District attorney offices evince equally sharp competition. DANY has six main trial bureaus. In a weekly rotation that advances through the days of the week, each bureau, for one 24-hour day a week, catches all complaints brought by police officers to DANY's Early Case Assessment Bureau (ECAB) at 100 Centre Street. But police officers and detectives, with prosecutors' collaboration, often hold cases until they get the bureau—and sometimes the specific prosecutor—they want. Even after a bureau, and an individual prosecutor within a bureau, catches a case through ECAB, the case can be stolen by bosses, by more experienced prosecutors, or by specialized bureaus such as the Sex Crimes Bureau, which claims, and gets, special jurisdiction over all high-profile rape cases in Manhattan. Historically, little love has been lost among the district attorneys of New York, Kings, Bronx, Queens, and Richmond counties, each of whom shapes his office to suit his own style, meet his own political agenda, and respond to the peculiarities of his constituency. Each has, and fosters in his subordinates, a bailiwick mentality that jealously guards big cases when crimes cross jurisdictional boundaries. Only interrelated crimes that pose marked threats to public safety prompt inter-jurisdictional cooperation.[2]

Federal authorities regularly steal cases from state prosecutors and investigators. For example, Assistant District Attorney Dan M. Rather, chief of the Gun Trafficking Unit at DANY, together with Detectives Ray Brennan, John Capers, Jim Killen, and other criminal investigators, labored long and hard to build a case against the Preacher Crew, headed by Clarence Heatley aka Preacher. The Preacher Crew was a Harlem-based drug and extortion ring with cult-like overtones, responsible, the state investigators thought, for at least forty murders over ten years. The crew's typical *modus operandi* was to pick up murder contracts in the drug trade for the standard fee of $5,000 per murder, fulfill the contracts, but then turn on those who took out the contracts, whom they identified as weak for relying on outsiders for their "wet work." The Preacher Crew then extorted fees from these hapless dealers for use of the street corners where they plied their trade. If the dealers did not listen to reason, the Preacher Crew killed them and took over their spots. Such killings were usually brutal, sometimes preceded by torture and sexual humiliation through ritual sodomy.

Just as Rather and his investigators were on the verge of securing indictments against the entire crew, FBI agents and federal prosecutors from the Southern District of New York swooped in and placed the case under federal jurisdiction. Ignoring intricacies of the case that involved murders in Georgia as well as New York, the federal prosecutors offered a plea to Heatley, arguably the worst of the lot, in order to turn him against his chief lieutenant, John Cuff, a former housing police officer, and several other key players in the ring, all subsequently arraigned in death penalty cases.[3] The federal intervention brought the state investigation to a grinding halt. The needs of large organizations and of individuals' careers within them regularly trump the hard work on the streets that is essential for public safety.

Detectives are prototypical men and women of action, but they work in a world of intricate procedures that curb their efforts to identify, find, and apprehend criminals. Criminal law consists in part of elaborate rules of conduct constructed over time to check state and police power. In order to resolve disputes, fix liability, or determine guilt, these procedures are interpreted and applied by judges, who themselves are trained as lawyers, in response to arguments from still other lawyers. In this sense, the law as an institution resembles the self-enclosure of other intellectual milieux. At any given moment, the law means what a particular judge says it means. The outcome of any legal process, and therefore of detectives' work on the streets, is always hostage to the vagaries of jurisdiction, judicial perceptions, whims, ideologies, or even the time of day.

From detectives' standpoint, legal procedures provide deep thickets in which suspects and seasoned criminals alike can hide. Detectives see criminals' entire recorded criminal histories, and they spend a great deal of down time with suspects, talking for hours on end. The judgments they form about a suspect's responsibility for specific crimes often emerge out of their assessments of the subject's character. The reliability of these assessments vary, of course, and sometimes even the best detectives are dead wrong. Partly to guard against such errors of judgment, the law deliberately ignores individuals' criminal histories and allows no consideration at trial of suspects' moral characters as discerned by state officials. The law insists that people be tried and judged for each separate individual criminal action with which they are charged, a bedrock assumption of the common-law notion of responsibility. Prosecutors are not allowed to mention previous crimes in front of a jury, even if these crimes are related to one another, unless there is a direct "narrative" relationship between a previous, even uncharged, crime and the particular crime at issue, and then only when the prosecutor can successfully argue the connection to a judge who knows the law. Unless detectives find a prosecutor willing to press a complicated case against a known criminal on marginal evidence ("He's guilty of *something*"), they regularly face the tension between their own moral certainty about someone's responsibility for a crime and their inability to fix guilt legally.

Sometimes legal procedures overturn detectives' work completely. In cases that depend wholly on eyewitness testimony, there obviously can be no cases if witnesses will not or cannot testify. Moreover, when the state is unable to produce witnesses, the rule is to grant bail pending trial, or even dismiss the charges outright. For instance, a drug-gang member named Pasqualito, arrested in late 1989 in the Bronx's 40th precinct by Detective Mark Tebbens for a double homicide committed the same year, was released on bail pending trial because the Bronx DA's office could not produce the witnesses to that shooting. While out on bail, Pasqualito went to Brooklyn with Lenny Sepulveda, the gang's leader, and shot in the face, but did not kill, one of the eyewitnesses to a notorious 1991 Bronx quadruple homicide. On the wounded witness's testimony, detectives arrested Pasqualito for that assault. But although he was on bail for the double homicide, and although the Brooklyn assault aimed to silence a witness to a quadruple homicide, Pasqualito was released on $25,000 bail that he posted in five minutes.

The Brooklyn night court district attorney and judge may not have known about the pending trial for the double homicide. In any event, given Pasqualito's bail on the double-homicide charge in lieu of available witnesses, the Brooklyn court could not hold him for that as yet legally unproved offense. Further, Pasqualito was not accused of participating in the quadruple homicide. Therefore the Brooklyn assault, though an attack on a witness to a slaughter, was seen and treated as just another serious assault, not as part of a pattern of gang criminal activity. Even Pasqualito was surprised when he received bail. By the grace of the system, he went on to commit several other violent crimes, including at least one murder, before he was finally captured, convicted of several crimes, and incarcerated.[4] From detectives' standpoint, the rationality of the law and the rational bureaucracy of the criminal justice system often produce irrationality.

Clever criminals understand and exploit the fragmentation of knowledge that inevitably flows from the bureaucratization of the system. In a world constructed of discrete "cases," the smart criminal follows required procedures to dispose of minor offenses that might trigger unwanted warrants, knowing that police, prosecutors, and courts alike are focused on their particular work tasks and are unlikely to connect a minor case to more serious crimes. Thus, Lenny Sepulveda, who was wanted by the police for murder, strode into Manhattan Criminal Court with his lawyer to "get rid of a gun case," counting on an overburdened clerk and judge not to check pending warrants for his arrest. He was apprehended only because Terry Quinn, DANY's supervising rackets investigator, anticipated such boldness.

Because criminal investigation depends on identification, the construction and maintenance of false identities comprise an industry in criminal underworlds. Most self-respecting drug dealers have multiple identities ready at hand, complete with validating documents. For about $400, one can buy an identity kit in

New York City that includes a social security number, pay stubs to demonstrate employment, letters of recommendation from previous employers, and tax records. With these phony documents one can get a welfare benefits card that provides a legitimate governmental validation of a fictitious identity. One can then not only enjoy monthly electronic deposits from the city's overwhelmed welfare apparatus but also use one's identity as a welfare recipient to obtain a driver's license, the de facto identity card of the United States. Some crime groups take identity creation to high levels. For example, the ferocious Jamaican posses that for years terrorized Edgecomb Avenue in the 30th precinct brought in "rude boys" from Kingston to assume the already fake identities of their young predecessors whose incompetence or bad luck had caused them to be "jointed," that is, chopped into manageable pieces and scattered throughout the metropolis.

Until around the turn of the millennium, there was no coordination of criminal records even between neighboring states. Therefore, an actual arrest in one state jurisdiction did not guarantee that someone could be identified in another state. Fingerprint identification depended entirely on the care of the officer taking the original set of inked prints, and on the care of other officers, or civilian examiners, in submitting fresh prints to computers and then manually comparing new to old prints. Digital photo-imaging technology, which can take fingerprints or, since April 2003, whole palm prints and transmit them electronically for comparison, has begun to aid law enforcement officers in identifying criminals. The terrorist attacks of September 11, 2001, initiated a national debate on the legitimacy of various identification techniques and spurred federal agencies to begin the tediously slow process of integrating separate databases to help detect and apprehend terrorists. When new techniques are adopted and databases meshed, police will use them to pursue run-of-the-mill criminals, although such efforts will be vigorously opposed by the defense bar and civil libertarians.[5]

In late 2003 the NYPD became the first police department in the United States to gain access to Interpol's heavily encrypted I-24/7 identification system, which makes criminal suspects' fingerprints, passports, photographs, and entire criminal records immediately available to police.[6] However, the general rule is that all such rational attempts adopted by individual detectives or by the criminal justice system as a whole to monitor, identify, apprehend, and control criminals merely serve as starting points for new forms of deception designed to thwart law enforcement and threaten public safety. Detectives see their work as never done.

Detectives' work regularly takes them behind respectable public faces, where they glimpse messy, sometimes tumultuous, sometimes sad, sometimes ironic, sometimes tragic, sometimes comic, sometimes despairing, sometimes vice-filled private lives. Detectives in the 34th precinct find that a homicide victim who always dressed immaculately in public and led a perfectly orderly, punctual, dutiful professional life as a schoolteacher, complemented by regular church-going, lived in an apartment thigh-deep in literally tons of trash, papers, magazines, old mail, advertisements, garbage, and rubble systematically collected over a period of years.[7] They encounter a widely acclaimed feminist professor regularly battered by her I-can't-give-him-up boyfriend. They meet an elderly couple bound together in a suicide pact. One day the old man wakes up from his

afternoon snooze and runs into the kitchen yelling "It's time!"
He attacks his surprised wife with a knife, cuts her throat, and
then slits his own wrists. But they both survive, and the woman
refuses to press charges, arguing gently: "I don't mind dyin, but I
just don't want him sneakin up on me."

Detectives from the 34th squad find a heart-attack victim in
his apartment, called to his maker while coupling with his white
Samoyed, found crushed beneath him ("Oh well, at least it was a
female dog"). They discover a solitary pleasure-seeker impaled
on a giant wooden phallus. They watch as a man walks into the
squad room clad only in his undershorts, claiming that he and a
friend were just having a quiet conversation in his parked car
when a robber reached through the window and snatched their
clothes. They come across otherwise respectable professionals
toying with narcotics or slavishly addicted to them. While investigating the murder of a man dressed up in women's clothes,
they uncover a genteel "butterfly society" of established professional men who cross-dress for Friday evening cocktails. They
encounter black professional men leading double lives on the
DL—the Down Low, a hypermasculine world where latent violence tinges sexuality—and white professional men stoned blind
on crystal meth (methamphetamine) to fuel serial conquests in
gay bathhouses and clubs.[8] They meet celebrities mired in uncontrollable, compulsive desires, some hungering to be tied up,
sand-papered, and whipped by a dominatrix, others longing to
humiliate and violate, and still others who, satiated with the
pleasures that fame and money can buy, delight in brushing
their wings against the flame of the raw and coarse vitality of
criminals in New York's vibrant downtown club scene, one small
piece of what detectives see as the elbow-rubbing alliance between American society's elites and its outlaws. To detectives, the

world is never as it seems. Beneath bright, tailored appearances lurk dark secrets.

Sometimes businesses are out-and-out fronts for criminal commerce, such as the famous chicken shack in the Bronx that, until its break-up by police and prosecutors in 1995, sold surplus Soviet AK-47s along with fried drumsticks. Even famous, respectable Wall Street financial houses are used as drug-money laundries by wily crooks who understand the profound vulnerabilities of highly bureaucratized systems.[9]

Detectives regularly cross paths with professionals who flirt with collaboration with criminals. Protected by a high-minded occupational ideology sanctioned by law, criminal defense attorneys are regularly afforded blind-eye professional courtesy by prosecutors and judges alike. As self-styled advocates for the persecuted, they maintain a publicly respectable distance from the depredations of their clients, even as they feed off the rich bottom land of criminal profits, sometimes delivered to them in paper bags filled with cash. While they invent fantastic fictions to protect their clients, sometimes suborning perjury in the process, they excoriate the police in court for procedural infractions and especially for the ruses they use to obtain statements, admissions, or confessions, morally equating such deceptions with bloody murder.[10] Plaintiffs' attorneys, working on contingency fees, file lawsuits against police officers and against the city and state of New York which, however frivolous, require inordinate amounts of time and money to resolve.

An imprisoned drug dealer serving time in prison sues former narcotics officers and the NYPD for loss of occupation and income, arguing that his yearly earnings slipped from $250,000 a year to $30 a month because of his arrest and conviction for narcotics trafficking. A young man fleeing from a just-completed subway robbery pivots to face two officers, points a pistol at them, which turns out to be a starter pistol, is shot by the police, and sustains severe internal injuries. He successfully sues the police and the city for unjustified use of force. Detective Joe Montuori receives a call from an old lady who tells him that a valuable antique brooch is missing from her apartment. She is convinced that the postman took it from her hall table when delivering a package because he is the only person besides herself who has been in the apartment. Montuori interviews the postman, who denies the theft. Then, the old lady calls Montuori to tell him that she has, miraculously, found the brooch. Montuori informs the postman, who promptly sues him and the city for mental anguish and humiliation.[11]

Plaintiffs' lawyers, in league with community activists and using accusations of racism to distract compliant judges and gullible juries from their clients' depredations, bring lawsuits against authors and publishers who have the nerve to point out the ugly realities of the drug trade and the moral confusion of a political and legal system that cannot decide what kind of public order it wants. Many politicians, protected by the knowing-and-not-knowing stance now firmly instituted as an excuse in our society, pander to whoever can commandeer the apparatus of advocacy and, perhaps, make substantial contributions to campaign chests. Prominent Washington Heights politicians frequent a restaurant that doubles as a drug-dealers' hangout, one politician even mugging for the camera alongside a notorious violent criminal.[12]

These journeys behind respectable public façades stir prurient interests in some detectives, but profound class resentments in most. Police officers come overwhelmingly from the working class. They are the sons and daughters of policemen, firemen,

craftsmen, laborers, bus and truck drivers, and factory and service workers of every sort. They see themselves and their brother
and sister officers constantly being blamed for the sorry outcomes of miserable social conditions they had no hand in fashioning, states of affairs that indeed were created, and are now
sustained, by elites of far higher social station than they. As detectives see it, the principal critics of the police are members, al-
lies, or servants of those same elites, the very people who benefit
most from the order that police maintain but who, far removed
from the savagery of the streets, can indulge in the indignation
that simultaneously guards and announces a cherished sense of
moral probity.

As visible symbols of authority, detectives and other police
find themselves regularly opposed and excoriated by intellectuals of various sorts. Of these, journalists exert the greatest day-
to-day influence on police work. Many journalists see their work
as a vocation to "comfort the afflicted and afflict the comfortable."[13] They frequently romanticize community activists, un-
critically taking their word that police use "excessive force" in
attempting to curb the violent crime that disproportionately
afflicts black and Hispanic residential areas of the city. Since its
founding in 1851, the *New York Times* has been the principal cru-
sader against police brutality and corruption in New York, as
part of the larger political struggle between key city elites and
emerging ethnic groups over what kind of order should prevail
and who controls crucial municipal bureaucracies such as the
police.[14]

As it happens, the police have given the *Times* plenty of material to work with over the years. NYPD officers as well as officers
in departments across the country do at times use excessive force
and abuse their authority in other ways. In Northern Manhat-

tan, one detective carried a sledge hammer in the trunk of his car, using it to destroy the apartments of culprits who ran from him on the street. The very mention of "Jason," the street handle of one notorious uptown "murderer" of a cop, turned many culprits' and law-abiding citizens' knees to jelly.

Some officers use their authority to hunt for sexual favors. Bystanders, witnesses, victims, their relatives, associates of criminals, and occasionally even culprits themselves become sexual prey for cops who pride themselves on being great ladies' men. In Washington Heights, with its multiple cross-cutting kinship and social network ties typical of the city's Hispanic areas, coupled with its entrenched Latino code of honor, one sows one's seed always at the risk of reaping whirlwinds. One officer started a torrid affair with a lovely woman connected to major neighborhood drug dealers. In time, the tensions of her divided loyalties proved too draining. She seized her police lover's gun and committed suicide. Every police officer who seeks sexual solace with a Latina and then breaks it off can expect a civilian complaint to be filed, on principle.

Some officers cross the line into thievery. Usually, thefts are petty, as when corpses wearing $100 sneakers at crime scenes show up barefooted at the city morgue—the kind of expropriation of property that police equate with firemen's "liberation" of stock in fire-damaged stores ("Whadda find if you look inside a fireman's coat? Big fuckin pockets"). But the drug trade multiplies opportunities for theft and enlarges the stakes.

On a rainy night in September 1992, two Colombian drug couriers were murdered in their automobile in uptown Manhattan. Citing the ugly weather, the police garage refused to tow the vehicle back to the station house for examination. Two detectives responded to the crime scene. One drove the blood-drenched vehicle to the house. The other returned in the squad car. In the dead of night, only the two detectives and a uniformed desk sergeant were awake in the house. A routine search of the death car's trunk in the station house garage revealed $70,000 in cash, according to the later statement of one detective, who admitted stealing about $23,000 and who accused the two other officers of splitting the balance. The matter came to light only by coincidence. A female colleague, visiting a male strip club thought that a nude dancer resembled that detective, so she photographed the dancer in all his glory. She left the photo and a note on the detective's desk the next morning, saying: "I know what you've been UP to!!" The detective panicked and gave up the theft to his union representative and to the Internal Affairs Bureau.[15]

Police officers sometimes organize themselves into rings or crews and cross the line into robbery. They "boom" known drug apartments, seize guns, narcotics, and money, hand over enough of the take to ensure prison time for the drug dealers, and then tuck away the rest of the proceeds for themselves. The river of cash flowing through the drug trade invites waywardness, venality, and the coarse brutality that police are sworn to eradicate.[16]

In recent years the *New York Times* has seemed to cling to its self-chosen vocation as indignant scourge of the police even when overwhelming evidence favors the police's version of events. In 1998 the *Times* ran a retrospective story on the aftermath of the early-1990s drug wars in Washington Heights, the bloodiest precinct in the city, state, and nation during that period.[17] The *Times* pointed to long-standing tensions between police and the Dominican immigrant community that dominates the area. It attributed these tensions, in part, to "a police officer's fatal shooting of an unarmed man in 1992" and to "the April 1997 death of Kevin Cedeno, shot in the back by an officer who was named 'cop of the month' by his colleagues soon after." The article quoted a community activist: "At least the drug dealers are not here to hurt you—they're here to make a profit." The *Times* article provided no further details on the shootings and failed to mention other well-documented versions of both events that contrasted sharply with what it chose to report.

As it happens, DANY led exhaustive grand jury investigations into both shootings to determine whether criminal charges should be brought against the officers involved, issuing reports readily available to the public and to the *New York Times*. Based on a great deal of evidence presented to the grand jury in the first case, including a taped police radio transmission that recorded the hand-to-hand, life-threatening combat between Kiko Garcia and PO Michael O'Keefe on the night of July 3, 1992, the grand jury determined that O'Keefe was justified in shooting Garcia, who resisted being disarmed of his .38-caliber revolver, recovered at the scene and later subjected to exhaustive traces on its history. Garcia turned out to be a local, small-time drug dealer. The grand jury investigation turned up a homemade videotape of Garcia and others in the street juggling bags that they themselves said contained cocaine. The tape had been seized by police in a warranted search of Garcia's boss's drug location on March 25, 1992.

Another grand jury exonerated PO Anthony Pellegrini in Kevin Cedeno's death of April 6, 1997. In its report of that investigation, DANY noted that, in the 0330 darkness on Amsterdam Avenue between 163rd and 164th streets, Officer Pellegrini's partner, PO Michael Garcia (no relation to Kiko), mistook the fleeing Cedeno's two-foot-long machete with black metal blade and black handle for a sawed-off shotgun. Cedeno had retrieved the machete from the nearby apartment of a close friend because, after attending a party, he and his friends had had an altercation on the street with a group of Hispanic youths near 162nd Street and Amsterdam. After getting the weapon, according to DANY's report, "Cedeno was visibly agitated, and though witnesses differ with respect to some details of his behavior, it is clear that he pulled out and brandished the machete at least once, and that he announced that he wanted to 'cut somebody.' Witnesses describe him as pacing. One states that when he and one of his antagonists began to argue, Cedeno started to pull the machete on him, and would have done so had the witness himself not intervened."

All the while, the street crowd grew in size and volubility, with some throwing bottles at rival factions. A woman alerted two plainclothes police officers in the 163rd subway station of the brewing street violence. These transit division officers radioed two uniformed transit officers upstairs in a radio car. These officers responded to the scene. Cedeno fled from the approaching officers at the urging of at least three companions because he was illegally armed while on parole for the armed robbery of a 57-year-old man. Around the same time, a local resident on 162nd Street called 911 with a "shots fired" report. Central quickly relayed that emergency call to the 33rd precinct, along with the dispatcher's warning that there was a "large dispute" in the street involving a knife.

Two uniformed officers responded to the scene. Pellegrini and Garcia, accompanied by two rookies just out of the police academy, also answered the dispatcher's call. By the time all of these police officers reached the scene, Cedeno was running up the east side of Amsterdam away from the direction of the reported gunfire. From the way he was holding the machete close to his body, it appeared that he was clutching a firearm. Garcia yelled:

"Oh, shit, Tony, he's got a gun." Pellegrini got out of the radio car and repeatedly ordered Cedeno to drop his weapon. He fired a single shot only when Cedeno, who had his back to the officers, dropped his shoulder, began to turn, and appeared to be swinging the object in his hands to point it toward Pellegrini, with Garcia all the while shouting that the man was armed.

That Pellegrini's shot hit Cedeno in the lower back was the object of close scrutiny by the grand jury, but the jury determined that the location of the fatal shot was a function of the street positioning of the actors. The grand jury based its findings on the testimony of thirty-six witnesses, including, as DANY's report said, no fewer than ten civilian witnesses "almost all friendly to Cedeno." Cedeno's autopsy revealed a .14 blood alcohol level (at the time, the New York standard for legal intoxication was .10), which, along with his evident agitation because of the earlier street altercation, seems to have slowed his responses considerably.

The kind of willingness to act decisively in the midst of perceived danger that Pellegrini displayed is celebrated by urban police officers as an occupational necessity and virtue. But in Washington Heights, and in many other quarters as well, the grand juries' findings and DANY's reports of them were dismissed outright as official propaganda covering up police brutality. In failing even to mention the existence of these detailed public records, which, in the O'Keefe case at least, another *Times* staffer had accurately reported almost six years earlier, and in recycling community sentiments as though they were settled facts, the editors and reporter of the *Times* countenanced and bolstered those perceptions.[18]

Whatever their own predilections for wish-news (a story so good that journalists wish it were true) or however much they are under the thumbs of their editors to write from certain angles, journalists have the opportunity to confront and interpret actual social reality. Journalists do their best work when they eyewitness and report extreme situations, such as war, terrorism, epidemics, mob violence, or catastrophic accidents, calamities with intrinsic drama that place a premium on physical courage and clearheaded narrative reportage.

Some intellectuals seem principally guided by ideological and moral commitments. Even when they are engaged in important work such as understanding the marginal phenomenon of false confessions, the troubling anomaly of increasing incarceration during periods of declining crime, or conflicting claims about the role of aggressive policing to control disorder amidst the stratification of all advanced industrial societies, they devote themselves with single-minded enthusiasm to ideological advocacy. Much of this advocacy is directed against police, the front-line soldiers in taking statements from criminals or suspects and arresting those involved in or accused of crimes.[19] These intellectuals demonstrate themselves as men and women in permanent opposition to authority of any sort, simultaneously asserting claims to rebelliousness and probity. Police regularly bear the brunt of the indignation that marks their discourse.

Other intellectuals find the moral ambiguities of criminal investigation troubling but nonetheless fascinating. An April 16–17, 1993, symposium at Williams College brought together judges, prosecutors, and detectives from both the transit and city police with professors from several colleges to discuss the crisis of the criminal justice system.[20] The moral anomalies of detectives' work framed the discussions. The exchange between these two occupational communities, both committed to the search for truth, often resembled a meeting in the Tower of Babel. Sometimes the dialogue became barbed, and at other times both groups found the conversation hilarious.

The conference started with graphic presentations of violent crimes in the hole and on the streets. The transit detectives showed a surveillance video that caught the terror of a feeble 63-year-old man trapped in the subway beneath a housing project in Brooklyn as he was stalked, toyed with, thrown down a set of stairs, stomped, and finally hit on the head several times with a gun by a cluster of seven youths, all of whom turned out to live directly upstairs. They also spoke about numbingly routine visits to parentless homes to pick up suspects, and hostile receptions from beleaguered grandmothers. The 34th precinct detectives talked about drug-related violence in uptown Manhattan, including their encounters with ravenous pit-bull watch dogs in drug-sale buildings, with victims who were tortured for information about drug stashes, and with the bodies of slain couriers, customers, robbers, dealers, or informants. They noted the typical refusal of Washington Heights residents to cooperate with the police and the vast sums of drug money laundered annually to the Dominican Republic through beauty parlors, *casas de cambio,* restaurants, and car garages. Both the transit and city police pointed out criminals' callousness, wanton disregard for others, and the increasing racial polarization of New York City. Their presentations evinced the dry, matter-of-fact manner of men whose work makes surprise a stranger.

At this point in the symposium, Detective Gennaro Giorgio gave a lengthy presentation of his investigation into the February 14, 1990, murder of Selma Fabisch at 20 Magaw Place in the upper end of the 34th precinct, once known as "Frankfurt-on-the-Hudson" because of its large community of German Jews, many of them escapees from Nazi Europe. Mrs. Fabisch's building's superintendent discovered the 84-year-old woman in her pajamas and robe, soaking wet in an empty bathtub, with a wet towel over her face. Detectives Jack Collich and Giorgio arrived at the scene at 1950 hours and quickly determined that Mrs. Fabisch had been tortured. Certain rooms of her apartment were in disarray with papers strewn everywhere, though the rest of the apartment was orderly. Medical technicians estimated the time of her death to be between 0600 and 0800 hours.

A canvass of the building turned up a witness who had seen a young man in front of Mrs. Fabisch's apartment at 0730 hours. Family members told detectives that Mrs. Fabisch had recently employed a health-care attendant through a local community council to care for her ailing husband, who eventually died on January 30, 1990. A call to the director of the community council revealed that the council had sent a man named Reginald Petty to Mrs. Fabisch in early January. Mrs. Fabisch's phone book listed the name Reginald Petty with a Brooklyn phone number. A search of criminal records determined that Petty had a record of seven arrests between 1982 and 1984, including arrests for assault, robbery, reckless endangerment, and attempted murder. But he had successfully pleaded all charges down and had served little prison time.

The phone number came back to Petty's girlfriend, "Sassy White." When the police interviewed Sassy, she said that Reggie had told her that he had given a ride to his friend, "Joe," who wanted to go to Manhattan. Reggie went to see Mrs. Fabisch, Sassy said, to apologize for using her bankcard several times. According to Sassy, Reggie went up to Mrs. Fabisch's apartment, spoke to her, and she gave him three dollars. He returned to the car where Joe was waiting, but then Joe went upstairs to Mrs. Fabisch's apartment. When he returned, Joe told Reggie that he had tied Mrs. Fabisch up and put her in the bathtub. Joe then gave Reggie several bankbooks recording certificates of deposit, which, Reggie told Sassy, he still had in his possession.

The detectives quickly traced Selma Fabisch's bank accounts back to a local branch office of Citibank. There, a bank officer said that a few weeks earlier a young man had tried to cash a $3,000 check on Mrs. Fabisch's account, a transaction that the bank refused. When the detectives asked the bank officer to view a photo array, she picked out Reggie Petty as the man who had tried to cash the check. A survey of Mrs. Fabisch's ATM card with Apple Bank revealed several transactions, all of which had been videotaped. The man using the ATM card was Reggie Petty. In the meantime, a search of traffic summonses on Magaw Place yielded two tickets that were written at 0542 hours in the early morning of February 14, 1990, to a car owned by Reginald Petty for double-parking in front of 15 Magaw Place.

Detectives Giorgio and Collich went to Brooklyn looking for Petty but found his brother instead. Petty's brother asked suspiciously what the police wanted with Reggie. Giorgio gave the brother a card with the 34th squad's phone number on it, but otherwise blank. On the card, Giorgio wrote BANK SQUAD in bold letters. Giorgio said that he needed to talk with Reggie to clear up some issues about checks. He assured Petty's brother that there was no rush on the matter. A few days later Petty did call Giorgio, who asked if Petty could stop by the 34th precinct to talk about the checks. Petty said that he was busy with work, a typical testing of the waters to see how anxious the police were to see him. Giorgio told him that he was also quite busy, that there was no rush, and that the matter was minor. Petty agreed to come to the station house in a few days. When Petty came, Giorgio greeted him casually, got him some coffee, and asked him to wait a few minutes while he finished up some other work on bank matters.

By the time the interviews with Petty began at 1100 hours on February 20, 1990, Petty was relaxed and ready to help Giorgio clear up the matter of the checks. Petty said that, when he worked for Mrs. Fabisch in January, he accompanied her to the bank several times to cash checks in order to buy groceries. He claimed that she gave him her ATM card and its code to withdraw money for her. He acknowledged keeping the card overnight and using it for himself, withdrawing a total of $400. He said that he owed a lot of money to loan sharks. Indeed, he felt that his life was in danger because he had been shot at only the month before. He said that he had not gone into Manhattan since mid-January. He agreed to take a lie detector test to verify his story. During that test, he was asked if he had killed Mrs. Fabisch. He emphatically denied doing so. The polygraph expert determined that Petty was lying.

When Petty was told that he had failed the polygraph, Giorgio immediately read him the *Miranda* warnings. He then said to Detectives Giorgio and Collich: "What if I was with her, and then left, and another person killed her?" He said that he could be a witness, because his friend, Joe, had killed Mrs. Fabisch. When the detectives asked Petty for an explanation, he gave another version of the story that Sassy White had reported. He said that he, accompanied by Joe, had gone to see Mrs. Fabisch. He had asked her for fifty dollars to help him out, but she gave him only three dollars. He returned to the car where Joe was waiting. Then Joe asked the number of Mrs. Fabisch's apartment and marched into the building. Reggie insisted that he had no idea what Joe was going to do. When Joe returned, he told Reggie that he had tied up the woman and thrown her in the tub. Reggie told the detectives that he did not want to get involved, but Joe had given him Fabisch's bankbooks, which he had kept in a safe at his father's house in Brooklyn.

The detectives interviewed Reggie again at 2350 hours. Reggie told the same story as before, but in much more elaborate detail. He said that it was 10:30 A.M. on February 14 when he went with Joe to Magaw Place. He had coffee and toast with Mrs. Fabisch in her kitchen and discussed her husband's death before asking her for a loan of fifty dollars. But now he said that he left Joe in upper Manhattan and drove back to Brooklyn. Later, Reggie reiterated, Joe gave him Fabisch's bankbooks and financial materials, and these were still in the safe.

Armed with a search warrant, the detectives went to Brooklyn and retrieved Mrs. Fabisch's bankbooks and other financial documents from the safe in Petty's father's apartment. When they returned, they placed Petty under arrest for the murder of Mrs. Fabisch and advised him once again of his rights. At 0320 hours on February 21, 1990, Giorgio and Collich interviewed Reggie yet again. Giorgio pointed out some discrepancies between things that Reggie had said, specifically that he had visited Mrs. Fabisch in mid-morning, and known facts, such as the early morning time of the traffic summons and the medical technicians' estimated time of Mrs. Fabisch's death. Giorgio also noted that certain of Reggie's statements just did not ring true. Did Reggie have anything else to say?

Then Reggie said that he and Joe had discussed robbing Mrs. Fabisch of her bankbooks and financial statements late in January. On February 14, Reggie and Joe went uptown to Magaw

Place. After spending time with Mrs. Fabisch, Reggie said that he left the door unlocked on his way out of her apartment. He said that he met Joe on the stairs, and Joe went into Fabisch's apartment. An hour later, Reggie said, he returned to the apartment and found Mrs. Fabisch in the water-filled bathtub with tape on her wrists, ankles, and mouth. Reggie claimed that he removed the tape and threw it near the toilet. Mrs. Fabisch was unconscious but alive. Joe stayed in the apartment, but Reggie went back to the car and discovered two traffic summonses on the vehicle. Reggie said that he drove Joe to Brooklyn, attended to some business, and then went back to Joe's apartment, where Joe had several financial documents from Mrs. Fabisch's apartment spread out on the table. Joe told Reggie to get some ready cash from the financial materials. Then Reggie took everything and put it in the safe in his father's apartment.

Reggie made the same statement, now videotaped, to the district attorney, who validated Giorgio's arrest of Reggie. The next day, Giorgio and Collich tracked down Joe, who turned out to be one "Joe Farmer." Joe said that on February 14, 1990, Reggie had approached him with financial documents, some of which were in German, asking his help in moving funds from Fabisch's accounts to his own. Reggie explained that he kept encountering a security block. Joe said that Reggie offered him twenty, and then fifty percent of the take, if he could bypass the security block and successfully transfer the funds. Joe admitted that he had looked into moving the funds but could find no way past the security block, so he turned down Reggie's proposition. Joe said that one of Reggie's schemes involved Joe's girlfriend, "Arielle," and that the police should check with her.

When the police interviewed Arielle, she said that she remembered February 14 well, because it was Valentine's Day. She had arisen early to go look for work. She had had a fight with Joe the night before, so she sneaked into his bedroom at 7:30 A.M. and left a Valentine's Day card on his bed where he was sleeping soundly. Arielle recalled seeing Reggie later that afternoon. But she saw no papers or bankbooks that day, or later that same week. She did recall seeing Reggie in Joe's apartment several weeks earlier and noticed that Reggie had several bank receipts in his possession at that time.

Further investigation could not shake Joe's alibi. At the very least, Reggie was guilty of felony murder on the basis of his own statements. The case went to trial. The jury did not believe Reggie's accounts and found him guilty of Mrs. Fabisch's murder. The judge sentenced him to life imprisonment.

Giorgio presented the case against the backdrop of the grisly crime scene photographs of Mrs. Fabisch, bound and tortured in her bathtub. He focused in particular on the trickery that he had used to lure Petty into the squad room and his meandering interviews that led Petty to contradict himself several times and to admit his intention to participate in the robbery of Mrs. Fabisch, a crime that resulted in her murder. Thus Petty essentially confessed to felony murder. Later, Giorgio described the case of Herman Myers and the murder of Guadalupe Diaz. He focused in particular on how, under the guise of investigating a series of burglaries, he mouse-trapped Myers into saying that he never went above 190th Street in Manhattan, a statement contradicted by the eight latent fingerprints Myers had left behind in Mrs. Diaz's apartment on West 204th Street.

The professors at the conference were appalled at the violence that the detectives described, particularly its heedless, extravagant viciousness.[21] They were fascinated by detectives' immersion in a world where such violence is routine. But they quickly pointed out that the criminals in the detectives' stories did exhibit some values—they sent money back to the Dominican Republic to support their families there; they bragged about the success they achieved in America to appreciative audiences, suggesting dense layers of community organization that the police didn't acknowledge; they took their young children out to the streets to see bodies riddled with bullets, clearly, one professor argued, giving parental warnings to the youngsters to be careful. The prosecutors and judges on the panel were taken aback by the professors' insistence on value systems among criminals, in the morally relativistic sense that the professors meant. But the police readily acknowledged that many criminals whom they encountered were indeed religious people, at least in a superstitious sense. They cited the *santería* shrines festooned with saints' pictures, burning candles, and dollar bills that police regularly find in drug-sale apartments in Washington Heights. A judge then recalled pre-sentencing statements made by several drug dealers convicted of multiple murders, who told the court that God, at least, loved them and knew that they had done nothing wrong. The professors argued that such evidence suggests that New York City serves as criminals' workplace, a world of rational calculation and occupationally specific moralities that provides the criminals with the wherewithal to pursue family lives closely resembling those of other people. The panelists pointed out that the same people who were shipping drug money to their families in the Dominican Republic were also building lavish mansions on the island for themselves.

Several professors then honed in on detectives' use of subterfuge in investigations, focusing on the cases of Reginald Petty and Herman Myers. They noted that the detectives' interactions with criminals, as detectives described them, resembled con-

games, with criminals trying to outwit the police and police trying to trick criminals into foolish admissions or contradictory
statements. "The way you're describing the relations between police and . . . perpetrators is almost as a game. The police play the
game and the perpetrators play the game." The police accepted
the characterization, with the proviso that the game was serious.
"The bad guys, I've found, they wanna talk because they wanna
convince you they didn't do somethin. A bad guy doesn't wanna
say I'm not talkin to you, screw you, you want me, come and get
me, because he's thinkin: 'I'll con him. I'll convince him I was
never there and I'm gone. I'm home free and I cooperated.' So
should *we* be con men? Yes. Should we be good liars? Not to our
wives, but to the bad guys? Absolutely."

One professor pressed Detective Giorgio on his interrogation
of Herman Myers, whose exculpatory statements, contradicted
by latent fingerprint evidence, convinced the jury of his guilt.

> *Professor:* So, in a sense, your evidence was actually that he lied to
> you?
> *Giorgio:* Yes.
> *Professor:* So, you're lying to him, but when he lies to you . . .
> *Giorgio:* I'm lyin to him to get him to talk.
> *Professor:* Okay, well, maybe he's lying to you because he has a differ-
> ent point. He never had the obligation to tell you the truth. Given
> he is not under arrest, given that he hasn't even been told what he
> is under suspicion of committing, why shouldn't he lie? . . . His
> lying is already set up by a set of lies, so why shouldn't he lie?
> Under what legal obligation does he have to tell you the truth,
> when to his knowledge he is not a suspect? That is, I could lie
> to you till I'm blue in the face . . . I always lie to the police and I
> had two parents and a dog and played baseball. It's the entry [of

lying] into the court of law as evidence . . . that is distressing. Aren't you counting on this person's lack of education and his lack of . . .

Giorgio: Sure, why not? I'm dealin with a felon. Are you tellin me I can't con him? Are you tellin me I shouldn't con him? Are you tellin me I shouldn't lie to him? Are you tellin me to play by the rules in every way? Get off that line.

Professor: I think if the context is one of subterfuge . . . against the alleged perp [that] is to me just kind of unfair . . . If I were not white and I were not educated, my relationships [with the police] would be much more problematic. I would be more likely to have a rap sheet. I mean you pick up a rap sheet as a consequence of decisions to arrest, prosecute, convict, and then the rap sheet in turn becomes further evidence in court if the defendant testifies. So . . . what constitutes evidence in a system in which, first of all, there is an understanding . . . that everybody is lying and cheating and trying to be clever and, secondly, that the system is unfair in the first place because it is racist and because it is class conscious . . . I believe I live in a racist country.

Giorgio pointed to the macabre picture of Mrs. Fabisch in death and asked: "What about her?" But another professor intervened, arguing that the question went beyond fairness to any particular person such as Reginald Petty or Herman Myers. At issue was the fairness of the system as a whole. Giorgio continued: "But I pick suspects for a reason. I don't pick them out of thin air. I can't seem to get through to you that if we have to use trickery or deception to try and get to the truth, I'm gonna try whatever it takes without breakin the law, every skill, every trick, every lie that I possess. If I think the guy's a killer, and [in both the Petty and Myers cases] I'm lookin at guys who were killers,

I'm gonna use it." At this point another professor, echoing many
academics' settled beliefs in the asymmetrical relationship between police and criminals—powerful versus powerless; good up-
bringings versus deprivation; establishment versus underdog—
suggested that all custodial interrogations are ipso facto coer-
cive. The detectives' advantages and skill were such, he thought,
that they could get anyone, even anyone in that conference
room, to admit to anything, even to crimes they hadn't thought
of, let alone committed. The use of subterfuge by the police
strengthened their already dominant hand.

Then another professor addressed the panel: "What strikes me
is how sophisticated the notions of interpretation are that in-
form the work you all do. I don't find anything extraordinary or
out of line about these kinds of strategies. To the contrary, it
seems to me like what you have done and what you do is to
deploy strategies that we use all the time in social interactions
with extraordinary sophistication. What's remarkable is . . . the
hermeneutical . . . skill it seems to take in employing those strategies in ways that I think are quite remarkable. A person becomes a kind of text that doesn't want to be read and so one has
to devise various kinds of strategies to allow one to read them."
Other professors immediately chimed in: "What's interesting is
that they have to be read and reread. It isn't simply one reading,
it's a number of different readings." The rereadings, they added,
happen in the multiple layers of checks, balances, and appeals in
the criminal justice system.

Panel members were startled by the notion that persons were
"texts" to be read and reread. Thinking of countless violent
crime scenes he had visited and the sour stench of interrogation
rooms, one detective later said, "Blood and sweat smell different
than ink." Police, prosecutors, and judges alike were flummoxed
by the word "hermeneutical" and inquired what it meant. A

professor explained that hermeneutics comes from the Greek *hermeneuein:* to expound, interpret, translate, and explain. *Hermeneus,* the noun, means herald, interpreter, or expounder. Both, he noted, are derived from the name of Hermes, the son of Zeus and Maia. Hermes was the herald and messenger of the gods, and also the god of science, eloquence, trickery, and theft. Hermeneutics, the professor concluded, is the science (perhaps the art?) of interpretation.

Panel members reacted uncertainly to this exegesis. One professor elaborated: "From my point of view, the whole system is rhetorical. The whole system involves different levels of persuasion, all the way from the detectives' discussion of a kind of game plan and gathering evidence to the final place the whole legal system ends up, which is in the jury box where you are trying to persuade these people." One of the judges queried the use of the word "rhetorical." The professor continued: "What it means is that the trial is aimed at persuasion and . . . that kind of persuasion is not simply going to be a function of assessing evidence rationally." Another professor ventured: "One of the things involved here is how people construct narratives. [So] police figure out [what's happened] by whatever means they have and they construct a plausible narrative, which can be used by the District Attorney, and can be used by the next level up . . . [And] defendants . . . learn how they [can] possibly construct narratives . . . to get them off . . . If there is a strong narrative which gets across to the jury then that is going to be believed. And even the physical evidence that convinces you because you've been trained to use it in a certain way by the rules of the game in court is not going to convince anybody. So what one has—whether you realize it or not—you've been using hermeneutics in talking about narratives. The narratives of deception against people."

A panelist asked again for an explanation of "hermanonics." Several professors fairly shouted in unison: "Hermeneutics!!" A detective then asked: "Who was Herman then?"

After the ensuing uproar abated, another professor said: "Where I see the two worlds clashing . . . is around precisely these notions we've been talking about: of truth; [the idea that] lying is okay in the service of the truth; [of] fact, evidence, and reality. And many of us in this room spend a lot of time thinking about how those notions are not as simple as they often appear to be in the trenches. How notions of fact, of what counts, are inextricably bound up with—I've got to use the word—hermeneutics, frameworks of interpretation that are established. What counts as evidence in one culture and in one situation may not count as evidence in another culture and another situation. So, there are complicated codes that operate here. When there is all this talk about getting to fact, all that to our ears, I mean, we spend our lives thinking about how problematic that is."

A panelist suggested that the sharp differences in perspective between the detectives, prosecutors, and judges, on one hand, and the professors, on the other, stemmed from the age-old conflict between the theoretical and the practical, between men and women immersed in abstract ideas and those in the hurly-burly of affairs, where one must make decisions that affect the lives, fortunes, and freedom of others. But several professors rejected this distinction. One said: "I want to take issue with your opposition between the theoretical and the real . . . I can't think that it is an unreal thing for any citizen to want to speculate about the nature of justice in our society or to speculate about the nature of lying or truth."

Finally, a professor tried to pull together several strands of the conversation. He said: "When we use terms like narrative or story or rhetoric, we're not saying narrative opposed to truth.

We're not saying rhetoric as opposed to common sense or interpretation as opposed to reality. What we're saying is that everyone is involved in this. Everyone in the system is involved in particular ways of constructing cases so as to make them as persuasive as possible to the particular audiences they have to address. And whether it's Detective Giorgio making a case to Warren Murray so that Warren can make his case in front of Judge Lowe, you're engaged in a process of presenting the truth as you see it in a way that's going to be persuasive. And . . . different kinds of techniques are involved in every step of the way. Different kinds of standards are used for evaluating those stories each step of the way. And what we're trying to figure out is how you guys go about constructing those stories and interpreting those stories. We're not saying that you are departing from the truth. We're not saying that there is no common sense involved. What we're saying is that it's a hermeneutic process."

In the end, the professors had explained to their own satisfaction the work of detectives. Their appraisals of the crucial importance of interpretive and story-telling skills in the detectives' world were on the mark. But, for their own part, the members of the panel wondered if the professors understood that, while words are deeds, deeds are also deeds, and that violent criminal deeds, in particular, shatter the moral and social order and must be righted. And the exigencies of righting wrong often clash with illusions of virtue.

The panelists memorialized the conference by instituting "The Hermeneutics Lunch," held annually for years at Forlini's Restaurant near the criminal courts building in Manhattan. The lunch provided an occasion to discuss the wonders of differing occupational rules-in-use; the rule of proportionality in evaluating means and ends, action and consequences; and the contrasting interpretive stances that work experiences produce.

The dreams that haunt detectives' nights reveal somber inner-worlds. One detective dreams of drawing his gun in combat, a defining moment in every policeman's life, but the barrel turns around and points back at himself. Another dreams of pulling the trigger in a lethal struggle, but his gun will not fire. Yet another dreams of shooting his gun at a deadly assailant, but the bullets go off in wild, aimless trajectories. And yet another dreams of making an arrest when, suddenly, out of nowhere, a gun appears magically in the culprit's hands.

One detective dreams of his prisoner escaping from custody during transport, precipitating a grueling Internal Affairs investigation. Another dreams of losing a prisoner, like Shorty, in the labyrinths of the criminal justice system. Another dreams of getting trapped in a prison revolt when delivering a prisoner and being subjected to humiliation, torture, and death. And yet another dreams of getting busted on a minor beef and getting housed at Rikers Island, lying awake all night in a dormitory with fifty inmates, terrified, as daylight gradually begins to break and light streams slowly through the high caged windows, and the prisoners begin to stir with the new day, all the while knowing that it's only a matter of time until the inmates discover the new man in their midst, recognize him as police, and come after him.

Another detective dreams of being demeaned in the courtroom by a defense attorney who is skilled in making detectives, and indeed all police, into villains. Another dreams of losing a big case and having to tell witnesses to whom he has promised safety that a feared predator is back on the streets. Another dreams of acting bravely and decisively in the midst of danger, only to end up being pilloried and prosecuted. And yet another dreams of being part of a defeated army in retreat, struggling to hold tattered standards high, even as enemy shells explode and splinter in his path.

Detectives' work carries other burdens. Detectives know that their regular interactions with criminals coarsen their views of human nature and create a general suspiciousness of others' actions and motives. They recognize that their constant encounters with the results of predatory violence numb them and, indeed, sometimes horrify them. They cannot openly reveal such sentiments to their fellow officers. The whole construction of the police world depends on maintaining the appearance of a rugged emotional distance, especially from the most emotion-laden and draining experiences of their work. Only the murder of a completely innocent victim, such as a child or a fellow officer killed in the line of duty, allows police to reveal to their brothers and sisters forlorn patches at the centers of their souls. Detectives try to compartmentalize their work from their families, hiding from those they love the commonplace savageries that they encounter on duty.

But the Job has a long arm. Its rhythms, its language, its images, its ugliness, its secrecy, its corrosive cynicism, its systematic demeaning of even the best officers reach under tightly closed doors and shape the relationships between husbands and wives, or between lovers, or between detectives and their children, sometimes subtly threatening, sometimes eroding the delicate frameworks that sustain intimacy and trust.

In the end, detectives live in a world of their own, one from which they exclude outsiders who have not shared their experiences. Detectives work together, eat and drink together, attend each other's children's marriages, and mourn with each other at funerals. Alcoholism becomes an occupational hazard for some. Every year a few police officers, including detectives, "eat their guns," a sobering reminder of the fragility of this seemingly robust world. Some detectives, dedicated men and women with vast street experience and profound understanding of how to work the interstices of interlocking bureaucracies, cannot imagine anything more satisfying than criminal investigation. But the great majority of detectives long for the day when they mark twenty years on the Job and, with a generous pension, can leave it behind forever.

Yet even their parting reminds detectives of their functionary status. When detectives put in their retirement papers and hand in their gold shields, the coveted symbols of occupational identity, their shield numbers as dear to them as their own names, clerks at One Police Plaza toss the tin into an overcrowded bin and tell officers who have spent their lives trying, often in vain, to keep the city safe: "Have a great life." The Job does nothing to retain its best detectives or to harness their knowledge and experiences for a new generation of investigators. Retirement rackets, amidst corny jokes, sendups of personal idiosyncrasies of partners, exchanges of gag gifts, memories of hard times, dangerous times, and good times, Frank Sinatra music, and heartfelt embraces, always have an undertone of resentment against the organizational behemoth in which detectives have spent the best years of their lives, directed in particular at the casual shrugging off of hard-won occupational wisdom.

Leaving the Job behind is not easy. Detective work can be exhilarating. Criminal investigation poses intriguing intellectual puzzles of agency, motive, and strategy. In trying to piece those puzzles together, detectives have the license to cross the checkerboard of little worlds that comprise modern society. The best detectives can place themselves easily on the level of any person of any social station. This access to social strata high and low fosters cosmopolitan habits of mind and conversational abilities rare in men and women of the working class, even as their necessary assertion of authority makes them intriguing companions in social gatherings. Their storytelling skills always shine there, especially when they talk about the truth-is-stranger-than-fiction lore of the underworld, an undeniably fascinating arena of action, vice, lust, greed, aggression, and violence. Among themselves, in a society dominated by an apparatus of advocacy that reduces social reality to varnished black-and-white caricatures, the company of colleagues who know the realities behind the public fictions, who see the world in unfiltered, riotous color, and who feel completely unburdened by the cautions, disclaimers, qualifications, and outright lies that mark all public discourse is profoundly comforting.

And when detectives make cases, they have the singular satisfaction of knowing that, often through their dogged persistence alone, they take violent predators off the streets and, at least for a time, help ensure public safety, a bedrock condition of any democratic social order. Detectives share a remarkable sense of occupational solidarity that binds them one to another, and to other police officers, in a brotherhood of secret knowledge, duty, risk, and, sometimes, death.

NOTES

1. In the Field

1. Police use military time, which proceeds on a 24-hour clock. 2340 hours, spoken as "twenty-three forty hours," is 11:40 P.M.; 0500 hours, spoken "oh five hundred hours," is 5 A.M.

2. This line of thinking forms part of a long-term examination of the social, institutional, cultural, moral, and epistemological foundations of modern American society, seen through ethnographic studies of paradigmatic occupations and professions. This larger project has explored issues such as how bureaucracy shapes moral consciousness through a study of corporate managers' occupational rules-in-use and how the distinctive habits of mind of interpretive experts that mark the mighty apparatus of advocacy have migrated into every nook and cranny of modern social structure. See Robert Jackall, *Workers in a Labyrinth* (Montclair, NJ: Allenheld, Osmun, & Co., 1978); *Moral Mazes: The World of Corporate Managers* (New York: Oxford University Press, 1988); Robert Jackall, ed., *Propaganda* (New York: New York University Press, 1995); and, with Janice M. Hirota, *Image Makers: Advertising, Public Relations, and the Ethos of Advocacy* (Chicago: University of Chicago Press, 2000).

3. The phrase comes from NYPD Captain Alexander "Clubber" Williams. When he was transferred in 1876 from an outer borough precinct to midtown, where brothels and gambling flourished under police protection, Williams said: "I've been having chuck steak since I've been on the force and now I'm going to have a bit of tenderloin." The Tenderloin originally extended from about 23rd Street to the upper 30s, from Madison Square to between Seventh and Eighth Avenues. As theatres, nightclubs, and bordellos moved uptown, so did the Tenderloin, eventually incorporating the entire area of Times Square, that is, Seventh Avenue and Broadway between 42nd and 53rd streets, east to Fifth Avenue, west to Eighth Avenue.

4. To prepare for the work with detectives, I attended two extensive training sessions for detectives in the New York Police Department's Homicide Investigation Course during the summers of 1989 and 1990, as well as a great many social functions for detectives and other police officers. I also took a lengthy Criminal Investigation Course with "white-shield" (rookie) detectives of the New York City Transit Police in summer 1991 and yet another Homicide Investigation Course with senior transit police detectives in November 1991.

5. The New York court system instituted the lobster shift in 1982 as crime rates in the city and arrests by the police outpaced the courts' abilities to process criminals during daylight hours. The lobster shift ended, for the time being, in April 2003 with the precipitous decline in crimes and arrests that followed September 11, 2001. My thanks to Detective Sergeant Brian McCabe for introducing me to the wonderland of the night arraignment court.

6. See, for instance, New York City Police Department, Drug Enforcement Task Force, "Dominican Narco-Traffickers: Emerging Dominican Semi-Organized Crime," February 18, 1992; United States House of Representatives, Hearing before the Select Committee on Narcotics Abuse and Control, 103rd Congress, First Session, "Dominican Drug Trafficking," March 24, 1993; United States Department of Justice, National Drug Intelligence Center, "The Dominican Threat: A Strategic Assessment of Dominican Drug Trafficking," June 1997; Drug Enforcement Administration, DEA Briefing Book, "Traffickers

from the Dominican Republic," July 1, 2000. To become familiar with the intricacies of money laundering, I attended two conferences conducted by a private organization that offers compliance symposia for the financial industry. I interviewed experts on money laundering in New York City and in Washington, DC, some working for the American Banking Association and others with the United States Treasury. The interviews followed a systematic review of all available congressional hearings on the subject.

7. For more on these events, see Chapter 10.

8. In fall 1992 I began an examination of all drug-related homicides committed in the 34th precinct beginning in 1987, the year that homicides in the jurisdiction began to soar. This work continued through the summer of 1994, with ongoing updates as various cases closed. The work consisted of a systematic review of the case files of all of the 638 homicides committed in the precinct during 1987–1993 and determined, using conservative criteria developed independently from those used by police, that 379 (59 percent) of these murders were drug-related, a percentage that police and prosecutors both considered too low. I interviewed the catching detectives about their investigations for all but a few of the 379 homicides determined to be drug-related.

9. Cambridge: Harvard University Press, 1997.

2. Looking for Shorty

1. I base this broad outline of the intertwined history of robbery on the subway and the growth of the New York City Transit Police Department on a systematic analysis of news articles appearing in the *New York Times* from 1936 to 1995 and on scores of interviews with police officers and detectives in the New York City Transit Police. The outline is also indebted to a brief, unpublished history of the Transit Police by Al O'Leary, Metropolitan Transit Authority spokesman, and to Jan M. Chaiken, Michael W. Lawless, and Keith A. Stevenson, *The Impact of Police Activity on Crime: Robberies on the New York City Subway System* (Santa Monica: Rand Corporation, 1974).

2. The privately-owned Independent (IND) subway hired six unarmed New York State Police in November 1933 to safeguard passengers and property on its lines. Two years later the IND added "station supervisors" with powers of arrest, but only on IND lines.

3. See Fred C. Shapiro and James W. Sullivan, *Race Riots, New York 1964* (New York: Crowell, 1964). After an extensive investigation into the incident, Lieutenant Gilligan was exonerated of any wrongdoing.

4. On May 19, 1971, POs Nicholas Binetti and Thomas Curry, assigned to guard District Attorney Frank Hogan's residence, were lured into a short high-speed chase down Riverside Drive by occupants of a blue Maverick automobile. When the officers pulled alongside the Maverick at 106th Street in the 24th precinct, devastating machine-gun fire from the vehicle blasted their patrol car. Although critically wounded, neither officer died. Two days later, materials delivered to both WLIB radio station and to the *New York Times* claimed responsibility for the assault on behalf of the Black Liberation Army.

   On May 21, 1971, POs Joseph A. Piagentini and Waverly M. Jones responded to an unfounded domestic assault call at the Colonial Park Apartments built on the site of the old Polo Grounds in the 32nd precinct. As the officers started to return to their cars, several assailants attacked them from behind, shooting PO Jones four times, killing him instantly. PO Piagentini took thirteen bullets, many from his own weapon, and died on the way to Harlem Hospital. Herman Bell, Anthony Bottom, and Albert Washington, all members of the Black Liberation Army, were eventually convicted of the murders.

   On January 27, 1972, at least three assailants assassinated POs Gregory Foster and Rocco Laurie at 11th Street and Avenue B in the 9th precinct. Foster was shot eight times, Laurie six, all from behind. Foster died instantly, Laurie the next morning on the operating table. The police received a message the next day from the George Jackson Squad of the Black Liberation Army, claiming responsibility for the shootings and announcing the start of a spring offensive. Police in St. Louis, Missouri, ended up in a shootout after a routine stop of a van, which turned out to be a mobile arsenal. In the vehicle was Laurie's 38-caliber Smith & Wesson service revolver. Ronald Carter died in that shootout, apparently killed accidentally by his partner, who was tried, but acquitted, for the murders of Foster and Laurie. The van itself had been rented by Joanne Chesimard aka Assata Shakur, one among eleven people linked to the assaults on POs Binetti and Curry and later convicted of the murder of New Jersey State Trooper Werner Foerster. Chesimard escaped from prison in 1979 and is now living in Cuba. The Foster-Laurie case remains open and active.

5. "Reputed Black Army Member Guilty in Police Murder," *New York Times,* March 29, 1974.

6. See Fox Butterfield, *All God's Children: The Bosket Family and the American Tradition of Violence* (New York: Alfred A. Knopf, 1995).

7. For a contemporary account of the graffiti wars, see Caryl S. Stern and Robert W. Stock, "Graffiti: The Plague Years," *New York Times,* October 19, 1980 (thanks to Sarah R. Hack). See also the accounts entirely sympathetic to the graffiti artists: Craig Castleman, *Getting Up: Subway Graffiti in New York* (Cambridge: MIT Press, 1982), and the film by Tony Silver, *Style Wars* (New York: Public Arts Films, 1983). See also the hagiographic film *Basquiat* by Julian Schnabel (Burbank, CA: Miramax, 1997), celebrating the art and short life of one of the most talented subway artists, Jean-Michel Basquiat.

In June and July 2001 the 35th annual Smithsonian Folklife Festival made New York City its main theme and brought to the Washington Mall members of Tats Cru, a South Bronx graffiti organization founded by former subway writers now charging $18 to $30 a square foot for commissioned work. The aging writers expressed themselves with a mural on a specially erected wall. For recent advocacy of graffiti artistry as "perhaps the most important art movement of the late 20th century," see Joseph Austin, *Taking the Train: How Graffiti Art Became an Urban Crisis in New York City* (New York: Columbia University Press, 2002). Austin lectures New Yorkers for

their provincialism in curbing graffiti artists even as the city's anar-
chy was attracting foreign writers whose countries' authorities forbade the destruction of public property in the name of art. See also
the on-line history of the graffiti movement in New York City http://
www.at149st.com/history.html.

One's abhorrence of or support for graffiti is a touchstone in
America's on-going culture wars. For polar opposite views, see Colin
Moynihan, "A Stirring Icon That Shook Things Up at 20," *New York
Times,* April 29, 2002, a panegyric to Peter Missing, whose symbol of
an upside-down martini glass, which originated as a protest against
drunk-driving checkpoints, has graced walls, phone booths, sidewalks, glass doors, and other surfaces throughout the city for two
decades, contrasted with Heather MacDonald, "Graffiti Is Metasta-
sizing Again in New York, and Guess Who's Applauding?" *New York
Sun,* July 17, 2002. See also Ivor L. Miller, *Aerosol Kingdom: Subway
Painters of New York City* (Jackson: University of Mississippi Press,
2002), a paean to the subway painters, one that locates their rebellion
in grievous class tensions emerging out of the devastation of contig-
uous communities in the South Bronx, home to many of the key
subway writers.

The Mayor's Anti-Graffiti Task Force, a coalition of nearly a score
of city agencies, was established by Executive Order 24 in 1995. The
task force has removed several million square feet of graffiti from
thousands of sites in all five boroughs. The Vandal Squad of the
Transit Division of the NYPD makes hundreds of arrests each year
for graffiti writing. See Craig McGuire, "Graffiti 2004," *Gotham Gazette,* January 12, 2004. http://www.gothamgazette.com/article/fea-
ture-commentary/20040112/202/832. But the battle over the significance of graffiti continues. The mid-2004 arrest and trial of
James De La Vega for painting a fish leaping between a bowl and a
glass on a brick wall in the South Bronx resurrected the 1980s debates about graffiti as art or vandalism. "Free De La Vega" shirts appeared all over the city; one of De La Vega's pieces fetched $2,500 at
an auction; and Soho shopkeepers hired street artists to festoon

West Broadway with edgy pictures. See Michelle Garcia, "N.Y. Artist's Brush with the Law," *Washington Post,* June 7, 2004.

8. *Untitled* (Defacement) 1984, date frequently given as 1983.

9. For an insider's account of the Bernhard Goetz trial, see Mark Lesly (with Charles Shuttleworth), *Subway Gunman: A Juror's Account of the Bernhard Goetz Trial* (Latham, NY: British American Publishers, 1988), esp. 97–123 and 137–156, which recount eyewitnesses' testimony of Goetz's shooting and Goetz's original videotaped statement to New Hampshire police, to whom he surrendered. For a legal analysis of New York's law on self-defense, see George P. Fletcher, *A Crime of Self-Defense: Bernhard Goetz and the Law on Trial* (New York: Free Press, 1988). For one account of the cultural and political significance of the Goetz case during a period of rapidly rising predatory crime and middle-class fear of crime, see Lillian Rubin, *Quiet Rage: Bernie Goetz in a Time of Madness* (Berkeley: University of California Press, 1988).

10. See Richard Emery, "The Even Sadder New York Police Saga," *New York Times,* December 12, 1987. Emery cites a report done for the New York City Transit Police's Internal Affairs Bureau in 1984. He says the report documents DANY's acceptance of guilty pleas from defendants caught by police in decoy stings even where their supposed victims told prosecutors that no crime had happened.

11. The advent of the MetroCard in New York City beginning in 1994 and its widespread adoption by subway users in the following few years made subway "token" booths much less desirable robbery targets. But the universal adoption of the MetroCard and the elimination of the subway token in May 2003 has generated other opportunities for crime. Now a "swiper" bends a MetroCard a certain way, swipes it through a turnstile three times in rapid succession, and ends up with a credit for $2. Or "swipers" purchase multiple unlimited ride cards (each card mandates an eighteen-minute wait before reuse in a given station), sabotage the MetroCard dispensing machines in a station with paper or other implements, and then sell individual rides to frustrated strap-hangers for the normal $2 fare, making hundreds of dollars before beleaguered MTA repairmen ar-

rive to fix the damage. On any given day in early 2004, between a third and a half of the 1,600+ MetroCard dispensing machines in the city were out of commission, due mostly to tampering. See Michael Luo, "Subway Headache: MetroCard Devices Often Need Repairs," *New York Times,* February 3, 2004.

12. Transit police detectives attributed the low rate of reporting crime to widespread public perceptions of the inefficiency and especially the arbitrary unfairness of the criminal justice system at least in New York City, where one's fate depends on the vagaries of the application of sometimes radically different standards by police, prosecutors, and judges in the city's several different jurisdictions.

13. The NYPD often closes cases, including many homicides, after a first arrest is made or an exceptional clearance is taken (in an exceptional clearance, the detective has enough to arrest and knows where the culprit is, but for some reason beyond his control is unable to effect the arrest). Upon taking office in January 1994, Mayor Rudolph Giuliani appointed William Bratton as commissioner of the New York City Police Department. Bratton quickly appointed Maple as his right-hand man. A year later, in April 1995, Bratton and Maple led the merger of the New York City Transit Police into the New York City Police Department. Bratton and his men initiated a great many changes in policing in New York during Bratton's three years as commissioner, including instituting zero-tolerance policing, with the privately stated aim of driving all crime in New York City into New Jersey and Connecticut.

Crime rates in each and every precinct were carefully monitored through a computerized CompStat system, and individual commanders were held accountable for spikes in crime in their jurisdictions. By pinpointing criminal activity with some exactitude, CompStat certainly helped commanders make rational decisions about deployment of officers and aided greatly in making targeted areas of the city, and to some extent the entire city, inhospitable to crime. Perversely, the CompStat system probably accelerated the premature closing of cases committed by multiple culprits with a single arrest, long the practice in the NYPD. I observed the CompStat session on November 3, 1995.

Both Bratton and Maple published popular books on their widely acclaimed success in reducing crime in New York City. See William Bratton (with Peter Knobler), *Turnaround: How America's Top Cop Reversed the Crime Epidemic* (New York: Random House, 1998). Bratton was named chief of the Los Angeles Police Department in fall 2002, largely on the strength of his claim to success in New York. See also Jack Maple (with Chris Mitchell), *The Crime Fighter: Putting the Bad Guys Out of Business* (New York: Doubleday, 1999). For a fictional account of Maple's transit-police career and life before his rise to fame, see Michael Daly, *Under Ground* (Boston: Little, Brown and Co., 1995).

14. All quotes in this section are from interviews with subway robbers in transit-police custody or from robbers' handwritten statements in transit-police case files.

15. The Decepticons, a street gang that gained notoriety in the late 1980s in New York City, in considerable disproportion to its actual numbers, mostly for targeting other youngsters as they left school, continued a long New York City tradition of youthful street gangs dating back at least to the mid-nineteenth century. See, for instance, Herbert Asbury, *The Gangs of New York: An Informal History of the Underworld* (New York: Knopf, 1928); Harrison E. Salisbury, *The Shook-up Generation* (New York: Harper & Row, 1958); Ira Henry Freeman, *Out of the Burning* (New York: Crown Publishers, 1960); New York City Youth Board, *Reaching the Fighting Gang* (New York, 1960); Lewis Yablonsky, *The Violent Gang* (New York: Macmillan, 1962); New York State Legislature, Assembly, Committee on Cities, Subcommittee on the Family Court, *The Resurgence of Youth Gangs in New York City* (Albany, 1974); Gary Hoenig, *Reaper: The Story of a Gang Leader* (Indianapolis: Bobbs-Merrill, 1975); Betty Lee Sung, *Gangs in New York's Chinatown* (New York: Office of Child Development, Department of Health, Education, and Welfare, City College, City University of New York, 1977); Anne Campbell, *The Girls in the Gang: A Report from New York City* (Oxford: Basil Blackwell, 1984); T. J. English, *Born to Kill:*

America's Most Notorious Vietnamese Gang, and the Changing Face of Organized Crime (New York: Morrow, 1995); Ko-lin Chin, *Chinatown Gangs: Extortion, Enterprise, and Ethnicity* (New York: Oxford University Press, 1996); Eric Schneider, *Vampires, Dragons, and Egyptian Kings: Gangs in Postwar New York* (Princeton: Princeton University Press, 1999); Douglas Century, *Street Kingdom: Five Years Inside the Franklin Street Posse* (New York: Warner Books, 1999); and Bruce Davidson, *Brooklyn Gang* (Santa Fe: Twin Palms, 1998). For some fictional treatments, see James DeJongh, *City Cool: A Ritual of Belonging* (New York: Random House, 1978); Richard Wright, *Rite of Passage* (New York: Harper-Collins Publishers, 1994); Phillip Baker, *Blood Posse* (New York: St. Martin's Griffin, 1995); and Hubert Selby, Jr., *The Willow Tree* (New York: Marion Boyars, 1998). In 1997 New York City experienced a sudden upsurge in the membership of the Crips and the Bloods, famous Los Angeles gangs that had migrated east. Police attributed more than 135 random initiation-rite slashings to these gangs, many of which happened in the subways.

16. The statement is based on a careful examination of all 1,002 descriptions of assailants by victims in Central Robbery's 1990 logbook for multiple-perpetrator-robbery complaints. When queried, Central Robbery squad members, who in 1991 were 65 percent black and Hispanic, saw nothing whatsoever surprising about the overwhelmingly one-sided cast of victims' descriptions of their assailants. Black detectives in particular, such as Detectives Kelvin Miles, Billy Carter, Sonny Archer, and Zack Jackson, all superior investigators who regularly get called Uncle Toms by black culprits and vilified mercilessly by them in other ways, simply accept the racial composition of subway predators and distance themselves as much as possible from the black culprits whom they arrest. However, despite such widespread, shared understanding about which groups are most likely to commit subway robberies, it is fatal to a police officer's career if one turns overwhelmingly empirical evidence into a voiced assumption about the race or ethnicity of the culprits of the very next case that one catches.

17. According to victims' descriptions of their assailants, older subway robbers are also overwhelmingly black and Hispanic. Victims do, however, describe their assailants as white in about 10 percent of reported subway robberies committed by individuals.

18. Jack Katz, *Seductions of Crime: Moral and Sensual Attractions in Doing Evil* (New York: Basic Books, 1988), is the best treatment of robbers' occupational virtue of "hardness." See, in particular, "Doing Stickup."

19. Ten out the sixteen complainants who stayed at CROB for lineups positively identified Shorty as the man who had robbed them, bringing the total of positive identifications to twenty-one. To detectives' astonishment, Shorty was offered and readily took a ten-year guilty plea to armed robbery.

3. When the Ball Fell

1. This sect declares, according to one of its regular street preachers, that "the so-called white Jews are not the true biblical Jews. They are imposters. The true Hebrew Israelites are those whose fathers are of Indian and Negro descent throughout North, Central, and South America. The only real Jews are the Negroes brought to North America in slavery." Thus, the Lost Tribes of Israel, in hierarchical order of importance, are: Negroes, West Indians, Haitians, Dominicans, Guatemalans, Panamanians, Cubans, North American Indians, Seminole Indians, Argentineans and Chileans, Mexicans, and, finally, Puerto Ricans.

2. Here are some other common Brooklyn street phrases collected during fieldwork. Some are criminal argot; some sendups of standard nomenclature; others verselike word play; others transvaluations of technical terms from another world; others dialectlike, metaphor-rich approximations of standard usage, in which street experience trumps standard usage; and several others, particularly the medical terms, simply phonological errors that suggest the state of the public schools in New York City. My thanks to Michael Erard for his linguistic advice on these phrases.

Phrase	Meaning
Acting in concrete	Acting in concert
Athletic flips with conversions	Epileptic fit with convulsions
Colossal bag	Colostomy
Diabolic	Diabetic
Electrocution school	Electrician school
Get indicated	Get indicted
Getting paid	Doing robberies
Fireballs in eucharist	Fibrosis of the uterus
Leg Iron Street	Legion Street
Lincoln Townhouse	Lincoln Town Car
Mongo Merry Street	Montgomery Street
Monogrammed headache	Migraine headache
Onions	Bunions
Persecuted	Prosecuted
Provoked	Revoked
Roaches of the liver	Cirrhosis of the liver
Singing merry Jesus	Spinal meningitis
Smoke insulation	Smoke inhalation
Statue of liberties	Statute of limitations
Streeticide	Outdoors homicide
Subway farez	Savoir faire
Throwing asparagus	Casting aspersions
Veranda rights	*Miranda* rights
Very closed veins	Varicose veins
Virginia	Vagina

3. See Mercer Sullivan, *"Getting Paid": Youth Crime and Work in the Inner City* (Ithaca: Cornell University Press, 1989).

4. The old Latin Quarter, the famed nightclub on Broadway at its junction with Seventh Avenue between 47th and 48th streets, opened in 1942 and catered to a demimonde of international playboys, Hollywood actors, big-name performers, and celebrities of every variety

who jammed the club nightly to see, and later mingle with, its glamorous feathered showgirls. The new Latin Quarter opened in 1984 at 200 West 48th Street near Broadway. The club catered to mostly black and Hispanic youngsters from all five boroughs and specialized in hip-hop music. It developed a reputation for dance-floor fisticuffs and on-the-street melees, including at least two drive-by shootings in 1987 aimed at youngsters who had just left the club.

5. There are centralized CATCH units in each of New York City's five boroughs. Each CATCH unit houses photographs of everyone arrested in that particular borough, going back decades. For further notes on this ancient system, see Chapter 8.

6. *People v. Eric Smokes and David Warren*, 00249/87, transcript pp. 1545–1546. My thanks to ADA Susan Axelrod of the Appeals Bureau of DANY and to Warren Murray, Chief of Trial Bureau 50, for allowing me to work with Ms. Alexrod's copy of the trial transcript.

7. Marcela Rojas, "A Day of Pride at Sing Sing: 20 Beat the Odds to Get Degrees," *Journal News,* June 3, 2004. The Mercy College program at Sing Sing is made possible through Hudson Link for Higher Education in Prison. My thanks to Carol Hagglund of Hudson Link for her assistance.

4. *The Girl in the Park*

1. Many Dominican community activists and politicians claim that references to Dominican involvement in the drug trade are racist. As it happens, the journalistic, governmental, and scholarly literature documenting Dominican involvement in narcotics trafficking is vast. See, for instance, Clifford Kraus and Larry Rohter, "Dominicans Allow Drugs Easy Sailing," *New York Times,* May 10, 1998, and "Dominican Drug Traffickers Tighten Grip on the Northeast," *New York Times,* May 11, 1998; United States Department of Justice, National Drug Intelligence Center, *The Dominican Threat: A Strategic Assessment of Dominican Drug Trafficking* (Washington, DC, 1997); "Dominican Drug Trafficking," Hearing before the Select Committee on Nar-

cotics Abuse and Control, House of Representatives, 103rd Congress, March 24, 1993; New York City Police Department, Drug Enforcement Task Force, "Dominican Narco-Traffickers: Emerging Dominican Semi-Organized Crime," February 18, 1992; Joseph Michael Rogers, "Political Economy of Caribbean Drug Trafficking: The Case of the Dominican Republic" (Ph.D. Diss., Florida Atlantic University, 1999); and Robert Jackall, *Wild Cowboys: Urban Marauders & the Forces of Order* (Cambridge: Harvard University Press, 1997).

2. For an account of the Westies, see T. J. English, *The Westies: Inside the Hell's Kitchen Irish Mob* (New York: Putnam, 1990). According to several detectives who grew up in Hell's Kitchen, the Westies kept some deceased rivals' penises pickled in jars on bar shelves for reminiscing about good old days over rounds of Jameson's whiskey. They also tried (to no avail, because of a lack of body oils) to use dismembered frozen hands to plant fingerprints on murder weapons.

3. *Santería* is an Afro-Caribbean syncretistic religion that melds ancient Yoruba gods with Catholic saints. It is widely, and variously, practiced in the Caribbean immigrant communities of New York. Washington Heights is dotted with *botánicas*, specialty shops where *santería* devotees and *brujos* (priests, but with the connotation of magical skills) can purchase literature, oils, statues of saints, herbs, and other paraphernalia necessary for the religion's rituals. The rituals include animal sacrifice. See, for instance, Miguel F. Santiago, *Dancing with the Saints* (Puerto Rico: Inter American University Press, 1993); George Brandon, *Santería from Africa to the New World: The Dead Sell Memories* (Bloomington: Indiana University Press, 1993); and Migene Gonzalez-Wippler, *Santería, the Religion: A Legacy of Faith, Rites, and Magic* (New York: Harmony Books, 1989).

5. Squad Work

1. The Manhattan South Homicide Squad (MSHS) assists detectives in precincts south of 59th Street. Similar homicide units work the Bronx, Queens, Brooklyn, and Staten Island, assisting precinct detectives in murder investigations.

2. Formerly, the desk officers were lieutenants. The old-time desk lieutenant was the master of his station house. Nobody came or went without his notice and permission. Everything that happened in the precinct was logged into his huge police blotter, which constituted the official record of reported crime and police response for his precinct. Moreover, the old-time lieutenant's ferocious protectiveness toward his men was legendary. Detective Matty Fallon tells the story of a rookie cop dragging a street person before such a desk lieutenant in Brooklyn to have the lieutenant validate his arrest. The lieutenant asked the rookie if the culprit had been searched. The rookie said that he hadn't yet done so. The lieutenant asked the culprit: "Do you have anything in your pockets that might endanger this officer?" The culprit replied no. The rookie proceeded to search the culprit's pockets and, lo and behold, came up with three unguarded spikes (uncapped needles). The desk lieutenant picked up his blotter and smashed the culprit over the head, knocking him out cold, and then ordered him transported to the detoxification unit to be stripped and searched. Desk lieutenants were replaced by sergeants in the late 1960s. Despite occasional vestigial traces of an expansive exercise of the desk officer role, the relentless bureaucratization of the Job has turned most current desk sergeants into clerks.

3. The latter detective eventually apprehended the gunman, who was convicted of the triple murder at trial.

4. This system inevitably produces low-level obfuscation to which district attorneys and courts usually turn a blind eye. When Detectives "White," "Black," and "Brown" all aid Detective "Green" on a homicide investigation, they usually provide Detective Green with verbal investigative reports of much, though not all, necessary work in the investigation. Unless Detective Green specifically asks for written reports on certain aspects of the investigation, he usually passes off work reported to him as his own in official reports in order to save his bosses the necessity of authorizing endless, often needless, hours of court time later.

5. See Robert Jackall and Janice M. Hirota, *Image Makers: Advertising, Public Relations, and the Ethos of Advocacy* (Chicago: University of Chi-

cago Press, 2000), 179–182, for an account of the Central Park Jogger case. Two sensational trials sent five youngsters to prison for the assault on the jogger, on the basis of their videotaped statements admitting the crime. More than a decade later, Matias Reyes, who was not previously charged in the case and who was serving time for raping and murdering a pregnant woman in 1991, came forward to confess his own (solitary, he claimed) attack on the woman investment banker. DNA evidence found at the crime scene in 1989 did in fact conclusively link Reyes to the attack on the jogger.

Reyes's admission and the newly reevaluated DNA evidence precipitated a massive reinvestigation of the case by DANY. On December 5, 2002, that office, in *Affirmation in Response to Motion to Vacate Judgment of Conviction,* Indictment No. 4762/89, asked the court to vacate the convictions of the five men for the assault on the jogger as well as their convictions for other assaults committed in Central Park on the same night, in light of the probable effect that the new evidence would have had in the original trials. On December 19, 2002, State Supreme Court Justice Charles J. Tejada did indeed vacate the convictions. A great deal of the highly racialized public debate surrounding the reevaluation of the case centered on whether police detectives had coerced the convicted youngsters' statements. At least two pieces of city and state legislation were submitted by Harlem-based politicians demanding the videotaping of all interaction between police and felony suspects from the moment custody begins.

On January 27, 2003, the NYPD released a *Central Park Jogger Panel Report* prepared by two prominent New York City attorneys and the police department commissioner of legal affairs. That report argued that the five youngsters convicted in 1990 had "most likely" participated in the beating and rape of the jogger. The panel said: "We adopt the view that the most likely scenario for the events of April 19, 1989 was that the defendants came upon the jogger and subjected her to the same kind of attack, albeit with sexual overtones, that they inflicted upon other victims in the park that night. Perhaps attracted to the scene by the jogger's screams, Reyes either joined in the attack as it was ending or waited until the defendants had moved on to their next victims before descending upon her himself, raping her and inflicting upon her the brutal injuries that almost caused her death. On this theory of the facts, there is no reason to believe that the defendants were prompted into making erroneous statements." A spokesman for DANY promptly disputed the police panel's interpretation.

6. The NYPD's Mounted Unit named one of its horses Keith Levine in honor of Sergeant Levine. Keith Levine, a 15-hand bay gelding, served the department and the city from 1992 to 2002. He was ridden for almost that entire period by Police Officer Joseph L. Perno, who had served on patrol in Midtown North when Sergeant Levine was murdered and who originally suggested naming the horse for his slain brother officer.

7. I spent the entire day of October 21, 1991, and parts of several other days during the same month on the street with PO Ryan.

8. See, for example, the famous case of Lieutenant Charles Becker, the head of the notorious Strong Arm squad in 1912 that kept gambling establishments faithful to their obligated tithes to Tammany Hall. Herman Rosenthal, a petty gambler who kicked against Becker's discipline, gave an account of the Tammany graft system enforced by the police department to journalist Herbert Bayard Swope of the *New York World*. Rosenthal was gunned down shortly afterward by four hitmen. The Manhattan district attorney, Charles Whitman, assisted greatly by Swope, who regularly printed releases of presumably secret grand jury testimony leaked to him by Whitman's office, blamed Becker for setting up Rosenthal's murder. And Becker was in fact convicted in two trials (the first trial was overturned by the Appellate Division for procedural irregularities by the trial judge) and executed, as were the four gunmen.

Although there is every reason to believe that Becker wanted Rosenthal dead, the evidence against the lieutenant for actually plotting the murder was thin indeed. In all likelihood, the murder was arranged by the very men who became the chief witnesses against Becker in order to protect Tammany Hall's big stake in gambling, a cash-cow bonanza overseen by Big Tim Sullivan. See Andy Logan, *Against the Evidence: The Becker-Rosenthal Affair* (New York: McCall, 1971); Henry H. Klein, *Sacrificed: The Story of Police Lieutenant Charles Becker* (New York: Isaac Goldman, 1927); and Jonathan Root, *One Night in July: The True Story of the Rosenthal-Becker Murder Case* (New York: Coward-McCann, 1961).

9. This hierarchical *structure* persists over time. In November 2002, for instance, the NYPD had 6,726 detectives, of which 681 were second-grade detectives and 211 were first-grade, with all the rest at third-grade. As in all years since 1995, the 5,939 detectives in September 2004 include those in the Transit Bureau and the Housing Bureau. Of this number, 5,203 were male and 736 were female. My thanks to Sam Katz of the New York City Detectives' Endowment Association for providing these figures.

10. For a particularly egregious example of similar judicial capriciousness, see Judge M. Langhorne Keith's dismissal of the murder indictment against John Muhammad for his wanton sniper murder of FBI analyst Linda Franklin in Falls Church, Virginia, in 2002. On January 6, 2004, a Fairfax county police detective sent a facsimile followed by a teletype to a jail in Manassas, Virginia, where Mr. Muhammad was being held, as a "detainer," an order to hold Muhammad for trial. Normally, a detainer simply expresses the interest of one jurisdiction in a prisoner being held by another jurisdiction to prevent an unwarranted release from custody. On May 27, 2004, after the Fairfax police department completed its own investigation, a Fairfax detective personally served Muhammad with an arrest warrant. On October 1, 2004, Judge Keith dismissed charges against Muhammad, saying that the January 6 detainer amounted to an arrest and that his "statutory right to a speedy trial was violated." See Thomas Crampton, "One Murder Charge Dismissed in a Sniper Attack in Virginia," *New York Times,* October 2, 2004.

11. For what most prosecutors consider an especially obnoxious exam-

ple of such moralistic self-righteousness, see David Heilbroner, *Rough Justice: Days and Nights of a Young D.A.* (New York: Pantheon, 1990).

12. The impartiality of the courts, in New York City at least, is a subject of constant dispute. In New York State, justices are both elected and appointed to the New York State Supreme Court, the state's trial court for felonies and major civil matters. As the *New York Times* ("New York's Farcical Judicial Elections," November 2, 2002, editorial) points out, scarce spots on the ballot "within a generally mediocre pool of candidates" are the property of "lucrative [Democratic Party] clubhouse patronage." In New York City, the main route to the Supreme Court bench is through mayoral appointment to the Criminal Court, which adjudicates misdemeanors and petty civil matters. "Acting" Supreme Court justices are then typically appointed from the ranks of Criminal Court judges, for terms that last fourteen years. Thus, the quality of most judges elected or appointed to the bench, with the greatest effect on the everyday life of New Yorkers, is seen by all the main players in the criminal justice system—police, prosecutors, defense attorneys, and judges themselves—to be wholly dependent on New York City's quasi-tribal politics. As Jeffrey A. Segal and Harold J. Spaeth argue convincingly about the United States Supreme Court in *The Supreme Court and the Attitudinal Model* (Cambridge: Cambridge University Press, 1993), so in the New York courts do judges' ideologies regularly seem to trump dispassionate adherence to judicial norms.

Moreover, the 2002–2003 revelations about corruption on the Brooklyn bench—judges soliciting fees from attorneys for favorable rulings in civil cases, administrative judges assigning particular cases to political-clubhouse cronies on the bench, and other judges turning a blind eye to their colleagues' involvement in bribery—fueled more debate about the state of the New York bench. For preliminary accounts, see William Glaberson and William K. Rashbaum, "Indictment of a Brooklyn Judge Provides Details of Seemingly Routine Corruption," *New York Times,* January 25, 2002; William K. Rashbaum,

"Another Brooklyn Judge Said To Be Reassigned," *New York Times,* February 23, 2002; and Andy Newman, "Judge in Corruption Plans a Guilty Plea, Court Papers Say," *New York Times,* July 2, 2002. For an analysis of the judicial patronage system in the Bronx and the incestuous relationships it inevitably produces between judges and lawyers who appear before them, see Clifford J. Levy, Kevin Flynn, Leslie Eaton, and Andy Newman, "A Bronx Judiciary Awash in Patronage, All Legal," *New York Times*, January 3, 2003. A similar system obtains in Manhattan. Despite their highly selective character, the annual reports of the New York State Commission on Judicial Conduct, which reviews the 1,000+ formal complaints made against New York judges every year, make lively reading. See www.scjc.state.ny.us.

13. For a sustained treatment of the social roots of the epistemological Hobbesianism that marks all public discourse in American society, see Jackall and Hirota, *Image Makers*, esp. 207–228. For excellent historical and institutional treatments of the transformation of the notion of responsibility in American society, see James L. Nolan, Jr., *The Therapeutic State: Justifying Government at Century's End* (New York: New York University Press, 1998), and *Reinventing Justice: The American Drug Court Movement* (Princeton: Princeton University Press, 2001).

14. See Robert Jackall, *Wild Cowboys: Urban Marauders & the Forces of Order* (Cambridge: Harvard University Press, 1997), 4–32.

6. Street Work

1. See Robert Jackall, *Wild Cowboys: Urban Marauders & the Forces of Order* (Cambridge: Harvard University Press, 1997), 270–284.

2. In 1970 Angela Yvonne Davis was accused of murder and kidnapping in what prosecutors argued was a conspiracy to free the Soledad Brothers, three black inmates of Soledad prison in California, themselves accused of murdering a white prison guard there. The most famous of the three was writer, revolutionary, and convicted armed robber George Jackson. According to the state indictment as reported in the *New York Times,* November 12, 1970, police discovered

that the guns used in an armed takeover and hostage-taking in the Hall of Justice in San Rafael, Marin County, California, on August 7, 1970, led by George Jackson's younger brother, Jonathan, had been purchased by Davis and were registered to her. At one point during the seizure of the hostages, according to some accounts, Jonathan Jackson had demanded the release of the Soledad Brothers. Prosecutors argued that Ms. Davis and Jonathan Jackson had been observed in each other's company several times in the days before the raid on the courthouse. The hostage-taking produced a wild shootout outside the courthouse in which Judge Harold J. Haley, one of five hostages taken, had his head blown off with a single-barrel sawed-off Spanish shotgun owned, according to the indictment, by Ms. Davis. Jonathan Jackson, along with James McClain, a San Quentin inmate on trial for assaulting a guard with a knife, and William Christmas, a fellow San Quentin inmate present to testify on McClain's behalf, were killed by police. Ruchell Magee, another San Quentin inmate in the courtroom to testify for McClain, was gravely wounded in the chest. Gary W. Thomas, the prosecuting attorney, was paralyzed for life. Three women jurors taken as hostages escaped alive.

After a nationwide hunt, during which Ms. Davis received succor from supporters across the country, she and a companion were apprehended by FBI agents in a Howard Johnson Motor Lodge on Eighth Avenue at 51st Street. The trial became a prototype of politicized legal proceedings involving black defendants in the following thirty years, and a paradigm of how to mobilize public opinion through racially-tinged propaganda to thwart the procedural rationality of the criminal justice system. Davis was acquitted of all charges, most specifically that she supplied the weapons used in the courthouse raid to Jonathan Jackson. She did not explain with any specificity why she fled California the day after the catastrophe and remained a fugitive for more than two months.

In 1995 Ms. Davis was appointed to a Presidential Chair in the university system that had fired her decades before for her Communist Party activities. She teaches in the History of Consciousness program at the University of California, Santa Cruz. She is a featured speaker at universities across the country on the abolition of prisons, on feminism, and on the plight of oppressed peoples everywhere. In October 2003 Professor Davis accepted the City of Paris's award of citizenship to Wesley Cook aka Mumia Abu-Jamal, who was unable to attend the ceremony because he is imprisoned for life after being convicted for the December 9, 1981, murder of Philadelphia Police Officer Danny Faulkner.

3. See Jackall, *Wild Cowboys*, 34–58, esp. 46–47.

4. More than thirty states have statutes that, in various ways, recognize unborn children who are victims of violent crime as members of the human family. However, New York State has contradictory statutes on this issue. The killing of an "unborn child" after twenty-four weeks of pregnancy constitutes a homicide (N.Y. Penal Law 125.00). However, a "person" is defined as "a human being who has been born and is alive" (N.Y. Penal Law 125.05). Versions of an Unborn Victims of Violence Act passed the United States House of Representatives in 1999 and again in 2001 but were not acted upon by the United States Senate. In January 2004 essentially the same act (H.R. 1997) was again passed by the House. In March 2004 the Senate (S. 1019) passed its version of the same bill. The law allows federal authorities to prosecute a culprit for injury or death sustained by a pregnant woman's unborn child, if inflicted in the course of committing any of the already-defined 68 federal crimes of violence against the woman herself. The long-run implications for "fetal rights" of the United States Supreme Court's refusal to reconsider the conviction of a woman found guilty of the homicide of her nearly-full-term still-born daughter because of her cocaine use during pregnancy are as yet unclear. See *State v. McKnight,* 576 S.E.2d 168 (S.C. 2003), cert. denied, *Regina D. McKnight v. South Carolina,* 124 S. Ct. 101 (U.S. 2003).

7. *Waiting for Chocoláte*

1. Street "runners," sometimes called "steerers," do not themselves hold drugs for sale but instead direct customers to drug dealers who

typically ply their trade in sale apartments off the streets. Runners take a commission on sales, either in cash or in kind. In some operations, they are expected to screen or vouch for customers to the dealers, but no respectable dealer relies solely on such assurances. The instituted occupational role of runner only marginally insulates dealers from the main hazards of their trade: undercover police and especially robbers.

2. Robert Jackall, *Wild Cowboys: Urban Marauders & the Forces of Order* (Cambridge: Harvard University Press, 1997), 60–100.

3. By 2001 the labor-saving wanted card system had changed into a labor-intensive program. By then, every time a detective filed a wanted card he had to create a folder detailing his weekly routine checks into a standardized list of search items. Wanted card units were created and disbanded, depending on manpower needs.

4. For accounts of the Brinks robbery, see John Castellucci, *The Big Dance: The Untold Story of Weatherman Kathy Boudin and the Terrorist Family That Committed the Brinks Robbery Murders* (New York: Dodd Mead, 1986), and Susan Braudy, *Family Circle: The Boudins and the Aristocracy of the Left* (New York: Alfred A. Knopf, 2003).

5. This account follows closely Cowell's statement to Detectives Austin Francis Muldoon and Joel Potter taken in the 34th precinct on August 5–6, 1991, written down by Detective Muldoon and signed by Julian Cowell and the two detectives. I also viewed Cowell's videotaped statement taken the next day by an assistant district attorney. These two statements differ only in minor details.

6. *People v. Yukl,* 25 N.Y.2d 585 at page 589. See also *Hicks v. United States,* 382 F.2d 158.

7. *People v. Balint,* 92 A.D.2d 348; 460 N.Y.S.2d 563; 1983 N.Y. App. Div.

8. See Harold J. Rothwax, *Guilty: The Collapse of the Criminal Justice System* (New York: Warner Books, 1997). Rothwax's anecdotal polemic against the system that he served for more than a quarter of a century, both as a defense attorney and a judge, particularly against the system's elevation of procedural over substantive justice, elicited both lavish praise and extreme condemnation. Rothwax's main point—that criminal proceedings should aim to discover "truth," in

the sense of who committed specific illegal actions, instead of being the juggling of legal "filters" designed to keep relevant information from juries—was deemed to be a radical idea.

9. On December 7, 1999, Cowell made a motion to vacate his conviction on the grounds that his trial counsel "coerced him into pleading guilty by threatening to withdraw from the case on the eve of trial if appellant did not plead guilty." Cowell further alleged that his attorney had promised that he wouldn't serve more than ten years of his sentence. The New York State Supreme Court denied Cowell's motion and the Appellate Division denied his application to appeal. Cowell filed two other motions with the same claims and received the same results.

10. *People v. Cowell,* 782 N.Y.S.2d 458 (1st Dept. 2004). See also Defendant-Appellant's Brief and Respondent's Brief. My thanks to ADAs Marc Frazier Scholl and Meredith Boylan of DANY for their assistance in locating the Appellate Division materials.

11. My thanks to ADAs Steve Saracco and Stacey Mitchell of DANY for providing me with access to their copy of the original trial transcript of *People v. Rodriguez,* 281 A.D.2d 375 (1st Dept. 2001).

8. Tracing the Past

1. In one of its endless reorganizations, the NYPD created "homicide zones" in the early 1970s, each with its own homicide squad. Manhattan had five such zones. Homicide squad detectives caught cases in their own zones. Sometimes the Manhattan Homicide Task Force assisted zone detectives in their investigations. The 5th Homicide Zone extended from 86th Street and the Hudson River to 165th Street and the Harlem River, and then from 165th Street to the upper tip of Manhattan, from river to river. The 5th Homicide Zone squad turned out of the 24th precinct in 1973, and then moved to the 26th precinct in 1974.

2. Parkhurst, an 1866 graduate of Amherst College, led a crusade against police graft and tolerance of debauchery and vice, especially

in lower Manhattan's infamous Five Corners area. Parkhurst's moral outrage was the prelude to the Lexow Committee, the first great inquisition into police corruption in the city. See Charles W. Gardner, *The Doctor and the Devil, or Midnight Adventures of Dr. Parkhurst* (New York: Gardner & Co., 1894). See Gabriel Chin, *New York City Police Corruption Commissions, 1894–1994*, vol. 1 (Buffalo: W. S. Hein, 1997).

3. Thomas F. Byrnes, *Professional Criminals of America* (New York: Cassell & Co. Ltd., 1886).

4. There is a vast and growing literature on eyewitness identification and its purported problems, emanating mostly from experimental psychologists. Perhaps the leading figure in the field is Professor Gary L. Wells of Iowa State University. For a state-of-the-art overview of this area of inquiry and a comprehensive bibliography, see Gary L. Wells and Elizabeth A. Olson, "Eyewitness Identification," *Annual Review of Psychology* 2002. Wells aided the United States Department of Justice, National Institute for Justice, in formulating the guidelines in *Eyewitness Evidence: A Guide for Law Enforcement* (Washington, DC: Government Printing Office, 1999), developed in response to the DNA-technology overturning of the convictions of more than sixty prisoners in the 1990s, most of whom had been incarcerated on the basis of eyewitness evidence. These rules are still being revised. Now sequential-photo identifications are replacing photo arrays; sequential lineups are replacing panel lineups; and soon all identification procedures will be double-blind to prevent even the accusation of police influence on witnesses. Joining Wells in efforts with the Justice Department and local police are other important scholars in this area, including Roy S. Malpass. See, for example, G. L. Wells, R. S. Malpass, R. C. L. Lindsay, R. P. Fisher, J. W. Turtle, and S. M. Fulero, "From the Lab to the Police Station: A Successful Application of Eyewitness Research," *American Psychologist* 55, no. 6 (2000): 581–598.

5. Simon A. Cole, *Suspect Identities: A History of Fingerprinting and Criminal Identification* (Cambridge: Harvard University Press, 2001). Cole's book argues that the colonial origins of using fingerprints for identi-

fication have tainted all subsequent uses by authorities, a colorful interpretation that one need not accept to appreciate the solid research presented in his volume. See my review of Cole's book in "Tales Told by Loops, Whorls, and Ridges," *Science* 293 (2001): 1771–1772. Some of the material in this section was originally published in the *Science* review and is used again in different form with the permission of the AAAS.

6. All the while, another identification technique competed with fingerprinting for ascendancy. Anthropometry, invented by Alphonse Bertillon, mandated detailed measurements of skulls, feet, and other body parts, reduced to highly standardized *portraits parlés,* which purportedly enabled authorities to ascertain identities. Anthropometric measurements, sometimes in conjunction with fingerprinting, continued in use well into the twentieth century. Eventually fingerprinting came to be seen as a convenient and efficient alternative to the cumbersome *Bertillonage* system.

7. Cole, *Suspect Identities,* 73–88 in particular.

8. Ibid., 88–89, 168–189. See also the lively, first-hand account of the history of fingerprints and their use as evidence in the United Kingdom and its colonies by Gerald Lambourne, *The Fingerprint Story* (London: Harrap, 1984).

9. See Colin Beavan, *Fingerprints: The Origins of Crime Detection and the Murder Case that Launched Forensic Science* (New York: Hyperion, 2001).

10. *Hamilton v. H.M. Advocate,* Court of Justiciary, Scotland, October 19, 1933. Opinion of Lord Justice-General Clyde. My thanks to Yuki A. Hirose of Frankfurt Kurnit Klein & Selz, PC, for help in locating this decision.

11. See United States Navy Department, Bureau of Navigation, *Finger-Print Evidence* (Washington, DC: Government Printing Office, 1922); United States Department of Justice, Federal Bureau of Investigation, *The Science of Fingerprints: Classification and Uses* (Washington, DC: Government Printing Office, 1984). For an example of testimony by a fingerprint examiner the scientific basis of which was challenged by the court, see *United States v. Parks,* Central District of Cali-

fornia, CR-91-358-JSL, testimony of fingerprint examiner Diane Castro, transcript, 585–607, cited in Cole, *Suspect Identities,* 272–273.

12. The finger numbering begins with the right thumb (No. 1) to the right little finger (No.5), then to the left thumb (No. 6) to the left little finger (No. 10).

13. *People v. Myers,* 150–177.

14. *People v. Myers, Appellant,* Supreme Court of New York, Appellate Division, First Department, 220 A.D. 2d 272; 632 N.Y.S. 2d 111; 1995 N.Y. App. Div. Lexis 10052. The Appellate Division's opinion also stated: "The expert testimony regarding the institution and mechanics of the Statewide Automated Fingerprint System did not constitute improper bolstering, but rather was properly admitted to explain why the police apprehended defendant after a lapse of seventeen years . . . Nor was there any error in admitting the testimony of the expert witness that based upon the number of comparison points and the quality of the latent fingerprints taken from the crime scene, the witness had no doubt that the fingerprints in question were those of the defendant."

15. An evidence-planting scandal with profound impact on police throughout New York State involved several members of the New York State Troopers in 1992, five of whom were eventually convicted. See "Former State Trooper Explains Ways That He Fabricated Evidence," *New York Times,* April 16, 1993; "Ex-Trooper Gets Prison for Faking Evidence," *New York Times,* June 12, 1993; and "Police Investigation Supervisor Admits Faking Fingerprints," *New York Times,* July 30, 1993. The most famous alleged evidence-planting occurred in the O. J. Simpson trial, where Simpson's attorneys successfully accused the Los Angeles police department of planting evidence against their client.

16. Simon Cole, *Suspect Identities,* 281. Cole notes that only 44 percent of 156 examiners who took an "external proficiency" test conducted by the Collaborative Testing Service under the aegis of the International Association for Identification made no mistakes at all in matching prints. Twenty-two percent of those tested wrongly re-

ported as positive matches fingerprints from completely different people.

17. *The Sokal Hoax: The Sham That Shook the Academy* (Lincoln: University of Nebraska Press, 2000), by the editors of *Lingua Franca,* graphically illustrates the profound differences between those who adhere to an epistemology based on empirical realities and those for whom all knowledge is either projected onto the world or spun in such a way as to make it unverifiable.

18. 509 U.S. 579 (1993). For a survey of the significance of *Daubert* and its offspring, see www.daubertontheweb.com.

19. The origin of the "general acceptance" test was *Frye v. United States,* 293 F. 1013, 1014 (App. D.C. 1923).

20. 509 U.S. 579 at 593–594.

21. *Kumho Tire Co., Ltd. v. Carmichael,* 526 U.S. 137 (1999).

22. In the federal courts, the issue has been the admissibility of fingerprint evidence under Federal Rule of Evidence 702 in light of the *Daubert* decision as modified by *Kumho.* In the state courts, the issue has been the admissibility of fingerprint evidence under particular state criminal procedures, usually but not always controlled by *Daubert/Kumho.* For a complete list of citations from 1999–2001, see *United States v. Llera-Plaza,* Cr. No. 98-362-10, 11, 12, U.S. District Court, Eastern District of Pennsylvania, *Government's Combined Motion in Limine to Admit Latent Print Evidence and Response to Defendant Acosta's Motion to Preclude the Introduction of Latent Fingerprint Evidence,* Appendix B, "Court Challenges to Fingerprint Evidence." My thanks to Richard Manieri of the United States Attorney's office of the Eastern District of Pennsylvania for providing me with the briefs submitted by AUSAs Thomas R. Perricone and Paul A. Sarmousakis on behalf of United States Attorney Michael L. Levy. My special thanks to Duffy Graham for his help on the legal issues discussed in this chapter.

23. *United States v. Mitchell,* Cr. No. 96-407, Eastern District of Pennsylvania, and *State of Georgia v. McGee,* Indictment No. 99-CR-277, Superior Court of Carroll County, Georgia.

24. "Will Fingerprinting Stand Up in Court?" *New York Times,* March 9, 2002, A15.

25. See *United States v. Llera-Plaza,* Eastern District of Pennsylvania, decisions of January 7, 2002, and March 13, 2002. The opinions are available on the Web at: http://www.paed.uscourts.gov/documents/opinions/02D0046P.pdf and http://www.paed.uscourts.gov/documents/opinions/02D0182P.pdf. Judge Pollak's decisions were widely reported in the press.

26. *U.S. v. Llera-Plaza,* March 13, 2002, 24. For case citations since Judge Pollak's January 7, 2002, decision, see "Legal Challenges to Fingerprints," http://onin.com/fp/daubert_links. In particular, see *U.S. v. Mitchell,* 2004 WL 908359 (3rd Cir. April 29, 2004). The public perception of the reliability of fingerprint identification experts suffered grievously from the bungled case of Brandon Mayfield of Portland, Oregon. Three FBI fingerprint experts identified a partial latent fingerprint found on a bag of explosive detonators near the site of the Al Qaeda terrorist attack in Madrid, Spain, on March 11, 2004, as belonging to Mr. Mayfield, a Muslim convert and once a defense attorney for a Muslim radical in a domestic matter. The FBI experts stated that there were fifteen matching points between Mayfield's print and the latent print discovered at the crime scene. Spanish authorities initially cited eight matching points, but they later retreated completely and said that the latent print belonged to one Ouhnane Daoud, an Algerian. FBI experts admitted the misidentification, and the Oregon district court released Mayfield with apologies. See Jennifer Mnookin, "The Achilles' Heel of Fingerprints," *Washington Post,* May 29, 2004.

Government officials claimed that they had reason to suspect Mayfield once the mismatched fingerprints brought him to the FBI's attention. According to briefs signed by the government, warranted searches of Mayfield's home, office, and safe deposit box revealed that Mayfield had accessed on his computer: airplane schedules to Madrid; "a website apparently sponsored by the Spanish national passenger rail system—the target of the March 11, 2004, bombings"; and "virulently anti-Semitic articles . . . which appeared to blame Jewish people for various world problems." See *Reply Memorandum in Support of Motion to Amend Order Requiring Destruction of Seized Items* (Misc. No. 04-MC-9071) filed by United States Attorney Karin Immergut and her assistants in the United States District Court in the District of Oregon on September 13, 2004. The United States filed the motion with a request to the court that it be sealed; but in opposing the motion Mayfield's attorneys also opposed the request for sealing, and the court unsealed the memorandum.

The Mayfield mismatch of fingerprints colored other stories as well. Benjamin Weiser of the *New York Times* wrote an excellent article about the Kafkaesque case of Rene Ramon Sanchez, who was arrested several times for the crimes of one Leo Rosario because police department clerks mixed up Sanchez's fingerprints, taken after a 1995 traffic violation, with those of Rosario, arrested the night before on a drug charge. Although the two men bore no physical resemblance whatsoever to each other, the clerical mistake repeatedly made Sanchez into Rosario in the eyes of the criminal justice system whenever Sanchez was stopped by police because Rosario's prints popped up and were taken as a more valid form of identification than photographs. As it happens, Rosario's occupation led him into a lot of trouble, which became trouble for Sanchez.

When Mr. Weiser wrote his article, in which he cites the Mayfield case, his editors titled the piece "Can Prints Lie? Yes, Man Finds to His Dismay," *New York Times,* May 31, 2004. But the point was not the misidentification of prints, as in the Mayfield case, but a cautionary tale about the consequences of a simple misfiling in a vast bureaucracy.

9. A Death in the Field

1. Horseweed *(Conyza canadensis)*, also known as mare's tail, is a composite, semicosmopolitan weed of the aster family. Horseweed bears

yellowish flowers and, with its great hardiness, range, and genetic plasticity, often grows to eight-to-ten-foot heights.

2. Translation: "Through this holy oil, and through the great goodness of His mercy, may the Lord pardon thee whatever sins thou hast committed."

3. On August 25, 1997, during a walkthrough reconstruction of PO Lozada's murder in Bushwick, Detective Vincent Carrera and I encountered one of the same two women sitting on the same stoop, along with her daughter. The other woman witness was long dead.

4. Compare the hesitancy about tape-recording suspects' statements in 1984 with demands for universal tape-recording of all encounters between criminal investigators and suspects fifteen years later. See, for instance, Jim Dwyer, "Cornered Minds, False Confessions," *New York Times,* December 9, 2001.

5. Jeter's attorneys later contended that Justice Thaddeus Owens, the New York State Supreme Court judge who presided over Jeter's trial, erred in admitting expert testimony on voice spectrographic analysis without a preliminary hearing on its scientific status and reliability. New York lower courts had previously split on the reliability of such evidence, and there had been no ruling on the issue by higher courts. The Court of Appeals of New York agreed with this claim, but called the admission of that expert testimony a harmless error in light of the other evidence against Jeter, including the two statements that Carrera and Cardi had taped. *People of the State of New York, Respondent v. Darryl Jeter, Appellant,* Court of Appeals of New York, 80 N.Y.2d 818; 600 N.E.2d 214, June 11, 1992.

6. *New York Times,* September 29, 1984.

7. While cross-examining Detective Carrera, however, Mr. Foard called his own client "Gerald." Carrera quickly pointed out that Foard had made the mistake while fully alert instead of at the end of a grueling night of work.

8. Indeed, when word of the findings of trace amounts of opiates and cocaine in PO Lozada's bile and urine was leaked to the press, Justice Thaddeus Owens immediately forbade counsel on both sides to

speak to the media. When Foard did give interviews to broadcast journalists after the order, Justice Owens found him in summary contempt of court. None other than William Kunstler, the dean of radical defense attorneys, represented Foard in this dispute. The Appellate Division, Second Department of the Supreme Court of New York, later reversed the contempt order. For details, see *People of the State of New York, Plaintiff, v. Darryl Jeter, Defendant, Richard Foard 3d, Appellant; Thaddeus E. Owens, as Justice of the Supreme Court, Kings County, Respondent.* 116 A.D. 2d 558; 497 N.Y.S.2d 414, January 13, 1986.

9. Vincent Carrera died on May 7, 2004, of natural causes. A friend sang Frank Sinatra's "My Way" at his funeral mass, and his daughter's husband played "Call to the Post" on the trumpet, a tribute to Carrera's avid interest in horse racing.

10. On September 11, 2001, NYPD Police Officer Moira Smith of the 13th precinct, formerly of the New York City Transit Police, and Captain Kathy Mazza of the Port Authority of New York and New Jersey Police died in the line of duty while helping civilians escape from the collapsing twin towers of the World Trade Center.

10. The Long Arm of the Job

1. INS had four law-enforcement divisions—Border Patrol, Investigations, Inspections, and Detention and Removal. ICE combines the U.S. Customs Office of Investigations, the INS Investigations and Detention and Removal divisions, the Federal Protective Service, and the Federal Air Marshals Service.

2. See Robert Jackall, *Wild Cowboys: Urban Marauders & the Forces of Order* (Cambridge: Harvard University Press, 1997), for the only example up to 2004 in New York City of such interborough cooperation. In *People v. Rincon,* DANY's Homicide Investigation Unit, borrowing investigators and prosecutors from Bronx and Kings counties, put together a tri-borough investigation and prosecution of members of a drug gang responsible for narcotics and arms trafficking, murder, and mayhem throughout the city.

3. See *United States v. Heatley,* SDNY S11 96 Cr. 515, which documents

some, but by no means all, of the crew's crimes. Heatley pleaded guilty to thirteen murders, conspiracy to commit murder, robbery, extortion, and other crimes on February 5, 1999, in the U.S. District Court in Manhattan. Before the final deal with Heatley was reached, the federal prosecutors refused to accept Heatley's initial "proffer" (the so-called Queen for a Day arrangement whereby defendants reveal to prosecutors crimes done by others to which they will testify even as they review their own criminal histories) on the grounds that they already had the same information from other sources. But in the end, the prosecutors closed the deal with Heatley in order to hammer Cuff. John Cuff pleaded guilty to ten murders and other crimes on March 22, 1999.

4. See Jackall, *Wild Cowboys,* for the complete story of the double and quadruple homicides and for a sketch of Pasqualito's criminal career.

5. On the issue of identification, see Shaila K. Dawan, "Elementary, Watson: Scan a Palm, Find a Clue," *New York Times,* November 21, 2003. By the end of 2003, the NYPD's computerized palm print database contained more than 100,000 palm prints.

Just as local police have begun using tools developed for the national "war on terror" for their own purposes, so too have district attorneys begun to use post-9/11 state anti-terror laws to prosecute street criminals. On May 13, 2004, Robert T. Johnson, district attorney of the Bronx, announced a 70-page indictment (2210/2004) against Edgar Morales aka Puebla and seventeen other members of "The St. James Gang" for conspiracy to "intimidate or coerce a civilian population," pursuant to New York State's Anti-Terrorism Statute (NYS Penal Law 490) enacted on September 17, 2001. The indictment names "numerous acts of violence and destructive behavior" that furthered the conspiracy. My thanks to Assistant District Attorney Edward Friedenthal for providing me with the indictment.

6. See Daryl Khan, "NYPD To Have Access to Interpol Data," *Newsday,* November 19, 2003.

7. Similar cases of obsessive hoarding come to light from time to time. See, for instance, Robert D. McFadden, "Bronx Man Rescued from

His Own Paper Prison," *New York Times,* December 30, 2003. But normally hoarders, such as the famous Collyer brothers, who saved over 180 tons of newspapers, magazines, and trash in their Harlem apartment, stay reclusive. See Franz Lidz, "The Paper Chase," *New York Times,* October 26, 2003. The 34th precinct homicide victim lived a full, outwardly normal professional, middle-class life. Her murder remains a mystery.

8. Benoit Denizet-Lewis, "Double Lives on the Down Low," *New York Times Magazine,* August 3, 2003; Andrew Jacobs, "The Beast in the Bathhouse," *New York Times,* January 12, 2004. See also Frank Owen, "No Man Is a Crystal Meth User to Himself," *New York Times,* August 29, 2004.

9. See, for instance, *United States v. Madrid,* Southern District of New York, 01 CR 21 and S4 02 CR 416. The indictment in the latter case alleges that Consuelo Marquez, then an employee of Lehman Brothers, created fronts, shell companies, and offshore accounts all over the world that funneled wire transfers and checks to launder millions of dollars of drug money for Mario Ernesto Villanueva Madrid, a former Mexican state governor, and his son Luis Ernesto Villanueva Tenorio. After lawyers hired by Lehman Brothers persuaded United States attorneys that Marquez had simply outwitted her superiors with fake documents, the indictment did not accuse the firm of criminal activity. See Bill Berkeley, "A Glimpse into a Recess of International Finance," *New York Times* November 12, 2002. My thanks to Richard Sullivan, Assistant United States Attorney, Southern District of New York, for providing me with the indictments and *United States v. Madrid, Government's Memorandum of Law in Opposition to the Pre-Trial Motions of Defendant Consuelo Marquez.* Marquez pleaded guilty to one count of bank and wire fraud and fifteen counts of wire fraud. United States Attorney, Southern District of New York, "Former Lehman Brothers Broker Pleads Guilty to Ban and Wire Fraud Charges," Public Information Office release, September 8, 2004. As of late September 2004, Marquez still faced trial for "laundering millions of dollars in narcotics proceeds through Lehman accounts."

10. Every once in a while outright criminal activities of members of the defense bar come to public attention. See, for instance, the case of Pat V. Stiso aka Gabriella, whose offices were in the Bronx, who pleaded guilty in the Southern District of New York to obstruction of justice and conspiracy to distribute heroin. Prosecutors argued that Stiso was the "house counsel" to the Maisonet drug gang from the Hunts Point section of the Bronx. In his allocution, Stiso acknowledged that he received money from gang members in his offices, knowing that the money was the "proceeds of narcotics activity," "for the purposes of preserving and concealing these funds to further the conspiracy." Stiso also acknowledged making false statements to one court and a false bail application to another court, using still other monies that he knew were the "proceeds of narcotics activity" to procure property that could be posted as security for his client. Stiso also admitted divulging information that he knew he was specifically prohibited from disclosing. On questioning from the judge, the prosecutors pointed out that this last offense was very serious indeed. Stiso provided Francisco Maisonet, the boss of the drug gang, the name of the government witness who had secretly testified against Maisonet. *United States v. Pat Stiso,* S5 CR 817, August 12, 1998. My thanks to Richard Sullivan, Assistant United States Attorney, Southern District of New York, for providing me with a copy of the plea proceedings.

11. Several attorneys general of different states made political hay out of *pro se* prisoner lawsuits in the 1990s, most notably Dennis C. Vacco of New York, who published during his term at least three Top Ten lists of frivolous inmate claims. Vacco's 1998 list had the story of a young man who sued New York State for $15 million for forcing him into a life of crime by denying him a driver's license after he failed to respond to several traffic tickets, and the story of another young man, in prison for burglary, who, because the commissary sold him a box of stale Pop Tarts, sued the state for $35,000 for mental anguish, pain, and suffering and a new box of the taste treats. "News from Attorney General Dennis C. Vacco," *The Capitol,* Albany, New York, September 10, 1998. Vacco was pressing for state legislation to penalize

New York State prisoners who filed such meritless suits. In 1996 the United States Congress had passed the Prison Litigation Reform Act, which sharply restricted federal prisoners' access to federal courts. See Margo Schlanger, "Inmate Litigation," *Harvard Law Review* 116 (April 2003), for a review of federal inmate claims before and after that legislation. Schlanger argues that the federal legislation has thwarted even constitutionally meritorious claims. She also notes the propagandistic use of egregious claims by attorneys general.

12. The photograph was discovered by Detectives Garry Dugan and John Bourges in the criminal's mother's apartment while the detectives were investigating a shooting.

13. This well-known phrase, now a motto of journalists everywhere, seems to have originated with Finley Peter Dunne's *Observations by Mr. Dooley* (New York: R. H. Russell, 1906), 240. "Th' newspaper does ivrything f'r us. It runs th' polis foorce an' th' banks, commands th' milishy, conthrols th' ligislachure, baptizes th' young, marries th' foolish, comforts th' afflicted, afflicts th' comfortable, buries th' dead an' roasts thim afterward."

14. The history of police corruption in New York City and investigations into it is extensive. The Knapp Commission Report on Police Corruption (1972), the landmark descendant of the Lexow Committee Report (1895), the Curran Commission Report (1913), the Seabury Investigation Report (1932), and the Hefland Investigation Report (1955) noted not only the prototypical form of police corruption, that is, the graft normally associated with any bureaucracy whose officials have authority to enforce or not to enforce regulations, but also warned of the growing corrupting influence of narcotics money on police officers. Major organizational reforms followed the Knapp Commission's report, including the institution of separate narcotics units to do narcotics investigations and arrests. The Mollen Commission Report (1994), instituted by the Dinkins administration, documents the startling extent to which the Knapp Commission's warning about narcotics-money-induced corruption had become re-

ality. The Giuliani administration instituted a zero-tolerance polic-
ing strategy that allowed uniformed police to make narcotics arrests.
For an overview of the history of police corruption in New York City
and the many efforts to reform it, see Gabriel J. Chin, *New York City
Police Corruption Investigation Commissions, 1894–1994* (Buffalo, NY: W. S.
Hein, 1997), 6 vols.

The stratification of police organizations by talent and especially
by hard work resembles that of all large organizations, particularly
that of other large civil-service municipal bureaucracies. In my estimate, based on wide comparative study of and experience in different bureaucracies, including corporations and the academy, about 20
percent of the employees of most large organizations have retired
while still on the job. Excepting favoritism in its various forms (the
universal plague of all bureaucracies), moral entrepreneurs' contin-
ual focus on certain types of police corruption—perjury, extortion,
and especially excessive or unwarranted use of force, inexcusable
crimes but committed by a relatively small percentage of officers—
causes them to miss the most widespread form of police corruption,
namely, laziness.

15. The officer who accused his fellows did, in fact, surrender what was
left of "his share" of the money, about $18,000. But, despite extensive
searches by the Internal Affairs Bureau, no trace of money was ever
found on his fellow officers' properties. Both were acquitted in departmental trials, and the district attorney declined to prosecute
them. Both retired from the department. The detective who made
the accusations remains on the force, though in a different precinct.

16. See, for instance, Shaila K. Dewan and William K. Rashbaum, "Arrests Jolt the Police, but Some See a Pattern," *New York Times*, December 14, 2003, and William K. Rashbaum, "Stolen Drug Money Is
Found in Investigation of Detectives," *New York Times*, December 30,
2003. But it is also the case that drug dealers, when busted, routinely
claim that police have stolen money from them, knowing that such
accusations will cast doubt on the validity of their arrests. Jamaican and Dominican drug dealers in particular understand that the

blame-the-police-first intricacies of police and judicial bureaucracies always help obscure their own crimes.

17. See Jackall, *Wild Cowboys*, 65–73, for a detailed treatment of drug-related violence in Washington Heights during this period.

18. See David M. Halbfinger, "In Washington Heights, Drug War Survivors Reclaim Their Stoops," *New York Times,* May 18, 1998. A later edition of this article was entitled "Where Fear Lingers: A Special Report. A Neighborhood Gives Peace a Wary Look." The DANY reports, as is practice, are in letter form from DANY to the commissioner of the NYPD. See Robert M. Morgenthau, district attorney of New York, to Raymond W. Kelly, acting police commissioner, September 10, 1992. See also Robert M. Morgenthau to Howard Safir, commissioner, July 1, 1997. Both are available from the Office of Public Information of the District Attorney of New York. The *New York Times* did not print a letter to the editor pointing out the reporter's oversight of these public reports. The earlier report in the *Times* was Robert D. MacFadden, "In Police Shooting, a Preponderance of Evidence Indicated Self-Defense," *New York Times,* September 12, 1992. Part of the account given here of these important police shootings was published in a different form in Robert Jackall, "What Kind of Order?" *Criminal Justice Ethics,* Summer/Fall 2003, 54–67, and is reprinted here with the permission of the editors of *Criminal Justice Ethics.*

19. On false confessions, see, for example, Richard J. Ofshe and Richard A. Leo, "The Decision to Confess Falsely: Rational Choice and Irrational Action," *Denver University Law Review* 74, no. 4 (December 1997), and "The Consequences of False Confessions: Deprivations of Liberty and Miscarriages of Justice in the Age of Psychological Interrogation," *Journal of Criminal Law and Criminology* 88 (1998), 429–496. Both Ofshe and Leo regularly serve as expert witnesses on the issue of "police coercion" of confessions.

On the issue of the relationship between aggressive policing and civil order, see Bernard E. Harcourt, *Illusion of Order: The False Promises of Broken-Windows Policing* (Cambridge: Harvard University Press, 2001); David Garland, *The Culture of Control: Crime and Social Order in*

Contemporary Society (Chicago: University of Chicago Press, 2001); Loïc Wacquant, *Les Prisons de la Misère* (Paris: Raisons D'Agir Éditions, 1999); Jackall, "What Kind of Order?," 54–67. On prison reform, see Marc Maurer, *Americans behind Bars: U.S. and International Rates of Incarceration* (Washington, DC: The Sentencing Project, 1995), and Jenni Gainsborough and Marc Maurer, *Diminishing Returns: Crime and Incarceration in the 1990s* (Washington, DC: The Sentencing Project, 2000), for examples of advocacy scholarship.

20. Participants included: New York Supreme Court Justices Richard B. Lowe III and Leslie Crocker Snyder; Assistant District Attorneys Linda Fairstein and Warren Murray; Detectives Louie Bauza, Gennaro Giorgio, and Joseph Montuori from the NYPD; Detective Sergeant Edward Keitel and Detective Jeremiah Lyons from the New York City Transit Police Department; and John Miller, then a news reporter for a New York television station. About fifty professors took part in the meetings. The conference was sponsored by the Oakley Center for the Humanities and Social Sciences. My thanks to Professor Jean-Bernard Bucky, then director of the Center, for sponsoring and funding the event, and to Rosemary Lane, assistant to the director, for her invaluable help in organizing it.

21. All quotes here are drawn verbatim from the taped and transcribed proceedings of the three two-hour sessions of the 1993 conference. The transcript runs 221 double-spaced typed pages. In addition to sharply condensing the material, I have taken the liberty of editing and reordering it for clarity.

ACKNOWLEDGMENTS

I am grateful to several institutions for indispensable support in writing this book. I want to express my deep appreciation to the New York City Police Department, the former New York City Transit Police Department, and the District Attorney of New York for permitting me to do fieldwork that extended over many years. Resources from the Willmott Family Professorship and later from the Class of 1956 Professorship at Williams College greatly aided me in my work. A Fellowship for College Teachers from the National Endowment for the Humanities partially funded the intensive phase of the fieldwork in 1991–1993, and a grant from the Harry Frank Guggenheim Foundation made possible my work on drug-related violence in the 34th precinct.

I am especially indebted to the great number of men and women in law enforcement without whose help I could not have done this work. For my fieldwork in the Midtown North precinct of the NYPD, my special thanks to Detectives Pete Castillo, Robert Chung, George Delgrosso, Pete Panuccio, Alex Renow, and Brian McCabe, then a detective in the precinct's robbery squad. Kenny Ryan, then a uniformed police officer, schooled me in street policing. Later, Detectives Harry Bridgwood and Danny Rizzo were invaluable in reconstructing the investigation into the murder of Sergeant Keith Levine.

For my fieldwork in the 34th precinct of the NYPD, I extend my great appreciation to the detective squad's commanding officers: then Detective Lieutenant Joseph Reznick and Detective Sergeants Julio Alicea, Angel Flores, and Bobby Maas, as well as to Detective Captain Sal Blando, and, later, then Detective Lieutenant Kenny Lindhal. My special thanks to then Detective Louie Bauza, Detectives John Bourges, José Caban, Judy Chandon, Eddie Cruz, Mike Davis, Hughie Drain, Matty Fallon, Bobby Geis, Gennaro Giorgio, Dave Gonzalez, Mark Gowrie, Ronnie Hicks, Tony Imperato, Mike Janazzo, Charlie Lungaro, Danny Medina, Joe Montuori, Angel Morales, Pete Moro, Tim Muldoon, Joe Neenan, Gil Ortiz, Annie Peters, Hank Primus, Frankie Rosado, Marta Rosario, Billy Seimer, Bobby Small, and Kevin Walla. Police Administrative Assistant Marina Amiaga was always extremely helpful to me.

The 34th squad was a crossroads for detectives from other squads. Among those officers who aided me greatly were: Detectives Humberto Arroyo, Mike Burke, Garry Dugan, John Hickey, John Lafferty, Tommy McCabe, Tommy McKenna, Dennis O'Sullivan, Ray Pierce, Joel Potter, Danny Rodriguez, John Schlagler, Mike Sheehan, Tommy Sullivan, and Mark Tebbens. The homicide investigation course that I took with Detective Lieutenant Vernon Geberth gave me many insights into investigative work. Al Marini, Jack Healy, and Tommy Scotto of the Detectives Endowment Association supported my work from its beginnings.

For my fieldwork with the New York City Transit Police, my special thanks to Colonel Dean Esserman of the Providence Police Department, then general counsel to the NYCTP, and to Detective Sergeant Tommy Burke for arranging my fieldwork with Central Robbery. My great appreciation to the commanding officers of Central Robbery in fall 1991, especially Detective Sergeant Ed Keitel, as well as to Lieutenant John Walsh, Detective Sergeants Dennis Boodle, John Dove, Vertel Martin, and Diane Pinetti. My special thanks to Detectives Jeff Aiello, Sonny Archer, Terry Benson, Imani Booker, Billy Carter, Elias Conde, John Cornicello, Billy Courtney, Bobby Dwayne, Roger Fanti, Tony Gonzalez, Paddy Hannigan, Billy Hartwicke, Zack Jackson, Debbie Lawless, Zeke Lopez, Kelvin Miles, Don Mounts, Bob Nardi, Jimmy Nuciforo, Carl Nuñez, Toya Pizzaro, Vinnie Romeo, Steve Schumacher, Jack Thomson, Kenny Tunkel, Vinnie

Valerio, Emilio Velez, Felix Vigo, Ed Vreeland, Lee Winters, and Stacy Weiss. Central Robbery was also a crossroads for detectives from other squads. Detectives Jerry Lyons, Mike Sapraicone, and Joe Quirk of the Major Case Squad and Detective Ed Delahunt of Transit District 2 were of special importance to the fieldwork. Detective Vinnie Carrera was the chief instructor of the course on criminal investigation that I took with white shield detectives in June 1991, and he was invaluable later in reconstructing the investigation into the murder of PO Irma Lozada. Eric Seidel, formerly of the Kings County District Attorney Office, also aided me in documenting that investigation.

I spent most of my time at the District Attorney of New York in Trial Bureau 50, and I owe its chief, Assistant District Attorney Warren Murray, a great deal. Special thanks to other ADAs in that bureau who assisted me in the Early Case Assessment Bureau phase of my fieldwork: ADAs Jenny Bliss, Linda Ford, David Futter, Nancy Guess, Kevin Hynes, Sherry Holland, Kate Jones, Mike Jones, Doreen Klein, Christie Moran, Joel Seidemann, and Chuck Quackenbush. Later, ADAs Stacey Mitchell and Steve Saracco helped in reconstructing the investigation into the Chocoláte case, and ADA Dan Bibb aided me on several other matters. ADA Elizabeth Lederer, Chief of Trial Bureau 80, assisted me greatly in reconstructing the investigation into the murder of Sergeant Keith Levine; ADA Dan M. Rather, Chief of Gun-Trafficking, provided me with important information about the Preacher Crew; and ADA Susan Alexrod of DANY's Appeals Bureau helped me document the investigation into the murder of Jean Casse. The time I spent in DANY's Homicide Investigation Unit was crucial to my understanding of criminal investigation. My special thanks to ADA Walter Arsenault, then chief of HIU, and to ADA Daniel Brownell, then Deputy Chief, and to ADA Deborah Hickey. My thanks as well to Senior Rackets Investigator César Aponte and especially to Supervising Rackets Investigator Terry Quinn.

I owe special thanks to my own attorneys, Edward H. Rosenthal and Yuki A. Hirose of Frankfurt Kurnit Klein & Selz, PC, both for their excellent legal advice and for their support and friendship. My thanks also to Lori Silver of the Office of the General Counsel at Harvard University for her careful reading of the manuscript.

Duffy Graham and Arthur J. Vidich read the manuscript with the greatest care and improved it immeasurably with their penetrating critiques. Michael Erard and Lizette Klussmann also offered many incisive suggestions. At Harvard University Press, William Sisler and Michael Aronson supported the project from its beginnings; Tim Jones designed the book with great flair; and Susan Wallace Boehmer brought her remarkable imaginativeness to its editing.

My thanks to Manuel Martínez for our years-long conversation about food, baseball, and the human comedy. Thanks also to colleagues at Williams College for their invaluable assistance: Donna Chenail, Robin Haley, Linda Saharczewski, and Margaret Weyers of the Faculty Secretarial Office; Barbara Agostini and Gail Rondeau, the one and only Calendar Girls; and research librarians Lori DuBois, Christine Menárd, and Rebecca Ohm.

Janice M. Hirota and Yuriko Hirota Jackall have listened to all the stories in this book more than once. And they have sustained me in every way imaginable.

Finally, I want to thank my students at Williams College for the unbroken dialogue that we have shared for three decades. The book originated in a course entitled "Ways of Knowing." It grew over the years through countless conversations and reflections with these remarkable young men and women. I am delighted to dedicate this work to them.

INDEX

This index does not include all the pseudonymous names used in the text. Those pseudonymous names that are indexed appear in quotation marks.

Index

425

Wolfpacks, robberies by, 45–59, 83, 269, 342

Women, as police officers, 136, 305, 329, 334, 404n10

Written statements, 55, 108–111, 161, 212–214, 240, 317, 381n5, 387n4, 395n5

Zeins, Wally, 76, 81

Zero-tolerance policing, 380n14, 408n14